DAM FAILURE MECHANISMS AND RISK ASSESSMENT

DAM FAILURE MECHANISMS AND RISK ASSESSMENT

Limin Zhang
The Hong Kong University of Science and Technology, Hong Kong, China

Ming Peng
Tongji University, Shanghai, China

Dongsheng Chang
AECOM Asia Company Ltd, Hong Kong, China

Yao Xu
China Institute of Water Resources and Hydropower Research, Beijing, China

WILEY

This edition first published 2016

Registered Office
John Wiley & Sons Singapore Pte. Ltd., 1 Fusionopolis Walk, #07-01 Solaris South Tower, Singapore 138628.

For details of our global editorial offices, for customer services and for information about how to apply for permission to reuse the copyright material in this book please see our website at www.wiley.com.

Library of Congress Cataloging-in-Publication Data

Names: Zhang, L. M. (Limin), author. | Peng, Ming, 1981– author. | Chang, Dongsheng, 1983– author. |
Xu, Yao, 1982– author.
Title: Dam failure mechanisms and risk assessment / Limin Zhang, Ming Peng, Dongsheng Chang, Yao Xu.
Description: Singapore ; Hoboken, NJ : John Wiley & Sons, 2016. | Includes bibliographical references and index.
Identifiers: LCCN 2016010573| ISBN 9781118558515 (cloth) | ISBN 9781118558546 (epub)
Subjects: LCSH: Dam failures.
Classification: LCC TC550.2 .Z44 2016 | DDC 627/.8–dc23
LC record available at https://lccn.loc.gov/2016010573

A catalogue record for this book is available from the British Library.

Set in 10/12pt Times by SPi Global, Pondicherry, India

1 2016

To Linda Li, Yan Zhu, Xin Wang, and Jing Bai

Contents

Foreword

I felt privileged to write the foreword for the book by Desmond Hartford and Gregory Baecher (2004) *Risk and Uncertainty in Dam Safety*, published by Thomas Telford Ltd in 2004. In that book, the authors described probabilistic analysis tools for dam risk analysis and decision-making, including guiding principles for risk analysis, methods for reliability analyses, and decision-making tools such as event tree and fault tree analyses.

This new book by Zhang, Peng, Chang, and Xu, *Dam Failure Mechanisms and Risk Assessment*, published by John Wiley & Sons, Ltd in 2016, presents the subjects in more detail by emphasizing practical applications of the analyses. The book describes the causes, processes, and consequences of dam failures. It covers up-to-date statistics of past dam failures and near-failures, mechanisms of dam failures, dam breaching process modeling, flood routing and inundation analyses, flood consequence analyses, and dam-breaching emergency management decisions. The authors integrate the physical processes of dam breaching and the mathematical aspects of risk assessment and management and describe methodologies for achieving optimal decision-making under uncertainty. The book emphasizes the two most common failure mechanisms for embankment dams: internal erosion, which has received increased attention in recent years, and overtopping. Empirical and numerical methods are used to determine dam breaching parameters such as breach geometry and peak flow rate, and for analyzing the dam breaching flood routing downstream.

The methodologies described by the authors may be used by government dam regulatory agencies for evaluating risks, and by dam owners to evaluate dam safety and the planning and prioritizing of remedial actions. I strongly recommend this up-to-date book, as it represents a most valuable contribution to the state of the art, paving the way for practical applications of probabilistic analysis tools to dam risk assessment and management.

Kaare Høeg
Professor Emeritus, University of Oslo, Norway
Expert Adviser, Norwegian Geotechnical Institute (NGI)
Honorary President, International Commission on Large Dams (ICOLD)
Formerly, President of ICOLD (1997–2000)

Foreword

As of 2015, the International Commission on Large Dams (ICOLD) has registered more than 60,000 large dams higher than 15 m around the world. Among these, 38,000 are in China. With functions of flood control, irrigation, hydropower, water supply, etc., dams contribute significantly to social-economic development and prosperity. On the other hand, dam failures do occur sometimes and can result in huge loss of life and property. Accordingly dam safety is of great importance to society. China alone has reported more than 3500 cases of failures of constructed dams. In the past 15 years or so, China also faced the mitigation of risks of large landslide dams, particularly those triggered by the 2008 Wenchuan earthquake.

Dam risk management requires not only a good understanding of dam failure mechanisms and probability, but also rapid evaluation of flood routing time and potential flooding areas. For instance, in the mitigation of the risks of the Tangjiashan landslide dam in June 2008, three likely overtopping failure modes were considered, the dam breaching impact area for each failure mode was evaluated and the flooding routing time was forecast. Consequently, approximately 250,000 people downstream of the dam were evacuated before the breaching of the large landslide dam. The book *Dam Failure Mechanisms and Risk Assessment* by Zhang, Peng, Chang, and Xu covers the wide spectrum of knowledge required for such a complex dam risk analysis and management case. This book is unique in that:

1. It is the first book that introduces the causes, processes, consequences of dam failures and possible risk mitigation measures in one nutshell;
2. It integrates the physical processes of dam failures and the mathematical aspects of risk assessment in a concise manner;
3. It emphasizes integrating theory and practice to better demonstrate the application of risk assessment and decision methodologies to real cases;
4. It intends to formulate dam-failure emergency management steps in a scientific structure.

ICOLD published statistics of dam failures in 1995, which have not been updated in the past 20 years. This book publishes three of the most updated and largest databases: a database of 1443 cases of constructed dam failures, a database of 1044 cases of landslide dam failures, and a database of 1004 cases of dike failures. The latest statistics of failures of constructed dams, landslide dams and dikes are reported accordingly. I consider the compilation of these latest databases one of the most important contributions to dam safety in the past 20 years.

I am confident this book will assist dam or dike safety agencies in evaluating the risks of dams, making decisions for risk mitigation, and planning emergency actions.

Jinsheng Jia
Professor, China Institute of Water Resources and Hydropower Research, Beijing
Honorary President, International Commission on Large Dams (ICOLD)
Formerly, President of ICOLD (2009–2012)
Vice President and Secretary General, Chinese National Committee on Large Dams

Preface

Every dam or dike failure touches the nerve of the public, as in the cases of the Banqiao dam failure in China in August 1975, the New Orleans dike failures during Hurricane Katrina in August 2005 and the Tangjiashan landslide dam breach in China in June 2008. The Banqiao dam failure caused the inundation of an area of 12,000 km^2 and the loss of more than 26,000 lives. The dike failures in New Orleans resulted in a death count of approximately 1,600 and an economic loss of US$100-200 billion, making it the single most costly catastrophic failure of an engineered system in history. The failure of the Tangjiashan landslide dam in June 2008 prompted the evacuation of 250,000 residents downstream the dam for two weeks.

Dam or dike risk analysis involves not only the calculation of probability of failure, but also the simulation of the failure process, the flood routing downstream the dam or dike, and the evaluation of flood severity, elements at risk, the vulnerability of the elements at risk to the dam-breaching flood and the flood risks. Once the risk is analyzed, it must be assessed against risk tolerance criteria. If the risk level is deemed too high, proper risk mitigation measures, either engineering or non-engineering, should be taken to lower the risk level. The effectiveness of any risk mitigation measures and the impact of any mitigation measures on the overall risk profile should also be evaluated. Non-engineering risk mitigation measures such as warning and evacuation are often the most effective. When a dam or dike failure is imminent, a dynamic assessment of hazard propagation and scientific decisions for risk mitigation are preferred. The worldwide trend is to make accountable decisions by quantitatively expressing the dam-failure risks.

The aforementioned dam risk analysis and management process involves physical aspects of dam failure mechanisms, failure processes, flood routing and flood damage, as well as risk assessment and management methodologies. Several excellent books are available on selected topics of dam safety. For instance, Hartford and Baecher (2004) describe uncertainties in dam safety and present probability theory and techniques for dam risk assessment; Singh (1996) introduces hydraulics of dam breaching modelling. In this book, we intend to introduce in one nutshell the essential components that enable a quantitative dam risk assessment. The mechanisms, processes and consequences of dam failures as well as risk assessment and decision methodologies for dam emergency management are introduced.

This book consists of three parts, with Part I devoted to dam and dike failure databases and statistics, Part II to dam failure mechanisms and breaching process modeling and Part III to dam failure risk assessment and management.

Part I (Chapters 1–5) presents three latest databases of the failure of 1443 constructed dams, 1044 landslide dams and 1004 dikes. The statistical analyses of failures of constructed embankment dams, landslide dams, concrete dams and dikes are presented separately. International Commission on Large Dams (ICOLD) released a statistical analysis of dam failures in 1995 and an updated analysis is long-awaited. In this book, the statistics for the failure of various types of dams are updated including the latest failure cases around the world and failure cases in China that were not included in the ICOLD analysis in 1995. The detailed failure cases are presented in Appendices A and B, which are of retention value to the dam safety community.

Part II (Chapters 6–9) presents two most common dam failure mechanisms (i.e. internal erosion in dams and their foundations and overtopping erosion of dams) and dam breaching modeling. The initiation, continuation and progression of concentrated leak erosion, backward erosion, contact erosion and suffusion are described separately in Chapter 6. The mechanics of overtopping erosion, methods for determining soil erodibility parameters and classification of soil erodibility are presented in Chapter 7. These two chapters lay the foundation for understanding and simulating the process of dam failure. Subsequently, we present methodologies of dam breaching process modeling and flood routing analysis following the time sequence of a dam failure: dam breach modeling and determination of dam breaching parameters such as breach geometry and peak flow rate (Chapter 8), and analysis of dam-breach flood routing downstream the dam (Chapter 9).

Part III (Chapters 10–14) presents key components in assessing the risks of a specific dam. This part begins with the introduction of several methods for analyzing the probability of failure of dams (Chapter 10). Subsequently, we present methodologies for the evaluation of inundation zones and vulnerability to dam-breaching floods (Chapter 11), the assessment of dam failure risks (Chapter 12), and dam breach contingency risk management and optimal decision making under uncertainty (Chapter 13). Finally, risk-based decision making is illustrated in the case study of the Tangjiashan landslide dam failure (Chapter 14).

Limin Zhang

Acknowledgements

Many individuals have contributed to the methodologies presented in Dam Failure Mechanisms and Risk Assessment. Several graduate students and post-doctoral research fellows, over the years, developed numerical methods for simulating dam and dike failure processes and flood routing, and assessing dam-failure risks. Special thanks go to, in alphabetical order, Kit Chan, Chen Chen, Hongxin Chen, Qun Chen, Jozsef Danka, Liang Gao, Mingzi Jiang, Jinhui Li, Yi Liu, Tianhua Xu, Jie Zhang, Lulu Zhang, Shuai Zhang, and Hongfen Zhao.

In the past decade, we collaborated closely with several research teams on contemporary dam safety issues, particularly with Prof. Jinsheng Jia of China Institute of Water Resources and Hydropower Research (IWHR) on compilation of a database of dam failures and distresses, with Prof. Jianmin Zhang of Tsinghua University on dam safety under extreme seismic and blasting loading conditions, with Prof. Runqiu Huang of State Key Laboratory for Geohazard Prevention and Environmental Protection and Prof. Yong You of Institute of Mountain Hazards and Environment of the Chinese Academy of Sciences on mitigating the risks of the landslide dams triggered by the Wenchuan earthquake, with Prof. Dianqing Li and Prof. Chuangbing Zhou of Wuhan University on the life-cycle safety of dam abutment slopes, and with Sichuan Department of Transportation on risk-based decision for mitigating the risks of debris flow dams for highway reconstruction near the epicenter of the Wenchuan earthquake. We were fortunate to have had so many opportunities to solve contemporary practical dam safety problems. The research collaborators are gratefully acknowledged.

The late Prof. Wilson Tang is fondly remembered by all the co-authors of this book. He offered enthusiastic encouragement for us to initiate the book project. We sincerely thank Professors Alfred Ang, Hongwei Huang, Bas Jonkman, Suzanne Lacasse, Chack Fan Lee and Farrokh Nadim who provided critical comments on the PhD theses supervised by the first author. These theses form part of this book. We also appreciate the efforts of Nithya Sechin, Maggie Zhang, Adalfin Jayasingh and Paul Beverley of John Wiley & Sons, Ltd, who edited the book, and those who reviewed the book proposal.

We are grateful to Natural Science Foundation of China for their financial support under grant Nos. 50828901, 51129902 and 41402257, to the Research Grants Council of the Hong Kong Special Administrative Region under grant Nos. C6012-15G and 16212514, to Sichuan Department of Transportation under contract No. SCXS01-13Z00110/11PN and to the Ministry of Science and Technology under grant No. 2011CB013506.

About the Authors

Limin Zhang, Department of Civil and Environmental Engineering, The Hong Kong University of Science and Technology (HKUST), Hong Kong, China. Dr Zhang is currently Professor of Civil Engineering at HKUST. In the past 10 years, Dr Zhang graduated five PhD students on the topic of dam or dike risks. A large part of this book is based on these theses. Dr Zhang's research areas cover embankment dams and slopes, geotechnical risk assessment, and foundation engineering. He has published over 190 refereed international journal papers and over 150 international conference papers. He is a Fellow of ASCE, Past Chair of the Executive Board of GEOSNET (Geotechnical Safety Network), Vice-Chairman of ASCE Geo-Institute's Risk Assessment and Management Committee, Editor-in-Chief of International Journal *Georisk*, Associate Editor of ASCE's *Journal of Geotechnical and Geoenvironmental Engineering*, and a member of the editorial board of *Soils and Foundations*, *Computers and Geotechnics*, *Journal of Mountain Sciences, International Journal Geomechanics and Engineering*, and *Geomechanics and Geoengineering*.

Ming Peng, Department of Geotechnical Engineering, College of Civil Engineering, Key Laboratory of Geotechnical and Underground Engineering of Ministry of Education, Tongji University, Shanghai, China. Dr Peng received his PhD degree in June 2012 from the Hong Kong University of Science and Technology and is currently an assistant professor in Tongji University. His research areas include risk analysis methodologies, flood vulnerability analysis, and decision theory. He is an author of more than 10 referred international journal papers on dam safety.

Dongsheng Chang, AECOM Asia Company Ltd., Shatin, Hong Kong, China. Dr Chang received his PhD degree in January 2012 from the Hong Kong University of Science and Technology and is currently a civil engineer. He is an expert in internal erosion and overtopping erosion of dams. He invented a laboratory device for testing the internal erodibility of soils under complex stress conditions. Dr Chang is a winner of the HKIE Outstanding Paper Award for Young Engineers.

Yao Xu, China Institute of Water Resources and Hydropower Research (IWHR), Beijing, China. Dr Xu received his PhD degree in January 2010 from the Hong Kong University of Science and Technology. He is a senior engineer at Chinese National Committee on Large Dams in Beijing. His areas of expertise include dam safety evaluation, diagnosis of dam distresses, dam breaching analysis, risk analysis methodologies, and sustainable development of hydropower.

Part I

Dam and Dike Failure Databases

1

Dams and Their Components

1.1 Classification of Dams

A dam is a barrier that impounds water. Dams have been an essential infrastructure in society that contributes to socioeconomic development and prosperity. They are built for a number of purposes, including flood control, irrigation, hydropower, water supply, and recreation. Dams can be classified in many ways, depending on their size, materials, structural types, construction methods, etc. According to the definition of the International Commission on Large Dams (ICOLD, 1998a), a reference dam height for distinguishing large dams from small dams is 15 m. Based on the materials used, dams can be classified as earthfill or rockfill dams, concrete dams, masonry dams, cemented sand and gravel dams, and others. Dams of earthfill or rockfill materials are generally called embankment dams. Based on the structural types adopted, dams can be divided into gravity dams, arch dams, buttress dams, and others. Very often dams are constructed with a combination of two or more structural forms or materials. Of the various types of dam, embankment dams are the most common.

ICOLD (1998) has published a world register of dams, which gives some facts regarding the numbers of different types of dam throughout the world. There are 25,410 dams over 15 m high, of which 12,000 were built for irrigation, 6500 for hydropower and 5500 for water supply, although many of them serve more than one purpose. Embankment dams of earthfill predominate over the others, comprising about 64% of all reported dams, while those of rockfill comprise 8%. Masonry or concrete gravity dams represent 19%, arch dams 4%, and buttress dams 1.4%. Dams lower than 30 m form 62% of the reported dams, while those lower than 60 m comprise 90% and those higher than 100 m just over 2% of the total number of dams.

Topography and geology are the two primary factors in weighing the merits of dam types. These interrelated characteristics of the dam site influence the loading distribution on the foundation and the seepage patterns through the reservoir margins. Embankment dams can be built on a variety of foundations, ranging from weak deposits to strong rocks, which is one of the most important reasons for their wide use in the world. A dam project usually comprises several components, including a water-retaining structure (e.g. the dam), a water-releasing structure (e.g. the spillway), a water-conveying structure (e.g. conduits), and others (e.g. power plants). In addition to the main

Dam Failure Mechanisms and Risk Assessment, First Edition. Limin Zhang, Ming Peng, Dongsheng Chang, and Yao Xu.

structure of the dam, there are appurtenant structures such as the spillway, conduit, and power plant around a dam that are necessary for the operation of the whole dam system. Failures of, or accidents involving dams may be attributed to defects in either the dams themselves or their appurtenant structures.

Landslide dams are natural dams caused by rapid deposition of landslides, debris flows, or rockfall materials. The formation of most landslide dams is trigged by rainfall or earthquakes. Earthquake is the most important cause. For instance, the Wenchuan earthquake in 2008 triggered 257 sizable landslide dams. Landslide dams and constructed embankment dams are similar in materials but different in geometry, soil components, and soil parameters. The differences largely influence the failure modes and breaching mechanisms of these two types of dam.

Dikes are a special type of dam. Although the height of a dike is typically small compared with that of a dam, a dike often protects a significant worth of property. Hence the failure statistics and mechanisms of dikes are also introduced in this book.

Here in Chapter 1, the structures of constructed embankment dams, natural landslide dams, concrete gravity dams, concrete arch dams, and dikes are introduced briefly.

1.2 Constructed Embankment Dams

Commonly constructed embankment dams can be divided into homogeneous dams, earth and rockfill dams with cores, and concrete-faced rockfill dams. A homogeneous dam (Figure 1.1) consists mainly of one single type of material. Such a dam is often constructed for soil and water conservation purposes, and many dams can be constructed along a gully in which soil erosion is serious. A conduit or other type of water passage facility may be installed inside a homogeneous dam. The interface between the conduit and the surrounding soils may easily become a channel of concentrated leak erosion.

An embankment dam can be constructed with earthfill or rockfill. Any dam which relies on fragmented rock materials as a major structural element is called a rockfill dam (Singh and Varshney, 1995). High quality rockfill is ideal for high-rise dams because it provides high shear strength and good drainage. A rockfill dam often has a vertical earth core or inclined earth core for seepage control. When a vertical core is adopted, the dam is zoned, with rockfill zones on both sides, a low-permeability zone (i.e. the earth core) in the middle, and transition and filter zones in between the core and the rockfill zones (Figure 1.2). The filters protect the earth core from internal erosion. They must be much more permeable than the core material and not be clogged by particles migrated from the core. The function of the transition zones is to coordinate the deformations of the core and the rockfill to minimize the stress arch effect and differential settlements.

Figure 1.2 shows a typical section of the Shuangjiangkou dam with a vertical core. Located on the upper reach of the Dadu River in Sichuan, China, the dam is 314 high, one of the highest dams in the world. The overburden at the dam site is relatively shallow (48–57 m) and was excavated so that the vertical core could be constructed on the bedrock.

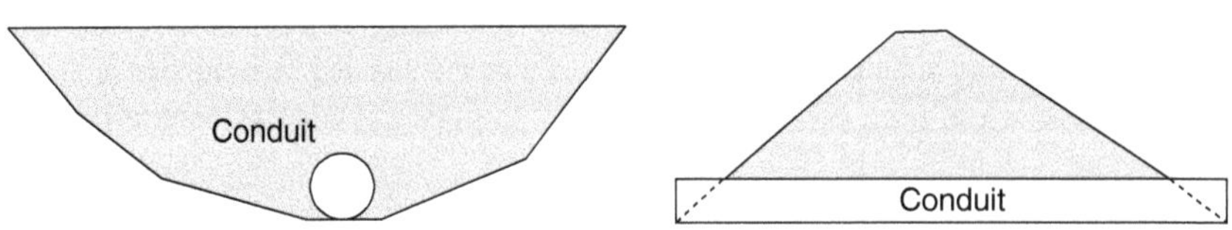

Figure 1.1 A homogeneous dam with a conduit

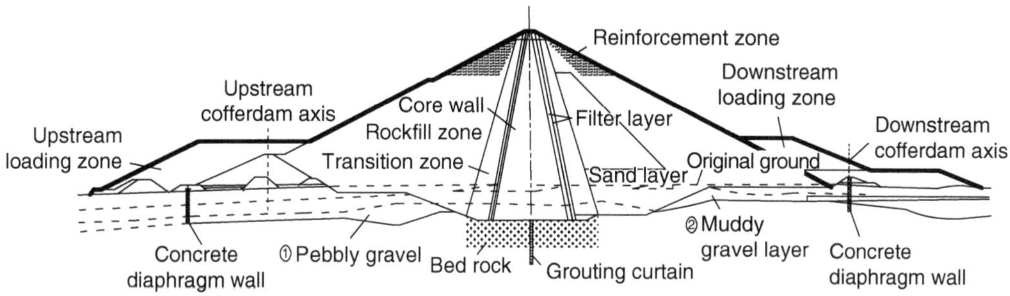

Figure 1.2 A section of the 314 m high Shuangjiangkou rockfill dam

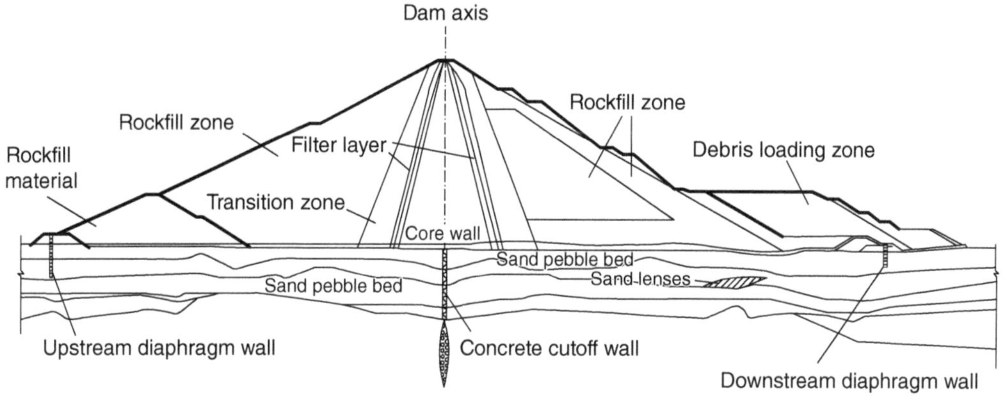

Figure 1.3 **A** section of the Pubugou rockfill dam with a vertical earth core

When the overburden is thick and pervious, cut-off walls may be required to minimize the seepage through the foundation and seepage-related problems in the foundation and the abutments. The cut-off walls for the 186 m high Pubugou rockfill dam are shown in Figure 1.3 as an example. This dam is also situated on the Dadu River in Sichuan, China. The maximum overburden thickness is 78 m. The overburden alluvium materials are gap-graded, and highly heterogeneous spatially. Hence two concrete cut-off walls were constructed through the overburden. The connection between the core and the walls was carefully designed to avoid stress concentrations and to alleviate the unfavorable influence of the large differential settlement between the walls and the surrounding soil. A highly plastic clay zone was constructed; one cut-off wall was embedded in the plastic clay and the other wall was connected to a drainage gallery, which was in turn embedded in the plastic clay.

A sloping upstream earth core may be adopted (Figure 1.4) when weather conditions do not allow the construction of a central vertical core all year round. The sloping core and the filters can be placed after the construction of the downstream rockfill. In this way, during staged construction, the rockfill can be placed all year round while the sloping core is placed during the dry season when mixing of the clayey core materials is practical. Figure 1.4 shows a section of the 160 m high Xiaolangdi dam with a sloping core. The bottom width of the core is 102 m and the core is extended to the upstream cofferdam through an impervious blanket. The overburden exceeds 70 m and two concrete cut-off walls were constructed, one beneath the sloping core and the other beneath the upstream cofferdam.

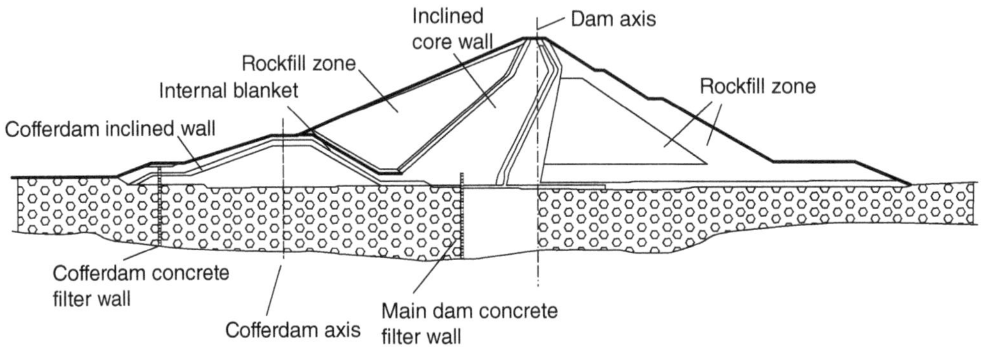

Figure 1.4 A section of the Xiaolangdi dam with an inclined earth core

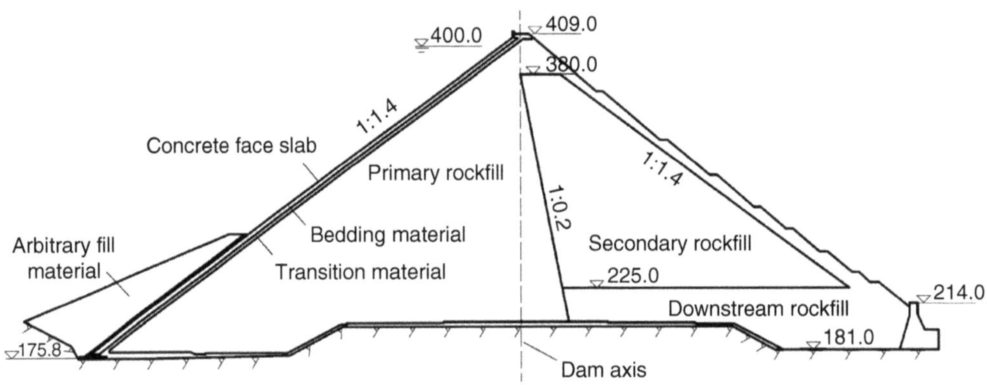

Figure 1.5 A section of Shuibuya concrete-faced rockfill dam

The seepage through a rockfill dam can also be controlled by placing an impervious reinforced concrete plate at the upstream face of the dam. Such a dam is termed as a concrete-faced rockfill dam (CFRD). Figure 1.5 shows a section of the 233 m high Shuibuya CFRD located on the Qingjiang River in Hubei, China. A CRFD consists of the rockfill, the face plate, and transition and filter zones. The rockfill provides free drainage and high shear strength so that the profile of the dam can be smaller than that of a cored dam. The reinforced concrete face plate is cast with longitudinal and transverse joints and waterstops to allow for differential movements of the plate. The transition zone serves as a cushion to support the face plate. When the joints leak or the plate cracks, the transition and filter zones also limit the leakage, and the filter between the transition zone and the rockfill prevents internal erosion at the interface. CFRDs have been the most common type of high rise rockfill dams over the past decade. Since 1985, more than 80 CFRDs higher than 100 m have been constructed or are under construction in China (Jia et al., 2014). However, separation of the face plate from the cushion and extruding rupture of the face plate have occurred in several CFRDs.

Figure 1.6 The Hongshiyan landslide dam formed on 3 August 2014 in Yunnan Province, China

1.3 Landslide Dams

A landslide dam is part of a natural landslide deposit that blocks a river and causes damming of the river. Once the river is blocked, the lake water level may rise quickly in the flood season since there is no flood control facility for such a natural dam. The dam may therefore be overtopped within a short period after the formation of the dam. For this reason, the risks posed by a landslide dam are rather high.

Figure 1.6 shows a recent landslide dam, at Hongshiyan, which was triggered by an earthquake on 3 August 2014 in north-eastern Yunnan Province, China. Landslides occurred on both sides of the Niulan River at Hongshiyan and formed a large landslide dam with a height of 83 m, a width of 750 m and a dam volume of $12 \times 10^6\,m^3$. The large lake with a capacity of $260 \times 10^6\,m^3$ threatened more than 30,000 people both upstream and downstream of the landslide dam.

1.4 Concrete Gravity Dams

A gravity dam is an essentially solid concrete structure that resists imposed forces principally by its own weight (Jansen, 1988) against sliding and overturning. Gravity dams are often straight in plan although they may sometimes be curved to accommodate site conditions. Along the dam axis, the dam can be divided into an overflow section and a non-overflow section. The dam is constructed in blocks separated by monolith joints (i.e. transverse contraction joints) and waterstops. The monolith joints are vertical and normal to the dam axis, cutting through the entire dam section. Therefore, when analyzing the safety of a concrete gravity dam, one or several blocks may be assumed to fail simultaneously. Since the stability is of primary concern in designing a gravity

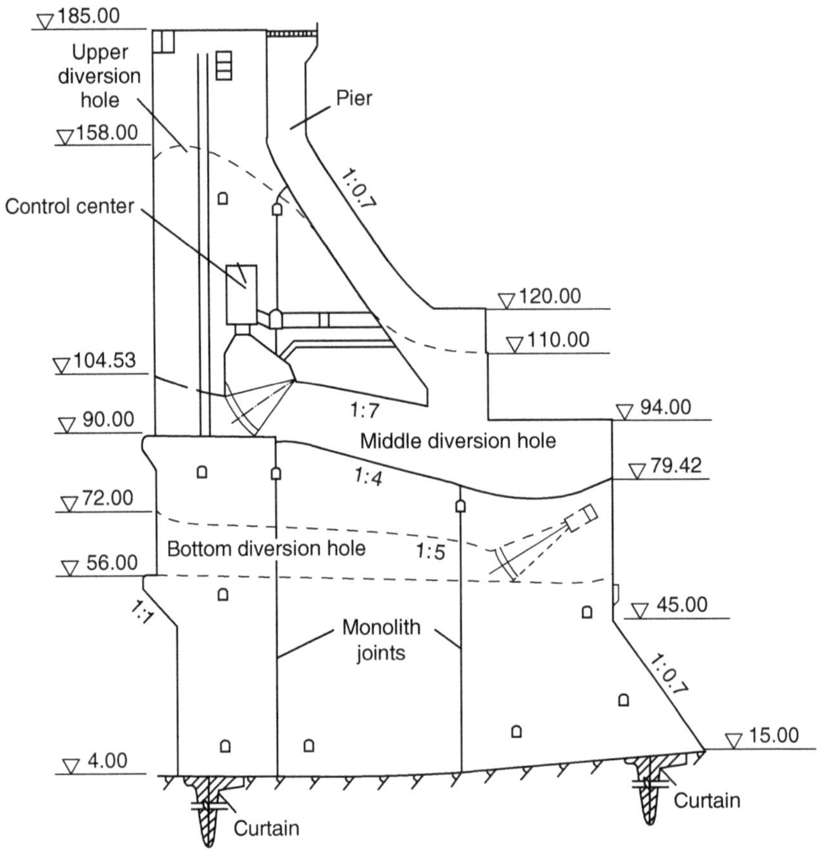

Figure 1.7 An overflow section of the Three-Gorges concrete gravity dam

dam, the treatment of its foundation is of paramount importance. In addition, grouted curtains are preferred to cut off the underseepage and minimize the uplift force on the base of the dam.

Figure 1.7 shows an overflow section of the 181 m high Three-Gorges concrete gravity dam. The dam has an axial length of 2309.47 m and a crest width of 15 m. Two curtain walls were grouted in the foundation, one near the upstream and the other near the downstream of the dam.

1.5 Concrete Arch Dams

A concrete arch dam is a shell structure that is curved both longitudinally and transversely. The water pressure is transferred to the abutments through the arch effects, and the primary load in the arch dam is compressive. Since concrete has high compressive strength, the cross-section of an arch dam can be significantly smaller than that of a concrete gravity dam. Due to the very large thrust loads, the stability of the abutments is critical to the success of an arch dam.

Figure 1.8 shows the construction of the 305 m high Jinping I arch dam in China. The widths of the crest and the base of the arch are 13 m and 58 m, respectively. Fractured rocks and deep tension cracks were found on the left abutment. Hence a deep key was excavated into the left abutment and a 155 m tall concrete foundation was constructed to ensure the safety of the abutment.

Figure 1.8 The Jinping I arch dam and its left abutment foundation

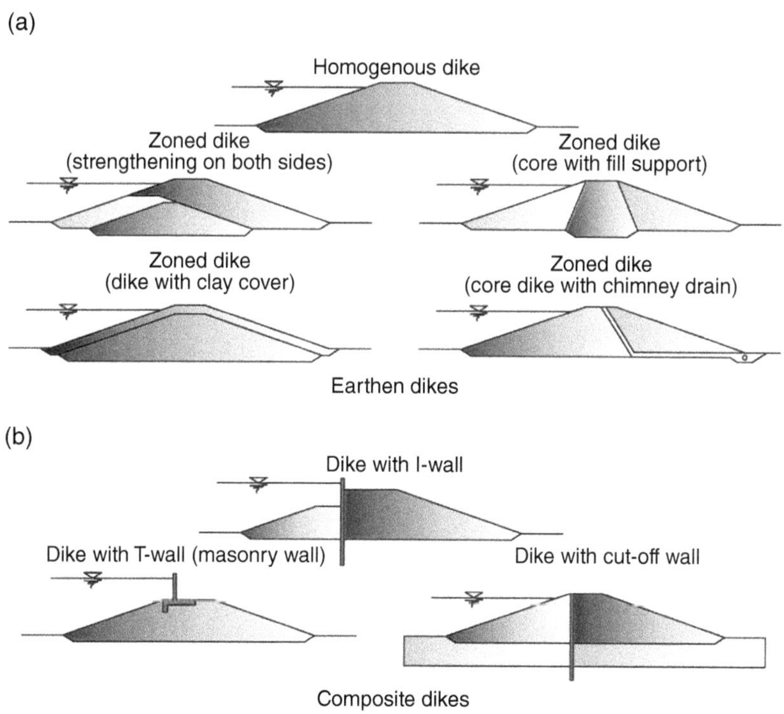

Figure 1.9 Classification of dikes. *Source:* Danka (2015). With permission from ASCE

1.6 Dikes

Two main categories of dikes – earthen dikes and composite dikes – are differentiated as shown in Figure 1.9. The earthen dikes are constructed only of granular or cohesive soils, with the materials satisfying requirements of water retention and stability. There are two different types of earthen dikes: homogeneous and zoned dikes. Homogeneous dikes are built in places rich in fine-grained soils, which is favorable as it has low permeability and high resistance to erosion. Zoned dikes are preferred when there is a lack of good quality fine-grained soils. This type of structure is popular, since the application of coarse fills can lead to savings on the use of fines. Two main types of zoned dikes are common: those with an impermeable core supported by fills, and those with an impermeable mask on the top of a fill material. In Figure 1.9, only one configuration of impermeable core is illustrated, but other layouts are also used. A special type of earthen dike is formed due to historical strengthening. This multi-layer structure is formed piecemeal as the safety standards evolve or as social needs lead to the heightening of a former section. The characterization of such dikes is challenging due to the lack of accurate information about historic geometries, and differences in construction and quality management methods.

Composite dikes include additional superstructures or embedded structures such as T-wall, I-wall, masonry wall, cut-off wall, gateway, spillway, and discharge pipes. The common layouts are presented in Figure 1.9b. T-walls or I-walls can increase the elevation of the crest hence lower the probability of overtopping. The use of such walls can also lead to a smaller dike width compared with earthen dikes of the same elevation, thus saving space. Application of retaining structures on the water side can also serve the purpose of enhancing surface protection against erosion. Cut-off walls are used to prevent seepage through the subsoil or the embankment.

2

Statistical Analysis of Failures of Constructed Embankment Dams

2.1 Database of Failures of Constructed Embankment Dams

To study failures of constructed embankment dams, 1443 cases of failure of constructed embankment dams have been collected from the literature (e.g. USCOLD, 1975, 1988; Vogel, 1980; Stanford University, 1994; Singh, 1996; Xu and Zhang, 2009) and compiled into a database. The failure cases are from over 50 countries, including the USA, India, the UK, and China. Details of the characteristics of the embankment dams and the failure information are presented in Appendix A. Details of 75 failure cases that have sufficient information on dam breach parameters are presented in Table 2.1.

Table 2.2 shows the reservoir capacities of 1065 failed embankment dams. Among the cases with known capacities, most reservoirs have capacities less than $1 \times 10^8\,m^3$. Table 2.3 shows the heights of the failed embankment dams, of which more than half are lower than 15 m. Table 2.4 presents the construction times of the failed embankment dams. The embankment dams constructed during two periods, 1910–1919 and 1960–1969, appear to have suffered the highest rate of failure. Table 2.5 shows the ages of the embankment dams at the time of failure. The data indicates that an embankment dam is most likely to fail within its first five years of service, especially during the initial impounding during the first year after construction.

2.2 Failure Modes and Processes

The breach of embankment dams is a process of erosion of the embankment materials by flow of water either over or through the dam or its foundation. The former leads to overtopping of the dam and the subsequent external erosion; the latter gives rise to internal erosion. In the following sections, the breaching characteristics of embankment dams by overtopping and internal erosion will be discussed separately.

Dam Failure Mechanisms and Risk Assessment, First Edition. Limin Zhang, Ming Peng, Dongsheng Chang, and Yao Xu.

Table 2.1 Summary of information of 75 dam failure cases (O = overtopping; P = internal erosion; S = sliding; HD = homogeneous dams; DC = dams with corewalls; FD = concrete faced dams; ZD = zoned-fill dams; HE = high erodibility; ME = medium erodibility; LE = low erodibility)

No.	Dam name	Location	Dam type	Dam height H_d (m)	Reservoir capacity V_d ($\times 10^6$ m^3)	Dam erodibility	Failure mode	Volume of water above breach invert V_w ($\times 10^6$ m^3)	Depth of water above breach invert H_w (m)	Breach geometry: Height H_b (m)	Breach geometry: Top width B_t (m)	Breach geometry: Bottom width B_b (m)	Breach geometry: Average width B_{ave} (m)	Breach geometry: Side slope Z Z:1(h:v)	Peak outflow rate Q_p (m^3/s)	Failure time T_f (hour)
1	Apishapa	USA	HD	34.1	22.5	HE	P	22.2	28.0	31.1	91.5	81.5	86.5	0.44	6850	2.5
2	Baimiku	China	HD	8.0	0.2	ME	O	0.2	8.0	8.0	40.0	—	—	—	—	—
3	Banqiao	China	DC	24.5	492.0	HE	O	607.5	31.0	29.5	372.0	210.0	291.0	0.3	78100	5.5
4	Bayi	China	HD	30.0	30.0	ME	P	23.0	28.0	30.0	45.0	35.0	40.0	0.17	5000	—
5	Castlewood	USA	DC	21.3	4.23	ME	O	6.17	21.6	21.3	54.9	33.5	44.2	0.5	3570	—
6	Chenying	China	HD	12.0	4.25	ME	O	5.0	>12.0	12.0	—	—	—	—	1200	1.83
7	Coedty	UK	DC	11.0	0.31	HE	O	0.311	>11.0	11.0	67.0	18.2	42.7	2.22	—	—
8	Dalizhuang	China	HD	12.0	0.6	ME	O	0.6	12.0	12.0	40.0	—	—	—	—	—
9	Danghe	China	DC	46.0	15.6	LE	O	10.7	24.5	25.0	96.0	20.0	58.0	0.66	2500	3.0
10	Davis Reservoir	USA	FD	11.9	58.0	ME	P	58.0	11.58	11.9	21.3	15.4	18.3	0.25	510	—
11	Dells	USA	—	18.3	13.0	HE	O	13.0	18.3	18.3	112.8	—	—	—	5440	0.67
12	Dongchuankou	China	HD	31.0	27.0	HE	O	27.0	31.0	31.0	—	—	—	—	21000	—
13	Dushan	China	HD	17.7	0.67	ME	O	0.67	17.7	17.7	70.0	—	—	—	—	—
14	Elk City	USA	DC	9.1	0.74	ME	O	1.18	9.44	9.14	45.5	27.7	36.6	1.0	—	0.83
15	Erlangmiao	China	HD	12.1	0.196	ME	O	0.196	9.0	9.0	36.0	1.6	18.8	1.9	—	—
16	Fengzhuang	China	HD	10.0	0.625	ME	O	0.625	>8.0	8.0	40.0	30.0	35.0	0.63	—	—
17	Frankfurt	Germany	HD	9.8	0.35	LE	P	0.352	8.23	9.75	9.2	4.6	6.9	0.4	79	2.5
18	French Landing	USA	HD	12.2	—	HE	P	3.87	8.53	14.2	41.0	13.8	27.4	0.97	929	0.58
19	Frenchman Dam	USA	HD	12.5	21.0	ME	P	16.0	10.8	12.5	67.0	54.4	54.6	0.5	1420	3.0
20	Frias	Argentina	FD	15.0	0.25	ME	O	0.25	15.0	15.0	62.0	—	—	—	400	0.25
21	Gouhou	China	FD	71.0	3.3	LE	P	3.18	44.0	48.0	138.0	61.0	99.5	1.27	2050	2.33
22	Grand Rapids	USA	DC	7.6	0.22	ME	O	0.255	7.5	7.5	12.2	9.1	10.7	—	—	0.5
23	Hart	USA	HD	—	—	HE	P	6.35	10.7	10.8	106.6	41.2	73.9	3.03	—	—
24	Hatfield	USA	—	6.8	12.3	HE	O	12.3	6.8	6.8	—	6.1	91.5	—	3400	2.0

25	Hell Hole	USA	—	67.1	30.6	ME	P	30.6	35.1	56.4	175.1	66.9	121.0	0.96	7360	—
26	Horse Creek	USA	FD	12.2	21.0	ME	P	12.8	7.01	12.8	76.2	70.0	73.1	0.83	3890	3.0
27	Hougou	China	HD	8.0	0.24	ME	O	0.24	8.0	8.0	—	—	20.0	—	—	—
28	Huoshishan	China	HD	13.0	0.22	HE	O	0.22	16.0	16.0	45.0	15.0	30.0	0.94	—	—
29	Huqitang	China	HD	9.9	0.734	LE	P	0.424	5.1	9.0	12.0	3.0	7.5	0.5	50	4.0
30	Jiahezi	China	HD	18.0	80.0	HE	P	42.0	12.0	18.0	181.0	—	—	—	—	—
31	Johnstown	USA	ZD	38.1	18.9	ME	O	18.9	24.6	24.4	128.0	61.0	94.5	1.38	8500	—
32	Kelly Barnes	USA	HD	11.6	0.505	HE	O	0.777	11.3	12.8	35.0	18.0	27.3	0.85	680	0.5
33	Kodaganar	India	HD	11.5	12.3	ME	O	12.3	11.5	11.5	—	—	—	—	1280	—
34	Lake Frances	USA	HD	15.2	0.865	ME	P	0.789	14.0	17.1	30.0	10.4	18.9	0.65	—	1.0
35	Lake Latonka	USA	HD	13.0	4.59	ME	P	4.09	6.25	8.69	49.5	28.9	39.2	1.18	290	3.0
36	Lijiaju	China	HD	25.0	1.14	ME	O	1.14	25.0	25.0	—	—	—	—	2950	—
37	Little Deer Creek	USA	HD	26.2	1.73	HE	P	1.36	22.9	27.1	49.9	9.3	29.6	0.75	1330	0.33
38	Liujiatai	China	DC	35.9	40.54	ME	O	40.54	35.9	35.9	—	—	—	—	28000	—
39	Longtun	China	DC	9.5	30.0	HE	O	30.0	>9.5	9.5	181.0	—	—	—	—	—
40	Lower Otay	USA	DC	41.2	49.3	ME	O	49.3	>39.6	39.6	172.0	93.8	133.0	1.0	—	—
41	Lower Two Medicine	USA	HD	11.3	19.6	HE	O	19.6	11.3	11.3	84.0	50.0	67.0	1.5	1800	—
42	Lyman	USA	ZD	19.8	49.5	HE	P	35.8	16.2	19.8	107.0	87.0	97.0	1.0	—	—
43	Lynde Brook	USA	DC	12.5	2.52	ME	P	2.88	11.6	12.5	45.7	15.3	30.5	1.22	—	—
44	Mahe	China	HD	19.5	23.4	HE	O	23.4	>19.5	19.5	—	—	—	—	4950	—
45	Mammoth	USA	DC	21.3	13.6	ME	O	13.6	21.3	21.3	—	—	—	—	2520	3.0
46	Martin Cooling Pond Dike	USA	—	10.4	136.0	HE	P	136.0	8.53	12.8	—	—	186.0	—	3115	—
47	Niujiaoyu	China	DC	10.0	0.16	LE	P	0.144	7.2	7.2	20.0	6.0	13.0	0.93	—	3.0
48	Oros	Brazil	ZD	35.4	650.0	LE	O	660.0	35.8	35.5	200.0	130.0	165.0	1.0	9630	—

(Continued)

Table 2.1 (*Continued*)

No.	Dam name	Location	Dam type	Dam height H_d (m)	Reservoir capacity V_d ($\times10^6$ m^3)	Dam erodibility	Failure mode	Volume of water above breach invert V_w ($\times10^6$ m^3)	Depth of water above breach invert H_w (m)	Breach geometry					Peak outflow rate Q_p (m^3/s)	Failure time T_f (hour)
										Height H_b (m)	Top width B_t (m)	Bottom width B_b (m)	Average width B_{ave} (m)	Side slope Z Z:1(h:v)		
49	Otter Lake	USA	HD	6.1	0.15	ME	P	0.109	5.0	6.1	17.1	1.5	9.3	1.28	—	—
50	Potato Hill Lake	USA	HD	—	—	ME	O	0.105	>7.77	7.77	26.2	6.8	16.5	1.25	—	—
51	Prospect	USA	HD	—	—	HE	P	3.54	1.68	4.42	91.4	85.4	88.4	0.69	116	—
52	Qielinggou	China	HD	18.0	0.7	HE	O	0.7	18.0	18.0	—	—	—	—	2000	0.17
53	Quail Creek	USA	HD	24.0	50.0	ME	P	30.8	16.7	21.3	72.1	67.9	70.0	0.1	3110	—
54	Rito Manzanares	USA	HD	7.3	0.0247	HE	P	0.0247	4.57	7.32	19.0	7.6	13.3	0.77	—	—
55	Schaeffer Reservoir	USA	DC	30.5	3.92	HE	O	4.44	>30.5	30.5	210.0	64.0	137.0	2.25	4500	0.5
56	Shangliuzhuang	China	HD	14.0	0.11	ME	O	0.11	14.0	14.0	30.0	—	—	—	—	—
57	Shanhu	China	HD	11.5	2.15	HE	P	1.78	12.5	13.0	58.0	24.0	41.0	1.31	—	—
58	Sheep Creek	USA	HD	17.1	1.43	ME	P	0.91	14.02	17.1	30.5	13.5	22.0	0.5	—	—
59	Shilongshan	China	HD	14.0	2.06	ME	O	2.06	14.0	14.0	50.0	—	—	—	—	—
60	Shimantan	China	HD	25.0	94.4	HE	O	117.0	27.4	25.8	446.0	288.0	367.0	0.32	30000	5.5
61	Spring Lake	USA	HD	5.5	0.135	HE	P	0.136	5.49	5.49	20.0	9.0	14.5	1.0	—	—
62	Statham Lake	USA	HD	5.5	—	ME	O	0.564	5.55	5.12	23.8	18.2	21.0	0.54	—	—
63	Swift	USA	FD	57.6	37.0	ME	O	37.0	47.85	57.6	225.0	225.0	225.0	0.0	24947	—
64	Teton	USA	ZD	93.0	356.0	ME	P	310.0	77.4	86.9	237.9	64.1	151.0	1.0	65120	4.0
65	Tiemusi	China	HD	12.0	0.11	HE	O	0.11	12.0	12.0	60.0	—	—	—	—	—
66	Tongshuyuan	China	HD	13.0	0.4	ME	O	0.4	>10.0	10.0	30.0	—	—	—	—	—
67	Trial Lake	USA	—	—	—	ME	P	1.48	5.18	5.18	25.2	16.8	21.0	0.82	—	—
68	Upper Pond	USA	—	5.2	0.22	ME	O	0.222	5.18	5.18	25.4	7.6	16.5	1.71	—	—
69	Wanshangang	China	HD	13.0	1.5	ME	O	1.5	>12.0	12.0	50.0	30.0	40.0	0.83	—	—
70	Wilkinson Lake	USA	DC	3.2	—	HE	P	0.533	3.57	3.72	35.5	22.5	29.0	1.74	—	—
71	Winston	USA	DC	7.3	0.664	LE	O	0.662	6.4	6.1	21.3	18.3	19.8	0.2	—	5.0
72	Yuanmen	China	HD	19.2	6.4	HE	O	6.4	19.2	19.2	—	—	—	—	—	0.5
73	Zhonghuaju	China	HD	16.0	0.14	HE	O	0.14	16.0	16.0	—	—	—	—	—	0.4
74	Zhugou	China	DC	23.5	15.4	HE	O	18.43	23.5	23.5	159.0	110.0	135.0	0.98	11200	0.43
75	Zuocun	China	DC	35.0	40.0	HE	O	40.0	35.0	35.0	—	—	—	—	23600	1.0

Table 2.2 Reservoir capacities of the failed embankment dams

Reservoir capacity ($\times 10^6\,m^3$)	Number of cases	Percentage (%)
≥1000	7	0.7
1000–100	20	1.9
100–10	71	6.7
10–1	59	5.5
<1	112	10.5
Unknown	796	74.7
Sum	1065	100.0

Table 2.3 Heights of the failed embankment dams

Dam height (m)	Number of cases	Percentage (%)
≥100	3	0.3
100–60	10	0.9
60–30	53	5.0
30–15	186	17.5
<15	659	61.9
Unknown	154	14.5
Sum	1065	100.0

Table 2.4 Construction times of the failed embankment dams

Construction year	Number of cases	Percentage (%)
Before 1800	8	0.8
1800–1849	10	0.9
1850–1859	10	0.9
1860–1869	15	1.4
1870–1879	7	0.7
1880–1889	22	2.1
1890–1899	32	3.0
1900–1909	43	4.0
1910–1919	55	5.2
1920–1929	41	3.8
1930–1939	35	3.3
1940–1949	23	2.2
1950–1959	42	3.9
1960–1969	59	5.5
1970–1979	37	3.5
1980–1989	9	0.8
After 1990	2	0.2
Unknown	615	57.7
Sum	1065	100.0

Table 2.5 Ages of the embankment dams at failure

Dam age	Number of cases	Percentage (%)
Under construction	32	3.0
0–1	89	8.4
1–5	80	7.5
5–10	38	3.6
10–15	38	3.6
15–20	26	2.4
20–30	36	3.4
30–40	21	2.0
40–60	36	3.4
60–80	22	2.1
80–100	10	0.9
100–150	10	0.9
>150	4	0.4
Unknown	623	58.5
Sum	1065	100.0

2.2.1 *Overtopping*

Overtopping occurs as a result of insufficient spillway capacity or an extreme flood exceeding design criteria. The water flow over the embankment from overtopping introduces tractive shear stress on the downstream surface. The erosion process begins at a weak spot where the tractive shear stress exceeds a critical resistance that keeps the material in place. Then the process will proceed under the action of flowing water while the eroded material is transported downstream. The initial breach is often V-shaped and becomes larger as the erosion progresses, and finally has an inverse trapezoidal shape. The extent of breaching depends on the duration of overtopping and the properties and structure of the embankment dam itself. The mechanics of overtopping erosion will be introduced in Chapter 7 in detail.

The erosion characteristics are different for granular and cohesive embankments. For granular embankments, surface slips take place quickly due to the overtopped water flow on the downstream slope; hence granular materials are removed rapidly layer by layer. Towards the end of the erosion process, the breach tends to flatten, depending upon the material and the longitudinal slope of the foundation surface. Figure 2.1 shows the breaching process of a granular embankment dam by overtopping. According to the definitions by Wahl (1998), the breach initiation phase involves steps 1–2, while the breach development phase involves steps 3–7 in Figure 2.1. In the breach initiation phase, the breach develops from the downstream to the upstream face; in the breach development phase, the breach widens along the longitudinal direction, hence the breach flow section increases rapidly.

For cohesive embankments, erosion often begins at the embankment toe and advances upstream, undercutting the slope and in turn causing the removal of large chunks of materials due to tensile or shear failure of the soil on the over-steepened slope. In some cases, a series of stair-step headcuts form on the downstream face of the dam. Figure 2.2 shows the breaching process of cohesive embankments by overtopping, of which the breach initiation phase involves steps 1–4, while the breach development phase involves steps 5–7.

If a granular embankment contains a vertical cohesive core, the core may be eroded in a manner similar to that for a cohesive embankment. If the core is inclined with structural support from the downstream shell, structural failure of the core can occur as the downstream shell erodes.

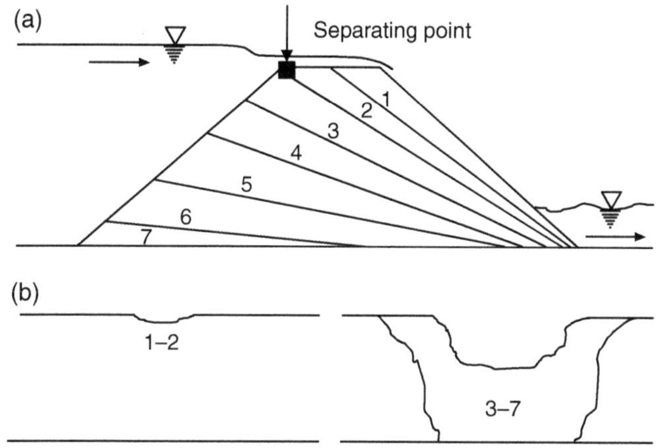

Figure 2.1 Breaching process of a granular embankment dam by overtopping: (a) side view, (b) downstream view (breach initiation phase, steps 1–2; breach development phase, steps 3–7)

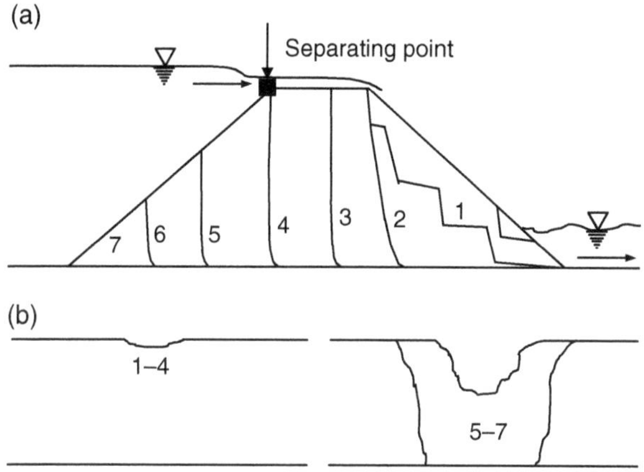

Figure 2.2 Breaching process of a cohesive embankment dam by overtopping: (a) side view, (b) downstream view (breach initiation phase, steps 1–4; breach development phase, steps 5–7)

2.2.2 *Internal Erosion*

Another common failure mode of embankment dams is internal erosion. This occurs when soil particles are carried away locally, usually in suspension, by the hydrodynamic forces of the flowing water in the embankment or the foundation, from the embankment to the foundation, or around and into conduits through the embankment. Internal erosion can occur in both granular and cohesive embankments. The mechanics of internal erosion will be introduced in Chapter 6 in detail.

Similar to overtopping, the breaching characteristics by internal erosion are also different for granular and cohesive embankment dams. At the original stage of internal erosion, soil particles are often removed slowly by the seeping water. Then a pipe is formed in the downstream slope after a significant amount of material has been carried away. Once the pipe is formed, the erosion process becomes rapid and the pipe advances upstream. For granular embankment dams, the portion of the materials above the pipe becomes unstable and collapses, which is followed by a

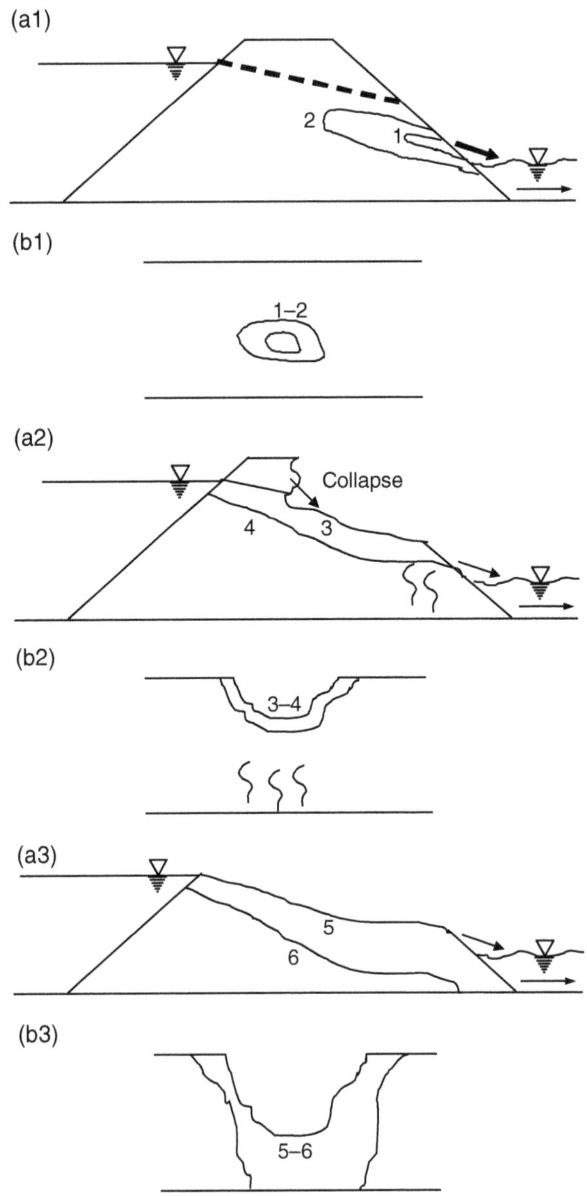

Figure 2.3 Breaching process of a granular embankment dam by internal erosion: (a) side view, and (b) downstream view (breach initiation phase, steps 1–4; breach development phase, steps 5 and 6)

similar subsequent process as in the case of overtopping. Figure 2.3 shows the whole breaching process of a granular embankment dam by internal erosion, of which the breach initiation phase involves steps 1–4, while the breach development phase involves steps 5 and 6. For cohesive embankments, the portion of the materials above the pipe may keep stable; hence the initial pipe in the downstream slope continues to advance till the upstream slope face. A penetrated pipe is finally formed through the embankment. The shape and dimensions of the pipe can vary greatly in

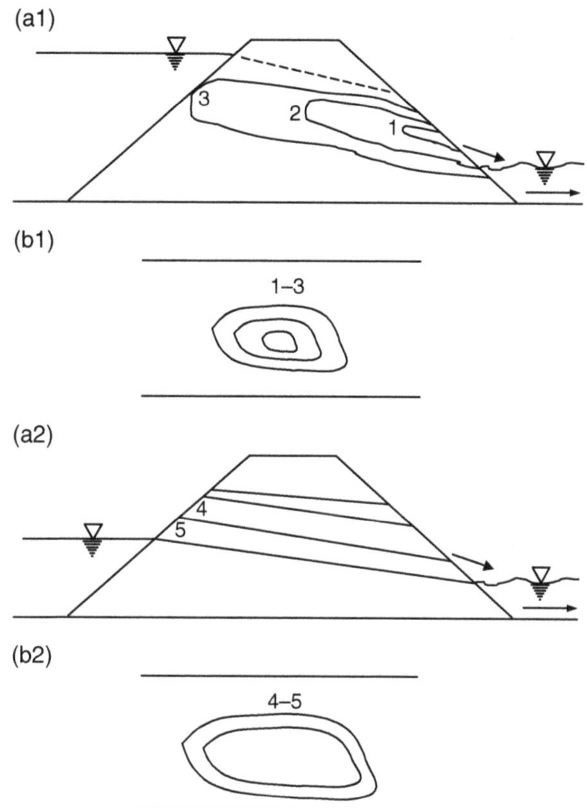

Figure 2.4 Breaching process of a cohesive embankment dam by internal erosion: (a) side view, and (b) downstream view (breach initiation phase, steps 1–3; breach development phase, steps 4 and 5)

specific cases. Figure 2.4 shows the whole breaching process of a cohesive embankment dam by internal erosion, of which the breach initiation phase involves steps 1–3, while the breach development phase involves steps 4 and 5.

2.3 Common Causes of Embankment Dam Failures

This section follows a classification of failure causes similar to that proposed by the Ministry of Water Resources of the People's Republic of China, as shown in Table 2.6 (MWR, 1993). In Table 2.6, internal erosion is a general term that describes internal failure mechanisms related to seepage. Figure 2.5 shows the percentages of causes of embankment dam failures. Most of the cases are caused by either overtopping or quality problems. These two causes led to over 80% of failures. Figure 2.6 shows specific percentages of quality problems leading to dam failures, indicating that 64% of quality problems are associated with internal erosion. Overall, the most common causes of embankment dam failures are overtopping and internal erosion. The principal influencing factor on overtopping is insufficient spillway capacity. The single most adverse factor for internal erosion is cracks, which can be caused by differential settlement, shrinkage, interfaces with abutment or embedded structures, or hydraulic fracturing.

Table 2.6 Categories of dam failure causes involved in the dam failure database in Appendix A (based on MWR, 1993)

No.	Cause
1	Overtopping 1. Insufficient spillway capacity 2. Extreme flood exceeding design criteria
2	Quality problems 1. Internal erosion in dam 2. Sliding of dam 3. Internal erosion in foundation 4. Internal erosion around spillway 5. Quality issues in spillway 6. Internal erosion around culverts and other embedded structures 7. Quality issues with culverts and other embedded structures
3	Poor management 1. Loss of reservoir capacity for flood control due to over storage prior to flood season 2. Poor maintenance and operation 3. Temporary heightening of spillway crest not removed in time 4. Organization issue: unclear responsibility for dam management
4	Disasters 1. Earthquakes 2. Wars and terrorist attacks 3. Breaching of upstream dam 4. Reservoir landslides 5. Rodent den
5	Others 1. Spillway blockage due to bank slide in reservoir 2. Excavation on dam for discharging 3. Poor design options 4. Poor planning of project layout
6	Unknown

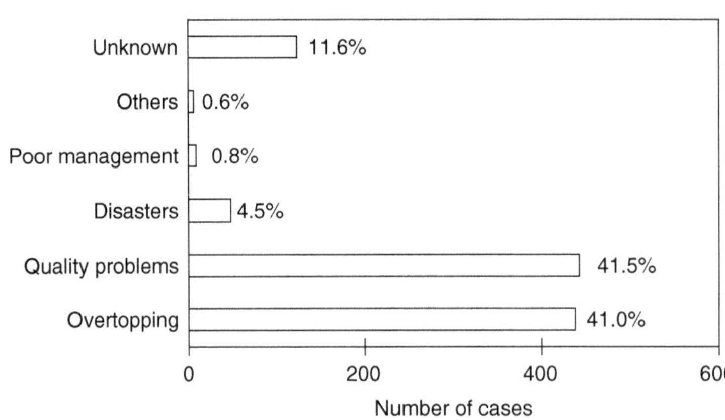

Figure 2.5 Percentages of causes for embankment dam failures. *Source:* Zhang et al. (2009). With permission from Taylor & Francis

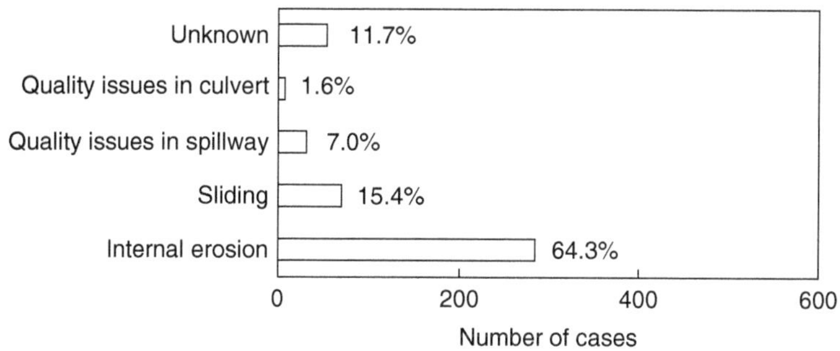

Figure 2.6 Percentages of quality problems for earth dam failures. *Source:* Zhang et al. (2009). With permission from Taylor & Francis

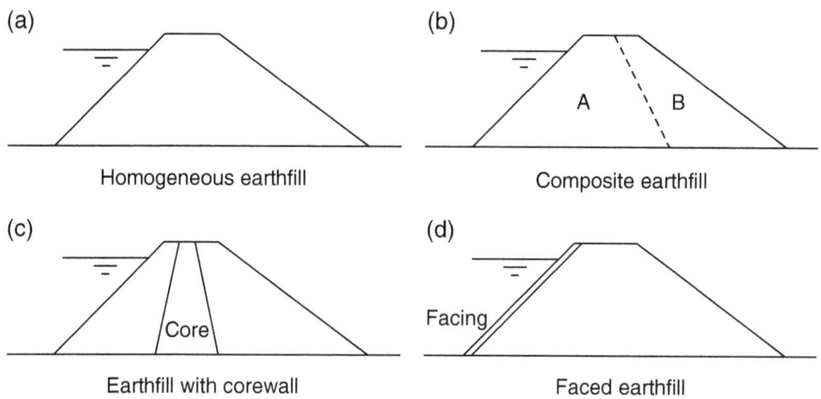

Figure 2.7 Sketches of four typical types of embankment dams

2.4 Failure of Different Types of Embankment Dams

According to dam zoning, embankment dams can be classified into four categories as sketched in Figure 2.7: (1) homogeneous earthfill dams, (2) composite earthfill dams, (3) earthfill dams with corewalls, and (4) concrete faced earthfill dams. Homogeneous earthfill dams are composed almost of the same material throughout the cross-section. Composite earthfill dams are composed of several materials, each for one zone (e.g. zones A and B consist of two different types of soil in Figure 2.7b). An earthfill dam with a corewall contains earth fills and a low-permeability wall (e.g. clay or concrete), which is often built vertically or inclined towards the upstream of the dam. A concrete faced earthfill dam is composed of earth fills and an impermeable concrete facing on the upstream slope. Table 2.7 shows the subdivision of the failed embankment dams in terms of the dam types in Figure 2.7. Note that the information available only allows the identification of the specific types of 129 embankment dams.

Table 2.8 summarizes the failure causes for the four types of embankment dams based on 129 cases. For the homogeneous earthfill dams and the composite earthfill dams, overtopping and internal erosion are two dominant failure causes. In contrast, for the earthfill dams with corewalls, failures are mostly attributed to overtopping; internal erosion appears to be less common. It is evident from the comparison that the corewall plays a key role in preventing internal erosion in the dam. In practice, one or two layers of filters are usually constructed on the downstream side of the

Table 2.7 Subdivision of failed embankment dams in terms of dam type

Dam type	Number of cases	Percentage (%)
Homogeneous earthfill	61	47.3
Composite earthfill	45	34.9
Earthfill with corewall	21	16.3
Concrete-faced earthfill	2	1.5
Sum	129	100.0

Table 2.8 Summary of failure causes for four types of embankment dams

Dam type	Overtopping	Piping	Other quality problems	Poor management	Disasters	Sum
Homogeneous earthfill dams	23	26	5	1	3	58
Composite earthfill dams	18	17	6	—	1	42
Earthfill dams with corewall	13	6	1	—	1	21
Concrete faced rockfill dams	—	2	—	—	—	2

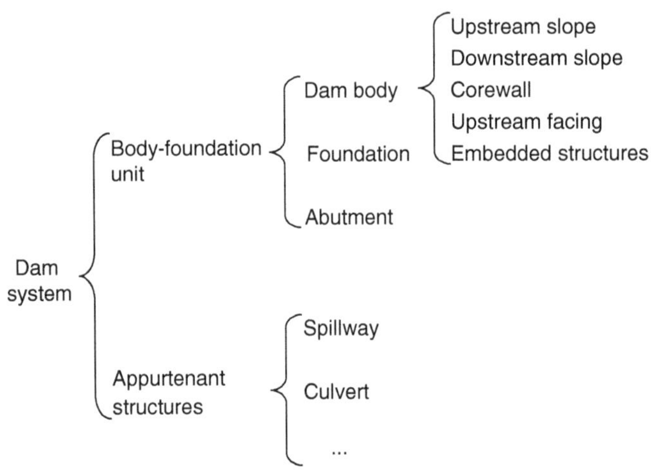

Figure 2.8 Potential locations at risk in a dam system. *Source:* Zhang et al. (2009). With permission from Taylor & Francis

corewall, which help prevent internal erosion failure in the core. Owing to insufficient data in the database (only two cases), the failure causes for concrete faced rockfill dams are not conclusive.

In order to better characterize the failure modes and causes, a dam system can be divided into several parts, termed 'potential locations at risk', as shown in Figure 2.8. Note that 'at risk' here means the possibility of a failure mode associated with a specific potential location. For instance, overtopping may be due to insufficient capacity of the spillway in a dam system. The two primary failure causes, overtopping and internal erosion, can be discussed in relation to these potential locations at risk. Homogeneous and composite earthfill dams are grouped together owing to their similar failure causes, as indicated in Table 2.8.

2.4.1 Analysis of Homogeneous and Composite Earthfill Dams

Dam failures due to overtopping often occur in the flood season. As presented before, insufficient spillway capacity is a major reason for overtopping. Therefore, the most important potential location at risk is the spillway. The second potential location at risk is the downstream slope. The erodibility of the downstream slope material is one of the controlling factors for the erosion process over time, and hence determines whether a dam eventually fails or not. The foundation may be another potential location at risk. This is because the settlement of a dam is often attributed to a weak foundation, and the foundation settlement will cause the settlement of the dam crest and reduce the freeboard of the dam. In summary, spillway, foundation, and downstream slope are believed to be potential locations at risk in terms of overtopping failure. Given these potential weaknesses of a dam, however, whether overtopping occurs or not depends on the loading (i.e. floods).

For dam failures owing to internal erosion, the impact of inflow floods does not seem to be extremely significant, although floods do increase the possibility of internal erosion by imposing larger hydraulic gradients. Whether internal erosion occurs or not primarily depends on the dam system itself, including the configuration, filter design, construction quality of the dam, as well as geological conditions and connections of the dam with the abutment slopes or any embedded structures.

Compared to overtopping failures, more potential locations at risk are found for internal erosion failures, which can be any part of the dam system, as shown in Figure 2.8. For instance, a faulty foundation can lead to internal erosion failure in several ways. In one scenario, if the abutment bedrock is too steep or contains weak planes not adequately treated, an adverse seepage through the foundation can develop. In a second scenario, soft soils in the foundation can lead to differential settlement cracks in the dam, resulting in internal erosion through the dam. In a third scenario, weak seams left at the interface between the dam and the foundation, or between segregated fill layers, can result in contact seepage along the interface (e.g. Zhang and Chen, 2006). Any of the above scenarios may be accompanied by other scenarios. Therefore, it is rather difficult to detect the internal erosion paths. However, a dam is more likely to survive the internal erosion condition if timely remediation actions are taken.

2.4.2 Analysis of Earthfill Dams with Corewalls

Failures of earthfill dams with corewalls due to overtopping are similar to those for homogeneous/composite earthfill dams (Section 2.4.1), except at the corewall. The corewall often consists of less erodible materials, such as clays or concrete. The erosion process of the corewall zone is relatively slow, allowing more time for warning and taking actions to limit the extent of dam failure.

For dam failures caused by internal erosion, appreciable differences exist between homogeneous or composite earthfill dams and earthfill dams with corewalls. The corewall controls the seepage through the dam body; the internal erosion paths are more likely to be associated with the foundation, the abutment, or their interfaces with the dam. This does not mean that the dam body is absolutely free from internal erosion. Instead, hydraulic fracturing is a common phenomenon that creates preferential flow paths in the clay corewall (e.g. Zhang and Du, 1997). As a result, the corewall clays erode as water flows along the hydraulic fractures. It may be concluded that foundation, abutment, or their interfaces with the dam body are notable potential locations at risk for earthfill dams with corewalls, while internal erosion may develop through the dam, owing to either hydraulic fracturing of the corewall or poor contact with any embedded structures.

Table 2.9 Summary of potential locations at risk for typical types of embankment dams

Dam type	Overtopping		Internal erosion	
	High risk	Low risk	High risk	Low risk
Homogeneous/composite earthfill dams	Spillway; downstream slope	Foundation	Dam body; foundation; abutment	—
Earthfill dams with corewall	Spillway; downstream slope	Foundation; corewall	Foundation; abutment	Dam body

The locations at risk in homogeneous/composite earthfill dams with or without corewalls are summarized in Table 2.9. These potential locations at risk are ranked in two levels: 'high risk' and 'low risk'. The classification is a little arbitrary, being simply based on frequency of occurrence in the dam failure database.

3

Statistical Analysis of Failures of Landslide Dams

3.1 Database of Failures of Landslide Dams

Several databases of landslide dam failures have been reported previously, as summarized in Table 3.1. However, the breaching parameters can be found only in a dozen cases, which hinders the investigation of landslide dam failures. Recently, the authors have compiled a large database of 1267 landslide dam failures, of which 257 cases were formed during the Wenchuan earthquake in 2008 in China. The detailed information on 1044 cases in the new database is presented in Appendix B, including 34 cases of large landslide dams due to the Wenchuan earthquake. The information of the locations of the 257 landslide dam cases caused by the Wenchuan earthquake is reported by Cui et al. (2009). The details of 52 landslide dam failures with sufficient information of breaching parameters are presented in Table 3.2. Based on the information in Appendix B and Table 3.2, the formation, triggers, types of landslides, dam heights, lake volumes, longevity, and failure modes of these landslide dams are studied in this section.

3.1.1 Locations of Landslide Dams

Landslide dams form most frequently where narrow steep valleys are bordered by high rugged mountains (Costa and Schuster, 1988). Areas rich of landslide source materials such as sheared and fractured or hydrothermally altered bedrock, like south-west China, west of North America, Japan, Italy and New Zealand, are prone to the formation of landslide dams (Costa and Schuster, 1988, 1991; Korup, 2002). As shown in Figure 3.1, out of the 1267 landslide dam cases in the database, 58.4% (740 cases) occurred in China, 7.9% (100 cases) in Japan, 7.2% (91 cases) in USA, 6.1% (77 cases) in Italy, 3.3% (42 cases) in Canada, 2.4% (30 cases) in New Zealand, 1.8% (23 cases) in Switzerland and 17% (164 cases) in other places.

Dam Failure Mechanisms and Risk Assessment, First Edition. Limin Zhang, Ming Peng, Dongsheng Chang, and Yao Xu.

Table 3.1 Review of datasets of landslide dams

ID	Number of cases	Case location	Reference
1	1267	Throughout the world	Chapter 3 in this book; Peng and Zhang (2012a)
2	463	Throughout the world	Costa and Schuster (1991)
3	38	Canadian Cordillera	Clague and Evans (1994)
4	147	China	Chai et al. (1998)
5	68	Northern Apennines	Casagli and Ermini (1999)
6	>130	New Zealand	Korup (2002)
7	84	Throughout the world	Ermini and Casagli (2003)
8	388 (dataset 3 included)	China	Yan (2006)
9	257	Areas affected by Wenchuan Earthquake, China	Cui et al. (2009)
10	32	Areas affected by Wenchuan Earthquake, China	Xu et al. (2009)

3.1.2 Formation Times of Landslide Dams

Landslide dams are common events frequently recorded in historical literature. The database in Appendix B registers cases formed from the Quaternary Period to 2010, as shown in Figure 3.2. Most of the cases were formed in the recent decades, with 13.2% (167 cases) in 1900–1949, 35.7% (452 cases) in 1950–1999, and 23.7% (300 cases, including the 257 cases caused by the Wenchuan earthquake in 2008 in China) in 2000–2010. Besides, some ancient cases still exist, including 60 cases that were formed before 0 BCE. The formation times of those cases are obtained approximately by C^{14} identification. For example, a rock slide with a volume of $14\times10^6\,m^3$ blocked an unnamed tributary (watershed: $6.4\,km^2$) of the Fiume Irminio in south-eastern Sicily, Italy, 5000 years ago (Nicoletti and Parise, 2001). This landslide dam, with a maximum thickness of 60 m, still survives and has not experienced significant overtopping.

3.1.3 Triggers of Landslide Dams

As shown in Figure 3.3, the formation of most landslide dams (91.6%) was trigged by earthquakes and rainfall. Earthquake is the most important cause, inducing 48.2% (525 cases) of the landslide dams, among which 257 cases were caused by the 2008 Wenchuan earthquake. Rainfall is the second most likely cause, which accounts for 43.4% (471 cases) of the landslide dams. For example, in 1889, a heavy storm produced 53 landslide dams in the Totsu River basin in Japan with a water watershed of $1100\,km^2$ (Swanson et al., 1986).

3.1.4 Types of Landslide

Most of the landslide dams were caused by slides, slumps, flows and avalanches. As shown in Figure 3.4, out of the 621 landslide dam cases with known types of landslides, 258 (41.5%) are caused by rock and soil slides or slumps; 176 (28.3%) by mud, debris, or earth flows; 162 (26.1%) by rock and debris avalanches; 14 (2.3%) by rock and debris falls; 11 (1.8%) by other reasons such as lateral spreads. More than one half (646 cases) of the landslide dams are recorded without information of landslide type.

Table 3.2 Information of 52 landslide dam failures

No.	Name	Location	Data of formation	Dam height H_d (m)	Dam width W_d (m)	Dam volume V_d ($10^6 \times m^3$)	Lake volume V_l ($10^6 \times m^3$)	Dam erodibility	Breach geometry			Peak outflow rate Q_p (m^3/s)	Breach time T_b (hour)	References
									Breach depth H_b (m)	Top width W_t (m)	Bottom width W_b (m)			
1	Arida River, Hanazono Village	Japan	July 18, 1953	10	150	0.18	0.047	High	—	—	—	890[3]	—	Costa and Schuster, 1991; Tung, 2008
2	Arida River, Hanazono Village	Japan	July 20, 1953	60	500	2.6	17	Low	—	—	—	750[3]	—	Costa and Schuster, 1991; Tung, 2008
3	Bairaman River	Papua New Guinea	May 11, 1985	200	3000	200	50	High	70	—	—	8000	3	King et al., 1989
4	Bireh-Ganga River	India	Sep 22, 1983	274	2750	286	460	High	97.5	—	—	56650	—	Costa, 1985; Walder and O'Connor, 1997
5	Buonamico River	Italy	Jan 3–4, 1973	90	700	21	7.5	High	50	—	—	—	—	Costa and Schuster, 1991
6	Cedar Creek	USA	1966	3	150	1.7	0.053	High	2	10	—	—	—	Costa and Schuster, 1991
7	Donghekou	China	May 12, 2008	20	750	12	6	Medium	10	25	15	800–1000	—	Chang et al., 2009
8	East Fork Hood River	USA	Dec 25, 1980	10.7	225	0.07–0.1	0.105	Medium	—	—	—	850	—	Costa, 1985;Costa and Schuster, 1991
9	Hime River	Japan	Aug 9, 1911	60	500	1.9	16	Medium	—	—	—	1800[3]	—	Costa and Schuster, 1991; Tung, 2008
10	Hongshi River	China	May 12, 2008	50	500	18	4	Medium	10	—	8–10	400–600	—	Chang et al., 2009
11	Iketsu River	Japan	Aug 19, 1889	140	180	3.4	26	Medium	—	—	—	480[3]	—	Costa and Schuster, 1991; Tung, 2008

(Continued)

Table 3.2 (*Continued*)

No.	Name	Location	Data of formation	Dam height H_d (m)	Dam width W_d (m)	Dam volume V_d ($10^6 \times m^3$)	Lake volume V_l ($10^6 \times m^3$)	Dam erodibility	Breach geometry: Breach depth H_b (m)	Breach geometry: Top width W_t (m)	Breach geometry: Bottom width W_b (m)	Peak outflow rate Q_p (m^3/s)	Breach time T_b (hour)	References
12	Jackson Creek Lake	USA	May 18, 1980	4.5	180–425	0.77[1]	2.47	Low	—	—	—	477	—	Costa and Schuster, 1991
13	Kaminirau River	Japan	July 26, 1788	50	500	2	2.2	Medium	—	—	—	440[3]	—	Costa and Schuster, 1991; Tung, 2008
14	Kano River, Kitatotsugawa Village	Japan	Aug 20, 1889	15	130	0.094	1.3	Medium	—	—	—	1600[3]	—	Costa and Schuster, 1991; Tung, 2008
15	Kano River, Totsugawa Village	Japan	Aug 20, 1889	20	120	0.1	0.6	Medium	—	—	—	1300[3]	—	Costa and Schuster, 1991; Tung, 2008
16	Kawarabitsu River	Japan	Aug 21, 1889	80	750	13	40	Medium	—	—	—	2000[3]	—	Costa and Schuster, 1991; Tong, 2008
17	La Josefina	Ecuador	Mar 29, 1993	100	1100	20	200	Medium	43	—	—	10000	6	Plaza-Nieto and Zevallos, 1993; Walder and O'Connor, 1997
18	Mantaro River[2]	Peru	Aug 16, 1945	133	580	3.5	301	High	56	—	—	35400	—	Snow, 1964
19	Mantaro River	Peru	Apr 25, 1974	175	3800	1300	670	Medium	107	243	30	10000	12	Lee and Duncan, 1975
20	Mt Adams	New Zealand	Oct 6, 1999	80–100	700	10–15	5–7	High	40–50	100–150	30–60	2000 – 3000	5.5	Becker et al., 2007; Hancox et al., 1999
21	Naka River, Kaminaka Town	Japan	July 25, 1892	80	330	3.3	75	Medium	—	—	—	5600[3]	—	Costa and Schuster, 1991; Tung, 2008
22	Nishi River, Totsugawa Village	Japan	Aug 20, 1889	20	200	0.6	1.3	Medium	—	—	—	980[3]	—	Costa and Schuster, 1991; Tung, 2008

23	Nishi River, Totsugawa Village	Japan	Aug 20, 1889	25	250	0.63	1.8	Medium	—	—	—	1200^{3}	—	Costa and Schuster, 1991; Tung, 2008
24	Nishi River, Totsugawa Village	Japan	Aug 20, 1889	20	120	0.63	0.4	High	—	—	—	1100^{3}	—	Costa and Schuster, 1991; Tung, 2008
25	Ojika River	Japan	Sep 1, 1683	70	400	3.8	64	Low	—	—	—	620^{3}	—	Costa and Schuster, 1991; Tung, 2008
26	Pisque River	Ecuador	Jan 2, 1990	58	450	1	2.5	Medium	30	50	—	700	3.5	Plaza-Nieto et al., 1990; Walder and O'Connor, 1997
27	Ram Creek	New Zealand	May 24, 1968	40	1200	2.8	1.1	Medium	30	—	30	1000	—	Nash et al., 2008
28	Rio Paute	Ecuador	Mar 29, 1993	112	800	25	210	Medium	—	—	—	8250	4–6	Canuti et al., 1993
29	Rio Toro River	Costa Rica	June 13, 1992	70–100	600	3	0.5	Medium	30–50	40–80	—	400	—	Mora et al., 1993
30	Sai River	Japan	Mar 24, 1847	65–100	1000	21	350	Low	—	—	—	3700^{3}	—	Costa and Schuster, 1991; Tung, 2008
31	Shiratani River, Totsugawa Village	Japan	Aug 21, 1889	190	500	10	38	Low	—	—	—	580^{3}	—	Costa and Schuster, 1991; Tung, 2008
32	Sho River	Japan	Nov 29, 1586	100	600	19	150	Low	—	—	—	1900^{3}	—	Costa and Schuster, 1991; Tung, 2008
33	Susobana River	Japan	Mar 24, 1847	54	250	1.2	16	Medium	—	—	—	510^{3}	—	Costa and Schuster, 1991; Tung, 2008

(Continued)

Table 3.2 (*Continued*)

No.	Name	Location	Data of formation	Dam height H_d (m)	Dam width W_d (m)	Dam volume V_d ($10^6 \times m^3$)	Lake volume V_l ($10^6 \times m^3$)	Dam erodibility	Breach geometry			Peak outflow rate Q_p (m^3/s)	Breach time T_b (hour)	References
									Breach depth H_b (m)	Top width W_t (m)	Bottom width W_b (m)			
34	Tanggudong	China	June 8, 1967	175	3000	68	680	High	—	—	55	53000	6	Li et al., 1986
35	Tangjiashan	China	May 12, 2008	82	611.8	20.37	246.6	Medium	30	145–235	80–100	6500	14	Huang et al., 2008; Hu et al., 2009
36	Tegermach River	USS.R	1853	120	60	20	6.6	High	—	280–340	50–60	4960	—	Glazyrin and Reyzvikh, 1968; Costa, 1985
37	Totsu River, Amakawa Village	Japan	Aug 20, 1889	80	350	2.5	17	Medium	—	—	—	2400[3]	—	Costa and Schuster, 1991; Tung, 2008
38	Totsu River, Daito Village	Japan	Aug 20, 1889	18	450	0.036	0.78	High	—	—	—	3400[3]	—	Costa and Schuster, 1991; Tung, 2008
39	Totsu River, Daito Village	Japan	Aug 20, 1889	10	380	0.23	0.93	High	—	—	—	3500[3]	—	Costa and Schuster, 1991; Tung, 2008
40	Totsu River, Kitatotsugawa Village	Japan	Aug 20, 1889	110	690	3.1	42	Medium	—	—	—	4800[3]	—	Costa and Schuster, 1991; Tung, 2008
41	Totsu River, Nakatotsugawa Village	Japan	Aug 20, 1889	7	250	0.073	0.65	High	—	—	—	6900[3]	—	Costa and Schuster, 1991; Tung, 2008
42	Tsao-Ling	Taiwan	Aug 10, 1942	217	1600	~ 191[1]	157	Medium	—	—	—	—	4.5	Costa and Schuster, 1991; Chen, 2000
43	Tsao-Ling	Taiwan	Aug 15, 1979	90	—	5	40	Medium	—	—	—	—	3	Costa and Schuster, 1991; Chen, 2000

44	Tsatichhu	Bhutan	Sep 10, 2003	110	700	5	1.5	High	—	—	—	6900	—	Dunning et al., 2006
45	Tunawaea landslide dam	New Zealand	Aug 17–18, 1991	70	550	4	0.9	Medium	—	—	—	250	—	Jennings et al., 1993; Walder and O'Connor, 1997
46	Wetlina River	Poland	July 25, 1980	6–10	60	0.15	—	Medium	3–5	10–15	—	—	—	Costa and Schuster, 1991
47	Wyoming	USA	June 23, 1925	70–75	3000	38	80	Medium	15	—	—	—	—	Costa and Schuster, 1991
48	Xiaogangjian	China	May 12, 2008	70–120	300	2	12	Medium	30	80	—	3950	—	Liu et al., 2009a
49	Xiaojiaqiao	China	May 12, 2008	57–67	200	2.42	20	Medium	37.3	131.6	8	1000	6.5	Liang, 2009a
50	Yibadao	China	May 12, 2008	25	140–180	0.15	3.79	Medium	8	—	—	—	—	Jiang et al., 2009
51	Yigong	China	Apr 9, 2000	60–100	2200–2500	300	3000	High	58.39	—	128	124000	9.25	Shang et al., 2003
52	Zepozhu, Dong River	China	Oct 27, 1965	51	650	29	2.7	Medium	—	—	—	560	4	Costa, 1985; Li et al., 1986

Notes: [1]The dam volume with a sign of "~" is estimated as one half of the product of dam width, dam length, and dam height. [2]Only the Mantaro River landslide dam failed by internal erosion in 1945; the others failed by overtopping. [3]The peak outflow rates of the cases in Japan were estimated values based on Tung (2008).

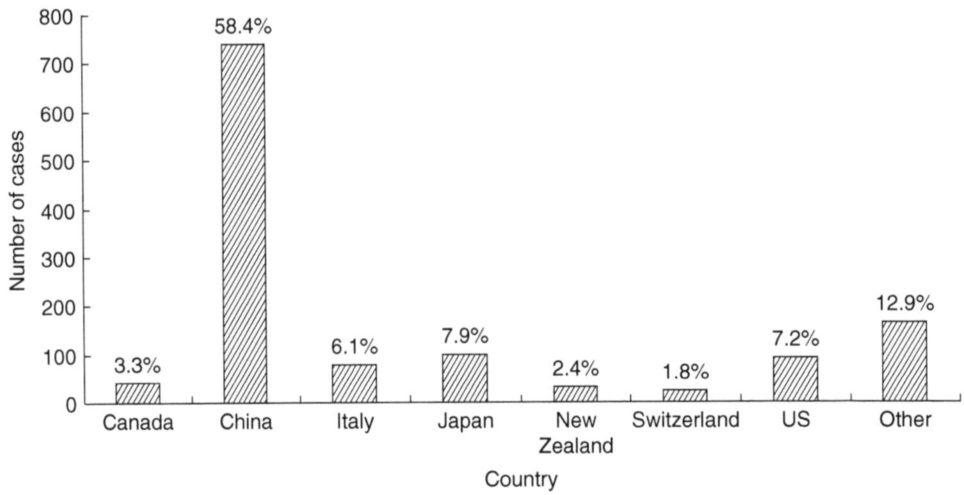

Figure 3.1 Locations of landslide dams (1267 cases)

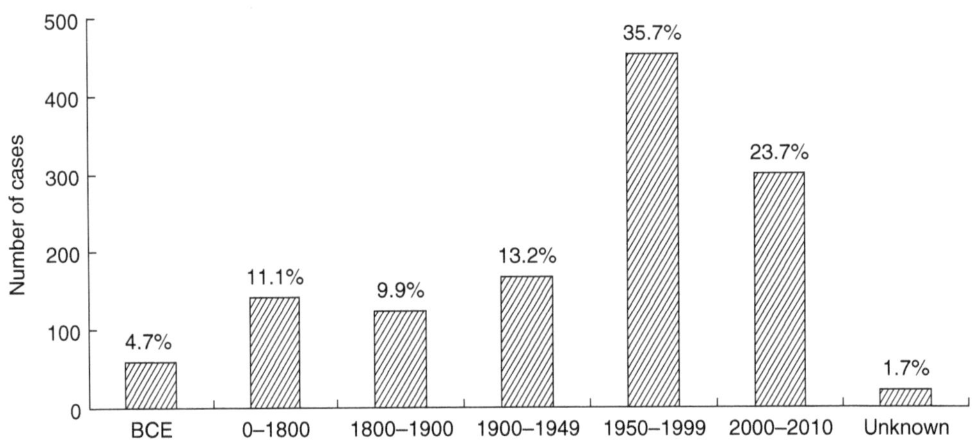

Figure 3.2 Formation time of landslide dams (1267 cases)

3.1.5 Dam Heights and Lake Volumes

Landslide dams are formed by rapid deposition of landslide, debris flow, or rockfall materials. Therefore, the dam height and lake volume are more random than constructed dams. Out of 451 cases with known information of dam height, 84 (18.6%) have heights of 0–10 m, 89 (19.7%) of 10–20 m, 71 (15.7%) of 20–30 m, 68 (15.1%) of 30–50 m, 81 (18.0%) of 50–100 m, 44 (9.8%) of 100–200 m, and 14 (3.1%) of greater than 200 m, as shown in Figure 3.5. Information of lake volume is available in 228 cases. Most of them (78.1%) have lake volumes of $0.1–100 \times 10^6 m^3$, as shown in Figure 3.6. There may be more cases with smaller lake volumes. However, detailed information of these cases is often difficult to obtain due to their short life spans, as some of these dams often existed only from

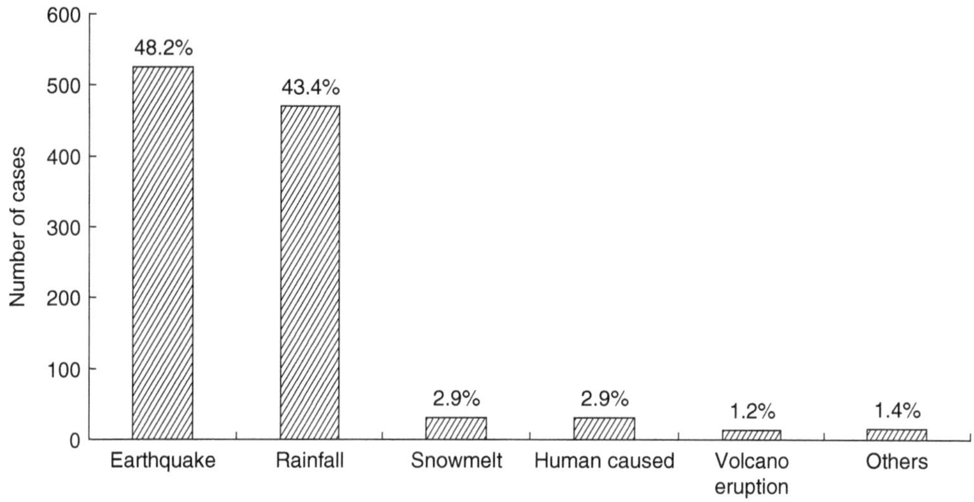

Figure 3.3 Triggers of landslide dams (1086 cases)

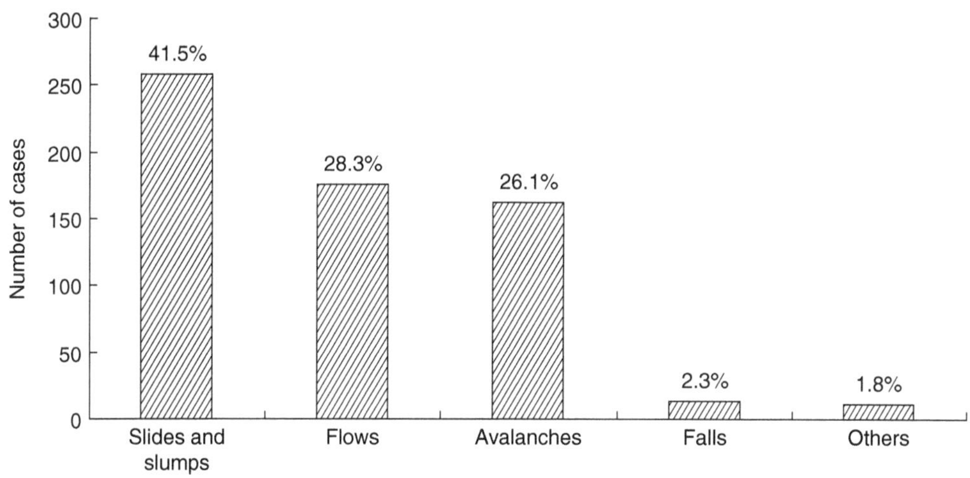

Figure 3.4 Types of landslides (621 cases)

minutes to hours. The recorded largest landslide dam in the world is Lake Sarez formed on 18 February 1911 in Tajikistan. The dam had a height of 800 m, a dam volume of $2\times10^9\,m^3$ and a lake volume of $16\times10^9\,m^3$. The rockslide dam still exists (Costa and Schuster, 1991).

3.2 Stability, Longevity, and Failure Modes of Landslide Dams

3.2.1 Stability of Landslide Dams

Of the 1267 landslide dams in the database, 101 remain stable against water erosion. However, it is inappropriate to conclude that 8% of the landslide dams in history are stable, since the existing cases are often easier to record than those that failed soon after formation.

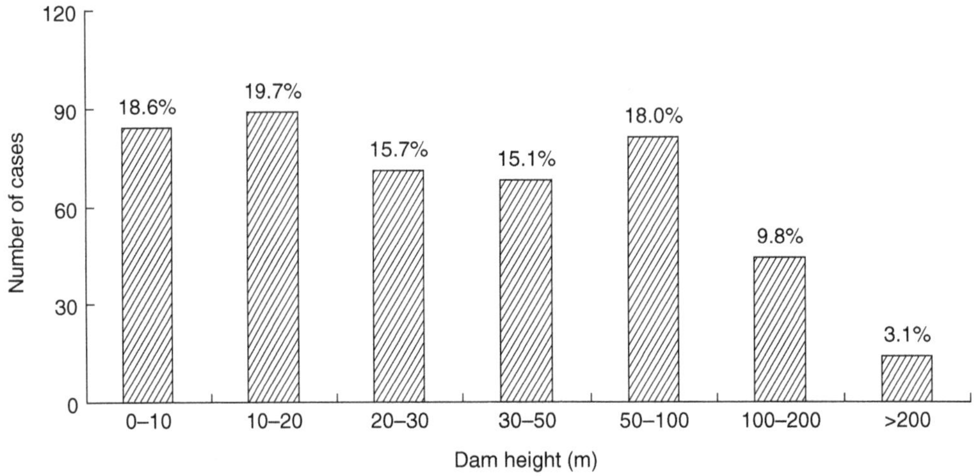

Figure 3.5 Heights of landslide dams (451 cases)

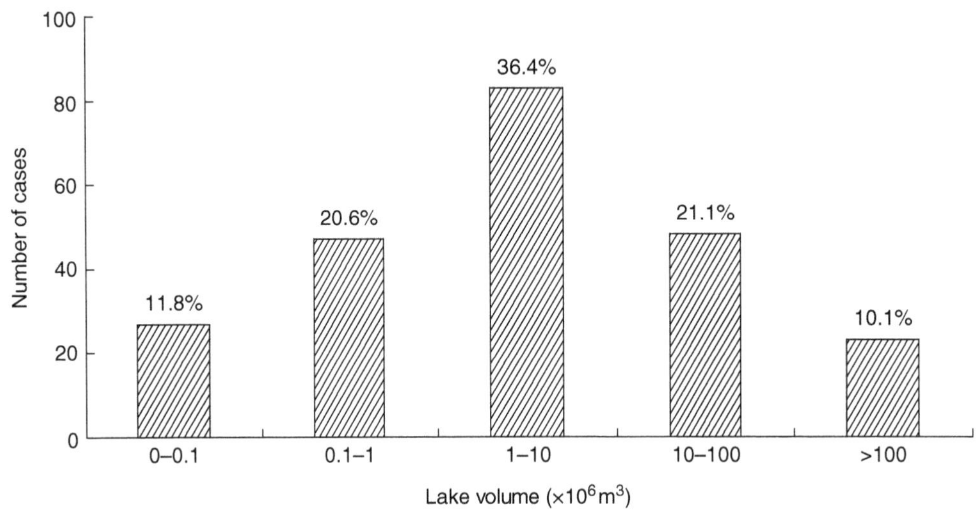

Figure 3.6 Lake volumes of landslide dams (228 cases)

The existing landslide dams may be naturally stable or protected with engineered measures. Schuster and Costa (1986) summarized the types of landslide dams that may be stable:

1. With large size. Landslide dams are typically much wider than constructed earth dams and involve larger volumes of materials. A massive blockage provides some degree of protection against failure by any mechanism. An extreme example is the debris-avalanche blockage of Sprit Lake at Mount St Helens, which extends 24 km down the lake (Schuster and Costa, 1986).
2. With erosion-resistant bedrock and well-graded materials. The most important characteristic of a dam in preventing failure is its resistance to erosion, either at the surface of the dam due to overtopping or internally due to seepage.

3. With small inflow rate. In some cases, the reservoir inflow rate may not exceed the rate of water losses from seepage through the dam and evaporation from the reservoir surface. The Bitang Lake in 1961 in Gansu Province falls into this category, having a dam height of 65–155 m and a lake volume of $4.2 \times 10^6\,m^3$ (Li et al., 1986).
4. With bedrock at the reservoir outlet. In a few cases, landslide-dammed lakes have formed natural spillways across their adjacent bedrock abutments, which prevent overtopping of the blockages. A well-known example of this type is Lake San Cristobal in south-western Colorado, which has a bedrock spillway that has existed for 700 years (Schuster, 1985).
5. Engineering measures. Some landslide dams have been stabilized by engineering measures, such as construction of channeled spillways, pipe and tunnel outlets, and reinforcement of the spillway with concrete. The details of engineering measures will be presented in Section 3.3.

Canuti et al. (1998) and Casagli and Ermini (1999) defined a blockage index (BI), and used this parameter for a preliminary forecast of landslide stability:

$$BI = \log\left(\frac{V_d}{A_b}\right) \tag{3.1}$$

where V_d is the landslide dam volume (m^3) and A_b is the catchment area (km^2). Normally, landslide dams with complete formation has $BI \geq 3$; an upper boundary for failed dams is given by $BI = 5$, while an approximate lower boundary for stable dams is given by $BI = 4$.

Ermini and Casagli (2003) utilized the geomorphic index (DBI) by combining three important variables (dam height H_d, dam volume V_d, and catchment area A_c) to evaluate the stability of a landslide dam:

$$DBI = \log\left(\frac{H_d \cdot A_c}{V_d}\right) \tag{3.2}$$

A dam with $DBI < 2.75$ will be classified as a stable dam, and one with $DBI > 3.08$ will be classified as an unstable dam.

Korup (2004) used discriminant analysis to more objectively classify stable and unstable landslide dams with log-transformed geomorphic variables. Dong et al. (2009) utilized discriminant analysis, based on a Japanese dataset consisting of 43 well-documented landslide dams, to determine the significant variables, including log-transformed peak flow (or catchment area), and log-transformed dam height, width, and length in hierarchical order.

The ability to evaluate the stability of a landslide dam has increased significantly in the past decade. Rational analysis based on mechanics can be routinely conducted. Detailed breaching processes and analyses of the failures caused by internal erosion and overtopping will be introduced in Chapters 6, 7, and 8.

3.2.2 Longevity of Landslide Dams

The destiny of landslide dams differs from case to case. Some fail soon after their formation, like the Hsiaolin landslide dam triggered by an intense rainfall in southern Taiwan in 2009 (Dong et al., 2011). The landslide dam, with a height of 44–68 m and a lake volume of $9.9 \times 10^6\,m^3$, only existed for around an hour, and was overtopped in 8.4 minutes. Some cases can last for a long time, such

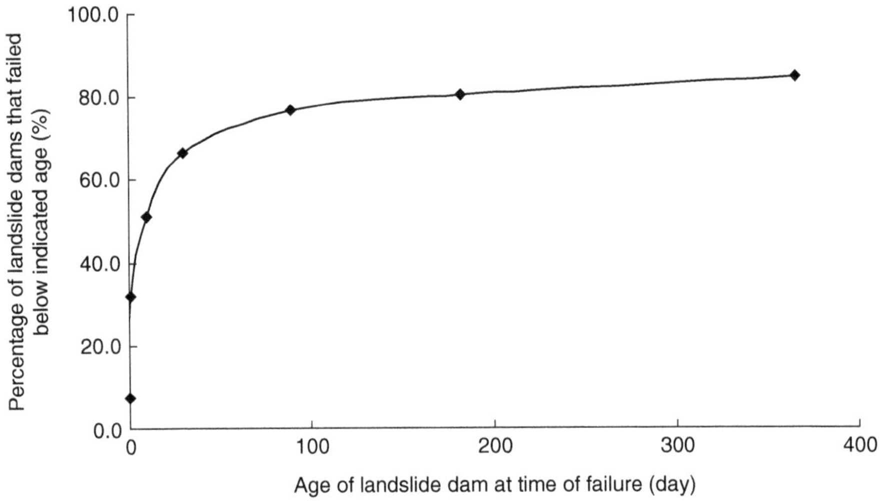

Figure 3.7 Age of landslide dams at time of failure (275 cases)

as the case formed in an unnamed tributary of the Fiume Irminio, which has existed for around 5000 years without significant overtopping. The longevity of some cases may be dependent on the safety of other dams. For example, a series of landslide dams along the Jianjiang River was triggered by the Wenchuan earthquake in 2008, China. Four relatively small landslide dams – Kuzhuba, Xinjie Village, Baiguo Village and Yanyangtan Landslide Dams – were located downstream of a larger landslide dam at Tangjiashan. The four smaller dams were overtopped by the flood from the breaching of the Tangjiashan landslide dam (Cui et al., 2009; Xu et al., 2009).

The longevity of 275 landslide dam cases in the database covers a wide range from several minutes to 20,000 years. Figure 3.7 shows the longevity distribution for the failed landslide dam cases. Some 84.7% of the cases lasted shorter than a year; 80.4% shorter than 6 months; 66.5% shorter than a month; 51.3% shorter than a week and 32.0% less than a day. Note that the corresponding percentages reported by Costa and Schuster (1988) are 85%, 80%, 56%, 41%, and 27%, respectively. The longevity of landslide-dammed lakes depends on factors such as volume, size, geometry, and sorting of blockage materials; rate of seepage through the blockage; rate of sedimentation, and inflow into the newly formed lake (Costa and Schuster, 1988).

Most of the landslide dams had a relatively short life, because they were made up of unconsolidated or poorly consolidated earth material and had no engineered control measures, such as an impervious zone for water barrier, filter zones to prevent internal erosion, drain zones to control pore pressures, or a spillway or other protected outlet to release water. Some landslide dams tend to have short life whenever small landslides block a stream with a large drainage area or inflow rate. Besides, landslide dams of predominantly soft, low density, fine-grained or liquefiable sediment lack resistance to erosion and may not last long.

3.2.3 Failure Modes

Without channelized spillways or other protected outlets, landslide dams commonly fail by overtopping. Figure 3.8 compares the failure modes of landslide dams and constructed man-made earth and rockfill dams. Out of the 171 landslide dam failure cases with known failure modes, 91.8% (157

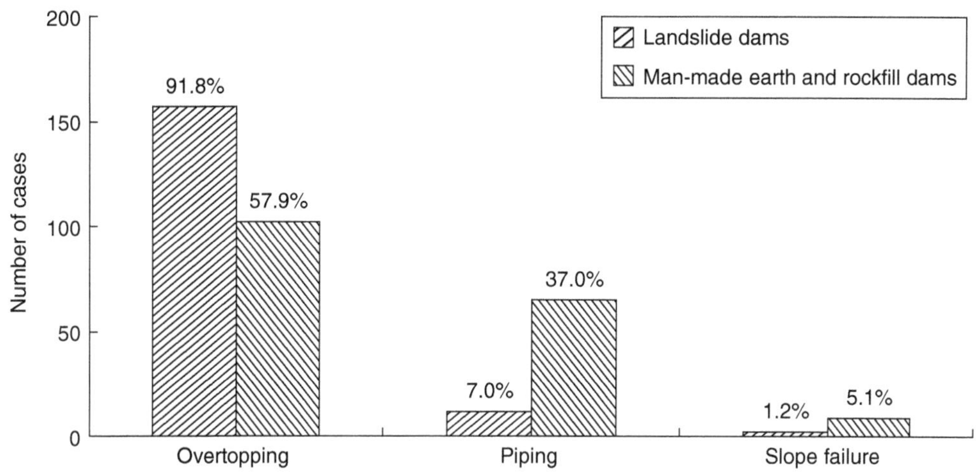

Figure 3.8 Comparison of failure modes of the failed landslide dams (171 cases) and man-made earth and rockfill dams (176 cases)

cases) failed by overtopping, 7.0% (12 cases) by internal erosion, and 1.2% (2 cases) by slope failure. The distribution of the failure modes of landslide dams is different from that of man-made earth and rockfill dams. The corresponding percentages for 176 man-made earth and rockfill dam failure cases (Xu and Zhang, 2009) are 58% by overtopping, 37% by internal erosion, and 5% by slope failure.

Comparing with constructed embankment dams, landslide dams have not undergone systematic compaction and contain large pores. Despite a large number of landslide dams with seepage problems, internal erosion failure is uncommon (7.0%, or 12 cases, with evidence of internal erosion failure), while a large percentage of man-made earth and rockfill dams failed in this mode (37%). Table 3.3 shows the landslide dams that failed by piping. The main reason for the lower chance of internal erosion failure is that landslide dams are often massive, with a large width (parallel to the river direction). The hydraulic gradient is typically too small to initialize internal erosion. Slope failures seldom occur in landslide dams (only two cases), because the dams formed in a dynamic manner with gentle slope angles.

3.3 Mitigation Measures for Landslide Dams

Landslide dams cause flooding both upstream with backwater and downstream when dam breaching occurs. For example, the Tangjiashan landslide dam, which was the largest induced by the Wenchuan earthquake, forced approximately 5000 people in 13 villages upstream of the dam to leave their homes until the dam breached a month later. The breaching flood of this dam caused another quarter of a million people downstream of the dam to evacuate for two weeks. When the potential floods threaten people or property, mitigation measures must be taken to reduce the flood risks.

Two types of mitigation measures are frequently adopted: engineering measures and non-engineering measures. Non-engineering measures (e.g. warning and evacuation) mitigate risks by reducing the elements at risk (i.e. people and both movable and immovable properties); for

Table 3.3 Landslide dams that failed in internal erosion

Dam name	Country	Formation time	Dam volume ($\times 10^6\,m^3$)	Dam height (m)	Dam width (m)	Lake volume ($\times 10^6\,m^3$)	Reference
Allpacoma landslide dam	Bolivia	2005	21	90	700	7.5	Costa and Schuster, 1991; Ermini and Casagli, 2003
Buonamico River	Italy	1973	0.55	32	240	14.9	Costa and Schuster, 1991; Ermini and Casagli, 2003
Cache Creek	United States	1906	—	70–75	3000	80	Costa and Schuster, 1991; Schuster and Costa, 1986
Gros Ventre River	United States	1925	—	60	150	—	Costa and Schuster, 1991
Labu Khola	Nepal	1968	3.5	133	580	301	Costa and Schuster, 1991
Mantaro River*	Peru	1945	—	7	—	—	Costa and Schuster, 1991
Mt Ruapehu Tephra	New Zealand	2007	3	70–100	600	0.5	Massey et al., 2010
Rio Toro River	Costa Rica	1992	20	120	60	6.60	Mora et al., 1993; Costa and Schuster, 1991
Tegermach River (Lake Yashinkul)	USSR	1835	—	—	—	—	Costa and Schuster, 1991
Trinity River	United States	1890	7–12	110	700	4–7	Costa and Schuster, 1991
Tsatichuu River*	Bhutan	2003	0.02	12	—	—	Dunning et al., 2006
Vallucciole Creek	Italy	1992	21	90	700	7.5	Costa and Schuster, 1991; Ermini and Casagli, 2003

Note: * The two cases on Mantaro River and Tsatichuu River failed first in slope failure, then in internal erosion.

instance, transferring people and movable properties out of the flooded areas. Engineering measures not only reduce the elements at risk by limiting flooded areas but also reduce dam failure probability, flood severity, and the vulnerability of the elements at risk. The studies on risk analysis and decision-making for non-engineering measures refer to Chapters 11 and 13, respectively.

3.3.1 Stages of Landslide Dam Risk Mitigation

Mitigation measures for landslide dams may be applied in three stages: before the formation of dam (stage 1), before dam breaching (stage 2), and after dam breaching (stage 3).

1. In stage 1, some measures may be taken to "prepare" for a potential landslide dam before the formation of the dam. For example, in 1961, a 2.3 km tunnel was constructed as a precautionary measure for the Vaiont Reservoir in northern Italy because of concern that a landslide might block the river. At that time, cracks in the slope indicated a potential landslide. In October 1963, a $270 \times 10^6\,m^3$ rock slide blocked the reservoir and dammed the Vaiont River. The tunnel served as a perfect diversion structure as neither the inlet nor the outlet of

the diversion tunnel was buried by the landslide (Schuster and Evans, 2011). Other measures include reinforcing riverbank slopes and constructing check dams to prevent debris flows from blocking rivers and forming debris flow dams.

2. The mitigation measures in stage 2 are of vast significance and urgency, which are the major part introduced in this section. The aim of these measures is to reduce the flood risks both upstream and downstream of the dam.
3. Landslide dams seldom are breached completely because of the massive body and the presence of low-erodibility materials such as large rocks. The threats of residual landslide dams should not be ignored, as the secondary breaching floods may still be large enough to impact people downstream. Normally, the safety of a residual landslide dam should be evaluated considering possible floods in the future. A residual dam which cannot resist certain floods may be reinforced or removed. For instance, a major part of the Ganderberg landslide dam, which formed in 1404 and blocked the Passer River in Italy, failed and drowned 400 people downstream. A reinforced outlet channel and flood gates were constructed in 1773 in order to control runoff from rainstorms or snowmelt. However, on 22 October 1774, the residual landslide dam still failed completely by overtopping (Schuster and Evans, 2011).

3.3.2 Engineering Mitigation Measures for Landslide Dams

When a landslide dam forms and threatens people either upstream or downstream of the dam, immediate engineering measures must be taken to stabilize the blockage, prevent the lake water from reaching dangerous levels, or control the erosion rate before breaching. These engineering measures can be divided into two types: short-term and long-term measures as summarized in Table 3.4 (Schuster, 2006; Liu, 2008a; Yang et al., 2010; Schuster and Evans, 2011; Cui et al., 2012).

Short-term measures include diversion of inflow, drainage by pumps or siphons and control of erosion rate during breaching. Diversion of inflow can be used when man-made hydraulic facilities (e.g. reservoir or irrigation systems) are available upstream of the dam. A good example is the Randa landslide dam formed in 1991 in Vispa River, Switzerland. Part of the river flow was diverted by the Grand Dixence hydroelectric facility. This allowed time to construct an open-channel spillway across the dam. The dam was still stable thus far (Bonnard, 2004). For landslide dams without diversion conditions, pumps or siphons can also be used to control the rising reservoir level. For example, the Higashi Takezawa landslide dam was triggered by the M6.8 Mid-Niigata Prefecture earthquake in 2004 (Sattar and Konagai, 2012). Twelve pumps were installed to keep the lake water level below the overflow point. After that, a 280 m long concrete-faced spillway was constructed. The landslide dam is still stable and under the supervision by CCTV.

Some short-term measures may be applied to accelerate the erosion rate to avoid persistent rising of water level in the reservoir when low erodibility materials exist. For example, blasting was applied to the Xiaogangjian-upstream landslide dam induced by the 2008 Wenchuan earthquake to clear large rocks that hindered further breaching of the dam (Yang et al., 2010). It is also possible to slow down the erosion rate by means of end-tipping man-made structures into the breach as for river closure in dam engineering. Chen et al. (2011) conducted model experiments and studied the effects of man-made structures on controlling the erosion rate during dam breaching. The peak outflow rates were reduced by 27–62% by end-tipping man-made structures.

Long-term measures include drainage tunnels through abutments, drainage conduits, and spillways through dams. Construction of drainage tunnels is an expensive and time-consuming measure. This often requires the presence of bedrock abutments for tunnel excavation and a

Table 3.4 Overview of engineering risk mitigation measures for landslide dams

Type	Measure	Target	Applicable conditions	Typical case
Short-term measures	Diversion of inflow water upstream	Slowing down the filling of the reservoir for long term measures	Man-made hydraulic facilities upstream such as reservoir or irrigation systems	Randa landslide dam, Switzerland, 1991 (Bonnard, 2004)
	Drainage by pumps or siphons	Slowing down the filling of the reservoir for long term measures	Power supply for pumps	Spirit landslide dam, USA, 1980 (Schuster and Evans, 2011); Higashi Takezawa landslide dam, Japan, 2004 (Sattar and Konagai, 2012)
	Control of erosion rate during breaching	Keeping the erosion rate not too large or small	Blasting to loosen the low erodibility materials; man-made structures to slow down the erosion rate	Xiaogangjian landslide dam, China, 2008 (Yang et al., 2010); Model experiments (Chen et al., 2011)
Long-term measures	Drainage tunnel through a abutment	Draining the reservoir water	Access road to dam site; water rising can be controlled in long term; presence of bedrock abutments; normally for large landslide dams	Spirit landslide dam, USA, 1980 (Schuster and Evans, 2011)
	Drainage conduit through the dam	Draining the reservoir water	Access road to the dam site; water rising can be controlled in relative long term	Higashi Takezawa landslide dam, Japan, 2004 (Sattar and Konagai, 2012)
	Spillway	Draining the reservoir water or reducing the peak outflow rate during breaching	Commonly used on various conditions	Madison River landslide dam, USA, 1959 (Schuster and Evans, 2011); Tangjiashan landslide dam, China, 2008 (Peng and Zhang, 2013b).

temporary road to the dam site for transporting equipment and materials. The Spirit Lake landslide dam in the state of Washington, USA, was a good example, which was induced by the 1980 eruption of Mount St Helens. A tunnel 2.59 km in length and 3.4 m in diameter was constructed using a tunnel-boring machine. A cofferdam was built before the completion of the tunnel as the intake was below the lake level (Schuster and Evans, 2011).

Installing drainage conduits (e.g. culverts or pipes) is another way to limit the elevation of the lake. Compared with drainage tunnels, drainage conduits are cheaper and relatively easier to construct. However, the water level of the intake for a conduit is often higher than that for a tunnel, and the discharge capacity of a conduit is smaller than that of a tunnel. Thus, the water volume in the landslide lake with conduits is often larger than that with a tunnel, and the residual risks of the landslide dam with conduits may be higher than those with a tunnel. This measure was adopted to dispose of the Higashi Takezawa landslide dam in Japan in 2004. Five pipes and an open channel were constructed with sufficient drainage capability for water discharge including snowmelt during early spring (Sattar and Konagai, 2012).

Construction of a spillway across a dam or an adjacent abutment is the most common long-term measure for landslide dam risk mitigation. A spillway may be designed to be erosion-resistant for preventing rapid breaching with a large flood. For instance, the Madison River landslide dam in south-western Montana, USA, in 1959 was properly controlled with a 75 m wide spillway across the crest. The lake outlet has only eroded 6–7 m during the 50 years since the construction of the spillway (Schuster and Evans, 2011). However, not all spillways are successful in preventing rapid breaching and downstream flooding, especially when the available time before overflow is limited and no access is available in remote areas. In these cases, constructed spillways are meant to lower the water level and the breaching flood discharge rather than to serve as a permanent structure. The Tangjiashan landslide dam comes into this category, which will be introduced in Section 3.3.4.

3.3.3 Engineering Measures for the Landslide Dams Induced by the Wenchuan Earthquake

Figure 3.9 shows the locations of 37 landslide dams induced by the Wenchuan earthquake. The detailed parameters of those landslide dams are shown in Table 3.5. The mitigation measures taken for the landslide dams induced by the Wenchuan earthquake can be divided into three categories: enlargement of the overflow spillway, construction of a spillway before overflow, and non-engineering measures.

3.3.3.1 Enlarging the Overflow Spillway

The M_S 8.0 Wenchuan earthquake that occurred on 12 May 2008 in Sichuan, China, induced approximately 50,000 landslides covering a wide area of 285 × 20 km (Sato and Harp, 2009; Zhang, 2009b). At least 257 of these landslides blocked rivers and formed landslide lakes (Cui et al., 2009). A large number of landslide dams could not be properly disposed before overflow. It is necessary to enlarge the overflow channels and prevent water level rising as soon as possible. For the landslide dams that could be accessed, equipment such as backhoes, bulldozers, and draglines were used to widen and deepen the overflow spillways in order to increase the outflow rate and prevent persistent rising of the water level. Four cases fell into this category: the Donghekou, Guanzipu, Tangjiashan, and Suijiayuanzi landslide dams, as shown in Table 3.5. When the dam materials were hard to excavate (e.g. consisting of large rocks), blasting was used to break up the dam materials. Figure 3.10(a) shows the construction of a spillway for the Xiaogangjian-downstream landslide dam with backhoes after blasting. Seven other landslide dams such as the Hongshihe, Jiangdanwan, and Laohuzui landslide dams were mitigated with a combination of blasting and backhoe excavation (Table 3.5).

Some landslide dams were hard to access and large equipment could not be transported to the dam sites. In these cases, the overflow spillways were enlarged by blasting and manual excavation. Thirteen landslide dams in Table 3.5, such as the Shibangou, Xiaogangjian-upstream, and Heidongya landslide dams were disposed in this manner. Figure 3.10(b) shows the blasting of the Shibangou landslide dam. The dam breached soon after the blasting with a breach depth of 8 m and a relatively small peak outflow rate. However, blasting could not be perfectly controlled in some cases. The Xiaogangjian-upstream landslide dam breached after four rounds of blasting, with a large peak outflow rate of 3900 m^3/s that caused damage to Hanwang Town.

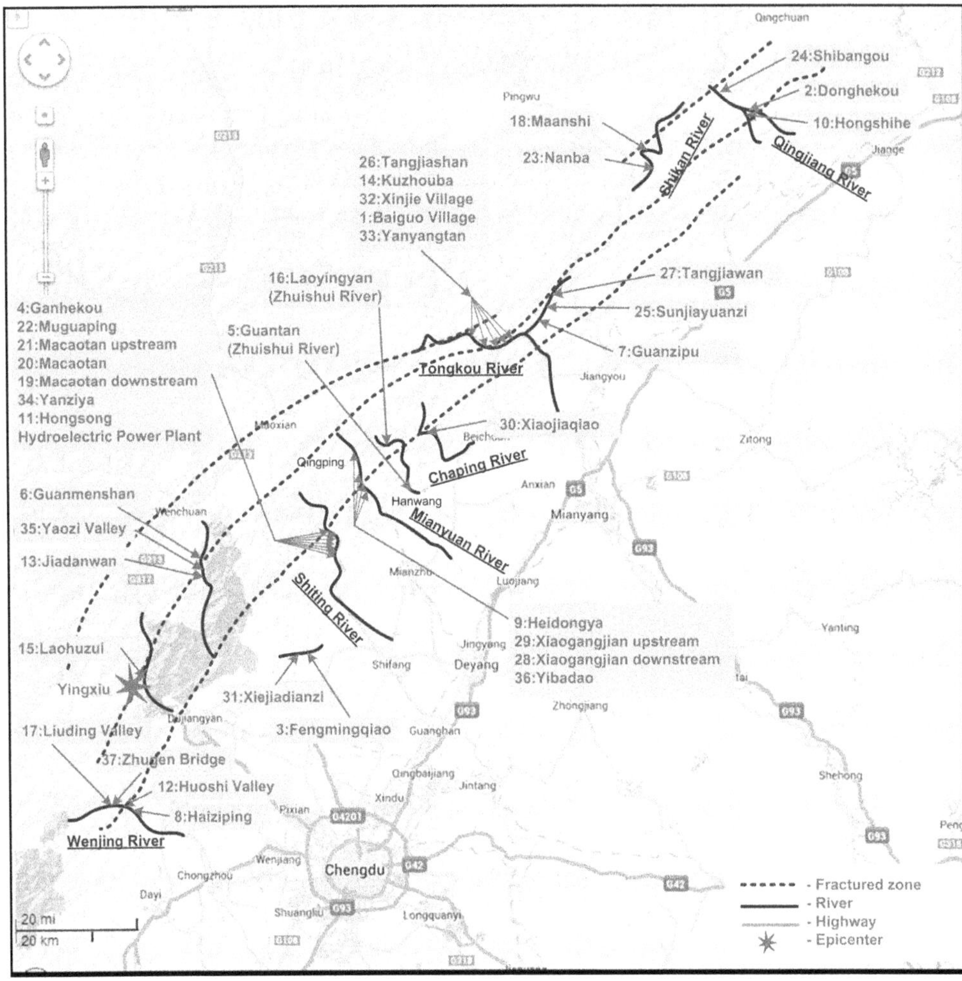

Figure 3.9 Locations of 37 landslide dams induced by the Wenchuan earthquake (Based on Google Maps). *Note*: the numbers of the landslide dam cases refer to Table 3.5

3.3.3.2 Construction of a Spillway before Overflow

If conditions permit, mitigation measures should be taken before overflow of the dams. In the Xiaojiaqiao landslide dam case, as shown in Figure 3.11(a), which had a dam height of 57–67 m and a reservoir capacity of $20 \times 10^6\,m^3$, a spillway 18 m in depth, 40 m in width, and 350 m in length was constructed within 10 days using backhoes before overflow. The peak discharge of the dam breaching through the spillway was about $1000\,m^3/s$, which was largely reduced from that without the spillway (He et al., 2009). The situation of the Ma'anshi landslide dam as shown in Figure 3.11(b) was different. As there was no access to the dam, only a relatively small spillway 3 m in depth, 3 m in bottom width, and 280 m in length was excavated manually assisted by blasting. The peak discharge of the breaching flood was estimated as $2200\,m^3/s$. Fortunately, there were no fatalities because the people downstream were evacuated in advance.

For landslide dams with extremely high risks, large excavation equipment had to be transported to the dam sites, even without access roads. The road to the Tangjiashan landslide dam, as shown

Table 3.5 List of high-risk landslide dams caused by the Wenchuan earthquake in 2008, and their disposal measures

No.	Name	Landslide dam			Catchment area (km^2)	Lake capacity ($10^6 m^3$)	Engineering measure (Spillway dimensions in meters: D = depth W_b = bottom width W_t = top width L = length)	Estimated influenced people downstream	Risk level	Failure mode (Q_p = peak outflow rate, B_d = breach depth)	References
		Height (m)	Width (m)	Length (m)							
1	Baiguo Village (Tongkou River)	10–20	200	100	3565	0.8	No engineering measure	2000	III	Overtopped by the flood of the Tangjiashan landslide dam	Li et al. (2008); MWR (2009)
2	Donghekou (Qingjiang River)	20	750	500	1220	6	Enlarging the overflow spillway with backhoes. D = 5, W_b = 30 and L = 400	50,000	III	Overtopped	Ren and Dang (2008); IWHR (2008)
3	Fengming Bridge (Shajin River)	10	300	100	—	1.8	No engineering measure	8000	III	Overtopped. $Q_p = 500\ m^3/s$	Li et al. (2008); IWHR (2008)
4	Ganhekou (Shiting River)	10	—	—	—	0.5	No engineering measure	5000	IV	Overtopped	Li et al. (2008)
5	Guantan (Zhuishui River)	60	120	200	215	10	Enlarging the overflow spillway with blasting and backhoes	4370	II	Overtopped	Li et al. (2008); He et al. (2010); MWR (2009)
6	Guanmenshan (Baisha River)	80	150	230	—	3.7	Enlarging the overflow spillway with blasting	<1000	III	Overtopped	Zhou et al. (2009)
7	Guanzipu (Tongkou River)	60	400	450	—	5.85	Enlarging the overflow spillway with backhoes. D = 3.3, W_b = 30 and L = 300	2000	III	Overtopped	Rao and Tang (2008)

(*Continued*)

Table 3.5 (*Continued*)

No.	Name	Landslide dam			Catchment area (km^2)	Lake capacity ($10^6 m^3$)	Engineering measure (Spillway dimensions in meters: D = depth W_b = bottom width W_t = top width L = length)	Estimated influenced people downstream	Risk level	Failure mode (Q_p = peak outflow rate, B_d = breach depth)	References
		Height (m)	Width (m)	Length (m)							
8	Haiziping (Wenjing River)	8	500	50	—	—	Enlarging the overflow spillway with blasting	1200	IV	Overtopped	Li et al. (2008); IWHR (2008)
9	Heidongya (Mianyuan River)	50–80	700	120	—	1.8	Enlarging the overflow spillway with blasting	70,000	II	Overtopped	Li et al. (2008); IWHR (2008); MWR (2009)
10	Hongshihe (Hongshi River)	50	500	400	70	4	Enlarging the overflow spillway with blasting and backhoes. D = 5, W_b = 5 and W_t = 25	50,000	II	Overtopped	Ren and Dang (2008); IWHR (2008)
11	Hongsong Hydroelectric Power Plant (Shiting River)	37	100	60	398	0.8	Enlarging the overflow spillway with blasting and backhoes. D = 6, W_b = 15 and L = 100	2000	II	Overtopped	Yang et al. (2010); MWR (2009)
12	Huoshigou (Wenjing River)	120	500	40	—	1.5	Enlarging the overflow spillway with blasting	150	III	Overtopped	Li et al. (2008); IWHR (2008)
13	Jiadanwan (Baisha River)	60	220	200	164.5	6.1	Enlarging the overflow spillway with blasting and backhoes. D = 6, W_b = 6–15 and L = 540	<1000	III	Overtopped. B_d = 20 m	Yang et al. (2010); Zhou et al. (2009)
14	Kuzhuba (Tongkou River)	60	200	300	3555	2	No engineering measure	70,000	II	Overtopped by the flood of the Tangjiashan landslide dam	Li et al. (2008); IWHR (2008)

15	Laohuzui (Minjiang River)	20	400	270	22564	>10	Enlarging the overflow spillway with blasting and backhoes	>100,000	II	Overtopped	Yang et al. (2010)
16	Laoyingyan (Jushui River)	106–140	500	300	58.3	5	Construction of a spillway with blasting and backhoes. D = 5, W_b = 12 and W_t = 20	60,000	I	Overtopped	He (2009); MWR (2009)
17	Liuding valley (Wenjing River)	60	500	50	—	3	Widen and deepen the overflow spillway with explosives	1200	III	Overtopped	Li et al. (2008); IWHR (2008)
18	Ma'anshi (Shikan River)	67.6	275	950	—	11.5	Construction of a spillway with explosives: D = 3, W_b = 3 and L = 280	35,000	II	Overtopped. $Q_p = 2200\ m^3/s$,	Luo (2008); Liu (2010)
19	Macaotan-Downstream (Shiting River)	30	—	100	—	0.1	Enlarging the overflow spillway with blasting. D = 4, W_b = 20 and L = 94	4000	II	Overtopped	Li et al. (2008); MWR (2009)
20	Macaotan (Shiting River)	40–50	80	90	—	0.25	Enlarging the overflow spillway with blasting. D = 4.5, W_b = 50 and L = 160	4500	II	Overtopped	Li et al. (2008); MWR (2009)

(Continued)

Table 3.5 (*Continued*)

No.	Name	Landslide dam			Catchment area (km^2)	Lake capacity ($10^6 m^3$)	Engineering measure (Spillway dimensions in meters: D=depth W_b=bottom width W_t=top width L=length)	Estimated influenced people downstream	Risk level	Failure mode (Q_p=peak outflow rate, B_d=breach depth)	References
		Height (m)	Width (m)	Length (m)							
21	Macaotan- Upstream (Shiting River)	40–50	300	160	—	0.6	Enlarging the overflow spillway with blasting. D=8, W_b=50 and L=160	5000	II	Overtopped	Li et al. (2008); MWR (2009)
22	Muguaping (Shiting River)	15	—	100	—	0.04	Enlarging the overflow spillway with blasting. D=6, W_b=20 and L=120	2000	III	Overtopped	Li et al. (2008); MWR (2009)
23	Nanba (Shikan River)	25–50	200	625	156.2	6.86	Construction of a spillway with backhoes. D=2, W_b=4 and W_t=11	5000	II	Overtopped	Li et al., 2008; Dai et al. (2008); MWR (2009)
24	Shibangou (Qiang River)	30–75	800	450	1140	11	Enlarging the overflow spillway with blasting and backhoes	150,000	II	Overtopped. B_d=8 m	Ren and Dang (2008); MWR (2009)
25	Sunjiayuanzi (Tongkou River)	50	180	—	—	5.6	Construction of a spillway with backhoes. D=3, W_b=20 and L=300	2000	III	Overtopped	Rao and Tang (2008); MWR (2009)
26	Tangjiashan (Tongkou River)	82	802	611.8	3550	316	Construction of a spillway with backhoes and bulldozers. D=12, W_b=8, W_t=44 and L=475	1.3 M	I	Overtopped. Q_p=6500 m^3/s, B_d=42 m	Peng and Zhang (2012)

27	Tangjiashan (Tongkou River)	30	—	300	—	15.2	Construction of a spillway with backhoes: $D=4$, $W_b=26$, $W_t=36$ and $L=300$	2000	II	Overtopped	Rao and Tang (2008); MWR (2009)
28	Xiaogangjian Downstream (Mianyuan River)	30	150	150	346	2	Enlarging the overflow spillway with blasting	47,190	II	Overtopped by the flood of the Xiaogangjian upstream landslide dam	Li et al. (2008); He et al. (2010); Xu et al. (2009)
29	Xiaogangjian Upstream (Mianyuan River)	70	300	250	346	12	Enlarging the overflow spillway with explosives. $D=3$, $W_b=30$ and $W_t=66$	47,190	II	Overtopped. $Q_p=3900\,m^3/s$; $B_d=30\,m$	Yang et al. (2010); MWR (2009)
30	Xiaojiaqiao (Chaping River)	57–67	200	200	154.8	20	Construction of a spillway with backhoes. $D=18$, $W_b=40$ and $L=350$	114,000	I	Overtopped. $Q_p=1000\,m^3/s$; $B_d=37\,m$	He et al. (2009); He et al. (2010); MWR (2009)
31	Xiejiadianzi (Shajin River)	10	250	70	—	1	No engineering measure	8000	III	Overtopped	Li et al. (2008); IWHR (2008);
32	Xinjie Village (Tongkou River)	20	200	350	3560	2	No engineering measure	32,000	III	Overtopped by the flood of the Tangjiashan landslide dam	Li et al. (2008); MWR (2009)

(Continued)

Table 3.5 (*Continued*)

No.	Name	Landslide dam			Catchment area (km^2)	Lake capacity ($10^6 m^3$)	Engineering measure (Spillway dimensions in meters: D = depth W_b = bottom width W_t = top width L = length)	Estimated influenced people downstream	Risk level	Failure mode (Q_p = peak outflow rate, B_d = breach depth)	References
		Height (m)	Width (m)	Length (m)							
33	Yanyangtan (Tongkou River)	20–30	150	—	—	4	No engineering measure	2000	II	Overtopped by the flood of the Tangjiashan landslide dam	Li et al. (2008); MWR (2009)
34	Yanziyan (Shiting River)	<10	20	40	—	0.03	Enlarging the overflow spillway with blasting. D = 6, W_b = 20 and L = 80	2000	IV	Overtopped	Li et al. (2008);Wang and He, 2008
35	Yaozi Valley (Baisha River)	60	250	180	—	6.2	Enlarging the overflow spillway with blasting and backhoes. W_b = 25 and L = 260	<1000	III	Overtopped	Zhou et al. (2009)
36	Yibadao (Mianyuan River)	25	140–180	120	400	3	Enlarging the overflow spillway with blasting and backhoes. D = 2, W_b = 15 and L = 60	47,190	III	Overtopped. B_d = 8 m	Yang et al. (2010); Zhou et al. (2009)
37	Zhugen Bridge (Wenjing River)	90	500	68	—	4.5	Enlarging the overflow spillway with blasting	1600	III	Overtopped	Li et al. (2008); IWHR (2008)

Figure 3.10 Enlarging the overflow spillways: (a) the Xiaogangjian-downstream landslide dam with backhoes (based on Liu and Du, 2010); (b) the Shibangou landslide dam with blasting (based on Xinhuanet.com, 2008)

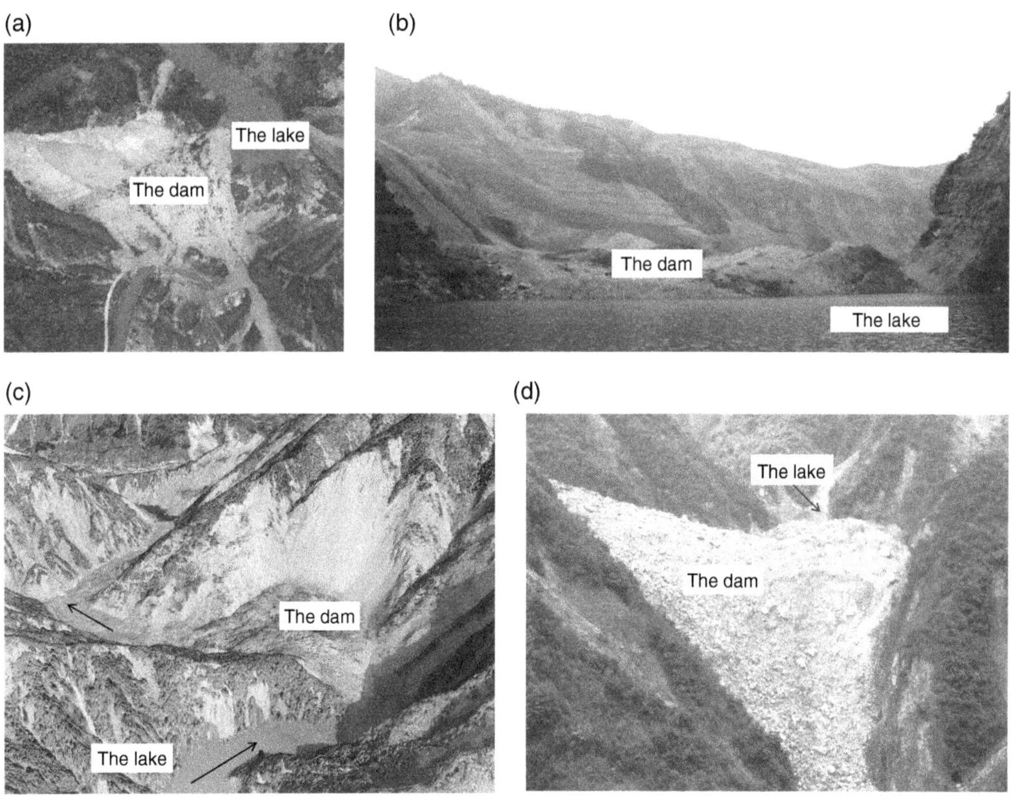

Figure 3.11 Landslide dams mitigated with constructed spillways before overflow: (a) the Xiaojiaqiao landslide dam (based on Liang, 2009a); (b) the Ma'anshi landslide dam (based on Liang, 2009b); (c) the Tangjiashan landslide dam (based on Liu, 2008); (d) the Laoyingyan landslide dam (based on Liang, 2009c)

in Figure 3.11(c) was completely blocked by a series of landslides. The engineers and excavation equipment were transported to the dam site by heavy-duty helicopters. The details of the mitigation measures for the dam will be described in the next section. The Laoyingyan landslide dam, as shown in Figure 3.11(d), with a dam height of 106 m and a lake volume of $5\times10^6\,m^3$, had no area for landing of helicopters at the dam site. A temporary landing apron was constructed 6 km away from the dam site. Engineers and small excavation machines were transported to this temporary landing apron and then to the dam site. A spillway with a depth of 5 m, a bottom width of 12 m and a top width of 20 m was constructed before overflow. The dam was partially overtopped with a relatively small peak outflow rate.

3.3.3.3 Non-engineering Measures

Some landslide dams were not disposed through engineering measures because they were too small to cause flood damage. A large number of landslide dams without detailed records were of this category. Some had overtopped before any actions could be taken, such as the Fengming Bridge, Ganhekou, and Xiejiadianzi landslide dams. In these cases, non-engineering measures such as warning and evacuation were implemented before dam breaching. Some were located at the downstream of a large landslide dam and were overtopped by the flood from the breaching of the large dam. The Baiguo Village, Kuzhuba, Xinjian Village, and Yanyangtan landslide dams were four landslide dams downstream of the Tangjiashan landslide dam, which were overtopped following the breach of the Tangjiashan dam with a peak discharge of $6500\,m^3/s$. Figure 3.12 shows the locations of the Tangjiashan landslide dam and two smaller landslide dams downstream, namely the Kuzhuba and Xinjian Village landslide dams (Cui et al., 2009).

Figure 3.12 The Tangjiashan landslide dam and two small landslide dams downstream: the Kuzhuba and Xinjian village landslide dams (based on Cui et al., 2009)

3.3.4 *Mitigation Measures for the Tangjiashan Landslide Dam*

Triggered by the Wenchuan earthquake on 12 May 2008, a slope of Tangjiashan Hill slid, blocked the Tongkou River, and formed a large landslide dam. The dam was found on 14 May through an aerial investigation. From the analysis of aerial photos and existing hydraulic and topography information, the dam was estimated to have a height of 82 m, a width of 802 m, a length of 611 m, a dam volume of $20.4 \times 10^6\,m^3$ and a lake capacity of $316 \times 10^6\,m^3$ (Cui et al., 2009; Hu et al., 2009; Liu et al., 2010). The estimated population at risk downstream was 1.2 million.

The landslide dam could not be accessed due to blockage of roads by landslides until 21 May when experts managed to arrive at the dam site by helicopter. The geological conditions of the dam were obtained two days later through in-situ investigations. The dam mainly consisted of three layers (Figure 3.13): the upper layer of gravely soils with a thickness of 5–15 m, the middle layer

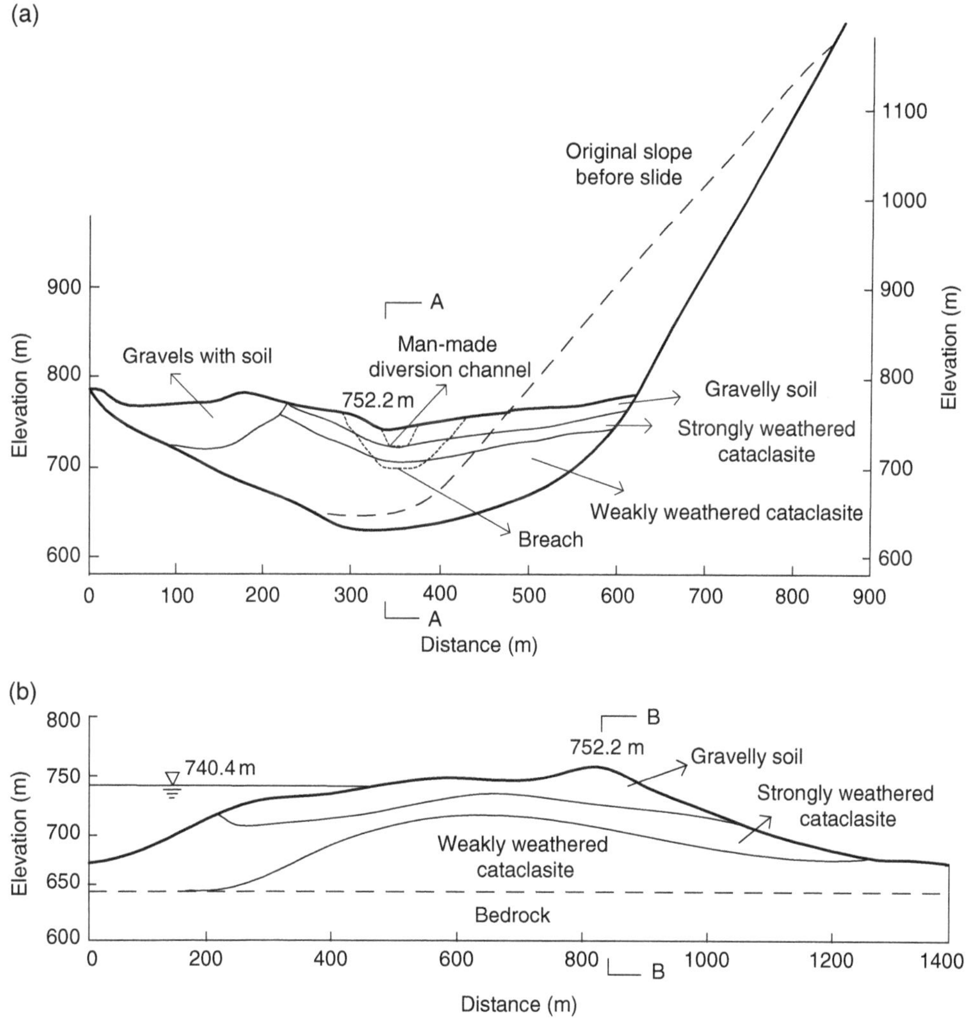

Figure 3.13 Cross-sections of the Tangjiashan landslide dam: (a) across the river; (b) along the river

Table 3.6 Parameters of Tangjiashan landslide dam

Landslide dam		Diversion channel (spillway)	
Parameter	Value	Parameter	Value
Dam height	82 m	Depth	12 m
Dam width	802 m	Top width	44 m
Dam length	611 m	Bottom width	8 m
Dam volume	20.4×10^6 m^3	Length	475 m
Lake capacity	316×10^6 m^3	Lake capacity after the construction of the spillway	247×10^6 m^3

of strongly weathered cataclasite with a thickness of 10–15 m, and the bottom layer of weakly weathered cataclasite with a thickness of 50–80 m (Hu et al., 2009; Liu et al., 2010). The coefficient of erodibility of the dam varies from 120 mm^3/N-s for the top layer to 10 mm^3/N-s for the bottom layer (Chang and Zhang, 2010). The dam-breaching peak outflow rate was estimated as 33,000–70,000 m^3/s with the corresponding population at risk (PAR) downstream of 0.2–1.2 million (Huang et al., 2008). The predicted floods were very large. The government decided to reduce the dam-break risks by excavating a diversion channel.

By 25 May 2008, 24 bulldozers, 16 backhoes and 4 trucks were shipped to the dam site by helicopters (Liu, 2010). After intensive excavation with bulldozers and backhoes from 26 May to 31 May, a spillway with a depth of 12 m, a bottom width of 8 m, a top width of 44 m and a length of 475 m was completed. The lake capacity was lowered from 316×10^6 m^3 to 247×10^6 m^3, as shown in Table 3.6. More than 0.25 million people were evacuated from their homes by 31 May. The lake water overflowed through the spillway at 7:00 on 7 June (Liu et al., 2010). The flow rate in the first two days was rather small and the lake volume kept rising. To accelerate the dam breaching process and reduce the evacuation cost, blasting was conducted on 8 June to loosen the rock and soil materials. The breaching of the dam started at 6:00 on 10 June. The peak outflow rate was 6500 m^3/s at 12:30 pm. Beichuan Town, which is located 4.5 km downstream of the dam was severely impacted by the flood. No people died because everyone had been evacuated 2 weeks before the dam breaching. Although nearly 0.2 million people were evacuated from Mianyang City, the second largest City in Sichuan Province 85 km downstream of the dam, no flood occurred in that city as the peak discharge (7110 m^3/s) was lower than the defense flood of 12,000 m^3/s (Peng and Zhang, 2012b, 2012c, 2013b). Four landslide dams – the Baiguo Village, Kuzhuba, Xinjian Village, and Yanyangtan landslide dams – were overtopped by the large flood from the breaching of the Tangjiashan landslide dam.

4

Statistical Analysis of Failures of Concrete Dams

4.1 Database of Failures of Concrete Dams

To study failures of concrete dams, 100 dam failure cases throughout the world, excluding China, have been collected from the literature (e.g. USCOLD, 1975, 1988; Vogel, 1980; Stanford University, 1994; Singh, 1996) and compiled into a database. Details of the characteristics of the concrete dams and the failure information are shown in Appendix A.

Table 4.1 shows the reservoir capacities of the failed concrete dams. Among the cases with known capacities, most reservoirs have capacities less than $1 \times 10^8\,m^3$. Table 4.2 shows the heights of the failed concrete dams, of which more than one half are less than 15 m. Table 4.3 presents the construction time of the failed concrete dams. The concrete dams constructed during the period of 1900–1929 appear to have suffered the highest rate of failure. Table 4.4 further shows the ages of the concrete dams at the time of failure. Similar to embankment dams, a concrete dam is most likely to fail within its first five years of service, especially during the first year after construction.

4.2 Failure Modes and Processes

Concrete dams often fail all of a sudden. The failure of a gravity dam may take a short time on the order of 0.1–0.5 h while the failure of an arch dam is usually assumed to range from instantaneous to 0.1 h (e.g. FERC, 1988; ICOLD, 1998b). Due to the very short failure time, the breaching floods of concrete dams are more serious than those of embankment dams with similar heights and reservoir capacities.

Concrete gravity dams are typically constructed from numerous concrete monoliths over a wide valley. When a concrete gravity dam fails, one or several monoliths are washed away. Concrete arch dams are usually constructed into a whole structure at a deep canyon with steep abutment slopes. When a concrete arch fails, the entire structure or a large segment of it gives way all at once.

Dam Failure Mechanisms and Risk Assessment, First Edition. Limin Zhang, Ming Peng, Dongsheng Chang, and Yao Xu.

Table 4.1 Reservoir capacities of failed concrete dams

Reservoir capacity ($\times 10^6$ m^3)	Number of cases	Percentage (%)
≥100	2	2.0
100–10	6	6.0
10–1	13	13.0
<1	9	9.0
Unknown	70	70.0
Sum	100	100.0

Table 4.2 Heights of failed concrete dams

Dam height (m)	Number of cases	Percentage (%)
≥100	1	1.0
100–60	7	7.0
60–30	9	9.0
30–15	20	20.0
<15	54	54.0
Unknown	9	9.0
Sum	100	100.0

Table 4.3 Construction time of failed concrete dams

Construction year	Number of cases	Percentage (%)
Before 1850	0	0.0
1850–1859	1	1.0
1860–1869	0	0.0
1870–1879	1	1.0
1880–1889	0	0.0
1890–1899	0	0.0
1900–1909	10	10.0
1910–1919	12	12.0
1920–1929	12	12.0
1930–1939	5	5.0
1940–1949	2	2.0
1950–1959	5	5.0
1960–1969	3	3.0
1970–1979	3	3.0
After 1980	0	0.0
Unknown	46	46.0
Sum	100	100.0

In general, the breaching process of failure of concrete dams, either for gravity dams or arch dams, is simple; hence the prediction of breaching parameters appears to be straightforward since the breach is often multiple parts of the removed structure. The ruptured dam section becomes the pivotal point for the mass of water to be released. The most adverse condition is that the whole structure may be removed, which allows for prescribing an upper bound for the outflow hydrograph.

Table 4.4 Ages of failed concrete dams

Dam age	Number of cases	Percentage (%)
Under construction	5	5.0
0–1	12	12.0
1–5	11	11.0
5–10	3	3.0
10–15	4	4.0
15–20	1	1.0
20–30	5	5.0
30–40	3	3.0
40–60	5	5.0
60–80	3	3.0
80–100	0	0.0
>100	0	0.0
Unknown	48	48.0
Sum	100	100.0

Table 4.5 Causes of concrete dam failures

Cause	Number of cases	Percentage (%)
Overtopping	30	30.0
Quality problems	48	48.0
Disasters	5	5.0
Poor management	2	2.0
Unknown	15	15.0
Sum	100	100.0

4.3 Common Causes of Concrete Dam Failures

Table 4.5 shows the causes of the concrete dam failures. The most common cause of failure is quality problems, which is followed by overtopping. These two causes led to about 80% of all failures. Table 4.6 shows the detailed quality problems leading to concrete dam failures. The most important quality problem is internal erosion in the foundation. In Table 4.6, 13 out of the 19 internal erosion failures occurred at the foundation. The second most important quality problem is sliding or overturning, which is also closely related to the foundation. Therefore, quality problems at the foundation in the form of internal erosion and structural instability are very important for concrete dams. This confirms the fact that the foundation plays a key role for concrete dams that rely on gravity, arch, or buttress resistances. It is consistent with the finding of ICOLD (1995) that foundation problems with internal erosion (21%) and insufficient shear strength (21%) are the two most common causes of concrete dam failures.

Table 4.7 shows the subdivision of the failed concrete dams in terms of structure type. Among the 100 concrete dams, 71 cases are gravity dams, which are the most common type of concrete dam. Table 4.8 shows the causes of the gravity dam failures. Most of the cases are caused by either quality problems or overtopping. These two causes led to nearly 80% of all the failures. Table 4.9 shows the sub-causes of quality problems leading to gravity dam failures. Internal erosion and sliding or overturning are the two main quality problems, of which most are associated with the foundation.

Table 4.6 Sub-causes of quality problems for concrete dam failures

Sub-cause	Number of cases	Percentage (%)
Internal erosion	19	39.6
Sliding/overturning	12	25.0
Quality issues in spillway	2	4.2
Quality issues in culvert and other embedded structures	2	4.2
Unknown	13	27.0
Sum	48	100.0

Table 4.7 Subdivision of failed concrete dams in terms of structure type

Structure type	Number of cases	Percentage (%)
Gravity	71	71.0
Arch	10	10.0
Buttress	9	9.0
Unknown	10	10.0
Sum	100	100.0

Table 4.8 Causes of gravity dam failures

Cause	Number of cases	Percentage (%)
Overtopping	23	32.4
Quality problems	29	40.8
Disasters	4	5.6
Poor management	2	2.8
Unknown	13	18.4
Sum	71	100.0

Table 4.9 Sub-causes of quality problems for gravity dam failures

Cause	Number of cases	Percentage (%)
Internal erosion	9	31.0
Sliding/overturning	7	24.2
Quality issues in spillway	2	6.9
Quality issues in culvert and other embedded structures	2	6.9
Unknown	9	31.0
Sum	29	100.0

In Table 4.7, there are ten cases of arch dam failures and nine cases of buttress dam failures. Among the ten arch dam failures, seven failures were caused by quality problems, of these five cases occurred at the foundation. Among the nine buttress dam failure cases, eight failures were caused by quality problems, of which six failures occurred at the foundation. Therefore, quality problems at the foundation (internal erosion or structural instability) are the most common cause of failure for arch and buttress dams.

5

Statistical Analysis of Failures of Dikes

5.1 Introduction

Chapters 2 to 4 presented databases and statistical analysis of failures of constructed embankment dams, landslide dams, and concrete dams respectively. This chapter presents a database of 1004 dike failure cases, with information on pre-breach dike geometry, materials, types of dike, failure mechanisms, breaching length, depth, and peak discharge. Frequent failure mechanisms are also evaluated.

Dike breaching cases have been reported by Nagy and Tóth (2005), NSF (2006), Li et al. (2003), and Danka and Zhang (2015) among others. A few researchers have studied mechanisms of historical breaches (van Baars and van Kempen, 2009; Nagy, 2007); the past breaches were most frequently caused by external erosion (overtopping, wave erosion, etc.), internal erosion, and slope failures. The physical process of dike breaching by overtopping has been detailed in three PhD theses (Visser, 1998; Zhu, 2006; Danka, 2015) related to homogeneous sand and clay dikes. Earthen sea dikes with an impermeable mask were studied by D'Eliso (2007), while Howard et al. (2009) illustrated the stages of dike breaching caused by internal erosion. Large-scale experiments on the breaching mechanisms caused by overtopping and concentrated leak were conducted in Norway (Morris et al., 2005). Other important failure mechanisms, such as micro instability and horizontal shear failure, have also been studied (e.g. TAW, 1998, 1999; CIRIA et al., 2013).

5.2 Database of Dike Breaching Cases

Danka (2015) compiled a database of 1004 dike breaching cases from the literature (e.g. Li et al., 2003; Nagy and Tóth, 2005; NSF, 2006; Nagy, 2007; US Army Corps of Engineers, 2007; URS, 2008; van Baars and van Kempen, 2009; Bayoumi and Meguid, 2011). The majority of the

Dam Failure Mechanisms and Risk Assessment, First Edition. Limin Zhang, Ming Peng, Dongsheng Chang, and Yao Xu.

cases were reported from Hungary, Germany, China, USA, and the Netherlands (Figure 5.1a). Many of these cases were reported by Nagy and Tóth (2005) who studied dike breaches in the Carpathian Basin in Europe. The database provides information on the breach location, date, length, depth, and development time; data about the pre-breach dike geometry (height, width, slope inclination, etc.), dike material, and likely failure mechanism; and additional available records (casualties, flooded area, peak outflow, etc.). The collected cases mainly represent river or canal dike breaches; only a few Dutch and New Orleans cases are related to seawater dikes.

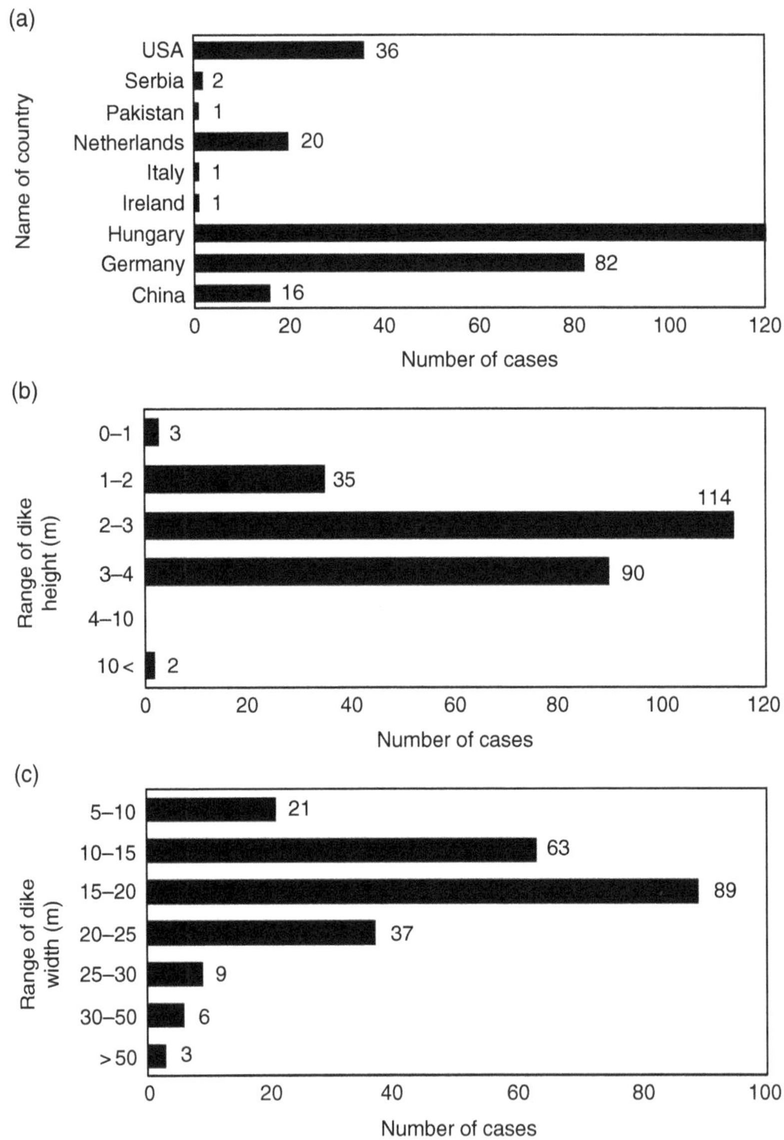

Figure 5.1 Statistics of the locations, heights and widths of the breached dikes. *Source:* Danka and Zhang (2015). With permission from ASCE

Most of the dike breaching cases lack detail. The dike height is available in 268 cases (Figure 5.1b) and the width in 228 cases (Figure 5.1c). Statistics show that the heights of the breached dikes are mostly in the range of 2–4 m and only in two cases are above 10 m. Though dikes often protect significant economic values, the International Commission on Large Dams categorizes most dikes as small dams, as the height of a large dam must be 15 m or above. Peak discharge is rarely available although it is important for estimating the inundation zone and supporting the risk management decisions. Therefore an attempt was made to estimate the peak discharge using Visser's formula (1998), which calculates the peak discharge in early stages of breaching as

$$Q = m\left(\frac{2}{3}\right)^{3/2} \sqrt{g}L\left(H_w - Z_{br}\right)^{3/2} \tag{5.1}$$

where m is the discharge coefficient ($m \approx 1$), g is the gravitational acceleration, L is the length of the breach, H_w is the floodwater elevation, and Z_{br} is the elevation of the bottom of the breach. In Visser's study (Visser, 1998), the difference between the floodwater level and the bottom of the breach is set as the dike height, and the breach length is defined as the final breach length, thus the peak discharge could be overestimated. Revision of the peak discharge values could be required on a case by case basis.

Part of the database (42 cases) is presented in Table 5.1, which lists detailed information on dike geometry, material, and observed failure mechanism. The database distinguishes three different soil materials based on their erosion and strength characteristics: coarse-grained, fine-grained, and organic soils. Since Danube and Tisza dikes are mainly constructed of clay, silty-clay, and different types of silt materials, these dikes are assumed to be constructed of fines when information is unavailable. The estimated discharges using Equation (5.1) are also presented in the last column of Table 5.1.

5.3 Evaluation of Dike Failure Mechanisms

5.3.1 *Most Relevant Failure Mechanisms*

The most frequent failure mechanisms observed in the dike failure dataset are external erosion, internal erosion, slope failure, and embedded structural failures. These failures are defined briefly here.

Two types of external erosion were observed. The first type, overtopping, initiates when the elevation of the dike crest is not sufficient, and the floodwater level exceeds it. While floodwater is flowing through the weir, the dike material erodes. This process is controlled by hydraulic conditions and dike material properties. Visser (1998) defined the breaching stages of homogeneous sand dikes, and proposed a mathematical model to describe the development when a sand dike is overtopped. Zhu (2006) used shear stress based governing equations to describe the erosion process, and applied a jet-erosion model to simulate homogeneous clay dike breaching. Recently more realistic heterogeneous dike and dam erosion models have become available (e.g. D'Eliso, 2007; Morris, 2011). The mechanics of overtopping erosion will be introduced in Chapter 7.

The second type of external erosion develops when the outer part of a dike section erodes due to high flow velocity or wave activities along the river direction or on shores. If the area of the cross-section decreases, horizontal shear failure might be triggered, or when the outer slope angle becomes too steep, it may lose stability. The erosion process is controlled by the erosion resistance

Table 5.1 Detailed data of dike breaches (ext. erosion = external erosion; int. erosion = internal erosion; L = left bank; R = right bank)

Country	Location	River – bank	Failure mechanism, *f*	Date of failure (y.m.d)	Height, *h* (m)	Dike Width, *w* (m)	Dike type, *t*	Dike material, m	Breaching length, *L* (m)	Breaching depth, *D* (m)	Peak discharge, Q_{peak} (m^3)
HUN	Szeremi f.	Duna – L	Int. erosion	1956.03.14	3.6	20.2	Earthen	Fine	5	6.8	58
HUN	Tolna	Duna – R	Ext. erosion	1956.03.11	2.0	14.0	Earthen	Fine	23	3.6	111
HUN	Tolna	Duna – R	Ext. erosion	1956.03.11	2.2	15.0	Earthen	Fine	22	4.2	122
HUN	Tolna	Duna – R	Ext. erosion	1956.03.11	1.5	11.5	Earthen	Fine	65	2.9	204
HUN	Bogyiszló	Duna – R	Ext. erosion	1956.03.11	2.3	15.5	Earthen	Fine	38	5.1	226
HUN	Margittasziget, Dunafalva	Duna – L	Int. erosion	1956.03.14	2.8	18.0	Earthen	Fine	40	7.7	320
HUN	Tolna	Duna – R	Ext. erosion	1956.03.11	2.1	14.5	Earthen	Fine	70	4.7	363
HUN	Kisbodak	Duna – R	Int. erosion	1954.06	2.5	16.5	Earthen	Coarse	58	10.5	391
HUN	Dunakiliti	Duna – R	Slope failure	1954.06	3.0	19.0	Earthen	Coarse	53	12.0	470
HUN	Bogyiszló	Duna – R	Ext. erosion	1956.03.11	2.4	16.0	Earthen	Coarse	75	4.8	475
HUN	Szeremi f.	Duna – L	Int. erosion	1956.03.14	3.2	20.0	Earthen	Fine	55	6.7	537
HUN	Bogyiszló	Duna – R	Ext. erosion	1956.03.11	2.7	17.5	Earthen	Fine	77	20.9	582
HUN	Szeremi f.	Duna – L	Int. erosion	1956.03.14	4.0	24.0	Earthen	Fine	45	8.0	614
HUN	Bogyiszló	Duna – R	Ext. erosion	1956.03.11	2.2	15.0	Earthen	Coarse	127	5.0	707
HUN	Szeremi f.	Duna – L	Ext. erosion	1956.03.13	2.8	18.0	Earthen	Fine	90	17.7	719
HUN	Bogyiszló	Duna – R	Ext. erosion	1956.03.11	2.2	15.0	Earthen	Coarse	139	5.2	773
HUN	Vöröshíd-Baja	Duna – L	Slope failure	1956.03.13	5.7	32.5	Earthen	Fine	35	11.6	812
HUN	Margittasziget, Dunafalva	Duna – L	Ext. erosion	1956.03.13	2.6	17.0	Earthen	Fine	120	7.5	858
HUN	Pőrböly	Duna – R	Slope failure	1956.03.13	3.0	19.0	Earthen	Coarse	100	6.6	886
HUN	Tivadar	Tisza – R	Ext. erosion	1947.12.31	3.2	15.8	Earthen	Fine	100	11.4	976
HUN	Margittasziget, Dunafalva	Duna – L	Ext. erosion	1956.03.13	3.5	21.5	Earthen	Fine	90	8.6	1005
HUN	Ásványráró	Duna – R	Int. erosion	1954.06	3.5	21.5	Earthen	Coarse	95	16.8	1061
HUN	Szeremi f.	Duna – L	Int. erosion	1956.03.14	4.2	25.0	Earthen	Fine	75	8.2	1101

HUN	Margittasziget, Dunafalva	Duna – L	Int. erosion	1956.03.13	3.5	21.5	Earthen	Fine	100	8.6	1116
HUN	Margittasziget, Dunafalva	Duna – L	Int. erosion	1956.03.13	3.5	21.5	Earthen	Fine	100	8.8	1116
HUN	Bogyiszló	Duna – R	Ext. erosion	1956.03.11	3.0	19.0	Earthen	Coarse	128	6.1	1134
HUN	Margittasziget, Dunafalva	Duna – L	Ext. erosion	1956.03.13	3.4	21.0	Earthen	Fine	110	13.2	1176
HUN	Vajastorok	Duna – L	Slope failure	1956.03.13	3.7	22.5	Earthen	Fine	100	17.4	1213
HUN	Margittasziget, Dunafalva	Duna – L	Int. erosion	1956.03.13	3.7	22.5	Earthen	Fine	110	11.9	1335
HUN	Margittasziget, Dunafalva	Duna – L	Int. erosion	1956.03.13	4.0	24.0	Earthen	Fine	100	12.2	1364
CH	Jiujiang	Yangtze	Int. erosion	1998.08.07	4.5	19.1	Composite	Fine	61.1	8.5	1500
HUN	Tass	Duna – L	Int. erosion	1956.03.08	4.0	24.0	Earthen	Fine	110	19.0	1500
HUN	Vajastorok	Duna – L	Ext. erosion	1956.03.13	3.9	23.5	Earthen	Fine	120	7.6	1576
HUN	Ásványráró	Duna – R	Ext. erosion	1954.06	3.0	19.0	Earthen	Fine	184	9.8	1630
HUN	Dombori	Duna – R	Int. erosion	1956.03.11	4.5	26.5	Earthen	Fine	114	22.0	1855
HUN		Duna – R	Int. erosion	1956.03.10	3.0	19.0	Earthen	Fine	250	18.0	2215
HUN	Békavári-Baja	Duna – L	Other	1956.03.14	5.2	30.0	Earthen	Fine	110	17.6	2224
NL	Wilnis	Canal	Sliding	2003.08.26	4.0	28.0	Composite	Organic	50	9.1	
HUN	Szekszárd	Sió – R	Ext. erosion	1956.03.10	2.1	11.4	Earthen	Fine	11	4.0	
HUN	Tiszabecs	Batár – L	Ext. erosion	1947.12.30	3.0	15.0	Earthen	Fine	66	6.0	
CH	Lizhou Polder	Yangtze	Ext. erosion	1998.07.24	9.5	65.5	Earthen	Fine	390	17.0	
USA	New Orleans	17th str. Can.	Sliding	2005.08.29	3.5	25.0	Composite	Fine (org.)	137	8.0	
HUN	Dunafalva	Duna – L	Ext. erosion	1956.03.13	3.5	21.5	Earthen	Fine	90	8.6	1005

of the dike material and the wave dynamics, which is governed by meteorological conditions and dike geometry. The breaching process of sea dikes due to waves, which erode the waterside slope, has been studied for instance in Stanczak and Oumeraci (2012).

Internal erosion is witnessed when the energy of seepage water through or beneath the dike is sufficient to detach particles from the soil matrix inside the dike, which increases permeability and deteriorates the soil structure. Four internal erosion mechanisms can take place: concentrated leak, backward erosion, contact erosion, and suffusion (ICOLD, 2013), which will be introduced in detail in Chapter 6.

Dike slope failure could also lead to breaching. Seepage forces and reduction of the shear strength of dike materials must be considered under flood conditions. At high water levels, horizontal sliding might be a potential hazard, especially if organic subsoil or dike material is present.

Rapid breaching caused by embedded structure failures has been observed in several cases, for instance in Wilnis (van Baars, 2005) and New Orleans (NSF, 2006). In Wilnis, a peat dike, supported with a timber sheet pile wall, lost its stability under dry weather conditions; the horizontal sliding opened a 50 m breach. Following the destruction caused by Hurricane Katrina, extensive research was carried out on I-wall structures including numerical analyses (e.g. Yuan and Whittle, 2013; Song et al., 2014) and centrifuge tests (e.g. Sasanakul et al., 2008; Ubilla et al., 2008; Steedman and Sharp, 2011). The formation of a gap between the I-wall and the surrounding soil media was highlighted.

5.3.2 Statistics of Observed Failure Mechanisms

The failure mechanisms for 503 breaching cases are identified (Figure 5.2). The failure mechanisms are observed in the following order (Figure 5.3): external erosion (68.6%), internal erosion (14.3%), slope failure (7.5%), embedded structure related failure (3.2%), horizontal sliding (0.6%), and other mechanisms (5.8%: distresses caused by human or animal activities, subsoil failure, ice drifting, micro instability, etc.).

The three most important failure mechanisms of dikes, embankment dams, and landslide dams are compared in Figure 5.3. The breaching mechanisms of dikes and embankment dams differ in several aspects:

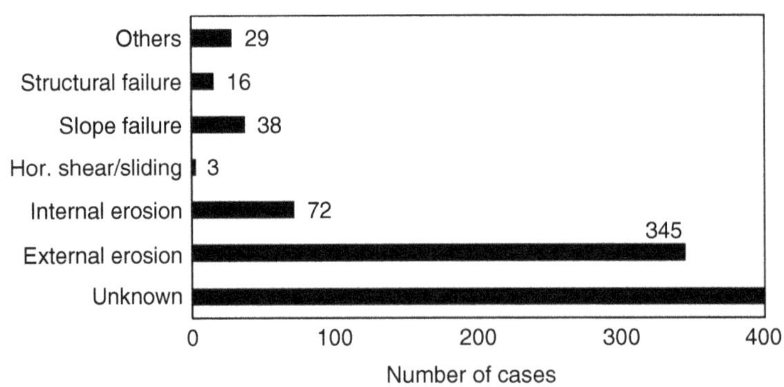

Figure 5.2 Statistics of observed failure mechanisms. *Source:* Danka and Zhang (2015). With permission from ASCE

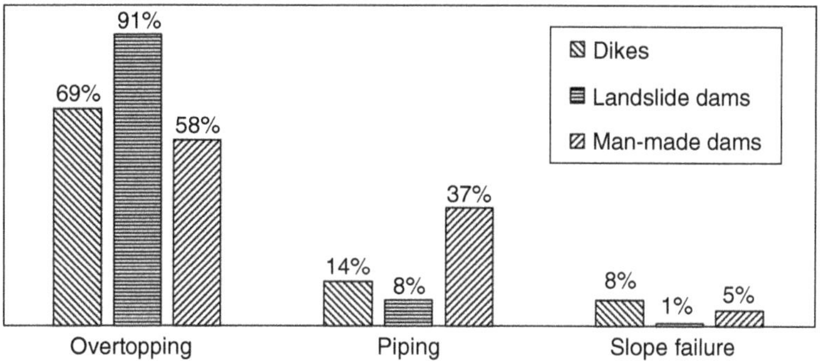

Figure 5.3 Frequency of failure mechanisms of dikes, landslide dams and constructed dams. *Source:* Danka and Zhang (2015). With permission from ASCE

1. The breach water from a dam is limited by the amount of stored water, while the breach water from a dike is governed by meteorological and hydrological conditions. Additionally, constructed dams have better floodwater control measures (e.g. spillways). Hence overtopping is more likely for dikes than for constructed dams.
2. Constructed dams are usually founded on solid bedrock so that foundation scour is less likely, while dikes are often founded on soils, and the breach depth can be larger than the dike height due to the foundation scour.
3. A dike is a long distributed system; the breach is longer than that of a dam.
4. Constructed dams are usually higher than dikes and the dam slopes are usually steeper. Accordingly, the hydraulic gradients in constructed dams are often larger and hence internal erosion related failures are more likely in constructed dams.

Similarly, the breaching mechanisms of dikes and landslide dams are also different:

1. Similar to breaching of constructed dams, the reservoir volume controls the hydraulic conditions of landslide dam breaching.
2. Limited flood control measures are available for dikes while on-site flood defense is often not available for landslide dams. Hence overtopping is more likely for landslide dams than for dikes.
3. Material properties of landslide dams (e.g. erodibility, permeability, and shear strength) are highly variable, owing to natural deposition and lack of compaction and quality control.
4. The crests of landslide dams are wider and their slopes are gentler than those of dikes. Under flood conditions the hydraulic gradients in dikes are larger. Therefore, dikes are more likely to fail in internal erosion and slope instability than landslide dams.

The breach length is a critical parameter for dikes along major rivers. The relationship between breaching length and failure mechanism is presented in Figure 5.4. Based on the statistics, external and internal erosion might lead to extremely long breaches; several breaches in the database are longer than 500 m. While slope failure or horizontal sliding caused shorter breaches; the longest one in the database is only 300 m. This is not surprising because slope failures or sliding are of local nature.

The evolution of observed breaching mechanisms over time is also evaluated (Figure 5.5). The breaches before 1920 are difficult to study due to the lack of detail. However, some conclusions

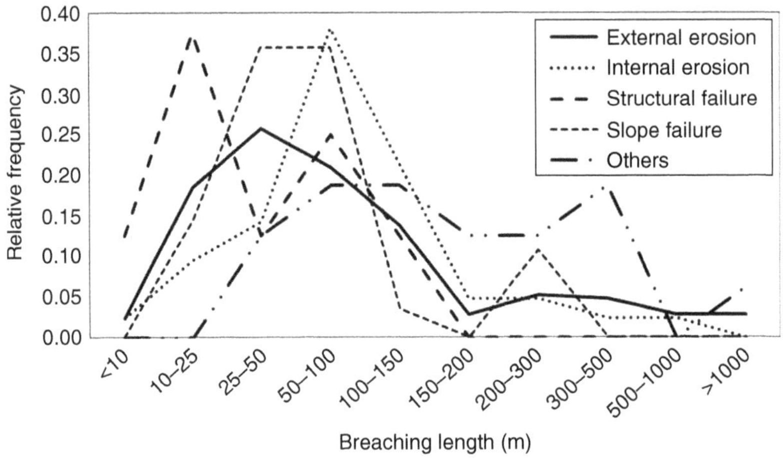

Figure 5.4 Breaching length as a function of breaching mechanism. *Source:* Danka and Zhang (2015). With permission from ASCE

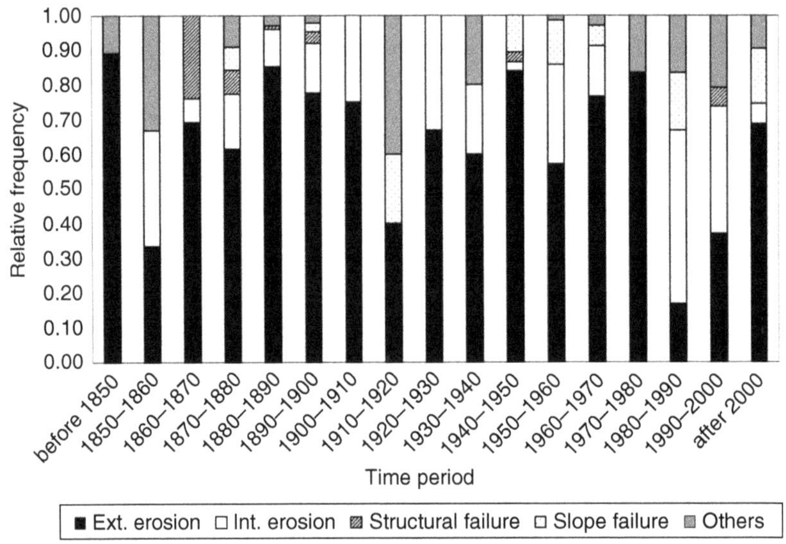

Figure 5.5 Evaluation of the frequency of breaching mechanisms on time scale. *Source:* Danka and Zhang (2015). With permission from ASCE

can be made. For instance, structural failure-induced breaches became less common over time owing to the increasingly stringent technical standards, better compaction technology, quality control, and understanding of the structural behavior. In addition, the percentage of breaches caused by animal activity or vegetation (under the category of 'others') is lower after World War I, which indicates improvements in dike maintenance.

New dikes were built and old dikes were reinforced to protect urban and agricultural developments in flood zones. The effect of new dike construction on an existing system is that the maximum floodwater level increases as a result of the narrowing down of the channel under flood conditions.

Therefore the likelihood of overtopping-triggered breach may become higher over time. The strengthening of a dike system, which usually includes heightening, has contradictory effects. With a higher dike, the likelihood of overtopping decreases but at the same time the increased hydraulic head difference may trigger breaches via internal erosion related failure mechanisms. According to Figure 5.5, in the periods of 1920–1950 and 1950–1980 new dikes were constructed extensively, while dike strengthening took place around 1950, after the end of World War II.

Results also show that, from 1980 to 2000, overtopping was not the most common cause of dike breaches, but internal erosion, structural failures, or horizontal sliding were observed more frequently. As mentioned above, the underlying reasons could be dike reinforcement and improvement in floodwater level forecast with the aid of more reliable meteorological and hydraulic predictions. Such improvements support flood defense activities; priorities and potential weak links can be identified more reliably. Since the millennium, extreme flood events with record-breaking flood levels have been observed all around the world. The peak flows on several European rivers set new records and more dikes have been overtopped. Hence the ratio of external erosion has increased again.

Part II

Dam Failure Mechanisms and Breaching Process Modeling

6

Internal Erosion in Dams and Their Foundations

6.1 Concepts of Internal Erosion

Constructed dams and dikes are a critical infrastructure for flood control, irrigation, and water and clean energy supply. However, due to problems such as substandard construction materials, poor quality control during construction, poor maintenance, extreme inflow, etc., an embankment dam may fail by internal erosion. Internal erosion occurs when soil particles are carried away locally, usually in suspension, by the hydrodynamic forces of the flowing water in the embankment or the foundation, from the embankment to the foundation, or around and into conduits through the embankment and adjacent walls supporting the embankment. According to the failure cases and distresses of embankment dams associated with internal erosion incidents (Fell et al., 2003; Foster et al., 2000a; Zhang et al., 2009), internal erosion has historically resulted in about 37% embankment dam failures. Therefore, it is crucial to gain fundamental understandings of the triggering mechanism and process of internal erosion.

Studies on the incidents of internal erosion have shown that there are different outflow kinetics and erosion kinetics that correspond to different levels of danger. To distinguish different initiation mechanisms or to discern the boundary conditions of the outflow that form the basis of these manifestations of internal erosion, four initiation mechanisms are classified (Fell and Fry, 2007): concentrated leak erosion, backward erosion, contact erosion, and suffusion (Figure 6.1). They are all related to the seepage within dams or dikes. The details of the four mechanisms are presented in Section 6.2. When internal erosion occurs within an embankment dam, it does not mean that the dam will definitely fail eventually. Whether the erosion propagates depends on hydraulic conditions and material properties. For instance, if the dam filters successfully function, the erosion may not continue to develop. Therefore it is necessary to divide the entire internal erosion process into several representative stages. Based on extensive laboratory and field internal erosion tests and analysis of past failure/incident cases of embankment dams triggered by internal erosion, the internal erosion process can be divided into four phases: initiation, continuation, progression to form a pipe, and formation of a breach.

Dam Failure Mechanisms and Risk Assessment, First Edition. Limin Zhang, Ming Peng, Dongsheng Chang, and Yao Xu.

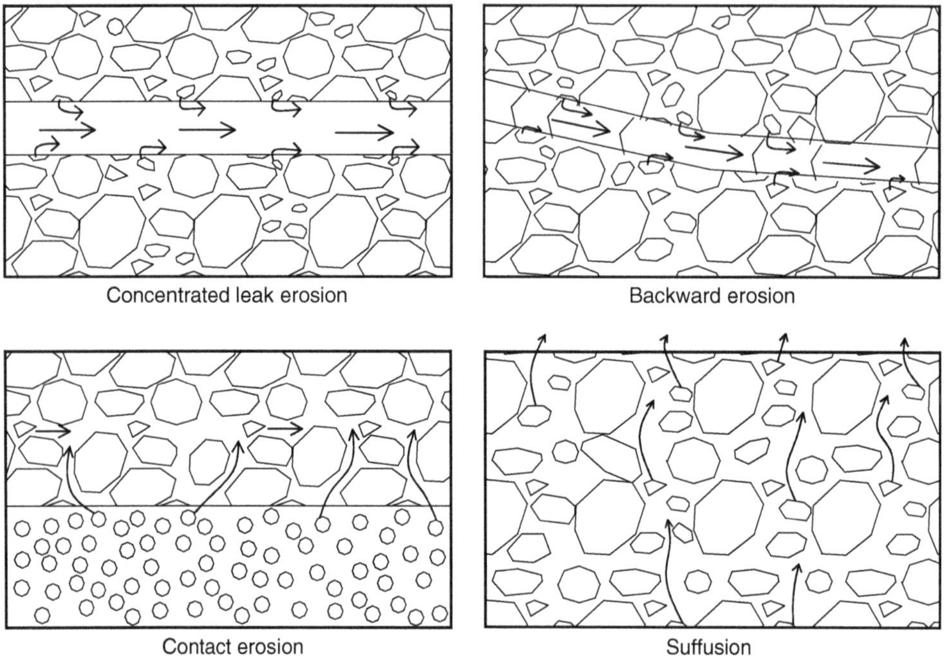

Figure 6.1 Four types of initiation mechanisms of internal erosion

The potential for internal erosion within an embankment dam depends both on the violence of seepage flow and the erosion resistance of the materials. The seepage flow is usually measured by hydraulic gradient in the analysis of the onset of internal erosion. The material properties and the mechanical conditions of the materials control the erosion resistance of the dam. Three classes of erosion resistance can be distinguished (ICOLD, 2013):

1. Non-plastic soils such as silts, sands, and gravels. These soils may collapse when saturated under flooding, will not sustain a crack when saturated, and are relatively easily eroded. As cohesionless soils become coarser due to the loss of fine particles, they progressively require more energy to initiate erosion. These soils are subject to backward erosion, contact erosion, or suffusion depending on their particle size distribution.
2. Plastic soils, such as clays, clayey sands, and clayey sandy gravels are generally more resistant to erosion than cohesionless soils. These soils are subject to concentrated leak erosion and contact erosion. Clay soils can hold a crack even when saturated. Higher energy is generally required to detach particles from cracks within a cohesive fill but the particles removed are small and easily carried through the crack. Backward erosion and suffusion will not occur in plastic soils under the gradients normally experienced in dams and their foundations.
3. Dispersive soils are plastic (clay) soils in which the chemistry of the seeping water causes dispersion (deflocculation) of the clay flocculation, breaking them down into smaller, more easily eroded particles.

A soil that fails to satisfy a geometric criterion does not imply high susceptibility to internal erosion, because the forces at the points of contact of the fine grains hinder their movement and, therefore, a relatively high gradient is necessary to initiate movement (Kovács, 1981). Accordingly, some embankment and hydraulic structures constructed using potentially unstable materials may not experience internal erosion in practice. This raises a challenge of confidently assessing the relative safety of soil structures built with potentially unstable materials. Currently, a typical measure of hydraulic conditions that control the onset of internal erosion in dams is critical hydraulic gradient. When the hydraulic gradient reaches a critical value, the fine particles in the dam body or its foundation may erode. The mechanical conditions of the dam material (i.e. stress state, relative compaction) also influence the initiation of internal erosion. A soil under high confining stress and high relative compaction is more resistance to erosion. As shown in Figure 6.2, the geometric conditions may influence the hydraulic conditions. For instance, the more susceptible to internal erosion of a soil is, the lower the critical hydraulic gradient. The mechanical condition also has an effect on hydraulic conditions. For instance, the critical hydraulic gradients that trigger internal erosion of a given soil are not the same under different stress states.

This chapter presents the fundamental mechanical behavior of internal erosion first from the geotechnical perspective. Then four initiation mechanisms of internal erosion (i.e. concentrated leak erosion, backward erosion, contact erosion, and suffusion) are discussed in detail. The causes, laboratory/field tests, and assessment methodologies for each initiation mode are presented

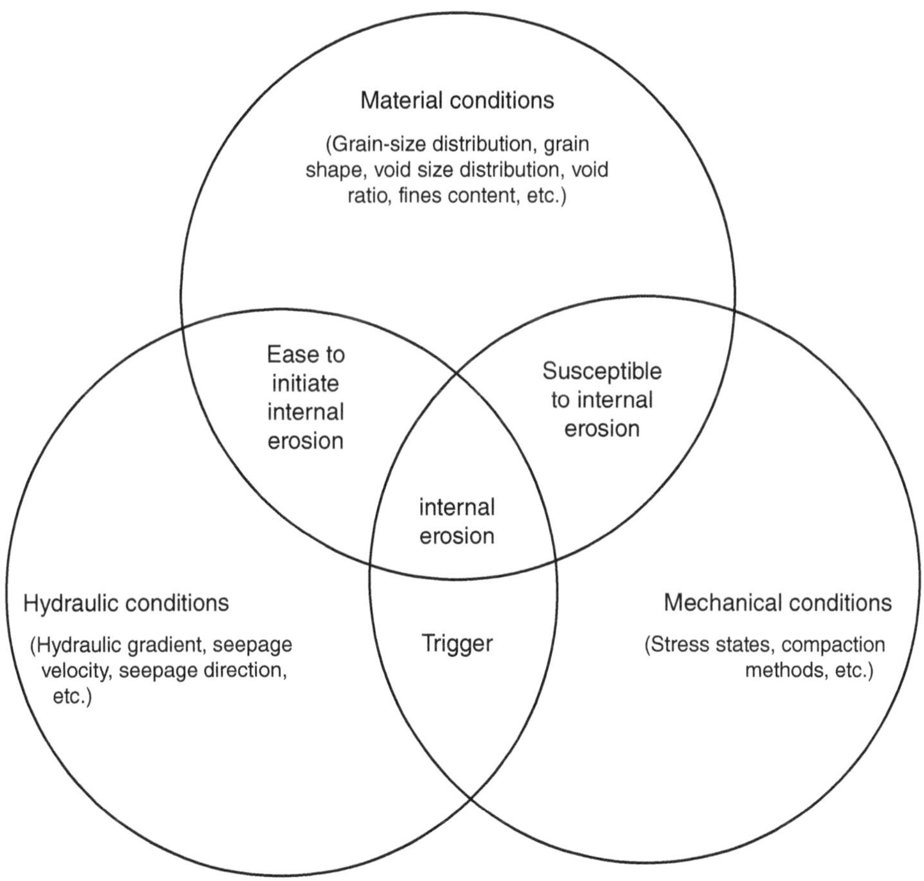

Figure 6.2 Coupling of three kinds of conditions (geometric, hydraulic and mechanical conditions) that trigger internal erosion

subsequently. Some commonly used filter criteria in practice are highlighted. Finally, some valuable topics in internal erosion are suggested for further study.

6.2 Mechanisms of Initiation of Internal Erosion

Internal erosion refers to the loss of soil particles within an embankment dam or its foundation by seepage forces. The entire internal erosion process of a dam can be generally divided into four phases: initiation, continuation, progression, and failure. Internal erosion can be initiated by four mechanisms: concentrated leak erosion, backward erosion, soil contact erosion, or suffusion. Figure 6.3 illustrates these four initiation modes for internal erosion within an embankment dam with a corewall. Each of these mechanisms is introduced in detail in the following sections.

6.2.1 Concentrated Leak Erosion

Erosion in a concentrated leak may occur in preferential paths (i.e. tensile cracks) in an embankment dam. Cracks can be caused by differential settlement, hydraulic fracture in clayey corewalls, or desiccation in voids adjacent to a concrete wall or conduits, in a continuous permeable zone containing coarse and/or poorly compacted materials which form an interconnected voids system, or in animal burrows into dikes and small dams, and along tree roots rotting in dams and forming holes. Where there is an opening in plastic soil, and sometimes in unsaturated silt, silty sand, or silty sandy gravel, through which concentrated leakage occurs, the sides of the opening may be eroded gradually by the leaking water. The diameter of the 'pipe' increases gradually in response to the erosion. Concentrated leak erosion is the most common initiation mode of internal erosion according to past failure cases, which can occur either during the first reservoir filling or in its operation.

Wister Dam is a homogeneous clay dam. A large storm caused the reservoir water to rise rapidly for the first time to a maximum depth of about 18 m on 28 January 1949. Then, with the large bottom outlets open, the reservoir water level started to drop slowly. On 30 January, seepage carrying eroded embankment material was first noticed on the downstream slope with an estimated quantity of about 150 L/s. This increased to more than 500 L/s in the following several days. By 3 February, with bottom outlets discharging at full capacity, the reservoir level had dropped about 4 m, exposing several erosion tunnels with diameter of about 0.6 m entering the upstream slope, all at about the

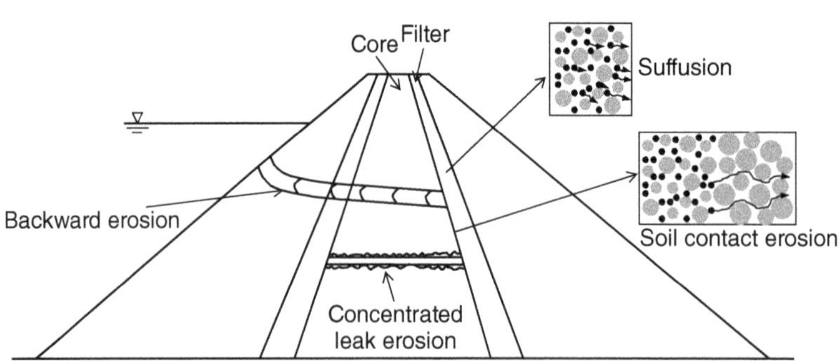

Figure 6.3 Illustration of initiation of internal erosion by four modes: backward erosion, concentrated leak erosion, soil contact erosion, and suffusion

same elevation, over a length of about 90 m. The erosion was initiated by hydraulic fracture (Sherard, 1986). Subsequently, the reservoir water level was lowered below the tunnels and the dam was saved.

6.2.2 Backward Erosion

Backward erosion involves the erosion of soil particles at the exit end of a seepage path due to a high exit velocity there. It can be simply divided into two types: backward erosion piping and global backward erosion (ICOLD, 2013). For the backward erosion piping, the exit of the seepage path may situate at a free surface, such as the ground surface downstream of a soil foundation or the downstream face of a homogeneous embankment. The detached particles are carried away by the seepage flow, and the erosion process gradually propagates towards the upstream side, following the critical seepage paths. The presence of backward erosion piping is often exhibited by the presence of 'sand boils' at the downstream side of the dam or dike. Backward erosion piping in the foundation of the Xiangjiang dike in China occurred on 22 September 2002. Several backward erosion piping points were found in the downstream foundation of the Xiangjiang dike during extreme rainfall. Due to the relatively permeable foundation materials (i.e. usually sandy soils), the sand particles can be easily picked up at the seepage exits of the downstream side under critical hydraulic conditions. A connected pipe can be formed if no mitigation measures are taken. Such an erosion mechanism often occurs in small dams and dikes founded on permeable layers. For instance, backward erosion piping has occurred several times in the North Pearl River Dike System in Guangdong Province in China during its operation (Zhang et al., 2011a).

For the global backward erosion, the exit of the seepage path may situate at the interface between the clay core and the downstream filter of a central core earth or rockfill embankment dam. The progression of the erosion process is assisted by gravity, and there is no need for a cohesive soil layer to form the roof for the pipe. If the core materials are not properly protected by the filter or the transition materials, backward erosion may initiate near the core–filter interface. Eventually, a sinkhole may form due to the global backward erosion.

6.2.3 Contact Erosion

Soil contact erosion involves selective erosion of fine particles in the interface of two different types of soils under seepage, for instance, at the base-filter interface in embankment dams. On one hand, if the pores formed by the coarse layer are too large, the fine particles in the finer layer can easily go through under seepage flow. On the other hand, if the coarse layer is not coarse enough to dissipate the possible pore water pressure generated within the fine layer, the coarse layer will not function successfully.

The natural soils of the alluvial plains are generally composed of successive layers of sediments of different origins, forming numerous interfaces between materials that have different grain size distributions. Moreover, the structure of embankment dams and dikes requires placing different layers of materials (i.e. core, filter, drains, transition, and shoulder), each having a specific function and thus different properties. A significant number of interfaces are thus found within the dam body and its foundation.

In order for contact erosion to occur, there are two conditions that need to be met: the pores of the coarse soils must permit the passage of the fine particles through their constrictions (the geometric condition), and the hydraulic constraint induced by the flow must be enough to wash out the fine particles when in contact with the coarse layer (the hydraulic condition). In terms of safety of hydraulic structures, these two conditions (geometric conditions and hydraulic conditions) have to be satisfied.

6.2.4 Suffusion

Suffusion involves selective erosion of fine particles within the matrix of coarse soil particles in a soil whose particle size distribution does not satisfy self-filtering conditions under seepage flow. The coarser particles are not transported and the effective stresses are largely transferred through the matrix of the coarser particles. For a soil susceptible to suffusion, once some fine particles are washed away, the microstructure of the soil will change accordingly, which may induce a mutation of hydraulic and mechanical conditions in the soil. In severe cases, the loss of fine particles could induce concentrated flow and eventually lead to backward erosion failure. If the hydraulic conditions are not too severe, stable situations may be reached with some of the finer fraction remaining in equilibrium with the seepage stresses. Suffusion may commence during cyclic periods of water loads or during high reservoir or river water level. Soil susceptible to suffusion is considered internally unstable. Broadly graded soils with a flat tail of fines and gap-graded soils are often internally unstable. Glacial tills are sometimes vulnerable to suffusion.

Internal erosion initiated by suffusion is a progressive process and can it induce the following incidents: (a) slope failure at the downstream face; (b) excessive settlement of the dam with loss of freeboard; and (c) formation of sinkholes at the crest and upstream of the dam, as illustrated in Figure 6.4. In (a), as the loss of the fine particles occurs at the seepage exist of the downstream slope, the local pore water pressure increases in addition to the decrease of shear strength of the soil. The slope stability is affected. Local slope failure may occur and the process then continues backward to the upstream. In (b), due to the loss of fine particles within the core or filters, the coarse matrix may rearrange, and the total volume decrease and then settlement occurs. In (c), due to the loss of fine particles within the much permeable flow paths, the fine particles are washed away along these paths. Therefore, a change of the volume of the coarse matrix occurs along these paths and a local sinkhole forms. The El Batan Dam in Mexico suffered suffusion, and two sinkholes near the crest were observed along the outlet axis (Flores-Berrones et al., 2011).

6.3 Initiation of Concentrated Leak Erosion Through Cracks

Concentrated leak erosion is the most common initiation mode of internal erosion in embankment dams. The concentrated leak is a "preferential path" in an embankment dam caused in various ways as mentioned in Section 6.2.1. Whether or not concentrated leak erosion occurs depends on whether a "preferential path" can be formed within the dam body or its foundation, which is eventually controlled by the stress conditions in the dam, the material properties, and the hydraulic conditions within the "preferential path". In the absence of filters or where filters are not satisfactory, the concentrated leak erosion will progress to form a pipe depending on the following (ICOLD, 2013):

1. whether the water flowing through the crack will cause the soil on the sides of the crack to swell, closing the crack or reducing the width so the forces imposed on the sides of the crack by the water flowing through the crack are insufficient for erosion to progress

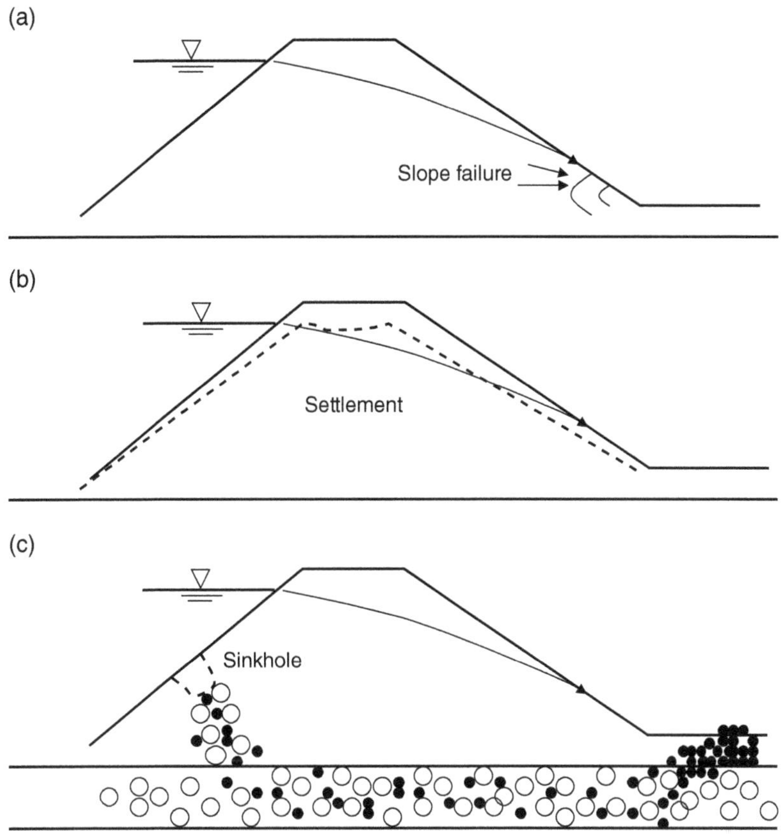

Figure 6.4 Consequences of internal erosion initiated by suffusion: (a) downstream slope failure; (b) settlement of dam with loss of freeboard; and (c) sinkhole in upstream face

2. whether the soil will hold the pipe roof
3. whether upstream zones will limit the erosion process or stop crack developments.

Once the progression of formation of pipe cannot be stopped properly, the pipe could be enlarged due to the pipe flow. The rate of the enlargement of the pipe depends on the soil erodibility and the violence of the pipe flow. In this section, the causes of concentrated leak, the testing of soil erodibility under pipe flow, and the physical modeling tests of concentrated leak erosion are presented.

6.3.1 Causes of Concentrated Leak

Concentrated leak occurs in cracks, hydraulic fractures, a gap adjacent to a conduit, a continuous permeable zone, animal burrows, and holes caused by tree roots in an embankment dam and its foundation. The causes of each kind of concentrated leak are as follows.

6.3.1.1 Cracks

Cracks in an embankment dam can be divided into three categories: desiccation cracks, transverse cracks, and longitudinal cracks. Desiccation cracking in embankment dams is caused by the drying and shrinking of certain types of embankment soils (e.g. high plasticity clay), particularly during an extended period of low reservoir levels. Desiccation cracks do not persist to a great depth so they only become an issue for reservoir levels near the crest level. Desiccation cracks usually develop in a random, honeycomb pattern. Typically, desiccation cracks occur in the crest and the downstream slope. Desiccation cracks may also occur on the surface of the first stage of dams which are built in two stages. The desiccation cracks may evolve to "preferential path" for seepage flow, which may eventually trigger concentrated leak erosion. The failure of Holland Dam Site A in 1997 was possibly due to the concentrated leak erosion in desiccation cracks (Richards and Reddy, 2007).

Transverse cracks appear approximately perpendicular to the axis of the dam. This type of crack is mainly caused by differential settlement. The differential settlement can frequently develop in an embankment where the abutments are steep and/or have steps in the profile (Figures 6.5a and b). The differential settlement can also occur if the foundation includes compressible materials, or small-scale irregularities are present in the foundation of the core (Figure 6.5c). During the construction of the dam, the partially saturated compacted soil in the embankment consolidates and settlement occurs. If the abutments are steep or have steps in the profile, differential settlements occur due to the variations in the height of the embankment. If the foundation contains compressible materials locally, large settlement may occur in the dam body above the compressible materials due to the high compressibility and high stress conditions. Subsequently, differential settlement may occur due to the non-synchronized deformation. For cracks to occur the small-scale irregularities need to be persistent over all or most of the distance across the core, and have steps greater than approximately 3–5% of the embankment height (Bonelli, 2013). Transverse cracks could also be generated by ground motion caused by a strong earthquake or dynamic vessel impact. The details of dynamic impact on transverse crack are out of the scope of this book. Transverse cracks may provide a path for seepage through the embankment. When the depth of the crack extends below the reservoir level, very rapid erosion of the dam may occur, eventually leading to breaching of the dam.

Longitudinal cracks occur in the direction roughly parallel to the axis of the dam. This type of crack mainly occurs in zoned dams. For instance, in an embankment dam with corewall and filter layers, uneven settlement between adjacent embankment zones may occur due to the different compressibility of materials in different zones. Eventually, longitudinal cracks could be generated because of the uneven settlement. The longitudinal cracks can also occur on the scarp of an unstable slope in earthfill dams or dikes. If progressive failure occurs on the downstream slope of the dam, tension cracks occur on the scarp of the slope in longitudinal direction and seepage water may concentrate in the cracks in turn. Eventually, the slope may fail due to the decreased soil shear strength. In this case, the cracks may be arc-shaped. Moreover, when embankment dams experience a large earthquake they often settle and spread in the upstream–downstream direction, leading to longitudinal cracks. Austrian Dam (Harder, et al. 1991) is a typical case where longitudinal and transverse cracks occurred on the downstream slope and the abutment, respectively.

6.3.1.2 Hydraulic Fracturing

Hydraulic fracturing is another common triggering factor for concentrated leak erosion. It refers to fracturing or cracking induced in a soil when hydraulic pressures in the soil exceed the minor principal stresses and tensile strength of the soil. Owing to the different stiffnesses of soils in

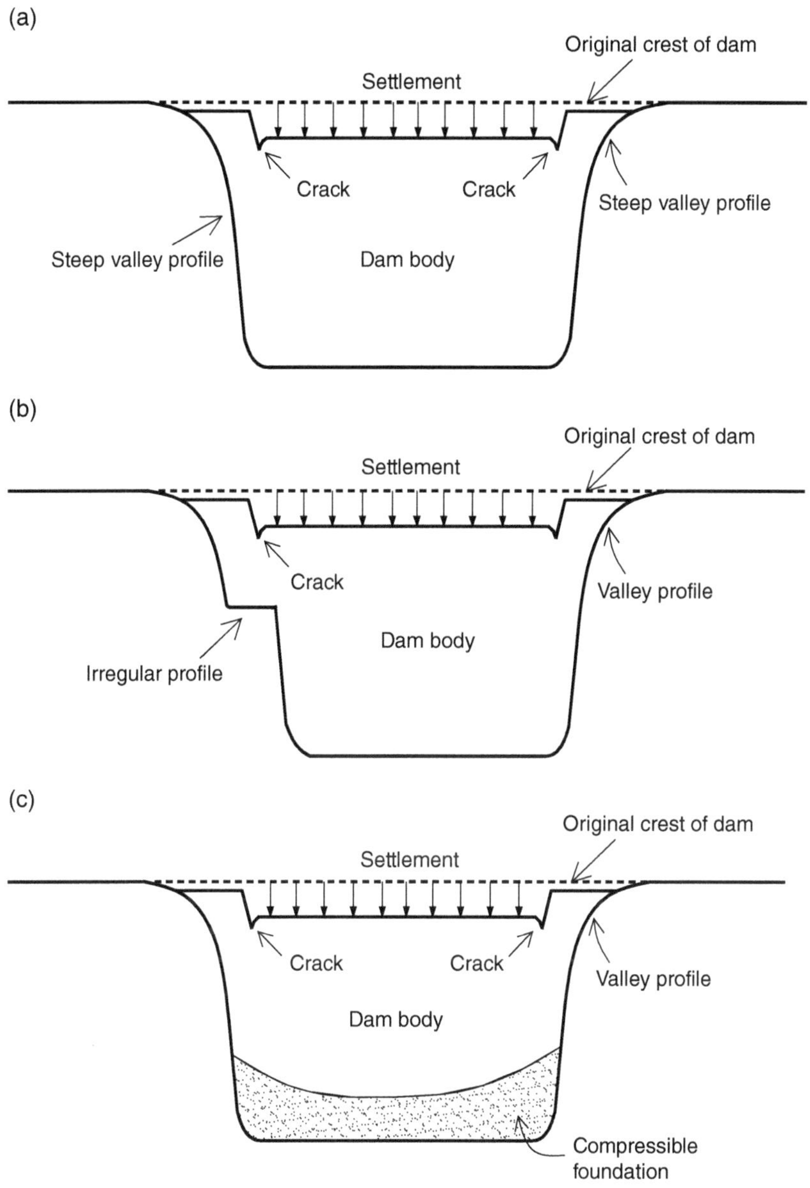

Figure 6.5 Illustration of transverse cracks in an embankment dam due to (a) steep abutment slope; (b) irregular abutment surface; and (c) compressible foundation material

different zones within an embankment dam or its foundation, differential settlement usually occurs. The differential settlement creates strains in the compacted embankment, changing the internal stress distribution. In this action the internal pressure is increased at some locations and decreased at other locations. In localized, but fairly large, zones within the dam the minor principal stress is reduced to near zero, or to even tensile stress if the dam has sufficient true cohesion to withstand tension. When the reservoir level rises above one of these surfaces with low stresses in the impervious section of the dam, water can enter along the surface from the upstream face in a concentrated thin layer. When the reservoir water pressure becomes slightly greater than the embankment stress, there is no resistance to the entry of water into the embankment in a concentrated leak. As the reservoir level continues to rise, increasing the pressure in the water-filled crack, the stress conditions in the embankment are changed, the impervious embankment material deforms, and the crack is opened wider.

In any area of low stresses in an embankment dam, particularly areas where "arching" may lead to portions of the embankment not experiencing the full weight of the overlying soils, hydraulic fracturing may be easily triggered. Evidence provided by Sherard et al. (1972) and Sherard (1986) indicate that hydraulic fracturing in relatively low homogeneous dams can occur, particularly when the embankment material is dispersive. For high central core dams, with rockfill or gravel shells, the impervious core width is generally less than the head of water retained, and the conditions for stress transfer are much more severe. Therefore, if hydraulic fracturing is common in homogeneous dams with their relatively wide impervious sections, even without unusually large settlements, it should be expected even more commonly in thin central core dams. The evidence from records of dam performance is in two main categories (Sherard, 1986): (1) Dams with inadequate filters which have developed erosive leaks through the core on the first reservoir filling; and (2) dams without erosive leaks in which exploratory boring in the core has encountered water-filled cracks.

An obvious example of this condition would be in a narrow, steep-sided excavation, perhaps a cut-off/key trench or an excavation for a conduit. For these types of feature, the overlying soil is able to bridge over the narrow excavation, resulting in relatively low stresses in the backfilled trenches. If the reservoir water has access to these areas, high water pressures could lead to hydraulic fracturing of the low stress soils. This mechanism has been suggested to have played a role in the failure of Teton Dam, which featured a very narrow, steep-sided key trench backfilled with low plasticity soil (USBR, 2011). Moreover, hydraulic fracturing can also occur within an embankment during improper drilling or grouting operations. Since pressurized fluids are used in these operations, it is possible to create a situation where high-pressure fluid exists in a drill hole, with pressures that exceed the stresses in the dam, which are a function of the height of the fill. Thus, particularly vulnerable locations within a dam are where existing confining stresses are anomalously low. The high hydraulic pressures force open a crack in the embankment, which could theoretically propagate for a significant distance. Although in many cases the crack may be filled with the drilling mud or grout used in the triggering operation, there is a potential that an untreated defect could be formed within the embankment core.

The above-mentioned hydraulic fracturing in embankment dams has been reflected from many field evidences, such as Balderhead Dam in the UK (Vaughan et al., 1970), Hyttejuvet Dam in Norway (Wood et al., 1976), Viddalsuatu Dam in Norway (Vestad, 1976), Wister Dam and Teton Dam in the US (Sherard, 1986), and Manicouagan Dam in Canada (Lo and Kaniaru, 1990).

6.3.1.3 Gap/Cracks Adjacent to Conduit and a Continuous Permeable Zone

Conduits convey water from a reservoir through, under, or around an embankment dam in a controlled manner. Most conduits are part of the outlet works. However, owing to problems such as inadequate compaction of the material under the haunches of the conduit, lack of effective filter collar/diaphragm, and excessive energy used to compact the embankment material adjacent to the conduit, concentrated leak erosion could occur around the conduit. Figure 6.6(a) shows the formation of a low density zone at the haunches of the conduit. Owing to the use of circular shaped conduit, a low density zone could be formed during the compaction of the materials adjacent to the conduit. Compacting soil adjacent to a conduit is difficult, and compaction efforts can dislodge the conduit and create pathways for future concentrated flow (USBR, 2012). Both seepage flow and flow due to leakage of the defective conduit can generate hydraulic fracture in the low density zone. Moreover, drying of the soil in the side walls of the trench in which the culvert is placed during construction can also cause cracks around the conduit. Concentrated leak erosion can occur if the hydraulic force overcomes the erosion resistance of the material within the crack, particularly if the conduit is not properly protected by filter collars. This is more often observed in some old dams. However, recent failure cases indicate that many failures occurred even though filter collars were installed correctly, and the collars could be seen to be intact after the failure. A portion of the flow path may have been away from the interface between the conduit and the surrounding soil in some dam failures.

In some cases, excessive compaction efforts are made to densify the materials adjacent to the conduit. This can lead to a gap forming below the conduit owing to the lifting of the conduit under the high compaction energy. Figure 6.6(b) shows the formation of a gap under the conduit. Under certain hydraulic conditions, the erosion of the soils around the gap can be easily initiated. The eroded failure path can enlarge to a size that can empty a reservoir rapidly.

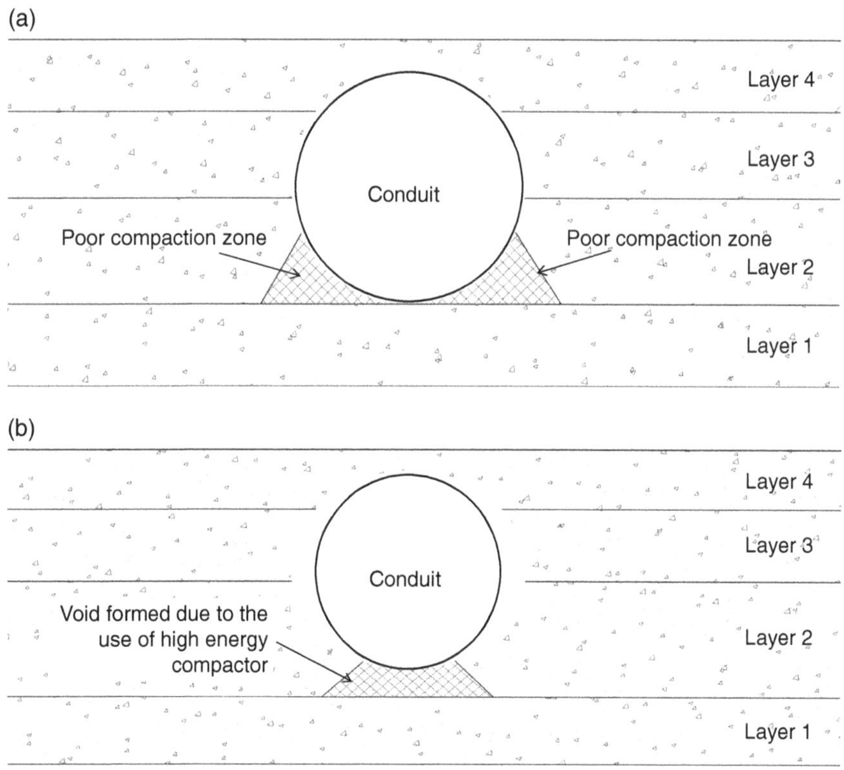

Figure 6.6 (a) Low density zones subjected to hydraulic fracture and (b) formation of a void below the conduit due to high energy compaction

If internal erosion occurs in soils surrounding a conduit defect, the potential for eventual failure is high, because all soils when subjected to continued flow along a preferential flow path are erosive over time. Highly erodible soils, such as non-plastic silts, broadly graded silty coarse-grained soils, and dispersive clays, could develop erosion features more quickly (FEMA, 2005a).

A continuous permeable zone within an embankment dam could also be one of the causes of concentrated leak erosion. Harmful segregation of widely graded materials during the placement in the field is inevitable, particularly if the materials are internally unstable. As a result of the segregation, hydraulic conditions change in the continuous permeable zone, and the shear stress generated by the seepage flow could easily overcome the erosion resistance of the soil in the permeable zone. If the filter cannot operate properly, concentrated leak erosion can be induced. Moreover, embankment dams may be constructed in several stages, and the interfaces between different stages could be a weak zone. Thus, concentrated leak erosion could also occur at the interfaces.

6.3.1.4 Animal Burrows and Holes Caused by Tree Roots

Animal burrows in an embankment or dike can lead to nearly continuous holes through the embankment or limited length of holes (FEMA, 2005b). The presence of animal burrows, either on the upstream or downstream slope, can distort the established phreatic surface and impact the flow net. The upstream burrows can allow the normal pool elevation to extend into the dam embankment, forcing the phreatic surface further into the embankment. Likewise, downstream intrusions can raise the phreatic surface on the downstream slope. The overall effect can be a significant alteration to the phreatic surface. Dramatic changes to the phreatic surface can shorten the seepage paths, increase seepage volumes, increase hydraulic gradient, and cause internal erosion in the embankment materials.

Vegetation growing on dams and dikes may also lead to a potential concentrated leak and initiation of erosion, which can be a result of (FEMA, 2005c):

1. uprooted trees that produce large voids and reduced freeboard
2. decaying roots that create seepage paths and internal erosion problems
3. cracking, uplifting, or displacing concrete structures and other facilities
4. allowing roots to wedge into open joints and cracks in the foundation rock along abutment groins and the embankment toe, thus increasing piping and leakage potential
5. root penetration of conduit joints and joints in concrete structures.

6.3.2 Need for Studying Soil Erodibility for Concentrated Leak Erosion

Improved methods for modeling concentrated leak erosion process incorporate quantifiable erodibility parameters that can be estimated from basic soil properties or measured in the laboratory or in situ. Soil erodibility is a very complex fundamental parameter, which reflects the intrinsic property of soil. Once a concentrated leak forms, whether the erosion can be generated or not or the erosion rate if it can be initiated is governed by soil erodibility.

Soil erosion is a complex process that involves both hydraulic and mechanical conditions. Over the years, efforts have been made to characterize various aspects of soil erodibility, such as the influence of pore fluid chemistry on the erodibility of clay and the influence of soil mechanical properties: relative density, water content, and shear strength.

Several surface erosion tests such as the jet index test, the rotating cylinder test, and the flume test have been carried out to investigate erosion within a soil mass. As different test methods involve different particle transport mechanisms, and are applicable to different types of soils, their results are difficult to compare. For instance, after conducting a large number of tests using the hole erosion method and the jet erosion index method, Wahl et al. (2008) concluded that the most important factors that can contribute to the differences in erodibility parameters obtained from the two kinds of tests are the accuracy of the stress descriptions used in each analysis, inherent differences in the nature of the hydraulic attack upon the eroding surface, the way that the flow exploits different weaknesses in the soil structure, and differences in the geometry of the exposed soil surface.

In order to quantify the relative erodibility of various types of soils commonly found in embankment dams, a sufficiently wide range of soils should be investigated by the same erosion test method. The selected test should be able to provide adequate control over the various mechanical properties that influence soil erodibility.

6.3.3 Laboratory Tests on Concentrated Leak Erosion

Currently, one of the most common test methods for concentrated leak erosion is the hole erosion test (HET). It was developed at the University of New South Wales (UNSW) as an index test to model internal erosion in concentrated leaks in earth dams (Wan and Fell, 2004a, b; Wan, 2006) as shown in Figure 6.7. In the HET, the soil specimen is compacted inside a standard mold used for the standard compaction test. A 6 mm diameter hole is drilled along the longitudinal axis of the soil sample to simulate a concentrated leak. The hydraulic head difference is increased stepwise until progressive erosion of the hole is produced. Once progressive erosion is produced, the upstream head remains constant until the end of the test. The HET uses flow rate as an indirect measurement of the diameter of the pre-formed hole. The flow within the hole generates shear stresses on the hole face. The soil will erode once the generated shear stress exceeds the critical shear stress of the soil.

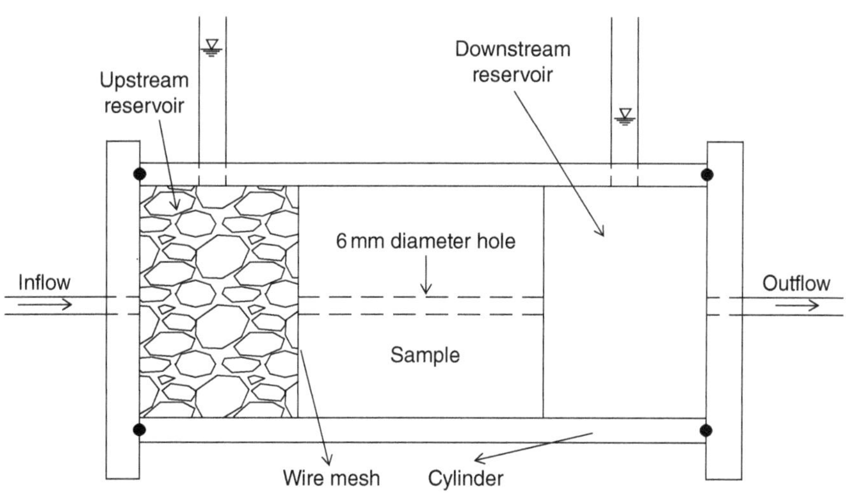

Figure 6.7 Schematic of hole erosion test (HET) apparatus (modified from Wan and Fell, 2004a)

In the HET, the hydraulic shear stress along the pre-formed hole can be expressed as

$$\tau = \rho_w g i_t \frac{d\phi_t}{dt} \tag{6.1}$$

where i_t is the hydraulic gradient across the soil sample at time t and ϕ_t is the diameter of the pre-formed hole, assumed to remain circular at time t.

In calculating the hydraulic shear stress in the hole, the following assumptions are made:

1. Flow through the soil matrix is negligible.
2. Only the soil surface along the pre-formed hole provides shear resistance.
3. Energy losses due to the eroding fluid entering and exiting the pre-formed hole are negligible.
4. The cross-section along the length of the soil specimen is uniform and circular.
5. The flow profile in the pre-formed hole throughout the length of the soil specimen is fully developed.
6. The hydraulic head difference across the soil specimen equals the total friction head loss.

The rate of erosion per unit surface area of the hole at time t, denoted by $\dot{\varepsilon}_t$, is given by

$$\dot{\varepsilon}_t = \frac{\rho_d}{2} \frac{d\phi_t}{dt} \tag{6.2}$$

where ρ_d is the dry density of the soil.

The diameter ϕ_t of the pipe at time t can therefore be estimated as

$$\phi_t = \left(\frac{16 Q_t f_{Lt}}{\pi \rho_w g i_t} \right)^{1/3} \quad \text{for laminar flow} \tag{6.3}$$

$$\phi_t = \left(\frac{64 Q_t^2 f_{Tt}}{\pi^2 \rho_w g i_t} \right)^{1/5} \quad \text{for turbulent flow} \tag{6.4}$$

where f_{Lt} and f_{Tt} are the friction factors relating shear stress to the mean flow velocity, and Q_t is the flow rate at time t (m^3/s).

The soil erodibility can be estimated by measuring the flow rate, the hydraulic gradient, and the final hole diameter.

Lim (2006) observed two limitations in the HET:

1. The pre-formed hole by drilling may densify the soil at the hole surface, which may influence its erosion resistance.
2. The effects of slaking may be present, where soil particles detach from the specimen due to the presence of water (hydrostatic conditions), rather than the applied shear stress.

Table 6.1 Summary of factors affecting concentrated leak erosion

Category	Description
Material properties	Soil type Plasticity index Water content Degree of saturation Fines content Soil mineralogy Dispersivity Relative compaction etc.
Hydraulic conditions	Hydraulic gradient Salt concentration Fluid viscosity etc.

6.3.4 Factors Affecting Concentrated Leak Erosion

Concentrated leak erosion can occur in cracks, hydraulic fractures, gap/cracks adjacent to conduit, continuous permeable zone, animal burrows, and tree roots, etc. The factors influencing the formation of these defects can be divided into two categories: (1) material properties and (2) hydraulic conditions. Table 6.1 summarizes the common factors that influence concentrated leak erosion.

6.3.4.1 Material Properties

The material properties are considered to be the intrinsic factor that influences the initiation of concentrated leak erosion. This includes soil type, relative compaction, plasticity index, water content, degree of saturation, fines content, soil mineralogy, and dispersivity. All these factors can be reflected in soil erodibility. Critical erosive shear stress indicates the erosion threshold, whereas the coefficient of erodibility expresses how fast the erosion is. Bonelli and Benahmed (2011) argued that evaluating the erodibility of soil, both in terms of erosion threshold and erosion rate, is critical to evaluating the safety of a water retaining structure. Although the initiation mechanisms for surface erosion and concentrated leak erosion are not the same, the main control variables for soil erodibility, both for surface erosion and concentrated leak erosion, are similar. The main factors affecting soil erodibility in surface erosion are presented in Chapter 7.

6.3.4.2 Hydraulic Conditions

Once a concentrated leak is generated within a dam, concentrated water flow within the leak can introduce shear stress on the soils around the leak (i.e. crack, hydraulic fracture, soil structure interface, etc.). Ideally, the shear stress induced by the pipe flow within a circular pipe can be expressed as (i.e. Reddi et al., 2000; Wan and Fell, 2004a):

$$\tau = \gamma_w i \frac{\phi_t}{4} \tag{6.5}$$

where γ_w is the unit weight of water (N/m^3); i is the hydraulic gradient along the pipe; and ϕ_t is the diameter of the pipe (m).

Equation (6.5) indicates that the hydraulic gradient is an important parameter influencing the concentrated leak erosion. The hydraulic gradient can be expressed as

$$i = \frac{\Delta h}{L} \tag{6.6}$$

where Δh is the total head difference between the inlet and outlet of the pipe and L is the total length of the pipe.

This indicates that the location of the concentrated leak and the length of the pipe are also important control variables for the ease of initiation of concentrated leak erosion. Particularly when the water table has a rapid increase during an unexpected heavy rainfall, the concentrated leak along the conduit or transverse cracks can be more easily triggered. Transverse cracks through the core of a dam are particularly dangerous because the cracks provide ready paths for concentrated seepage.

The hydraulic gradient that leads to initiation of concentrated leak erosion could be lower than the theoretical value for "sand boil". For example, the hydraulic gradient at the AV Watkins Dam incident was calculated to be 0.08, and the hydraulic gradient at USACE's Wister Dam was reported to be 0.02 (USBR, 2012). However, little research has been conducted to investigate the critical hydraulic gradient that leads to concentrated leak erosion. The ease of initiation of concentrated leak erosion is also influenced by the properties of the seepage flow, for instance, the salt concentration and the viscosity of fluid. The HET conducted by Lim (2006) using both tap water and distilled water, indicated that most testing soils showed a greater resistance against erosion when tap water was used as the eroding fluid due to the salt contained in the tap water. Sinco et al. (2010) carried out similar HETs on sand using deionized water, deionized water with 6% bentonite, and deionized water with 1% fines as permeating fluid. The results indicate that a higher density of a permeating fluid introduces a lower shear stress on the soil.

6.3.5 Soil Dispersivity

Soil dispersivity is mainly due to the presence of exchangeable sodium present in the composition of clay, where the attractive forces are weaker than the repulsive forces. A dispersive clay has a preponderance of sodium, whereas an ordinary clay has a preponderance of calcium and magnesium cations in the pore water. Sherard et al. (1976a) found that soils with percent dispersion greater than 50% are susceptible to dispersion and concentrated leak erosion, whereas those with a percent dispersion less than 15% are not susceptible. The commonly used laboratory tests for determining the dispersivity of a soil are the pinhole erosion test (Sherard et al., 1976b; ASTM, 2006), the double-hydrometer test (Volk, 1937; ASTM, 2007), and the chemical test (Sherard et al., 1976a). The basic theory and procedures for the pinhole test and the double-hydrometer test are now presented briefly.

6.3.5.1 Pinhole Test

The pinhole test was developed by Sherard et al. (1976b) to directly measure the dispersibility of compacted fine soils in which water is made to flow through a small hole in a soil specimen to simulate water flow through a crack or other concentrated leakage path in the impervious core of

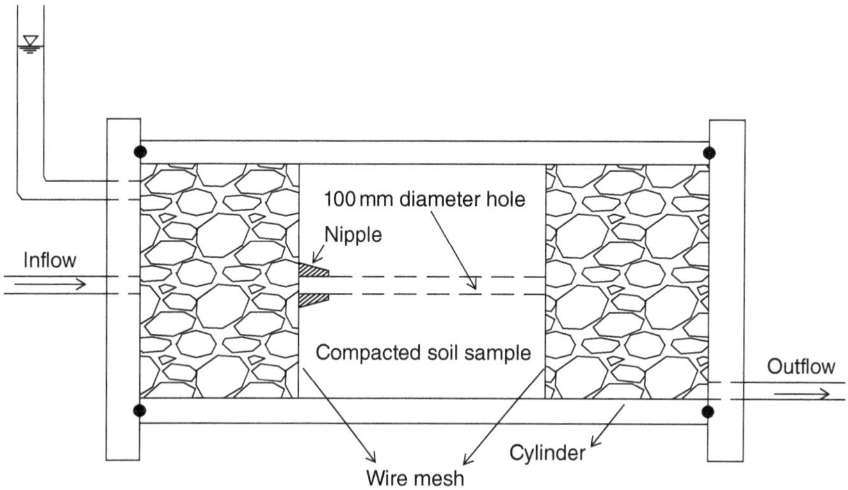

Figure 6.8 Schematic of pinhole test apparatus (modified from Sherard et al., 1976a)

a dam or other structure. The test has become a widely used physical test and recommended by ASTM as a standard test (ASTM, 2006).

A 1.0 mm diameter hole is drilled through cylindrical soil specimen 25 mm in length and 35 mm in diameter (Figure 6.8). Distilled water is percolated through the pinhole under heads of 50, 180, 380 and 1020 mm for 5 min each. The measuring variables include the flow rate, the final hole size, and the effluent turbidity. The test result is evaluated from the appearance of water, the rate of flow, and the final size of the hole in the specimen. The general test procedure is as follows (Sherard et al., 1976b):

1. The soil specimen must be preserved at the natural water content by shipping and storing before testing in airtight bags or containers.
2. Any coarse sand or gravel particles in the test material should be removed by sieving the material through a No. 10 sieve (2 mm).
3. The natural water content is measured, and the water content is brought to near the plastic limit by adding the required distilled water.
4. The 38 mm long clay specimen is compacted in the cylinder on top of pea gravel and wire screen in five layers to the target degree of compaction.
5. A plastic nipple is pushed into the top of the specimen and the hole is punched through the nipple with a 1.0 mm diameter stiff steel wire.
6. After the apparatus is assembled, distilled water is percolated through the hole in the sample under heads of 50, 180, 380, and 1020 mm for time periods of 5 min to 10 min at each head. The outflow is measured continuously using a flask, and the color of the outflow is observed through the side of the flask and vertically through the column of fluid in the flask.
7. At the end of the successive flow tests, the apparatus is dismantled and the soil specimen is extruded from the cylinder and broken open to examine the size of the hole.

Sherard et al. (1976b) divided the soil dispersivity into six categories. The principal differentiation between dispersive and non-dispersive soils is given by the test results under

50 mm of head. If the flow under 50 mm of head is visibly cloudy and does not become clear with time, the specimen is classified as highly dispersive (D1). If the flow at 50 mm of head has a slight but easily visible color as seen from the side of the flask at the end of 5 min, continue the test for a total of 10 min. If flow continues colored, stop the test. If the rate of flow at the end of 10 min has not exceeded 0.8 ml/s, and the hole diameter exceeds two needle diameters, classify the soil as intermediate (ND4). If the flow at the end of 10 min exceeds 1.0 ml/s, and the hole diameter exceeds two needle diameters, classify the specimen as dispersive (D2). If the flow under 50 mm of head is clear at the end of 5 min, raise the head to 180 mm and continue the test. At 180 mm of head if the water continues to flow clear, or has only a trace of color as seen from the side of the flask, after 5 min raise the head to 380 mm and continue the test. If the water has color and the rate of flow increases rapidly, stop the test. The dispersivity of the soil is classified as intermediate (ND3). If the flow is completely clear as seen from the top of the measuring flask, raise the head to 1020 mm after 5 min. If the flow has a slight color or exceeds 3.5 ml/s, stop the test and classify the soil as non-dispersive (ND2). If the flow continues completely clear at rate usually less than 4.0 ml/s, classify the soil as non-dispersive (ND1). If the flow continues has a bare trace of color or exceeds 5.0 ml/s, classify the soil as non-dispersive (ND2).

6.3.5.2 Double-hydrometer Test

The Soil Conservation Service (SCS) double hydrometer test is one of the most commonly used methods for classifying dispersive clay soils. The test evaluates the dispersibility of a soil by measuring the natural tendency of the clay fraction to go into suspension in water (Volk, 1937). The procedure involves the determination of the percentage of particles in the soil that are finer than 0.005 mm, using the standard hydrometer test with strong mechanical agitation and a chemical dispersant. Meanwhile, a parallel test is also carried out, in which no chemical dispersant is added and the solution is not mechanically agitated. The dispersion ratio is defined as the ratio of the percentage of particles smaller than 0.005 mm in the method without using dispersion agent and the percentage of particles smaller than 0.005 mm in the standard method. Dispersion ratios greater than 50% are considered dispersive, those of 30–50% are moderately dispersive, and those less than 30% are non-dispersive. The double-hydrometer test is also a standard testing method adopted in ASTM (2007).

6.3.5.3 Chemical Test

Ingles and Wood (1964) recognized the presence of exchangeable sodium as a main contributing chemical factor in dispersive clay behavior. The dispersion potential of the soil is evaluated with respect to the type and quantity of cations present in the pore water of the soil specimen. Sherard et al. (1976b) defined sodium absorption ratio (SAR) of the pore water as an indicator for soil dispersivity

$$SAR = \frac{Na^{+}}{0.5\left(Ca^{2+} + Mg^{2+}\right)} \tag{6.7}$$

The testing procedure is as follows:

1. The soil is mixed with distilled water to a consistency near the Atterberg liquid limit.
2. Saturation extract is prepared from the soil sample using a filter.
3. Four main cations – calcium, magnesium, sodium, and potassium – are determined in terms of milliequivalents per liter.

The percentage of sodium can then be calculated using Equation (6.7). The use of the SAR is based on the fact that soils in nature are in equilibrium with their environment. In particular, there is a relationship between electrolyte concentration of the soil pore water and the exchangeable ions in the clay adsorbed layer.

6.4 Initiation of Backward Erosion

Backward erosion is a form of internal erosion that mainly occurs in non-plastic silts or sands of a foundation or the core of a dam or dike. It involves the detachment of soil particles when the seepage exits to a free unfiltered surface, such as the ground surface downstream of a soil foundation, the downstream face of a homogeneous embankment, and a coarse rockfill zone immediately downstream from the fine-grained core. The detached particles are carried away by the seepage flow, and the process gradually works its way towards the upstream side of the embankment or its foundation until a continuous pipe is formed. A distinction is made between pipe formation in a near-horizontal direction below a cohesive layer (known as backward erosion piping) and in a near-vertical pipe formed in non-plastic embankment cores (known as global backward erosion) (ICOLD, 2013). In this section, the conditions that trigger the backward erosion are presented first, then several commonly used methods to assess the initiation of backward erosion are reviewed. Finally, the process of formation of a pipe due to backward erosion is described.

6.4.1 Susceptibility of a Dam or Dike to Backward Erosion

For backward erosion piping to occur within a dam or dike, the following criteria should be met (ICOLD, 2013):

1. The soil where pipes develop would need to be non-plastic. For instance, backward erosion piping was observed in silty sand, sandy silt, and clean sand. Even in the same soil, the susceptibility of the soil to backward erosion piping is also influenced by its relative density, seepage direction and stress state, etc. Richards and Reddy (2012) conducted extensive backward erosion piping tests on clean sand under true triaxial stress conditions and found that the resistance to initiation of backward erosion in a low-plastic soil increases with the increase of major principal stress and the increase of relative compaction. The effect of specific gravity, angularity, and gradation of the soil was investigated by Fleshman and Rice (2014). Their test results indicate that soils with higher specific gravity, angular soils, and graded soils showed greater backward erosion piping resistance.
2. There would have to be a free surface at the downstream of the dam core at which erosion could initiate; for example if there is no filter or an inadequately filtered contact between the core and a transition zone.

3. The hydraulic conditions (e.g. flow path, source of water, critical hydraulic gradient) should reach critical conditions.
4. Continuous stable roof would have to form for the pipe to progress. This might occur at the phreatic surface if the partially saturated soil above the surface was silty, or if more cohesive strata were layered in the core.

Backward erosion piping is dominant in areas where the conditions described above are met. This is often the case for river dikes, such as those found along the main rivers in the Netherlands, the Mississippi river in the USA, and the Yangtze rivers in China. In these river systems, backward erosion piping is known to have led to the occurrence of sand boils and dike failures during floods.

In the Netherlands, dike overtopping and backward erosion piping occurred in 1995, when 100,000 acres of land were flooded and 240,000 people had to be evacuated (Ammerlaan, 2007). During these floods, approximately 120 sand boils were observed along rivers Rhine, Waal, IJssel, and Maas, indicating the susceptibility of Dutch dikes to backward erosion piping (Bonelli, 2013). From the first part of the FLORIS research (Floris, 2005) it is clear that backward erosion piping forms a realistic threat for the stability of the Dutch dikes. Several failures in the past attributed to backward erosion piping were also recorded, such as the failures near Zalk, Nieuwkuijk, and Tholen (Vrijling, 2010).

Along the Mississippi river in the USA, the St Louis District authorities documented excessive underseepage (seepage through the aquifer in the foundation of the dike) and the formation of sand boils during the floods of 1973, 1993, and 1995 (USACE, 2014). Despite continued investigations, analyses, and the addition of control measures since the 1950s, excessive underseepage and the formation of sand boils were still observed. It is found that the occurrence of sand boils increases with subsequent flood events (Glynn and Kussmaul, 2004). For instance, higher frequencies of sand boils during the 1995 flood than during the record flood of 1993 have been documented. During the flood of 1993, the stage of the Mississippi river equaled or exceeded the 50-year design stage. Dike failure initiated by backward erosion piping during the 1993 flood was found in the Kaskaskia Island Levee District because no relief wells had been installed in the Kaskaskia Levee District at the time of the 1993 high water. During the Hurricane Katrina in New Orleans, the north breach on the Inner Harbor Navigation Channel (IHNC) at the Lower Ninth Ward was deduced due to the underseepage-induced erosion and backward erosion piping based on the understanding of the geological profile of the foundations, detailed numerical analysis, and the observed timing and geometry of this failure (Seed et al., 2008). The US Army Corps of Engineers (USACE) has carried out extensive studies of backward erosion piping over many years (USACE, 2014). Either a history of significant flood fight effort being required to maintain system integrity during high water events or a seepage analysis indicating the need for significant seepage remediation that had not yet been constructed could have resulted in a negative evaluation finding.

In China, backward erosion piping has occasionally occurred in the Jiujiang Dike and the Jingjiang Dike along the Yangtze River, the North Pearl River Dike along the North Pearl River, and the Yellow River Dike along the Yellow River. For instance, during the 1998 flood, several dike breaches occurred along the Yangtze River, some of which were caused by backward erosion piping (Liu et al., 2008). Depending on the geological information of the foundation materials, the location of sand boil could occur at a long distance from the dike toe. Chen (2003) summarized the locations of backward erosion piping during the 1998 flood. Nearly 50% of the failures occurred within 50 m, and the longest distance for the piping to occur could be up to 500 m. Wang and Cao

(1991) pointed out that the piping locations can extend from 10 m to 400 m in the Jingjiang Dike. According to numerical analysis and field investigation, backward erosion piping in several segments of the North Pearl River dike was also found to be a main failure mode (Zhang et al., 2011a). According to the statistics by Cao (1994), dike failures caused by backward erosion piping accounted for 90% of the total number of failures of dikes in China.

One of the common features in these dike failures is that their foundation materials are usually sandy materials, and in some cases a thin cohesive layer underlain by a sandy layer. Particularly during a heavy flood, the water level in the river increases quickly and the pore water pressure in turn builds up dramatically, especially in the downstream toe of the dike. This introduces large hydraulic gradients at the seepage exit, which in turn generate sand boil. For a cohesive layer underlain by a sandy layer, cracking of the top layer as a result of high water pressure can be triggered. Eventually, backward erosion piping can occur along the cracks. In many areas, seepage is detected by the presence of wet areas downstream before the formation of sand boil. Backward erosion piping more likely occurs in dikes than in embankment dams. Because the length of a dike could be up to several kilometers, the huge amount of foundation materials susceptible to piping may not be replaced completely by good materials. Therefore, the reliable solutions for such water barrier systems could be: (1) proper installation of water relief wells in the downstream side, (2) implementation of impermeable layer either vertically or horizontally with proper compaction to increase the seepage paths, or (3) adopting effective land berms (i.e. sufficient depth and width).

6.4.2 Methods for Assessing Backward Erosion

Prediction tools for backward erosion have been developed since early the 1900s, and improvements are still being made (Van Beek et al., 2014). Most of the methods rely to a large extent on case histories or laboratory and full-scale experiments, and they are empirical or semi-empirical in nature. In this section, several commonly cited methods are presented.

6.4.2.1 Bligh's Method (Bligh, 1910)

Bligh (1910) was among the first to establish a relationship between the average hydraulic gradient across the structure, the soil type, and the occurrence of internal erosion on the basis of data derived from a large number of field piping studies referring to Van Beek et al. (2014). This resulted in an empirical formulation for estimating the critical gradient. The critical gradient (H_c/L) is defined as the ratio between the critical head (H_c, the head that leads to ongoing erosion) and seepage length (L, the shortest distance between the upstream and downstream levels). The value of the ratio depends on the type of material and is given for four different types of materials as shown in Table 6.2. It is assumed that the seepage path was mainly concentrated near the base of the structure and thus the seepage length refers to the base length of the dike.

The critical hydraulic gradient proposed by Bligh (1910) does not take into account the possible presence of dispersive soils in the case histories used to develop the safe margin. Based on the equation, the piping potential is directly proportional to the hydraulic head and inversely proportional to the length of seepage path. It is still in common use today although it is an empirical equation.

Table 6.2 Bligh's criteria for critical hydraulic gradient (after Richards and Reddy, 2007)

Type of foundation material	Critical hydraulic gradient, L/H_c
Riverbeds of light sandy sand	18
Fine micaceous sand	15
Coarse-grained sand	12
Boulders or shingle and gravel and sand	5–9

6.4.2.2 Terzaghi's Method (Terzaghi, 1939, 1943; Terzaghi and Peck, 1996)

Terzaghi (1939) argued that Bligh's rule is based on the tacit assumption that the critical hydraulic gradient should depend on nothing but the grain size of the subsoil. After a detailed study, it was found that the critical head depends not on the average gradient along the line of shortest flow net, but on the greatest hydraulic gradient at which the water flows out of the ground. For a perfectly uniform subsoil and for a horizontal discharge surface, the danger zone is located at the downstream edge of the structure. Terzaghi and Peck (1996) showed that backward erosion piping will initiate when a heave or zero effective stress condition occurs in cohesionless soils at the downstream toe of a dike or a dam. The condition can be expressed as

$$\gamma' h = i h \gamma_w \tag{6.8}$$

$$i_c = (G_s - 1)(1 - n) \tag{6.9}$$

where h is the height of the soil element considered.

Terzaghi (1943) introduced the concept of factor of safety against backward erosion, which is simply the effective weight of a prism of soil in the area of expected heave, divided by the excess hydrostatic pressure beneath it. The critical factor of safety was found by trial and error using a number of potential depths for the prism. The factor of safety, F_s, is expressed as:

$$F_s = W' / U_e \tag{6.10}$$

where W' is the effective weight of the most critical sand prism and U_e is the uplift pressure at the base of the prism.

Terzaghi's concept of critical hydraulic gradient and F_s against backward erosion has been the dominating influence in the design of dikes and dams, particularly in USA and China (e.g. The Ministry of Water Resources of People's Republic of China, 1999; USACE, 2000a).

Kovács (1981) extended Terzaghi's critical hydraulic gradient to a general case as shown in Figure 6.9. Based on limit equilibrium of the forces acting on a rectangular soil element at the sloping exit face including the soil weight reduced by the uplift force, the hydrodynamic force created by the percolating water on the soil element, and the frictional force along the base of the soil element, the critical gradient is expressed as

$$i_c = \left(\frac{\gamma_b}{\gamma_w}\right)\left(\frac{\tan\phi' \cos\beta - \sin\beta}{\cos(\beta - \alpha) + \tan\phi' \sin(\beta - \alpha)}\right) \tag{6.11}$$

where ϕ' is the internal friction angle, β is the slope angle, and α is the seepage angle defined in Figure 6.9.

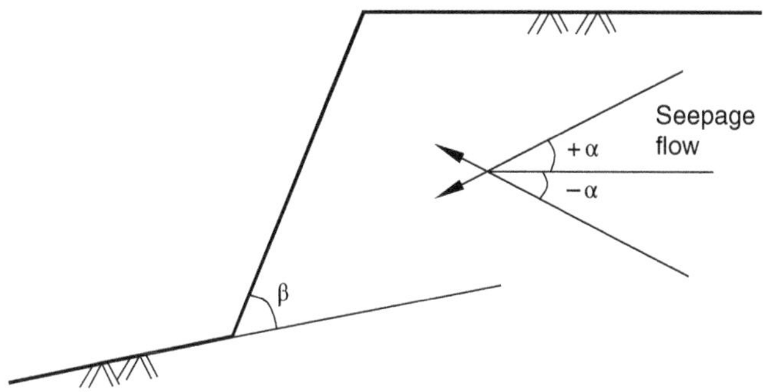

Figure 6.9 Schematic of observed processes

6.4.2.3 Sellmeijer's method (Sellmeijer, 1988)

Sellmeijer and Koenders (1991) and Koenders and Sellmeijer (1992) developed a mathematical model for backward erosion piping, based on the large-scale experiments carried out at Delft Laboratories in the 1980s. The model is a solution of a linear groundwater flow problem under rather unusual boundary conditions. The bottom of the structure is impermeable. Outside the structure, the pressure equals the equivalent water level. The boundary conditions in the sand boil and the piping channel depend on the conditions of limiting equilibrium. The model covers three essential areas (Sellmeijer et al., 2011): (1) groundwater flow through the subsoil, (2) pipe flow through the erosion channel, and (3) limit equilibrium of sand particles at the bottom of the channel. For the single particle force balance for a grain at the bottom of the erosion channel, four distinct forces are considered. The horizontal ones are the drag force due to the channel flow and the horizontal flow force. The vertical ones are the weight of a particle and the vertical flow force. At limit equilibrium, Sellmeijer et al. (2011) assumed that the flow forces do not affect the coarse particles' movement. Consequently, a two-force approach is selected where the drag force and weight of the sand particle are applied:

$$\frac{H_c}{L} = F_R F_S F_G \tag{6.12}$$

$$F_R = \eta \frac{\gamma_p{}'}{\gamma_w} \tan\vartheta \left(\frac{RD}{RD_m}\right)^{0.35} \left(\frac{U}{U_m}\right)^{0.13} \left(\frac{KAS}{KAS_m}\right)^{-0.02} \tag{6.13}$$

$$F_S = \frac{d_{70}}{\sqrt[3]{\kappa L}} \left(\frac{d_{70m}}{d_{70}}\right)^{0.6} \tag{6.14}$$

$$F_G = 0.91 \left(\frac{D}{L}\right)^{\frac{0.28}{\left(\frac{D}{L}\right)^{2.8} - 1} + 0.04} \tag{6.15}$$

where F_R is the resistance factor, F_S is the scale factor, relating particle size and seepage size, F_G is the geometrical shape factor, L is the seepage length, H_c is the critical hydraulic head, η is the

White's constant, γ_p is the unit weight of particle, γ_w is the unit weight of water, ϑ is the bending angle of sand, *RD* is the relative density, *U* is coefficient of uniformity, *KAS* is the roundness of particles, *k* is the coefficient of permeability, and *D* is the height of sand layer.

Sellmeijer et al. (2011) pointed out that Equations (6.12)–(6.15) are based on small-scale tests. It is not clear if there may be a less prominent effect in larger structures. The piping rule may be only applied within the limits of the involved parameters during testing.

6.4.3 Formation of a Pipe due to Backward Erosion

Van Beek et al. (2010), Sellmeijer et al. (2011), and Van Beek et al. (2011) divided the process leading to failure due to backward erosion piping into four phases: seepage, backward erosion, widening, and failure. Due to complex ground conditions, the erosion process could, in practice, be much more complicated than that explained in this section based on a simple ground profile. However, the basic principles of backward erosion can be explained by using the simple model.

Phase 1: Seepage

When a hydraulic head difference exists in the foundation, seepage can be generated. If there is a relatively impermeable layer on a permeable stratum, an open exit is required for the occurrence of seepage leading to backward erosion. The open exit can be generated by cracking, ditching, hydraulic fracture, etc. If the foundation only contains a permeable layer, seepage can be directly introduced to the surface of the downstream foundation.

Phase 2: Initiation and Progression of Backward Erosion

Once the seepage is generated, the hydraulic gradient at the seepage exit determines whether erosion can initiate. The initiation of backward erosion requires fluidization of sand near the exit point. Fluidization occurs when the pore pressures in the sand exceed the effective stresses, such that the sand will turn into a fluid state. There is a concentration of flow lines near the exit point, and therefore the local gradient near the exit point will be higher than the average gradient under the dike (Bonelli, 2013).

At the beginning of the process there is a rearrangement of grains, individual grain movements, and formation of small channels (ICOLD, 2013). The process reaches equilibrium under the hydraulic head applied. Very small amounts of sand are transported. With an increase in the hydraulic head to the critical value, sand is transported continuously. There will be a concentration of flow lines near the exit point that leads to higher water flow velocities near the exit point, and this increases the probability of backward erosion piping. The pipe will develop at the interface of the sand layer and a cohesive layer, as the clay layer will form a roof to the pipe (Van Beek et al., 2011). If a clay layer lies on the sand layer within the foundation, a sand boil appears at the open exit. Sand boils are sand mounds, which may or may not deposit sand at their edges (Sellmeijer et al., 2011). If no clay layer exits, a sand boil may be introduced at the most critical seepage exit, and the pipe will propagate along the most critical seepage path. In Sellmeijer's model, it is assumed that the stability of grains on the bottom of the pipe is the most important control variable for the erosion process. The critical shear stress exerted by the water flow on the grains on the bottom of the pipe will determine the likelihood of the grains being transported. It is assumed that the pipe cannot progress when the grains are in equilibrium, whereas the pipe will increase in length when the critical shear stress is exceeded, causing deepening of the pipe. The flow barely increases in this phase of the process. When the critical head is exceeded, the pipe grows towards the upstream side of the dike.

Phase 3: Widening of Erosion Pipe

As soon as the pipe progresses across the entire base of the water-retaining structure and reaches the upstream side, a pressure surge will occur in the pipe. The flow in the pipe is increased enormously because the flow resistance of the sand disappears. The resulting flow surge causes a large quantity of sand to be eroded near the upstream inlet of the pipe (Bonelli, 2013). The widening pipe develops from upstream to downstream. The widened pipe has a large flow capacity, causing further scouring of the pipe.

During the widening process, the flow and sand transport at the exit point do not increase significantly (Van Beek et al., 2011). When the pipe is widened from the upstream side to the downstream side, the sand transport and flow increase significantly.

Phase 4: Failure and Breakthrough

Van Beek et al. (2011) found that once the widening process is completed, the increase of flow through the pipe results in further scouring of the pipe, both in the sand and in the clay, such that the dike deforms and cracks. The scour in the sand is driven by water flow in a similar way as during pipe formation, whereas the erosion of the cohesive base of the dike shows similarities to concentrated leak erosion.

Failure takes place shortly after the first increase of sand transport, but can also be delayed due to settlement of the dike, which causes the pipe to close. The connection between the upstream and downstream side must be re-established for sand transport and flow to increase again. Several phases of reconnection and deformation can take place before the dike fails. The period from increase of sand transport to failure may therefore be as short as 20 minutes, but it could also take a few days, if the widening process needs to be restarted due to settlement. Failure and breakthrough are accompanied by a violent flow of sand and water, mud fountains, cracks in the dike and subsidence of the dike.

6.5 Initiation of Contact Erosion

Contact erosion appears at the interface of two soil layers (e.g. a coarse soil and a fine soil layer) subject to a seepage flow. Particles of the finer layer may be destabilized by the seepage flow and transported through the pores of the coarser layer. The natural soils of the alluvial plains are generally layered, composed of successive layers of sediments of different origins and grain size distributions. Moreover, a zoned embankment dam requires placing multiple layers of materials, each having a specific function and thus different properties. A significant number of interfaces are thus naturally formed within the body of the structure as well as its foundation. The seepage flow could be parallel to the contact in the coarse layer, perpendicular to the interface towards the coarse

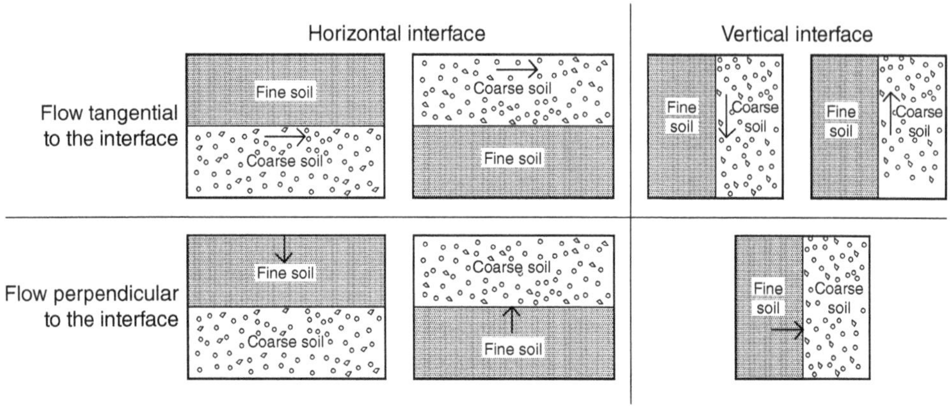

Figure 6.10 Elementary configurations of contact erosion depending on the orientation of the interface and of the seepage flow

layer, or inclined to the interface towards the coarse layer as shown in Figure 6.10. Bonelli (2013) pointed out that the sensitive interfaces are most frequently horizontal, since they are the result of the successive deposit of sediment layers or of layers made up of the materials that build up a structure. However, certain interfaces, such as the corewall–filter interface can frequently be inclined or even vertical.

In this section, the fundamentals of contact erosion are presented in detail first. Then several common test methods for contact erosion are introduced. Finally the critical hydraulic gradient that initiates contact erosion is introduced.

6.5.1 Fundamental Aspects of Contact Erosion Process

6.5.1.1 Typical Conditions for Contact Erosion to Occur

Contact erosion usually occurs at the interface between a coarse layer and a fine layer. Typically the fine layer can be made of sand, silt, clay, or a mixture of these, whereas the coarse layer is mostly composed of gravel, possibly mixed with sand and silt. In the case of contact erosion, the coarse layer is systematically a granular type of soil, whereas the layer of fine soil can have highly variable characteristics: granular and moderately fine in the case of a type of sand, fine and granular in the case of fine sand, and very fine and cohesive in the case of silt or clay (Bonelli, 2013).

Along rivers, the soil is often made up of pebbles, sand, and silt, all transported by the water. The presence of a gap in the grain size distribution of these alluvia, which is typically 1–5 mm, is commonly observed. The presence of interfaces between layers of distinct grain size distributions is thus highly probable. In zoned embankment dams, a central core is usually made of silt or compacted clay. Besides this core, upstream and downstream shoulders are added, which are meant to ensure the mechanical stability of the structure and are created with a more permeable coarse material. In this situation, contact erosion most likely occurs at the base–filter and filter–shoulder interfaces. Particularly for the embankment dams constructed in the early 20th century, the filters could be coarser than the modern filter criteria (Foster and Fell, 2001). This means that the potential of contact erosion for such zoned dams is relatively high.

6.5.1.2 Loads and Resistance to Contact Erosion

Contact erosion requires two conditions to initiate: (1) geometric condition – large pores should be formed geometrically to allow the fine particles to pass through; in particular, the sizes of the constrictions within the void network should allow the fine particles to pass, and (2) hydraulic condition – the hydraulic force should be large enough to detach and transport the fine particles. The permeability of the coarse layer is much larger than that of the fine layer, and herein the seepage flow is much more significant than in the fine layer. Therefore, if the pores in the coarse soil are large enough and the outflow is sufficiently intense, the fine particles in the fine layer could be detached by the flow and transported through the pores. The detailed description for the two conditions for the initiation of contact erosion will be presented in Sections 6.5.3 and 6.7.

6.5.1.3 Consequences of Contact Erosion

If contact erosion occurs in the configuration of fine soil below coarse soil, the layer of fine soil is eroded up to the point where cavities appear between the coarse grains. The local porosity increases progressively. The development of this cavity will lead to the local collapse of the grains of the coarse layer. These grains will fill in the cavity. As a consequence, it causes a progressive subsidence of the grains in the coarse layer and eventually settlements propagate within the structure above the zone of contact erosion. It can have consequences regarding the geotechnical properties of the soils and of the structure (Bonelli, 2013):

1. A sinkhole may be created eventually at the top of the dike.
2. Degraded areas could be generated, with low density and poor mechanical resistance, or even cavities. This can introduce a loss of slope stability of the structure or initiation of a concentrated leak erosion, and cause failure of the entire structure.
3. Clogging of granular filters or of drains located in the upstream of the erosion zone can also occur. The particles eroded and transported will then accumulate in the areas where the hydraulic and geometric conditions no longer allow their transport. Due to the clogging of filter layer, the increase of pore water pressure may happen. Local instability can be triggered.

If contact erosion occurs in the configuration of fine soil below coarse soil, during the development of contact erosion, a significant loss in the materials can induce settlements, which propagate through the structure in relation to the soil mechanical properties. Consequently, sinkholes can form at the surface or degraded areas, causing preferential outflows or stability issues (Bonelli, 2013).

In zoned embankment dams, the cavities created by contact erosion at the core–filter interface may not collapse but can be the start of backward erosion piping if the core materials are very cohesive. This process has been identified in hydraulic flume tests, where it appeared that the downstream part of the opening was held open by hydraulic fracture (ICOLD, 2013). Similarly to the case where fine soil lays below coarse soil, a weak zone with lower density and lower shear strength due to the development of contact erosion can cause loss of stability. Finally, the eroded fine particles may clog the filter layer if the constriction sizes of the filter layer do not allow most of the fine particles to pass. The pore water pressure in the filter layer may increase subsequently which may result in instability of the dam body.

6.5.2 Laboratory Investigation on Contact Erosion

Over several decades, a number of contact erosion tests have been conducted to investigate the initiation of contact erosion, the transportation of the fine particles, and the hydraulic regime that initiates contact erosion. The testing apparatus is illustrated in Figure 6.11. A large rectangular/cylindrical cell in which two layers of soil are installed successively is connected to a water supply system that permits hydraulic flow either parallel to the coarse–fine interface or perpendicular to it. The instrumentation used permits at least measuring either the pressure drop or the flow rate imposed in order to deduce the hydraulic gradient across the specimen. The outflow rate can also be measured by weighing, or using a turbidity meter. The experimental procedure usually consists of imposing a constant hydraulic head for a period of time, depending on the soil type, then determining whether contact erosion has been initiated. Due to the different flow directions, the response of the erosion process may not be the same, thus the cases of different orientations of seepage flow are presented separately in the following section.

6.5.2.1 Fine Soil Below a Coarse Soil Layer

This configuration of fine soil under a coarse soil layer has been widely studied (Istomina, 1957; Brauns, 1985; Bezuijen et al., 1987; Wörman and Olafsdottir, 1992; Den Adel et al., 1994; Guidoux et al., 2010; Béguin, 2011). When the fine soil mainly consists of sand, erosion particles are mainly transported in the form of bed-load. The initiation and transportation of fine particles can be modelled by classical surface erosion criteria.

Bezuijen et al. (1987) carried out a series of tests to validate their proposed models in predicting block and/or gravel revetment stability. They proposed a predictive law for critical velocity and gradient based on Shields's criterion. In particular, it was highlighted that it is not sufficient to consider the external hydraulic load alone. It is also necessary to analyze the internal flow regime within the coarse soil layer. Finally, they concluded that, for filter stability associated with the tested materials, the hydraulic gradients are relatively low, and so an approach based on geometrically sand tight filter criteria is too strict.

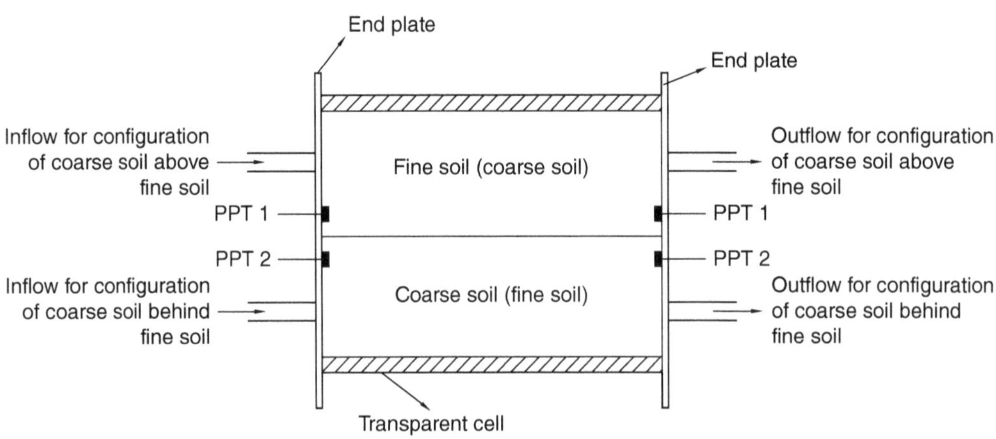

Figure 6.11 Illustration of testing apparatus for contact erosion under horizontal flow

Wörman and Olafsdottir (1992) conducted extensive flume tests to verify their developed constitutive transport equation relating interfacial transport rate with state variables for the flow. The following equation was deduced from basic physical considerations:

$$a\left(G^{*}/n_{1}\right)^{b}=\mathrm{Re}\left(d\right)^{c}\left(\frac{e_{1}}{8}\frac{1}{\left(G_{s}-1\right)}i\frac{D_{H}}{d_{85}}\right) \tag{6.16}$$

where G^* is the dimensionless transport parameter, D_H is the characteristic grain size, and a, b, and c are empirical coefficients.

In the tests, a horizontal interface was formed between a 0.07 m thick layer of fine quartz sand and an upper layer of granite gravel. Both materials were uniform, with a ratio $d_{100}/d_{10}=1.41$. The empirical constants, a, b, and c in Equation (6.16), were determined from a regression analysis based on the experimental data. On a 90% confidence level, the coefficients were found to be

$$a=0.374,\ b=0.084\pm0.016,\ c=0.140\pm0.038 \tag{6.17}$$

Den Adel et al. (1994) conducted small-scale laboratory tests of contact erosion on non-cohesive base soils with a median grain size d_{50} greater than 0.15 mm. The following observations were reported:

1. Base particles are dislodged separately; there is no large-scale fluidization mechanism.
2. Base particles remain near the filter interface; the material is not dispersed throughout the filter; transport takes place in a thin layer near the interface.
3. At low water velocities in the filter, i.e. less than 1.3 times the critical filter velocity, the layer of moving base material has a thickness of one base particle.
4. At higher water velocities in the filter the layer of moving particles is thicker, but never more than five base particles.
5. At a constant water velocity the accumulated mass of the transported material increases continuously as a function of time.

Based on the results of the series of tests, a simple model was derived for horizontal flow:

$$T\left(L,t\right)=\frac{\mu^{0}v}{c_{11}}\left(t-\frac{1}{c_{11}v'}\right)\text{ when }\frac{c_{11}v'L}{v}\text{ is large} \tag{6.18}$$

$$T\left(L,t\right)=\mu^{0}v'Lt-\frac{1}{2}\mu^{0}v'\frac{L^{2}}{v}\text{ when }\frac{c_{11}v'L}{v}\text{ is small} \tag{6.19}$$

where μ^0 is volume density of particles at rest at the interface between filter and base, v is velocity, v' is frequency to dislodge particles, and c_{11} is a coefficient.

Three describing constants are determined based on the experimental results, each one pertaining to part of the physical process: (1) the frequency of dislodging particles from the base layer, (2) the velocity of the moving particles, and (3) the frequency of stopping particles in the filter.

Guidoux et al. (2010) carried out extensive laboratory tests on contact erosion, putting a particular emphasis on silts and sand–illite mixtures that have not been tested before. They concluded

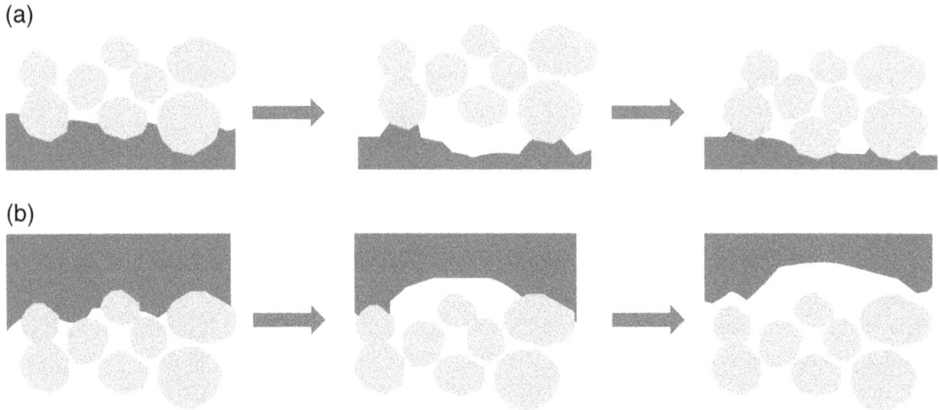

Figure 6.12 Local geometrical evolution in the two configurations of contact erosion: (a) fine soil below coarse soil; (b) coarse soil below fine soil (modified after Béguin, 2011)

that the erosion process can be divided into three phases. In the first phase, no erosion occurs and the turbidity remains zero, except for a few minutes when the flow rate is increasing to the next step. Indeed, the steep variation of the flow rate removes fluidized particles at the interface, but after a few minutes the turbidity returns to zero. In the second phase, when the flow velocity exceeds the critical velocity, the base soil begins to be continuously removed, and the turbidity seems to be almost constant during each step of this period. This process is named as 'steady erosion'. During this part of the experiment the base soil is significantly eroded, the coarse soil geometry changes; the coarse soil grains are moving and arches form inside the coarse soil. In the third phase, when the arches collapse, the contact area between coarse soil and base soil tends to increase and erosion tends to decrease. Particularly, Guidoux et al. (2010) pointed out that the critical velocities were almost independent of the coarse soils involved, and they also highlighted that, in order to estimate the contact erosion threshold, velocity is a more convenient criterion than hydraulic gradient. The works by Guidoux et al. (2010) and Béguin (2011) indicated that the critical velocity for contact erosion has a remarkable increase with the increase of fines content, as erosion no longer concerns an individual particle but the aggregate for strongly cohesive soils. The cavities that form in the configuration of the fine soil under the coarse soil are filled because of the successive collapses of coarse grains, thus causing locally somehow a regular cycle of erosion (Figure 6.12(a)).

6.5.2.2 Fine Soil on a Coarse Soil Layer

There is limited laboratory investigation on this configuration in the literature (i.e. Schmitz, 2007; Béguin, 2011). The phenomenon is complex and cannot be directly linked to river erosion. A safe hydraulic condition for contact erosion in this configuration should be based on the transport velocity of the particles, considering that detachments occur whatever the value of the flow velocity (ICOLD, 2013).

Schmitz (2007) carried out experimental tests of erosion of silt layers above coarse layers, which are more or less cohesive. The critical velocity of the initiation of erosion was defined by a sharp increase in the mass of transported soil. Starting from this level of hydraulic load, cavities appear in the fine soil layer, and erosion develops up to the point where it causes sinkholes at the top of the sample. It was also noticed that the soil erosion resistance increases with the increase of vertical stress on the specimen. This is considered to be due to the increase in the shear strength of the fine layer with the increase of vertical stress. Moreover, Schmitz's tests indicated that the grain size distribution of the coarse layer did not show any obvious influence on the critical velocity.

Béguin (2011) carried out six tests on silt embankments over gravel. The results indicated that, in the case of cohesive soils, contact erosion is similar to that in the configuration of fine material below a coarse soil layer. The direction of gravity is the only parameter that distinguishes them. It seems that, even in the silt above gravel configuration, where erosion might be expected to be initiated when silt particles "fall" into the gravel, the initiation of erosion is dependent on the transport of particles, not by detachment (ICOLD, 2013). However, Béguin (2011) pointed out that there is a discrepancy that appears between the two configurations when erosion lasts a long time. In the configuration of the fine soil on the coarse soil, the cavities are not necessarily filled by the coarse materials. Depending on the mechanical characteristics of the fine soil, these cavities can either collapse or grow to the point where they generate preferential outflows (Figure 6.12b).

6.5.2.3 Base–Filter Interface Erosion

One of the main purposes of contact erosion tests at the base–filter interface reported in the literature is to assess the capability of the filter to prevent the loss of fine particles in the base soil. In the early days, the filter materials mainly ranged between SP, SW, GP, and GW (i.e. USACE 1941; Karpoff, 1955). The tests were carried out on samples subjected to upward or downward gradients, the intensity of which varied between 1 and 30. Most of the tests were reported unstable when particle migration was observed visually through transparent permeameters and further confirmed by grain size analysis before and after flow. The division between stable and unstable combination was then used to propose geometric criteria for filters, based solely on grading.

To simulate the most representative conditions that are likely to occur at the core–filter interface, Lafleur (1984) carried out contact erosion tests in which a broadly graded cohesionless till was submitted to a steady unidirectional flow. The samples were compacted at densities that were 91–97% of the Proctor standard. The flow was downwards, implying that the viscous drag of the water on the particles was acting in the same direction as gravity. The hydraulic gradients were unidirectional and limited to 8. The results indicated that the critical combinations for the testing materials were obtained for a ratio (D_{15}/d_{85}) of 8.4. By contrast, Sherard et al. (1984) conducted extensive contact erosion tests on uniform sand base. They concluded that, for filters with D_{15} larger than about 1.0 mm, the main criterion for judging filter acceptability would be $D_{15}/d_{85} < 5$. For finer filters used for protecting silts and clays, many criteria are available as presented in Section 6.7.

Extensive laboratory contact erosion tests were conducted to investigate the effect of various factors (confining pressure, filter thickness, and the magnitude and rate of gradient increase) on erosion process. Tomlinson and Vaid (2000) summarized the following findings:

1. Confining pressure had a minor negative impact on stability. This is apparently due to the collapse of arches, which occurs in loose sand in the filtration zone with increase in stress. Confining pressure level did not influence the stability of soil–filter systems with larger grain size ratios.

2. A minimal filter thickness is necessary to establish a filtration zone. A filter thinner than this minimum value will allow contact erosion to occur more readily.
3. Imposing the gradient rapidly prevented a filtration zone from properly forming, thus triggering contact erosion at a smaller critical gradient.

6.5.3 *Threshold of Contact Erosion*

The hydraulic threshold from which contact erosion occurs when geometric conditions do not impede the transport of eroded particles has been developed analytically or based on experimental results. The hydraulic threshold can be expressed as a critical hydraulic gradient, a critical velocity, or a critical erosive shear stress from different angles.

6.5.3.1 Critical Hydraulic Gradient

Assuming that the pressure drop is constant along the flow path, Ergun (1952) developed an equation expressed as

$$i = aq + bq^2 \tag{6.20}$$

where q is the mean velocity; and a and b are coefficients that both depend on coarse soil porosity, n_F, and on coarse soil characteristic grain size D_H defined by Equation (6.30) applied to the coarse soil gradation curve.

$$Re = \begin{cases} [10^{-5}, 2.3] \\ [5, 80] \\ > 120 \end{cases} \quad a = \begin{cases} 192.24 \dfrac{(1-n_F)^2}{n_F^3} \dfrac{\nu}{gD_H^2} \\ 182 \dfrac{(1-n_F)^2}{n_F^3} \dfrac{\nu}{gD_H^2} \\ 225 \dfrac{(1-n_F)^2}{n_F^3} \dfrac{\nu}{gD_H^2} \end{cases} \quad b = \begin{cases} 0 \\ 1.92 \dfrac{(1-n_F)}{n_F^3} \dfrac{1}{gD_H} \\ 1.61 \dfrac{(1-n_F)}{n_F^3} \dfrac{1}{gD_H} \end{cases} \tag{6.21}$$

Where Re is the Reynolds number, d_j is the particle size of the fraction j of the coarse soil gradation curve, and F_j is the mass fraction of the fraction.

The two empirical expressions for a and b in Equation (6.21) are related to flow characteristics via the particle Reynolds number Re that depends itself on fluid properties and the mean flow velocity. The two coefficients have been determined by Fand et al. (1987) who proposed the following expression:

$$i = 182 \frac{(1-n_f)^2 \nu_w}{n_f{}^3 gD^2} U + 1.92 \frac{(1-n_f)}{n_f{}^3 gD} U^2 \tag{6.22}$$

where ν_w is the kinematic viscosity of water, n_f is the coarse soil porosity, and g is the gravitational acceleration.

Guidoux et al. (2010) conducted extensive contact erosion tests at the interface between coarse soils and base soils, with a special emphasis on gap-graded base soils and sand/illite mixtures. They concluded that the measured critical hydraulic gradients are similar to the theoretical values calculated based on Equation (6.21).

By assuming a model porous medium composed of capillary tubes of the same permeability k as the real medium, Reddi et al. (2000) developed a relationship between shear stress, water pressure difference, and permeability as

$$\tau = 1.414\left(\frac{\Delta p}{L}\right)\left(\frac{k}{n}\right)^{1/2} \tag{6.23}$$

By using the expression proposed by Kozeny and Carman (Bear, 1972) for intrinsic permeability k:

$$k = \frac{c_0 n^3 D_{50}^{\ 2}}{36(1-n)^2} \tag{6.24}$$

where Kozeny's parameter c_0 is usually taken as 0.2. The expression for shear stress is

$$\tau = \sqrt{\frac{2}{5}}\frac{n\rho_w g i D_{50}}{6(1-n)} \tag{6.25}$$

Substituting Equation (6.25) into Shields's equation, the critical hydraulic gradient for contact erosion can be expressed as

$$i_c = \frac{6(1-n)(G_s-1)d_{50}}{n\beta_i\beta_w D_{50}}\Theta_c \tag{6.26}$$

where G_s is the specific gravity of the fine particles, Θ_c is the critical Shields number deduced from the Shields diagram, and β_w and β_i are empirical coefficients.

6.5.3.2 Critical Velocity

For cohesionless base soil, Brauns (1985) proposed an expression of critical velocity q_c for contact erosion, which was validated with sand as the base soil. If $n_F D_{15}/d_{85} > 10$, q_c then is

$$q_c = 0.7 n_F \sqrt{(G_s-1) g d_{50}} \tag{6.27}$$

where n_F is the coarse soil porosity; and g is the gravity acceleration. In the Brauns model, critical velocity depends on the porosity of the coarse soil, the relative density of particles, and the diameter of the particles of fine soil d_{50}.

For cohesive base soil, the experimental results by Guidoux et al. (2010) indicated that d_{50} is not a relevant soil characteristic, and therefore should not be used to predict the critical velocity.

The effective grain diameter D_H defined by Koženy (1953) is a more representative grain size description for a base soil.

$$D_H = \left(\sum_{j=1}^{m} \frac{F_j}{d_j} \right)^{-1} \tag{6.28}$$

where d_j is the particle size of the fraction j of the base soil gradation curve and F_j is the mass fraction of the fraction j.

By assuming that adhesion can be estimated by a force proportional to the diameter of the particle, Guidoux et al. (2010) developed a critical velocity equation expressed as

$$q_c = 0.7 n_F \sqrt{(G_s - 1) g D_H \left(1 + \frac{\beta}{D_H^2} \right)} \tag{6.29}$$

where β is a coefficient function of base-soil compaction, particles shapes, electrodynamic characteristics of the particles, and chemical condition. By fitting the experimental data from Guidoux et al. (2010) β is estimated as $5.3 \times 10^{-9}\,\text{m}^2$.

6.6 Initiation of Suffusion

Suffusion involves selective erosion of fine particles within the matrix of coarse soil particles under seepage flow. For a soil susceptible to suffusion, once the fine particles are removed, the microstructure of the soil will change accordingly. This could induce a reduction of the soil shear strength and a mutation of hydraulic conditions in the soil. In this section, the control parameters for suffusion are introduced first. Then the most recent laboratory tests are presented. Finally the assessment criteria (geometric and hydraulic) are explained in detail.

6.6.1 Control Parameters for Likelihood of Suffusion

Suffusion of a soil is influenced by geometric conditions of soil, hydraulic conditions, and mechanical conditions. Suffusion usually occurs in gap-graded granular soils and broadly graded soils with a steep slope in the coarse fraction and a gentle slope in the fine fraction. From geometric point of view, grain size distribution and pore-size distribution are two most important factors. Grain shape and pore shape also affect the transport of the fine particles. Round-shape grains are easier to pass the constrictions. From mechanical standpoint, degree of compaction is the most important parameter. Obviously, compaction effort affects the pore size and its distribution and the constriction size. With regards to the drag forces that trigger the process of suffusion, the hydraulic gradient and seepage direction are two important parameters. Under a low hydraulic gradient, suffusion may not occur even the soil is internally-unstable as the seepage force may not overcome the resistance of the fines. Under upward flow conditions, the critical hydraulic gradient is larger comparing with those in downward and horizontal flow cases (Adel et al., 1988). The effective stress of the soil is also an important factor for the development of suffusion, because the stress influences the microstructure of the soil and in turn the hydraulic response.

6.6.2 *Laboratory Testing of Suffusion*

For several decades, a great number of laboratory suffusion tests on cohesionless soils (e.g. USACE, 1953; Kenney and Lau, 1985; Lafleur et al., 1989; Burenkova, 1993; Chapuis et al., 1996; Honjo et al., 1996; Liu, 2005, Moffat, 2005, Fannin and Moffat, 2006; Li, 2008; Cividini et al., 2009; Moffat et al., 2011) and on cohesive soils (e.g. Sun, 1989; Lafleur and Nguyen, 2007; Wan and Fell, 2008) have been conducted. Typically the soil specimen was under a downward seepage within a cylindrical seepage cell, with diameters of 25–580 mm. The hydraulic gradients were kept constant in some tests during the testing process (e.g. Aberg, 1993; Chapuis et al., 1996; Honjo et al., 1996; Lafleur and Nguyen, 2007; Bendahmane et al., 2008; Wan and Fell, 2008), but were increased gradually in other tests (e.g. Lafleur et al., 1989; Skempton and Brogan, 1994; Fannin and Moffat, 2006; Cividini et al., 2009). Most tests followed similar testing procedures, although there were differences in the magnitude of hydraulic gradient, flow direction, sample filter size, and sample vibration. The maximum hydraulic gradients applied in most of the tests were high to mobilize geometrically unstable particles. The soils ranged from clay to gravel, and most of them belong to broadly graded soils with values of coefficient of uniformity larger than 3. Nearly one third of the soils contained a significant amount of fines. The soils include crushed stone, moraine soil, fluvial soil, sandy gravel, completely decomposed granite (CDG), Loire sand, Leighton Buzzard sand, Fontainebleau sand, silica sand, concrete sand, clayey silt, kaolin, and bentonite. Moreover, the particle size distributions of the test soils represent well the actual base soils and filters for embankment dams. Based on the test results, many geometric criteria have been developed, which will be presented in Section 6.6.3.

The soils within embankment dams or dikes are usually under complex stress states, especially in embankment dams with corewalls (Zhang and Du, 1997). Figure 6.13 illustrates the major and minor principal stresses in different zones and the foundation of an embankment dam after reservoir filling. The flow direction is nearly perpendicular to the major principal stress direction in the upstream filter and parallel to the major principal direction in the downstream filter. Inside the corewall, the flow direction makes a certain angle with the principal stress direction. In the upstream and downstream foundations, the flow directions are nearly parallel to the respective major principal stress directions, downward in the upstream and upward in the downstream. Chang and Zhang (2011, 2013a) and Chang et al. (2012) conducted extensive laboratory investigations on suffusion, considering the effect of stress conditions. In this section, the testing apparatus, test soils and test results are presented briefly.

A stress-controlled internal erosion testing apparatus was developed to systematically investigate the initiation and development of internal erosion subjected to complex stress states. The general layout of the apparatus is shown in Figure 6.14. It consists of a triaxial system, a pressurized water supply system, a soil collection system, and a water collection system. The triaxial system allows the independent control of hydraulic gradient and stress state. The final stress state can be approached by setting a stress path through the control system. To accommodate high permeability soils, a high-capacity water supply system consisting of three water tanks 200 mm in diameter and 400 mm in height are used. The hydraulic gradient can be changed at an increment of 0.05 for a soil specimen 100 mm in height. The soil collection system includes three main components: a transparent subsidence funnel, a T-fitting, and a detachable container. The subsidence funnel helps identify the initiation of internal erosion. The T-fitting is used to separate the outflow water and eroded soil through a steel mesh screen at the inlet of the horizontal drainage hose. The detachable container is used to collect the eroded soil at a certain interval. The water collection system consists of a downstream reservoir, an acrylic cylindrical container, and an electronic balance. The downstream reservoir is used to maintain a constant water head at the bottom of the soil specimen. The outflow

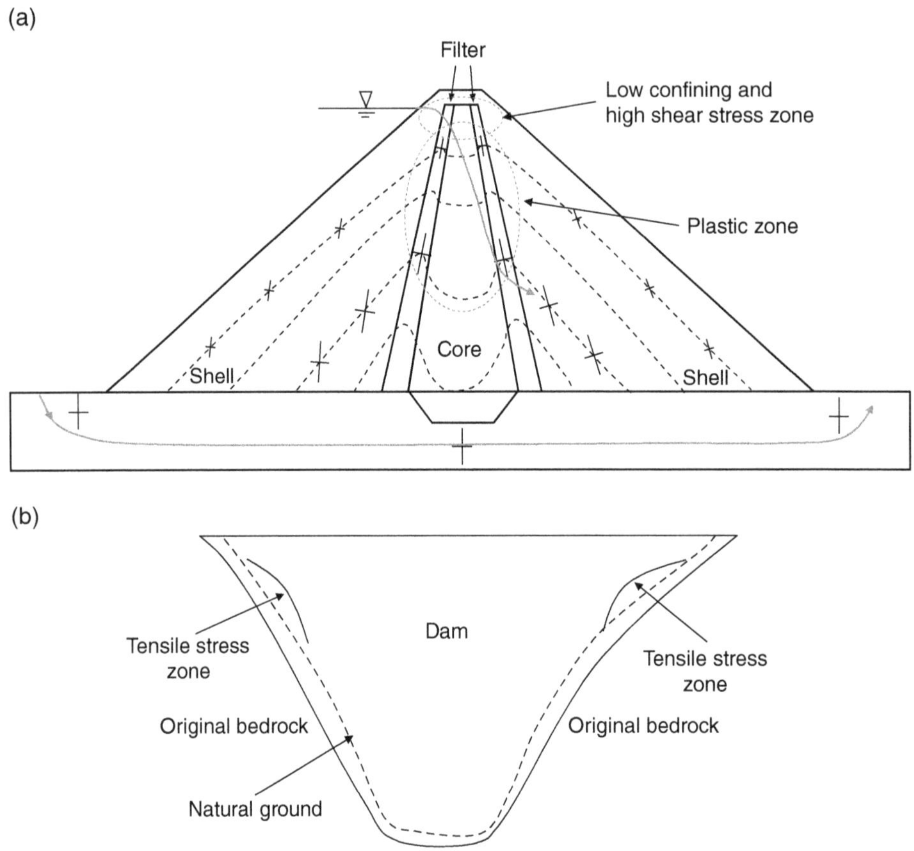

Figure 6.13 Illustration of (a) the principal stress contour and the orientation of the major and minor principal stresses in different zones after reservoir filling in a cross-section of an embankment dam with a corewall and (b) the tensile stress zones at the abutment of an embankment dam

is weighted at a certain time interval to measure the variations in soil permeability during the erosion process.

The soil was obtained by mixing two soils: commercial washed Leighton Buzzard sand (fraction E) and completely decomposed granite. The grain size distribution of the test soil is shown in Figure 6.15. The soil has a gap ratio of 7.9 and is described as poorly graded sand (SP) according to the Unified Soil Classification System (ASTM, 2000a). Standard soil compaction tests (ASTM, 2000b) were conducted to find its optimal water content. The physical properties of the test soil are summarized in Table 6.3.

To study the critical hydraulic gradients of internal erosion under such complex stress conditions, three series of erosion tests were carried out under three stress conditions: isotropic stress conditions, triaxial compression stress conditions (i.e. the major principal stress is along the seepage direction), and triaxial extension stress conditions (i.e. the major principal stress is perpendicular to the seepage direction). The stress states in the three series of tests are shown in Figure 6.16. Conventional drained triaxial compression and extension tests indicate that the soil shows dilative response during both the triaxial compression and extension tests. The 22 stress states in Figure 6.17

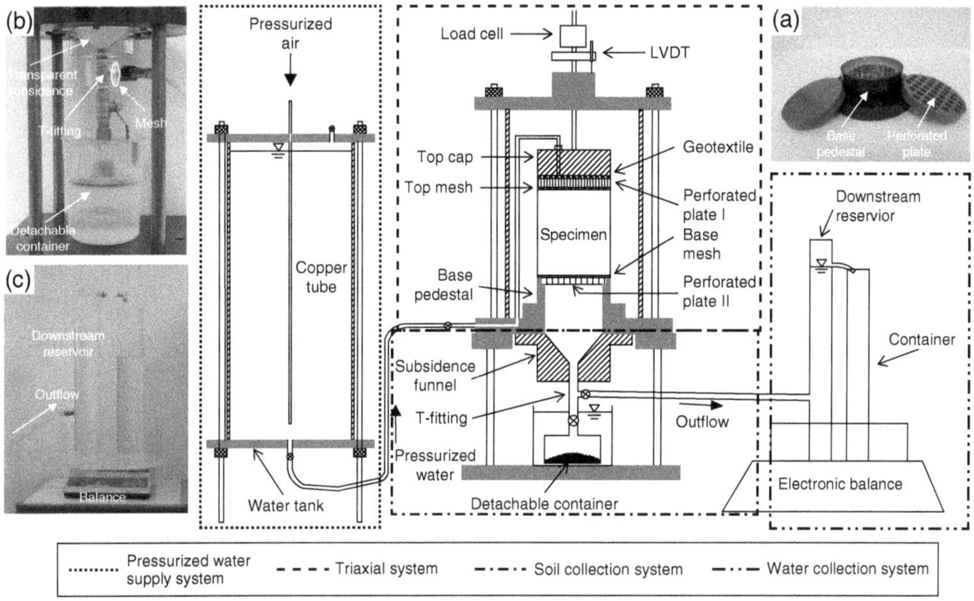

Figure 6.14 Schematic of the testing apparatus and components

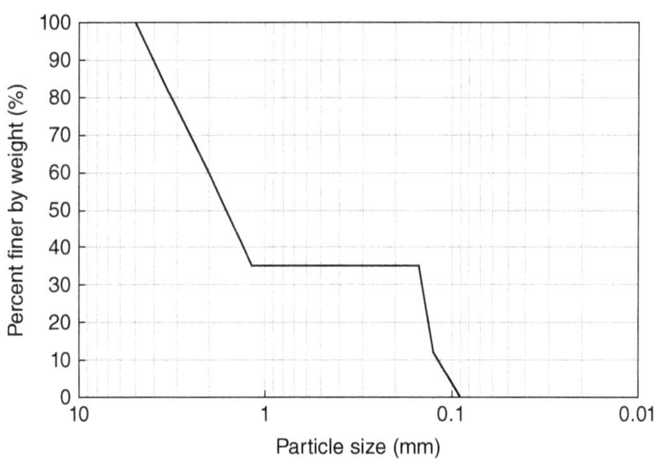

Figure 6.15 Grain size distribution of the test soil

fall on seven stress state lines of different shear stress ratios (i.e. $\eta = 1.50$, 1.20, 0.75, 0.00, −0.33, −0.60, and −0.92). Here, $\eta = q/p'$ is the shear stress ratio, q is the deviatoric stress, and p' is the mean effective stress. The conditions for each test are summarized in Table 6.4.

The erosion process can be divided into three phases (Figure 6.17): (1) initiation phase, in which some fine particles erode but the specimen deformation is limited, (2) development phase, in which a large amount of fine particles is washed out and large specimen deformation occurs, and (3) failure phase, in which the soil experiences shear failure due to large seepage forces and the loss of excessive particles. Take the progression of internal erosion in three specimens (i.e. GS-I-1,

Table 6.3 Physical properties of the test soil

Physical property	Value
Standard compaction tests	
Maximum dry density, ρ_{dmax} (kg/m^3)	1895
Optimum water content, w_{opt} (%)	6.5
Grain size distribution	
Percentage of gravel (%)	2
Percentage of sand (%)	98
Coefficient of uniformity, C_u	16.7
Coefficient of curvature, C_c	0.09
Specific gravity, G_s	2.61

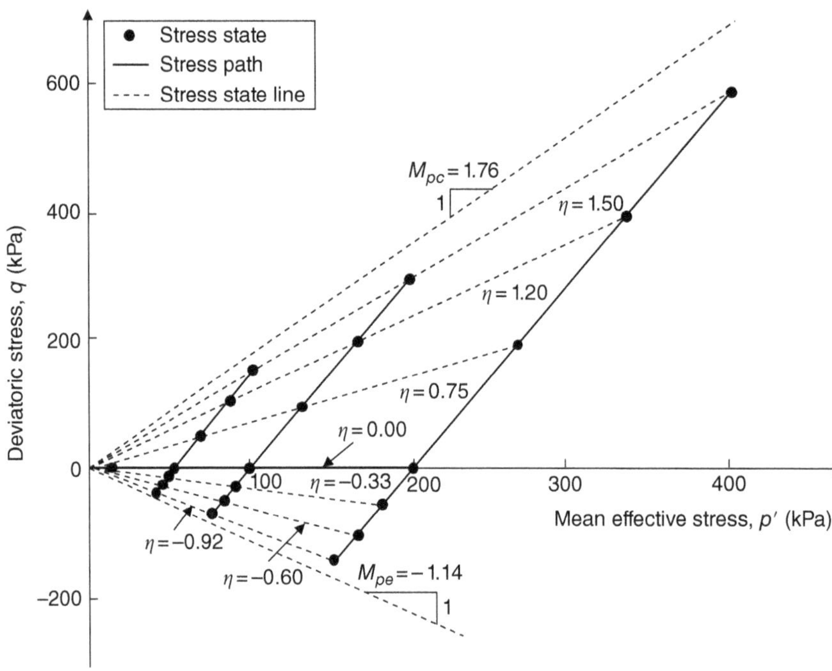

Figure 6.16 Stress states in the internal erosion tests

GS-C-6, and GS-E-6) under the same confining stress of 50kPa but different shear stresses as examples to elaborate the soil response in these three phases.

In the initiation phase, some fine particles start to erode, and the soil permeability increases in response to the loss of fine particles. The hydraulic gradients corresponding to this phase are $0.87 \leq i < 8.22$, $1.32 \leq i < 1.93$, and $0.85 \leq i < 1.00$ for specimens GS-I-1, GS-C-6, and GS-E-6, respectively. Figure 6.18(a) shows that the weights of the eroded soil are 2.5%, 0.30%, and 0.04% of the total soil for specimens GS-I-1, GS-C-6, and GS-E-6, respectively. The specimen deformations both in the axial and radial directions are very small in this phase (Figures 6.18b, c]. With the increase of seepage force, the fine particles with little contacts with the coarse particles will erode. Since such fine particles are mostly not in the strong force chains, the specimen deformations are still small.

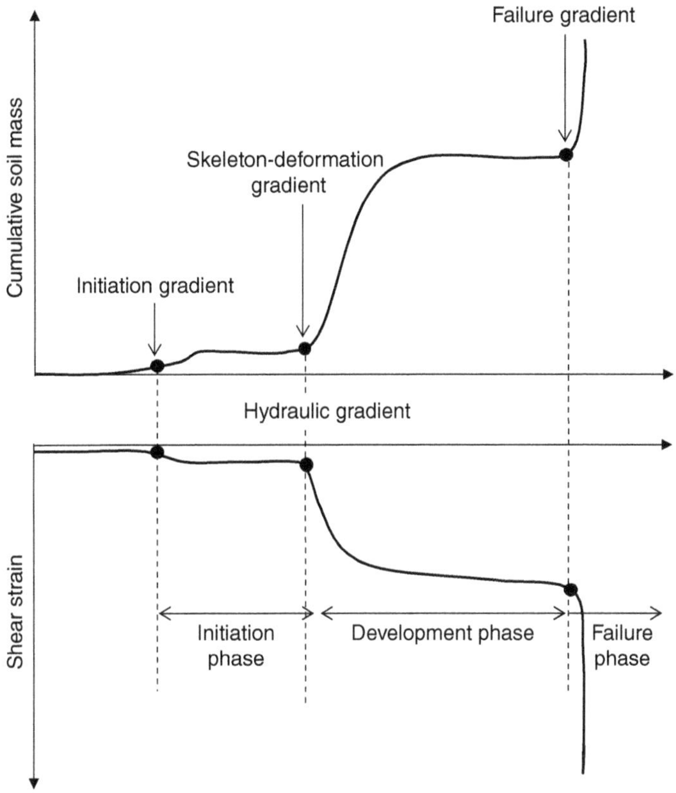

Figure 6.17 Phases of internal erosion and associated critical hydraulic gradients

In the development phase, sudden increases in eroded soil mass, permeability, and specimen deformation occur, but the soil finally reaches new equilibrium. Sudden increases in eroded soil mass, permeability, and specimen deformation are observed at an applied hydraulic gradient of 1.93 for specimen GS-C-6 and 1.00 for specimen GS-E-6 in Figure 6.18. For instance, the eroded soil mass increases nearly 16 times, and the axial and radial strains increase to 3.7% and −1.9%, respectively when the hydraulic gradient increases from 1.66 to 1.93 for specimen GS-C-6. For specimen GS-E-6 that is under extension, the eroded soil mass increases nearly 13 times, and the axial and radial strains increase to −1.7% and 1.1%, respectively when the hydraulic gradient increases from 0.85 to 1.00. In this phase, the seepage forces were so high that a large amount of fine particles within the coarse matrix were washed away. The force chains may buckle due to the loss of lateral support induced by the loss of fine particles.

In the failure phase, very large deformations occur and the soil fails due to shear failure. Specimen GS-E-6 experiences very large deformations and a significant loss of fine particles when the hydraulic gradient is incrementally increased to 6.80 (Figure 6.18). The shear failure was caused by the increased pore water pressure at the top of the specimen and the weakened soil structure due to the excessive loss of fine particles. The maximum applied hydraulic gradients for specimens GS-I-1 and GS-C-6 were 8.22 and 9.24, respectively. The two specimens did not fail at these gradients.

Table 6.4 Summary of the testing program

Specimen identity	Testing series	Initial porosity	Porosity at applied stress state	Applied stress state		
				Mean stress, p' (kPa)	Deviatoric stress, q (kPa)	Stress ratio, $\eta = q/p'$
GS-I-0	I	0.337	0.341	10	0	0.00
GS-I-1	I	0.336	0.330	50	0	0.00
GS-I-2	I	0.338	0.325	100	0	0.00
GS-I-3	I	0.335	0.320	200	0	0.00
GS-C-4	II	0.336	0.320	67	50	0.75
GS-C-5	II	0.337	0.314	83	100	1.20
GS-C-6	II	0.335	0.309	100	150	1.50
GS-C-7	II	0.338	0.311	133	100	0.75
GS-C-8	II	0.336	0.306	167	200	1.20
GS-C-9	II	0.337	0.302	200	300	1.50
GS-C-10	II	0.337	0.305	267	200	0.75
GS-C-11	II	0.338	0.298	333	400	1.20
GS-C-12	II	0.336	0.292	400	600	1.50
GS-E-4	III	0.335	0.328	45	−15	−0.33
GS-E-5	III	0.336	0.325	42	−25	−0.60
GS-E-6	III	0.337	0.320	38	−35	−0.92
GS-E-7	III	0.338	0.322	90	−30	−0.33
GS-E-8	III	0.335	0.318	83	−50	−0.60
GS-E-9	III	0.337	0.313	77	−70	−0.92
GS-E-10	III	0.336	0.318	180	−60	−0.33
GS-E-11	III	0.338	0.312	167	−100	−0.60
GS-E-12	III	0.336	0.306	153	−140	−0.92

6.6.3 Geometrical Criteria for Internal Stability of Soils

The potential of internal erosion is controlled by the geometric information of the soil. Gain size distribution is one of the most important geometric factors. A large number of laboratory tests were conducted to investigate the potential of internal erosion of different soils. The testing conditions for most previous internal erosion tests are summarized in Table 6.5. Based on the test results, several geometric criteria (Table 6.6) were developed to assess the internal stability of a soil. This section focuses on geometric criteria.

6.6.3.1 Gradation Based Internal Stability Criteria

Gradation based criteria can be divided into three groups based on: (1) the uniformity of the grain size distribution curve (e.g. Istomina, 1957), (2) the filtering capacity of the coarse particles to the fine particles within the soil (e.g. Kezdi, 1979; Kenney and Lau, 1985; Chapuis, 1992; Li and Fannin, 2008), and (3) the flatness of the grain size distribution curve at the coarse fraction and the fine fraction (e.g. Burenkova, 1993; Wan and Fell, 2008).

Istomina (1957) (see Kovács, 1981) developed a simple method for evaluating internal stability of sandy gravels using the coefficient of uniformity, C_u, as an indicator. Soils with C_u of less than 10 are internally stable, and those with C_u larger than 20 are likely to be internally unstable. Those with C_u between 10 and 20 fall into the transitional zone.

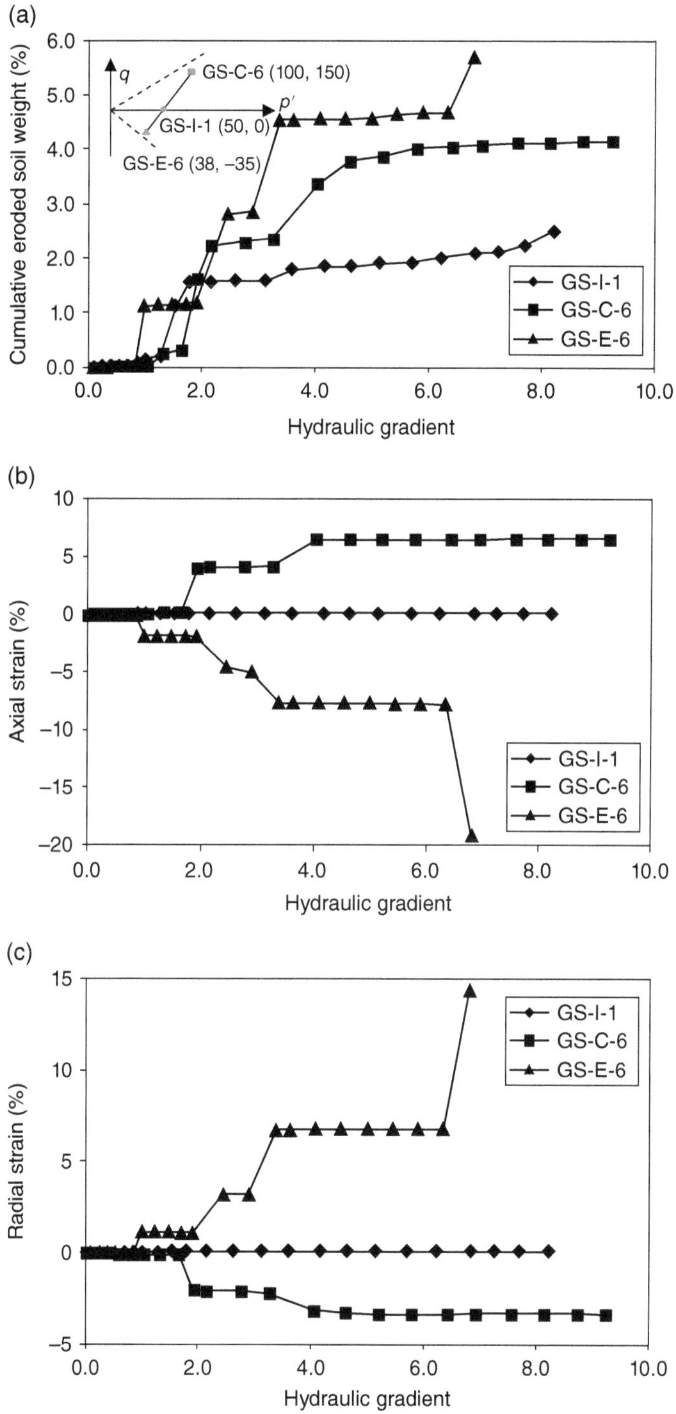

Figure 6.18 (a) Normalized cumulative eroded soil mass, (b) axial strain, and (c) radial strain of specimens GS-I-1, GS-C-6, and GS-E-6 during the internal erosion testing (the eroded soil mass is normalized by the total mass of the specimen)

Table 6.5 Summary of experimental investigation of internal stability of soils

References	Soil properties				Testing information				Test results	
	C_u	$(H/F)_{min}$	$(d_{15c}/d_{85f})_{max}$	P	i	Vibration	Flow direction	Saturated	No. of stable cases	No. of unstable cases
USACE (1953)	11–24	0.07–0.73	2.6–8.1	0	0.5–16.0	Y	D	Y	3	1
Kenney and Lau (1985)	3–71	0.49–3.08	1.7–51.0	0	—	Y	D	—	8	6
Adel et al. (1988)	17–57	0.25–1.02	6.2–41.0	0	0–0.7	N	H	—	1	1
Lafleur et al. (1989)	7–19	0.23–0.65	3.5–6.3	0	2.5–6.7	Y	D	—	2	1
Sun (1989)	3–7000	0–0.55	10.6–37.0	7–43	20.0	N	U	Y	6	10
Aberg (1993)	3–30	0.34–1.91	1.4–8.9	0–12	21.6	Y	D	—	6	2
Burenkova (1993)	3–109	0.17–0.77	8.1–82.0	0	2.5	N	D	N	4	4
Skempton and Brogan (1994)	4–24	0.14–2.70	4.0–10.9	0	0–1.0	N	U	Y	2	2
Chapuis et al. (1996)	18–32	0.35–0.99	2.3–10.6	0	9.8	Y	D	Y	2	1
Honjo et al. (1996)	2–13	0–2.75	1.2–7.4	0	9.8, 13.0	Y	D	Y	5	4
Liu (2005)	18–152	0.3–1.00	3.3–25.3	0	0–1.0	N	U	—	2	7
Mao (2005)	20–54	0–0.80	5.0–22.0	0	—	N	U	—	0	8
Moffat (2005)	73–85	0.30	13.7–14.3	0–4	1.0–62.0	N	D & U	Y	0	2
Fannin and Moffat (2006)	4–21	0–2.20	2.4–7.4	0	0.1–18.5	Y	D	Y	2	0
Lafleur and Nguyen (2007)	12–24	0.08–0.18	2.7–12.8	16–27	10.0–11.0	N	D	Y	3	0
Bendahmane et al. (2008)	42–173	0–0.03	19.3	10–30	5.0–140.0	N	D	Y	1	2
Li (2008)	6–36	0–1.30	2.8–19.0	0–19	0.08–31.0	N	D & U	Y	1	6
Wan and Fell (2008)	21–5709	0–0.62	5.0–99.0	3–40	8.0	N	D & U	—	10	9
Cividini et al. (2009)	32	0.07–0.56	8.6–11.3	25–31	0.2–0.99	N	U	Y	2	0
Total									60	66

C_u: coefficient of uniformity; H: mass fraction between grain size d and $4d$; F: mass fraction at any grain size d; d_{15c}: diameter of the 15% mass passing in the coarse part; d_{85f}: diameter of the 85% mass passing in the fine part; P: fines content; i: hydraulic gradient.

Table 6.6 Summary of commonly used geometric criteria for soil internal stability

References	Material description	Geometric criteria	Remarks
Istomina (1957) (see Kovács 1981)	Sandy gravel	$C_u \leq 10$: internally stable $10 \leq C_u \leq 20$: transitional $C_u \geq 20$: internally unstable	
Kezdi (1979)	All soils	$(d_{15c}/d_{85f})_{max} \leq 4$: internally stable	
Kenney and Lau (1985)	Granular soils	$(H/F)_{min} \geq 1.0$: internally stable	Medium to dense soils
Burenkova (1993)	Granular and graded soils	$0.76\lg(h'') + 1 < h' < 1.68\lg(h'') + 1$: internally stable	
Mao (2005)	Granular soils	$p_f \geq 100/4(1-n)$: internally stable	
Wan and Fell (2008)	Well-graded soils	$30/\log(d_{90}/d_{60}) < 80$, or $30/\log(d_{90}/d_{60}) < 80$ and $15/\log(d_{20}/d_5) > 22$ internally stable	
Li and Fannin (2008)	Granular soils	For $F < 15$, $(H/F)_{min} \geq 1.0$; internally stable For $F > 15$, $H \geq 15$; internally stable	H and F is obtained at $(H/F)_{min}$

h': d_{90}/d_{60}; h'': d_{90}/d_{15}; n: porosity; p_f: the finer fraction by weight (for gap-graded soils, p_f = mass passing at the gap location (%); for well graded soils, p_f = mass passing (%) at the division diameter $d_f = 1.3\sqrt{d_{15}d_{85}}$).

Following the idea of a filtration criterion (Terzaghi, 1939) (i.e. the base soil particles cannot pass through the pores formed by the filters), a soil can be considered to be composed of two components: coarse and fine components (Kezdi, 1979) as shown in Figure 6.19. If the coarse component can serve as a filter for the fine component, the soil is considered to be internally stable. In this method, the grain size distribution curve is split into two parts at a arbitrary grain diameter, if it satisfies $(D_{15c}/d_{85f})_{max} < 4$ (where D_{15c} is the diameter of the 15% mass passing in the coarse part, d_{85f} is the diameter of the 85% mass passing in the fine part), and the soil is internally stable. Sherard (1979) presented the same criterion but changed the value from 4 to 5. Fannin and Moffat (2006) suggested that a soil with $(D_{15c}/d_{85f})_{max}$ close to 4 is stable, whereas a soil with $(D_{15c}/d_{85f})_{max}$ close to 7 exhibits internal instability at relatively small hydraulic gradients.

Kenney and Lau (1985) proposed a geometric criterion for granular soils. The particles finer than grain size d (having a mass fraction F) will likely be washed out if there are not enough particles between grain size d and $4d$ (having a mass fraction H) as shown in Figure 6.20. The boundary $(H/F)_{min}$ was initially set to be 1.3, but it was eventually changed to 1.0 after a comprehensive discussion by Milligan (1986), Ripley (1986), and Sherard and Dunnigan (1986). In other words, a soil is internally unstable if $(H/F)_{min} < 1.0$. H/F can be obtained from the grain size distribution curve over a portion of its finer fraction given by $F \leq 20\%$ for soils with a widely graded primary fabric ($C_u > 3$), and by $F \leq 30\%$ for soils with a narrowly graded primary fabric ($C_u \leq 3$). The internal stability is independent of grain size and is only dependent on the geometrical form of the grading curve.

Kezdi's criterion and Kenney and Lau's criterion are similar in mathematical expressions where the secant slope of the grain size distribution curve can be used to indicate the internal stability of the soil (Chapuis, 1992). Kezdi's criterion starts from the consideration that the pores formed by the coarse particles should filter the fine particles. Li and Fannin (2008) proposed an internal stability criterion for granular soils by combining Kezdi's criterion and Kenney and Lau's criterion. At $F < 15\%$, Kenney and Lau's criterion is used, whereas Kezdi's criterion is adopted when F is larger than 15%.

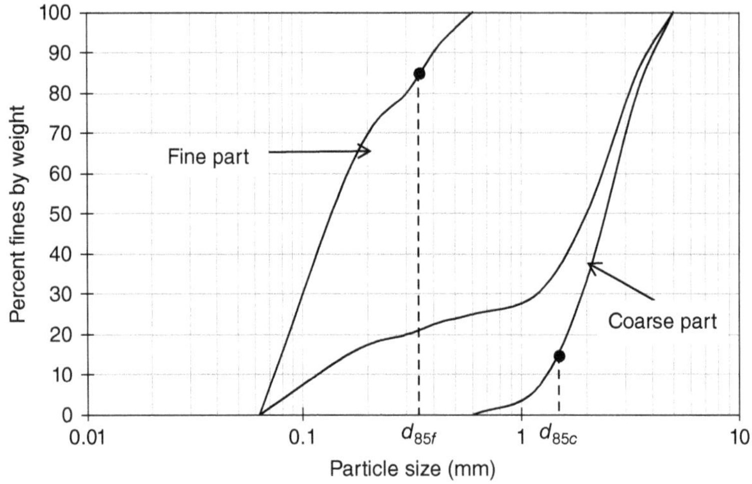

Figure 6.19 Kezdi's criterion for assessing internal stability of granular soils

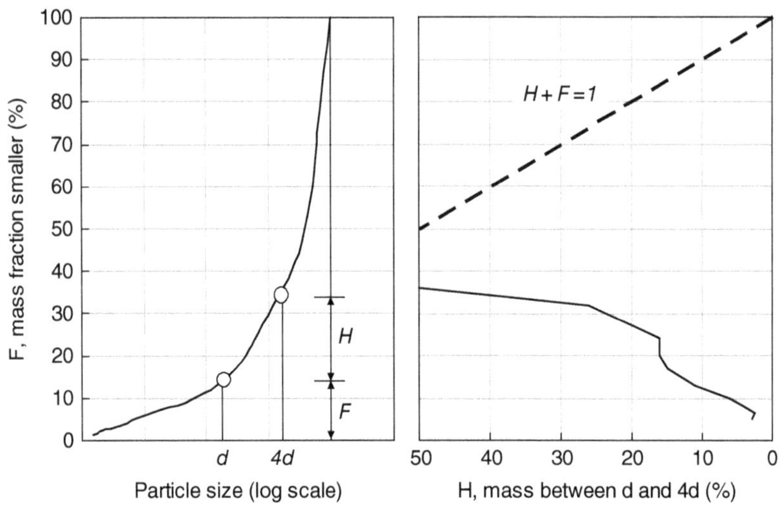

Figure 6.20 Kenney and Lau's criterion for assessing internal stability of granular soils

A soil with a grain size distribution curve of a gentle slope at the fine fraction part and a steeper slope at the coarse fraction part is more susceptible to internal erosion. Burenkova (1993) developed a criterion to assess the internal stability of granular soils based on the results of laboratory tests on 22 granular sand-gravel soils with maximum particle sizes up to 100 mm, and coefficients of uniformity up to 200. The internal stability depends on two conditional factors of uniformity:

$$h' = d_{90} / d_{60} \tag{6.30}$$

$$h'' = d_{90} / d_{15} \tag{6.31}$$

where d_{90}, d_{60}, and d_{15} are the diameters of particles for which 90%, 60%, and 15% are smaller by weight, respectively. If h' and h'' satisfy the following relation, the soil is considered internally stable:

$$0.76\lg(h'')+1 < h' < 1.86\lg(h'')+1 \tag{6.32}$$

Noticing that the control variables for different soils are not exactly the same, Chang and Zhang (2013b) divided the soil into two groups, well-graded and gap-graded, based on results of tests on 131 different soils reported in the literature. Each group is further divided into three categories: category I (with fines contents less than 5% for well-graded soils and 10% for gap-graded soils), category II (with fines contents larger than 20% for well-graded soils and 35% for gap-graded soils), and category III (with fines contents of 5–20% for well-graded soils and 10–35% for gap-graded soils).

The internal stability of a well-graded soil with a fines content less than 5% can be evaluated using $(H/F)_{\min}$ as a control variable, and the internal stability of a well-graded, low plasticity soil with a fines content of 5–20% can be assessed according to a relation between $(H/F)_{\min}$ and fines content. A well-graded soil with a fines content larger than 20% is deemed to be internally stable, as shown in Figure 6.21(a). The internal stability of a gap-graded soil with a fines content less than 10% can be assessed using the gap ratio as a control variable, and the internal stability of a gap-graded, medium plasticity soil with a fines content of 10–35% can be evaluated using gap ratio and fines content as two control variables. A gap-graded soil with a fines content larger than 35% is deemed to be internally stable as shown in Figure 6.21(b).

6.6.3.2 Controlling Constriction-size-based Internal Stability Criteria

Controlling constriction size is a characteristic of the void network, and this size is equal to the diameter of the largest particle that can possibly be moved through the material by seepage (Kenney et al., 1985; Indraratna et al., 2011). The constriction size distribution curve can be obtained once knowing the particle size distribution and relative density of a soil by using a probabilistic method. Within a broadly graded soil, the controlling constriction size of the coarse skeleton affects the potential of internal erosion of the soil. Thus by dividing a soil into a coarse component and a fine component, the controlling constriction size could be obtained from the coarse component. Indraratna et al. (2011) established a controlling constriction-size-based internal stability criterion. D_{c35} (35% constrictions of a given constriction size distribution curve are smaller) is chosen as the controlling constriction size of the coarse component, and d_{85f} is used to represent the particle size of the fine component. If a soil satisfies $D_{c35}/d_{85f} < 0.73$, it is considered as internally stable. If $D_{c35}/d_{85f} > 0.82$, the soil can be judged as internally unstable. Within these two boundaries, further laboratory testing is needed. Whether the controlling constriction size-based criterion can accurately predict the internal stability of a broadly graded soil depends on the accuracy of the calculation of constriction size distribution. The criterion proposed by Indraratna et al. (2011) is based on the assumption that the coarse particles contact with each other and the fine particles fill within the pores formed by the coarse particles. However, for soil specimens compacted by moisture tamping, which is a common compaction method in dam engineering practice, some fine particles could also form the matrix.

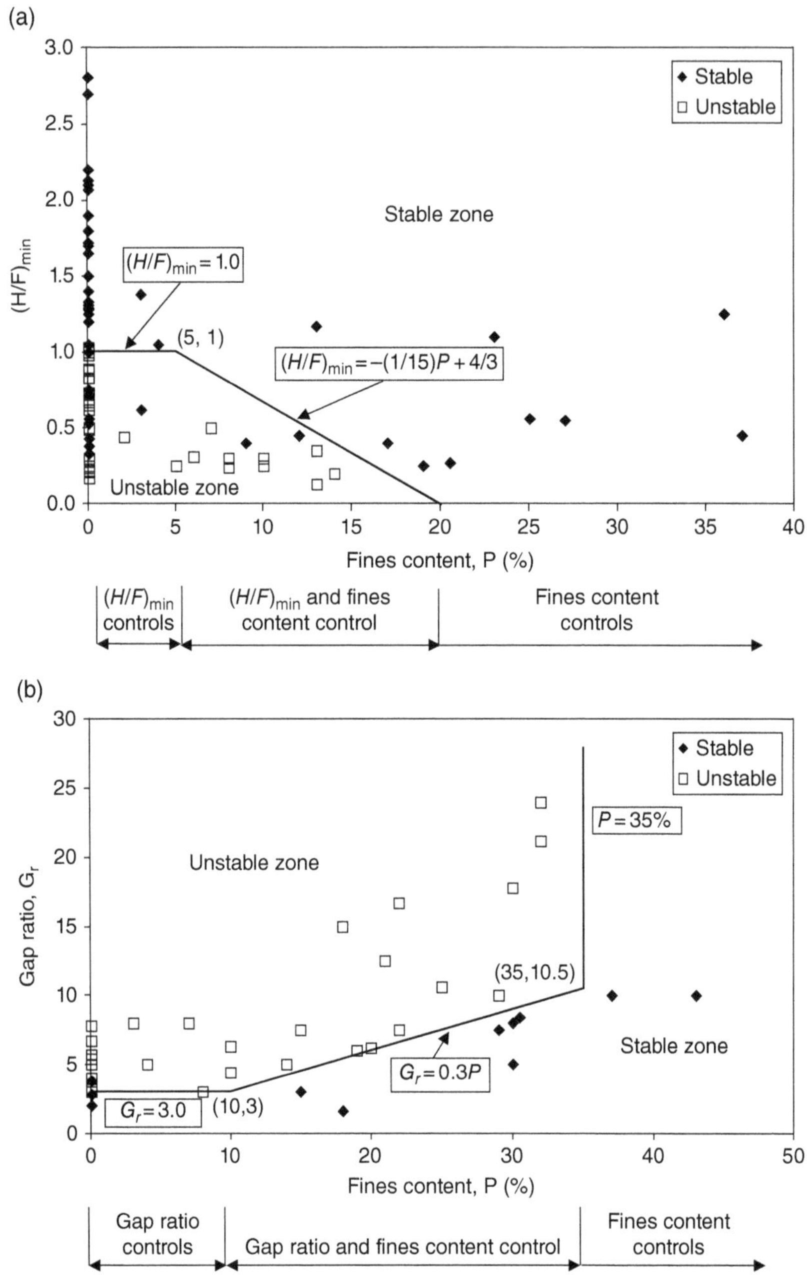

Figure 6.21 Composite geometric criteria for (a) well-graded soils and (b) gap-graded soils

Table 6.7 Summary of models of critical hydraulic gradient for granular soils

References	Equations
Terzaghi (1939)	$i_c = (G_s - 1)/(1+e)$
Wu (1980)	$i_c = (0.026 + 0.345D_r + 1.7P)(G_s - 1)\dfrac{d_i}{d_i + ed_{\Theta K}}$
Skempton and Brogan (1994)	$i_c = \alpha(G_s - 1)/(1+e)$
Indraratna and Radampola (2002)	$i_c = \dfrac{2}{3}\dfrac{d_i^2}{(d_i^2 + 0.375D_0^2)}(G_s - 1)[\cos\beta\tan\phi + \sin\beta]$
Liu (2005)	$i_c = \alpha_1(G_s - 1)/(1+e)\dfrac{d_i}{d_{\Theta K}}$
Li and Fannin (2011)	$i_c = \dfrac{\alpha}{1-0.5\alpha}(\sigma'_{vm} + 0.5(G_s - 1)/(1+e))$

G_s: specific gravity; e: void ratio; D_r: relative density; d_i: particle size to be washed out; $d_{\Theta K}$: effective particle size; D_0: minimum pore diameter; α: stress reduction factor; β: angle of the pore channel with horizontal axis; ϕ: angle of repose; σ'_{vm}: initial effective vertical stress. α_1: shape factor.

6.6.4 Critical Hydraulic Gradients for Suffusion

As mentioned in the previous section, the geometric conditions control the potential of internal erosion of a soil, whereas hydraulic conditions govern the onset of internal erosion. In other words, geometric conditions reflect the intrinsic properties of the soils, and hydraulic conditions express the external triggers. Critical hydraulic gradient is one of the most important factors of hydraulic conditions. Table 6.7 summarizes several critical hydraulic gradient equations.

6.6.4.1 Zero effective stress concept

Under upward seepage flow conditions, the effective vertical stress within a soil decreases with the increase of hydraulic gradient. Critical hydraulic gradient for sand boil (Terzaghi, 1939) is defined as when the effective stress within the soil becomes zero, which means that the pore water pressure at a certain depth of the soil is equal to its overburden stress. It corresponds to the soil failure and is developed for uniform sand. For internally unstable sandy gravels, the results presented by Skempton and Brogan (1994) indicate that the critical hydraulic gradient can be much lower than the theoretical values (at sand boil). This is explained from a stress transformation standpoint in that the overburden load is taken predominantly by the matrix structure of gravel particles, leaving the sand under relatively small stress. However, both Terzaghi's critical hydraulic gradient and Skempton and Brogan's critical hydraulic gradient do not consider the effect of overburden stress. Based on the consideration of the effective stress in the finer fraction of an internally unstable soil under overburden stress, Li and Fannin (2011) proposed a theoretical hydromechanic envelope to characterize the critical hydraulic gradient of a soil under overburden stress. The critical hydraulic gradient for an internally unstable soil subjected to an effective vertical stress can be expressed as

$$i_c = \frac{\alpha}{1-0.5\alpha}\left(\sigma'_{vm} + 0.5(G_s - 1)/(1+e)\right) \quad (6.33)$$

where α is a stress transformation coefficient and σ'_{vm} is the effective vertical stress of the soil. The critical hydraulic gradient based on zero effective stress can only be applied to cases with upward seepage.

6.6.4.2 Particle Level Considerations

Methods based on zero effective stress take the soil as a continuum. The critical hydraulic gradient can also be developed based on limit equilibrium of the moveable particles. Within a broadly graded granular soil, the moveable fine particles within coarse pore networks can be washed out if the hydraulic gradient reaches a critical condition (Wu, 1980; Kovács, 1981). Based on the forces acting on a fine particle within the coarse pore channels, the critical hydraulic gradients for suffusion (Wu, 1980; Liu, 2005) were developed. The expressions are shown in Table 6.7. The critical gradient of suffusion is associated with the initiation of internal erosion. The critical hydraulic gradients developed in particle level are based on the assumption that the pore network can be represented by parallel channels with different diameters.

6.6.4.3 Critical Hydraulic Gradients in Different Erosion Phases

Based on the internal erosion tests, corresponding to the three phases proposed by Chang and Zhang (2013a), three critical hydraulic gradients are defined as shown in Figure 6.17: initiation, skeleton-deformation, and failure hydraulic gradients. The initiation hydraulic gradient, i_{start}, is defined here as the critical gradient that initiates internal erosion without visible soil deformation. It corresponds to the minimum gradient in the initiation phase. The skeleton-deformation hydraulic gradient, i_{sd}, is defined as the gradient that induces significant soil deformation and increased eroded soil mass, which corresponds to the minimum gradient in the development phase. The failure hydraulic gradient, i_f is defined as the gradient that induces soil shear failure. It corresponds to the minimum gradient in the failure phase.

Initiation Hydraulic Gradient

A general observation is that the initiation hydraulic gradient increases with the decrease of soil porosity, no matter whether the soil is under isotropic, triaxial compression, or triaxial extension stress conditions as indicated in Figure 6.22. For instance, under isotropic stress states, i_{start} increases gradually from 0.82 to 1.56, as the soil porosity decreases from 0.330 to 0.320. It is also observed that the initiation hydraulic gradients under the extension stress states are generally higher than those under the compression stress states under the same porosity. The initiation gradient increases with the increase of mean effective stress under the same shear stress ratio, but at a reducing rate. At the initiation hydraulic gradient, some fine particles filling the large pores of the skeleton start to erode, whereas the skeleton structure of the soil remains stable. The initiation

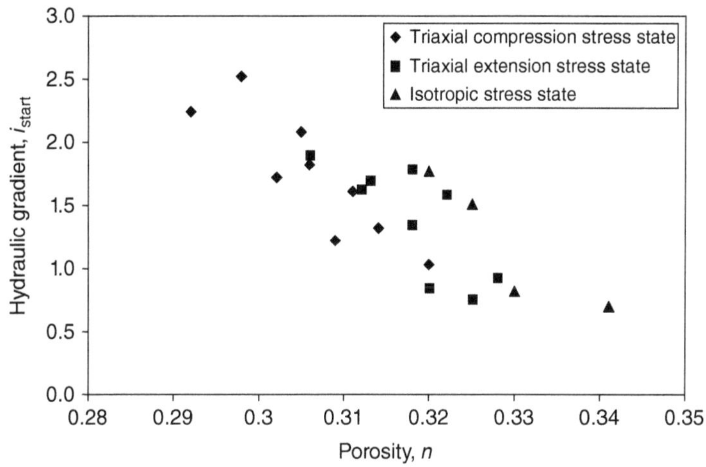

Figure 6.22 Variations in initiation hydraulic gradient with porosity under different stress states

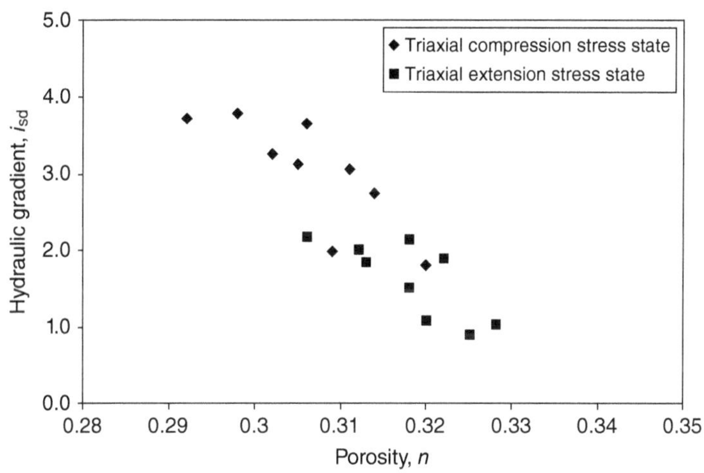

Figure 6.23 Variations in skeleton-deformation hydraulic gradient with porosity under different stress states

hydraulic gradient is mainly controlled by the constriction size and connectivity of the pores within the soil.

Skeleton-deformation Hydraulic Gradient

A general observation is that the skeleton-deformation hydraulic gradient increases with the decrease of soil porosity no matter whether the soil is under triaxial compression or triaxial extension stress conditions, as indicated in Figure 6.23. For instance, the soil porosity of specimens GS-C-6, GS-C-9, GS-C-12 (under the same shear stress ratio of 1.50 but different confining stresses) are 0.309, 0.302, and 0.292, respectively, and their corresponding skeleton-deformation hydraulic gradients are 1.66, 2.76, and 3.27, respectively. It should be noted that the deformations of the specimens under the isotropic stress conditions were small; the development phase was not observed within the applied hydraulic gradients up to 8.22 for specimen GS-I-1, 8.34 for specimen GS-I-2, and 8.32 for specimen GS-I-3. Under the same soil porosity, the skeleton-deformation hydraulic gradients under triaxial compression stress conditions are higher than those under triaxial extension stress conditions. For instance, the skeleton-deformation hydraulic gradient of specimen GS-C-4 is larger than that of specimen GS-E-6, although they are under the same soil porosity of 0.320.

After the initiation hydraulic gradient is reached, as the hydraulic gradient is further increased to the skeleton-deformation hydraulic gradient, the strong force chains formed by the coarse particles buckle due to the loss of radial support for the triaxial compression cases or axial support for the triaxial extension cases. The fine particles provide lateral support for the force chains and the loss of the fine particles undermines the stability of the strong force chains. Hence sudden erosion and deformation occur. After that, a new equilibrium state is reached if shear failure does not occur. The skeleton-deformation hydraulic gradient is mainly controlled by the buckling potential of the strong force chains within the soil.

Failure Hydraulic Gradient

Shear failure did not occur in most of the internal erosion tests in the range of applied hydraulic gradients except for specimens GS-E-6, GS-E-9, and GS-E-12. Their corresponding failure hydraulic gradients are 6.8, 1.98, and 2.43. The initial stress states of these three specimens are already close to shear failure, so a limited change in the stress state due to seepage would bring the specimens to failure.

To further study the effect of stress state on shear failure of soils, Chang (2012) and Zhang and Chang (2012) developed a theoretical formula for failure hydraulic gradient. Consider a soil specimen under triaxial stress state and downward seepage, as shown in Figure 6.24(a). The pore water pressure keeps constant at the bottom of the specimen and increases at the top of the specimen; hence a downward seepage flow is generated. The effective axial stress at the top of the specimen decreases, while the effective axial stress at the bottom of the specimen increases. The effective radial stress at the top of the specimen also decreases as the pore water pressure increases, whereas the effective radial stress at the bottom of the specimen keeps constant. In the p′–q space, the stress states at the top and the bottom of the specimen under downward seepage can be established as

$$\text{At top:} \quad \begin{aligned} p' &= \frac{1}{3}\left(\sigma'_{1t0} + 2\sigma'_{3t0}\right) - \Delta u \\ q &= \sigma'_{1t0} - \sigma'_{3t0} = \text{constant} \end{aligned} \tag{6.34}$$

$$\text{At bottom:} \quad \begin{aligned} p' &= \frac{1}{3}\left(\sigma'_{1b0} + 2\sigma'_{3b0}\right) + \frac{1}{3} i\gamma_w h \\ q &= \sigma'_{1b0} - \sigma'_{3b0} + i\gamma_w h \end{aligned} \tag{6.35}$$

where σ'_{1t0} and σ'_{3t0} are the initial effective axial stress and radial stress at the top of the specimen, respectively; σ'_{1b0} and σ'_{3b0} are the initial effective axial stress and radial stress at the bottom of the specimen, respectively; Δu is the increment of pore water pressure; i is the applied average hydraulic gradient through the specimen (take downward as positive); γ_w is the unit weight of water; and h is the height of the specimen.

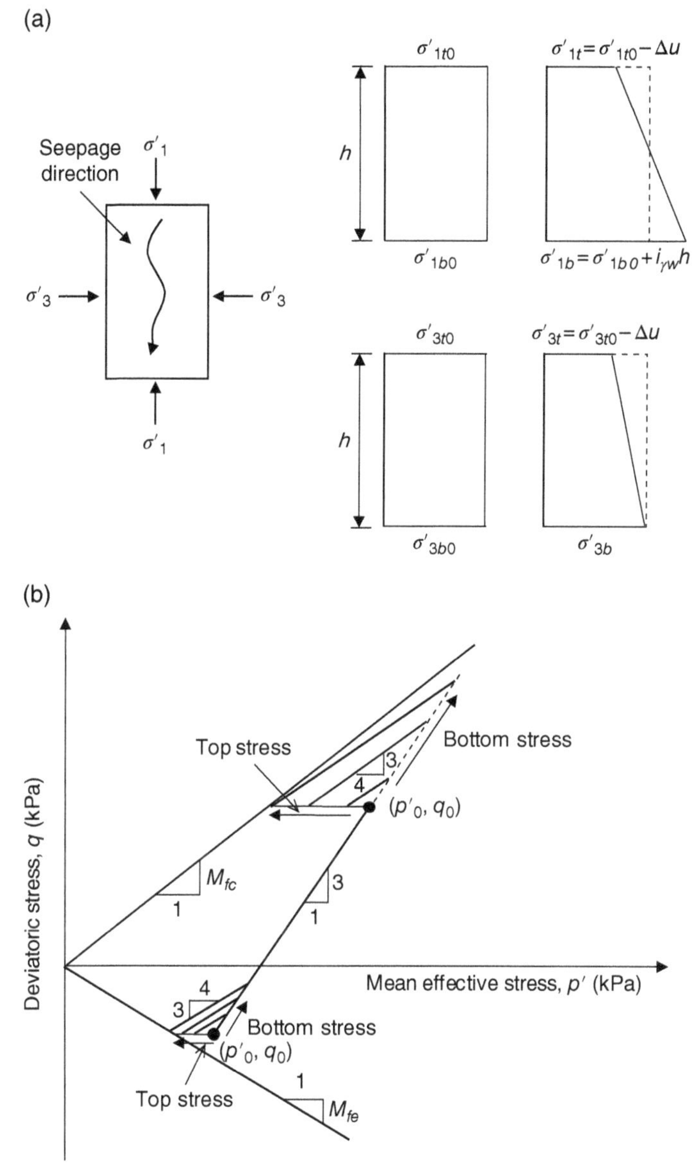

Figure 6.24 (a) Illustration of the stresses at the top and bottom of a specimen under downward seepage and (b) evolution of the stress states of the specimen under triaxial compression and extension stress conditions with increase of hydraulic gradient in the p'-q space

The effective mean stress at the top of the specimen decreases with increasing hydraulic gradient, whereas the deviatoric stress remains constant. Both the effective mean stress and the deviatoric stress at the bottom of the specimen increase with increasing hydraulic gradient. Thus the evolution of the stress state at the top of the specimen follows a constant q stress path, while the evolution of the stress state at the bottom of the specimen follows a drained triaxial stress path (3:1 line). Assuming the stresses are linearly distributed along the specimen, the evolution of the stress states of the specimen under increasing hydraulic gradients is shown in Figure 6.24(b). Shear failure of the specimen will occur when the stress path reaches the failure envelope. Under both triaxial compression and extension stress conditions, the stress state at the top of the specimen reaches shear failure slightly earlier than that at the bottom of the specimen under downward seepage. At the failure state, the mean stress and the deviatoric stress satisfy the following requirement:

$$q = M_f p' \tag{6.36}$$

where M_f is the failure shear stress ratio.

Substituting Equations (6.34) and (6.35) into Equation (6.36), the failure hydraulic gradient, i_f can be obtained as

$$i_f = \frac{p'_0}{\gamma_w h}\left(1 - \frac{\eta_0}{M_f}\right) \tag{6.37}$$

where p'_0 and η_0 are the initial mean stress and initial shear stress ratio, respectively.

Following the same theory, the failure hydraulic gradient for soils under upward seepage can also be developed, which is the same as Equation (6.37). It shows that the failure hydraulic gradient is a function of the initial stress state, shear strength, and drainage path of the soil element.

6.7 Filter Criteria

6.7.1 Functions of Filter

Filters are placed in the embankment body, its foundation, or other areas of hydraulic structures (i.e. conduit) to intercept water flowing through cracks or openings in a base soil and to block the movement of eroding soil particles into the filter and to intercept water flowing through the pores of the base soil, allowing passage of the water while preventing the movement of base soil particles. In other words, the pore size in the filter medium should be large enough to allow free passage of seeping water and at the same time small enough to retain soil particles of the base material.

The basic functions of filters that govern the capability of providing control for internal erosion are summarized by ICOLD (2013):

1. Retention criterion. The filter limits the transport of particles of the base soil it is protecting. Soil particles are caught at the filter face, reducing the flow of water through cracks or openings and preventing further erosion and enlargement of the cracks or openings.
2. Permeability criterion. The filter is sufficiently permeable to transport the water flowing through the base soil into drainage layers or to the toe of the dam to prevent the build-up of excess pore water pressure in the base soil.
3. Internal stability. The filter is internally stable and not subject to finer particles eroding from within the filter, and will not hold a crack due to cementation or the presence of plastic fines.
4. No cohesion. Fines within the filter material should be limited so that the filter is cohesionless and no cavities or cracks can be sustained within the filter.
5. Strength. The shear strength of the filter should be high enough to prevent crushing under the stresses within the dam.

In terms of the retention criterion, filters should provide total retention as suggested by Giroud (2010) in zoned dams on the downstream side of clay core. The function is to prevent loss of clay by stopping particles that may migrate as a result of backward erosion, concentrated leak erosion, etc. These filters become progressively clogged by the fine particles in the core. The opening of the filters should be sufficiently small to retain all fine particles likely to mitigate. The retention function of the filters is mainly controlled by their geometric conditions. The filter criteria are a reflection of the retention capability of the filter, which will be presented in Section 6.7.2. Moreover, to minimize the segregation of broadly graded filters during construction, the maximum range of filter particle size should be limited (Sherard et al., 1984; Milligan, 2003).

In terms of the permeability criterion, the permeability of the filter should be permeable enough to prevent the build-up of excessive pore water pressure at the core–filter interface. This is achieved by limiting the fines content and by controlling the grain size, d_{15f} of the filter. Sherard et al. (1984) pointed out that the soil permeability directly relates to the square of its particle size, d_{15}. Therefore, the permeability requirement can be satisfied if the filter satisfies the following relation according to Terzaghi (1939):

$$d_{15f} / d_{15b} > 4 \tag{6.38}$$

In terms of the no cohesion criterion, the filter should not hold a crack. This can be achieved by limiting the fines content and the plasticity of the fines. The fines should be non-plastic. In terms of strength of filters, the filter should have enough strength to prevent their crushing during the construction and their operation. This can be achieved by specifying durability and crushing strengths of the filter aggregates.

6.7.2 *Filter Criteria*

6.7.2.1 Gradation-based Filter Criteria

Design criteria in current engineering practice are mostly based on empirical relationships derived from tests carried out on different combinations of base and filter materials. The Terzaghi's filter retention criterion for granular filter is one of the commonly used empirical criteria expressed as

$$d_{15f} / d_{85b} < 4 \tag{6.39}$$

As the relationship of Equation (6.39) is based on experimental results, different researchers subsequently assigned different values, depending on the type of base soil, method of testing, and personal experience. For example, Sherard et al. (1984) suggested a value of 9 for well-graded base soil, and Indraratna et al. (1996) suggested a value 4–5 for a lateritic clay base soil.

Indraratna and Raut (2006) pointed out that although Equation (6.39) is valid for uniform base and filter materials, there are a few major limitations associated with this equation when used with well-graded materials: (1) It does not distinguish between uniform and well-graded filters if they have the same d_{15}, which invariably makes any design involving well-graded filters conservative, (2) the effect of filter compaction is ignored, and (3) the gradation condition of base soil (i.e. uniform and well-graded base soils) cannot be differentiated if they have the same d_{85}.

An enhanced method for identifying effective filters should address the limitations raised above. Further studies (e.g. Sherard and Dunnigan, 1985; Sherard and Dunnigan, 1989) divided the base soil into four groups (the gravel fraction should be ignored):

1. Group 1 – fine silts and clays with more than 85% passing the No. 200 sieve (0.075 mm).
2. Group 2 – silty and clayey sands and sandy silts and clays with 40–85% passing the No. 200 sieve.
3. Group 3 – silty and clayey sands and gravel with 15–39% passing the No. 200 sieve.
4. Group 4 – silty and clayey sands and gravelly sands with 15% or less passing the No. 200 sieve.

Based on the classification of different base soils and the test results, Sherard and Dunnigan (1989) proposed a widely accepted criterion for different groups of soil as summarized in Table 6.8. USDA (1994) and USBR (2004) adopted the Sherard and Dunnigan's criteria as the design rule for effective filters. Moreover, USDA (1994) added other constrains to prevent the use of possibly gap-graded filters and possible segregation during construction as shown in Table 6.8. USDA (1994) also presented the design of protective filters and examples for different categories of base soil.

Foster and Fell (2001) summarized the no erosion filter (NEF) tests conducted by previous researchers, and carried out additional tests particularly focusing on dispersive core materials.

Table 6.8 Filter design criteria

Base soil			Filter criteria $(d_{15f})_{max}$			
Category	Description	Fines content (< 0.075 mm) (%)	Sherard and Dunnigan (1989)	USDA (1994)	Foster and Fell (2001)	Locke and Indraratna (2002)
1	Fine silts and clays	≥85	$\leq 9d_{85b}$ but mot less 0.2 mm	$\leq 9d'_{85b}$ but mot less 0.2 mm	For non-dispersive base material: $\leq 9d_{85b}$ For dispersive base material: $\leq 6.4d_{85b}$	$\leq 12d'_{85b}$
2	Sands, silts, clays, and silty and clayey sands	40–85 (In Foster and Fell's criterion, 35–85)	≤ 0.7 mm	≤ 0.7 mm	For non-dispersive base material: ≤0.7 mm For dispersive base material: ≤0.5 mm	If plasticity index of base soil, PI > 10: $\leq 9d'_{85b}$ If plasticity index of base soil, PI < 10: $\leq 4d'_{85b}$
3	Silty and clayey sands and gravel	15–39 (In Foster and Fell's criterion, 15–35)	$\leq (40-A)/(40-15)(4d_{85b}-0.7\,mm)+0.7\,mm$	$\leq (40-A)/(40-15)(4d'_{85b}-0.7\,mm)+0.7\,mm$	$\leq 1.6\,(35-A)(4d_{85b}-0.7)/20+0.7$	—
4	Sands and gravel	≤15	$\leq 4d_{85b}$	$\leq 4d'_{85b}$	$\leq 7d_{85b}$	—

d'_{85b} represents the particle size of the reduced base soil with 85% passing rate. The reduced base soil means the gradation of the base soil is adjusted by removing the particles larger than 4.75 mm; *A* is the percentage of fines content.

For the design of critical filters for dispersive core materials, it is recommended that NEF tests be carried out to confirm the d_{15f} for no erosion and that the subdivision of soil groups 2 and 3 should be changed to a fines content of 35% instead of 40% as defined by Sherard and Dunnigan (1989). Moreover, they proposed new boundaries for the design of critical filters for dispersive core materials, as summarized in Table 6.8.

After a detailed investigation on the broadly graded base materials, Locke and Indraratna (2002) modified the Sherard and Dunnigan's criterion for group 1 and 2 soils (Table 6.8) to prevent the possible non-self-filtering properties of broadly graded base soil. Because the coarse fraction of the material is often too coarse to act as an effective filter for the fine fraction, the filters designed by traditional filter design criteria (e.g. Sherard and Dunnigan, 1989) to retain the coarse fraction of the base soil may allow continued erosion of the fine fraction.

6.7.2.2 Constriction-size-based Filter Criteria

To overcome the limitations in conventional filter criteria, great efforts have been made to investigate the constriction size of the filters. In particular, the constriction size can cover the effect of relative density on filter performance. Constriction-based filter criteria are all still under development.

Controlling constriction size provides a quantitative measure of the size of the channels within a granular filter. The controlling constriction size can be obtained from the constriction size distribution of a filter material. The constriction size distribution is dependent on the numbers and sizes of filter particles and on their spatial arrangement. Indraratna and Raut (2006) and Indraratna et al. (2007) developed an analytical computation procedure to determine the constriction size distribution (CSD) of a granular filter, based on the surface area of particles, the particle size distribution (PSD) and the anticipated relative density or porosity of the filter material. Figure 6.25 shows a typical CSD of a granular filter. D_{c35} and D_{c95} are then readily obtained from the CSD computed.

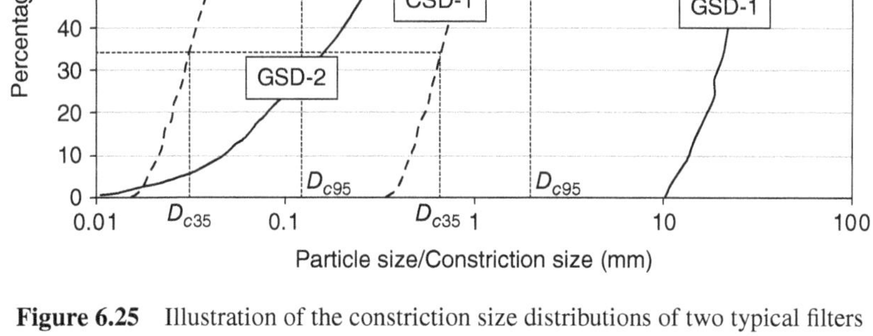

Figure 6.25 Illustration of the constriction size distributions of two typical filters

In effective filtration, the base particles smaller than D_{c35} are initially lost through the filter during filtration; however, these particles eroded between the constriction sizes D_{c35} and D_{c95} are still captured and retained at various depths in the filter depending upon the base particle sizes (Indraratna and Raut, 2006). It is the larger base particles that initiate self-filtration upon retention at the base–filter interface. Thus, D_{c95} forms a rational approach to separate self-filtering fraction of a base soil with respect to a given filter. The value of D_{c95} varies depending on the filter gradation, grain size, and compaction. The self-filtration size of the base soil is given by d_{85}* where $d_{85}{*} = d_{85}$ of the modified base soil grading curve by neglecting the base particles larger than the self-filtering constriction size, D_{c95}. This provides the following constriction-based filter design criterion for effective filtration (Raut and Indraratna, 2008):

$$D_{c35} / d_{85}{*} \leq 1 \tag{6.40}$$

6.8 Continuation of Internal Erosion

Whether internal erosion continues or not depends on whether the exit point is free or filtered. If it is filtered, the materials in the filters should satisfy certain filter criteria (e.g. Sherard and Dunnigan, 1989; GEO, 1993; ICOLD, 1994; USSD, 2011); otherwise the continuation of internal erosion may occur.

Given that erosion has been initiated, and the filters are absent or unable to stop erosion, internal erosion will progress to form a "pipe". For different initiation modes, the progression is not the same. For internal erosion initiated by backward erosion, the development of a pipe depends on the hydraulic gradient, the flow rate within the pipe, and the soil properties. For internal erosion initiated by concentrated leaks, the hydraulic shear stress and the soil erodibility within the hole will control the enlargement of the hole. For the internal erosion initiated by suffusion, the hydraulic and mechanical responses of the soil may mutate due to the loss of fine particles within the coarse matrix. In severe cases the loss of fine particles could induce concentrated flow and eventually lead to piping failure (Schuler, 1995).

Potential breach mechanisms for internal erosion involve the gross enlargement of pipe diameter, dam settlement, sinkhole on the crest leading to overtopping, and instability of the downstream slope (Foster et al., 2000b).

The initiation of internal erosion by suffusion requires a soil particle to detach from its parent material and become transported through sufficiently large voids and constrictions. Detachment and transport of soil particles requires a hydrodynamic force against the inter-granular friction. For internal erosion to continue, it requires large seepage forces to drag the relatively stable fine particles within the pores and large soil constrictions to permit further movement of the fine particles. During the suffusion process, the microstructure and the mechanical response of the soil change continuously.

In terms of practical dams or other hydraulic structures, whether continuation of internal erosion can occur or not depends on the dam conditions. According to ICOLD (2013), continuation of internal erosion is likely to occur in the following conditions:

1. a homogeneous embankment dam or dike
2. downstream shoulder of fine grained materials allowing a crack to occur
3. a zoned dam with filter layer or transition zone not satisfying the practical filter criteria

4. erosion into a crack or open joint (e.g. open joint or crack in a conduit or adjoining concrete structure), allowing the soil around the crack or open joint to pass
5. the critical erosion path exiting into a toe drain.

6.9 Progression of Internal Erosion

Internal erosion involves variations in erosion rate, permeability, and deformation during the entire erosion process. Some laboratory tests were conducted to study the progression of internal erosion by applying a constant large hydraulic gradient (e.g. Honjo et al., 1996; Tomlinson and Vaid, 2000; Bendahmane et al., 2008). Bendahmane et al. (2008) investigated the erosion rate of sand–clay mixture under several constant hydraulic gradients (20, 40, 60, and 100). The erosion rate increases suddenly to a maximum value and then decreases to a very small value for sand–clay mixture under a hydraulic gradient of 60. The maximum erosion rate is influenced by the applied hydraulic gradient and clay contents. It increases with the hydraulic gradient and decreases with the clay content. Tomlinson and Vaid (2000) studied the filtration failure process by applying a large constant hydraulic gradient of 14.7. Once the gradient was applied, large settlement (12 mm of the specimen 100 mm in height) and filtration failure occurred immediately. Honjo et al. (1996) further observed that the total settlement of the specimen that experiences internal erosion increases with the increase of coarse particle grain size ratio and gap ratio for well-graded and gap-graded soils, respectively.

The progression of internal erosion has also been investigated by applying hydraulic gradient in steps in the laboratory tests (e.g. Adel et al., 1988; Skempton and Brogan, 1994; Tomlinson and Vaid, 2000; Moffat et al., 2011). The progression of internal erosion was found to be a progressive process until failure. Skempton and Brogan (1994) investigated the internal erosion process of an internally unstable soil. When the hydraulic gradient is less than 0.10, the coefficient of permeability stays constant at a value of 4.5 cm/s. With further increase of hydraulic gradient, local erosion of fine particles was observed and the permeability increased slightly. When the hydraulic gradient was increased to 0.28, strong erosion was observed and the coefficient of permeability increased to 1.6 cm/s, which is approximately four times the initial value.

Tomlinson and Vaid (2000) studied the influence of the rate of gradient increase on filtration failure. The hydraulic gradient was applied in two ways: increasing in steps of 0.5 per 10 minutes and increasing in one step of 23 in 1 minute. The failure hydraulic gradient for the one-step case is only one third of that of the multi-step case. This indicates that the soil that may normally be stable at a given gradient may become unstable if the same gradient is imposed rapidly. Moreover, initiation of internal erosion of internally unstable soil is usually under low hydraulic gradient, thus the hydraulic gradient should be applied in multiple steps to better characterize the hydraulic information of initiation of internal erosion.

In general, whether progression of internal erosion can occur or not depends on hydraulic and mechanical conditions. The water seeping through the dam must be with sufficient velocity to provide sufficient energy or drag force to continue to transport particles along openings and off the external surfaces of the dam in a continuous process. The progression of internal erosion can stop if the flow in the concentrated leak pipe or in the soil experiencing concentrated leak, contact erosion, or suffusion is limited by head losses in the upstream or downstream zones. For erosion to progress, the cracks triggered by hydraulic fracture should be sustained. Moreover, the erosion pipe or the cavity through which eroded particles are being transported should not collapse. ICOLD (2013) indicated that plastic soils, when both saturated and partially saturated, can sustain openings. Non-plastic silts, sands, and gravels will in general collapse when they become saturated.

6.10 Suggested Topics for Further Research

Internal erosion has been studied extensively. However, owing to the complicated conditions within an embankment dam/dike in practice, many fundamental aspects of internal erosion should be paid more attention. Further research on the following topics is recommended:

1. Internal erosion is controlled by geometric, hydraulic, and mechanical conditions. It is a coupled process. For instance, the mechanical condition can influence the critical hydraulic condition for initiation of internal erosion. More efforts should be made to investigate the coupling effects of different controlling factors on internal erosion.
2. Most of the previous testing materials were reconstituted samples. However, the soil conditions within an embankment dam/dike and their foundations could be much more complicated than reconstituted conditions. Therefore, intact samples should be tested in laboratory to represent the site conditions.
3. In the previous studies on concentrated leak erosion, a fully opened man-made hole is usually generated before the erosion test. However, within an embankment dam/dike, the propagation of the cracks may also be influenced by the erosion process, the stress state, and the hydraulic regime. Even without erosion, the opening of the cracks is sensitive to changes in the water content in the soil. More research should be conducted to investigate the coupled process of propagation of cracks and the concentrated leak erosion.
4. As loss of some fine particles is usually inevitable, the mechanical responses of a soil after the loss of certain fine particles should be studied: the stress–strain relationship, the soil shear strength, and the soil stiffness.
5. The internal erosion process usually involves the initiation, continuation, and progression phases. More studies should be conducted to investigate the entire internal erosion process in a systematic manner.
6. Currently extensive laboratory and large-scale physical modeling tests have been conducted. However, more attention should be paid to how to relate findings from element tests and model tests.
7. The current design guidelines require that the filter materials be non-cohesive, to prevent formation of cracks in the filter layer. In this regard, the fines content in the filter should be less than 5%. However, during the filtration process of the core, some fine clay and silt particles are retained in the filter. The filter may evolve to cohesive features and may be susceptible to cracking. Thus more efforts should be made to investigate long-term behavior of filters (e.g. the maximum fines content that a filter can sustain to generate cracks).
8. The stress states within a dam with a corewall and filters are typically complex. The corewall is relatively thin and has relatively low permeability, which can generate high hydraulic gradients. In low-confining stress, high shear stress, and high hydraulic gradient conditions, high seepage force could be generated, which could induce shear failure of dams. However, there is a lack of fundamental understanding of the shear failure of soil induced by seepage forces at the element level.

7

Mechanics of Overtopping Erosion of Dams

Overtopping erosion of dams is one of the most common causes of dam failure. The overtopped water results in channel erosion over the crest and the downstream face driven by shear forces. The overtopping erosion process is often a progressive surface erosion process (i.e. sediment transport in dispersed particles) for non-cohesive embankments, and a headcut erosion process (i.e. formation and migration of a vertical or nearly vertical drop on the bed) for cohesive embankments. Understanding the mechanisms of surface erosion and the characterization of soil erodibility are the basis for modeling the overtopping erosion process of dams. In this chapter, the mechanics of surface erosion is presented first, then methods of characterizing soil erodibility are introduced. Finally, the characteristics of dam failure due to overtopping erosion are summarized.

7.1 Mechanics of Surface Erosion

The surface erosion in this chapter refers to the erosion of dam materials by the overtopped water. Surface erosion is a physical process by which the soil particles on the surface of a dam are removed by the flowing water. A hydraulic model, a geotechnical model, and an erosion model, or a model that couples the three elements, are required for modeling the overtopping erosion. The hydraulic model allows the prediction of the hydraulic parameters that control soil detachment and deposition (i.e. flow velocity, flow depth, and hydraulic roughness). These variables can then be combined in different ways to form composite predictor variables for a soil erosion model (Gimenez and Govers, 2002). The geotechnical model allows analysis of the soil stability during the breaching process. The erosion process depends on soil and flow conditions. The erosion model is used to deal with the detachment and deposition of dam materials, which is the interaction between the flowing water and the dam materials. This section focuses on the initiation, transport, and deposition process of sediment.

Dam Failure Mechanisms and Risk Assessment, First Edition. Limin Zhang, Ming Peng, Dongsheng Chang, and Yao Xu.

7.1.1 *Incipient Motion of Sediment*

Incipient motion refers to the initiation of surface erosion. As the properties of different kinds of materials vary significantly, the initiation mechanisms are not the same. In the following sections, the erosion mechanisms for granular soils and cohesive soils are presented separately.

7.1.1.1 Granular Soils

The forces and angular moments acting on a granular soil particle under surface flow conditions are shown in Figure 7.1. Forces independent of the flow are the weight, G, and the hydrostatic buoyancy, F_h, which can be represented by the submerged weight of the body, G'. The hydrodynamic force, F_D, has to overcome the submerged weight and the resistant force, F_R. The flow velocity at the top of the bed particle is assumed to be the same as the flow velocity near the bed, while the velocity at the bottom of the particle in the soil bed is the seepage velocity. Since the velocities of the flow at the top and bottom of the particle are different, a lift force, F_L, is introduced by the pressure difference.

The forces, G', F_D, and F_L, can be expressed as

$$G' = \frac{1}{6}\pi d^3 \left(\gamma_s - \gamma_w\right) \tag{7.1}$$

$$F_D = C_D A \frac{\rho_w u^2}{2} \tag{7.2}$$

$$F_L = C_L A \frac{\rho_w u^2}{2} \tag{7.3}$$

where C_D and C_L are the drag and lift coefficients, respectively, A is the surface area, and u is effective velocity near the bed particle.

The velocity distribution near the bed is characterized by a local mean velocity profile, $\bar{u}=f(y)$, as shown in Figure 7.1, and a turbulent fluctuation velocity, u'. If the flow near the boundary layer

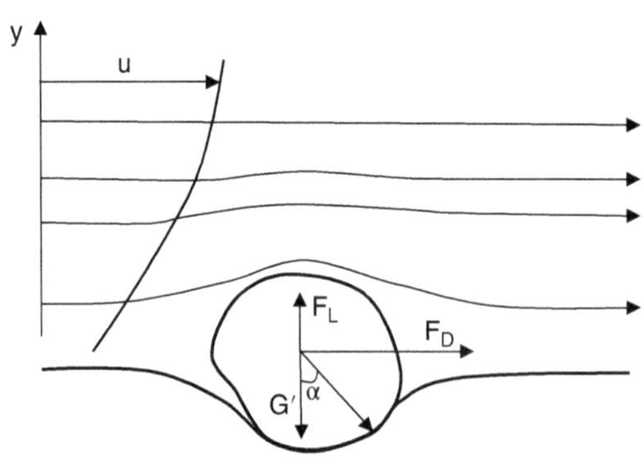

Figure 7.1 Forces balance on a particle at the surface of a bed (modified from Wiberg and Smith, 1987)

is laminar (i.e. the laminar sublayer thickness, $\delta' \geq 5d$ or the Reynolds number < 3.5), the individual particle will not shed eddies. If the water flow is laminar, the dominance of viscous forces results in the flowing water attempting to drag the assembly of particles at the surface along with it (Annandale, 2006).

As the flow velocity increases, the exposed particle sheds eddies, and a wake will be generated in the lee of the particle. The size of the wake depends on the size and shape of the particle and on the point of separation of the boundary layer formed on the particle. Annandale (2006) pointed out that when water flows over a matrix of sediment particles in a steady, time-invariant manner, it leads to the development of a pressure difference over the particles, which reduces to relating kinetic energy and pressure energy. The drag force, in this case, is the result of viscous shear and pressure differences in front of and behind the particle. The particle shape and local Reynolds number affect the drag force.

The incipient motion of the particle is controlled by rolling or sliding. Under critical conditions, the incipient motion of sediment can either be expressed in terms of critical average velocity or critical shear stress.

Under steady flow, the following elementary analytical solutions for the erosion of granular soil by White (1940) consider the balance of moments only:

$$M_{G'} = \frac{1}{6}\pi d^3 (\gamma_s - \gamma_w) \frac{d}{2} \sin\alpha \tag{7.4}$$

$$M_{F_D} = \beta \rho_w u^2 \frac{1}{4}\pi d^2 \frac{d}{2} \cos\alpha \tag{7.5}$$

where $M_{G'}$ is the moment by the submerged weight, M_{FD} is the moment by the drag force, β is a parameter which accounts for the effects of turbulent and the point of application of the drag force, and α is the angle of the slope.

Under moment equilibrium of the particle, the critical threshold velocity is

$$u_c = \left(\frac{2\tan\alpha}{3\beta}\right)\left[(G_s - 1)gd\right]^{1/2} \tag{7.6}$$

Shields (1936) defines the motion of the particle under sliding based on force equilibrium. The drag force can be replaced as

$$F_D = C_D A \frac{\rho_w u^2}{2} = f\left(a_1, \frac{ud}{\upsilon}\right)\rho_w d^2 u^2 = \tau_0 d^2 f\left(a_1, a_2, \frac{ud}{\upsilon}\right) \tag{7.7}$$

where a_1 and a_2, are the particle-shape factor and τ_0 is the shear stress applied on the surface of the particle.

The resistant force was assumed to depend only on the form of the bed and the submerged weight of the particle:

$$F_R = a_3 (G_s - 1)\rho_w g d^3 \tag{7.8}$$

where a_3 is a factor accounting for the bed form.

By force equilibrium, and assuming a level ground and uniform particle size, the critical condition yields

$$\frac{\rho_w u_c^2}{(G_s-1)\rho_w g d} = \frac{\tau_c}{(G_s-1)\rho_w g d} = f\left(\frac{du}{\upsilon}\right) = f(Re) \quad (7.9)$$

where Re is the Reynolds number, τ_c is the critical erosive shear stress, and u_c is the critical flow velocity.

Under turbulent flow, in addition to the pressure difference induced by the steady portion of the flowing water, additional fluctuating pressures are induced by vortices that develop in the turbulent flow, which introduces pressure differentials onto the bed (Hofland et al., 2005). The interaction between the vortex and the boundary introduces a reduction in the pressure on the boundary surface. The pressure reaches a minimum underneath the eddy when it interacts with the boundary, which indicates the development of varying low pressure on the top surface of the particle. This causes a differential distribution of pressure over the particle, resulting in a net force that tends to rotate the particle. The turbulence can affect the particle erosion in several ways:

1. The particle may be moved by the drag exerted by a passing eddy.
2. The eddy may lower the local pressure and the particle may be ejected from the bed by hydrostatic pressure.
3. The eroded particle may be put into suspension rather than moved along the bed.

The threshold of erosion is in principle in a probabilistic manner, since both the applied stresses in a turbulent flow and the response of a particle due to its position show randomness. The critical erosive shear stress of a particle of diameter d_i in the sediment mixture cannot be determined from the Shields diagram. Considering the fully developed turbulent flow and sediment mixture with different particle sizes, Egiazaroff (1965) derived the threshold of a particle of diameter d_i on the bed

$$\theta_{ci} = \frac{0.1}{\left[\log(19 d_i / d_{50})\right]^2} \quad (7.10)$$

where θ_{ci} is the critical dimensionless shear stress.

The critical shear stress depends on the grading of the mixture and surface structure. Hayashi et al. (1980) derived a formula for calculating the critical shear stress

$$\begin{aligned} \theta_{ci} &= \theta_{ca}\frac{d_i}{d_a} & \frac{d_i}{d_a} < 1.0 \\ \theta_{ci} &= \theta_{ca}\left(\frac{\log 8}{\log(8 d_i / d_a)}\right)^2, & \frac{d_i}{d_a} > 1.0 \end{aligned} \quad (7.11)$$

where d_a is the average particle size of the mixture and θ_{ca} is the critical dimensionless shear stress for particle size of d_a.

The formulas of Egiazaroff and of Hayashi et al. are only related to the grain size, and the Hayashi et al. formula uses the same critical shear stress for all the particles finer than d_a. By introducing a correction factor, η_i, Wu et al. (2000) defined the critical shear stress as

$$\frac{\tau_{ci}}{(G_s - 1)\rho_w g d_i} = \theta_c \eta_i \tag{7.12}$$

$$\eta_i = \left(\frac{p_{ei}}{p_{hi}}\right)^m \tag{7.13}$$

where p_{ei}, p_{hi} are the total exposed and hidden probabilities of particle d_i, respectively and m is an empirical parameter.

The correction factor used in Equation (7.13) is not only related to the grain size but also to the bed material gradation. Better prediction can be achieved with Equation (7.13), even when the bed-material gradation is very wide and highly irregular.

During the overtopping erosion process of dams, the erosion mainly occurs on the downstream slope. The incipient motion of particles on slopes is not only affected by the drag and lift forces, but also by the component of gravitational force along the slope. As shown in Figure 7.2, at the condition of incipient motion, the forces tangential (F_t) and normal (F_n) to the bed are related to the angle of repose of the sediment (θ), where

$$\tan\theta = \frac{F_t}{F_n} \tag{7.14}$$

Substituting the value of each of the terms in Figure 7.2, the drag force F_D, the lift force F_L, and the submerged weight force G' into Equation (7.14) and rearranging the equation, Chiew and Parker (1994) derived the critical shear stress/shear velocity of sediment on a steep slope as

$$\tau_c = \rho u_c^2 = \frac{2(G_s - 1)dc_1 f}{c_2} \frac{(\tan\theta\cos\phi - \sin\phi)}{C_D + C_L\tan\theta} \tag{7.15}$$

where ϕ is the slope angle, c_1 and c_2 are the correction factors for volume and area of particles, and f is a measure of friction.

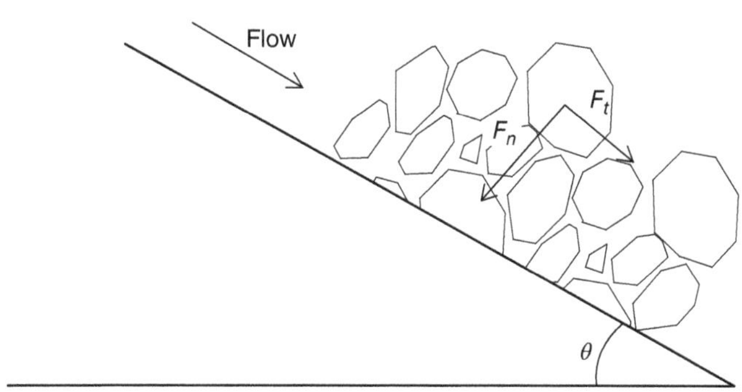

Figure 7.2 Forces acting on sediment particles lying on steep slope

7.1.1.2 Cohesive Soils

The detachment and subsequent entrainment of cohesive material can result from two processes: (1) hydraulic erosion – the lift and drag imposed by channel flow or (2) subaerial erosion – the weakening and weathering of bank material imposed by dynamic soil moisture conditions (Julian and Torres 2006). When the water starts flowing over a particle or a cluster of particles, three things happen (Briaud et al. 2008): (1) a drag force and associated shear stresses develop at the interface between the soil particles, (2) the normal stress on top of the soil particle decreases because of the water flow, and (3) the normal stresses and shear stresses applied at the boundaries fluctuate with time because of the turbulence. The combination of the mean value and the fluctuations around the mean of the drag force and uplift force can be large enough to pluck and drag the soil particle or soil particle cluster away and generate erosion.

The erosion process of cohesive soils has not been widely reported in the literature, with greater attention being given to the eroded surface features after testing. Moore and Masch (1962) observed that appreciable quantities of sediment become loose, and the water becomes cloudy when the shear stress applied reaches the critical shear stress. The features of the eroded surface depend strongly on the orientation of the clay aggregates. Where the aggregates are packed face to face, the erosion is a kind of peeling off loose patches where interlayer bonds have failed. The eroded surface of stiff clay is pitted. When the clay is soft, the randomly occurring pressure fluctuations remold the crater left after a lump of clay has been eroded from the surface.

Laboratory tests conducted by Kothyari and Jain (2008) on incipient motion of cohesive soils indicate three types of erosion:

1. Pothole erosion. At a lower percentage of clay in bed material (up to 30%), the detachment generally initiates in the form of individual particles detached from the bed leaving the formation of potholes over the bed surface at different locations.
2. Line erosion. At higher percentages of clay in bed material (greater than 30%), the detachment begins by removing thin flakes from the sediment surface and that removal continues and spreads both longitudinally and laterally, but more in the longitudinal direction.
3. Mass erosion. This type of condition was frequently observed in clay–sand–gravel mixtures, particularly for higher values of energy slope and for sediment beds having clay percentages of more than 30%. Here the detachment was seen to occur in the form of chunks of varying sizes. The pieces of sediment appeared to be ripped or torn off from the bed surface.

The incipient motion of cohesive material can also be described by critical shear stress. Van Rijn (2007) assumed that the critical shear stress for particle size smaller than 62 μm is affected by cohesive particle–particle interaction effects, including clay coating effects (ϕ_{cohesive}), by packing (or bulk density) effects (ϕ_{packing}), and by biological and organic material effects (ϕ_{bo}). The critical bed-shear stress is proposed to be represented as

$$\tau_c = \phi_{bo}\phi_{packing}\phi_{cohesive}\tau_{c,0} \tag{7.16}$$

where $\tau_{c,0}$ is the critical shear stress of cohesionless soil.

Currently, the critical shear stress of cohesive soil is mainly measured in the laboratory or in the field, or calculated based on empirical equations. These will be presented in Section 7.2.

7.1.2 Sediment Transport

Sediment transport is the movement of soil particles, typically due to a combination of the force of gravity acting on the sediment, and/or the movement of the fluid in which the sediment is entrained. Once the erosion of the sediment is initiated, the particles can be transported by the hydrodynamic forces. Usually, the transport of particles by rolling, sliding, and saltating is called the bed-load transport. Bed-load is generally thought to constitute 5–10% of the total sediment load in a stream, making it less important in terms of mass balance. Suspended load is carried in the lower to middle parts of the flow, and moves at a large fraction of the mean flow velocity in the stream. Bed-material load is the sum of bed-load and suspended load. It can be calculated either by separating the bed-load and transport load or directly using bed-material load transport equations. The sediment transport capacity is a function of flow conditions and sediment properties.

7.1.2.1 Bed-load Transport

Bed-load transport formulas are founded on the premise that a specific relation exists between hydraulic variables, sedimentological parameters, and the rate at which bed-load is transported. Four principal approaches have emerged to the design of bed-load transport formulas, based upon bed shear stress (e.g. du Boys, 1879; Meyer-Peter and Müller, 1948; Smart, 1984; Wu et al., 2000), stochastic functions for sediment movement (e.g. Einstein, 1942; Brown, 1950; Parker et al., 1982), work done (e.g. Bagnold, 1956, 1980; Yalin, 1963, 1972), and statistical and numerical analysis (e.g. van Rijn, 1984a), respectively. In this section, several most widely quoted bed-load formulas are introduced.

Formulas Based on Bed Shear Stress

The bed-load sediment transport equations based on the tractive shear stress theory postulate that the sediment transport is a function of nine variables:

$$q_{s,b} = f\left(\tau_c, \gamma', \gamma_w, \gamma_s, d, u, R_h, S, g\right) \tag{7.17}$$

where $q_{s,b}$ is the volume of sediment transported per second per unit channel width, τ_c is the bed shear stress, γ' is the specific weight of submerged dam material, γ_w is the specific weight of water, γ_s is the specific weight of dam material, d is the particle size, u is the mean flow rate, R_h is the hydraulic radius, S is the energy slope, and g is the gravitational acceleration.

Many of the bed-load transport formulas have a similar form to that developed by Du Boys (1879). The Du Boys formula takes a very simple form as

$$q_s = C_s \tau \left(\tau - \tau_c\right) \tag{7.18}$$

where τ is the total shear stress on the bed and C_s is the sediment parameter. In Equation (7.18), the shear stress, τ can be expressed as

$$\tau = \gamma_w R_h S \tag{7.19}$$

For a typical breach cross-section as shown in Figure 7.3, R_h can be calculated as

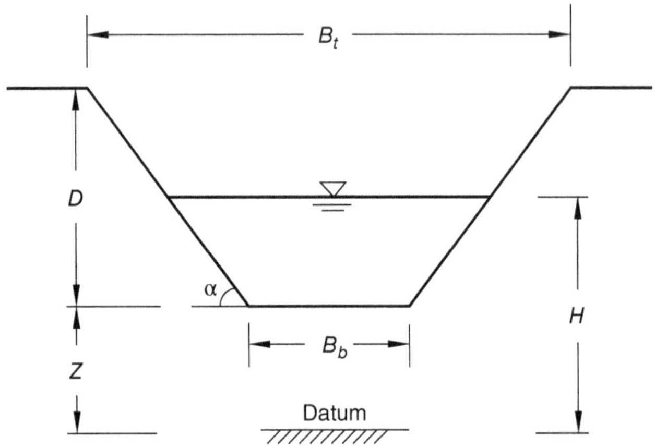

Figure 7.3 Geometry of a typical breach cross-section

$$R_h = \frac{(H-Z)\sin\alpha + B_b \cos\alpha}{2(H-Z)+B_b\cos\alpha}(H-Z) \tag{7.20}$$

where H is the elevation of the water surface, Z is the elevation of the breach bottom, B_b is the breach bottom width, and α is the angle of the side slope.

Based on the experiments undertaken at Swiss Federal Institute of Technology (ETH), Zurich, with uniform 5.05 and 28.6 mm diameter gravels, Meyer-Peter et al. (1934) developed a bed-load sediment transport formula, which is originally expressed as

$$q_{s,b} = \left(250q^{2/3}J - 42.5d\right)^{3/2} \tag{7.21}$$

where q is the water flow, J is the channel slope, and d is the characteristic particle size.

Based on a considerable number of additional flume tests on uniform, graded, and lightweight materials, the Meyer-Peter formula was further modified by Meyer-Peter and Müller (1948) as

$$q_{s,b} = 8g^{0.5}\frac{1}{\gamma'}\frac{1}{\gamma_w^{\ 0.5}}\left(\tau - \tau_c\right)^{1.5} \tag{7.22}$$

$$\tau = \gamma_w \left(\frac{k_s}{k_r}\right)^{1.5} R_h S \tag{7.23}$$

$$\tau_c = 0.047\gamma' d_{50} \tag{7.24}$$

where k_s is the total roughness of the channel bed and k_r is the grain roughness.

Substituting Equations (7.23) and (7.24) into Equation (7.22) yields

$$q_{s,b} = 8g^{0.5}\frac{\gamma_w}{\gamma'}\left[\left(\frac{k_s}{k_r}\right)^{1.5} R_h S - 0.047\frac{\gamma'}{\gamma_w}d_{50}\right]^{1.5} \tag{7.25}$$

Considering the non-uniform sediment conditions, Wu et al. (2000) developed a non-uniform bed-load sediment transport formula as

$$q_{si} = 0.0053\left[\left(\frac{n'}{n}\right)^{1.5}\frac{\tau}{\tau_{ci}} - 1\right]^{1.5} \tag{7.26}$$

where n is Manning's roughness coefficient for the channel bed, n' is Manning's roughness corresponding to the grain roughness, and τ_{ci} is the critical shear stress for particle d_i, in non-uniform sediment mixtures.

Formulas Based on Stochastic Functions for Sediment Movement

The Einstein–Brown formula (Einstein, 1942; Brown, 1950) follows a different line of analysis, adopting a probabilistic approach that takes no account of the critical erosive shear stress, which is expressed as

$$q_{s,b} = \Phi\gamma_s K\left[\left(G_s - 1\right)gd\right]^{0.5} \tag{7.27}$$

where K is a dimensionless constant, Φ is a dimensionless sediment transport rate function, and G_s is the specific gravity.

The Einstein–Brown formula indicates that the number of deposited grains in a unit area depends on the number of grains in motion and the probability that the hydrodynamic forces permit the grains to deposit. The number of eroded grains in the same unit area depends on the number of grains in that area and the probability that the hydrodynamic forces are strong enough to move them. At equilibrium, the number of grains deposited must be equal to the number of grains eroded.

In Equation (7.27), K is defined as

$$K = \left[\frac{2}{3} + \frac{36\nu^2}{\left(G_s - 1\right)gd^3}\right]^{0.5} - \left[\frac{36\nu^2}{\left(G_s - 1\right)gd^3}\right]^{0.5} \tag{7.28}$$

where ν is the kinematic viscosity of water.

Formulas Based on Work Done

Bagnold (1956) developed a bed-load sediment transport formula based on the concept that the sediment transport rate equals the work done in the channel. The useful work done can be expressed as

$$W_b = \frac{\gamma^* Q_b}{\gamma_s d}\left(\tan\alpha_0 - \tan\beta\right) \tag{7.29}$$

where γ^* is the unit stress, Q_b is the grain transport rate, $\tan\alpha_0$ is the grain-stress ratio, and β is the slope angle. The available flow energy rate is

$$w_b = A(\tau/\rho)^{1/2}(\tau - \tau_c) \tag{7.30}$$

where A is a function of Reynolds number ($A = 8.5$ for $Re > 70$ and $A = 5.5$ for $Re < 5$ with transition between these).

Assuming the sediment transport rate equals the work done in the channel, the bed-load transport formula can be expressed as

$$q_{s,b} = \frac{ABG_s}{(G_s - 1)g\cos\beta\sqrt{\rho_w}}(\tau - \tau_c)(\tau)^{1/2} \tag{7.31}$$

In Equation (7.31), constant B is the ratio of the efficiency of transport to the friction coefficient,

$$B = \frac{\left[2\tan\alpha/(3C_D)\right]^{1/2}}{1 - \tan\beta/\tan\alpha} \tag{7.32}$$

where C_D is a drag coefficient.

In a later study (Bagnold, 1980), the work done was assumed to relate to the stream power and the formula of bed-load sediment transport is given as

$$q_{s,b} = q_{sr}\frac{G_s}{G_s - 1}\left[\frac{\omega - \omega_0}{(\omega - \omega_0)_r}\right]^{3/2}(Y/Y_r)^{-2/3}(d/d_r)^{-1/2} \tag{7.33}$$

$$\omega = \gamma_w YJu \tag{7.34}$$

$$\omega_0 = 5.75\left[0.04(\gamma_s - \gamma_w)\rho_w\right]^{3/2}(g/\rho_w)^{1/2}d^{3/2}\log(Y/d) \tag{7.35}$$

$$q_{sr} = 0.01 \quad (\omega - \omega_0)_r = 0.5 \quad Y_r = 0.1 \quad d_r = 0.0011 \tag{7.36}$$

where Y is the flow depth and d is the particle size.

Yalin (1963, 1972) developed a bed-load sediment transport equation by incorporating both probabilistic reasoning and the rate of doing work. The increase of bed-load discharge is assumed to be due to the number of particles participating in the motion. The equation is expressed as

$$q_{s,b} = 0.635\gamma_s du_* s\left[1 - \frac{1}{as}\ln(1 + as)\right] \tag{7.37}$$

where

$$s = (\theta_0 - \theta_c)\theta_c \tag{7.38}$$

$$a = 2.45\sqrt{\theta_c}\left(\frac{\gamma_w}{\gamma_s}\right)^{0.4} \tag{7.39}$$

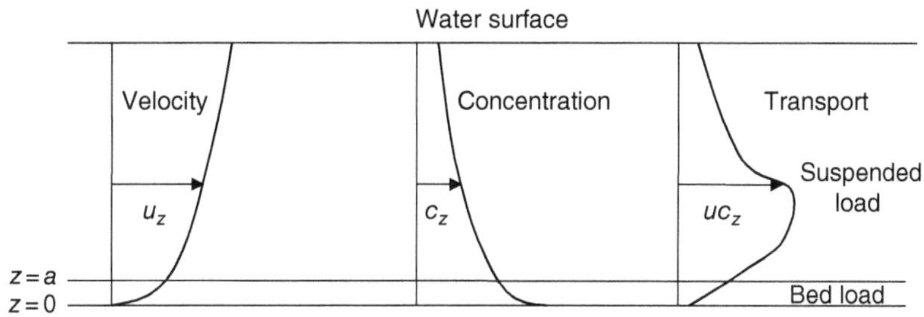

Figure 7.4 Schematic of suspended load transport

where θ is the Shields number. Yalin's bed-load formula was developed for uniform sediments on the basis of a theoretical consideration of the mechanics of the average motion of a saltating grain.

Formulas Based on Statistical and Numerical Analysis

Considering the bed-load transport as the transport of particles by rolling and saltating along the bed surface, van Rijn (1984a) developed a bed-load transport equation as the product of the saltation height, the particle velocity, and the bed-load concentration. The motion of the bed-load particles is assumed to be dominated by gravitational forces, while the effect of turbulence on the overall trajectory is supposed to be of minor importance. The equations of motions for a solitary particle are solved numerically to determine the saltation height and particle velocity. The sediment transport equation is expressed as

$$q_{s,b} = 0.053\left[\left(G_s - 1\right)g\right]^{0.5} d_{50}^{1.5} \gamma_s \frac{T^{2.1}}{D_*^{0.3}} \tag{7.40}$$

where T is the transport stage number, which can be expressed as $T = [(u_*')^2 - (u_{*,c})^2]/(u_{*,c})^2$, and D_* is the particle parameter, which can be expressed as $D_* = D_{50}[(G_s - 1)g/v^2]^{1/3}$.

7.1.2.2 Suspended-load Transport

If the shear stress generated by the water flow exceeds the critical shear stress of the soil to a certain level, the amount of turbulence generated at the bed and its upward diffusion reach a stage where the grains lose contact with the bed and are held in suspension. The upward diffusion of turbulence keeps the particles in suspension against gravity. The velocity of particles in suspension is almost equal to the fluid velocity. The quantity of sediment transported as suspended load is often much greater than that of the bed-load, which moves slowly. An essential part of morphological computations in the case of flow conditions with suspended sediment transport is the use of a reference sediment concentration as a bed boundary condition. The suspended sediment concentration decreases with distance up from the bed. The rate of decrease depends on the ratio of the fall velocity and the bed shear velocity.

Figure 7.4 shows a schematic graph of the flow velocity, sediment concentration, and suspended load in a typical sediment transport regime. In a steady and uniform flow, the vertical distribution of the sediment concentration profile can be described by:

$$(1-c)cw_{s,m}+\varepsilon_s\frac{dc}{dt}=0 \tag{7.41}$$

where c is the sediment concentration, $w_{s,m}$ is the particle fall velocity in a fluid–sediment mixture, and ε_s is the sediment diffusion coefficient.

Particle Fall Velocity

The settling velocity of a particle relates to particle size, shape, submerged specific weight, water velocity, sediment concentration, etc. It can be expressed as

$$w_s=\left[\frac{a_1}{a_2}\frac{2}{C_d}(G_s-1)gd\right] \tag{7.42}$$

where a_1, a_2 are the shape coefficient associated with volume and area and C_d is the drag coefficient.

In a laminar settling region, according to Stokes's law, without considering the inertia effect, the drag coefficient is

$$C_d=\frac{24}{\mathrm{Re}} \tag{7.43}$$

Substituting Equation (7.43) into Equation (7.42), the settling velocity of a spherical particle can be expressed as

$$w_s=\frac{1}{18}(G_s-1)g\frac{d^2}{\nu} \tag{7.44}$$

If the inertia effect is considered, the drag coefficient can be expressed as (Goldstein, 1929)

$$C_d=\frac{24}{Re}\left(1+\frac{3}{16}Re-\frac{19}{1280}Re^2+\frac{71}{20480}Re^3+\cdots\right) \tag{7.45}$$

Equation (7.45) is valid for Re up to 2.0.

According to van Rijn (1984b), in a clear, still fluid, the particle fall velocity of a solitary sand particle smaller than about 100 μm (Stokes-range) can be described by Equation (7.44). For suspended sand particles in the range 100–1000 μm, the following equation proposed by Zanke (1977) can be used:

$$w_s=10\frac{\nu}{d}\left\{\left[1+\frac{0.01(G_s-1)gd^3}{\nu^2}\right]^{0.5}-1\right\} \tag{7.46}$$

For particles larger than about 1000 μm the following simple equation can be used:

$$w_s=1.1\left[(G_s-1)gd\right]^{0.5} \tag{7.47}$$

In Equations (7.44), (7.46), and (7.47), the particle diameter expresses the representative particle diameter of the suspended sediment particles, which may be considerably smaller than d_{50} of the bed material.

Wu and Wang (2006) developed a generalized expression of settling velocity as

$$w_s = \frac{M\nu}{Nd}\left[\sqrt{\frac{1}{4}+\frac{4N}{3M^2}\frac{d^3}{\nu^2}(G_s-1)g}-\frac{1}{2}\right]^n \tag{7.48}$$

In Equation (7.48)

$$M = 153.5e^{-0.65S_p},\ N = 5.65e^{-2.5S_p},\ n = 0.7+0.9S_p \tag{7.49}$$

where S_p is the Corey shape factor.

Experiments with high sediment concentrations have shown a substantial reduction of the particle settling velocity due to the presence of the surrounding particles (van Rijn, 1984b). This effect, referred to as hindered settling, is caused by the return flow of water induced by the settling of sediment. For normal flow conditions with particles in the range 50–500 μm the reduced particle settling velocity can be described by the following equation (Richardson and Zaki, 1954):

$$w_{s,m} = (1-c)^4 w_s \tag{7.50}$$

Diffusion Coefficient

Diffusion coefficient represents the ability of sediment to diffuse into the flow. The diffusion of sediment particles is related to the diffusion of fluid momentum defined by van Rijn (1984b) as

$$\varepsilon_s = \beta\phi\varepsilon_f \tag{7.51}$$

where ε_f is the diffusion coefficient of fluid.

The β-factor describes the difference in the diffusion of a discrete sediment particle and the diffusion of a fluid 'particle' and is assumed to be constant over the flow depth. The ϕ-factor expresses the damping of the fluid turbulence by the sediment particles and is assumed to be dependent on the local sediment concentration. Usually, the diffusion of fluid momentum (ε_f) is described by a parabolic distribution over the flow depth (d):

$$\varepsilon_f = \frac{z}{d}\left(1-\frac{z}{d}\right)\kappa u_* d \tag{7.52}$$

where κ is the constant of Von Karman for clear fluid and u_* is the overall bed-shear velocity. The parabolic constant distribution is expressed as

$$\varepsilon_{f,\max} = 0.25\kappa u_* d \quad \text{for } z/d \ge 0.5 \tag{7.53}$$

$$\varepsilon_f = 4\frac{z}{d}\left(1-\frac{z}{d}\right)\varepsilon_{f,\max} \quad \text{for } z/d < 0.5 \tag{7.54}$$

Near-bed Concentration of Suspended Load

The suspended sediment concentration near the bed changes dramatically with elevation. The accurate measurement of the near-bed concentration of suspended sediment is difficult. A reference level (a) related to the bed-form height is introduced to solve the problem. Einstein (1950) assumed that the bed layer is two-grain-diameter thick and the near-bed concentration of suspended load relates to the bed-load transport rate as

$$c_a = \frac{q_{s,b}}{23.2u'_* d} \quad (7.55)$$

where u'_* is the skin friction velocity.

Van Rijn (1984b) related the reference level to the equivalent roughness height (k_s) or half the bed-form height. Below the reference level, the transport of all sediment particles is considered as bed-load transport. The reference concentration can be expressed as:

$$c_a = 0.015\frac{d_{50}}{a}\frac{T^{1.5}}{D_*^{0.3}} \quad (7.56)$$

where T and D_* are particle parameter and transport stage parameter, which are given as

$$T = \frac{(u'_*)^2 - (u'_{*,c})^2}{(u'_{*,c})^2} \quad (7.57)$$

$$D_* = d_{50}\left[\frac{(G_s - 1)g}{v^2}\right]^{1/3} \quad (7.58)$$

In principle, the suspended sediment transport is given by integrating the product of local sediment concentration and flow velocity over the suspended zone. Thus, for unit width of flow,

$$q_{s,s} = \int_a^d cu\,dy \quad (7.59)$$

Einstein (1950) developed a suspended load transport equation using the Rouse distribution sediment concentration and the logarithmic distribution of flow velocity, which yields

$$q_{s,s} = 11.6u_* c_a a\left[2.303\log\left(30.2\frac{d}{\Delta}\right)I_1 + I_2\right] \quad (7.60)$$

where

$$I_1 = \left[0.216\frac{A^{z-1}}{(1-A)^z}\right]\int_A^1\left(\frac{1-y}{y}\right)^z dy \quad (7.61)$$

$$I_2 = \left[0.216\frac{A^{z-1}}{(1-A)^z}\right]\int_A^1 \left(\frac{1-y}{y}\right)\ln y dy \tag{7.62}$$

Considering the near-bed sediment concentration, Equation (7.60) can be changed to

$$q_{s,s} = q_B\left[2.303\log\left(30.2\frac{d}{\Delta}\right)I_1 + I_2\right] \tag{7.63}$$

van Rijn (1984b) developed a suspended sediment transport equation as

$$q_{s,s} = \frac{u_* c_a}{\kappa}\left(\frac{a}{d-a}\right)^{z_1} \left\{\int_a^{0.5d}\left(\frac{d-y}{y}\right)^{z_1}\ln\left(\frac{y}{y_1}\right)dy + \int_{0.5d}^{d}\exp\left[-4z_1\left[\left(\frac{y}{d}\right)-0.5\right]\right]\ln\left(\frac{y}{y_1}\right)dy\right\} \tag{7.64}$$

The transport of sediment particles below the reference level (a) is considered as bed-load transport. Equation (7.60) can be represented with an inaccuracy of about 25% by ($0.3 \le z_1 \le 3$ and $0.01 < a/d < 0.1$):

$$q_{s,s} = F\bar{u}dc_a \tag{7.65}$$

$$F = \frac{(a/d)^{z_1} - (a/d)^{1.2}}{(1-a/d)^{z_1}(1.2-z_1)} \tag{7.66}$$

where $\bar{u}$ is the mean flow velocity and d is the flow depth.

7.1.2.3 Total Load Transport

Total load transport of bed material can be obtained by the summation of the bed-load and suspended load transport as follows:

$$q_{s,t} = q_{s,b} + q_{s,s} \tag{7.67}$$

The methods of Einstein (1950), van Rijn (1984a, b), Bagnold (1956), and Wu et al. (2000) follow this approach.

The total transport of bed material can also be directly calculated by using the energy approach (e.g. Engelund and Hansen, 1967; Yang, 1973). Using Bagnold's stream power concept and the similarity principle, Engelund and Hansen (1967) established the following sediment transport equation:

$$q_{s,t} = \frac{0.05\bar{u}^5}{(G_s-1)g^{0.5}d_{50}C} \tag{7.68}$$

where C is the Chezy coefficient.

Yang (1973) assumed that the sediment transport is related to the unit stream power, defined as $\bar{u}I$. The total load transport rate is given by:

$$q_{s,t} = 10^{-3} c_t \bar{u} h \tag{7.69}$$

The total sediment concentration defined as the ratio of the sediment and fluid discharge per unit width is expressed as:

$$\log(c_t) = M + N \log\left(\frac{\bar{u}J}{w_s} - \frac{\bar{u}_c J}{w_s}\right) \tag{7.70}$$

For sand,

$$M = 5.435 - 0.286 \log \frac{w_s d}{v} - 0.457 \log \frac{\bar{u}}{w_s} \tag{7.71}$$

$$N = 1.799 - 0.409 \log \frac{w_s d}{v} - 0.314 \log \frac{\bar{u}}{w_s} \tag{7.72}$$

For gravel,

$$M = 6.681 - 0.633 \log \frac{w_s d}{v} - 4.816 \log \frac{\bar{u}}{w_s} \tag{7.73}$$

$$N = 2.784 - 0.305 \log \frac{w_s d}{v} - 0.282 \log \frac{\bar{u}}{w_s} \tag{7.74}$$

Ackers and White (1973) presented that the transport of coarse sediments, which are mainly in bed-load, is attributed to the stream power corresponding to the grain shear stress, while the transport of fine sediments, which are mainly in suspended load, is related to the turbulent intensity, and in turn the total stream power. Based on an analysis of 925 sets of flume and field data, an empirical formula is developed:

$$q_{s,t} = K \bar{u} d_{35} \left(\frac{\bar{u}}{u_*}\right)^n \left(\frac{F_{gr} - F_{gr,c}}{F_{gr,c}}\right)^m \tag{7.75}$$

In Equation (7.75), K, m, and n are coefficients and F_{gr} is a mobility factor, which can be expressed as

$$F_{gr} = \frac{u_*^{\,n}}{\left((G_s - 1) g d_{35}\right)^{0.5}} \left(\frac{\bar{u}}{5.66 \log(10h / d_{35})}\right)^{1-n} \tag{7.76}$$

A comparison of the mentioned formulas in this section is presented by Wu et al. (2000) based on 1859 sets of uniform bed-material load data selected from Brownlie (1981). It shows that none of them well predicts the real conditions. Due to the fact that the possible discrepancy of any existing sediment transport formula may exceed two- or three-fold, proper selection of the sediment transport formula is very important. The study site should be similar to those involved in the tests used to develop the formula.

7.1.2.4 Sediment Transport on Steep Slopes

The above-mentioned bed-load sediment transport formulas are mainly applied for slopes in the range of 0.0004–0.02%. In a steep slope, which is common for dams and dikes, those formulas may not be applicable, since gravity plays an important role. Moreover, steep channels have grains of a wider distribution, composed of finer, more mobile sediment and large, rarely mobile grains. The large grains can bear a significant portion of the total shear stress and thereby reduce the stress available to move the finer sediment. Conventional bed-load transport equations often over-predict the sediment flux in steep channels by several orders of magnitude (Yager et al., 2007). The sediment transport equations over-predict the sediment flux probably because (1) they do not account for the stress borne by rarely mobile grains, (2) they do not differentiate between highly and rarely mobile sediment, and (3) they do not account for the limited availability of mobile sediment. However, the studies on sediment transport on steep slopes are relatively rare comparing to the gentle slope cases. This section introduces several sediment transport equations for steep slopes, which are developed based on laboratory testing results.

Smart (1984) carried out some sediment transport tests on a steep flume and finally developed a bed-load transport formula for steep channels, which is a modification of Meyer-Peter and Müller's formula expressed as

$$q_s = 4\left[\left(\frac{d_{90}}{d_{30}}\right)^{0.2} J^{0.6} C\tau^{0.5}\left(\tau - \tau_c\right)\right] \tag{7.77}$$

where C is the resistance factor.

Bed-load transport experiments were performed with clay-suspension flows in a steep channel by Rickenmann (1991). With increasing clay concentration, the bed-load transport capacity increased due to the decreasing solid/fluid ratio. Once the effect of the increasing viscosity of the fluid becomes important and the flow around the grains is laminar, the transport capacity decreases markedly with further increasing clay concentration. Based on the test results, a regression calculation yields the following transport equation, valid for the slope range $0.1\% < J < 20\%$:

$$\Phi_{s,b} = \frac{3.1}{\left(G_s - 1\right)^{0.5}}\left(\frac{d_{90}}{d_{30}}\right)^{0.2} \theta_m^{\ 0.5}\left(\theta_m - \theta_c\right) F^{1.1} \tag{7.78}$$

where F is the Froude number.

On the basis of experiments performed within the range of mild and steep slope angles, nearly to the angle of repose, and within the range of the relative Shields parameter of 2–6, Damgaard et al. (1997) concluded that for positive and small negative slopes the effect of slope on bed-load transport is described adequately by altering the critical Shields parameter for the slope. For large negative slopes, extra correction is needed. Finally, an empirical equation is developed for sediment transportation both on gentle and steep slope, which is expressed as

$$\Phi_{s,b} = 8(\theta - \theta_c)^{3/2} f_{slope} \tag{7.79}$$

$$f_{slope} = \begin{cases} 1 + 0.8\left(\dfrac{\theta_{c,0}}{\theta}\right)^{0.2}\left(1 - \dfrac{\theta_{c,\beta}}{\theta_{c,0}}\right)^{(1.5+\theta/\theta_c)} & \text{for } -\phi_s < \beta \le 0 \\ 1 & \text{for } 0 < \beta \le \phi_s \end{cases} \tag{7.80}$$

7.2 Determination of Erodibility of Soils

Soil erosion can be described by the tractive shear stress induced by the water flow and soil erodibility. Erodibility is a term describing the resistance of a soil subjected to water flow. Soil erodibility is an intrinsic property of soil, which reflects the erosion resistance of the soil and it can be described by two parameters: critical erosive shear stress and coefficient of erodibility. The critical erosive shear stress corresponds to the initiation of erosion, which means that a soil erodes when the shear stress exerted by the flow exceeds the critical erosive shear stress. The coefficient of erodibility indicates the amount of soil loss in a time interval under a given shear stress. The critical erosive shear stress reflects the ease of initiation of erosion in the soil, while the coefficient of erodibility represents how fast the soil erodes. A shear stress equation, which is similar to shear stress based sediment transport equation, is often used to describe soil erosion (e.g. Graf, 1984):

$$E = K_d(\tau - \tau_c) \tag{7.81}$$

where E is the erosion rate of the soil, K_d is the coefficient of erodibility, and τ_c is the critical shear stress of soil at initiation of erosion.

Although the use of flow shear stress as a soil detachment predictor can be contested, K_d and τ_c have been selected as the most "universal" parameters to describe erosion resistance to concentrated flow (Knapen et al., 2007).

7.2.1 *Critical Erosive Shear Stress*

Several testing methods have been devised to measure the critical erosion shear stress. For granular soils, Shields's chart (the Shields parameter is the ratio of the shear stress to a particle's submerged weight per unit of surface area at critical conditions) provides an estimate of critical erosive shear stress based on a particles size, using a representative particle diameter and assuming no inter-particle actions. Neill (1973) proposed a critical erosion shear stress equation (Table 7.1) by performing flume tests, which is a function of specific density, mean particle size, and water depth. However, the equation involves flow depth, which is not an intrinsic property of the testing materials. Annandale (2006) developed a critical erosive shear stress equation for uniform granular soils under laminar flow condition. The critical erosive shear stress depends on particle size, particle unit weight, and friction angle.

Table 7.1 Summary of empirical equations for calculating the critical erosive shear stress of soils

References	Expression	Soil type	Testing method
Smerdon and Beasley (1961)	$\tau_c = 0.16(PI)^{0.84}$ $\tau_c = 10.2(d_r)^{-0.63}$ $\tau_c = 3.54\times10^{-28.1d_{50}}$ $\tau_c = 0.493\times10^{0.0182P_c}$	Cohesive soil	Flume test
Neill (1973)	$\tau_c = 0.7609\gamma_w(G_s-1)d_{50}^{2/3}d_w^{\ 1/3}$	Gravel, glass balls	Flume test
Mitchener and Torfs (1996)	$\tau_c = 0.015(\rho_b - 1000)^{0.73}$	Mixture of mud and sand	Flume test
Julian and Torres (2006)	$\tau_c = 0.1 + 0.1779P + 0.0028P^2 - 2.34\times10^{-5}P^3$	Cohesive soil	Field monitoring
Annandale (2006)	$\tau_c = \frac{2}{3}\gamma_w d(G_s-1)\tan\varphi$	Homogeneous granular soil	—
Thoman and Niezgoda (2008)	$\tau_c = 77.28 + 2.20(Act) + 0.26D_r - 13.49G_s - 6.4pH + 0.12w$	Cohesive soil	Submerged jet
Chang and Zhang (2011)	$\tau_c = 6.8(PI)^{1.68}P^{-1.73}e^{-0.97}$	Landslide deposits	Submerged jet

PI: plasticity index (%); d_r: dispersion ratio (ratio of the total weight of silt and clay sized aggregates to the total weight of silt and clay sized particles); d_{50}: mean particle size (m); P_c: percent of clay by weight (%); γ_w: unit weight of water (kN/m³); d_w: depth of flow (m); ρ_b: bulk density of soil (kg/m³); φ: friction angle; *d*: particle diameter (m); *Act*: activity (plasticity index divided by the percent of clay); and *w*: water content (%).

For cohesive soils, laboratory tests are usually carried out to evaluate their critical erosive shear stress as the erosion mechanisms of cohesive soils are much more complex. Based on testing results, several empirical equations were developed to describe the critical erosive shear stress, such as Smerdon and Beasley (1961), Mitchener and Torfs (1996), Julian and Torres (2006), and Chang et al. (2011). The expressions of these equations are summarized in Table 7.1. Because the soil erodibility is affected by many factors, in-situ measurements of soil erodibility are preferred. Thoman and Niezgoda (2008) presented a predictive model to estimate the critical erosive shear stress from a selection of related cohesive soil properties by conducting in-situ jet index tests on 16 reaches. A linear relation was established between critical erosive shear stress and activity, dispersion ratio, soil pH, specific gravity, and water content. Knapen et al. (2007) collected 522 published critical erosive shear stress values, and variations of four orders of magnitude were found.

Empirical or analytical equations are valuable to quantitatively describe the initiation of surface erosion. However, due to the variation in experimental techniques, different criteria defining the initiation of erosion, and different assumptions, conflicting results are found. For instance, Smerdon and Beasley's equation shows that the critical erosive shear stress decreases with increasing dispersion ratio, whereas Thoman and Niezgoda's equation indicates that the critical erosive shear stress increases with increasing dispersion ratio. Moreover, the materials in previous tests were mainly narrowly graded alluvial, sand, and fines (clay + silt). Therefore, caution should be exercised when adopting an empirical equation.

7.2.2 Coefficient of Erodibility

Unlike critical erosive shear stress, few empirical or analytical equations for coefficients of erodibility are available based on basic soil properties. Hanson and Simon (2001) observed an inverse relationship between the critical erosive shear stress and the coefficient of erodibility for stream beds of highly erodible loess. They found that the coefficient of erodibility can be represented by the following equation:

$$K_d = 0.2\tau_c^{-0.5} \tag{7.82}$$

Based on results of in-situ jet index tests, Thoman and Niezgoda (2008) proposed a similar inverse relationship between K_d and τ_c as

$$K_d = 1.11\tau_c^{-0.37} \tag{7.83}$$

However, the parameters that have an influence on critical shear stress do not necessarily influence the coefficient of erodibility in a similar way. Chang et al. (2011) found that the fines content and soil plasticity have significant influence on critical erosive shear stress, but the influences on coefficient of erodibility are weak. The coefficient of erodibility shows six orders of magnitude variations according to Knapen et al.'s statistics on 470 published data of coefficient of erodibility. A reasonable approach might be to establish the relations between K_d and basic soil properties by conducting systematic laboratory and field tests. However, quantitative description of coefficient of erodibility is still limited.

Soil erodibility is a soil-specific parameter, which reflects the erosion resistance of a soil subjected to water flow. It is mainly controlled by the soil properties and influenced by the flow properties. For different soils, the factors may not be the same. For granular soils, the controlling factors are relatively simple, including gravitational force and soil parameters (grain size distribution, grain shape, and particle density). The critical erosive shear stress increases with the increase of mean particle size and specific gravity.

For cohesive soils, the most important parameters are clay content, grain size distribution, bulk density, clay type, plasticity index, dispersion ratio, and water chemical compositions. Soil texture (e.g. clay fraction, particle size) plays a dominant role in determining many of the other properties of a soil that might be used in the estimation of erodibility parameters. Moreover, soil texture is a relatively simple routine measurement to make. Therefore, soil texture should be treated as one of the most important parameters controlling soil erodibility. Raudkivi (1990) reported that approximately 10% of clay content will suffice to control the soil properties in erosion processes. Moreover, the compaction effort also has a significant influence on the soil erodibility. An increase in compaction effort enhances the erosion resistance and the increase is more pronounced for soils with high fines contents. This is due to the increased interlocking of soil particles and the subsequent increased number of inter-particle bonds when compacting. For cohesive soils with higher clay contents, compaction also changes the particle orientation whereby compacted clays tend to have more parallel-oriented particles (Grissinger, 1966) and are therefore more resistant than loose materials. The erosion resistance of clay is significantly affected by cohesion. Natural intact structured clays show a much higher erosion resistance than do destructed clays formed by complete remolding and reconsolidation (Lefebvre et al., 1985, 1986). Soil moisture content is also an important parameter for soil erodibility. When moisture content is high, soil particles are far apart, and opportunities for cohesion to develop are small. Conversely, when moisture content is low, soil particles are not free to reorient themselves to a position with low energy and high cohesion between particles (Shainberg et al., 1996). Under the optimum moisture content at which maximum cohesion forces can develop and soils will have the lowest erodibility. The shear strength of cohesive sediment is closely related

Table 7.2 Summary of factors affecting the erodibility of cohesive materials

Factors	
Soil composition	Grain size distribution Clay content Clay type Plasticity index Dispersion ratio Cohesion Water content Shear strength Organic matter
Soil structure	Density Stress history Anisotropy
Fluid composition	Salinity pH Sediment Concentration Temperature

to its resistance to erosion. An increase in the shear strength normally generates an increase in the critical shear stress for erosion and a decrease in the erosion rate. It is suggested that critical shear stress should not be modeled by a single dependent parameter as it is influenced by multiple soil parameters. The main factors controlling soil erodibility are summarized in Table 7.2.

7.2.3 Laboratory Tests

The best way to quantify the soil erosion resistance is to test the soil either in situ or in the laboratory. Several testing methods were developed in the past to quantitatively measure the coefficient of erodibility and the critical erosive shear stress of soils, such as rotating cylinder tests, erosion function apparatus tests, hole erosion tests, flume tests, and submerged jet tests. The rotating cylinder tests and the erosion function apparatus tests are mainly used in the laboratory, while the flume tests and the submerged jet tests can be implemented either in the field or in the laboratory.

7.2.3.1 Flume Test

The flume test is an effective method for investigating soil erosion resistance under controlled flow conditions. The test can be carried out either in large-scale or small-scale depending on the soil conditions and research objectives. The cross-section of the flume can be either a U-shape or an inverse trapezoidal shape. The soils are placed and compacted in the channel bed to the density anticipated in the field. Then the slope of the flume is adjusted to a suitable value, which can be used to represent the field conditions. After that, the flow is increased gradually until surface erosion is observed. The critical erosive shear stress and erosion rate can be obtained by measuring the flow rate and eroded soil volume. It can be employed to measure both remolded soils and undisturbed soils. The tractive stress, τ, exerted on the flume bottom can be estimated by measuring the velocity profiles and using the Prandtl–von Karman equation for smooth channels (Chow, 1959):

$$\frac{V}{V_*} = 5.5 + 5.57\log\left(\frac{V_* y}{\nu}\right) \tag{7.84}$$

where V_* ($= (\tau/\rho)^{1/2}$) is the shear velocity and V is the velocity at depth y in the turbulent region.

For small flow depths and high slopes where velocity profile measurement was not possible, the tractive stress was estimated by

$$\tau = \gamma R_h J \tag{7.85}$$

The measured quantities in the flume test include eroded depth or eroded soil weight and flow velocity. After manipulation, the erosion rate and tractive shear stress can be obtained at different stages.

Hanson (1990) carried out large-scale flume tests outdoor on four soils over a two-year period. It was found that the soil erodibility is a function of soil properties, such as gradation and plasticity. Kamphuis and Hall (1983) performed a series of flume tests on consolidated clay to determine the onset of erosion. Kamphuis et al. (1990) conducted erosion tests in a 9 m long laboratory flume for undisturbed soil samples from four different places. It was found that the erosion of intact natural cohesive soils was controlled by the discontinuities present rather than their geotechnical properties. Gaskin et al. (2003) carried out flume tests for undisturbed Champlain Sea clay to determine the important factors influencing the erosion rate and the mechanisms of erosion of the material. Cracks and fissures within the soil lead to rapid erosion, and the wetting and drying process has a significant influence on the erosion resistance.

The flume test can be applied to measure the erodibility of both fine materials (clay, silt) and coarse materials (sand, gravel). Both intact and remolded specimens can be measured. The channel slope can be easily adjusted to represent the field conditions. By regulating the flow rate, the critical erosive shear stress can be measured, and then the erodibility coefficient can be calculated according to Equation (7.85). The shear stress at the water–soil interface is not measured directly but calculated based on the measurement of the flow velocity and visual identification of initiation of erosion.

7.2.3.2 Rotating Cylinder Test

The rotating cylinder test apparatus was developed by Moore and Masch (1962) to measure soil erodibility. A cylinder of cohesive soil 76 mm in diameter and 76 mm in height is mounted coaxially inside a slightly larger transparent cylinder which can rotate at any desired speed. The annular space between the soil sample and the rotating cylinder is filled with a fluid to transmit shear from the rotating cylinder to the surface of the soil sample. The quality of the eroding fluid can be adequately controlled. The sample is stationary, but is mounted on flexure pivots so that the shear stress transmitted to its surface resulted in a slight rotation of the supporting tube. The shear stress can be calculated by measuring the torque. The eroded soil can be collected and measured at certain time intervals at a given rotating speed. Therefore, the critical erosive shear stress can be measured by increasing the rotating speed until soil particles are eroded. The rotating cylinder technique provides the following advantages:

1. It requires smaller volumes of water and clayey soil.
2. The quality of eroding fluid can be adequately controlled.

3. The rate of erosion can be measured.
4. The shear stress can be determined by a torque measurement.
5. The influence of physicochemical parameters can be studied.

Chapuis and Gatien (1986) improved the technique so that both remolded soil samples and intact soil samples could be tested with an improved rotation guidance, a better alignment, a lower inherent friction, and a damped influence of end supports on the annular flow regime. Lim and Khalili (2009) modified the improved device by incorporating a sample assembly module, a torque reading module, and an electrical control amplifier. The diameter of the soil sample was increased to 100 mm.

The sample setup and the testing procedure included the following steps:

1. Drill a 16 mm diameter hole through the centerline of the sample (100 mm in diameter and 100 mm in height) using an electric drill.
2. Place the sample between the top and bottom sample plates and suspend the sample block inside the outer rotating cylinder (123 mm in internal diameter).
3. Fill the outer cylinder with eroding fluid through a valve.
4. Connect the torquemeter to the sample shaft and to the electric reading unit.
5. Set the testing parameters: duration of cylinder rotation, data-logging interval, cylinder rotation speed, output data file name, and run the test.
6. Upon completion of the test, remove, dry, and weigh the eroded material from the valve located at the bottom of the outer cylinder, and calculate the erosion rate.
7. For tests involving several stages, repeat Steps 3, 5, and 6 as necessary.

The rotating cylinder method is applicable to cohesive soil samples that are either intact or remolded. The erosion rate is calculated based on the eroded soil weight (the difference of the specimen weight before and after a test stage); however, disturbances to the specimen cannot be avoided during the measuring process. The eroded soil mass measurement was modified by Chapuis and Gatien (1986) by withdrawing the eroding fluid at each test stage. However, both eroded soil mass measurement methods may introduce uncertainty in that the eroded fine soil within the eroding fluid may change the shear stress acting on the specimen surface. For granular soils and broadly graded soils, this method may not be suitable. Another limitation is the calculation of the shear stress acting on the specimen. As the specimen diameter decreases with the surface erosion, the calculated shear stress based on the measured torque may not reflect the real value since the decrease of the specimen diameter is not measured, especially when the erosion is not uniform along the whole specimen.

7.2.3.3 Erosion Function Apparatus

Briaud et al. (2001) developed a new device to measure the erosion resistance of soil and soft rock, the erosion function apparatus (EFA) (Figure 7.5). It is composed of a rectangular pipe, a flow rate control system, a piston pushing system, and a thin wall steel tube (drilling or compacting sample). The cross-section and the length of the rectangular pipe are 101.6×50.8 mm and 1.22 m, respectively. The 75 mm outside diameter sampling tube is placed through the bottom of the conduit where water flows at a constant velocity. The piston at the bottom of the sample can extend and push the sample out of the thin wall tube until it protrudes 1 mm into the rectangular pipe. By regulating the

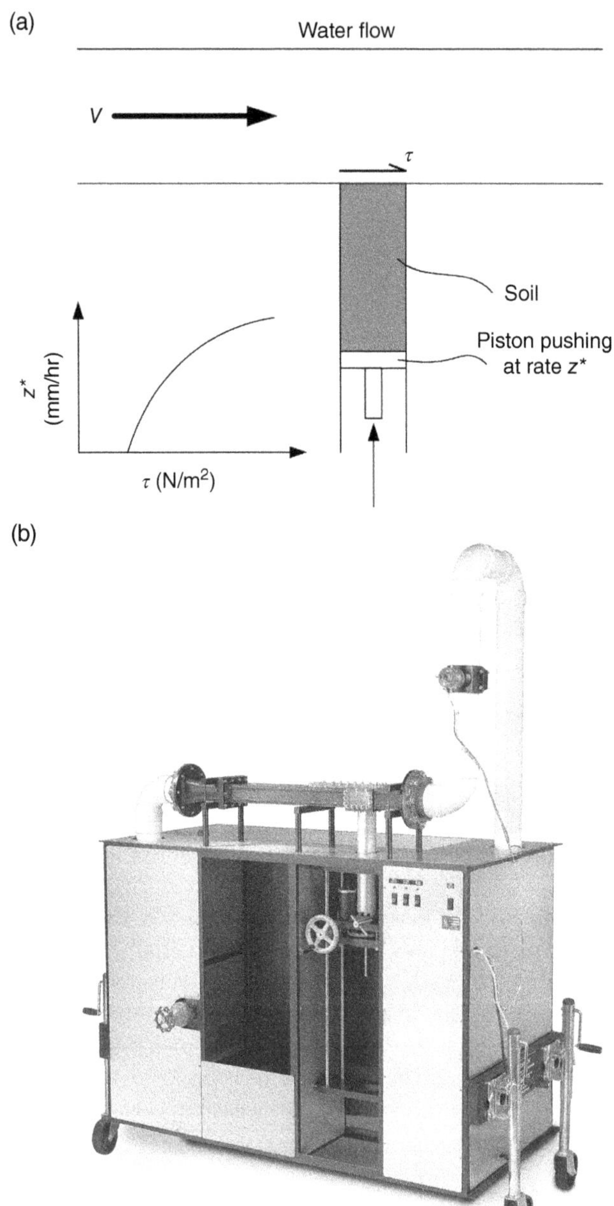

Figure 7.5 Schematic of the erosion function apparatus (EFA): (a) conceptual diagram and (b) testing section (reprinted with the permission of Prof. Jean-Louis Briaud)

flow rate, the time to erode the 1 mm protrusion can be obtained. The erosion rate and critical erosion shear stress can be obtained given the flow rate and erosion time.

The procedure for the EFA test is as follows:

1. Place the sample in the EFA, fill the pipe with water, and wait one hour.
2. Set the velocity to 0.3 m/s.

3. Push the soil 1 mm into the flow.
4. Record the amount of time it takes for the 1 mm soil to be eroded.
5. When the 1 mm of soils is eroded or after 1 hour of flow, whichever comes first, increase the velocity to 0.6 m/s and bring the soil back to a 1 mm protrusion.
6. Repeat step 4.
7. Next, repeat steps 5 and 6 for velocities equal to 1, 1.5, 2, 3, 4.5, and 6 m/s.

The erosion rate $\bar{z}$ at a given velocity v is

$$\bar{z} = \frac{h}{t} \tag{7.86}$$

where h is the length of soil sample eroded in a time t. The length h is 1 mm and t is the time required for the sample to be eroded flush with the bottom of the pipe (visual inspection through a Plexiglas window). The shear stress τ is

$$\tau = \frac{1}{8} f \rho v^2 \tag{7.87}$$

where τ is the shear stress on the wall of the pipe, f is the friction factor obtained from the Moody chart, ρ is the mass density of water, and v is the mean flow velocity in the pipe.

7.2.4 Field Tests

Common methods for soil erodibility measurement in the field are the jet index method and the flume test method. The flume test can be used either in the laboratory or in the field, and its working principle was presented in Section 7.2.3. This section focuses on the jet index method.

The jet index method was initially developed for in-situ erosion testing, as shown in Figure 7.6. It was adopted by the ASTM and recommended as a standard method (ASTM 2000). Hanson and Cook (2004) improved the device to a portable scale. Hanson and Hunt (2007) modified the apparatus for laboratory tests.

The apparatus includes a vertical submerged jet device, a pin profiler, a head tank, and a water delivery system. The vertical submerged jet device consists of three detachable parts: a base ring, a backwater tank, and an inner cylindrical liner. The height of the base ring is 65 mm. To handle the loose but relatively coarse fresh deposits encountered, the height of the base ring was increased to 125 mm to lessen the escape of water from the test area in Chang et al. (2011). The use of a water tank can eliminate the turbulent effect of water inflow. Water delivery may be accomplished by pumping directly from a body of water at the site, from a storage tank delivered to the site, or from a city water supply system if available. If the test water is delivered by a pump, a regulator should be used to control the flow rate, and the excess water could spill through an outlet. The apparatus uses a nozzle positioned above the center of the submerged specimen to produce a jet perpendicular to the soil surface. The jet, driven by a constant water head, applies a shear stress to the soil surface, which in turn causes scouring. The depth of the material removed increases and the shear stress exerted on the hole surface decreases with time. When the shear stress equals the critical erosive shear stress, the erosion process stops. The erosion rate can be calculated by measuring the erosion depth at several time intervals. Moreover, the critical erosive shear stress can be extrapolated from the measured data. This method has been successfully used to measure the soil erodibility of stream channels, stream banks, and landslide deposits.

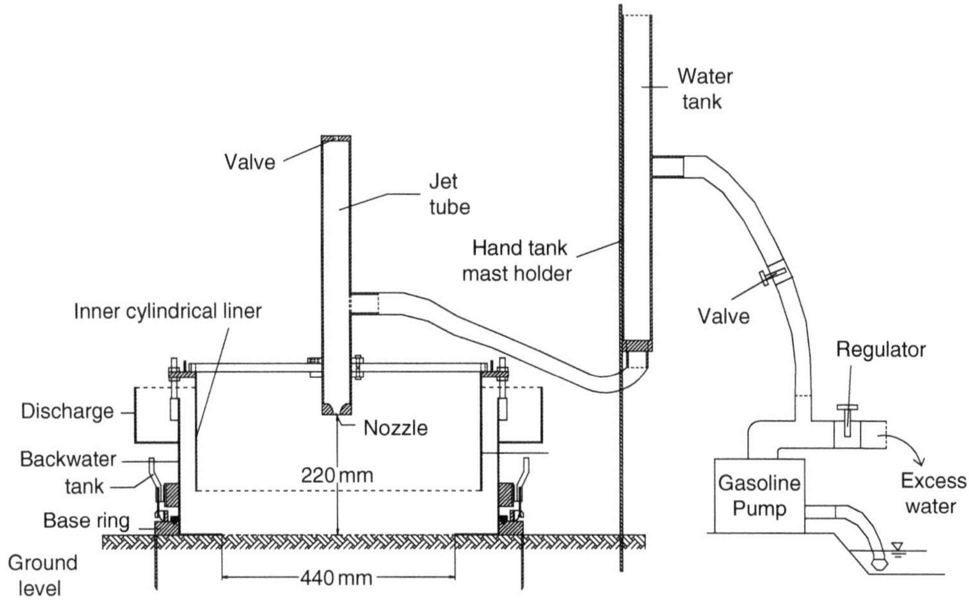

Figure 7.6 Schematic of the jet index apparatus

The test procedure follows ASTM designation D5882-00 and is summarized in Figure 7.7:

1. Level the ground.
2. Push the base ring into the soil by using wood cushion to protect the base ring from damage (Figure 7.7a).
3. Assemble other components including the backwater tank and the cylindrical liner (Figure 7.7b).
4. Use the pin profiler to determine the pre-testing soil elevation.
5. Remove the pin profiler and backfill the tank with water.
6. Place the jet apparatus on the backwater tank cylindrical liner (Figure 7.7c). Hold a flat 10–13 cm diameter disk approximately 3 cm from the nozzle outlet and make final adjustments to the head. The flat disk prevents discharge from the nozzle from impinging directly on the sample prior to testing.
7. Adjust the water head to be 0.91 m by regulating the valve (Figure 7.7d).
8. Remove the disk from the discharging jet to initiate testing.
9. Measure the scour depth using the pin profiler (Figure 7.7e) with a timing sequence of 600, 1200, 1800, and 3600 s. The final scour profile in a test is shown in Figure 7.7(f).

A dimensionless coefficient, jet index J_i, is used to quantify the soil erodibility that is determined from the results of a jet index test as follows:

$$J_i = \frac{D_s / t}{U_0 \left(t / t_0 \right)^{-0.931}} \tag{7.88}$$

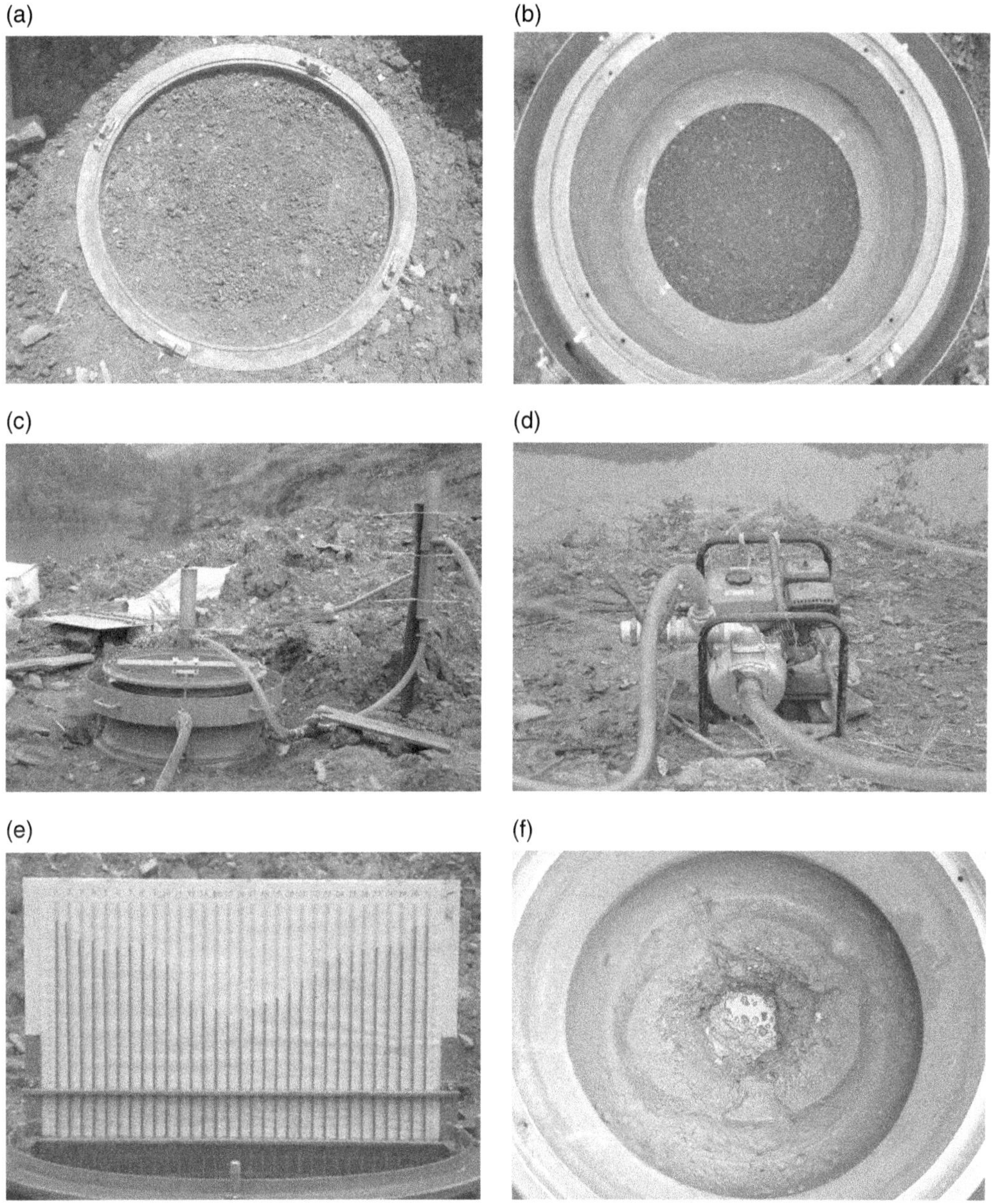

Figure 7.7 Field testing procedure: (a) levelling the ground and pushing the base ring, (b) assembling the backwater tank and the inner cylindrical liner, (c) connecting the jet tube and the water tank, (d) connecting the water tank to the gasoline pump, (e) measuring the scour depth using a pin profiler, (f) final scour profile

where D_s is the average maximum depth of scour determined from readings of the three central profiler pins – pins 15, 16, and 17 in Figure 7.9(e) – taken at each time interval; U_0 is the jet velocity at the nozzle; t is time; and t_0 is a reference time taken as 1 s.

The relationship between jet index, J_i and coefficient of erodibility, K_d (in mm³/N-s) is given as

$$K_d = 3e^{385J_i} \tag{7.89}$$

The critical erosive shear stress for a circular submerged jet is determined as

$$\tau_c = \tau_0 \left(\frac{H_p}{H_e} \right)^2 \tag{7.90}$$

where τ_0 is the maximum applied bed shear stress within the potential core of the jet, H_p is the potential core length from the origin of the jet (Figure 7.8), and H_e is the distance from the jet nozzle to the equilibrium depth of the scour (Figure 7.8). H_p is defined once the jet nozzle diameter, d_0, and the water head are selected: $H_p = C_d d_0$ where C_d is an orifice discharge coefficient with a typical value of 6.2. τ_0 is expressed as $\tau_0 = C_f \rho U_0^2$, where ρ is the density of water, and C_f is the bed coefficient of friction that is assumed to be 0.00416 (Hanson and Simon, 2001).

The jet index method is applicable to a variety of soils other than those that have a dominant soil structure of 7–8 cm or larger (that is, aggregate, clod, or particle size) (ASTM, 2000c). The jet erosion method can be performed on a relatively wide range of cohesive soils, but is not well suited for soils that include fractions of larger gravel, as the gravel may fail to wash out of the scour hole and could gradually armor the surface (Wahl et al., 2008).

Although great progress has been made on measuring soil erodibility, each method described above has some limitations. For instance, the shear stress is calculated based on the flow rate for

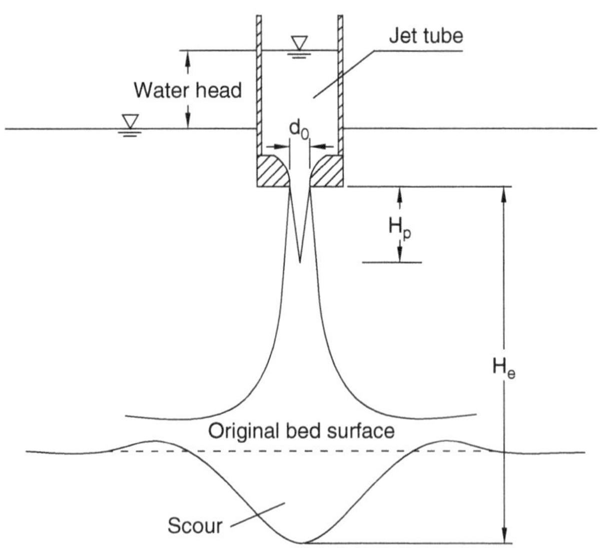

Figure 7.8 Schematic of jet scour parameters

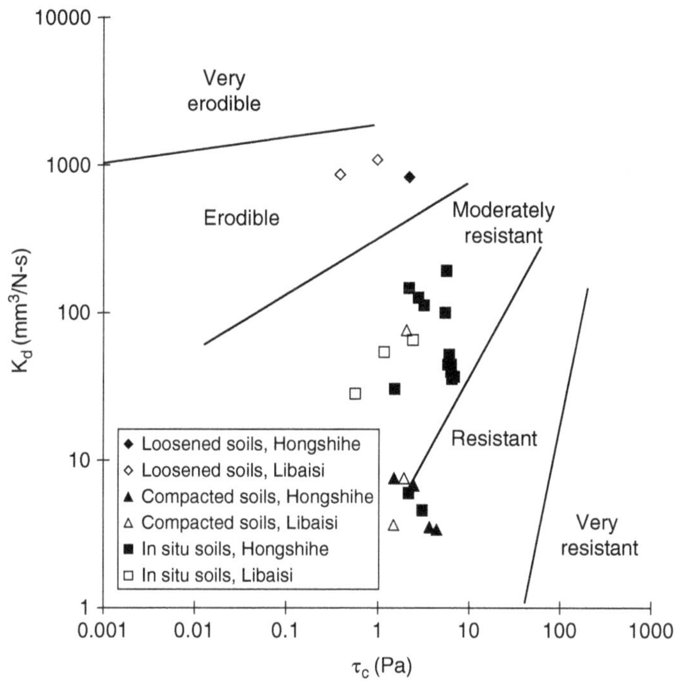

Figure 7.9 Classification of soil erodibility based on erosive shear stress and coefficient of erodibility

the flume test, thus it is an averaged value. However, erosion may occur along the weak zones (cracks) in intact soil samples. Thus the calculated shear stress is usually larger than the actual shear stress. The calculated shear stress of the rotating cylinder method is also an averaged value. Identification of the initiation of erosion for the flume test and the rotating cylinder test relies on visual observation of the surface of the specimen, which may be subjective, especially when the color of the testing soil is close to that of the surrounding fluid.

The rotating cylinder method and the hole erosion method are applicable to cohesive soils with no or little gravel content. The erosion function apparatus method can be applied to cohesive soils, sand and soft clay. Although the flume test works on a broad range of soils (clay, sand, and gravel), it is often limited to remolded soils or self-stable undisturbed soil samples. The jet index test method can accommodate a broader range of particle sizes, but the gravel fraction should be small.

7.2.5 Classification of Soil Erodibility

Two soil erodibility classification methods are available: Hanson and Simon's method (Hanson and Simon, 2001) and Briaud's method (Briaud, 2008). Hanson and Simon's method is based on critical erosive shear stress and the coefficient of erodibility, and Briaud's method is based on erosion rate, flow velocity, and hydraulic shear stress.

Based on results from 83 field jet erosion tests conducted in cohesive streambeds, Hanson and Simon (2001) developed erodibility categories based on the critical erosive shear stress and the coefficient of erodibility (Figure 7.9). There is a wide variation in the erosion resistance of the streambeds, spanning six orders of magnitude for critical erosion shear stress and four orders of magnitude for

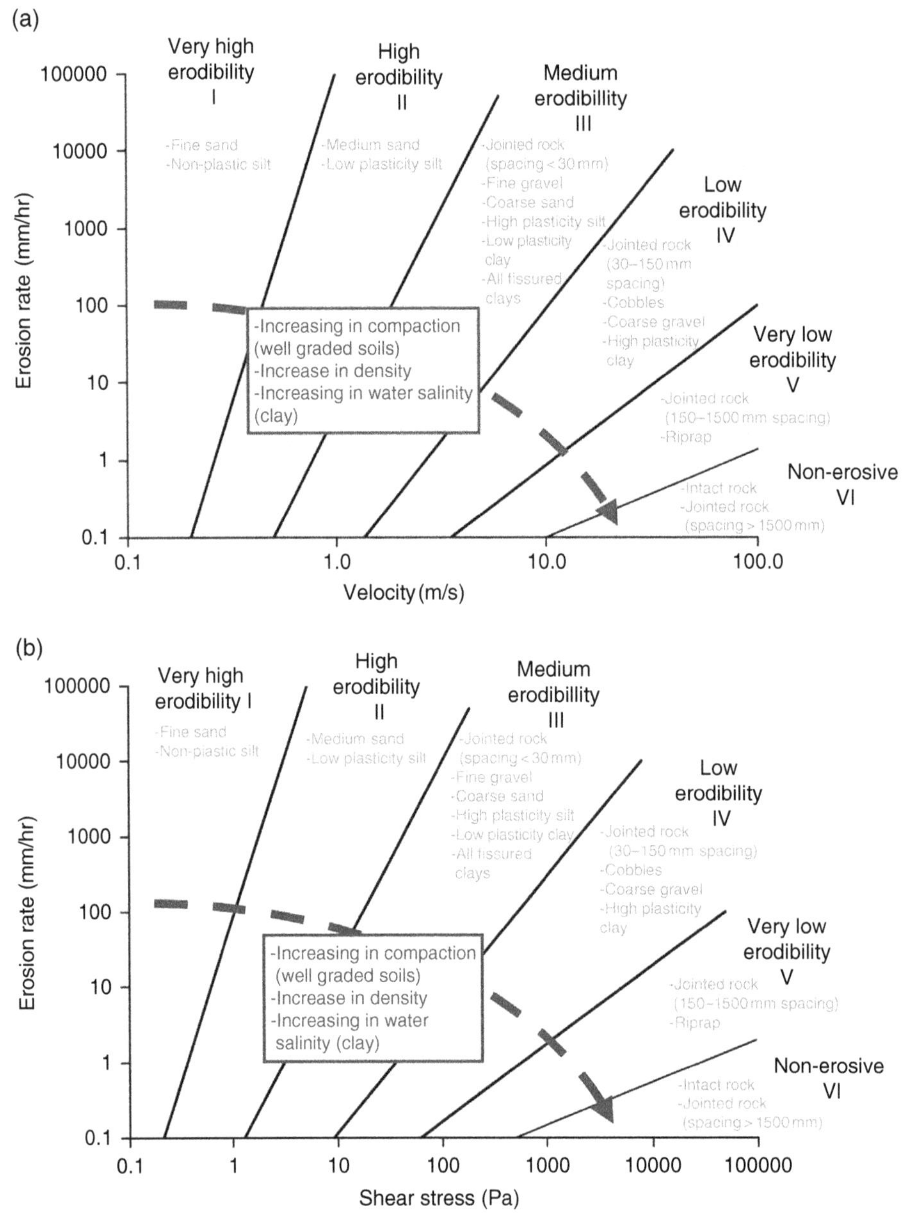

Figure 7.10 Classification of soil erodibility based on applied shear stress and erosion rate. *Source:* Briaud (2008). With permission from ASCE

coefficient of erodibility. Soil erodibility is classified into five categories (very erodible, erodible, moderately erodible, resistant, and very resistant). A soil with a higher critical erosive shear stress and lower coefficient of erodibility tends to be more erosion resistant. Among the soils tested, the most erodible bed materials were in south-eastern Nebraska, with τ_c close to 0.0 and K_d as high as 3.75 cm^3/N-s. These beds consisted of loess-derived alluvium originating from failed bank materials and deposited silts. These beds scoured rapidly even under low stresses from the jet. The most resistant bed materials were observed in the Yalobusha River Basin, Mississippi. The highest measured τ_c was 400 Pa and the lowest K_d was 0.001 cm^3/N-s.

Briaud (2008) and Briaud et al. (2008) proposed erosion categories using a plot of hydraulic shear stress against erosion rate/hydraulic shear stress based on results of over 15 years of tests for bridge foundations, river meanders, cliffs, and levees. Six erosion categories (very high erodibility, high erodibility, medium erodibility, low erodibility, very low erodibility, non-erosive) are proposed as shown in Figure 7.10. Generally, soil compaction leads to an increase in the erosion resistance.

7.3 Characteristics of Overtopping Erosion Failure of Dams

The overtopping failure of man-made dams occurs for several reasons: (1) insufficient design height in an earthfill dam or spillway crest, (2) reservoir storage capacity loss due to lack of maintenance, (3) malfunction of spillways, gates, or pipe outlets, (4) seiches induced by seismic activity and large displacements of water volumes in the reservoir caused by rock/soil slides or ice avalanches (Clague and Evans, 2000). Man-induced causes such as sabotage and terrorist attacks are gaining attention as another possible source of dam failure. When water flow overtops a dam, the flow could be turbulent. This turbulence imposes a rapid and significant pressure on the bed surface, which has an important effect on entrainment of soils (Raudkivi, 1990). Observations in laboratory experiments and case studies suggest a distinct difference between the erosion processes of non-cohesive, cohesive, and composite embankment dams. The characteristics of the overtopping erosion failures of different types are presented in the following sections.

7.3.1 Homogeneous Embankment Dams with Cohesionless Materials

The overtopping erosion mode is usually progressive surface erosion for non-cohesive embankments but headcut erosion for cohesive embankments. A large number large-scale field and laboratory dam failure experiments have been conducted. Some characteristics of overtopping failure of homogeneous embankment dams can be found from these tests.

Breach initiation phase: The breach initiation phase is the process whereby overtopping flow erodes material from the downstream face and crest to a point where it starts to affect flow over the embankment by eroding into the upstream face. Figure 2.1 in Chapter 2 shows the schematic of the breaching process of a granular embankment dam, of which the breach initiation phase involves stages 1–2. Once overtopping occurs, the seepage exiting on the downstream slope accelerates slope erosion. The phenomenon starts at or near the toe, where the seepage flow exits and turbulence of the impinging flow is greatest. Surface slips take place quickly due to the seepage existing on the downstream slope, and hence granular materials are removed rapidly layer by layer. While continuous erosion is likely to occur, lateral erosion is normally through failure of discrete sections of the embankment along the breach channel. The size of these discrete sections can vary from small to significant. Removal of the failed material is often very slow. Undermining of the breach channel side slopes causes large volumes of unsupported material to collapse into the center of the channel and be transported downstream. During the initiation phase, the outflow rate and the breach size are usually small, and the erosion at the crest of the dam is usually limited. With the retrograding of the erosion to the upstream side, the breach on the downstream slope develops gradually.

Breach Development Phase: When the scour finally reaches the upstream edge of the dam crest, the breaching process becomes rapid. The channel invert slope then starts to flatten due to the slope failure on the upstream slope. As erosion continues, the channel invert slope reduces to a terminal value, accompanied by lateral erosion and slope failures. The opening of the breach first mainly progresses down to the base of the dam before it expands laterally. In Figure 2.1, the breach development phase involves stages 3–7. Towards the end of the erosion process, the downstream slope may flatten, steepen, and erode back parallel, depending on the soil type and state and the general slope of the foundation surface. The foundation of the dam could also be eroded depending on its material properties. If the dam body includes some rocks, the breach may not erode to the original foundation. The peak outflow rate usually occurs during this stage.

7.3.2 Homogeneous Embankment Dams with Cohesive Materials

For dams with cohesive materials, when the water flow overtops the crest, impinging jet attacks the channel bed and creates a reverse roller. The impinging jet and flow circulation typically cause deepening of the bed and undermining of the near vertical slope (Hanson et al., 2005b). The overtopping erosion mode for cohesive embankments is usually headcut erosion. The overtopping erosion process of cohesive embankment dams can also be simply divided into the breach initiation phase and the breach development phase.

Breach initiation phase: Initial overtopping flow results in sheet and rill erosion on the downstream surface, with one or more master rills developing into a series of cascading overfalls. The cascading overfalls develop into a large overfall and finally headcut at the downstream face develops due to tensile or shear failure of the soil on the over-steepened slope. In some cases a series of stair-stepped headcuts form on the downstream face of the dam (Morris, 2005; Xu, 2010; Zhang et al., 2010). Instead of flowing down attaching to the headcut slope, the breach flow impinged like a jet upon the downstream flume bottom, imposing considerable erosive forces on the toe of the slope. Under the pressure imposed by the water flow and the erosive forces on the slope toe of the breach, the headcut retrograded from the downstream to the upstream crest of the embankment. Fluidization and consequent failure of the headcut surface layer due to the diffused falling water can speed up the retreat of the headcut. Owing to the loss of stability, parts of the materials in the side slopes of the breach are washed away, hence the breach is widened gradually. This phase ends when the headcut reaches the upstream crest. In this phase, the side slopes of the breach could be vertical due to the cohesive feature of the materials and soil suction. Moreover, the eroded soils can easily be washed away as the particle size is small in cohesive materials. The outflow rate is small because the erosion depth at the crest is limited due to the relatively high erosion resistance of the cohesive material. Figure 2.2 shows the schematic of the breaching process of a cohesive embankment dam, of which the breach initiation phase involves stages 1–4.

Breach development phase: Lowering of the crest occurs quickly when the headcut retrogrades to the upstream crest of the embankment. The breach is enlarged quickly both in depth and laterally. The downward erosion may stop when the erosion reaches the foundation, if the foundation of the dam is not erodible. Widening of the breach continues due to the lateral erosion and the resulting discrete side-slope instability. The peak outflow rate usually occurs in the development phase. The breaching process stops when the headwater is depleted or the tailwater elevation increases. Zhu (2006) pointed out that, along with the reduction of the height of the dike in the development phase, the role played by the headcut slope mass failure is about to decline; on the contrary, the role played by the flow shear erosion to the breach boundary is about to increase, due to the increase of the breach flow rate. Moreover, the cohesion of the dam materials also has a significant influence on the breaching process. Zhang et al. (2009) stated that the final width and depth of the breach, and the peak outflow become smaller, and the rate of the vertical cutting and horizontal expansion is lower when the cohesion of dam material is larger. Figure 2.2 shows the whole breaching process of a cohesive embankment by overtopping, of which the breach development phase involves stages 5–7.

7.3.3 Composite Embankment Dams

Where an embankment includes additional structures such as surface protection measures and concrete floodwall on the crest or clay-core, the breach process becomes more complex and harder to predict. When water overtops the crest of the dam, erosion starts on the downstream slope in the form of either surface erosion or headcut migration until the core or floodwall is reached. The rate of erosion can be increased or decreased by the presence of the additional structures and hence progression of the breach formation process becomes dependent upon structural failure processes. This erosion may affect the stability of the core and floodwall, and eventually lead to failures. The most common failure mechanisms of the core and floodwall include sliding, overturning, and bending (ASCE/EWRI Task Committee on Dam/Levee Breaching, 2011). For instance, for an embankment dam with a thin clay corewall, surface erosion will lead to the removal of the bulk embankment material, but the core will affect the rate and location of erosion. Failure of the core will then occur through a structural process due to a lack of support from the bulk fill material, allowing further erosion of the bulk fill material.

7.4 Suggested Topics for Further Research

Overtopping erosion failure of embankment dams and landslide dams is one of the most common failure modes. Although great effort has been made in predicting the dam failure process associated with overtopping erosion, engineering design aids and standard procedures have to be developed for dam breach analysis and risk management. The following issues require further research:

1. Bed-load equations are developed based on field and laboratory flume tests on gentle slopes with uniform-sized materials. In embankment dams or landslide dams, the materials are generally broadly graded and zoned. The processes of sediment transport under dam breaching conditions have to be better understood.
2. The current models for dam breaching prediction concern more the breach development phase. The breach initiation process is more difficult to predict; however, the breach initiation time is important for warning the downstream residents. Therefore, reasonable prediction of breach initiation is of great importance.
3. Erodibility of cohesive materials is as important a soil parameter as soil strength or permeability in embankment design. However, quantification of cohesive soil erosion and headcut migration is a weak point in all embankment breach models. The effects of soil type, moisture content, and compaction on cohesive soil erosion and headcut migration need to be better incorporated in modeling the breaching process of cohesive dams.
4. Erodibility testing of typical soils is needed so as to develop a database of 'indicative values of erodibility' for embankments made from different soils in different regions. A review and comparison of erodibility measuring testing techniques is required to improve understanding of soil erodibility and to recommend suitable techniques for typical soils.

5. It is essential to understand and integrate hydraulic, geotechnical, and structural processes when considering the dam breaching process. The geotechnical parameters and processes that significantly influence breaching are either oversimplified or, in some cases, neglected. Therefore, more efforts are needed to incorporate geotechnical issues such as slope stability during breaching, effect of unsaturated zone on slope stability and erosion, and erodibility of unsaturated materials.
6. During the dam breaching process, materials below the breach undergo stress relief due to the loss of the dam materials. How the unloading process affects the erodibility of the materials needs to be understood.
7. Many embankment dams comprise a combination of embedded structures – for example, flood embankments with concrete or sheet piled walls. The interaction between different components of the flood defense structures can significantly affect the rate of breach initiation and growth. Research is required to extend the capabilities of predictive breach models to simulate processes that occur within such structures.

8

Dam Breach Modeling

8.1 Methods for Dam Breach Modeling

Floods resulting from breaching of dams can lead to devastating disasters with tremendous loss of life and property, especially in densely populated areas. Past dam failure disasters showed that the loss of life is directly related to the evacuation time available, should failures occur. It is therefore very important to understand the process of dam breaching and, if possible, to obtain key breaching parameters required to quantitatively describe the dam breaching process. Earth and rockfill dams are widely used throughout the world, and most past dam failures have involved such dams.

Dam breaching parameters can be divided into two groups: geometric and hydrographic. An embankment dam breach often has a trapezoidal shape, with the geometric parameters of breach depth H_b, breach top width B_t, average breach width B_{ave}, breach bottom width B_b, and breach side slope factor Z, as shown in Figure 8.1. Any combination of three of the five geometric parameters determines the breach shape and size. Hydrographic parameters include peak outflow rate Q_p and failure time T_f. After the onset of breaching, the outflow through the breach increases, until it reaches a peak, Q_p, and then decreases until there is no longer any water in the reservoir or the breaching process ceases to develop (Chinnarasri et al., 2004). Failure time T_f is defined as the period from the inception to the completion of the breaching process (Singh and Snorrason, 1984). Wahl (1998) divided the whole breaching process into two phases: the breach initiation phase and the breach development phase. In the breach initiation phase, the outflow from the dam is small, consisting of a slight overtopping or a small flow through a developing pipe or seepage channel. In the breach development phase, the outflow and erosion develop rapidly. The breach development time is a parameter implied in most failure time prediction equations (Wahl, 2004). According to Fell et al. (2003), in most cases it has not been possible to identify the time of initiation of erosion, and the first signs of erosion tend to be at the progression phase. The time for initiation has been recorded where it is possible, for example from increased seepage flows. Therefore, from a practical standpoint, failure time is often recorded at the start of the breach development phase. Failure time T_f in this book is also regarded as the breach development time.

Dam Failure Mechanisms and Risk Assessment, First Edition. Limin Zhang, Ming Peng, Dongsheng Chang, and Yao Xu.

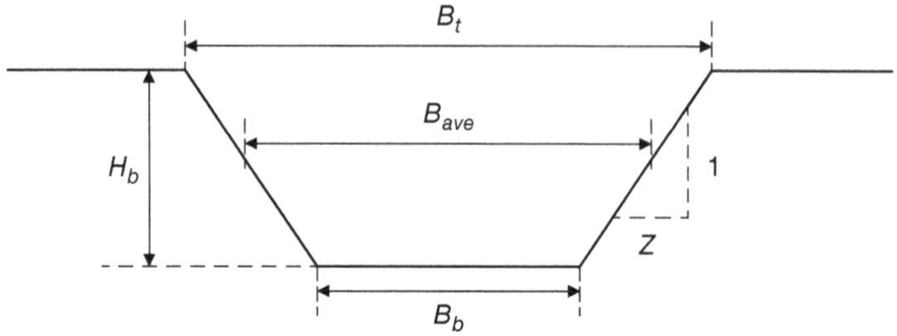

Figure 8.1 Geometric parameters of an idealized dam breach. *Source:* Xu and Zhang (2009). With permission from ASCE

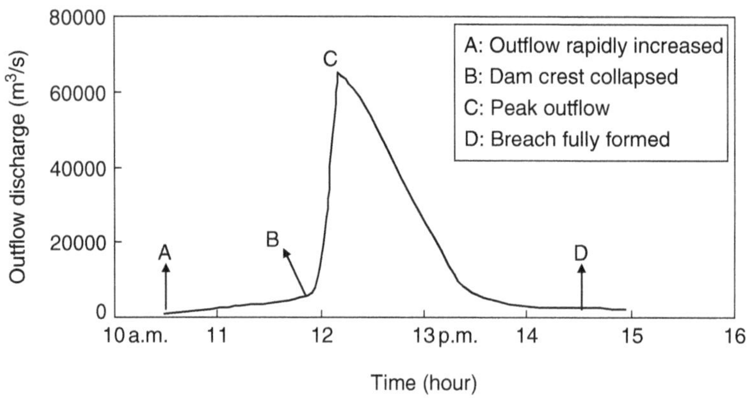

Figure 8.2 Observed outflow discharge during failure of Teton Dam on 5 June 1976 (based on Singh 1996). *Source:* Xu and Zhang (2009). With permission from ASCE

Figure 8.2 shows an example of the outflow hydrograph during the failure of Teton Dam. The time of 10:30 a.m. is a critical point to separate the breach initiation phase and the breach development phase. From that moment, the volume of water released increased rapidly as did the erosion of the embankment materials. After the collapse of the dam crest at 11:55 a.m., the outflow reached its peak discharge very soon. The failure time, T_f, in this case is approximately 4 hours.

A number of studies have been devoted to dam breach modeling. Wahl (1998) summarized existing prediction models for estimating embankment dam breaching parameters. In general, two sorts of prediction models were involved: (1) empirical formulas based on case study data (e.g. Kirkpatrick, 1977; Hagen, 1982; Bureau of Reclamation, 1982, 1988; MacDonald and Langridge-Monopolis, 1984; Singh and Snorrason, 1984; Costa, 1985; Von Thun and Gillette, 1990; Froehlich, 1995a, b), and (2) physically based prediction models (e.g. Brown and Rogers, 1977, 1981; Fread, 1977, 1988; Singh and Scarlatos, 1985). Most of the existing physically based prediction models rely on sediment transport formulas. These formulations may be appropriate for some stages of the breach process, but are not consistent with the mechanics of much of the breaching process as observed in the field and in the laboratory (Wahl, 1998). Recent developments on physically based dam breach models have been summarized by Wahl et al. (2008) and Morris et al. (2008). Three physically based models that have potential for further development draw great attention, including SIMBA (Temple et al., 2005; Hanson et al., 2005b), HR-BREACH (Mohamed, 2002), and Firebird BREACH (Wang and Kahawita, 2002). These methods are often parametric, making use of breach parameter estimates derived from regression equations (Wahl et al., 2008).

In the sense of being close to reality, empirical formulas based on case studies appear to be preferable when a database of numerous well-documented dam failure cases exists. The available empirical equations usually relate dam breaching parameters to one or two dam and reservoir properties, such as dam height (or depth of water above the breach invert at the time of failure) and reservoir storage (or volume of water stored above the breach invert at the time of failure). For instance, Froehlich (1995a) developed a prediction equation for the average breach width based on 60 cases of embankment dam failures and an equation for the breach formation time based on 21 cases of embankment dam failures. Froehlich (1995b) also proposed a regression equation for predicting the peak outflow rate based on data from 22 embankment dam failures.

The existing databases of dam failures are relatively small. In addition, about 75% of the cases had heights less than 15 m (Wahl, 1998), and hence there was a lack of data on failures of large dams. Note that according to the definition by ICOLD (1988), $H_r = 15$ m is a reference height for distinguishing large dams from small dams. Chapters 2 and 3 of this book present the largest databases of failures of constructed dams and landslide dams, especially failure cases from China that were often not available in previous studies. These failure cases provide a good basis for developing empirical formulas for determining the breaching parameters of large dams.

This chapter starts by presenting the available dam breaching data obtained from past dam failure cases. Then the empirical formulas based on case study data, numerical simulation of overtopping erosion, and numerical simulation of internal erosion are presented in three separate sections.

8.2 Dam Breaching Data

8.2.1 Embankment Dam Breaching Data

Chapter 2 presents a database of 1443 cases of constructed embankment dam failures. Most of these failure cases lack detailed quantitative information. In this section, 182 earth and rockfill dam failures that have some sort of quantitative information are extracted from the database. The failure cases are from eight countries, of which the USA and China are represented in 116 and 52 cases, respectively. Information on the characteristics of the dams, reservoirs, and failure information is collected. Complete information is not available in some of the 182 cases. Details of 75 failure cases that have sufficient information for developing breach parameter models are presented in Table 2.1. The cases comprise homogeneous dams, zoned-fill dams, dams with core-walls, and concrete faced dams. Figure 8.3(a) compares the percentages of the four types of dams and shows that homogeneous dams consist of half of all the cases. Figure 8.3(b) shows the heights of the failed dams. Among the dams with known heights, nearly half of them are large dams higher than 15 m. Figure 8.3(c) shows the reservoir capacities of the failed dams, of which most reservoirs have capacities less than 1.0×10^8 m^3. Figure 8.3(d) compares the failure modes. Most of the failures were caused by either overtopping or seepage erosion/piping, at 56.0% and 35.7% respectively. Dam failures caused by sliding consist of less than 5% of all the cases.

Note that the dam erodibility (high, medium, or low) in Table 2.1 is a relative measure of erodibility based on the information on material compositions and compaction conditions, dam cross-sectional geometry, construction time, and other relevant pieces of construction information. Based on the erosion resistance categories proposed by Briaud (2008), the rate of erosion can be very different for different soils. Specifically, rockfill and clay are often associated with medium to low erodibility while sand and silt are often associated with high to medium erodibility. Hence, material compositions are considered as a primary basis for classifying the dam erodibility. In addition, compaction conditions also play an important role in determining the dam erodibility, especially

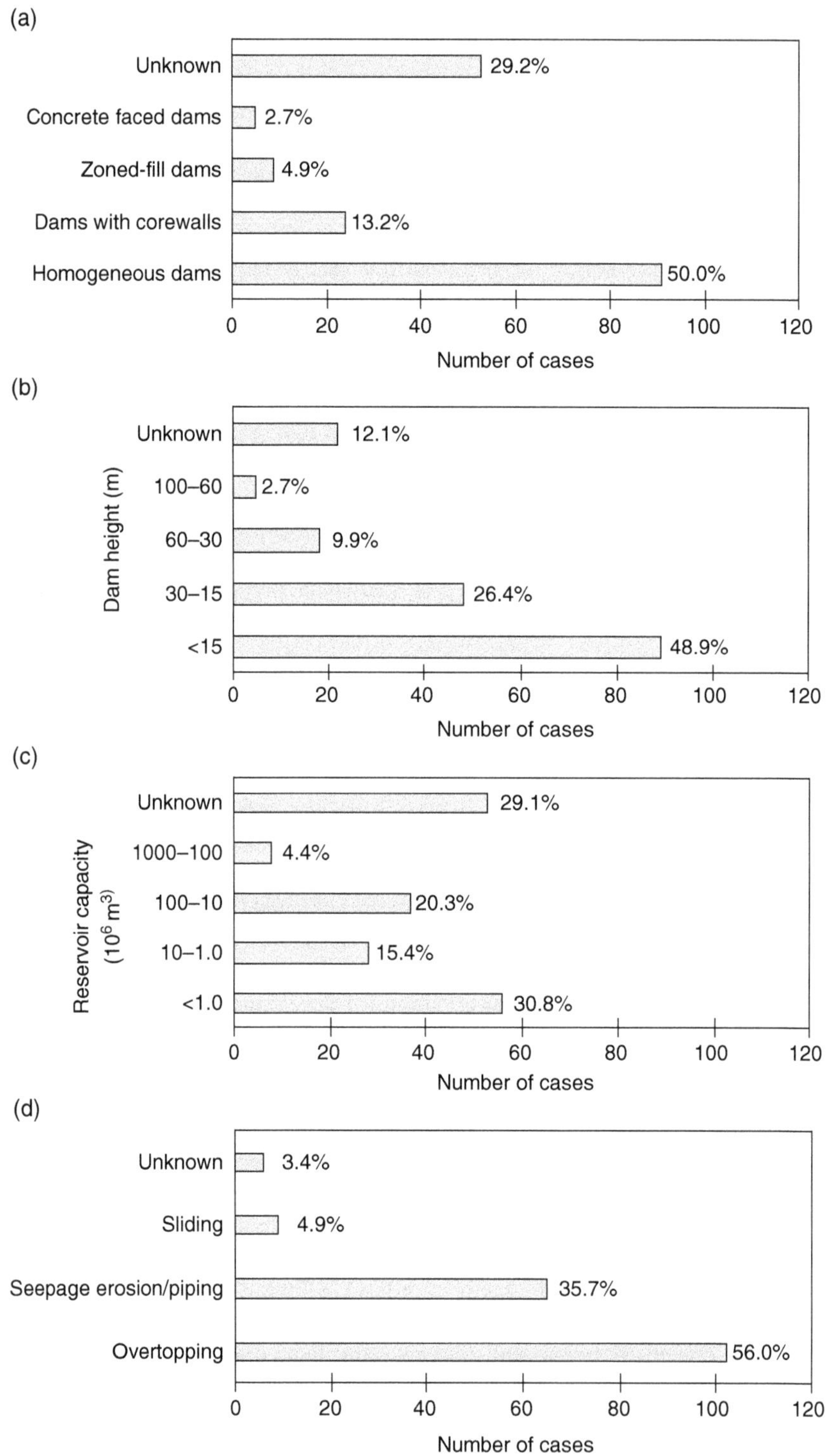

Figure 8.3 Statistics of (a) dam type, (b) dam height, (c) reservoir capacity, and (d) failure mode. *Source:* Xu and Zhang (2009). With permission from ASCE

for dams of fine soils. Briaud et al. (2008) reported that the erosion resistance increases with compaction effort and that the effect is more significant for some soils with high fines contents. It was also indicated by Wan and Fell (2004a) that the erosion rate index of a soil is influenced strongly by the degree of compaction. In this section, construction time and compaction method are important pieces of information for judging the compaction condition of a dam. For instance, dams built in China in the 1950s were often associated with poor compaction due to limitations of construction equipment and technology at that time, resulting in high dam erodibility. Other useful pieces of information such as dam cross-sectional geometry and slope surface protection are used as supplementary information for measuring the dam erodibility.

8.2.2 Landslide Dam Breaching Data

Landslide dams fail frequently. A database of landslide dam failures has been reported in Chapter 3. This database includes 1044 landslide dam cases from all over the world, with 52 of them having detailed breaching information (Table 3.2). Based on the database, the formation, triggers, longevity, and failure modes of these landslide dams have been studied in Chapter 3.

Based on information of ten failed landslide dam cases, Costa (1985) presented three empirical equations for estimating the peak outflow rate from breaching of landslide dams. Two control variables, dam height (H_d) and lake volume at time of failure (V_l) were taken into consideration. Three years later, Costa and Schuster (1988) presented a concept of potential energy (product of H_d, V_l, and the specific weight of water) for estimating the peak outflow rate based on information of 12 landslide dam breaching cases. Walder and O'Connor (1997) also suggested three equations for estimating the peak outflow rate based on drop in water level (d) or/and released water volume (V_0) using information from 18 landslide dam cases.

The existing datasets of landslide dam failures are lacking detailed breaching information. For example, out of the 463 cases in the database of Costa and Schuster (1991), only nine cases have records of final breach dimensions. Due to the limitation of data, the existing empirical models for estimating the peak outflow rate consider only one or two variables (dam height or drop of water level, and lake volume or released water volume), while other variables such as dam volume, dam width, and dam erodibility may help improve the prediction of peak outflow rate. In addition, no statistics based empirical models are available for estimating the breach size (depth and width) and breaching duration of landslide dam failures. These parameters provide key information for understanding possible failure mechanisms and can be used as key inputs for dam breaching and flood routing analyses (Fread, 1988; Chang and Zhang, 2010).

Empirical models for estimating the breaching parameters of man-made earthen dams are sometimes applied to landslide dams; the feasibility of such applications, however, has not yet been confirmed. Using the more extensive database described in Chapter 3, more accurate empirical models for estimating the breaching parameters of landslide dams can be developed, and differences in the breaching parameters of man-made earth and rockfill dams and those of landslide dams can be compared.

8.2.3 Dike Breaching Data

Dikes are a special form of embankment dams. Chapter 5 and Appendix C present a database of 1004 cases of dike failures. A list of 42 cases of dike failures that have detailed information on dike geometry, material, and observed failure mechanism is presented in Table 5.1.

Common failure modes of dikes have been summarized and compared with those of constructed embankment dams and landslide dams in Figure 5.3 in Chapter 5, based on the historic failure data.

In contrast to constructed embankment dams and landslide dams, only a limited effort has been made in the past to evaluate the breaching parameters for dikes, perhaps due to the scarcity of dike breaching data. Two empirical equations for dikes are published by Verheij (2002) and Nagy (2006). Nagy (2006) derived an equation for estimating dike breaching length as a function of dike height. Verheij (2002) proposed two equations for estimating the breaching length of dikes built of fine- and coarse-grained material, respectively, as a function of the breaching time. These equations will be introduced in Section 8.3.4. Due to the lack of dike specific equations, sometimes empirical models for dams are used to estimate the dike breaching parameters, which is shown to be inappropriate.

8.3 Empirical Analysis Methods

8.3.1 *Multivariable Regression*

Two functional forms are used to establish empirical relationships: additive form (linear) and multiplicative form (nonlinear):

$$\begin{aligned} Y_i &= b_0 + b_1X_1 + b_2X_2 + \left(b_{31}X_{31} + b_{32}X_{32} + b_{33}X_{33}\right) \\ &\quad + \left(b_{41}X_{41} + b_{42}X_{42}\right) + \left(b_{51}X_{51} + b_{52}X_{52} + b_{53}X_{53}\right) \end{aligned} \tag{8.1}$$

$$Y_i = b_0 X_1^{b_1} X_2^{b_2} \left(X_{31}^{b_{31}} X_{32}^{b_{32}} X_{33}^{b_{33}}\right)\left(X_{41}^{b_{41}} X_{42}^{b_{42}}\right)\left(X_{51}^{b_{51}} X_{52}^{b_{52}} X_{53}^{b_{53}}\right) \tag{8.2}$$

where Y_is ($i=1, 2, \ldots, 5$) are the five breaching parameters regarded as dependent variables (breach depth H_b, breach top width B_t, average breach width B_{ave}, peak outflow rate Q_p, and failure time T_f); X_is are control variables; and b_is are regression coefficients. Note that Equation (8.2) can be transformed into an additive function by taking a logarithmic transformation on both sides of the equation.

The five key breaching parameters are expressed in the dimensionless form shown in Figure 8.4. As mentioned before, most past research work took into account the depth of water above the breach invert at time of failure, H_w (or dam height H_d), and the volume of water stored above the breach invert at time of failure, V_w (or reservoir capacity V_d). For mathematical rigor, two non-dimensional control variables are used to represent H_w (or H_d) and V_w (or V_d): the relative dam height, $X_1=H_d/H_r$, and the reservoir shape coefficient, $X_2=V_w^{1/3}/H_w$. Three additional control variables, dam type X_3, failure mode X_4, and dam erodibility X_5, are discrete variables, which are expressed using virtual discrete variables that represent either presence or absence of an attribute. Note that homogeneous and zoned-fill dams are grouped together for dam type X_3 in the latter regression analysis due to their similar structure and failure characteristics (Zhang et al., 2009).

For any known values of X_is, Equation (8.1) or Equation (8.2) can be written as,

$$Y_i = g(X_1, X_2, X_{31}, X_{32}, X_{33}, X_{34}, X_{41}, X_{42}, X_{51}, X_{52}, X_{53}) + \varepsilon \tag{8.3}$$

Breaching parameters	
Breach depth	$Y_1 = H_b/H_d$
Breach top width	$Y_2 = B_t/H_b$
Average breach width	$Y_3 = B_{ave}/H_b$
Peak outflow rate	$Y_4 = Q_p/\sqrt{gV_w^{5/3}}$
Failure time	$Y_5 = T_f/T_r$

Control variables			
Dam height	$X_1 = H_d/H_r$		
Reservoir shape coefficient	$X_2 = V_w^{1/3}/H_w$		
Dam type	X_{31}	X_{32}	X_{33}
with corewalls	1[a](e[b])	0(1)	0(1)
concrete faced	0(1)	1(e)	0(1)
homogeneous/zoned-fill	0(1)	0(1)	1(e)
Failure mode	X_{41}	X_{42}	
overtopping	1(e)	0(1)	
seepage erosion/piping	0(1)	1(e)	
Dam erodibility	X_{51}	X_{52}	X_{53}
high erodibility	1(e)	0(1)	0(1)
medium erodibility	0(1)	1(e)	0(1)
low erodibility	0(1)	0(1)	1(e)

Note: [a]Values for additive regression analysis;
[b]Values for multiplicative regression analysis.
$H_r = 15$ m; $T_r = 1$ hour.

Figure 8.4 Summary of five breaching parameters and five control variables for constructed embankment dams. *Source:* Xu and Zhang (2009). With permission from ASCE

where ε is the residual with E(ε) = 0 and Var(ε) = σ_ε^2, where σ_ε is the standard deviation of the residual. The conditional expectation and variance of Y_i for given control variables X_is are respectively $E(Y| X_1, X_2,\ldots, X_{53})$ and $Var(Y| X_1, X_2,\ldots, X_{53}) = \sigma_\varepsilon^2$. The values of b_is and σ_ε^2 can be obtained by performing a regression analysis on a set of observed data of size n, namely, $(x_{j1}, x_{j2},\ldots, x_{j53}, y_j)$, $j = 1, 2, \ldots, n$. The value of σ_ε is assumed to be constant in the regression analysis.

The control variables and functional form are selected so that these control variables are best related to dam breaching parameters of interest. One indicator for the goodness-of-fit is related to the residual error, defined as $e_j = y_j - \overline{y}_j$, where y_j and $\overline{y}_j$ are the observed and predicted values for the jth dependent variable. Based on the least squares method, estimates of regression coefficients b_is are obtained by minimizing the sum of squared residuals,

$$\underset{b_0, b_1, \ldots, b_{43}}{Min} \sum_{j=1}^{n} e_j^2 = \sum_{j=1}^{n} (y_j - \overline{y}_j)^2 = \sum_{j=1}^{n} [y_j - g(x_j \mid b_0, b_1, \ldots, b_{53})]^2 \tag{8.4}$$

The coefficient of determination, R^2, is

$$R^2 = 1 - \frac{\Sigma\left(y_j - \overline{y}_j\right)^2}{\Sigma\left(y_j - y_{ave}\right)^2} = 1 - \frac{\Sigma e_j^2}{\Sigma\left(y_j - y_{ave}\right)^2} = 1 - \frac{SSE}{SST} \tag{8.5}$$

where *SSE* is the sum of the error squares, and *SST* is the sum of the squares about the mean of dependent variable *Y*, termed y_{ave}. The value of R^2 represents the variability in *Y* that is explained by the regression model under consideration. In other words, a model with a large R^2 indicates a high degree of goodness-of-fit to the dataset.

In developing regression models, it is necessary to strike a balance between the prediction accuracy and simplicity. To avoid using an excessive number of control variables, an adjusted coefficient of determination is proposed as,

$$R^2_{adj} = \frac{(n-1)R^2 - k}{n-(k+1)} \tag{8.6}$$

where k is the number of control variables used. The adjusted coefficient of determination R^2_{adj} with k variables is smaller than that with $(k-1)$ variables if the extra control variable does not significantly contribute to improving the explanation about the variation of the data. Therefore, by comparing the R^2_{adj} values, it is possible to choose a simplified prediction model that involves only a minimal number of key control variables.

The following procedure is used to establish the relationship between any one of the five breaching parameters and the five control variables:

1. *Data selection.* Choose the dam failure cases with known X_1, X_2, X_3, X_4, X_5, and Y values. The parameters of interest may not be available in some failure cases in the database.
2. *Regression analysis I.* Conduct additive (or multiplicative) regression analyses for Y by considering: (a) all five X_is, and (b) any combination of four X_is. Compare the coefficients of determination R^2 of the additive and multiplicative regression models considering all five X_is. Choose the one with the largest R^2, which is regarded as the best prediction model. Note that before conducting the multiplicative regression analysis, Equation (8.2) should be transformed into a linear function by taking the natural logarithm on both sides,

$$\ln Y = \ln b_0 + b_1 \ln X_1 + b_2 \ln X_2 + b_{31} \ln X_{31} + \ldots + b_{53} \ln X_{53} \tag{8.7}$$

3. *Ranking the importance of X_is to Y.* Determine the increment of the total squared error *SSE* associated with each prediction model neglecting one X_i over that associated with the prediction model considering all five X_is. Compare the increments of *SSE* due to the ignorance of X_i, and rank the importance of X_is according to their relative contributions to Y.
4. *Regression analysis II.* Conduct additive (or multiplicative) regression analyses for Y by considering (c) the combination of the three most important X_is, (d) the combination of the two most important X_is, and (e) only the most important X_i.
5. *Model selection.* Compare the adjusted coefficient of determination R^2_{adj} of each prediction model developed in Steps 2 and 4. Choose an additive (or multiplicative) regression model with the largest value of R^2_{adj}, and name the chosen model as the best additive (or multiplicative) regression model. Compare the values of R^2_{adj} associated with the best additive and multiplicative regression models. Choose the one with a largest R^2_{adj}, which is the best functional form with reasonably few control variables.
6. *Model refinement.* Check if there are any additional failure cases with the reduced number of X_is in the database when the best functional form involves fewer variables. If not, the model with the largest R^2_{adj} in Step 5 is the best-simplified prediction model. If yes, the additional cases are mixed with the failure cases utilized in Step 1. Based on the expanded dataset, the coefficients of the best regression model are updated and the model obtained is the best-simplified prediction model. Finally, both the best prediction model considering all five X_is and the best-simplified prediction model considering reasonably few X_is are obtained for each specific dam breaching parameter.

8.3.2 *Empirical Breaching Parameters for Constructed Embankment Dams*

8.3.2.1 Regression Equations

According to the analysis procedure presented earlier, the five dam breaching parameters (i.e. breach depth H_b, breach top width B_t, average breach width B_{ave}, peak outflow rate Q_p, and failure time T_f) are studied. Taking the analysis of breach depth H_b as an example, the corresponding results for Steps 1–6 in the analysis procedure are presented in detail:

1. Sixty-six cases of dam failures with known values of X_1, X_2, X_3, X_4, X_5, and H_b are selected from the database. H_b is represented by a dimensionless term, $y=H_b/H_d$, the ratio of breach depth to dam height.
2. Conduct additive (or multiplicative) regression analysis. Row 1 in Table 8.1 shows the additive regression results considering all five X_is while rows 2–6 show the results considering five combinations of any four X_is. Row 1 in Table 8.2 shows the additive regression results considering all five ln X_is while rows 2–6 show the results considering five combinations of any four ln X_is. The value of $R^2=0.35$ of the additive regression model is larger than that of $R^2=0.344$ of the multiplicative regression model, and hence the additive regression model considering all five X_is is the best prediction model,

$$\begin{aligned} Y = 0.453 - 0.025X_1 + 0.145X_{31} + 0.176X_{32} + 0.132X_{33} \\ +0.218X_{41} + 0.236X_{42} + 0.254X_{51} + 0.168X_{52} + 0.031X_{53} \end{aligned} \tag{8.8}$$

with a conditional standard deviation of $S_{Y|X}=0.109$. Using the original dam or reservoir variables, Equation (8.8) is expressed as,

$$\frac{H_b}{H_d} = 0.453 - 0.025\left(\frac{H_d}{H_r}\right) + B_1 \tag{8.9}$$

with $B_1=b_3+b_4+b_5$, in which $b_3=0.145$, 0.176, and 0.132 for dams with corewalls, concrete faced dams, and homogeneous/zoned-fill dams, respectively; $b_4=0.218$ and 0.236 for overtopping and seepage erosion/piping, respectively; $b_5=0.254$, 0.168, and 0.031 for high, medium, and low dam erodibility, respectively.
3. In Table 8.1, compared to the *SSE* value in row 1, the increases of *SSE* in rows 2–6 are 6.0%, 0.0%, 1.2%, 0.6%, and 41.6%, respectively. Therefore, the order of importance of the five variables influencing Y is $X_5 > X_1 > X_3 > X_4 > X_2$.
4. In Table 8.1, Row 7 shows the additive regression results considering X_1, X_3, and X_5; row 8 shows the results considering X_1 and X_5; row 9 shows the results considering X_5 only.
5. Rows 1–9 in Table 8.1 show that the additive regression model considering X_1 and X_5 has the largest value of $R^2_{adj}=0.2923$, which is regarded as the best additive regression model. Note that the nil value of R^2_{adj} in row 6 means that the calculated value of R_{adj} is negative, which suggests the corresponding regression model is meaningless. A similar multiplicative regression analysis is conducted, as shown in Table 8.2. The best multiplicative regression model considering X_1 and X_5 has the largest value of $R^2_{adj}=0.292$, which is slightly smaller than that for the additive regression model. Therefore, the additive regression model considering X_1 and X_5 can be regarded as the best functional form.

Table 8.1 Results of additive regression analyses for the dimensionless representation of breach depth h_b/h_d

No.	Case No.	Considered X_is	b_0	b_1	b_2	b_{31}	b_{32}	b_{33}	b_{41}	b_{42}	b_{51}	b_{52}	b_{53}	SSE	ΔSSE %	R^2	R^2_{adj}	$S^2_{Y\|X}$
1	66	$X_{1,2,3,4,5}$	0.453	−0.025	0.000	0.145	0.176	0.132	0.218	0.236	0.254	0.168	0.031	0.652	—	0.350	0.232	0.012
2	66	$X_{2,3,4,5}$	0.429	—	0.001	0.143	0.151	0.135	0.207	0.222	0.249	0.164	0.017	0.690	6.0	0.312	0.201	0.012
3	66	$X_{1,3,4,5}$	0.455	−0.026	—	0.145	0.177	0.132	0.218	0.237	0.255	0.169	0.031	0.652	0.0	0.350	0.246	0.012
4	66	$X_{1,2,4,5}$	0.524	−0.022	0.001	—	—	—	0.254	0.271	0.274	0.193	0.058	0.660	1.2	0.342	0.263	0.011
5	66	$X_{1,2,3,5}$	0.586	−0.025	0.000	0.185	0.223	0.178	—	—	0.297	0.210	0.079	0.656	0.6	0.346	0.255	0.012
6	66	$X_{1,2,3,4}$	0.543	−0.031	0.001	0.178	0.180	0.185	0.271	0.272	—	—	—	0.922	41.6	0.080	—	0.016
7	66	$X_{1,3,5}$	0.591	−0.026	—	0.186	0.227	0.178	—	—	0.299	0.212	0.080	0.656	0.7	0.346	0.267	0.011
8	66	$X_{1,5}$	0.726	−0.023	—	—	—	—	—	—	0.340	0.258	0.128	0.666	2.2	0.336	0.292	0.011
9	66	X_5	0.701	—	—	—	—	—	—	—	0.338	0.252	0.111	0.701	7.6	0.301	0.267	0.011
10	71	$X_{1,5}$	0.729	−0.025	—	—	—	—	—	—	0.343	0.257	0.129	0.705	—	0.353	0.314	0.011

Table 8.2 Results of multiplicative regression analyses for the dimensionless representation of breach depth h_b/h_d

No.	Case No.	Considered $\ln X_i$s	$\ln b_0$	b_1	b_2	b_{31}	b_{32}	b_{33}	b_{41}	b_{42}	b_{51}	b_{52}	b_{53}	SSE	ΔSSE %	R^2	R^2_{adj}	$S^2_{\ln Y\|\ln X}$
1	66	$\ln X_{1,2,3,4,5}$	−0.032	−0.047	0.002	−0.021	0.016	−0.028	−0.023	−0.010	0.109	0.015	−0.157	0.885	—	0.344	0.225	0.016
2	66	$\ln X_{2,3,4,5}$	−0.045	—	0.009	−0.018	−0.003	−0.023	−0.032	−0.013	0.106	0.015	−0.166	0.939	6.1	0.304	0.192	0.017
3	66	$\ln X_{1,3,4,5}$	−0.030	−0.047	—	−0.020	0.017	−0.027	−0.022	−0.008	0.110	0.016	−0.156	0.885	0.0	0.344	0.239	0.016
4	66	$\ln X_{1,2,4,5}$	−0.046	−0.043	0.003	—	—	—	−0.030	−0.015	0.101	0.012	−0.158	0.893	0.9	0.338	0.258	0.015
5	66	$\ln X_{1,2,3,5}$	−0.046	−0.048	0.004	−0.028	0.015	−0.032	—	—	0.104	0.009	−0.159	0.887	0.2	0.343	0.250	0.016
6	66	$\ln X_{1,2,3,4}$	−0.036	−0.052	0.011	−0.020	−0.014	−0.003	−0.014	−0.022	—	—	—	1.266	43.1	0.061	—	0.022
7	66	$\ln X_{1,3,5}$	−0.039	−0.049	—	−0.026	0.017	−0.030	—	—	0.106	0.011	−0.157	0.887	0.3	0.342	0.263	0.015
8	66	$\ln X_{1,5}$	−0.058	−0.045	—	—	—	—	—	—	0.096	0.006	−0.160	0.896	1.3	0.335	0.292	0.015
9	66	$\ln X_5$	−0.062	—	—	—	—	—	—	—	0.098	0.007	−0.167	0.949	7.3	0.296	0.262	0.015
10	71	$\ln X_{1,5}$	−0.058	−0.048	—	—	—	—	—	—	0.097	0.004	−0.160	0.929	—	0.351	0.312	0.014

6. It is found that five additional failure cases with known values of X_1, X_5, and Y are available (but values of X_2, X_3 or/and X_4 are unknown). These additional cases are mixed with the 66 cases chosen in Step 1. Based on the 71 cases, a better additive regression model considering X_1 and X_5 is obtained with R^2_{adj}=0.314 (see row 10 in Table 8.1), which is taken as the best-simplified prediction model:

$$Y = 0.729 - 0.025X_1 + 0.343X_{51} + 0.257X_{52} + 0.129X_{53} \quad (8.10)$$

with a conditional standard deviation of $S_{Y|X}$=0.103. Using the original dam or reservoir variables, Equation (8.10) is expressed as,

$$\frac{H_b}{H_d} = C_1 - 0.025\left(\frac{H_d}{H_r}\right) \quad (8.11)$$

where C_1 = 1.072, 0.986, and 0.858 for high, medium, and low dam erodibility, respectively. In this case, the best multiplicative functional form based on the results in row 10 in Table 8.2 is

$$\ln Y = -0.058 - 0.048\ln X_1 + 0.097\ln X_{51} + 0.004\ln X_{52} - 0.16\ln X_{53} \quad (8.12)$$

with a conditional standard deviation of $S_{\ln Y|\ln X}$=0.119. The corresponding equation using the original variables is

$$\frac{H_b}{H_d} = C_{11}\left(\frac{H_d}{H_r}\right)^{-0.048} \quad (8.13)$$

where C_{11} = 1.04, 0.947, and 0.804 for high, medium, and low dam erodibility, respectively. This equation is associated with R^2_{adj}=0.312, which is smaller than 0.314 for the best additive regression model.

Using the same analysis procedure, the regression results for all the five breaching parameters can be obtained, as summarized in Table 8.3 for the best prediction models and in Table 8.4 for the best-simplified prediction models. Table 8.5 summarizes the derived nonlinear equations using the original variables for breach top width B_t, average breach width B_{ave}, peak outflow rate Q_p, and failure time T_f. Most of the breaching parameters and the control variables tend to show nonlinear relationships. Therefore, the multiplicative regression model is in general more appropriate than the additive regression model.

The dam erodibility is an important factor influencing all five breaching parameters. Higher dam erodibility allows more rapid erosion development, and hence leads to larger breach depth and width values, a higher peak outflow rate, and a shorter failure time. It is also found that the reservoir shape coefficient, $V_w^{1/3}/H_w$, plays an important role in all prediction models except for breach depth. Generally, two typical reservoir shapes exist: a low dam with a relatively large reservoir capacity (resulting in a large $V_w^{1/3}/H_w$) and a high dam with a relatively small reservoir capacity (resulting in a small $V_w^{1/3}/H_w$). Consider two dams, a lower dam and a higher dam with the same reservoir capacity. The failure of the higher dam generates a larger peak outflow rate due to the higher gravitational potential energy. That is, a higher H_w leads to a lower $V_w^{1/3}/H_w$, V_w being constant, and results in a larger Q_p. In other words, $Q_p / \sqrt{gV_w^{5/3}}$ decreases with increasing $V_w^{1/3}/H_w$. In addition, T_f decreases with increasing Q_p if V_w is constant. Hence, T_f increases with $V_w^{1/3}/H_w$.

Table 8.3 Summary of the best prediction models for the five key dam breaching parameters

No.	Y (or ln$Y$)	Number of cases	b_0 (or lnb_0)	b_1	b_2	b_{31}	b_{32}	b_{33}	b_{41}	b_{42}	b_{51}	b_{52}	b_{53}	R^2	$S^2_{Y\|X}$ (or $S^2_{\ln Y\|\ln X}$)
1	H_b/H_d	66	0.453	−0.025	0.000	0.145	0.176	0.132	0.218	0.236	0.254	0.168	0.031	0.350	0.012
2	$\ln(B_t/H_b)$	54	0.060	0.092	0.508	0.061	0.088	−0.089	0.299	−0.239	0.411	−0.062	−0.289	0.620	0.169
3	$\ln(B_{ave}/H_b)$	45	−0.240	0.133	0.652	−0.041	0.026	−0.226	0.149	−0.389	0.291	−0.140	−0.391	0.648	0.184
4	$\ln(Q_p/\sqrt{gV_w^{5/3}})$	34	−1.744	0.199	−1.274	−0.503	−0.591	−0.649	−0.705	−1.039	−0.007	−0.375	−1.362	0.800	0.365
5	$\ln(T_f/T_r)$	28	−1.190	0.707	1.228	−0.327	−0.674	−0.189	−0.579	−0.611	−1.205	−0.564	0.579	0.793	0.365

Table 8.4 Summary of the best-simplified prediction models for the five key dam breaching parameters

No.	Y (or ln$Y$)	Number of cases	b_0 (or lnb_0)	b_1	b_2	b_{31}	b_{32}	b_{33}	b_{41}	b_{42}	b_{51}	b_{52}	b_{53}	R^2_{adj}	$S^2_{Y\|X}$ (or $S^2_{\ln Y\|\ln X}$)
1	H_b/H_d	71	0.729	−0.025	—	—	—	—	—	—	0.343	0.257	0.129	0.314	0.011
2	$\ln(B_t/H_b)$	61	−0.004	—	0.558	—	—	—	0.258	−0.262	0.377	−0.092	−0.288	0.584	0.159
3	$\ln(B_{ave}/H_b)$	53	1.713	—	0.739	—	—	—	−1.207	−1.747	−0.613	−1.073	−1.268	0.669	0.173
4	$\ln(Q_p/\sqrt{gV_w^{5/3}})$	39	−2.020	—	−1.276	—	—	—	−0.788	−1.232	−0.089	−0.498	−1.433	0.777	0.304
5	$\ln(T_f/T_r)$	30	−1.893	0.654	1.246	—	—	—	—	—	−1.375	−0.828	0.310	0.727	0.288

Table 8.5 Empirical equations for estimating the breaching parameters of constructed embankment dams

Parameters	Equations		Coefficients	No. of cases	R^2
H_b	Full-variable equation	$\frac{H_b}{H_d} = 0.453 - 0.025\left(\frac{H_d}{H_r}\right) + B_1$	$B_1 = b_3 + b_4 + b_5$ b_3=0.145, 0.176, and 0.132 for dams with corewalls, concrete-faced dams, and homogeneous/zoned-fill dams, respectively; b_4=0.218 and 0.236 for overtopping and seepage erosion/piping, respectively; b_5=0.254, 0.168, and 0.031 for high, medium, and low dam erodibility, respectively.	66	0.35
	Simplified equation	$\frac{H_b}{H_d} = C_1 - 0.025\left(\frac{H_d}{H_r}\right)$	C_1=1.072, 0.986, and 0.858 for high, medium, and low dam erodibility, respectively	71	0.31
B_t	Full-variable equation	$\frac{B_t}{H_b} = 1.062\left(\frac{H_d}{H_r}\right)^{0.092}\left(\frac{V_w^{1/3}}{H_w}\right)^{0.508} e^{B_2}$	$B_2 = b_3 + b_4 + b_5$ b_3=0.061, 0.088, and –0.089 for dams with corewalls, concrete faced dams, and homogeneous/zoned-fill dams, respectively; b_4=0.299 and –0.239 for overtopping and seepage erosion/piping, respectively; b_5=0.411, –0.062, and –0.289 for high, medium, and low dam erodibility, respectively;	54	0.62
	Simplified equation	$\frac{B_t}{H_b} = 0.996\left(\frac{V_w^{1/3}}{H_w}\right)^{0.558} e^{C_2}$	$C_2 = b_4 + b_5$ b_4=0.258 and –0.262 for overtopping and seepage erosion/piping, respectively; b_5=0.377, –0.092, and –0.288 for high, medium, and low dam erodibility, respectively	61	0.58
B_{ave}	Full-variable equation	$\frac{B_{ave}}{H_b} = 0.787\left(\frac{H_d}{H_r}\right)^{0.133}\left(\frac{V_w^{1/3}}{H_w}\right)^{0.652} e^{B_3}$	$B_3 = b_3 + b_4 + b_5$ b_3=–0.041, 0.026, and –0.226 for dams with corewalls, concrete faced dams, and homogeneous/zoned-fill dams, respectively; b_4=0.149 and –0.389 for overtopping and seepage erosion/piping, respectively b_5=0.291, –0.14, and –0.391 for high, medium, and low dam erodibility, respectively	45	0.65
	Simplified equation	$\frac{B_{ave}}{H_b} = 5.543\left(\frac{V_w^{1/3}}{H_w}\right)^{0.739} e^{C_3}$	$C_3 = b_4 + b_5$ b_4=–1.207 and –1.747 for overtopping and seepage erosion/piping, respectively; b_5=–0.613, –1.073, and –1.268 for high, medium, and low dam erodibility, respectively	53	0.67

(Continued)

Table 8.5 (*Continued*)

Parameters	Equations		Coefficients	No. of cases	R^2
Q_p	Full-variable equation	$\frac{Q_p}{\sqrt{gV_w^{5/3}}}=0.175\left(\frac{H_d}{H_r}\right)^{0.199}\left(\frac{V_w^{1/3}}{H_w}\right)^{-1.274}e^{B_4}$	$B_2=b_3+b_4+b_5$ b_3=−0.503, −0.591, and −0.649 for dams with corewalls, concrete faced dams, and homogeneous/zoned-fill dams, respectively; b_4=−0.705 and −1.039 for overtopping and seepage erosion/piping, respectively; b_5=−0.007, −0.375, and −1.362 for high, medium, and low dam erodibility, respectively	34	0.80
	Simplified equation	$\frac{Q_p}{\sqrt{gV_w^{5/3}}}=0.133\left(\frac{V_w^{1/3}}{H_w}\right)^{-1.276}e^{C_4}$	$C_4=b_4+b_5$ b_4=−0.788 and −1.232 for overtopping and seepage erosion/piping, respectively; b_5=−0.089, −0.498, and −1.433 for high, medium, and low dam erodibility, respectively	39	0.78
T_f	Full-variable equation	$\frac{T_f}{T_r}=0.304\left(\frac{H_d}{H_r}\right)^{0.707}\left(\frac{V_w^{1/3}}{H_w}\right)^{1.228}e^{B_5}$	$B_2=b_3+b_4+b_5$ b_3=−0.327, −0.674, and −0.189 for dams with corewalls, concrete faced dams, and homogeneous/zoned-fill dams, respectively; b_4=−0.579 and −0.611 for overtopping and seepage erosion/piping, respectively; b_5=−1.205, −0.564, and 0.579 for high, medium, and low dam erodibility, respectively	28	0.79
	Simplified equation	$\frac{T_f}{T_r}=C_5\left(\frac{H_d}{H_r}\right)^{0.654}\left(\frac{V_w^{1/3}}{H_w}\right)^{1.246}$	C_5=0.038, 0.066, and 0.205 for high, medium, and low dam erodibility, respectively.	30	0.73

Note: *No records are available for the low erodibility coefficient cases.

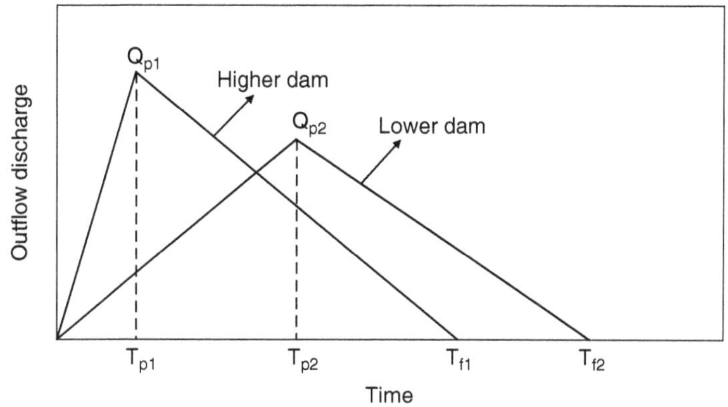

Figure 8.5 Sketch of outflow hydrographs resulting from the failures of a higher dam and a lower dam with the same reservoir capacity. *Source:* Xu and Zhang (2009). With permission from ASCE

Dewey and Gillette (1993) made a statement that the peak outflow may occur before the breach width is at its maximum, particularly for dams that are high relative to the reservoir area. Indeed, the peak outflow associated with a higher dam tends to occur earlier than that associated with a lower dam. Suppose T_p is the duration from the start of breaching to the peak outflow. The lower dam has a smaller Q_p, a larger T_p, and a larger ratio of T_p/T_f. Figure 8.5 shows a sketch of the different outflow hydrographs resulting from the failure of the two dams. The lower dam, with a higher value of $V_w^{1/3}/H_w$, allows a longer time of erosion, especially lateral erosion, resulting in a relatively larger breach width. Accordingly, the breach width increases with the reservoir shape coefficient. This is consistent with the observation by Wahl (1998) that for large reservoirs impounded by relatively small dams, the reservoir level does not change significantly during the early stages of the embankment breach, and the lateral erosion can take place while the reservoir level remains high.

Another distinct feature is that overtopping appears to cause wider breaches and larger peak outflow rates than does seepage erosion/piping. Dewey and Gillette (1993) provided two possible reasons: (1) overtopping initiates a breach over a length of the dam rather than at a localized point; (2) overtopping is often accompanied by continued inflow floods. In addition, in a seepage/piping failure, some water is released via a formed pipe through the dam, and therefore has no contribution to the later overtopping erosion after the collapse of the pipe.

The dam height has a minor influence on H_b/H_d based on Equation (8.11). A low dam tends to generate a relatively low discharge. Periods of relatively low discharge associated with non-aerated nappes can produce the most dramatic headcutting (Wahl, 1998), resulting in more complete breach height development for low dams. Hence, H_b/H_d decreases with increasing dam height. In contrast, the dam height does not affect the breach width significantly. This is because the breach width is mainly controlled by the duration of lateral erosion that is dependent of the reservoir shape coefficient, as mentioned earlier.

8.3.2.2 Comparison with Existing Prediction Equations

The Bureau of Reclamation (1982, 1988) proposed a set of formulas for B_{ave}, Q_p, and T_f:

$$B_{ave} = 3H_w \tag{8.14}$$

$$Q_p = 19.1 H_w^{1.85} \tag{8.15}$$

$$T_f = 0.011 B_{ave} \tag{8.16}$$

and Froehlich (1995a, b) proposed another set of empirical equations:

$$B_{ave} = 0.1803 K_0 V_w^{0.32} H_b^{0.19}, K_0 = 1.4 \text{ for overtopping, and 1.0 otherwise} \tag{8.17}$$

$$Q_p = 0.607 V_w^{0.295} H_w^{1.24} \tag{8.18}$$

$$T_f = 0.00254 V_w^{0.53} H_b^{-0.9} \tag{8.19}$$

The dataset of dam failures in Chapter 2 and Table 2.1 is utilized to evaluate model uncertainty. The model uncertainty is represented by a bias factor, λ, which is the ratio of measured value y_m to predicted value y_p. Table 8.6 summarizes the comparison of four sets of prediction models: the best prediction models and the best-simplified models derived in this section in Table 8.5, the Bureau of Reclamation models, and the Froehlich models. In general, the empirical formulas for predicting the breach geometry parameters have better performance than do those for predicting the peak outflow rate and the failure time. This is partially due to the relatively small measurement errors with the breach geometry and large errors with the estimates of peak flow rate and failure time. In addition, the best prediction models of Xu and Zhang (2009) are associated with bias factors closest to unity and smallest standard deviations. The dropping of dam type as a control variable only leads to slight increases in standard deviation. The Bureau of Reclamation equation (bias = 3.08) and the Froehlich equation (bias = 2.3) tend to severely underpredict the failure time.

It should be recognized that, in comparing the four sets of models, inherent advantages might exist for the models developed in this chapter. However, some case data in Table 2.1 was also used in deriving the Bureau of Reclamation equations and the Froehlich equations. This may be a fair basis for the comparison of these different prediction models.

8.3.2.3 Case Studies

Two case studies are presented to illustrate the application of the empirical prediction models in Table 8.5. One is the overtopping failure of Banqiao Dam, which occurred on 8 August 1975 in China. The other is the seepage erosion/piping failure of Teton Dam, which occurred on 5 June 1976 in the USA.

Failure of Banqiao Dam (HWRA 2005)

Banqiao Dam was built in the early 1950s on Ru River, as part of a project for flood control for the Huai River basin. It was located in Henan Province, the central part of China, which is about 700 km south-west of Beijing. In 1953, the clay-core earthfill dam was built with a height of 21.5 m and a storage capacity of $244 \times 10^6\,m^3$. Shortly after the first filling of the reservoir, excessive settlement of the dam was observed, resulting in a large number of cracks. This was due to the poor compaction of the materials in the dam. In 1956, the dam was strengthened and raised. A trench with dimensions of 1300 m × 30 m × 29.5 m was excavated on the dam and backfilled with compacted materials. The rebuilt dam was 24.5 m high and had a storage capacity of $492 \times 10^6\,m^3$ at the normal water level, with $375 \times 10^6\,m^3$ of it reserved for flood control. The dam crest was 6 m wide and 2020 m long. On 4–8 August 1975, an extreme storm occurred in Henan Province; the maximum 5-day rainfall

Table 8.6 Comparison of model biases for different prediction equations based on the failure cases in Table 8.1

Parameter	Full-variable equations in Table 8.5			Simplified equations in Table 8.5			Bureau of Reclamation (1982, 1988)			Froehlich (1995a, b)		
	Number of cases[a]	Mean	St. Dev.	Number of cases[a]	Mean	Sta. Dev.	Number of cases[a]	Mean	St. Dev.	Number of cases[a]	Mean	St. Dev.
H_b	62	1.01	0.08	66	1.01	0.08	—	—	—	—	—	—
B_t	52	1.03	0.32	59	1.03	0.32	—	—	—	—	—	—
B_{ave}	43	1.02	0.32	52	1.03	0.33	51	1.31	0.94	51	1.07	0.43
Q_p	33	1.01	0.34	38	1.03	0.39	36	0.95	0.80	36	1.30	0.84
T_f	27	1.02	0.43	28	1.02	0.44	27	3.08	4.00	27	2.30	2.05

[a] The number of cases after using the objective outlier-exclusion algorithm (Rousseeuw, 1998) at the 95% probability level.

Table 8.7 Predictions of breaching parameters for Banqiao dam

Breaching parameter	Observed value	Full-variable equations		Simplified equations	
		Mean prediction	95% prediction interval	Mean prediction	95% prediction interval
Breach depth (m)	29.5	25.2	19.9–30.6	25.3	20.2–30.3
Breach top width (m)	372	325	143–739	301	135–668
Breach average width (m)	291	273	116–643	262	114–601
Peak outflow (m^3/s)	78100	55268	16514–184971	52982	17566–159804
Failure time (hour)	5.5	3.0	0.9–10.1	3.2	1.1–9.5

reached a record of 1631 mm. Banqiao Dam failed from overtopping in the storm event. The measured values of the five key dam breaching parameters are presented in Table 8.7.

The prediction models in Table 8.5 are applied to estimate the five breaching parameters, including the mean value and the prediction interval at a significance level of 95%. Take the calculation of breach depth H_b using the best-simplified model as an example. The mean of H_b/H_d can be calculated given the value of H_d/H_r and the dam erodibility (Case No. 3 in Table 2.1). Based on the mean of H_b/H_d and the known conditional standard deviation of 0.103, the prediction interval of H_b/H_d can be obtained. To estimate the breach top width and the average breach width, the mean of H_b is used without considering its uncertainty. The predictions for the five breaching parameters are summarized in Table 8.7. The mean predictions indicate that the best prediction models give in general slightly better results than the best-simplified models do. The differences between the measurements and the mean predictions in Table 8.7 can be explained by two facts. First, the material of the Banqiao Dam was probably very erodible (very high erodibility) due to the poor compaction by foot traffic of the laborers who built the dam. Secondly, a volume of flood as much as $238.5 \times 10^6 m^3$ flowed into the reservoir after the initiation of overtopping, which was 40% of the original V_w value of $607.5 \times 10^6 m^3$. The poor compaction and the continued inflow floods had a major influence on the breaching parameters, which increased the breach size and the peak outflow rate. The poor compaction led to a decreased failure time while the continued inflow floods resulted in an increased failure time. In this case, the large inflow floods appear to have a greater influence as the failure time is under-predicted. All the measured values are within the 95% prediction intervals.

Failure of Teton Dam (USCOLD 1988)

Teton Dam was built on the Teton River in south-eastern Idaho in the USA. The dam was designed as a zoned earth dam with a maximum height of 93 m and a crest length of 945 m. It had a storage capacity of $355.6 \times 10^6 m^3$. The dam failed on 5 June 1976 during the initial filling of the reservoir. The failure started with an observed leak at about 7:45 a.m., near the toe of the dam on the right abutment. At about 10:30 a.m., loud noises were heard, followed by the sound of running water. From that moment, the outflow increased very rapidly as did the erosion of dam materials, which progressed to the breach of the dam at 11:57 a.m. (Figure 8.2). The measured values of the five key dam breaching parameters are shown in Table 8.8. The predictions for the five breaching parameters are also summarized in Table 8.8. All the measured values are within the 95% prediction intervals and the prediction errors are smaller than 20%.

Based on the statistics of the model biases, the equations for predicting the breach geometric parameters appear to be more accurate than those for predicting the peak outflow rate and the failure time. This is mainly because the measurements of dam-breach geometry in the field are more reliable.

Table 8.8 Predictions of breaching parameters for Teton dam

Breaching parameter	Observed value	Full-variable equations		Simplified equations	
		Mean prediction	95% prediction interval	Mean prediction	95% prediction interval
Breach depth (m)	87	78	57–98	77	58–96
Breach top width (m)	238	198	87–451	181	82–402
Breach average width (m)	151	150	64–354	127	55–291
Peak outflow (m^3/s)	65,120	75,220	22,475–25,1745	55,146	18,283–166,333
Failure time (hour)	4.0	4.1	1.2–13.6	3.2	1.1–9.5

Beware that individual cases should be expected to deviate from the mean model predictions. In addition, the prediction models for breaching parameters may not be used in non-evaluated conditions, such as dams with extremely large heights or with the sliding failure mode. Therefore, prediction models should be used with caution in a specific case study.

8.3.3 Empirical Breaching Parameters for Landslide Dams

8.3.3.1 Breaching Parameters and Influence Factors

A landslide dam, comparing to a man-made earth or rockfill dam, is usually in an irregular shape. In this chapter, the geometric parameters of a landslide dam are defined in Figure 8.6 following Costa and Schuster (1991):

1. *dam height* (H_d). the vertical altitude difference from the valley floor to the lowest point on the landslide dam (point of overflow in event of overtopping)
2. *dam length* (L_d). the crest length of the landslide dam measured perpendicular to the major valley axis
3. *dam width* (W_d). the base width of the landslide dam measured parallel to the main valley axis
4. *dam volume* (V_d). the part of landslide volume which blocks the river
5. *lake volume* (V_l). the volume of water ponded behind the landslide dam (usually the maximum volume).

The breaching parameters of landslide dams differ slightly from those of constructed embankment dams and are defined in Figure 8.7 as follows:

1. *breach depth* (H_b). the vertical distance from the lowest bottom of the breach to the original lowest point on the landslide dam crest
2. *breach top width* (W_t). the width of the breach on the crest of the landslide dam
3. *breach bottom width* (W_b). the width at the bottom of the breach
4. *peak outflow rate* (Q_p). the maximum water discharge rate during dam breaching
5. *breaching duration (time)* (T_b). the period from the inception to the completion of the breaching process (Singh and Snorrason, 1984).

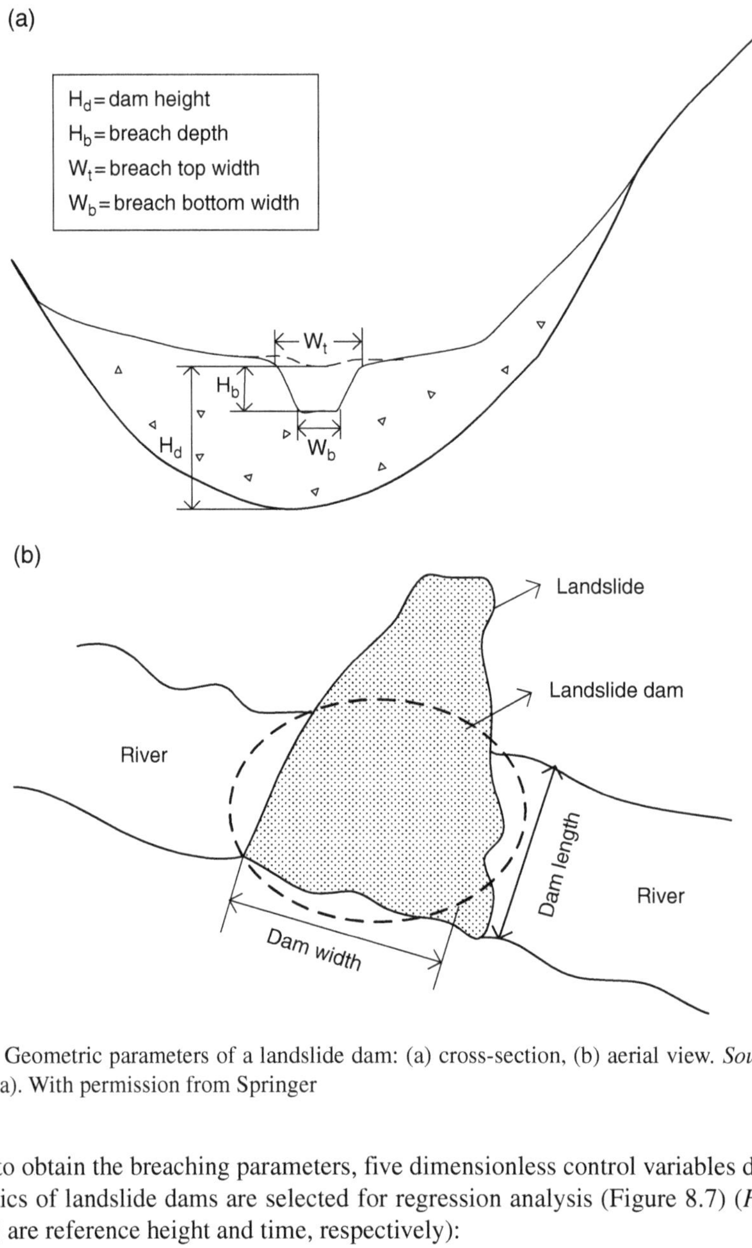

Figure 8.6 Geometric parameters of a landslide dam: (a) cross-section, (b) aerial view. *Source:* Peng and Zhang (2012a). With permission from Springer

In order to obtain the breaching parameters, five dimensionless control variables describing the characteristics of landslide dams are selected for regression analysis (Figure 8.7) (H_r [=1 m] and T_r [=1 hour] are reference height and time, respectively):

1. *dam height factor* (H_d/H_r), which describes the maximum water head or the potential water energy
2. *ratio of dam height to dam width* (H_d/W_d), which influences the slope of erosion and controls the velocity of water flow and the erosion rate
3. *dam shape coefficient* ($V_d^{1/3}/H_d$), which denotes the amount of rock and soil materials that may be taken away, and influences the breaching duration and the outflow rate
4. *lake shape coefficient* ($V_l^{1/3}/H_d$), which indicates the potential volume of water that may erode the dam and influences the breach size, the breaching duration, and the outflow rate

Breaching parameters	
Peak outflow rate	$Y_1 = (Q_p^2/g)^{1/5}/H_r$
Breach depth	$Y_2 = H_b/H_r$
Breach top width	$Y_3 = W_t/H_r$
Breach bottom width	$Y_4 = W_b/H_d$
Breaching duration	$Y_5 = T_b/T_r$

Control variables			
Dam height	$X_1 = H_d/H_r$		
Ratio of height and width	$X_2 = H_d/W_d$		
Dam shape coefficent	$X_3 = V_d^{1/3}/H_d$		
Lake shape coefficent	$X_4 = V_l^{1/3}/H_d$		
Dam erodibility	X_{51}	X_{52}	X_{53}
High erodibility	$1^a(e^b)$	0(1)	0(1)
Medium erodibility	0(1)	1(*e*)	0(1)
Low erodibility	0(1)	0(1)	1(*e*)

Note: [a]Values for additive regression analysis;
[b]Values for multiplicative regression analysis.
$H_r = 1$ m; $T_r = 1$ hour

Figure 8.7 Summary of five breaching parameters and five control variables for landslide dams. *Source:* Peng and Zhang (2012a). With permission from Springer

5. *dam erodibility,* which indicates the resistance of dam material to the erosive action of water flow. Erodibility is treated as a discrete variable with three values: high, medium, and low erodibility. Higher erodibility corresponds to a larger breach size and peak discharge but a shorter breaching duration (Knapen et al., 2007; Briaud, 2008; Chang et al., 2011). The judgment on dam erodibility may be subjective when the available information is limited.

8.3.3.2 Regression Equations for Estimating Breaching Parameters of Landslide Dams

The same procedure of regression analysis as in Section 8.3.1 is adopted to analyze the breaching parameters of landslide dams. Five breaching parameters shown in Figure 8.7 are analyzed based on different combinations of control variables. A group of equations with all the five control variables and a group of simplified equations with three control variables are obtained, which are presented in Table 8.9.

After comparing the importance of the five variables in terms of values of R^2 and ΔSSE as conducted for the peak flow rate, the significance of the control variables are ranked in Table 8.10. The most important control variables for each breaching parameter are different, but dam height factor X_1, lake shape coefficient X_4, and dam erodibility X_5 are generally more important than ratio of dam height to dam width X_2, and dam shape coefficient X_3. Important variables for each individual breaching parameter are summarized as follows.

1. *Peak outflow rate, Q_p.* The influences of dam height, dam erodibility, and lake volume are more substantial than those of dam volume and dam width. Dam height denotes the water energy to be released, dam erodibility denotes the breaching rate, and lake volume denotes both the water energy and the amount of water that may erode the dam.
2. *Breach depth, H_b.* The dam height is more important than the other four variables. H_bs are approximately proportional to dam height with an exponent close to 1.0.
3. *Breach top width, W_t.* The dam height and lake volume are much more important than the other three variables. W_t is not simply proportional to dam height or breach depth. This is because the breach can develop laterally after the breach reaches the final depth. Ceasing of vertical erosion can be caused by the presence of low erodibility soils or large rocks in deep layers.

Table 8.9 Empirical equations for estimating the breaching parameters of landslide dams

Parameters	Equations		Erodibility coefficient a		Num. of cases	R^2
Q_p	Full-variable equation	$\frac{Q_p}{g^{1/2}H_d^{5/2}}=\left(\frac{H_d}{H_r}\right)^{-1.417}\left(\frac{H_d}{W_d}\right)^{-0.265}\left(\frac{V_d^{1/3}}{H_d}\right)^{-0.471}\left(\frac{V_l^{1/3}}{H_d}\right)^{1.569}e^a$	High	1.276	45	0.946
			Med.	−0.336		
			Low	−1.532		
	Simplified equation	$\frac{Q_p}{g^{1/2}H_d^{5/2}}=\left(\frac{H_d}{H_r}\right)^{-1.371}\left(\frac{V_l^{1/3}}{H_d}\right)^{1.536}e^a$	High	1.236	45	0.939
			Med.	−0.380		
			Low	−1.615		
H_b	Full-variable equation	$\frac{H_b}{H_r}=\left(\frac{H_d}{H_r}\right)^{0.882}\left(\frac{H_d}{W_d}\right)^{-0.041}\left(\frac{V_d^{1/3}}{H_d}\right)^{-0.099}\left(\frac{V_l^{1/3}}{H_d}\right)^{0.139}e^a$	High	−0.316	21	0.871
			Med.	−0.520		
			Low	—*		
	Simplified equation	$\frac{H_b}{H_r}=\left(\frac{H_d}{H_r}\right)^{0.923}\left(\frac{V_l^{1/3}}{H_d}\right)^{0.118}e^a$	High	−0.500	21	0.870
			Med.	−0.673		
			Low	—*		
W_t	Full-variable equation	$\frac{W_t}{H_r}=\left(\frac{H_d}{H_r}\right)^{0.752}\left(\frac{H_d}{W_d}\right)^{0.315}\left(\frac{V_d^{1/3}}{H_d}\right)^{-0.243}\left(\frac{V_l^{1/3}}{H_d}\right)^{0.682}e^a$	High	1.683	13	0.855
			Med.	1.201		
			Low	—*		
	Simplified equation	$\frac{W_t}{H_r}=\left(\frac{H_d}{H_r}\right)^{0.911}\left(\frac{V_l^{1/3}}{H_d}\right)^{0.271}e^a$	High	0.588	13	0.780
			Med.	0.148		
			Low	—*		
W_b	Full-variable equation	$\frac{W_b}{H_d}=0.004\left(\frac{H_d}{H_r}\right)+0.050\left(\frac{H_d}{W_d}\right)-0.044\left(\frac{V_d^{1/3}}{H_d}\right)+0.088\left(\frac{V_l^{1/3}}{H_d}\right)+a$	High	0.775	10	0.841
			Med.	0.532		
			Low	—*		
	Simplified equation	$\frac{W_b}{H_d}=0.003\left(\frac{H_d}{H_r}\right)+0.070\left(\frac{V_l^{1/3}}{H_d}\right)+a$	High	0.624	10	0.822
			Med.	0.344		
			Low	—*		
T_b	Full-variable equation	$\frac{T_b}{T_r}=\left(\frac{H_d}{H_c}\right)^{0.262}\left(\frac{H_d}{W_d}\right)^{-0.024}\left(\frac{V_d^{1/3}}{H_d}\right)^{-0.103}\left(\frac{V_l^{1/3}}{H_d}\right)^{0.705}e^a$	High	−0.635	14	0.624
			Med.	−0.518		
			Low	—*		
	Simplified equation	$\frac{T_b}{T_r}=\left(\frac{H_d}{H_c}\right)^{0.293}\left(\frac{V_l^{1/3}}{H_d}\right)^{0.723}e^a$	High	−0.805	14	0.621
			Med.	−0.674		
			Low	—*		

Table 8.10 Order of importance of the control variables

Parameter	Order of importance	Most important factors
Q_p	$X_1 > X_5 > X_4 > X_3 \approx X_2$	X_1, X_5, X_4
H_b	$X_1 > X_5 \approx X_4 \approx X_3 \approx X_2$	X_1
W_t	$X_1 > X_4 > X_2 \approx X_5 > X_3$	X_1, X_4
W_b	$X_4 > X_3 > X_1 > X_5 > X_2$	X_4, X_3
T_b	$X_4 > X_1 \approx X_2 \approx X_3 \approx X_5$	X_4

4. *Breaching duration, T_b*. The lake volume is the most important control variable because the released water volume often controls the run-out time. The regression results for T_b are not as good as other parameters in terms of R^2. The reason may be that the breaching durations in the historical cases were recorded in different ways. In some cases, the T_b values were recorded from the start of breaching, while in other cases, the T_b values might be registered from a sudden increase in breach size.
5. *Breach bottom width, W_b*. The lake volume and dam volume are slightly more important, but the influence of the other three variables cannot be ignored. Similar to the analysis of W_t, W_b is not proportional to dam height or breach depth.

8.3.3.3 Comparison of Models for Landslide Dams

The accuracy of a particular prediction model is represented here by a bias factor, which is defined as the ratio of the measured value and the value estimated using the prediction model. Table 8.11 shows statistics of the bias factors for several empirical models for estimating the peak outflow rate from breaching of landslide dams. The models include Costa's model (1985) based on 10 failure cases, Costa and Schuster's model (1988) based on 12 cases, Walder and O'Connor (1997) based on 18 cases, and Peng and Zhang's model (2012a) in Table 8.9 based on 45 cases (with peak outflow rate data). The following observations can be made from the results in Table 8.11:

1. The lake volume (V_l) or released water volume (V_0) is more effective than dam height (H_d) or drop of water level (d) for use as a parameter in one-variable models.
2. The models with combined parameters – product of dam height and lake volume (or PE) or product of drop in water level and released water volume – are not necessarily more accurate than the one-variable (lake volume or released water volume) models.

Figure 8.8 compares the peak outflow rates estimated using three empirical models. The predictions are made by using Costa's equation [$Q_p = 181(H_d V_l)^{0.43}$], Walder and O'Connor's equation [$Q_p = 0.99(dV_0)^{1.59}$], and the full-variable model in Table 8.9. The dashed line (measured value = simulated value) is for comparison. The predictions from the full-variable models of Peng and Zhang (2012) are less scattered than the other two models. First, the models are developed with recent data with more accurate and detailed information. Second, in addition to dam height

Table 8.11 Comparison of the bias factors for the empirical models for predicting peak flow rate from failure of landslide dams

Authors	Costa (1985)			Costa and Schuster (1988)	Walder and O'Connor (1997)			Models in Table 8.9	
Equations*	$Q_p=6.3\ H_d^{1.59}$	$Q_p=672\ V_1^{0.56}$	$Q_p=181\ (H_dV_1)^{0.43}$	$Q_p=0.0158\ (PE)^{0.41}$	$Q_p=6.7\ d^{1.73}$	$Q_p=1.60\ V_0^{0.46}$	$Q_p=0.99\ (dV_0)^{0.40}$	Full-variable equations	Simplified equations
Number of cases	47	47	47	47	14	14	14	45	45
Mean bias factor	1.96	1.51	1.42	1.43	2.55	1.40	1.42	1.14	1.16
Standard deviation	4.72	1.87	1.77	1.70	4.10	1.13	1.08	0.55	0.60

Note: *H_d = dam height, V_1 = lake volume, PE (potential energy) = product of dam height, lake volume and specific weight of water, d = drop in water level, V_0 = released water volume.

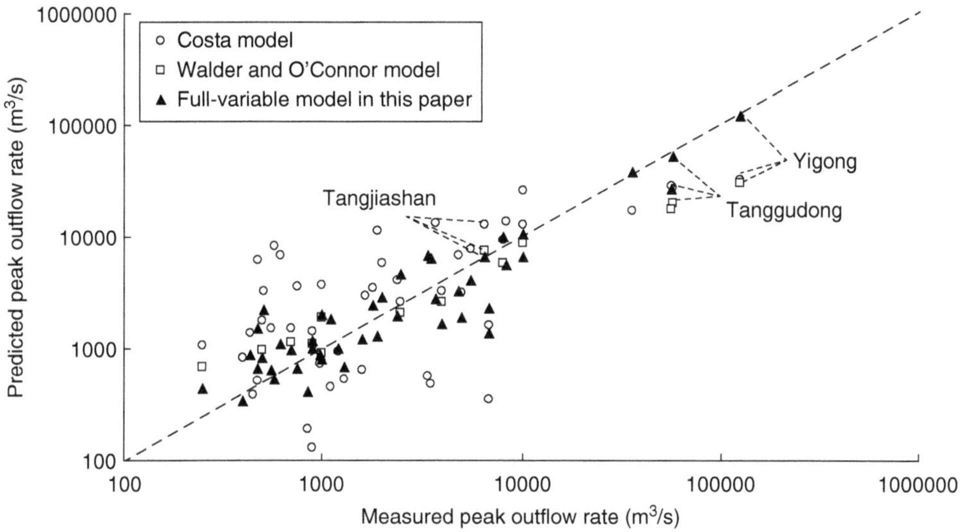

Figure 8.8 Comparison of three breaching models for peak outflow rates from failure of landslide dams. *Source:* Peng and Zhang (2012a). With permission from Springer

(or drop in water level) and lake volume (or released water volume), three additional variables (i.e. dam erodibility, dam volume, and dam width) are also included in the models, which influence the peak outflow rate to a certain extend.

8.3.3.4 Case Studies

Tangjiashan Landslide Dam (Liu et al., 2009a,b; Hu et al., 2009; Peng and Zhang, 2010)

Predictions of the five breaching parameters are made for the Tangjiashan landslide dam (Section 3.3.4), with both the models for landslide dams shown in Table 8.9 and those for constructed earth and rockfill dams in Table 8.5. The mean values of the predicted breach parameters and the prediction intervals at a significance level of 95% are presented in Table 8.12, together with the corresponding measured values. All the predicted values are within the 95% intervals. The last two predictions – breach bottom width and failure time – are not as good as the first three. This is because the bottom layer of the Tangjiashan landslide dam consists of weakly weathered cataclasite, which is less erodible and extends the breaching time. The constraint in the vertical erosion by the low erodibility layer also caused more erosion in the lateral direction, leading to a relatively small breach depth but a large bottom width. The predicted results with the models for earth and rockfill dams significantly overestimate the peak outflow rate and the breach size, but underestimate the breaching duration.

Mantaro River Landslide Dam (Lee and Duncan, 1975)

On 25 April 1974, a landslide with a volume of $1.3\times10^9\,m^3$ blocked the Mantaro River in Peru and formed a lake which reached a volume of $670\times10^6\,m^3$. The landslide dam had a height of 175 m, a width of 3800 m, and a soil volume of $1.3\times10^9\,m^3$. The dam material was a well-graded heterogeneous mixture of clay, silt, sand, gravel, and boulders. The material was quite impervious. No seepage was found from the downstream toe during its 42-day life span. Based on the

Table 8.12 Predictions of breaching parameters of Tangjiashan landslide dam

Breaching parameter	Measured values	Models for landslide dams in Table 8.9				Models for constructed embankment dams in Table 8.5			
		Full-variable		Simplified		Full-variable		Simplified	
		Mean	95% interval	Mean	95% interval	Mean	95% interval	Mean	95% interval
Peak outflow rate (m^3/s)	6500	6237	2149–18,101	7043	2325–21,338	23427	7000–78,406	19661	6518–59,301
Breach depth (m)	42	37.1	18.0–76.5	37.9	18.3–78.5	68.4	51.6–85.2	69.6	52.8–86.5
Breach top width (m)	145–235	165.1	107.3–238.8	123.3	60.1–252.8	263.4	115.8–599.6	248.6	112.0–551.8
Breach bottom width (m)	80–100	59.2	20.2–173.8	51.6	18.5–143.7	177.5	71.6–438.3	206.7	86.1–494.7
Breaching duration (hour)	14	8.0	3.5–16.7	8.1	3.7–17.6	5.9	1.8–19.8	7.4	2.5–21.8

Table 8.13 Predictions of breaching parameters of Mantaro River landslide dam

Breaching parameter	Measured values	Models for landslide dams in Table 8.9				Models for constructed embankment dams in Table 8.5			
		Full-variable		Simplified		Full-variable		Simplified	
		Mean	95% interval	Mean	95% interval	Mean	95% interval	Mean	95% interval
Peak outflow (m^3/s)	10,000	10323	3634–29,330	8951	2955–27,117	215153	64287–720,072	158356	52,502–477,636
Breach depth (m)	107	66.8	32.4–138.1	67.6	32.7–139.8	112.7	77.6–147.8	11.5.1	80.0–150.2
Breach top width (m)	243	196.5	131.7–293.1	209.7	102.2–430.1	453.4	199.3–1031.5	384.9	173.4–854.2
Breach bottom width (m)	30	31.7	10.8–92.9	38.1	13.7–106.1	232.6	92.2–582.7	280.2	116.0–674.3
Breaching duration (hour)	12	7.7	3.5–16.7	7.7	3.5–16.7	3.2	0.9–10.6	3.8	1.3–11.1

description, the soils are believed to be of medium erodibility. The dam breached on 6 June 1974. A volume of $500 \times 10^6\,m^3$ of water was released, leaving a breach 107 m in depth, 243 m in top width, and 30 m in bottom width. The peak outflow rate, $10{,}000\,m^3/s$, occurred on the afternoon of 8 June 1974.

The prediction results (Table 8.13) using both the full-variable models and the simplified models are generally within the 95% prediction intervals. The results from the best prediction models seem closer to the measured values than those from the simplified prediction models. The estimated results with Xu and Zhang's models for earth and rockfill dams in this case are not acceptable, particularly the peak outflow rate is grossly overestimated.

8.3.4 Empirical Breaching Parameters for Dikes

8.3.4.1 Breaching Parameters and Control Variables

Dike breaches are typically long compared with their depths, and the slopes of a dike breach are usually close to vertical. Therefore the breach can be approximated as a rectangular channel. In this section, the dike breaching parameters are taken as breaching length, breaching depth, and peak discharge. Information about peak discharge is limited in the database; the development time is available only for a few cases in a quantitative form.

A regression analysis procedure similar to that described in Section 8.3.2 is applied to derived empirical equations for predicting the breaching parameters of dikes based on the dike breaching data in Chapter 5. The control variables for dikes are selected via a study of the physical breaching process and availability of data. Five control variables are selected for regression analysis: dike height and dike width as continuous variables, and failure mechanism, dike type, and dike material as indicator variables. When an indicator variable is used in the logarithmic form, its value equals either e or 1. Since the structural failure mechanism implies a composite dike, either failure mechanism or dike type can be included in the analysis. The regression analyses involve four control variables at most. Note that the important variable for constructed or landslide dams, reservoir volume, is not applicable to dikes.

8.3.4.2 Regression Equations for Estimating Breaching Parameters of Dikes

Several empirical equations for dikes are presented in Table 8.14. Those based on the dike failure database in Appendix C are referred to Danka and Zhang (2015). The empirical equations of Verheij (2002) and Nagy (2006) are also included in this table.

The breaching length is available in 504 dike failure cases. However, not all these cases can be utilized in the regression analysis due to incomplete information of the control variables. Breaches larger than 1000 m or shorter than 10 m are considered as outliers. The longest breaches in the database were developed due to overtopping; most likely these dikes were not designed for the observed floodwater levels, and the dike system was overtopped extensively. In the case of the short breaches (breaching length shorter than 10 m), the breach was close to a failed structure (gate, spillway, etc.), which caused the wash-away of a small part of a dike. Sensitivity analysis shows that the importance of the control variables follows the order of $w > H_d > t > m$.

Information on breach depth is available in 155 cases. A model including dike height and failure mechanism is preferred, as it produces a similar error to that when more variables are involved. The models are recommended for calculating the peak flow rate. Sensitivity analysis indicates that the order of the relevance of the control variables is $m > t > H_d > w > f$.

Table 8.14 Empirical equations for estimating breaching parameters of dikes

Parameter	Equation	Values of indicator variables	Reference
Breach length	$L = 0.08\frac{w^{3.06}e^{m}}{H_d^{2.11}e^{t}}$	*m*: 0.38 for coarse-grained soils, 0.42 for fine-grained soils, and 0.35 for organic soils; *t*: 0.94 for composite dikes and 0 for earthen dikes.	Danka and Zhang (2015)
Breach length	$L = 67\log\left(\frac{T_b}{522}\right)$ (coarse grained – 'sandy')		Verheij (2002)
Breach length	$L = 20\log\left(\frac{T_b}{288}\right)$ (fine grained – 'clayey')		Verheij (2002)
Breach length	$L = 5.17e^{0.75H_d}$		Nagy (2006)
Beach width	$D = 0.91H_d^{1.02}e^{f}$	*f*: 0.74 for overtopping; 1.21 for piping; 0.81 for slope failure or horizontal sliding, and 1.15 for other failure mechanisms.	Danka and Zhang (2015)
Peal flow rate	$Q_p = 2.55\frac{H_d^{1.14}w^{0.58}e^{m}}{e^{t}}$	*m*: 2.84 for coarse-grained soils, 2.60 for fine-grained soils, and 1.47 for organic soils; *t*: 0.93 for composite dikes, 0 for earthen dikes.	Danka and Zhang (2015)
Peal flow rate (Simplified)	$Q_p = 4.15\frac{H_d^{1.67}e^{m}}{e^{t}}$	*m*: 3.47, 3.20 and 2.01 for coarse-grained soils, fine-grained soils, and soils with high organic content, respectively; *t*: 1.0 for composite dikes, and 0 for earthen dikes.	Danka and Zhang (2015)

Notes: H_d=Dike height; *w*=dike width, *t*=dike type, *m*=Material type, *f*=Failure mechanisms, T_f=breaching time.

8.3.4.3 Calibration of Empirical Models for Dikes

The empirical equations of Danka and Zhang (2015) and Nagy (2006) in Table 8.14 are calibrated against the cases in Appendix C. Models for man-made earth dams are sometimes used for the estimation of dike breaching parameters due to the lack of specific empirical formulas for dikes. For comparison, the equations of the Bureau of Reclamation (1982, 1988), Equations 8.14 and 8.15, are also included in the calibration for comparison purposes. Table 8.15 presents the bias factor and its standard deviation associated with each of the prediction models for dikes and the two Bureau of Reclamation equations for constructed dams. The mean bias factor denotes the mean value of the ratio of observed value to predicted value. The numbers of cases in Table 8.15 refer to those used during the regression analysis to determine the regression coefficients for these empirical equations. The bias factors of the Danka and Zhang models range between 0.95 and 1.02. However, the two empirical equations for constructed dams are associated with the bias factors of 12.53 and 4.46, respectively when applied to dikes, which vastly underestimate the final breaching length and peak discharge. Hence the empirical models for constructed dams should not be applied to dikes.

Table 8.15 Comparison of bias factors for prediction models for dikes

Parameters	Number of cases	Danka and Zhang (2015), Table 8.15		Nagy (2006), Table 8.15		Bureau of Reclamation's equations for constructed dams, Eqs. (8.14) and (8.15)	
		Mean	St. Dev.	Mean	St. Dev.	Mean	St. Dev.
Breach length, L	111	1.02	0.69	2.10	4.62	12.53	31.90
Breach width, D	113	0.96	0.53			—	—
Peak discharge, Q_p	90	0.95 0.99*	0.65 0.80*		—	4.46	4.62

*Simplified equation in Table 8.15.

8.3.5 *Comparison of Breaching Parameters for Landslide Dams and Constructed Embankment Dams*

Compared with landslide dams, more studies have been conducted on breaching parameters of constructed embankment dams. Breaching of landslide dams is sometimes studied by using empirical equations for man-made earthen dams (Kroup, 2002). This may not be appropriate as there are several distinct differences between these two kinds of dam (Costa, 1985; Costa and Schuster, 1988, 1991; Meyer et al., 1994; Kroup, 2002; Ermini and Casagli 2003; Chang and Zhang, 2010; Peng and Zhang, 2012a):

1. Landslide dams consist of heterogeneous and anisotropic rock and soil materials, so that their mechanical parameters, such as erodibility and permeability, vary significantly across the dams. The geometry of landslide dams is also irregular, bringing in more uncertainty in the control variables.
2. Soils in landslide dams are not compacted sufficiently; thus, their erodibility and permeability are generally higher than those of constructed embankment dams.
3. Given the same reservoir capacity, landslide dams are usually more massive. Therefore, more soil materials need to be eroded, which may increase the breaching duration and lower the peak discharge.
4. Landslide dams are usually wider and both their upstream and downstream slopes are gentler, which may reduce the likelihood of piping failure and slope failure.
5. Landslide dams often form in remote mountainous areas, which are difficult to monitor, so less performance data is available.
6. Limited flood control measures (e.g. spillways) and seepage control measures (e.g. corewalls and filters) are available in landslide dams. so most of them failed a short time after formation.

Figure 8.9 shows the lake volume to dam height relationship of failed constructed earth and rockfill dams (182 cases from Xu and Zhang, 2009a) and breached landslide dams (52 cases from Peng and Zhang 2012). The heights of these beached landslide dams cover a larger range, from several meters to nearly several hundred meters (e.g. the Bireh-Ganga River landslide dam with a dam height of 274 m). The heights of failed constructed earth and rockfill dams are mostly in a range of 5–50 m (95% of 160 cases with dam height information). With the same dam height, the volumes of the lakes retained by landslide dams are generally much smaller than those behind earth and rockfill dams because the landslide dams are usually in steep valleys.

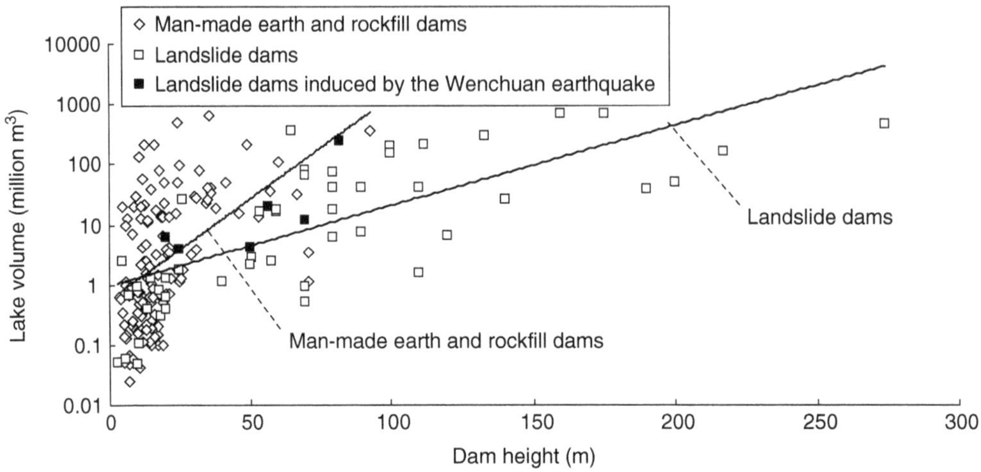

Figure 8.9 Comparison of dam height–lake volume relationships for constructed earth and rockfill dams and landslide dams. *Source:* Peng and Zhang (2012a). With permission from Springer

Figures 8.10–8.13 compare the influences of dam height and lake volume on the breaching parameters of landslide dams and man-made dams. Table 8.16 shows the biases of the models for constructed dams and landslide dams. The differences and similarities in the breaching characteristics of landslide dams and constructed dams are shown in these figures and the table.

1. *Peak outflow rate Q_p.* As shown in Figure 8.10, an increase in either dam height or lake volume generally causes an increase in peak outflow rate for both kinds of dam. The Q_p of landslide dams is less sensitive to dam height, as the larger width and volume of a landslide dam would consume more water energy. The sensitivity to the lake volume for both kinds of dams (shown in Figure 8.10b) is quite similar, which agrees with the results of Costa (1985). The models for constructed earth and rockfill dams will seriously overestimate the peak flow rate of landslide dams, with the mean bias factor of 0.34 as shown in Table 8.16. The mean bias factor with the models for landslide dams in Table 8.9 is 1.01, which is close to unity.
2. *Breach depth.* As shown in Figure 8.11, the breach depth of a landslide dam is less sensitive to the dam height, but more sensitive to the lake volume than the breach depth of constructed dams. As shown in Figure 8.9, with the same lake volume, the height of a landslide dam is often larger than that of a constructed dam, so the landslide dam is likely to breach to a larger depth. In the failure cases of constructed dams (Xu and Zhang, 2009), most of the high erodibility dams failed to "full breach depths" (i.e. the breach depth is close to the dam height). For landslide dams, however, very few cases failed to full breach depths. The equations for constructed dams can grossly overestimate the breach depth of landslide dams, with the mean bias factor of 0.58.
3. *Breaching duration.* As shown in Figure 8.12, the breaching duration of a landslide dam is generally longer than that of a constructed dam with the same dam height or lake volume. The main reason is that the volume and width of the landslide dam are typically larger, and the downstream slope is typically gentler than that of a constructed dam, resulting in reduced erosion velocity during breaching. Like the breach depth, the breaching duration of landslide dams is less sensitive to the dam height, but more sensitive to the lake volume. In fact, the breaching duration is statistically not related to the dam height, as shown in Figure 8.12(a). A larger dam

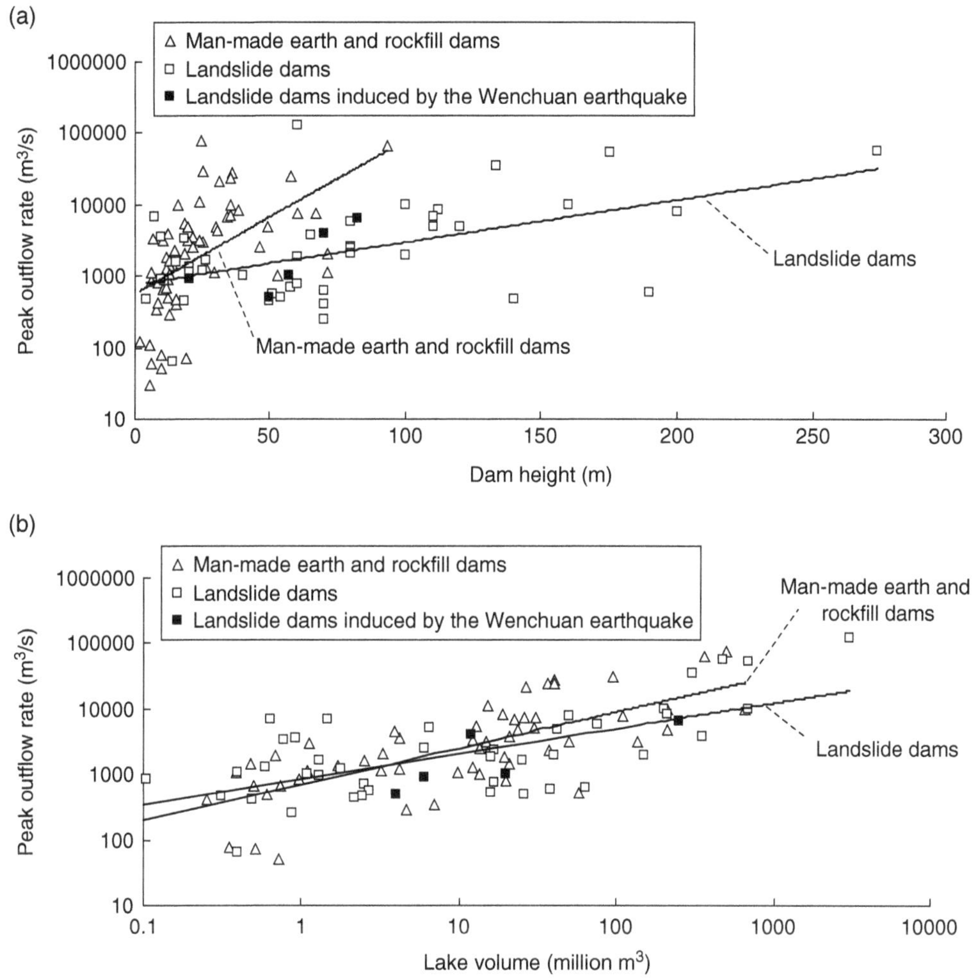

Figure 8.10 Comparison of peak outflow rates from failure of landslide dams and earth and rockfill dams: (a) influence of dam height, (b) influence of lake volume. *Source:* Peng and Zhang (2012a). With permission from Springer

height does not mean a longer breaching process, because the breach often does not develop to the bottom of the dam. The breach duration of a landslide dam could be, on average, underestimated by 2.76 times using the equations for constructed dams.

4. *Breach top width.* As shown in Figure 8.13, the breach top widths in constructed dams are more sensitive to dam height than in landslide dams. The results of Xu and Zhang (2009) show that for constructed dams, besides lake volume, dam type (with/without corewalls, concrete face), failure modes (overtopping or piping), and dam erodibility also influence the breach top width, which makes the statistical relation between breach top width and lake volume rather weak.

In summary, the direct application of the prediction models for constructed embankment dams to landslide dams would lead to overestimation of the breach size (by more than 60% using Xu and Zhang's models) and the peak outflow rate (by approximately 200% using Xu and Zhang's

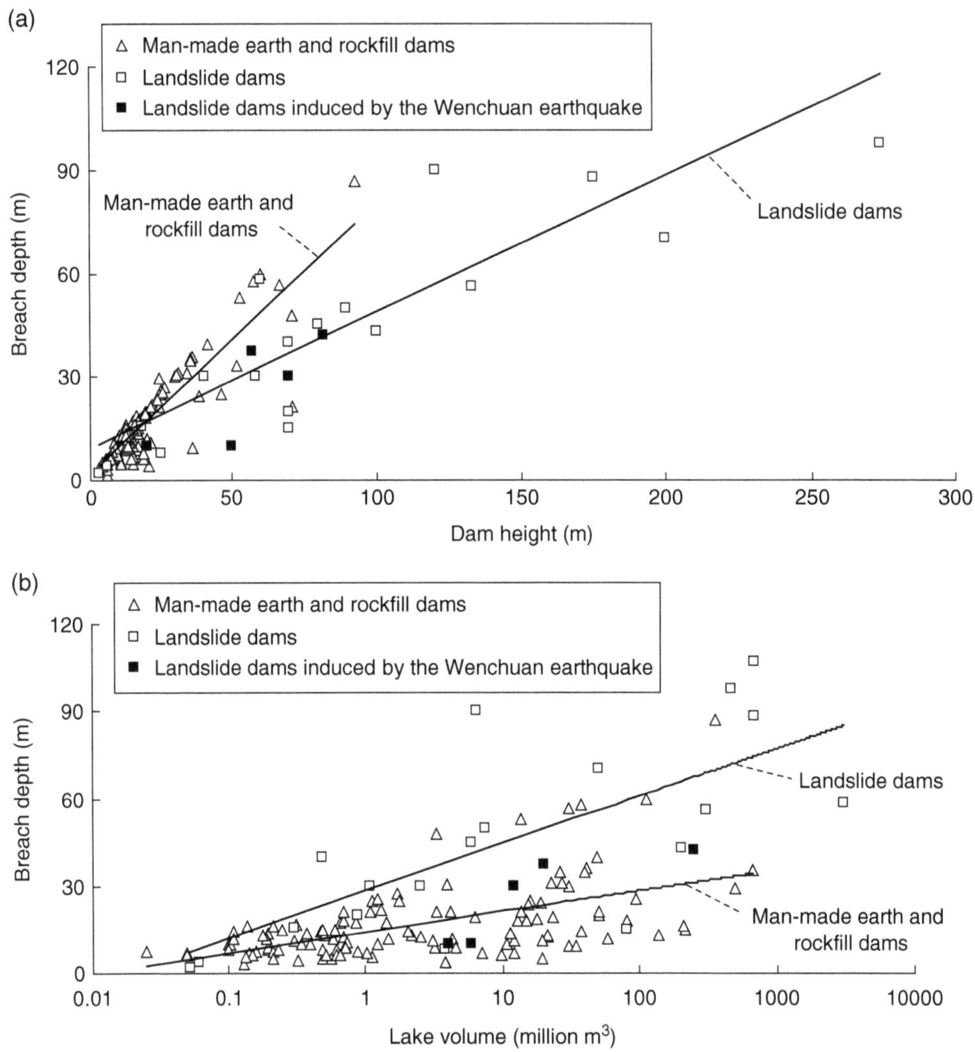

Figure 8.11 Comparison of breach depths of landslide dams and earth and rockfill dams: (a) influence of dam height, (b) influence of lake volume. *Source:* Peng and Zhang (2012a). With permission from Springer

models), and underestimation of the breaching duration (by approximately 50% using Xu and Zhang's models). Hence, separate models should be developed for predicting the breaching parameters of landslide dams.

8.4 Numerical Simulation of Overtopping Erosion

The overtopping failure of a dam is a complex phenomenon involving erosion, slope stability and the hydraulics of the overflow. Overtopping erosion is an interactive process between water flow and the resistance of the dam materials/structures. The water from the reservoir/lake erodes the

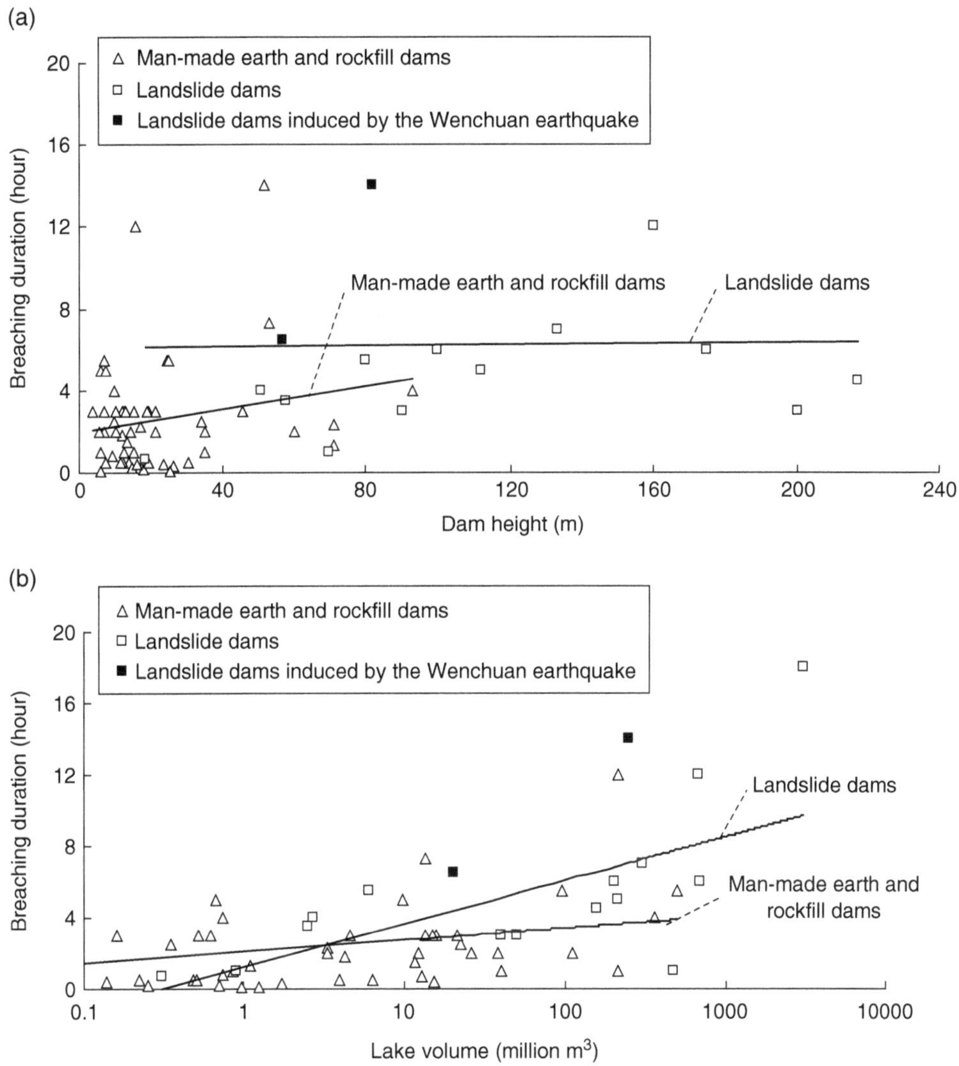

Figure 8.12 Comparison of breaching durations of landslide dams and earth and rockfill dams: (a) influence of dam height, (b) influence of lake volume. *Source:* Peng and Zhang (2012a). With permission from Springer

dam body, and a breach forms at the most erodible part of the crest. The breach develops with increasing vertical and lateral erosion along the channel and collapses of the side slopes. The empirical models for predicting overtopping breach parameters have been described in Section 8.3. This section describes numerical simulation of overtopping erosion.

Several physically based models have been developed to predict the overtopping process of dams, such as DAMBRK (Fread, 1977), BREACH (Fread, 1988), BEED (Singh and Quiroga, 1987), BRES (Visser, 1998), Tingsanchali and Chinnarasri's model (Tingsanchali and Chinnarasri, 2001), HR-BREACH (Mohamed, 2002), SIMBA (Temple et al., 2005; Hanson et al., 2005a), Wang and Bowles's model (Wang and Bowles, 2006), BRES-Zhu (Zhu, 2006), Faeh's model

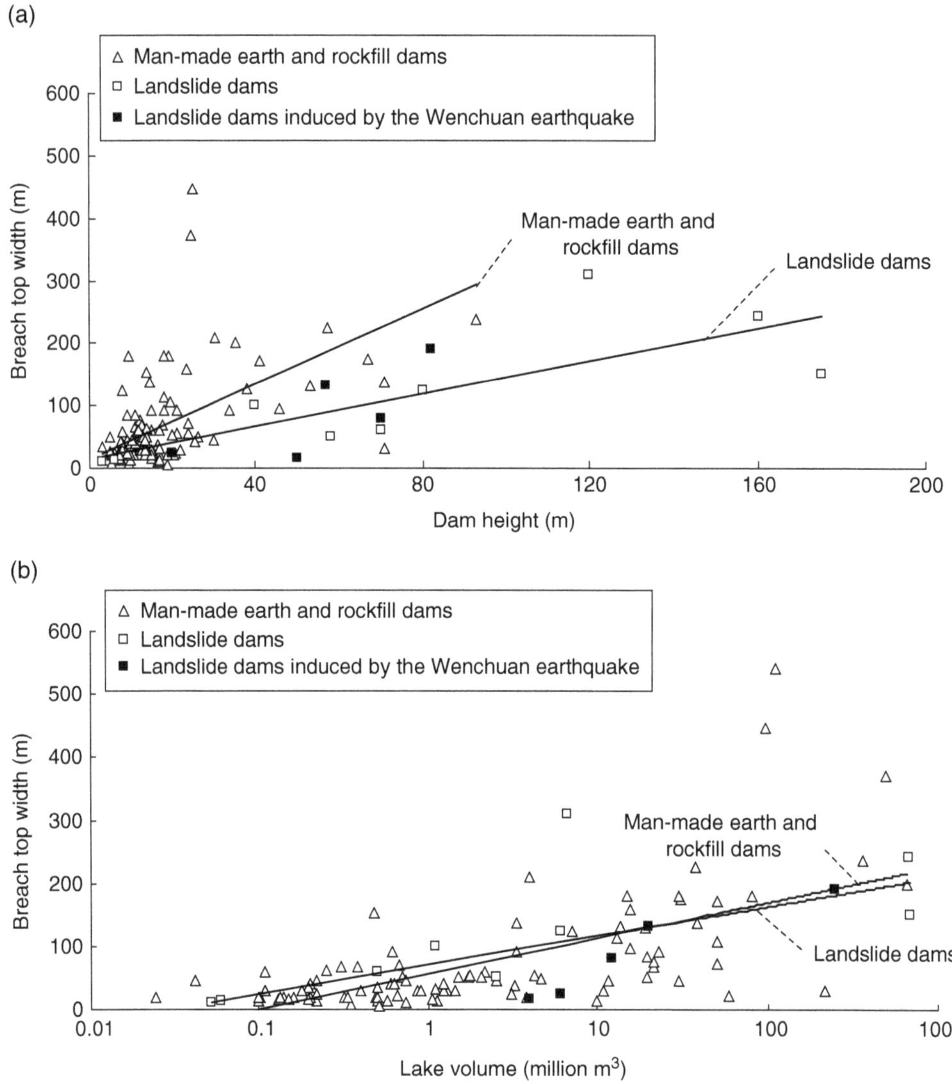

Figure 8.13 Comparison of breach top widths of landslide dams and earth and rockfill dams: (a) influence of dam height, (b) influence of lake volume. *Source:* Peng and Zhang (2012a). With permission from Springer

(Faeh, 2007), Wu and Wang's model (Wu and Wang, 2007), Wang et al.'s model (Wang et al., 2008), DABA (Chang and Zhang, 2010), and Cao et al.'s model (Cao et al., 2011). Table 8.17 summarizes the main characteristics of several simplified physically based models. A physically based dam breaching model should incorporate breach channel hydraulics, soil mechanics, and erosion mechanics as shown in Figure 8.14. These components might be modeled differently in specific models. The models can be classified into simplified physically based and erosion-mechanics based models. The basic theory and assumptions of these two kinds of models are introduced first, then the simulation of two dam failure cases is presented.

Table 8.16 Comparison of the bias factors for prediction models for landslide dams and constructed earth and rockfill dams

Parameter	Full-variable equations for landslide dams in Table 8.9			Simplified equations for landslide dams in Table 8.9			Full-variable equations for constructed earth and rockfill dams in Table 8.5			Simplified equations for constructed earth and rockfill dams in Table 8.5		
	No. of cases[1]	Mean[2]	St. Dev.	Number of cases[1]	Mean[2]	St. Dev.	No. of cases[1]	Mean[2]	St. Dev.	No. of cases[1]	Mean[2]	St. Dev.
Q_p	41	1.01	0.43	41	1.01	0.44	14	0.32	0.21	14	0.36	0.18
H_b	21	1.05	0.36	21	1.05	0.37	21	0.58	0.20	21	0.57	0.20
W_t	13	1.08	0.40	13	1.12	0.50	9	0.59	0.25	9	0.64	0.27
W_b	9	1.07	0.36	9	1.07	0.47	7	0.39	0.29	7	0.34	0.33
T_b	14	1.05	0.36	14	1.06	0.36	8	2.76	1.28	8	2.16	0.95

Note: [1]The number of cases after using the objective outlier-exclusion algorithm (Rousseeuw, 1998) at the 95% probability level.
[2]The mean value denotes the ratio of measured value to predicted value.

Table 8.17 Summary of simplified physically based dam breach models (after Singh 1996; Wahl 1998; and Morris 2000)

Models	References	Sediment transport	Breach shape	Breach development	Flow over the dam	Input parameters	Breach mode	Soil erodibility
DAMBRK	Fread (1977)	Assume linear erosion	Rectangular, triangular, trapezoidal	Linear or nonlinear; constant side slope angle	Broad-crest weir flow; orifice flow	Final breach shape and geometry; breach time; hydraulic conditions; others	Overtopping and piping	No
BEED	Singh and Quiroga (1987)	Einstein–Brown formula	Triangular, trapezoidal	Consider the stability of the side slope of the breach	Broad-crest weir flow	Initial breach shape and geometry; soil properties; hydraulic conditions	Overtopping	Constant
BREACH	Fread (1988)	Meyer-Peter and Müller formula; DuBoys formula	Rectangular, triangular,	Consider the stability of the side slope of the breach	Broad-crest weir flow, orifice flow	Dam geometry; soil properties; hydraulic conditions	Overtopping and piping	Constant

(Continued)

Table 8.17 (*Continued*)

Models	References	Sediment transport	Breach shape	Breach development	Flow over the dam	Input parameters	Breach mode	Soil erodibility
BRES	Visser (1998)	Bagnold-Visser, Engelund-Hansen and Van Rijn formula	Trapezoidal	Consider the stability of both the side slope of the breach and the downstream slope	Broad-crest weir flow,	Initial breach geometry; dam and foundation geometry; soil properties; hydraulic conditions; others	Overtopping	Constant
HR BREACH	Mohamed (2002)	Variety of sediment transport/erosion equations	Rectangular	Consider discrete growth using bending, shear, sliding and overturning failure of soil masses	Broad-crest weir flow; orifice flow	Initial breach; soil properties; others; hydraulic conditions	Overtopping and piping	Constant
SIMBA	Temple et al. (2005), Hanson et al. (2005)	Headcut erosion formula	Rectangular	Headcut formation, deepening, and upstream advancement; lateral widening	Broad-crest weir flow	Embankment geometry (slopes, height, crest width); soil properties; notch dimensions (bottom width, depth, side slope); hydraulic conditions	Overtopping	Constant
BRES-Zhu	Zhu (2006)	Headcut erosion formula	Trapezoidal	Consider the stability of both the side slope of the breach and the downstream slope; headcut formation and advancement	Broad-crest weir flow	Initial breach geometry; dam and foundation geometry; soil properties; hydraulic conditions; others	Overtopping	Constant
DABA	Chang and Zhang (2010)	Meyer-Peter and Müller formula	Trapezoidal	Consider the stability of both the side slope of the breach and the downstream slope of the breach	Broad-crest weir flow	Initial breach geometry; dam geometry; soil properties along the depth of the dam; hydraulic conditions;	Overtopping	Varied

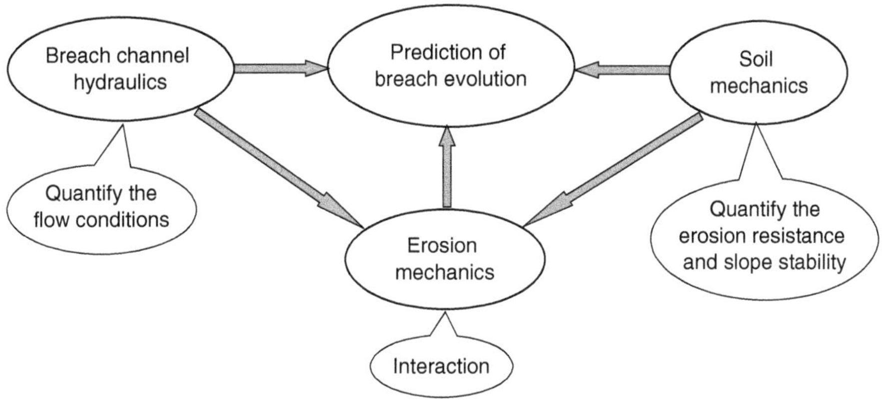

Figure 8.14 Components of a physically based dam breaching analysis model

8.4.1 Simplified Physically Based Methods

One of the main tasks of dam breach analysis is to predict the outflow hydrograph through the breach. Prediction of the outflow hydrograph can be further subdivided into prediction of the breach geometric characteristics (e.g. breach shape, breach width development, breach depth development) and outflow hydrograph through the breach. Breach formation involves a variety of processes, including sediment entrainment from the breach floor, mass movements from the breach sides into the flow, and mass movement on the downstream slope.

The actual breaching process is sophisticated. It involves not only water flow but also sediment transport and morphological changes, which in turn modify the breach flow. Moreover, the uncertainty in many parameters that govern dam breach morphology, especially those of a geotechnical nature, also complicates the breaching process. To capture the general breaching characteristics, many simplified physically based models are developed for embankment dams and landslide dams. Some assumptions are generally made for these models:

1. The initial breach geometry and the overtopping location need to be assumed beforehand.
2. The breach shape is often trapezoidal, rectangular, or triangular.
3. The breach cross-section is symmetrical.
4. The breach is straight along the flow direction.
5. The final breach depth is equal to the dam height.
6. The downstream slope is a constant or time-varying straight slope.
7. The relation between the reservoir water level and its surface area follows a linear or power function.
8. The flow through the breach is estimated using a broad-crested weir relation.
9. The erosion rate is often assumed to be a power function of the flow rate.
10. The soil erodibility does not vary along the depth of the dam.
11. The potentially strong eroding capability of transient flow is mostly not considered.
12. The dam materials are usually assumed to be fully saturated.

Simplified physically based models involve breach evolution, sediment transport, and breach hydraulics. The evolution of the breach, in relation to the collapse of side slopes and flow conditions, is often analyzed using simplified slope stability methods. In some cases, the side-slope failure is simplified by specifying a predefined "critical failure angle" which induces a slope failure once this critical angle is exceeded (e.g. Peviani, 1999; Faeh, 2007). Most of the models are based on local equilibrium bed-load sediment transport (Ponce and Tsivoglou, 1981; Fread, 1988; Singh and Scarlatos, 1988; Chang and Zhang, 2010) or non-equilibrium total-load sediment transport from clear-water in the reservoir to downstream turbid water (Visser, 1998; Wu et al., 2009). The hydrodynamics is modeled by solving the continuity equation of the reservoir/lake together with the breach outflow through a broad-crested weir. In the following sections, the key components of simplified physically based models are presented.

8.4.1.1 Evolution of Breach Geometry

A critical component of a simplified physically based model is the determination of the breach evolution. Here take the BREACH model (Fread, 1988) as an example to elaborate the analysis of the evolution of breach geometry.

The initial breach is assumed to be rectangular, as shown in Figure 8.15. The width of the breach (B_0) is governed by the following relation:

$$B_0 = B_r D_w \tag{8.20}$$

where B_r is a factor based on optimum channel hydraulics efficiency, and D_w is the depth of flow in the breach channel (m).

The initially rectangular-shaped channel changes to a trapezoidal one when the sides of the breach channel collapse due to slope instability. The collapse occurs when the depth of breach cut (H_c) reaches a critical value (H_k), which is a function of the dam's material properties as

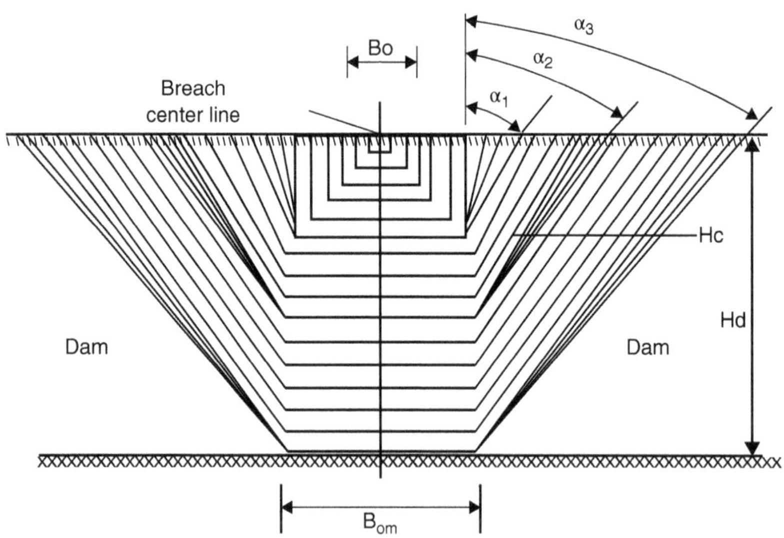

Figure 8.15 Evolution of breach geometry in the BREACH model: $\alpha_k=\pi/2-\theta$; θ=side slopes angle; B_0=width of the breach; B_{0m}=final width of the breach bottom; H_d=elevation of the dam; H_c=elevation of the breach bottom (modified from Fread 1988)

$$H_k = \frac{4c\cos\phi\sin\theta_{k-1}}{\gamma\left[1-\cos\left(\theta_{k-1}-\phi\right)\right]} \tag{8.21}$$

where c is the cohesion, kPa, γ is the unit weight of the material, N/m^{3,} ϕ is the internal friction angle, θ is the side slopes angle, and $k=1, 2, 3\ldots$

$$\theta_k = \frac{1}{2}\left(\theta_{k-1}+\phi\right) \tag{8.22}$$

Erosion is assumed to occur equally along the bottom and sides of the breach channel, except when the sides of the breach channel collapse. Thereupon, the erosion of the breach bottom will come to an end when the volume of collapsed material along the breach is removed at the rate of the sediment transport capacity of the breach channel at the instant of collapse. After this short pause, the breach bottom and sides continue to erode. When the breach has eroded downward to the original valley floor, further downward erosion is prohibited within the BREACH model; however, the side slopes of the breach can continue to erode.

The BREACH model also accounts for the breach enlargement by sudden collapses. During the breaching process, it is possible for the breach to enlarge by a sudden failure of the upper potions of a dam in the vicinity of the breach. The collapse is considered to be due to the pressure of the water on the upstream face of the dam exceeding the resistance of the dam structures. In the BREACH model, the downstream slope of the dam below the breach channel is assumed to be the same as the initial downstream slope of the dam. The BRES model (Visser, 1998) takes the steepening process of the downstream slope during the breaching process into account and further divides the entire breaching process into five stages:

Stage I. Steepening of the slope angle of the breach channel in the downstream slope from an initial value β_0 at t_0 up to a critical value at $t=t_1$. Generally, the erosion rate close to the toe of the downstream slope is larger than that on the upper part. This uneven rate of erosion induces steepening in time of the downstream slope. The outflow rate is relatively small during this stage, as shown in Figure 8.16.

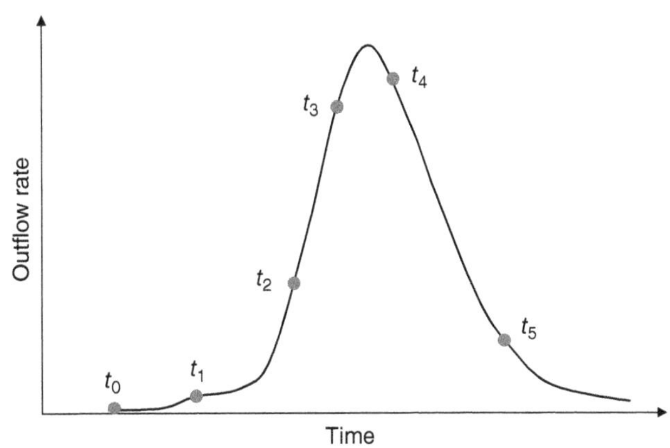

Figure 8.16 Outflow hydrograph associated with the stages defined by Visser (1998) during a dam failure

Stage II. Retrograde erosion of the downstream slope at constant angle β_1 for $t_1 < t < t_2$, yielding a decrease of the width of the crest of the dam/dike in the breach. During this process, the BRES-Zhu model (Zhu, 2006) and the DABA model (Chang and Zhang, 2010) also account for the erosion on the top of the breach. The erosion along the crest of dam/dike lowers the height of the dam/dike in the breach and increases the breach flow gradually. Due to the surface erosion on the crest of the breach and the increasing reservoir water level, the outflow rate during this stage increases significantly, as shown in Figure 8.16.

Stage III. Lowering of the top of the dam/dike in the breach channel, with constant angle of side slopes and equal to the critical value, resulting in an increase of the width of the breach for $t_2 < t < t_3$. The breach enlarges rapidly, accordingly so does the breach flow rate, which in turn accelerates the breach erosion process in the dam/dike. The outflow rate in this stage increases rapidly due to the enlargement of the breach, as shown in Figure 8.16.

Stage IV. Critical flow stage, where the breach flow is virtually critical throughout the breach for $t_3 < t < t_4$, and the breach continues to expand mainly laterally. The breach side slope angles remain at the critical value. The breach growth in the vertical direction depends in this stage on the erodibility of the foundation of the dike. Owing to the lateral erosion of the breach, the outflow rate can reach the peak value during this stage, as shown in Figure 8.16.

Stage V. The breach erodes mainly laterally similar to that in Stage IV, with the vertical erosion relying on the material conditions in the foundation of the dike for $t_4 < t < t_5$. Due to the growing obstructive effect of the rising reservoir water level, a gradual decrease of the breach flow velocity is expected. At t_5 the flow velocities in the breach become so small that the breach erosion stops. The flow through the breach continues and stops when the water level in the polder equals the outside water level. The outflow rate tends to decrease and approach the inflow rate eventually as most of the lake/reservoir water has been released during Stages II and IV, as shown in Figure 8.16.

The five-stage breaching process can be reflected by the evolution of the breach geometry in DABA model, as shown in Figure 8.17. When water erodes the soils at the side slopes below the water level, the side slopes above the water level will collapse. The soils at the breach channel bed can also erode during this process. During each time interval, the erosion depth at the bottom corner, A, is assumed to be the same along the vertical direction and the direction perpendicular to the previous side slope, as shown in Figure 8.17(a). This process will continue until the side slopes reach a critical value α_c. After that, the side slopes will recede, keeping the same angle, α_c, as erosion continues. During each time interval, the erosion depths at the bottom corner B are also the same along the vertical direction and the direction perpendicular to the side slope, as shown in Figure 8.17(b). This process will stop when the applied shear stress by the overtopped water flow is smaller than the erosion resistance of the soil at the channel bed. Once a hard layer in landslide dams or the foundation of embankment dams with its erosion resistance larger than the shear stress induced by the water flow is encountered, the breach cannot develop any further in the vertical direction. However, if the shear stress induced by the water flow at the side slopes is still larger than the erosion resistance of the side slope materials, the soils at the side slopes can still erode. In this case, the breach slopes will recede laterally keeping the same side slope angle as shown in Figure 8.17(c).

After water flows over the crest of the dam, the flow velocity increases gradually along the downstream slope until an energy loss equilibrium is reached. Before the slope reaches β_f, the downstream slope angle increases gradually, as shown in Figure 8.18. During each time interval, the increment in the downstream slope angle can be expressed as

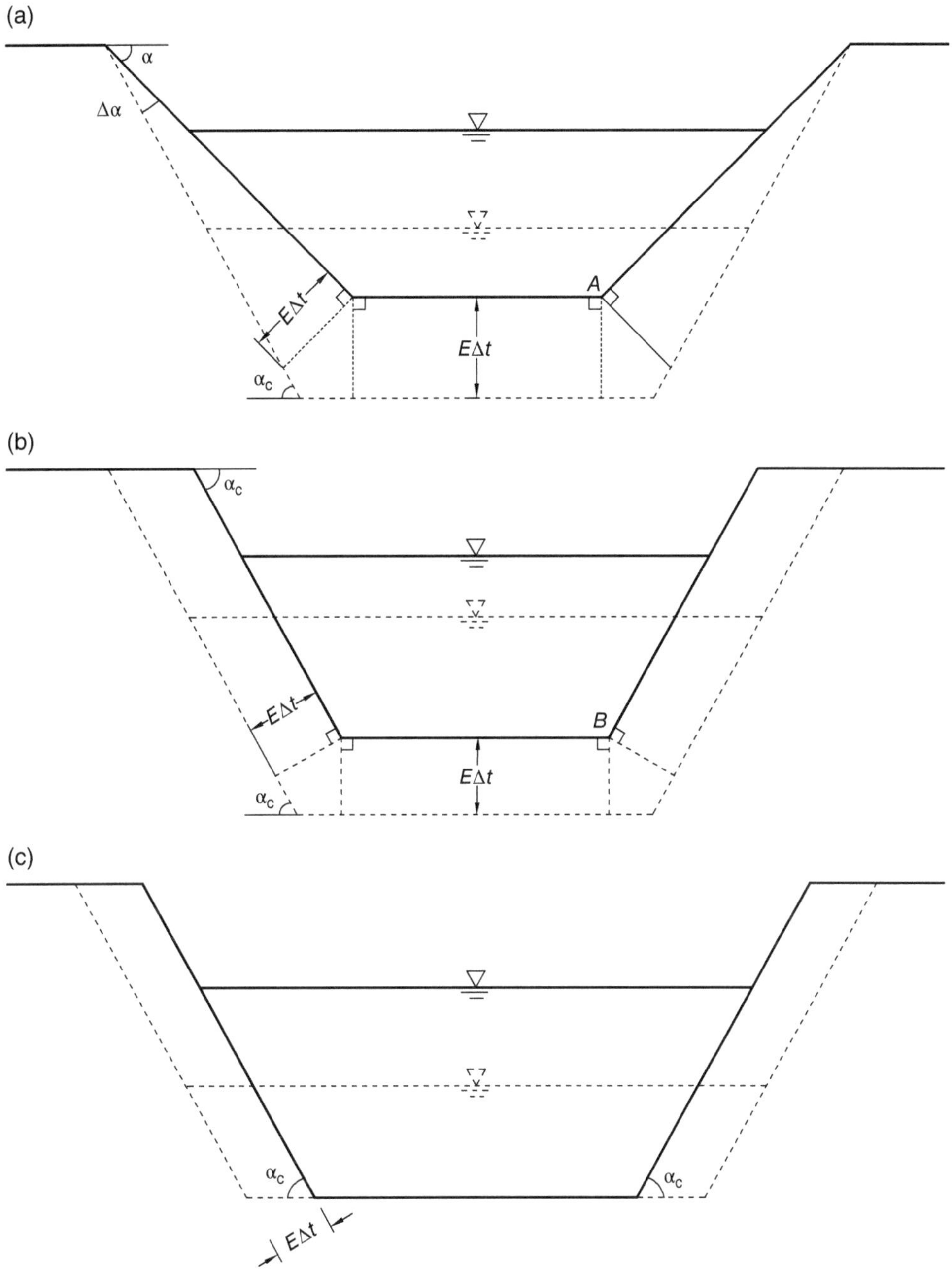

Figure 8.17 Breach enlargement process: (a) stage I, (b) stage II, (c) stage III

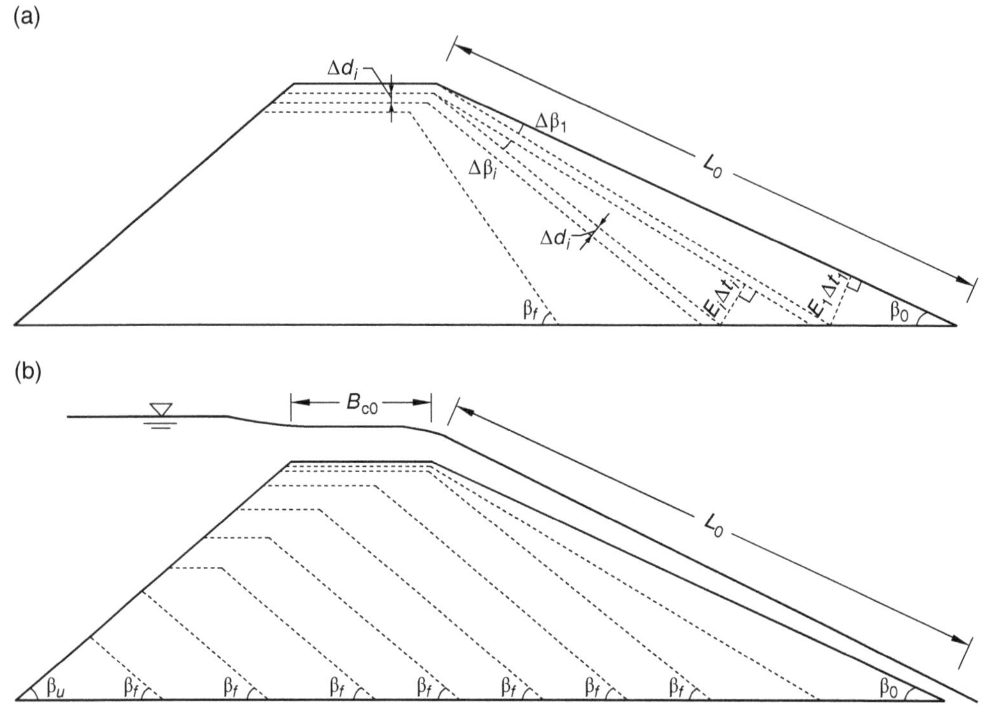

Figure 8.18 Longitudinal breach growth: (a) steepening of the downstream slope, (b) longitudinal breach development

$$\frac{1}{\tan \Delta\beta_i} = \frac{L_{i-1}}{E_i \Delta t_i} - \frac{1}{\tan \beta_{i-1}} \tag{8.23}$$

where $\Delta\beta_i$ is the increment of the downstream slope angle at the ith time interval, °, E_i is the erosion rate at the ith time interval, mm^3/m^2-s, L_{i-1} is the downstream slope length at the ith time interval, m, β_{i-1} is the downstream slope angle at the $i-1$th time interval, °, and Δt_i is the ith time interval, s.

To satisfy the consistency condition of erosion at the downstream crest, the same erosion depth will also occur at the top of the downstream slope (Figure 8.18a):

$$\Delta d_i = E_i \Delta t_i \tag{8.24}$$

After the downstream slope angle reaches β_f, the downstream slope is assumed to recede at the same slope angle.

Most cohesive earth dams breach by headcut erosion. Both the SIMBA model (Temple et al., 2005; Hanson et al., 2005a) and the BRES-Zhu model (Zhu, 2006) account for the breach formation by headcut erosion. Hanson et al. (2005a) divided the whole breaching process into four erosion stages: (I) headcut formation on the downstream slope, (II) headcut advancement through the embankment crest, (III) breach formation as the headcut enters the reservoir, and (IV) breach expansion during reservoir drawdown. Generally four types of erosion are likely to occur at the headcut in Stages II and III during the breach erosion process of a cohesive earth dam, as shown in Figure 8.19: (I) breach flow shear erosion along the approaching headcut top surface, gradually

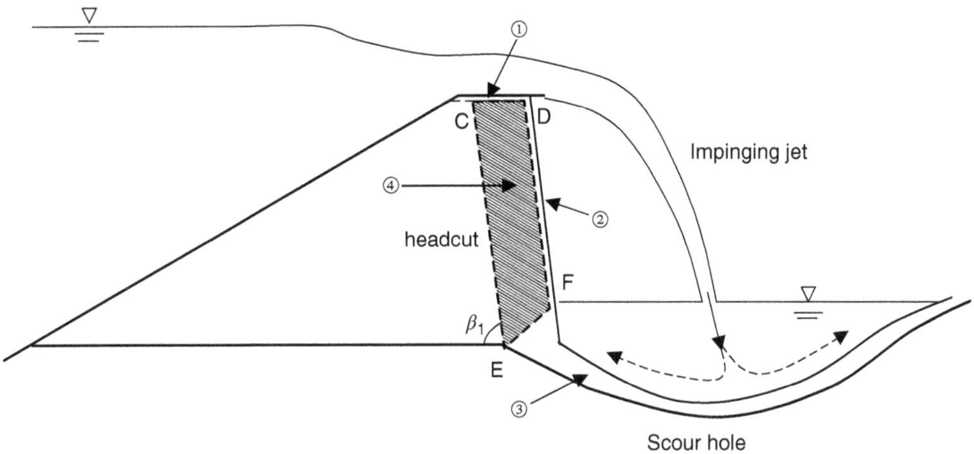

Figure 8.19 Sketch of headcut erosion of cohesive dam in Stages II and III. *Source:* Zhu (2006). Reprinted with permission of Dr. Yonghui Zhu

lowering the height of the dam/dike (see ① in Figure 8.19); (II) erosion along the headcut slope surface (② in Figure 8.19); (III) jet scour of the dam/dike foundation and the headcut undermining (③ in Figure 8.19); (IV) when the headcut undermining has developed to a certain critical degree, the headcut becomes unstable under the action of gravity and the seepage forces, and soil slope failure occurs from the headcut (④ in Figure 8.19). However, due to lack of detailed understanding of mechanisms of headcut erosion process, the models (i.e. SIMBA and BRES-Zhu) for simulating headcut erosion of earth dams are still under development. The headcut migration is the weakest point in the existing embankment breach models. Therefore, models for the headcut erosion failure of earth dams/levee will not be presented in this book.

8.4.1.2 Erosion and Sedimentation

Once an initial breach is formed on the downstream slope or crest of a dam, the hydrodynamic forces continue to enlarge the breach by removing the soil material. The eroded sediments are transported by the water flow. The extent of erosion and subsequent transport depend upon the extent of overtopping, material composition, downstream conditions, dam zoning, etc. Most of the simplified physically based breach models are based on bed-load sediment transport. Bed-load is the part of fluvial sediment load that moves immediately above the bed in a rolling or saltating mode. Its submerged weight is supported by a combination of fluid and solid reactive forces exerted at intermittent contacts with the breach channel bed (Gomez and Church, 1989). During the dam breaching, the bulk of sediment is transported as bed-load. The physics of bed-load transport in a dam breaching process remains incompletely understood. Currently, most of the simplified physically based models use a simple method for computing the bed-load discharge, for instance, the Du Boys formula (Du boys, 1879), and the Meyer-Peter–Müller formula (Meyer-Peter and Müller, 1948) in the BREACH model; the Einstein–Brown formula (Brown, 1950) in the BEED model; and a stress-based detachment rate formula (Hanson et al., 2005a) in the SIBA, BRES-Zhu, and DABA models. These sediment transport equations have been introduced in Chapter 7.

Application of bed-load transport formulas generally involves the following assumptions (Gomez and Church, 1989):

1. The flow and sediment properties, and hence the bed-load transport rate, for the period in question are invariant and can be described with reference to a steady state.
2. The bed-load transport rate is a unique function of tangible and comprehensive flow and sediment parameters.
3. The maximum possible amount of bed-load is being transported. In other word, all the mentioned sediment transport equations are derived for steady state subcritical flow conditions, for specific types of sediment, and for a certain range of sediment diameters. These conditions are likely to be violated during the dam breaching process since conditions are typically unsteady, the flow is supercritical, and a variety of soil types are used for embankment construction.

All the bed-load transport formulas adopted in physically based breach models are developed based on flume tests. In a real dam failure process, in addition to channel bed erosion, the erosion and collapse of the breach side slopes will also take place. The collapse obviously makes the sedimentation process complex. Uncertainties in predicted breach parameters and the flood hydrograph exist because of significant simplifications of the bed-load formula.

8.4.1.3 Hydraulics of the Flow over the Dams

Powledge et al. (1989) observed that flow over an embankment dam, with low or no tail water, goes through three zones as shown in Figure 8.20:

1. from a static energy head to a combination of static and dynamic head, proceeding from the calm reservoir to a subcritical velocity state over the upstream portion of the dam crest
2. through critical velocity on the crest and supercritical flow across the remainder of the dam crest to the downstream slope
3. rapidly accelerating turbulent supercritical flow on the steep downstream slope.

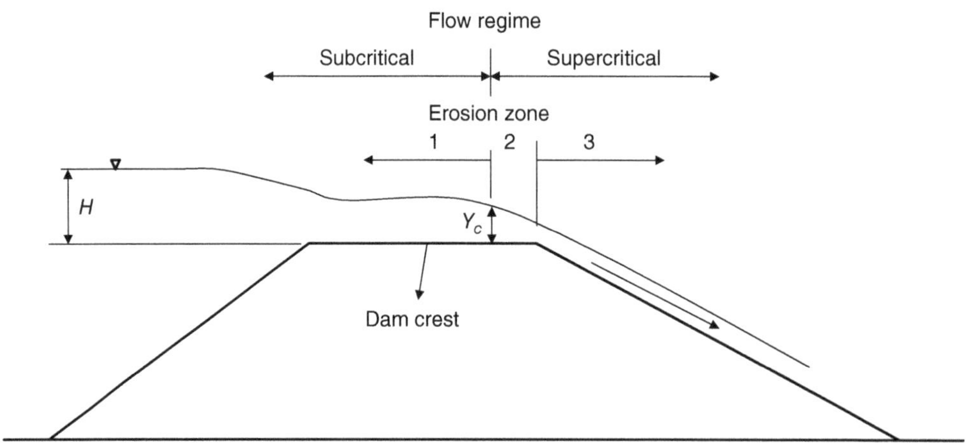

Figure 8.20 Hydraulic flow regime and erosion zones for overtopping erosion (modified after Powledge et al. 1989)

For the subcritical zone at the crest (zone 1 in Figure 8.20), the erosive stresses are low because the energy level is low and the energy slope is small; therefore, the flow velocities are small. The small energy slope imposes a low tractive shear stress even though the flow depth may be fairly large. Only when the dam crest consists of highly erodible material does erosion occur in the supercritical zone of the crest.

For the supercritical zone at the crest (zone 2 in Figure 8.20), the energy level is unchanged from that in zone 1, but the energy slopes can be increased significantly, which increases the tractive shear stress significantly. Erosion can occur at the downstream edge of the crest due to the high tractive shear stress.

For the supercritical zone on the downstream slope (zone 3 in Figure 8.20), the energy levels progressively increase as the flow proceeds down the downstream slope. The velocities can increase significantly due to the steep energy slope, until energy loss equilibrium is reached. Because of the steep energy slope, the tractive shear stresses are very large, resulting in high erosion potential.

Three elements are included in the hydraulics of flow over the dam: (1) flow over the crest, (2) flow through the breach, and (3) flow through the breach channel on the downstream face of the dam. The overtopped water triggers the formation of a breach. For man-made dams, the flow can be assumed to be a quasi-steady uniform flow as the reach of the breach channel is short and the channel slope on the downstream face is very steep (Fread, 1988). Cencetti et al. (2006) pointed out that the flow could be considered as a uniform flow along a landslide dam with a large channel length but a small slope. The crest of a landslide dam is often wider than that of an embankment dam; thus the flow at the crest can be assumed to be laminar or subcritical. The water flow over the crest as well as the breach can be assumed to be analogous to the flow over a broad-crest weir. Most simplified physically based breach models (DAMBRK, BREACH, BEED, BRES, HR-BREACH, SIMBA, BRES-Zhu, DABA) use the broad crested weir equation to simulate flow over the crest and on the downstream face of the embankment.

A general form of broad-crested weir flow equation through a breach is

$$Q = C_d B_a \left(H - Z\right)^{3/2} \tag{8.25}$$

where C_d is the breach discharge coefficient in the range 1.3–1.7 m$^{1/2}$/s (Coleman et al., 2002) and B_a is the average breach width.

For a rectangular breach with a breach width of B (Figure 8.21a), the broad-crested weir flow equation can be expressed as

$$Q_b = C_d B \left(H - Z\right)^{3/2} \tag{8.26}$$

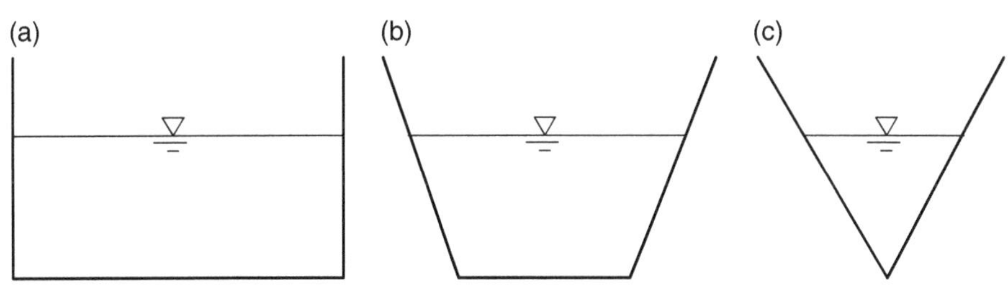

Figure 8.21 Three types of breach cross-sections: (a) rectangular, (b) trapezoidal, (c) triangular

For a trapezoidal breach with a breach bottom width of B_b and a side slope of α_b (Figure 8.21b), the broad-crested weir flow equation can be expressed as

$$Q_b = C_d\left[B_b + (H - Z)\tan\alpha_b\right](H - Z)^{3/2} \tag{8.27}$$

For a triangular breach with a breach width of B (Figure 8.21c), the broad-crested weir flow equation can be expressed as

$$Q_b = \frac{1}{2}C_d B(H - Z)^{3/2} \tag{8.28}$$

The breach outflow depends on the reservoir storage and water level. The inflow coming to the reservoir/lake increases the reservoir/lake storage. In order to simultaneously account for the effects of inflow and outflow, the reservoir/lake routing is performed by applying the mass balance equations:

$$A\frac{dH}{dt} = Q_{in} - Q_{out} \tag{8.29}$$

$$Q_{out} = Q_b + Q_s \tag{8.30}$$

where A is the reservoir/lake water surface area, m^2, Q_{in}, Q_{out}, Q_b, are the flow rates into the reservoir, out of the reservoir, and through the breach, respectively, m^3/s, and Q_s is the seepage discharge through the dam, m^3/s. Q_{in} can be obtained according to hydrological information and Q_s can be measured through field monitoring. The surface area can be calculated based on the relationship between lake water level and lake volume, assuming that the lake water level is horizontal or directly obtained from the relationship between the lake water level and the lake water surface area.

8.4.2 Detailed Physically Based Methods

Dam breaching is undoubtedly a complex process, which involves coupling of water flow, sediment transport, and morphological changes. For instance, dam breach flows are usually in mixed flow regimes with discontinuities, which could have strong eroding potential and could influence the morphological evolution of the channel bed. Owing to the strong erosion of the breach channel, the breach flows are applied on the movable beds. Dam break flows over movable beds differ from clear-water flows on fixed beds because they are significantly affected by high sediment concentration and rapid bed change. Detailed physically based modeling of dam breaching requires a thorough understanding of the interplay between these governing elements. Recent physically based models including Tingsanchali and Chinnarasri's model (Tingsanchali and Chinnarasri, 2001), Wang and Bowles's model (Wang and Bowles, 2006), Faeh's model (Faeh, 2007), Wu and Wang's model (Wu and Wang, 2007), Wang et al.'s model (Wang et al., 2008), and Cao et al.'s model (Cao et al., 2011) are summarized in Table 8.18.

One-dimensional, depth-averaged two-dimensional, and three-dimensional numerical models have been developed based on hydrodynamics, sediment transport, and slope stability. For a 2D model, the mass and momentum conservation equations for the water-sediment mixture and the

Table 8.18 Summary of detailed physically based dam breach models (after ASCE/EWRI Task Committee on Dam/levee Breaching 2011).
Source: ASCE/EWRI (2011). With permission from ASCE

Models	References	Breach morphology	Flow	Sediment transport	Slope stability	Numerical solution	Remarks
Tingsanchali and Chinnarasri's model	Tingsanchali and Chinnarasri (2001)	1D Exner equation	1D St Venant equations	Multiple formulas	Downstream slope stability	Finite difference	Overtopping, no lateral erosion
Wang and Bowles's model	Wang and Bowles (2006)	Clear-water scour	2D shallow water equations	Chen and Anderson's formula for erosion rate	3D slope stability	2D TVD finite difference	Cohesionless soil, overtopping
Faeh's model	Faeh (2007)	2D Exner equation	2D shallow water equations	Bed-load and suspended load formulas	Breach slope stability	2D finite volume	Overtopping of non-cohesive levee
Wu and Wang's model	Wu and Wang (2007)	1D Non-equilibrium total-load transport	Generalized shallow water equations	Wu et al. total-load capacity formula	Breach slope stability	Coupled 1D finite volume scheme	Overtopping of non-cohesive dam and levee
Wang et al.'s model	Wang et al. (2008)	2D non-equilibrium sediment transport	2D shallow water equations	Bed-load formula	Breach slope stability	Coupled 2D finite volume scheme	Cohesionless soil, overtopping
Cao et al.'s model	Cao et al. (2011)	2D non-equilibrium sediment transport	Generalized shallow water equations	Modified Meyer-Peter and Müller bed-load	Breach slope stability	Coupled finite volume scheme	Overtopping of landslide dam

mass conservation equations for the sediment and the bed material, respectively are generally expressed in a vector form as

$$\frac{\partial \mathbf{U}}{\partial t}+\frac{\partial \mathbf{F}}{\partial x}+\frac{\partial \mathbf{G}}{\partial y}+\mathbf{S}=0 \tag{8.31}$$

where **U** is the vector of conserved variables, **F**, **G** are the vectors of flux variables, **S** is the vector of source terms, t is the time, x is the streamwise coordinate, and y is the cross-channel coordinate.

Generally, the vectors **U**, **F**, and **G** in different models are the same, which can be expressed as

$$\mathbf{U}=\begin{bmatrix} h \\ hu \\ hv \\ hc \end{bmatrix} \tag{8.32}$$

$$\mathbf{F}=\begin{bmatrix} hu \\ hu^2+gh^2/2 \\ huv \\ huc \end{bmatrix} \tag{8.33}$$

$$\mathbf{G}=\begin{bmatrix} hv \\ huv \\ hv^2+gh^2/2 \\ hvc \end{bmatrix} \tag{8.34}$$

where h is the flow depth, m, u, v are the depth-averaged velocities in x and y directions, respectively, m/s, and c is the averaged sediment concentration in flow.

The vector **S** depends on how a model accounts for the depth-averaged flow velocity and moveable channel bed. Cao et al.'s model reflects the mechanism well, in which vector **S** is expressed as (Cao et al., 2011)

$$\mathbf{S}=\begin{bmatrix} (E-D)/(1-n_0) \\ gh\left(S_{bx}-S_{fx}\right)+\varepsilon\left(\frac{\partial^2 hu}{\partial x^2}+\frac{\partial^2 hu}{\partial y^2}\right)-\frac{\rho_w(G_s-1)gh^2}{2\rho}\frac{\partial c}{\partial x}-\frac{(\rho-\rho_0)(E-D)u}{\rho(1-n_0)} \\ gh\left(S_{by}-S_{fy}\right)+\varepsilon\left(\frac{\partial^2 hv}{\partial x^2}+\frac{\partial^2 hv}{\partial y^2}\right)-\frac{\rho_w(G_s-1)gh^2}{2\rho}\frac{\partial c}{\partial y}-\frac{(\rho-\rho_0)(E-D)v}{\rho(1-n_0)} \\ E-D \end{bmatrix} \tag{8.35}$$

$$\frac{\partial z}{\partial t}=\frac{D-E}{1-n_0} \tag{8.36}$$

where E is the sediment entrainment across the bottom boundary of the flow, m/s, D is the deposition fluxes across the bottom boundary of the flow, m/s, n_0 is the bed sediment porosity, S_{fx}, S_{fy} are the friction slopes in x and y directions, respectively, S_{bx}, S_{by} are the bed slopes in x and y directions, respectively, ρ is the density of water-sediment mixture, kg/m^3, ρ_0 is the density of saturated bed, kg/m^3, z is the bed elevation, m, and ε is the turbulent eddy viscosity.

The mass conservation for the water-sediment mixture differs from the traditional continuity equation for clear-water flows in shallow water hydrodynamics because of the mass exchange between the flow and the erodible bed. In dam break hydraulics, this could lead to highly concentrated debris flows. The momentum transfer due to sediment exchange between the water column and the erodible bottom boundary has to be considered. On the one hand, the deposition of sediment contained in the moving water column can transfer momentum to the flowing mixture, producing a positive source term related to the deposition flux. On the other hand, the entrainment of bed sediment, initially with vanishing momentum due to its original static state, will gain momentum from the flowing fluid, behaving as a sink (Cao et al., 2004). The bed shear stresses and turbulent shear stresses can be reflected in Equation (8.35).

The Meyer-Peter–Müller formula is also adopted to calculate the bed-load transport rate in Faeh's model (Faeh, 2007), Wu and Wang's model (Wu and Wang, 2007), Cao et al.'s model (Cao et al., 2011), and Van Emelen et al.'s model (Van Emelen et al., 2011). However, owing to a lack of understanding of sediment transport under embankment breach flow conditions, the conventional bed-load formula based on flume tests may encounter difficulties in calculating the erosion rate of dam breach (ASCE/EWRI Task Committee on Dam/Levee Breaching, 2011).

During the erosion process, local slope failure on side slopes and downstream slopes of a breach channel usually occurs due to the loss of stability. However, owing to the complex failure mechanism coupled with soil erosion, detailed slope stability analysis has not been considered in most of the detailed physically based models. For instance, the breach side slope failure is considered using a geometric approach in Faeh's model (Faeh, 2007), Wang et al.'s model (Wang et al., 2008), Cao et al.'s model (Cao et al., 2011), and Van Emelen et al.'s model (Van Emelen et al., 2011). As a result of vertical erosion, the slopes of the breach sides become steeper. If a certain critical slope value is exceeded, the slope collapses. As shown in Figure 8.22, under the action of the flow traction, the bank toe is scoured with a bank retreat distance of ΔB. The body of bank slope may then lose its stability under the action of gravity and failure may occur along the critical failure surface.

The retreat distance of bank toe during a time-step Δt is evaluated by Wang et al. (2008)

$$\Delta B = C_l \frac{\tau - \tau_c}{\gamma} e^{-1.3\tau_c} \Delta t \tag{8.37}$$

where C_l is the erodibility coefficient related to bank soil, γ is the bulk density of soil, N/m^3, and τ, τ_c are the flow shear stress acting on the bank toe and the critical shear stress of bank soil, respectively, Pa.

In the case of subsequent bank failure, since the preceding slide occurs along the critical failure surface, the bank slope α_0 will be equal to the angle α_c as shown in Figure 8.22(b). α_c can be evaluated as

$$\alpha_c = 0.5\left\{\arctan\left[\left(H_1 / H_2\right)\tan\alpha_0\right] + \phi\right\} \tag{8.38}$$

where ϕ is the internal friction angle of bank soil, °, H_1 is the initial bank height, m, and H_2 is the height of uneroded bank, m.

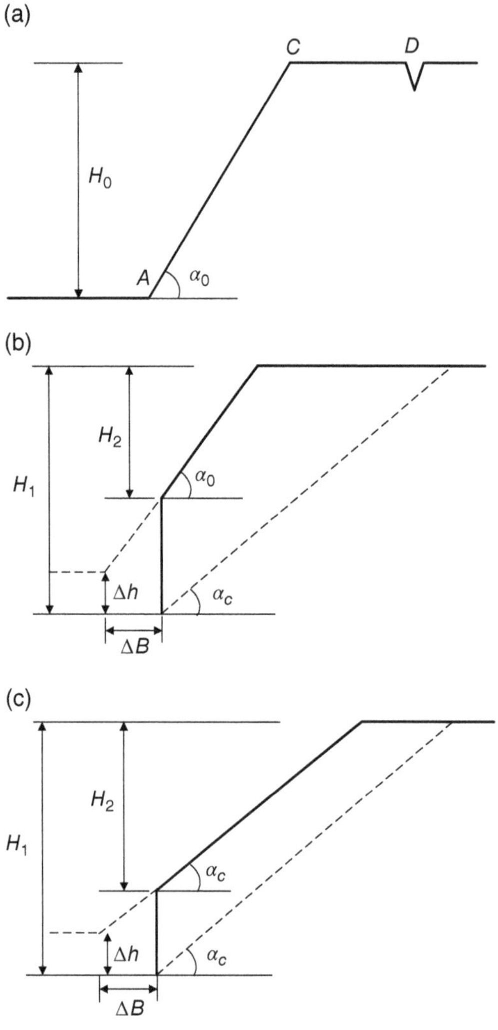

Figure 8.22 Sketch of bank collapse process: (a) initial breach, (b) initial bank failure, (c) subsequent bank failure (based on Osman and Thorne 1988; Wang et al. 2008)

The threshold of H_1/H_2 is evaluated to assess the stability of bank slope as

$$\left(\frac{H_1}{H_2}\right)_c = 0.5\left[\frac{\lambda_2}{\lambda_1} + \sqrt{\left(\frac{\lambda_2}{\lambda_1}\right)^2 - 4\left(\frac{\lambda_3}{\lambda_1}\right)}\right] \tag{8.39}$$

where

$$\lambda_1 = \sin\alpha_c \cos\alpha_c - \cos^2\alpha_c \tan\phi \tag{8.40}$$

$$\lambda_2 = 2c/(\gamma H) \quad (8.41)$$

$$\lambda_3 = \left(\sin\alpha_c \cos\alpha_c \tan\phi - \sin^2\alpha_c\right)/\tan\alpha_0 \quad (8.42)$$

where c is the cohesion of the soil, kPa. The bank slope will be stable when $(H_1/H_2) \le (H_1/H_2)_c$; otherwise, the bank slope will lose its stability and collapse along the failure surface.

The slope stability analysis in detailed physically based models is similar to the slope stability analysis in simplified physically based models. Downstream slope failure in the breach channel is considered in Tingsanchali and Chinnarasri's model (Tingsanchali and Chinnarasri, 2001) and Wang and Bowles's model (Wang and Bowles, 2006). In these two models, the downstream slope stability is modeled using the conventional method of slices.

8.4.3 Case Studies

The simulation of the overtopping erosion process of an embankment dam (i.e. South Fork Dam) and a landslide dam (i.e. Tangjiashan Landslide Dam) are presented in this section.

8.4.3.1 South Fork Dam (Quiroga and Singh, 1987)

The South Fork Dam, located near Johnstown, Pennsylvania, USA, failed on 31 May 1889, due to overtopping caused by extremely heavy rainfall. The dam was built in 1838–1853 by the Commonwealth of Pennsylvania as a reservoir for the state's Main Line of Public Works canal basin. The dam was built with earth and gravel, 22 m in height and 284 m in length across the river. On 31 May 1889, after several days of unprecedented rainfall, the dam finally failed. At the moment of failure, the reservoir water volume was approximately 18.9×10^6 m^3 with a surface area of 165 ha. The peak discharge through the breach was estimated to be 5600–8500 m^3/s. It took 3.5 h for the breach to develop completely (MacDonald and Langridge-Monopolis, 1984).

The BEED model was utilized to simulate the South Fork Dam failure using the following input parameters: Chezy's friction factor $= 80\sqrt{m}/s$, sediment diameter = 0.014 m, internal friction angle = 30°, and cohesion = 20 kPa. The parameters for Bagnold's equation were: $\tan(\alpha) = 0.24$ and $e_b = 0.10$. The outflow hydrograph, sediment graphs, and breach evolution sketch from the BEED model are shown in Figure 8.23. There is a close correspondence between the computed and observed breach peak outflow and breach formation time.

8.4.3.2 Tangjiashan Landslide Dam (Chang and Zhang, 2010)

Numerical simulation of the overtopping erosion process of the Tangjiashan landslide dam is described in this section. The general information of the landslide dam was introduced in Section 3.3.4. The profile of the Tangjiashan landslide dam along the breach channel is shown in Figure 8.24(a). The landslide dam mainly consists of two layers of materials: silty sandy gravel with cobbles in the top layer and rock with gravel and sand in the bottom layer. Several soil samples along the side slope of the breach at different depths were taken to measure the basic soil parameters in the laboratory.

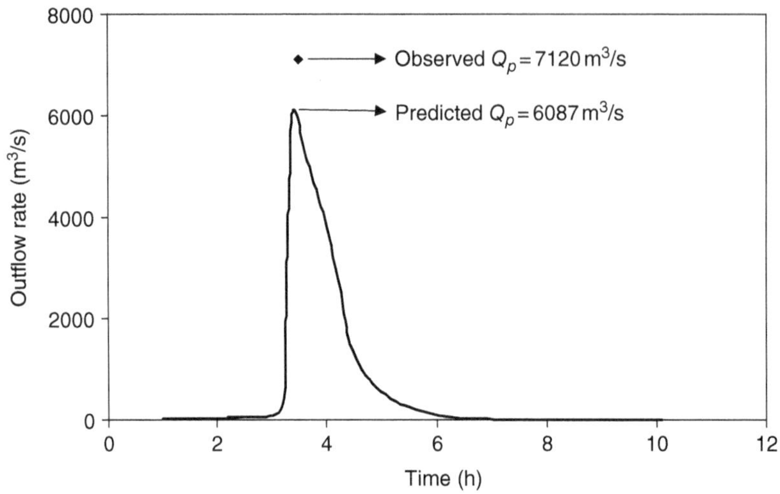

Figure 8.23 Breach hydrograph for the South Fork Dam (after Quiroga and Singh 1987)

According to the measured soil parameters, the soil erodibility at different depths can be calculated. The distributions of the coefficient of soil erodibility, K_d, and critical erosive shear stress, τ_c, along depth are shown in Figure 8.24(b). K_d decreases from 120 mm³/N-s to 10 mm³/N-s and τ_c increases from 10 Pa to 5500 Pa along the depth.

The initial discharge channel geometry is defined by the following parameters: $B_b = 8$ m, $B_t = 38$ m, $D = 10$ m, $B_{c0} = 350$ m, $\alpha_0 = 33.7°$, $\beta_0 = 13.5°$, $\beta_u = 20°$, $\alpha_c = 50°$, $\beta_f = 30°$. The input increment of erosion depth Δd_i is set to be 0.01 m.

The computational procedure is as follows:

1. Input the initial conditions of the breach geometry (B_{b0}, α_0, h_{w0}, B_{t0}, D_0, β_0, β_u), the dam geometry parameters (B_{c0}, L_0), the basic soil parameters along the depth of the landslide dam (e, C_u, d_{50}, PI, P, G_s, ϕ), the critical slope angles α_c and β_f, the lake volume and lake water surface area relationship, the inflow rate (Q_{in}), and the seepage discharge (Q_s).
2. Input the increment of erosion depth Δd_i at the crest.
3. Compute the erosion time Δt_i at the crest; then, based on the average water depth and time, compute the erosion depth along the downstream slope during this period ($\Delta d_i = E_i \times \Delta t_i$). Evaluate which stage the development of breach geometry is in and update the breach geometry.
   At stage I:

$$B_{bi} = B_{bi-1} + D_{i-1}\text{ctan}\,\alpha_{i-1} - \left(D_i + \Delta d_i\right)\text{ctan}\left(\alpha_{i-1} + \Delta\alpha\right); B_{ti} = B_{ti-1}; D_i = D_{i-1} + \Delta d_i;$$

   At stage II:

$$B_{bi} = B_{bi-1} + \Delta d_i \tan\left(\alpha_c / 2\right); B_{ti} = B_{ti-1} + \Delta d_i\left(1/\sin\alpha_c\right); D_i = D_{i-1} + \Delta d_i;$$

   At stage III:

$$B_{bi} = B_{bi-1} + \Delta d_i\left(1/\sin\alpha_c\right); B_{ti} = B_{ti-1} + \Delta d_i\left(1/\sin\alpha_c\right); D_i = D_{i-1}.$$

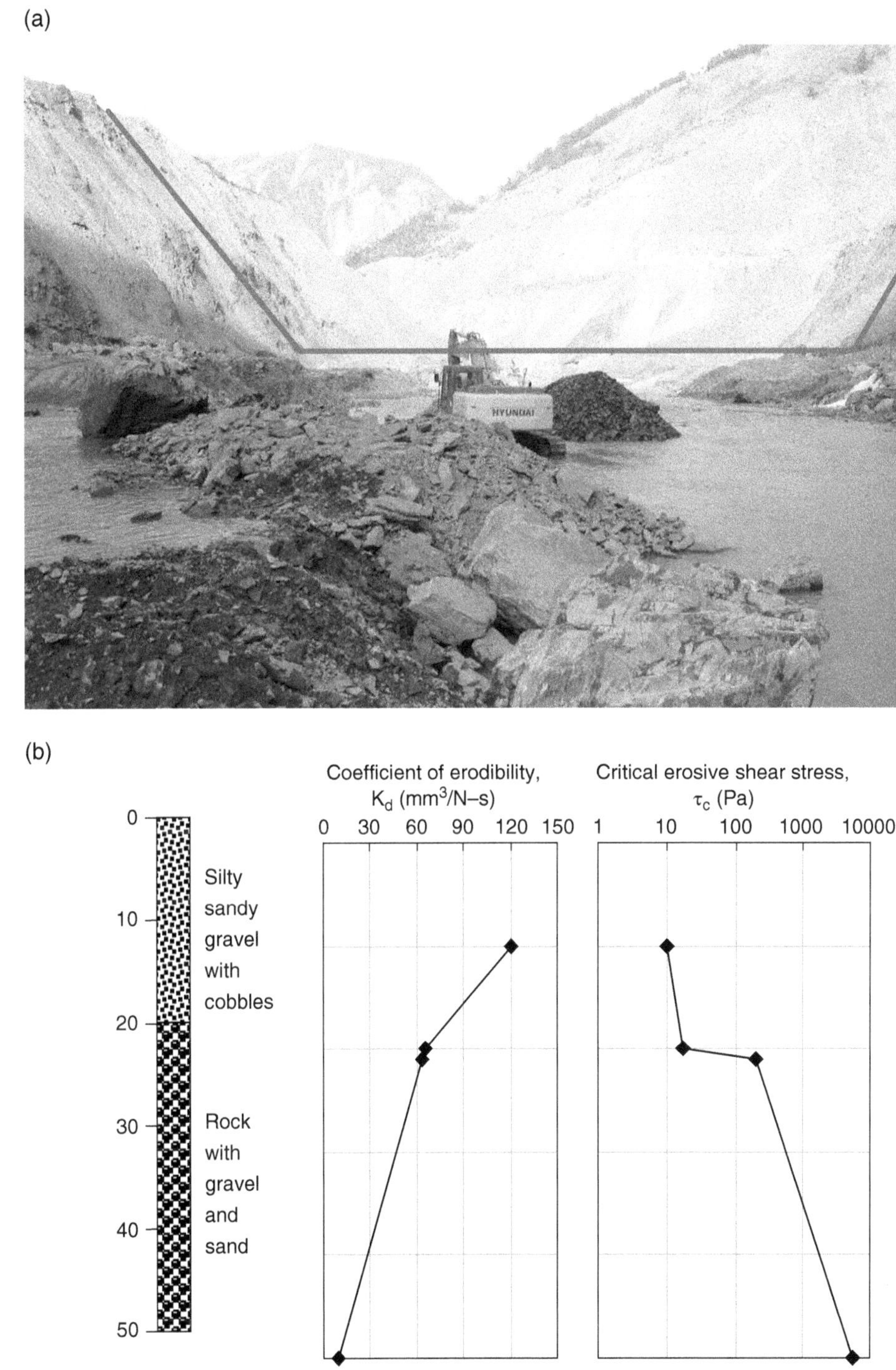

Figure 8.24 (a) Tangjiashan landslide dam after breaching, (b) soil profile and variations in soil erodibility along depth of Tangjiashan landslide dam

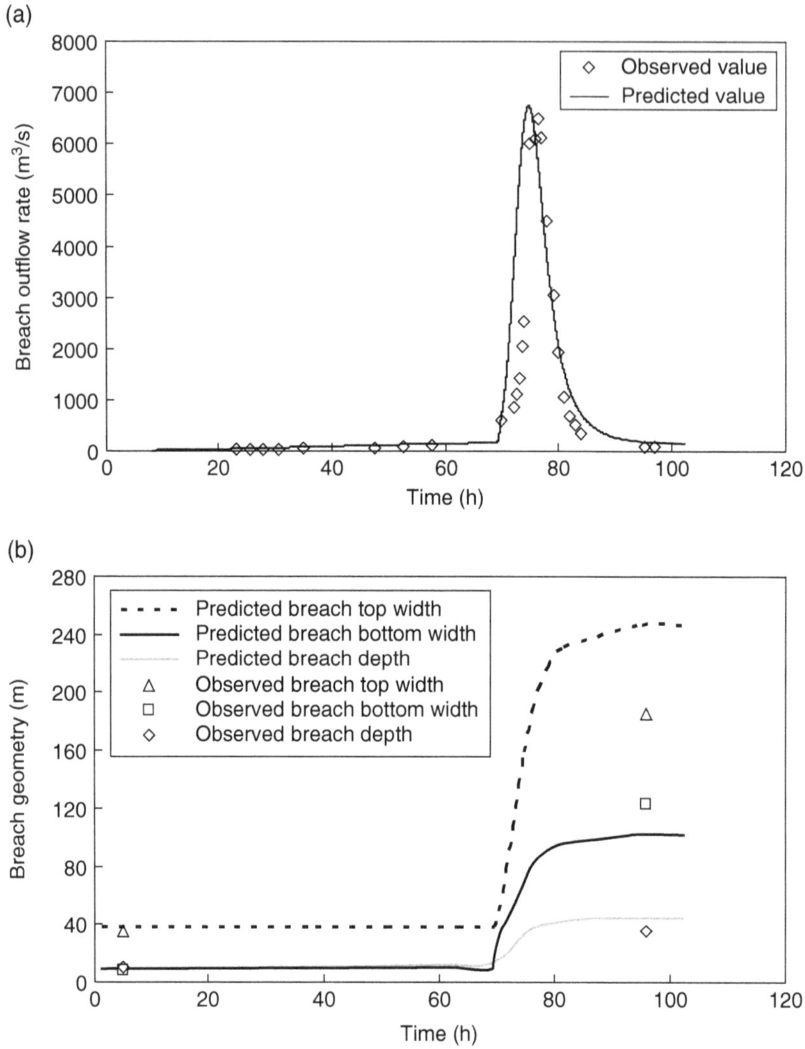

Figure 8.25 Predicted and observed breach parameters for the Tangjiashan landslide dam: (a) outflow hydrographs, (b) breach geometry

In these equations, the subscripts i, $i-1$, and $i+1$ represent the ith step, the (i – 1)th step, and the (i + 1)th step, respectively, D is the breach depth, and B_t is the breach top width.

4. Update the downstream slope angle, $\beta_i = \beta_{i-1} + \Delta\beta_i$, and check if β_i reaches β_f.
5. Compute discharge flow Q_{bi}.
6. Update the breach elevation, $Z_i = Z_{i-1} - \Delta d_i$, and water surface elevation.
7. Examine whether the crest has been eroded completely. If not, the erosion at both the crest and downstream slope should be simulated; otherwise, the erosion can only occur along the downstream slope.
8. Go to the next depth step.

9. Compute until the shear stress induced by the water flow is less than the critical erosive shear stress of the soils within the wetted area of the breach cross-section.
10. Plot the time histories of the breach outflow hydrograph, breach depth, breach top width, breach bottom width, water depth within the breach channel, and lake water level, respectively.

Figure 8.25(a) shows the predicted breach outflow hydrography along with the observed values. The predicted outflow hydrograph agrees well with the observed value. The predicted peak outflow rate is 6737 m^3/s, which is slightly larger than the observed value of 6500 m^3/s. The predicted breach initiation time is 69.4 hours, which is very close to the observed value of 69.9 hours. The evolution of the breach size (i.e. breach top width, breach bottom width, breach depth) is shown in Figure 8.25(b). The enlargement of the breach is slow in the first 69.4 hours, and then accelerates in the following 12 hours. After that the evolution rate slows down again until the final equilibrium state is reached. The predicted final breach top width, breach bottom width, and breach depth are 247, 101, and 45 m, respectively.

8.5 Numerical Simulation of Internal Erosion

It is possible to use numerical analysis to simulate the internal erosion process initiated by one of the four mechanisms (concentrated leak erosion, backward erosion, contact erosion, and suffusion) and the resulting outflow hydrograph. Currently, the models simulating dam breaching caused by internal erosion can be classified into two groups: continuum models and discrete-element models. Since surface and subsurface flows as well as pipe and open-channel flows all exist, detailed multidimensional simulation of dam breaching induced by piping is very challenging.

8.5.1 Continuum Methods

The existing models, such as DAMBRK (Fread, 1977) and BREACH (Fread, 1988) presented in Section 8.4, are able to predict the piping process after the formation of a connected "pipe" through the dam, whereas the internal erosion process prior to the formation of a connected pipe cannot be properly simulated. For instance, the formation process of a highly permeable zone due to the loss of fine particles within a dam cannot be predicted properly. However this is important to assess the long-term performance of a dam with internal erosion potential. In the literature, only limited study has been conducted on modeling the internal erosion process.

Piping models also involve breach evolution, sediment transport, and breach hydraulics. The breach evolution process is assumed to operate differently in different models. In the DAMBRK model, the initial and final breach geometry and the breach time are assumed. In the BREACH model, the pipe is assumed to be rectangular. The bottom of the pipe erodes vertically downward while its top erodes at the same rate vertically upwards. The width of the pipe is assumed constant until the top of the pipe collapses. After that, piping failure will turn to be an overtopping failure, and the evolution of the breach geometry follows Figure 8.15, which has been presented in Section 8.4.1. The sediment transport equations adopted in these two models for piping failure are the same as those for overtopping failure. The pipe flow during piping failure process is controlled by orifice flow

$$Q_b = A_b\left[2g\left(H-H_p\right)/\left(1+fL/B\right)\right]^{1/2} \tag{8.43}$$

where A_b is the cross-sectional area of the pipe channel, m^2, H_p is the elevation of the assumed centerline of the initially rectangular-shaped pipe channel, m, L is the length of the pipe channel, m, B is the pipe width, m, and f is the Darcy friction factor.

With the continued erosion of the pipe, the piping process may change to overtopping process once the soil above the pipe collapses. The transition is assumed to occur when the following inequality is satisfied (Fread, 1988):

$$H < H_p + 2\left(H_{pu} - H_p\right) \tag{8.44}$$

where H_{pu} is the elevation of the top of the pipe channel, m. After that, the pipe flow changes to weir flow and the overtopping failure process can be simulated based on Equation (8.28).

Obviously, the DAMBRK and BREACH models cannot simulate the initiation process of internal erosion. Recently, several researchers have explored numerical simulation of the initiation process of internal erosion by suffusion (Fujisawa et al., 2010) and backward erosion (Zhou et al., 2012). Bonelli et al. (2006), and Bonelli and Marot (2008) developed analytical solutions for piping erosion and internal erosion following mass and momentum balance and shear stress-based erosion law.

The numerical methods developed by Fujisawa et al. (2010) accounts for the saturated-unsaturated seepage flow of the pore liquid, the detachment of the soil particles from the soil fabric, and the migration of the eroded particles. The conservation of the liquid within an infinitesimal soil block is expressed by

$$\frac{\partial \theta}{\partial t}\Delta x_1 \Delta x_2 = \left(EA_c\right)\Delta x_1 \Delta x_2 - \frac{\partial v_1}{\partial x_1}\Delta x_1 \Delta x_2 - \frac{\partial v_2}{\partial x_2}\Delta x_1 \Delta x_2 \tag{8.45}$$

where θ is the volumetric water content, E is the erosion rate, m^3/s, A_c is the area of the erodible region per unit volume, m^2, and v_1, v_2 are the horizontal and vertical component of the flow velocity, m/s.

The internal erosion also decreases the volume of the soil skeleton by the loss of particles, which leads to the following equation:

$$\frac{\partial (1-n)}{\partial t}\Delta x_1 \Delta x_2 = -EA_c \Delta x_1 \Delta x_2 \tag{8.46}$$

The volume of the eroded soil particles within the pore fluid also changes with erosion, which can be expressed as

$$\frac{\partial C\theta}{\partial t}\Delta x_1 \Delta x_2 = \left(EA_c\right)\Delta x_1 \Delta x_2 - \frac{\partial C v_1}{\partial x_1}\Delta x_1 \Delta x_2 - \frac{\partial C v_2}{\partial x_2}\Delta x_1 \Delta x_2 \tag{8.47}$$

Substitute the relationship of $\theta = nS_r$ into Equations. 8.52 and 8.53, and combine these two equations, Equation (8.46) can be rewritten as

$$n\frac{\partial S_r}{\partial t} + \frac{\partial v_1}{\partial x_1} + \frac{\partial v_2}{\partial x_2} = \left(1 - S_r\right)EA_c \tag{8.48}$$

where S_r = degree of saturation.

Assuming internal erosion only occurs in the saturated zone (i.e. $S_r = 1$), and applying Darcy's law into Equation (8.48), gives

$$n\frac{\partial S_r}{\partial t}+\frac{\partial}{\partial x_i}\left(k\frac{\partial}{\partial x_i}\left(z+\frac{\mu_w}{\gamma_w}\right)\right)=0 \tag{8.49}$$

where k is the coefficient of permeability of the soil, m/s, z is the elevation of flow, m, and μ_w is the pore water pressure, kPa.

The Van Genuchten model (Van Genuchten, 1980) is used to describe the permeability and water retention characteristics of the soil as follows:

$$S_e=\left[\frac{1}{1+(\alpha\psi)^{n_G}}\right]^{m_G} \tag{8.50}$$

$$k=k_{sat}S_e^{\,1/2}\left[1-\left(1-S_e^{1/m_G}\right)^{m_G}\right]^2 \tag{8.51}$$

where S_e is the effective degree of water content, ψ is the suction, kPa, k_{sat} is the saturated permeability, m/s; α, m_G, n_G are the material constants.

The erosion of the fine particles is characterized by Equation 7.24. By assuming the soil pores assembly as a pore tube, as shown in Figure 8.26, the shear stress applied onto the pore wall can be estimated as

$$\tau=\gamma_w i\sqrt{2K/n} \tag{8.52}$$

where K is the intrinsic permeability of the soil, m^2, and i is the hydraulic gradient.

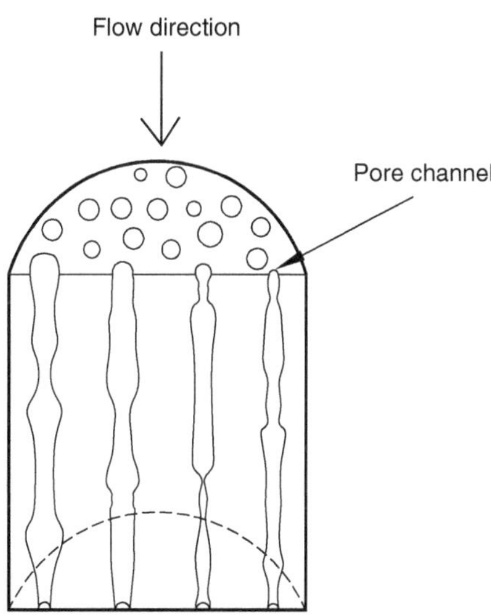

Figure 8.26 Illustration of pore channels within typical filter element

Continuum-based models can better simulate the seepage conditions within the dam during internal erosion, and the erosion zones within the dam satisfactorily. However, how the seepage flow affects the particle movement or rearrangement and how the change of void ratio affects the seepage flow is difficult to model. Moreover, piping is a sophisticated process involving not only soil–water interactions but also the effects of gravity. Arching effects and sudden collapse of pipe crown caused by the effect of gravity have been observed in several laboratory and field tests recently (Morris et al., 2007; Hanson et al., 2010a), which were found to play an important role during the piping process. These effects have to be properly considered in a piping model.

8.5.2 Particle Level Analysis

From discrete element standpoint, the modeling of piping failure includes modeling the fluid phase, fluid–particle interactions, and the particle motion. EI Shamy and Aydin (2008) stated that flow conditions that would result in piping encompass several issues that have to be accounted for when developing a computational model for fluid flow through deforming porous medium:

1. In the case of flood-induced piping, transient analysis has to be considered, since flooding of a river results in a rapidly increasing hydraulic head that a steady flow regime is unlikely to occur.
2. Soil particles may undergo large displacements leading to significant changes in porosity under such extreme flow conditions.
3. Variations in porosity affect the soil hydraulic conductivity and deformation characteristics.

In EI Shamy and Aydin's model, the porous soil medium is modeled as a mixture of two phases: the fluid phase (water) and the particulate solid phase. The fluid is idealized as a continuum by using an averaged form of the Navier–Stokes equations that account for the presence of the solid particles. A small-scale levee structure was simulated using this model.

Lominé et al. (2013) developed a method to simulate the concentrated leak erosion. This method is based on a coupling between two methods: the discrete element method (DEM) and the lattice Boltzmann method (LBM). The DEM is employed to simulate interactions between particles, whereas the LBM is used to describe an interstitial Newtonian fluid flow. The particles are detached under the hydraulic loading. Then, detached particles are transported by the flow toward the channel outlet. The irregular evolution of the leak can be well simulated.

The particle-level analysis of internal erosion can better reflect the response of particles during the erosion process. However, a common question for particle-level based modeling is its representation for the real problem. As the particle number is usually limited, the scale of the model is usually in the centimeter level. The practical application of such models is doubtful in the present time because dams usually have dimensions of tens to hundreds meters.

8.5.3 Case Studies

In this section, the simulations of the piping failure process of an embankment dam (i.e. Teton Dam) and a hypothetical piping failure of a landslide dam (i.e. Spirit Lake Landslide Dam) are presented.

Teton Dam (Independent Panel to Review Cause of Teton Dam Failure 1976, Seed and Duncan, 1987; Xu and Zhang, 2013)

The Teton Dam was a central-core zoned dam situated on the Teton River in Idaho, USA. The maximum height of the dam was approximately 92 m above the stream bed. The slopes of the upstream and downstream surfaces are 1:2.5 and 1:2, respectively. A typical cross-section of the dam can be found in Figure 8.27. The materials in Teton Dam can be divided into five zones. According to laboratory tests and in-situ drill hole test results, the materials in zone 1 can be classified as silt or silty clay (CL or CL-ML).

Teton Dam failed on 5 June 1976 during the first filling of the reservoir. The failure of the dam was probably initiated by piping. The exit of seepage water was found at 7:00 a.m. Before 10:00 a.m., turbid water was observed running out through the downstream surface, which was an indication of the initiation of piping failure. A large leakage flow rate was reported at 10:00 on the right abutment. Then the pipe kept enlarging and at about 12:00 noon, the soil mass above the pipe collapsed and the failure changed to an overtopping failure. The peak discharge occurred in the late afternoon during the overtopping process.

The piping failure process of Teton Dam is simulated using Xu and Zhang's piping model (Xu and Zhang, 2013). Only zone 1 and zone 2 materials are considered in the simulation for simplicity. Piping is set to be initiated from a small crack through the dam at an elevation of 1585 m, which is the elevation where a large leak was first observed. The initial crack width and height are assumed to be 0.02 m and 0.10 m, respectively. The overtopping failure process after all the soil mass above the pipe has collapsed is simulated using the DABA model developed by Chang and Zhang (2010).

The predicted evolution process of outflow hydrograph and water level during the failure process are shown in Figure 8.28. In the simulation, the piping process lasts about 10 hours before it finally turns into an overtopping process. The outflow rate of the pipe keeps increasing in the piping process and reaches a value of 4480 m^3/s before the transition of failure mode. After global collapse of soil mass happens, a sudden increase of dam failure discharge occurs. This is because of the assumption that the collapsed soil will be washed away immediately so that a broad-crested weir forms right after the collapse. The peak discharge occurs during the overtopping process in the afternoon.

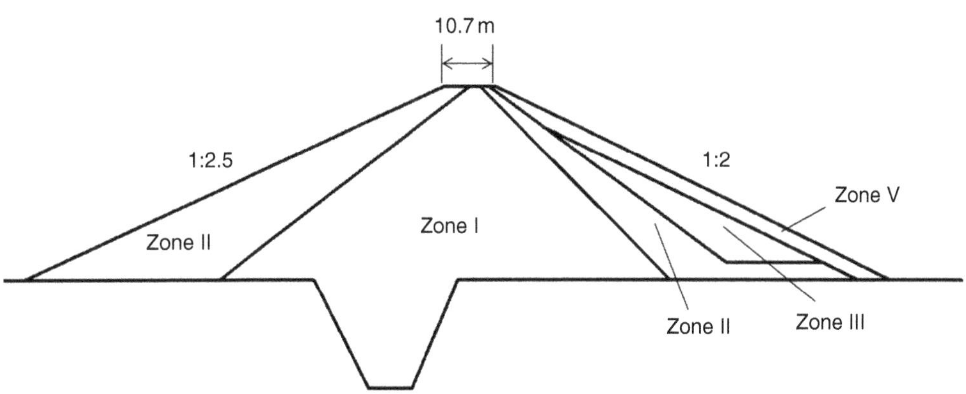

Figure 8.27 Typical cross-section of Teton Dam

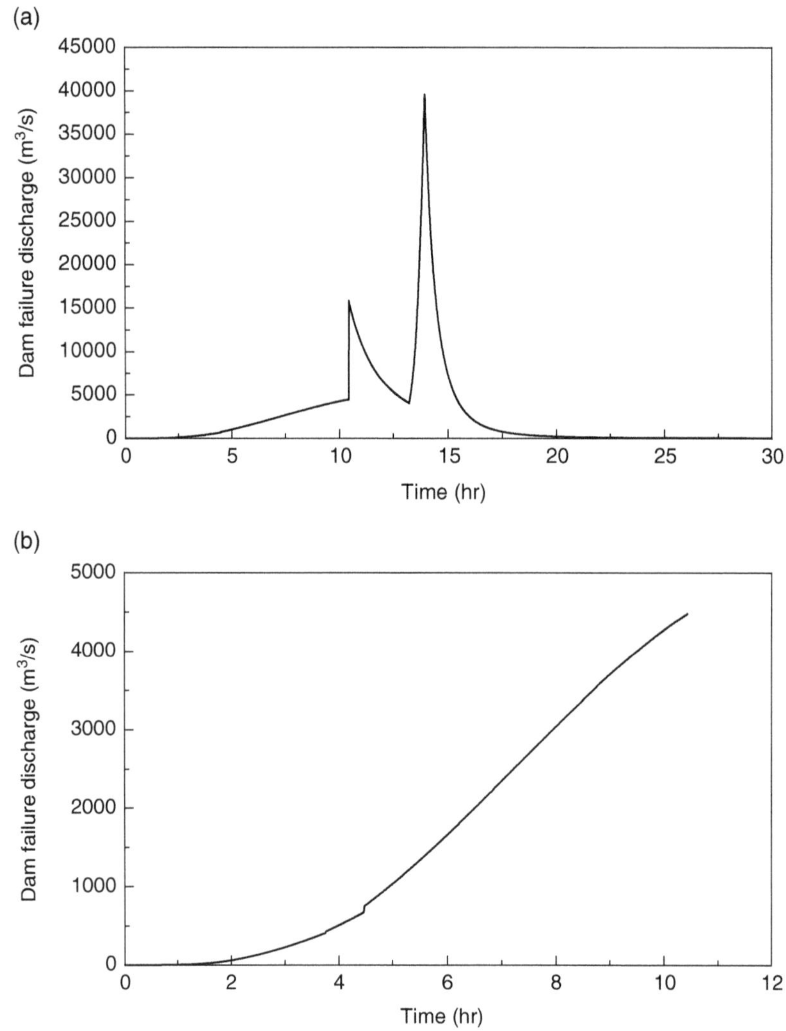

Figure 8.28 Predicted outflow hydrograph during the failure process of the Teton Dam: (a) whole failure process; and (b) piping process

Spirit Lake Landslide Dam (Sager and Chambers, 1986, Fread, 1988, Schuster and Evans, 2011)

The violent eruption of Mount St Helens on 18 May 1980 in Washington, triggered a massive debris avalanche which moved down the north side of the volcano deposition about $3\times10^9\,m^3$ of materials in the upper 17 km of the North Fork of the Toutle River valley. The debris avalanche dammed the North Fork and tributaries, impounding four landslide-dammed lakes, the largest being Spirit Lake with deposits of up to 152 m deep. Spirit Lake was dramatically changed by the avalanche. The lake has a maximum volume of $380\times10^6\,m^3$ at the elevation of 1059 m when breaching of the landslide dam was anticipated. A government panel concluded that the Spirit Lake dam could not hold water above elevation 1060 m without breaching, and estimated that the lake would reach this level by March 1983. By August 1982, $330\times10^6\,m^3$ of water had been impounded and the lake level had reached 1055 m. To avoid breaching, the Corps of Engineers installed temporary pumps to maintain the lake level at about elevation 1055 m and planned to construct a permanent outlet channel in the near future which would bypass the landslide dam and maintain safe lake levels. Four possible alternatives were considered for a permanent solution to the Spirit Lake hazard: (1) a permanent pumping facility, (2) an open-channel spillway, (3) a buried conduit, and (4) a diversion tunnel through the bedrock right abutment. Each alternative was evaluated in regard to (1) stability in relation to the debris avalanche, (2) stability in case of volcanic or seismic activity, (3) efficacy of construction, (4) environmental disturbance of the site, and (5) cost. Based on these considerations, it was decided to maintain a permanent lake level at elevation 1048 m by means of a tunnel 2.59 km in length and 3.4 m in diameter through the volcanic-ridge right abutment immediately west of Spirit Lake.

Greater than normal precipitation, failure of the pumping system, and/or addition of more avalanche material from another eruption of the volcano could cause the lake level to exceed elevation 1059 m and possibly cause the dam to fail. Such a hypothetical breach was simulated using the BREACH model (Fread, 1988). The landslide dam failure was assumed by piping, which was occurred at elevation of 1051 m. The input parameters are: geometry of the dam: $H_u = 1059$ m, $H_p = 1051$ m, $H_l = 1012$ m, ZD = 30, ZU = 22; material properties: $d_{50} = 7$ mm, $n = 0.32$, $\phi = 35°$, $c = 150$ kPa; and initial pipe width: 0.03 m

The simulated outflow has a peak of about 11200 m^3/s occurring 15 hours after the start of failure. The time of rise is about 2 hours. The final breach dimensions are $D = 47$ m, $W = 128$, and $\alpha = 50°$.

9

Analysis of Dam Breaching Flood Routing

Flood risk is closely related to the flood severity in flooded areas, specifically the water depth, flow velocity, and flood rising rate. The flood severity can be evaluated through a flood routing simulation. This chapter introduces the river hydraulics and numerical models for flood simulation. A case study is also presented to demonstrate the application of a numerical model.

9.1 River Hydraulics

The Navier–Stokes equations for an incompressible fluid are the core for most unsteady flow routing computer simulations (Gilles and Moore, 2010). The continuity of fluid motion is described as

$$\frac{\partial p}{\partial t}+\frac{\partial u}{\partial x}+\frac{\partial v}{\partial y}+\frac{\partial w}{\partial z}=0 \tag{9.1}$$

Using the equations of motion and continuity, the Navier–Stokes equations of fluid motion are as follows:

$$\rho\left(\frac{\partial u}{\partial t}+u\frac{\partial u}{\partial x}+v\frac{\partial u}{\partial y}+w\frac{\partial u}{\partial z}\right)=-\frac{\partial p}{\partial x}+\mu\nabla^2 u+\rho g_x \tag{9.2}$$

$$\rho\left(\frac{\partial v}{\partial t}+u\frac{\partial v}{\partial x}+v\frac{\partial v}{\partial y}+w\frac{\partial v}{\partial z}\right)=-\frac{\partial p}{\partial y}+\mu\nabla^2 v+\rho g_y \tag{9.3}$$

Dam Failure Mechanisms and Risk Assessment, First Edition. Limin Zhang, Ming Peng, Dongsheng Chang, and Yao Xu.

$$\rho\left(\frac{\partial w}{\partial t}+u\frac{\partial w}{\partial x}+v\frac{\partial w}{\partial y}+w\frac{\partial w}{\partial z}\right)=-\frac{\partial p}{\partial z}+\mu\nabla^2 w+\rho g_z \tag{9.4}$$

where ρ is the fluid density, x, y, and z are the Cartesian coordinates, t is time, u, v, and w are the velocity components in the x, y, and z directions, respectively, p is the pressure, μ is the viscosity, and g is the gravitational acceleration.

While these governing equations are applicable to almost all situations, computational constraints may dictate the degree of simulation detail that can be achieved.

9.1.1 One-dimensional Models

The governing equations of the most widely used one-dimensional (1D) models are Saint-Venant equations (Bates and De Roo, 2000). Several assumptions are often made for 1D analyses (Cunge et al., 1980):

1. The flow is one-dimensional.
2. The water level across the section is horizontal.
3. The streamline curvature is small and vertical accelerations are negligible.
4. The effects of boundary friction and turbulence can be accounted for using resistance laws analogous to those for steady flow conditions.
5. The average channel bed slope is small so the cosine of the angle can be replaced by unity.

Based on the conservation equations of mass and momentum for a control volume, the Saint-Venant equations are shown in differential form as

$$\frac{\partial A}{\partial t}+\frac{\partial Q}{\partial x}=0 \tag{9.5}$$

$$\frac{\partial Q}{\partial t}+\frac{\partial (uQ)}{\partial x}+gA\left(\frac{\partial h}{\partial x}-S_0\right)+gAS_f=0 \tag{9.6}$$

where Q is the discharge, A is the cross-sectional flow area, u is the longitudinal flow velocity, h is the flow depth, S_o is the bed slope, and S_f is the friction slope.

9.1.2 Two-dimensional Models

The 1D flow may not be valid in many situations, such as flow in a non-prismatic channel (i.e. a channel with varying cross-section and alignment), flow downstream of a partially breached dam, or lateral flow from a failed dike. Although flows in these situations are three-dimensional, we may simplify the analysis by considering them as two-dimensional flows by using vertically averaged quantities (Cook, 2008).

Most widely used two-dimensional codes utilize depth-averaged Navier–Stokes equations, commonly called the Saint-Venant shallow water equations:

$$\frac{\partial d}{\partial t}+\frac{\partial (dU)}{\partial x}+\frac{\partial (dV)}{\partial y}=0 \tag{9.7}$$

$$\frac{\partial dU}{\partial t}+\frac{\partial\left(dU^2\right)}{\partial x}+\frac{\partial\left(dUV\right)}{\partial y}=\frac{\partial\left(dT_{xx}\right)}{\partial x}+\frac{\partial\left(dT_{xy}\right)}{\partial y}-gh\frac{\partial z}{\partial x}-\frac{\tau_{bx}}{\rho} \tag{9.8}$$

$$\frac{\partial dV}{\partial t}+\frac{\partial\left(dUV\right)}{\partial x}+\frac{\partial\left(dV^2\right)}{\partial y}=\frac{\partial\left(dT_{xy}\right)}{\partial x}+\frac{\partial\left(dT_{yy}\right)}{\partial y}-gh\frac{\partial z}{\partial y}-\frac{\tau_{by}}{\rho} \tag{9.9}$$

where d is the flow depth, U and V are the average velocities in the x and y directions, T_{xx} and T_{xy}, and T_{yy} are the depth-averaged turbulent stresses, z is the water surface elevation, and τ_{bx} and τ_{by} are the bed shear stresses.

9.2 Numerical Models for Flood Routing Analysis

9.2.1 One-dimensional Numerical Models

A typical one-dimensional model for flood routing is HEC-RAS, developed by the US Army Corps of Engineers (HEC, 2008). In a HEC-RAS steady state simulation, water surface profiles are computed from one cross-section to the next by solving the energy equation with an iterative procedure called the standard step method. The energy equation is written as

$$Z_2+Y_2+\frac{\varepsilon_2 v_2^{\,2}}{2g}=Z_1+Y_1+\frac{\varepsilon_2 v_2^{\,2}}{2g}+h_e \tag{9.10}$$

where Z_1 and Z_2 are the elevation of the main channel inverts, Y_1 and Y_2 are the depth of water at cross-sections 1 and 2, V_1 and V_2 are the average velocities (total discharge/total flow area), ε_1 and ε_2 are the velocity weighting coefficients, g is the gravitational acceleration, and h_e is the energy head loss, as shown in Figure 9.1.

The energy head loss (h_e) between two cross-sections is comprised of friction losses and contraction or expansion losses. The equation of energy head loss is

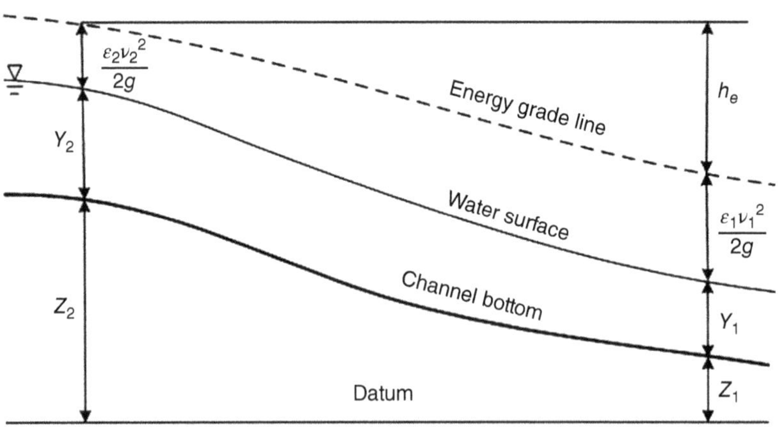

Figure 9.1 Water surface profile for energy equation (after HEC, 2008, public domain, http://www.hec.usace.army.mil/software/hec-ras/)

Table 9.1 Subcritical flow contraction and expansion coefficients (after HEC, 2008, public domain, http://www.hec.usace.army.mil/software/hec-ras/)

Conditions	Contraction	Expansion
No transition loss computed	0.0	0.0
Gradual transitions	0.1	0.3
Typical bridge sections	0.3	0.5
Abrupt transitions	0.6	0.8

$$h_e = L_R \bar{S}_f + C_L \left| \frac{\varepsilon_2 v_2^2}{2g} - \frac{\varepsilon_1 v_1^2}{2g} \right| \tag{9.11}$$

where L_R is the discharge weighted reach length, $\bar{S}_f$ is the representative friction slope between two sections, and C_L is the expansion or contraction loss coefficient.

Typical values for expansion or contraction loss coefficients for subcritical flows are shown in Table 9.1. A subcritical flow occurs when the flow velocity is smaller than the wave velocity. The weighted friction slope using the average conveyance equation and the distance weighted reach length are defined as

$$\bar{S}_f = \left(\frac{Q_1 + Q_2}{K_1 + K_2} \right)^2 \tag{9.12}$$

$$L = \frac{L_{lob}\bar{Q}_{lob} + L_{ch}\bar{Q}_{ch} + L_{rob}\bar{Q}_{rob}}{\bar{Q}_{lob} + \bar{Q}_{ch} + \bar{Q}_{rob}} \tag{9.13}$$

where K is the conveyance, L_{lob}, L_{ch} and L_{rob} are the cross-section reach lengths specified for flow in the left over-bank, main channel, and right over-bank, respectively, and $\bar{Q}_{lob}$, $\bar{Q}_{ch}$, and $\bar{Q}_{rob}$ are the arithmetic average of the flows between sections for the left over-bank, main channel, and right over-bank, respectively.

HEC-RAS subdivides the flow in the main channel from the over-banks. Conveyance is calculated for each subdivision as

$$Q = KS_f^{1/2} \tag{9.14}$$

$$K = \frac{AR^{2/3}}{n} \tag{9.15}$$

where K is the conveyance for the subdivision, n is Manning's roughness coefficient for the subdivision, A is the flow area for the subdivision, and R is the hydraulic radius for each subdivision.

The total conveyance for each subdivision is calculated as the sum of the conveyance from the left over-bank, main channel, and right over-bank. The flow in the main channel is subdivided only when Manning's roughness coefficient changes within the channel area. The Manning roughness coefficient for the composite main channel is defined as

$$n_c = \left[\frac{\sum_{i=1}^{N} \left(P_i n_i^{1.5} \right)}{P} \right]^{2/3} \quad (9.16)$$

where n_c is the composite or equivalent coefficient of roughness, P is the wetted perimeter of the entire main channel, P_i is the wetted perimeter of subdivision *I*, and n_i is the coefficient of roughness for subdivision *i*. Chow (1959) recommended preliminary values of Manning's roughness coefficient.

The physical laws that govern the unsteady flow of water are the principle of conservation of mass (continuity) and the principle of conservation of momentum. These laws are expressed mathematically in partial differential equations in HEC-RAS. Conservation of mass for a control volume states that the net rate of flow into the volume is equal to the rate of change of storage inside the volume. The final continuity equation is expressed (HEC, 2008) as

$$\frac{\partial A_T}{\partial t} + \frac{\partial Q}{\partial x} - q_l = 0 \quad (9.17)$$

where A_T is the total flow area, Q is the flow rate, x is the length of the channel, and q_l is the lateral inflow per unit length.

Conservation of momentum for a control volume states that the net rate of momentum entering the volume (momentum flux) plus the sum of all external forces acting on the volume is equal to the rate of accumulation of momentum. The momentum flux is the fluid mass times the velocity vector in the direction of flow. Three forces are considered: pressure, gravity, and boundary drag, or friction force. The final form of the momentum equation is given by

$$\frac{\partial Q}{\partial t} + \frac{\partial Qv}{\partial x} + gA\left(\frac{\partial z}{\partial x} + S_f \right) = 0 \quad (9.18)$$

where v is the flow velocity, A is the flow area, and z is the water surface.

The analysis with HEC-RAS 4.0 uses the following procedure:

1. Start a new project.
2. Enter geometric data.
3. Enter flow data and boundary conditions.
4. Perform the hydraulic calculations.
5. View and print results.

The first step is to establish a new project by entering its information. The geometric data is needed after the definition of the new project. The geometric data in flood routing analysis includes: map of reservoir and river reaches; geometric information of cross-sections and other structures such as dams, levees, and bridges; Manning's n values, and expansion or contraction loss coefficients for each cross-section.

Flow rate and boundary conditions have to be entered. These include inflow rate upstream the studied river, lateral inflow rate, gate openings in dams, and the boundary (e.g. friction slope) at the end of the river. Hydraulic calculations can be performed after the setting of the timescale and the outputs. The output results from HEC-RAS include the water level, flow velocity, flow rate, and water kinetic energy in each time interval.

Other examples of widely used one-dimensional models are DAMBRK/FLOWAV (BOSS, 1999) by US National Weather Service (NWS), MIKE11 (DHI, 2003) by Danish Hydraulic Institute (DHI), ISIS Flow (Lin et al., 2006) jointly developed by Halcrow and Wallingford, SOBEK 1D (Vanderkimpen et al., 2009) by Deltares and Delft Hydraulics, and NWS FLDWAV (Fread and Lewis, 1998) by National Weather Service (NWS) in the USA.

9.2.2 *Two-dimensional Numerical Models*

A typical two-dimensional model for flood routing is FLO-2D, which evolved from the diffusive hydrodynamic model (DHM) (Hromadka and Yen, 1987). The first version of FLO-2D was applied in 1988 in a flood insurance study by Federal Emergency Management Agency (FEMA). The two-dimensional constitutive equations include the continuity equation and two equations of motion:

$$\frac{\partial h}{\partial t}+\frac{\partial (hV_x)}{\partial x}+\frac{\partial (hV_y)}{\partial y}=i \tag{9.19}$$

$$S_{fx}=S_{ox}-\frac{\partial h}{\partial x}-\frac{V_x}{g}\frac{\partial V_x}{\partial x}-\frac{V_y}{g}\frac{\partial V_x}{\partial y}-\frac{1}{g}\frac{\partial V_x}{\partial t} \tag{9.20}$$

$$S_{fy}=S_{oy}-\frac{\partial h}{\partial y}-\frac{V_y}{g}\frac{\partial V_y}{\partial y}-\frac{V_x}{g}\frac{\partial V_y}{\partial x}-\frac{1}{g}\frac{\partial V_y}{\partial t} \tag{9.21}$$

where h is the flow depth, V_x and V_y are the depth-averaged velocities along the x and y directions, respectively, i is the excess rainfall intensity, S_{fx} and S_{fy} are the friction slopes along the x and y directions, respectively, expressed as function of the respective bed slope (i.e. S_{ox} and S_{oy}), the pressure gradient, the convective acceleration, and the local acceleration terms.

The total friction slope, S_f, is the sum of the yield slope, the viscous slope, and the turbulent dispersive slope. Based on a quadratic rheological model developed by O'Brien et al. (1993), S_f can be written as

$$S_f=\frac{\tau_y}{\gamma_m h}+\frac{K\eta V}{8\gamma_m h^2}+\frac{n_{td}^2 V^2}{h^{4/3}} \tag{9.22}$$

where τ_y, γ_m and η are the yield stress, unit weight, and viscosity of the debris flow mixture, respectively, K is a resistance parameter for laminar flow, V is the depth-averaged velocity, and n_{td} is the equivalent Manning coefficient, which accounts for both the boundary roughness and resistance arising from the solid particle contact. The value of n_{td} can be calculated (FLO-2D Software Inc., 2009) as

$$n_{td}=0.0538ne^{6.0896C_v} \tag{9.23}$$

where n is the Manning coefficient and C_v is the volumetric sediment concentration, which is the ratio of the volume of the solid materials to the volume of the water and the solid materials. When simulating flooding, only the third term (the equivalent Manning coefficient term) is considered.

If a debris flow is simulated, then the first term (yield stress term) and the second term (viscosity term) are also introduced.

In numerical simulation, the surface topography is discretized into a grid of uniform square zones first. Each zone is assigned a location on the grid system, an elevation, a roughness factor, and area and flow width reduction factors used to simulate flow blockage. The two-dimensional governing equations are reduced to a one-dimensional form

$$\frac{\partial h}{\partial t}+\frac{\partial(hV)}{\partial x}=i \tag{9.24}$$

$$S_f=S_o-\frac{\partial h}{\partial x}-\frac{V}{g}\frac{\partial V}{\partial x}-\frac{1}{g}\frac{\partial V}{\partial t} \tag{9.25}$$

There are eight potential flow directions for each grid: the four compass directions (north, east, south, and west) and the four diagonal directions (north-east, south-east, south-west, and north-west). Each velocity computation is essentially one-dimensional in nature and is solved independently of the other seven directions.

The differential form of the continuity and momentum equations in the FLO-2D model is solved with a central, finite difference numerical scheme. This explicit algorithm solves the momentum equation for the flow velocity across the grid element boundary, one element at a time. The solution algorithm incorporates the following steps:

1. The average flow geometry, roughness, and slope between two grid elements are computed.
2. The flow depth dx for computing the velocity across a grid boundary for the next time-step $(i+1)$ is estimated from the previous time-step i using a linear estimate (the average depth between two elements).
3. The first estimate of the velocity is computed using the diffusive wave equation. The only unknown variable in the diffusive wave equation is the velocity for overland, channel or street flow.
4. The predicted diffusive wave velocity for the current time-step is used as a seed in the Newton–Raphson solution to solve the full dynamic wave equation for the solution velocity. It should be noted that for hyperconcentrated sediment flows such as mud and debris flows, the velocity calculations include the additional viscous and yield stress terms.
5. The discharge Q across the boundary is computed by multiplying the velocity by the cross-sectional flow area. For overland flow, the flow width is adjusted by the width reduction factors.
6. The incremental discharge for the time-step across the eight boundaries (or upstream and downstream channel elements) is expressed as

$$\Delta Q_x^{i+1}=Q_n+Q_e+Q_s+Q_w+Q_{ne}+Q_{se}+Q_{sw}+Q_{nw} \tag{9.26}$$

   and the change in volume is distributed over the available storage area within the grid or channel element to determine an incremental increase in the flow depth

$$\Delta d_x^{i+1}=\Delta Q_x^{i+1}\Delta t\,/\,A_{surf} \tag{9.27}$$

   where Δt is the interval between time i and $i+1$.

7. The numerical stability criteria are then checked for the new grid element flow depth. If any of the stability criteria is exceeded, the simulation time is reset to the previous simulation time, the time-step increment is reduced, all the previous time-step computations are discarded and the velocity computations begin again.
8. The simulation progresses, with increasing time-steps, until the stability criteria are exceeded.

Other two-dimensional models include MIKE21 by DHI (2007), SOBEK 2-D (Vanderkimpen et al., 2009) by Deltares and Delft Hydraulics, FST2DH (FESWMS, 2002) by Federal Highway Administration, US Department of Transportation, and DIVAST by Lin et al. (2006).

9.2.3 Coupling of 1D/2D Numerical Models

Modeling of urban flooding has presented several challenges in using typical 1D/2D numerical codes (Patro et al., 2009). One-dimensional numerical models cannot accommodate complex floodplain flow fields or require post-processing to produce realistic flood extents. Two-dimensional numerical models are, on the other hand, hard to model structural elements that may produce supercritical or pressurized flow conditions (Gilles and Moore, 2010). These limitations may be overcome by coupling of 1D/2D numerical models. In a coupled model, a 1D numerical model of the river channel complemented by a 2D model of the floodplain is used to provide improvements in hydraulic modeling accuracy and computational efficiency. If an entire river reach is modeled using a 1D model, then computational nodes within that portion of the two-dimensional mesh will not become active, improving computational efficiency (Gilles and Moore, 2010).

Recently 1D/2D models have emerged, such as MOUSE–MIKE 21, which couples the 1D MOUSE sewer model with the 2D MIKE 21 overland model (Carr and Smith, 2006), Sobek Urban, which couples 1D SOBEK flow and 2D Delft FLS (Bolle et al., 2006), or TUFLOW (Phillips et al. 2005). Research studies have attempted to compare the performance of these two types of model, either by using commercial software (Kaushik, 2006) or by focusing on research applications such as SIPSON/UIM (Chen et al., 2007). Lin et al. (2006) coupled ISIS and DIVAST, Delft-FLS, LISFLOOD-FP, SOBEK 1D2D, and MIKE FLOOD.

9.3 Example – Tangjiashan Landslide Dam Failure

Tangjiashan Landslide Dam (Section 3.3.4), with a height of 82–124 m, a volume of $20.4 \times 10^6\,m^3$ and a reservoir volume of $316 \times 10^6\,m^3$, was the largest landslide dam induced by the Ms 8.0 Wenchuan earthquake. The dam posed tremendous threats to the people downstream and finally breached in the morning of 10 June 2008 with a peak flow rate of $6500\,m^3/s$. This section applies HEC-RAS (HEC, 2008) to simulate the dam breaching floods in this landslide dam failure case.

9.3.1 Geometric Information

The studied river reach downstream of the dam site is approximately 90 km long. Altogether, 5303 river cross-sections are selected, 26 of which have detailed elevation and distance data. Figure 9.2 shows the geometric locations of the landslide lake, the dam, and the towns and cities downstream. Figure 9.3 shows a cross-section of the Beichuan town, which is located 3.5 km downstream of the dam. Two river bank points are defined to divide the river and floodplains. The Manning's n values

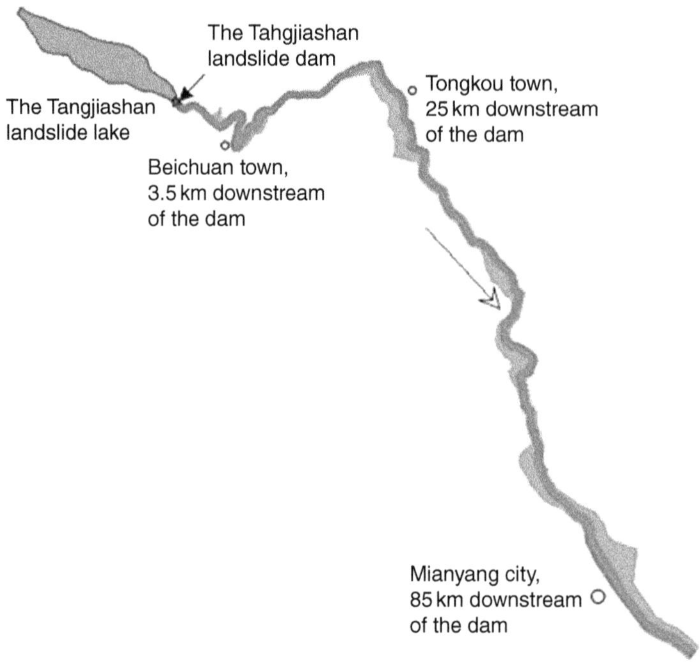

Figure 9.2 Geometric profiles of the Tangjiashan landslide dam and the river downstream

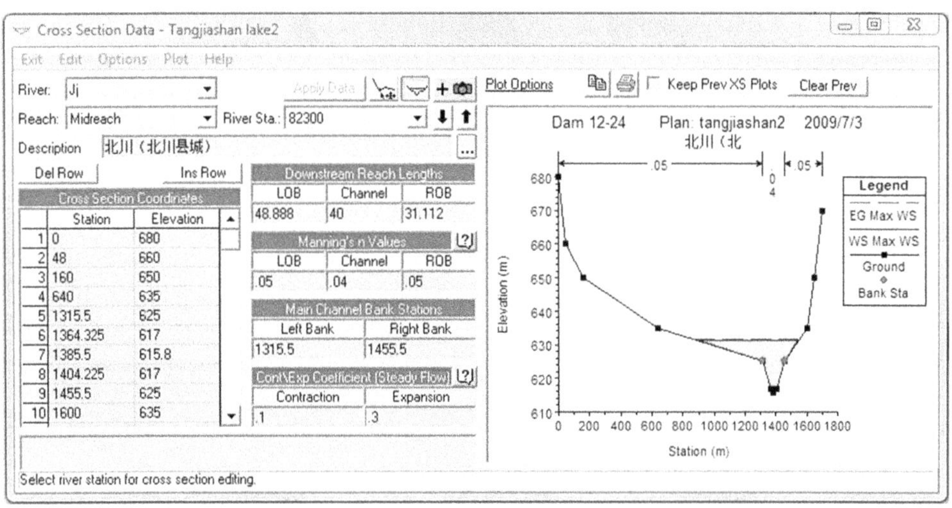

Figure 9.3 A cross-section of the river downstream of the dam

for the river channel and the floodplains are selected according to Chow (1959). Typical values for the contraction and expansion coefficients are selected following the HEC-RAS Reference Manual (HEC, 2008), as shown in Table 9.1. The diversion channel is simulated as a spillway as shown in Figure 9.4. With this spillway, the dam crest is lowered from 752 m to 740 m.

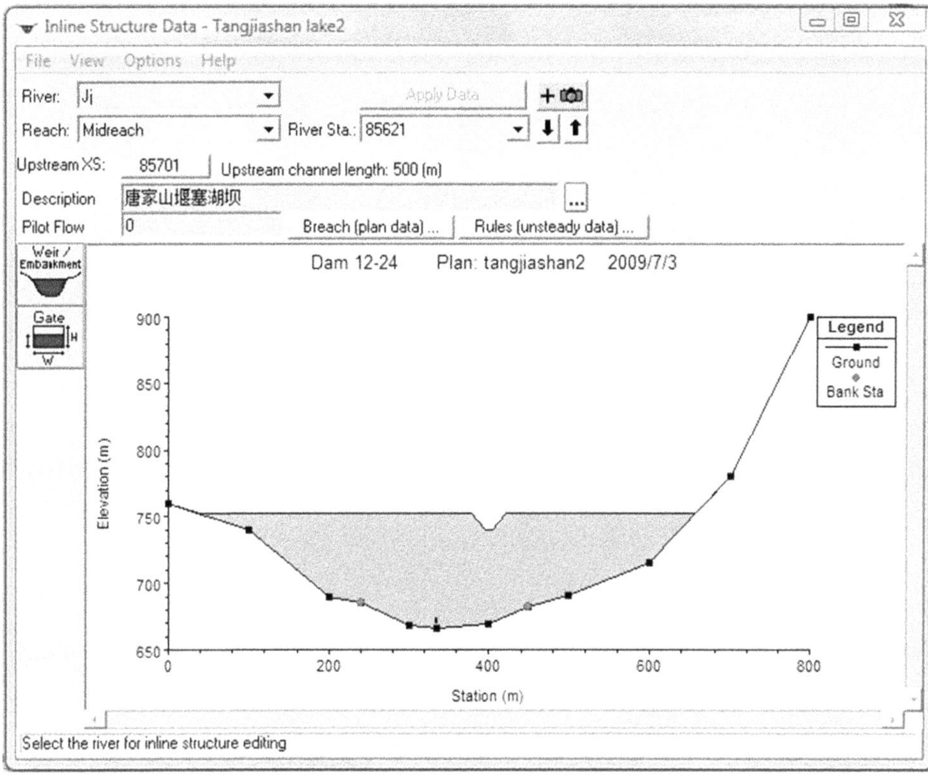

Figure 9.4 The cross-section of the dam

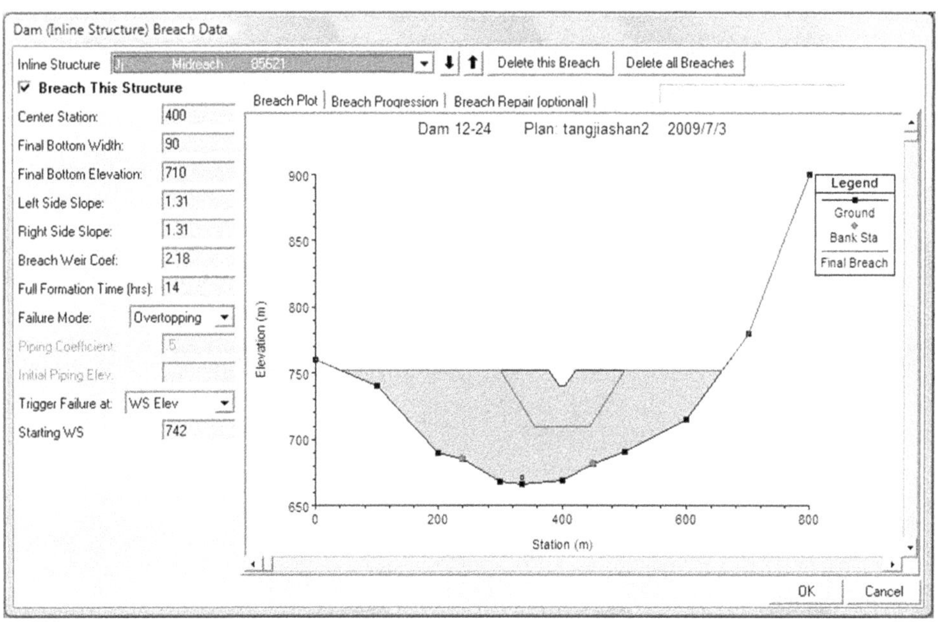

Figure 9.5 The dam breach dimensions

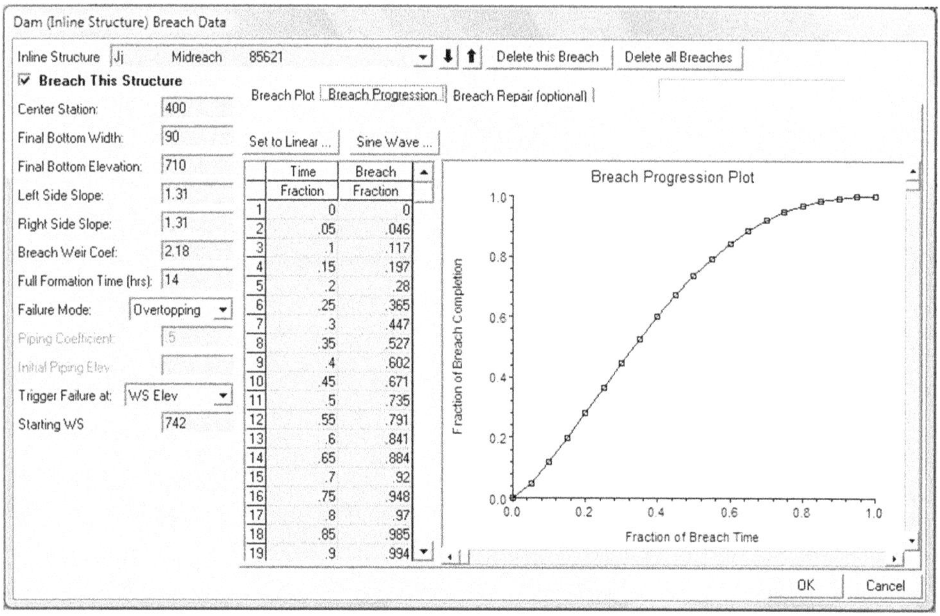

Figure 9.6 The dam breaching progression curve

9.3.2 *Dam Breaching Simulation*

After setting the geometric parameters, the next step is to input the breaching parameters. In HEC-RAS, the final breach size, breaching time, and the breach progression are needed before simulation. In this case, the final breach size is according to the real case, as shown in Figure 9.5. The final bottom width is 90 m, the breach depth is 42 m (final bottom elevation is 710 m), both the left and right side slopes are 1.31, the breach weir coefficient is 2.18, and the breaching time is 14 hours. Besides, the breaching is set to start when the water level is 2 m over the new crest elevation (742 m). Figure 9.6 shows the breach progression according to the records.

9.3.3 *Boundary and Initial Conditions*

Boundary conditions in this study are set for the seepage, lateral inflows, and the downstream boundary (Figure 9.7). The seepage is controlled by the opening of gates. The lateral inflow upstream of the Beichuan Town is set as 20 m^3/s to simulate the runoff generation (Figure 9.8). The friction slope of the downstream boundary is set as 0.001, as shown in Figure 9.9. The inflow rates into the lake are also set as lateral inflows, according to the records as shown in Figure 9.10. The initial conditions are set as Figure 9.11. Besides, the elevation of the lake is set as 738.5 m at the start of the simulation based on the records.

9.3.4 *Flood Routing Analysis and Results*

After inputting the geometric parameters, breach parameters, boundary conditions, and initial conditions, the flood from dam breaching and the flood routing can be simulated, as shown in Figure 9.12. Three options are to be set on this interface: simulation time settings, computation

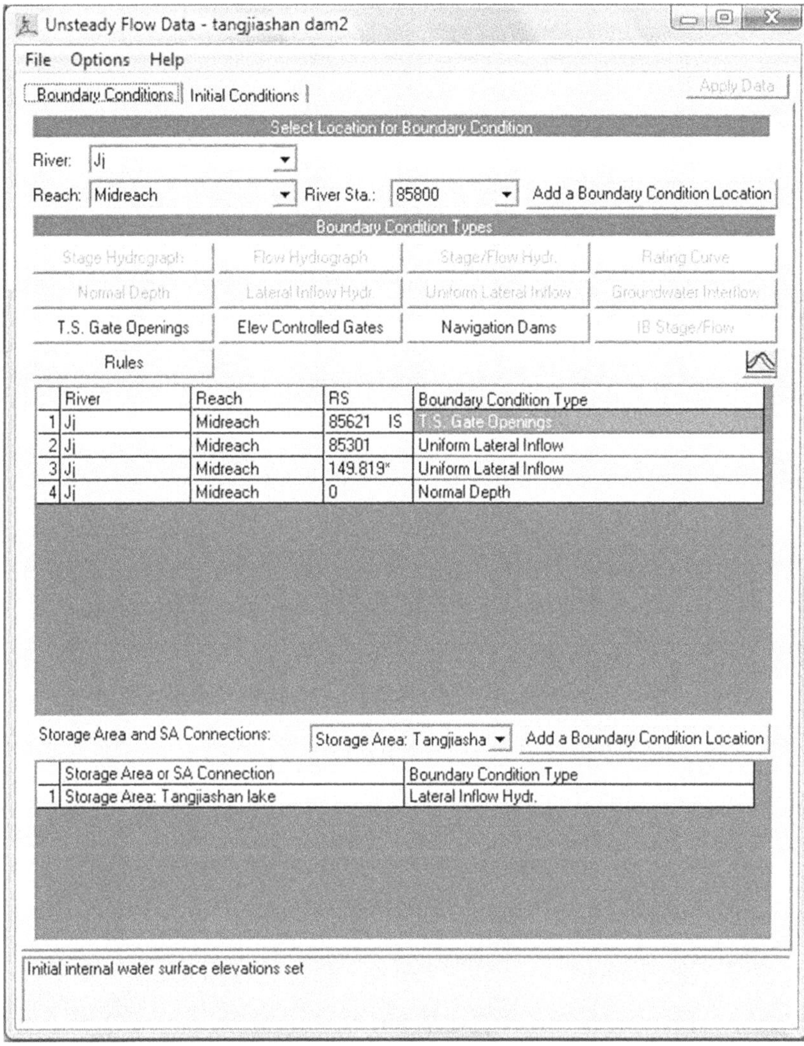

Figure 9.7 Setting boundary conditions

settings, and flow regime settings. The simulation time settings are to define the period of simulation. In this case, the period is from 22:00 on 6 June 2008 to 12:00 on 13 June 2008, which covers the beaching initiation period (from 7:00 on 7 June to 6:00 on 10 June) and the breach development period (from 6:00 on 10 June to 20:00 10 June). The computation setting includes the computation interval (or iteration time interval) and the output interval.

Figures 9.13 and 9.14 show the stages (water elevation) and flow hydrographs at the dam site and in the Beichuan town. The maximal river water elevations and the flow rates are shown in the upper-right corner in the two figures. For example, the maximal flow rate at the dam site is 6551 m^3/s. Figures 9.15 and 9.16 show the details of the outputs at the dam site and in the Beichuan town, including the geometric parameters, river water elevation, flow rate, and water volume. Figure 9.17 shows detailed outputs for selected river cross-sections.

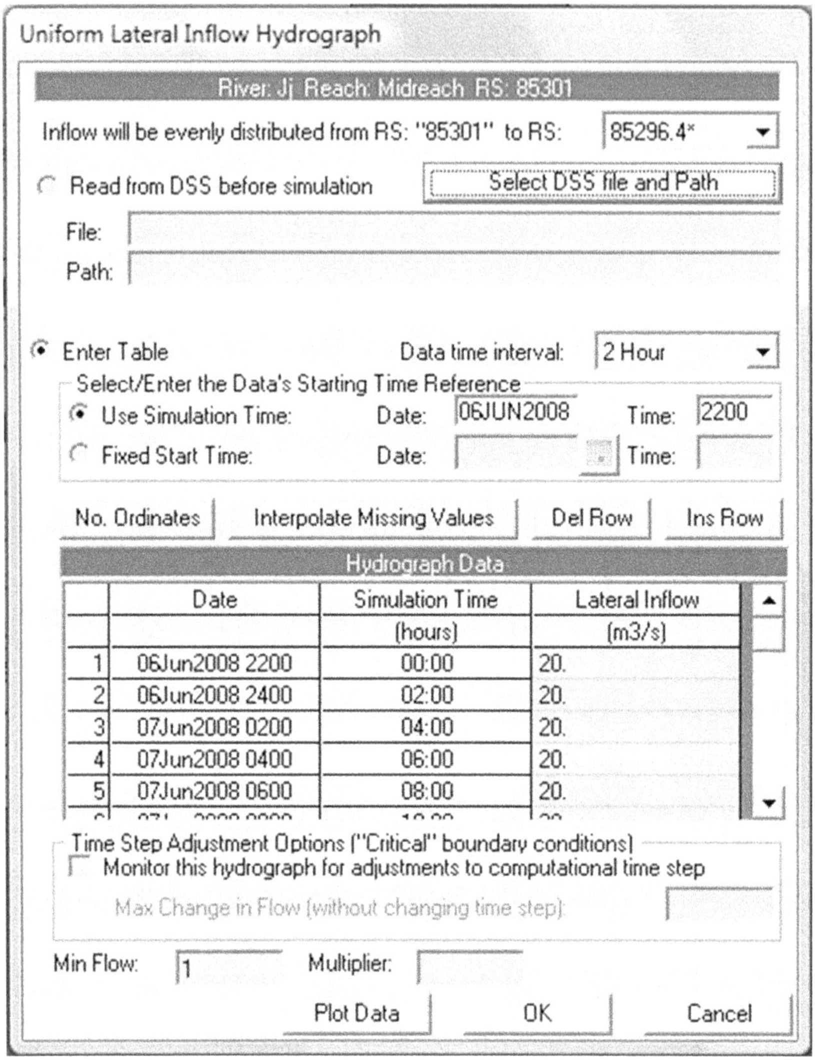

Figure 9.8 Lateral inflow rates upstream of Beichuan town

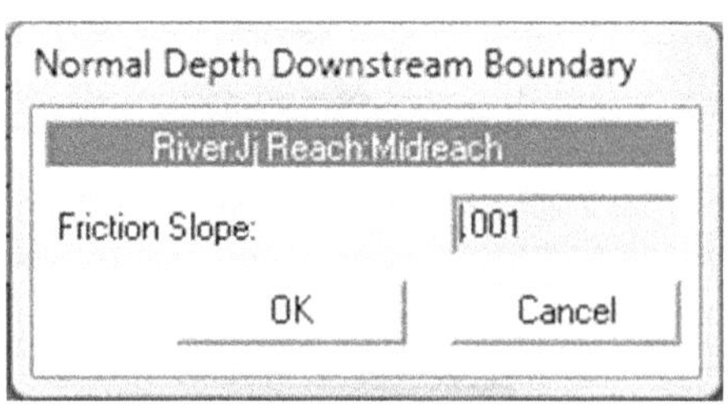

Figure 9.9 Downstream boundary

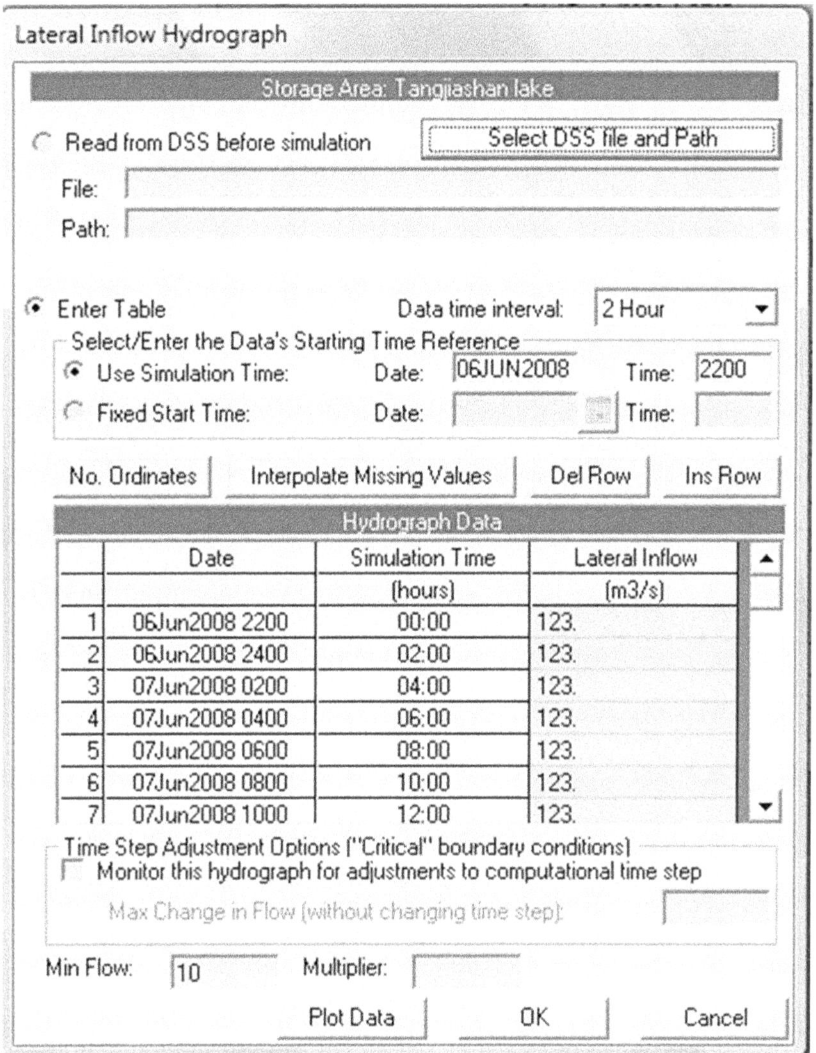

Figure 9.10 Inflow rates into the lake

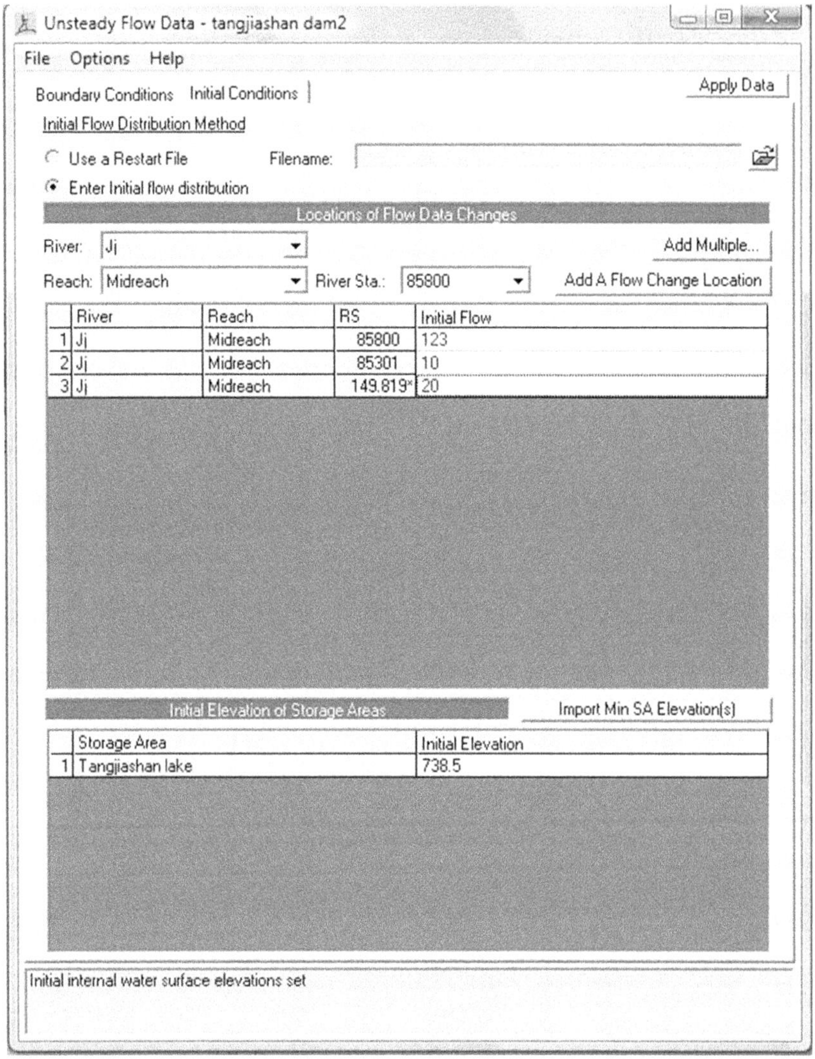

Figure 9.11 Setting initial conditions

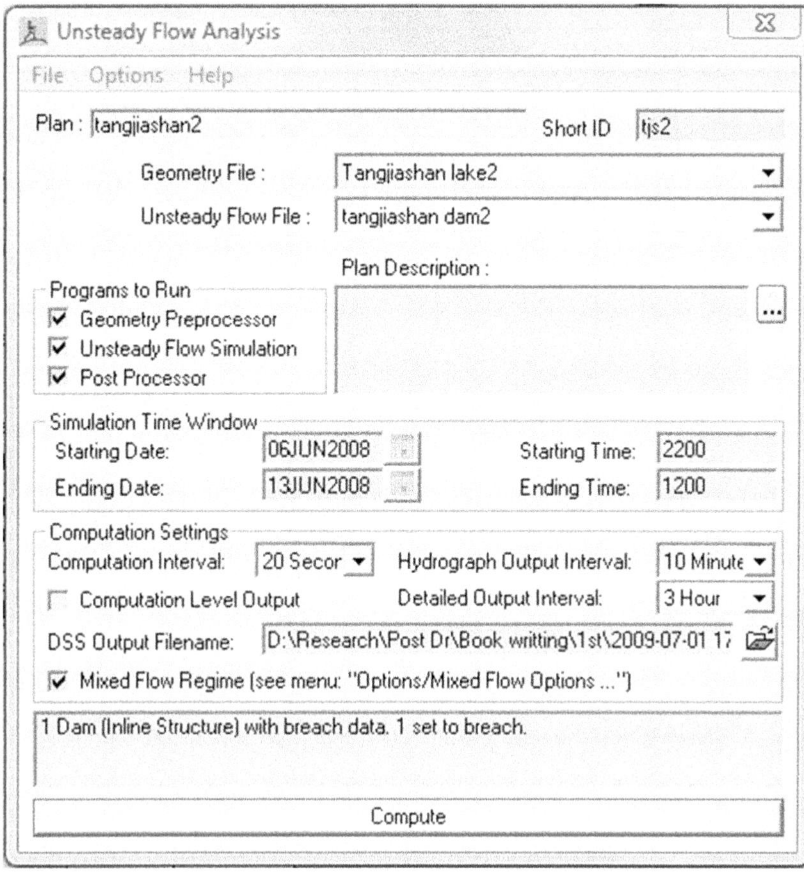

Figure 9.12 Flood simulation settings

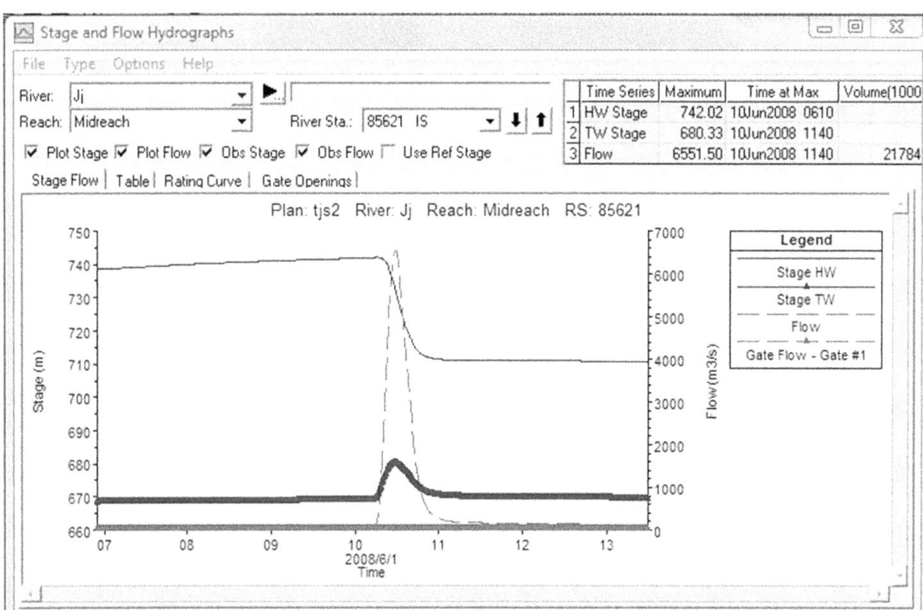

Figure 9.13 The stage and flow hydrographs at the dam site

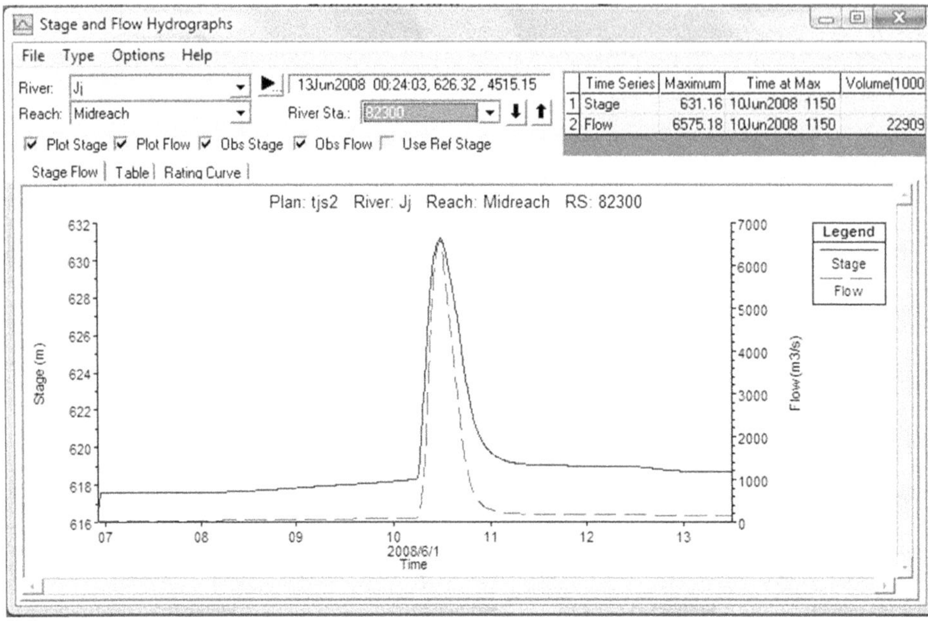

Figure 9.14 The stage and flow hydrographs in the Beichuan town

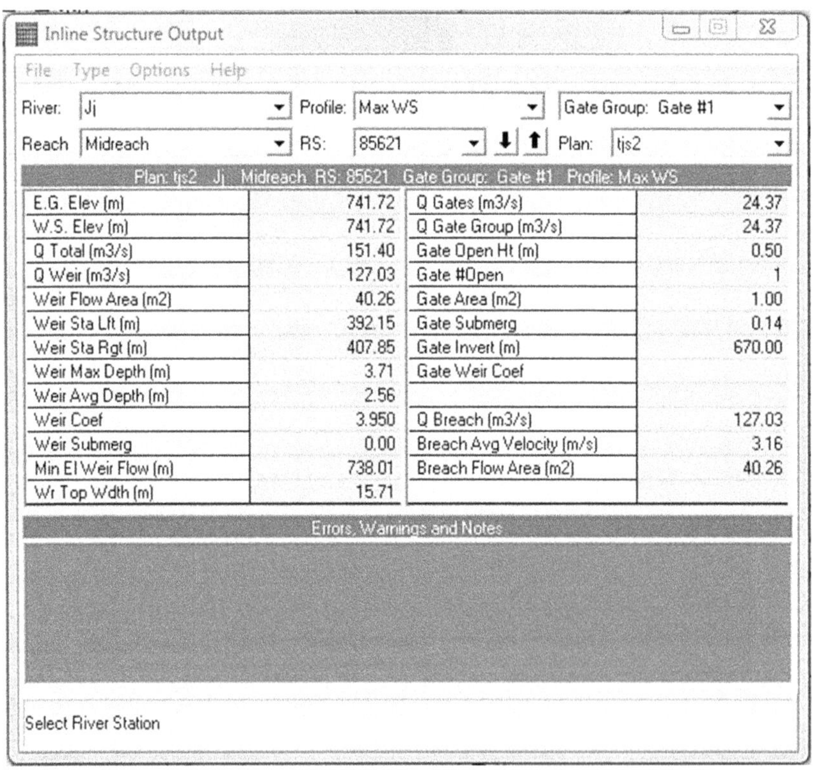

Figure 9.15 Details of the outputs at the dam site

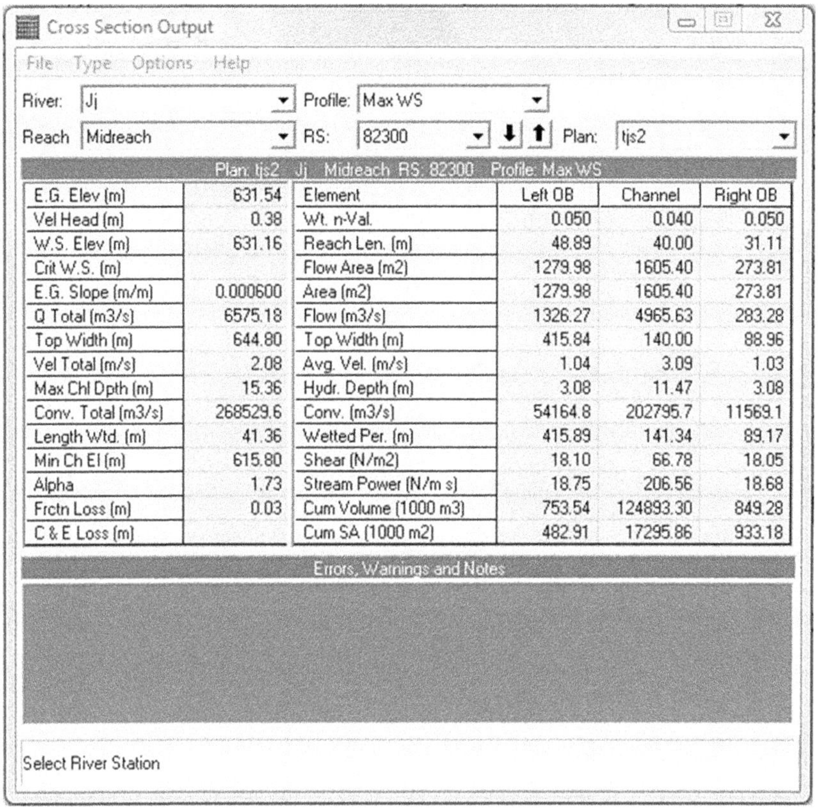

Cross Section Output

File Type Options Help

River: Jj | Profile: Max WS

Reach: Midreach | RS: 82300 | Plan: tjs2

Plan: tjs2 Jj Midreach RS: 82300 Profile: Max WS

E.G. Elev (m)	631.54	Element	Left OB	Channel	Right OB
Vel Head (m)	0.38	Wt. n-Val.	0.050	0.040	0.050
W.S. Elev (m)	631.16	Reach Len. (m)	48.89	40.00	31.11
Crit W.S. (m)		Flow Area (m2)	1279.98	1605.40	273.81
E.G. Slope (m/m)	0.000600	Area (m2)	1279.98	1605.40	273.81
Q Total (m3/s)	6575.18	Flow (m3/s)	1326.27	4965.63	283.28
Top Width (m)	644.80	Top Width (m)	415.84	140.00	88.96
Vel Total (m/s)	2.08	Avg. Vel. (m/s)	1.04	3.09	1.03
Max Chl Dpth (m)	15.36	Hydr. Depth (m)	3.08	11.47	3.08
Conv. Total (m3/s)	268529.6	Conv. (m3/s)	54164.8	202795.7	11569.1
Length Wtd. (m)	41.36	Wetted Per. (m)	415.89	141.34	89.17
Min Ch El (m)	615.80	Shear (N/m2)	18.10	66.78	18.05
Alpha	1.73	Stream Power (N/m s)	18.75	206.56	18.68
Frctn Loss (m)	0.03	Cum Volume (1000 m3)	753.54	124893.30	849.28
C & E Loss (m)		Cum SA (1000 m2)	482.91	17295.86	933.18

Errors, Warnings and Notes

Select River Station

Figure 9.16 Details of the outputs in the Beichuan town

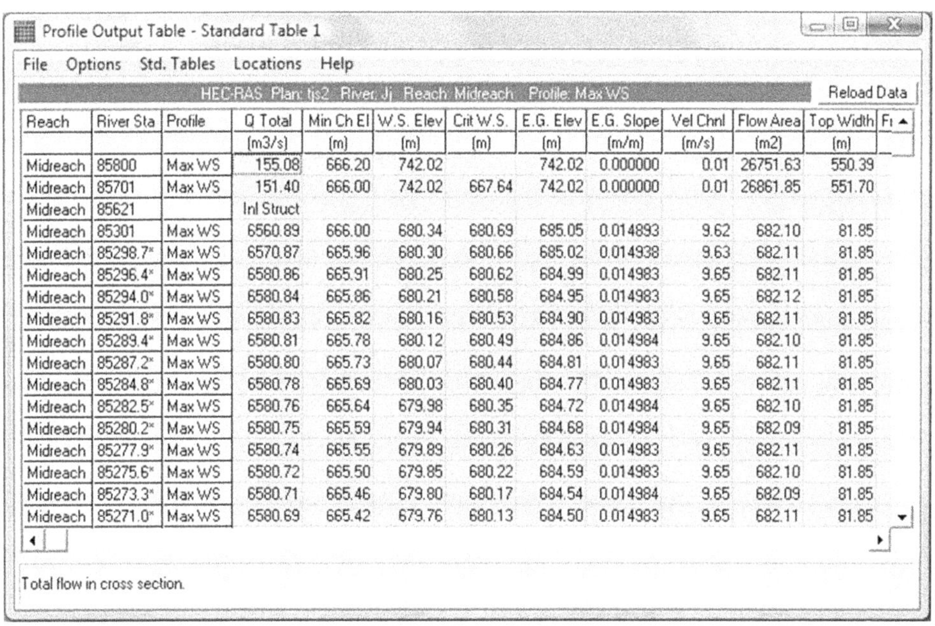

Profile Output Table - Standard Table 1

File Options Std. Tables Locations Help

HEC-RAS Plan: tjs2 River: Jj Reach: Midreach Profile: Max WS — Reload Data

Reach	River Sta	Profile	Q Total (m3/s)	Min Ch El (m)	W.S. Elev (m)	Crit W.S. (m)	E.G. Elev (m)	E.G. Slope (m/m)	Vel Chnl (m/s)	Flow Area (m2)	Top Width (m)	Fr
Midreach	85800	Max WS	155.08	666.20	742.02		742.02	0.000000	0.01	26751.63	550.39	
Midreach	85701	Max WS	151.40	666.00	742.02	667.64	742.02	0.000000	0.01	26861.85	551.70	
Midreach	85621		Inl Struct									
Midreach	85301	Max WS	6560.89	666.00	680.34	680.69	685.05	0.014893	9.62	682.10	81.85	
Midreach	85298.7*	Max WS	6570.87	665.96	680.30	680.66	685.02	0.014938	9.63	682.11	81.85	
Midreach	85296.4*	Max WS	6580.86	665.91	680.25	680.62	684.99	0.014983	9.65	682.11	81.85	
Midreach	85294.0*	Max WS	6580.84	665.86	680.21	680.58	684.95	0.014983	9.65	682.12	81.85	
Midreach	85291.8*	Max WS	6580.83	665.82	680.16	680.53	684.90	0.014983	9.65	682.11	81.85	
Midreach	85289.4*	Max WS	6580.81	665.78	680.12	680.49	684.86	0.014984	9.65	682.10	81.85	
Midreach	85287.2*	Max WS	6580.80	665.73	680.07	680.44	684.81	0.014983	9.65	682.11	81.85	
Midreach	85284.8*	Max WS	6580.78	665.69	680.03	680.40	684.77	0.014983	9.65	682.11	81.85	
Midreach	85282.5*	Max WS	6580.76	665.64	679.98	680.35	684.72	0.014984	9.65	682.10	81.85	
Midreach	85280.2*	Max WS	6580.75	665.59	679.94	680.31	684.68	0.014984	9.65	682.09	81.85	
Midreach	85277.9*	Max WS	6580.74	665.55	679.89	680.26	684.63	0.014983	9.65	682.11	81.85	
Midreach	85275.6*	Max WS	6580.72	665.50	679.85	680.22	684.59	0.014983	9.65	682.10	81.85	
Midreach	85273.3*	Max WS	6580.71	665.46	679.80	680.17	684.54	0.014984	9.65	682.09	81.85	
Midreach	85271.0*	Max WS	6580.69	665.42	679.76	680.13	684.50	0.014983	9.65	682.11	81.85	

Total flow in cross section.

Figure 9.17 Details of the outputs in selected cross-sections

Part III

Dam Failure Risk Assessment and Management

10

Analysis of Probability of Failure of Dams

10.1 Introduction

An excellent book *Risk and uncertainty in dam safety* (Hartford and Baecher, 2004) presents probabilistic analysis tools for dam risk analysis and decision-making. Several more general books – Baecher and Christian (2003), Ang and Tang (2007), Fenton and Griffiths (2008), and Phoon and Ching (2015) – introduce in detail probabilistic analysis methods in geotechnical engineering. This chapter is only meant to summarize five common tools in dam reliability and risk analysis, including failure modes and effects analysis (FMEA), event trees, fault trees, first-order reliability method (FORM)/first-order second-moment method (FOSM), and Monte Carlo Simulation. FMEA is a qualitative inductive method for analyzing potential failures in a system. Event trees are designed to model the possible event sequence following an initiating event. Fault trees are used to identify all possible sequences of causes leading to an undesired event. FORM and FOSM are approximation methods that only use the first term or both the first and second terms of a Taylor series expansion of the performance function to calculate the reliability of a system. Monte Carlo simulation creates a large number of sets of randomly generated values for input random variables and computes the performance function for each set, resulting in the probability distribution of an output variable. In addition to the above five traditional approaches, an emerging method named Bayesian networks is introduced. Bayesian networks, the combination of graph theory and probability theory, are well suited for studying a complex system (e.g. a dam) with multiple elements and their interactive influences in a systematic way.

10.2 Analysis Methods

10.2.1 Failure Modes and Effects Analysis

Failure modes and effects analysis (FMEA) was introduced in the late 1940s for military usage by the US armed forces. Later it was used for aerospace/rocket development to avoid errors in small sample sizes of costly rocket technology (US Department of Defense, 1980). Although initially

Dam Failure Mechanisms and Risk Assessment, First Edition. Limin Zhang, Ming Peng, Dongsheng Chang, and Yao Xu.

developed by the military, the FMEA methodology is now extensively used in a variety of industries including semiconductor processing, food service, plastics, software, healthcare, and civil (e.g. International Electrotechnical Commission, 1985; British Standards Institution, 1991; Canadian Standards Association, 1993; Langford, 1995).

FMEA is a form of reliability analysis that is used to map the consequences of specific events that can occur during the operation of an engineered system, and to use this information to identify and prioritize necessary actions (Hartford and Baecher, 2004). It can be a freestanding technique to give a structured understanding of the failure modes applicable to the components of an engineered system, or it can be an integral part of a more comprehensive probabilistic analysis of the risks associated with a multiple integrated system.

FMEA techniques are structured, logical frameworks that allow informed specialists to use available knowledge and information in a systematic way to lead to an understanding of the sources of risk in the system. The primary skill required is to understand the functional nature of the system being analyzed. This requires knowledge of how the system was designed and built, how the system function is achieved, and how it has been and is being operated. This gives a basic appreciation of the potential weaknesses within the system and forms the basis for carrying out the analysis.

FMEA can be carried out to varying depths within a system, making it essential to define the level in the system for which the analysis will be conducted. Systems can be broken down into subsystems, sub-subsystems, etc., and then to an elemental level. Complex systems can be broken down in a hierarchical manner whereby the failure effects at a lower level in the system become the failure modes at the next highest level in the system. In general, the system is broken down to a level where there is adequate understanding of the failure modes of the elements and the system.

FMEA can provide a systematic procedure for analyzing failure modes of dam systems by isolating each component and describing the effects of individual component failure modes on the system as a whole. It is conceptually straightforward and readily applied to discrete-component systems with little or no component redundancy. It clearly relates component failure modes to their effects on the performance of the system and presents them in an easily readable form. Figure 10.1 shows a general sequential failure diagram from FMEA.

FMEA cannot be used for quantitative analysis. A major disadvantage of FMEA is the difficulty of modeling and analyzing redundancy, as the individual component failure modes do not always result in observable system effects. Another significant disadvantage of FMEA is the difficulty of applying the method to systems where the components can exhibit varying degrees of failure, that is, the components do not necessarily go from a functional state to a non-functional state instantaneously.

In addition, FMEA may require a great deal of time and a very significant resource commitment for dealing with a complex system where the effects of multiple failures are often taken into account. These disadvantages become especially apparent when FMEA is used in dam safety applications where failures or distresses and their corresponding causes are varied, multiple, and interrelated.

10.2.2 Event Tree

Usually, a number of possible consequences may result from the occurrence of an initiating event. Event tree analysis (ETA) is a logic method, either qualitative or quantitative, that is used to identify the possible outcomes and, if required, their probabilities given the occurrence of an initiating event (Fell et al., 2000). Application of ETA to technological systems became common in the 1970s (Rasmussen, 1975) and has continued to increase since then.

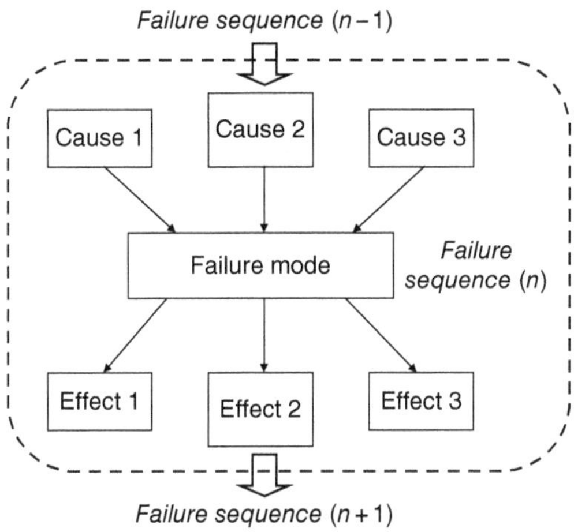

Figure 10.1 Sequential failure diagram from FMEA

ETA is widely used for facilities provided with engineered accident mitigating features to identify the sequence of events that follow the initiating event and produce specified consequences. A typical use is a post-accident analysis that identifies consequences of an accident sequence. This is the most common application of ETA for dam safety, where accident initiation is postulated and the event tree is used to illustrate how the various subsequent events and scenarios evolve. Applications of ETA to dam safety management include Whitman (1984), Lee et al. (1998), and Li et al. (2006).

One of the attractive features of ETA is the fact that it is a graphic display of the sequence of system states that lead to the occurrence of a specified undesired scenario. Typically, event trees illustrate known states that the system can reside in, given certain precursor conditions. When used in a qualitative way, ETA provides a logical way of describing system performance in terms of a "what happens if …" type of analysis (Hartford and Baecher, 2004). Usually, the event tree is in a binary form, where the various system states either occur or not occur, with the outcome of each event sequence described at the end.

The steps for a qualitative or quantitative ETA are straightforward in situations where the tree is precisely defined and the probabilities of event occurrences are available. The probability of occurrence of an event is often conditional only on the occurrence of the precursor event. The process of enumerating the event tree is simply a matter of multiplying the probabilities along individual branches. In general, ETA involves three typical attributes: (1) graphic tool, (2) qualitative analysis, and (3) quantitative assessment. Figure 10.2 shows a simple event tree with an initiating event and a number of possible consequences. It can be observed that a particular consequence depends on the subsequent events following an initiating event; in other words, for a given consequence to occur, a sequence of subsequent events, or path in the event tree, must occur.

ETA provides a logical and graphical means of illustrating the sequence of events from an initiating event to the complete set of possible outcomes, even for large systems. It is oriented to illustrating necessary system conditions or states that have a bearing on the undesired consequence, such as system failure. Once the tree is defined and the probabilities assigned, the enumeration of the tree is a straightforward matter, as there are no logical gates.

Like any analysis technique, ETA has its disadvantages. It becomes large for a complex system and hence inefficient. Also, it is difficult to use ETA to identify multiple initiating events, which is a common situation in dam safety applications.

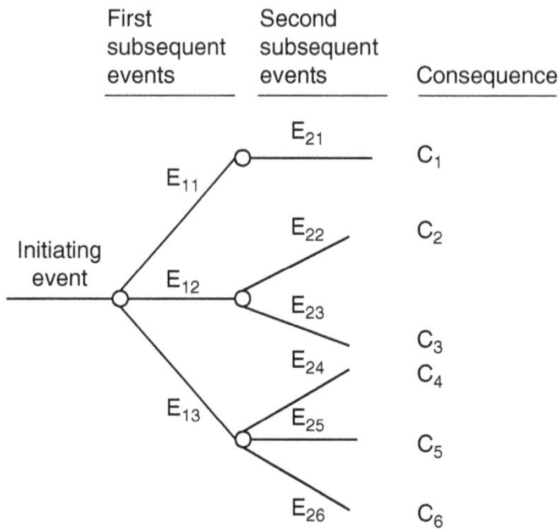

Figure 10.2 Event tree model

10.2.3 Fault Tree

For a complex system, the adverse conditions or faults that could lead to the occurrence of one of the potential modes of failure may be quite involved, such that some systematic means of identifying the various faults and their interactive effects on the failure event would become necessary. Fault tree analysis (FTA) serves exactly this purpose. FTA was developed in 1961 by H. A. Watson at Bell Telephone Laboratories. By the 1970s, the range of FTA applications had expanded well to various fields of engineering (e.g. Rasmussen, 1975; Cornell and Newmark, 1978; Lambe et al., 1981).

FTA is a deductive analysis method that is used to obtain information about a defined undesired event. In a top-down manner, FTA works from the general to the specific. One of the early steps in FTA is to specify a particular but general undesired event, such as failure of a system. Then the analysis proceeds to determine what the specific causes or modes of system failure are. Similar to ETA, FTA also involves the three attributes: (1) graphic tool, (2) qualitative analysis, and (3) quantitative assessment.

When FTA is conducted for a large system, such as a dam system, it is important to clearly define the bounds. Usually, multiple subsystems exist in a large system. FTA is used to model a number of different subsystems, of which many are functionally linked. Consequently, it is important that each subsystem and its constituent components should be clearly defined, avoiding overlap or oversight.

FTA may include a quantitative evaluation of the probabilities of the various faults or failure events leading eventually to the calculation of the probability of the main failure event. Aside from identifying all potential paths that could lead to failure, it may serve also to single out the critical events that contribute significantly to the likelihood of failure of a system and reveal the weak links, if any, in the system. Figure 10.3 shows a simple fault tree diagram for instance. It can be observed that FTA is a graphical decomposition of a top event into the union and/or intersection of subevents. The alternative subevents that could lead to the top event are logically related to the top event by "or" gates and "and" gates, as shown in Figure 10.3.

FTA provides a logical and graphic means of modeling and analyzing system failure modes. It is oriented to identifying system faults that have a bearing on the undesired event, such as system failure. As a modeling technique for assessing the reliability of systems, it is well developed and

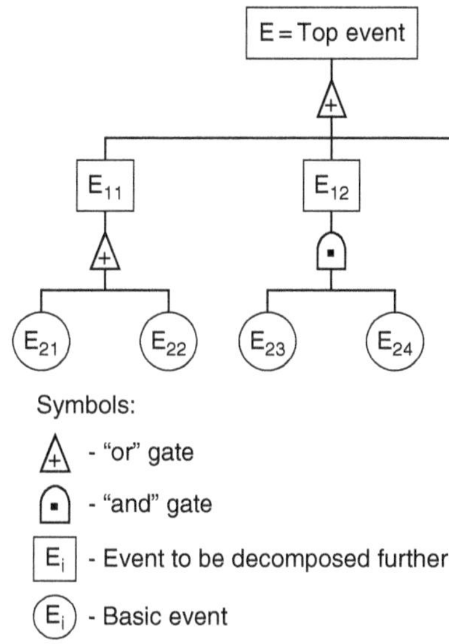

Figure 10.3 Fault tree diagram

accepted. However, a large fault tree can be difficult to understand. From the graphic display, it can be difficult to identify the system failure modes and to ascertain what are the important events among those that cause system failure. In addition, fault trees are not unique. They can be constructed in different ways, and therefore have different appearances.

10.2.4 First-order Reliability Method/First-order Second-moment Method

The reliability of a system is defined as the probabilistic measure of assurance of performance (e.g. Ang and Tang, 2007). The problems of reliability of engineering systems may be cast essentially as a problem of supply versus demand. Due to the presence of uncertainties, the available supply and required demand are random variables. Therefore, the reliability of a system is often measured in terms of probability. For this purpose, two basic random variables are defined: X_s is the supply capacity; Y_d is the demand requirement. A quantitative measure of safety is the probability of survival, P_s, given by

$$P_s = P(X_s \geq Y_d) \tag{10.1}$$

The complement of the probability of survival is the probability of failure, P_f, which can be expressed as

$$P_f = 1 - P_s = P(X_s < Y_d) \tag{10.2}$$

To evaluate the probability of failure, P_f, the above supply–demand problem may be formulated in terms of the safety margin, $M = X_s - Y_d$. As X_s and Y_d are random variables, M is also a random variable. In this case, failure is clearly the event ($M < 0$), and thus the probability of failure, P_f, is

$$P_f = P(M < 0) \tag{10.3}$$

The reliability of an engineering system may involve multiple variables. In particular, the supply and the demand may be functions of several other variables. For such cases, a generalized formulation in terms of the basic design variables is given by

$$g(\mathbf{X}) = g(x_1, x_2, \ldots, x_n) \tag{10.4}$$

where $\mathbf{X} = (x_1, x_2, \ldots, x_n)$ is a vector of basic variables of the system, and the function $g(\mathbf{X})$ determines the performance of the system. The limiting performance requirement is defined as $g(\mathbf{X}) = 0$, which is the "limit-state" of the system. The probability of failure means the probability that the system cannot perform its intended function. It is given by the integral

$$P_f = \int \int_{g(\mathbf{X})<0} \ldots \int f_{\mathbf{X}}(x_1, x_2, \ldots, x_n) dx_1 \, dx_2 \ldots dx_n \tag{10.5}$$

where $f_{\mathbf{X}}(x_1, x_2, \ldots, x_n)$ is the joint probability density function of $\mathbf{X}$. The integration is performed over the failure region of $g(\mathbf{X}) < 0$.

A simple method of expressing the probability of failure of an engineering system is to use the reliability index, β. It is defined as the ratio of the mean to the standard deviation of the performance function, $g(\mathbf{X})$,

$$\beta = \frac{\mu_g}{\sigma_g} \tag{10.6}$$

where μ_g = mean of $g(\mathbf{X})$ and σ_g = standard deviation of $g(\mathbf{X})$. If the distribution of $g(\mathbf{X})$ is normal, the probability of failure, P_f, is

$$P_f = \Phi(-\beta) \tag{10.7}$$

where $\Phi(\cdot)$ is the standard normal cumulative distribution function.

In practice, the joint probability density function of X, $f_X(x_1, x_2, \ldots, x_n)$, is often unavailable or difficult to obtain, due to insufficient data. The evaluation of the multiple integral in Equation (10.5) is extremely complicated. Therefore, approximation methods are proposed as a practical alternative. Various reliability analysis methods have been developed corresponding to the types and complexity of performance functions. Two commonly used methods, first-order reliability method (FORM) and first-order second-moment method (FOSM), are respectively introduced in the following.

Equation (10.4) can be written by using the Taylor series for multiple variables as follows:

$$\begin{aligned} g(\mathbf{X}) = g(\boldsymbol{\mu}) &+ \frac{1}{1!}\sum_{i=1}^{n}\frac{\partial g}{\partial x_i}(x_i - \mu_{x_i}) + \frac{1}{2!}\sum_{i=1}^{n}\sum_{j=1}^{n}\frac{\partial g}{\partial x_i}\frac{\partial g}{\partial x_j}(x_i - \mu_{x_i})(x_i - \mu_{x_j}) \\ &+ \frac{1}{3!}\sum_{i=1}^{n}\sum_{j=1}^{n}\sum_{k=1}^{n}\frac{\partial g}{\partial x_i}\frac{\partial g}{\partial x_j}\frac{\partial g}{\partial x_k}(x_i - \mu_{x_i})(x_i - \mu_{x_j})(x_k - \mu_{x_k}) + \ldots \end{aligned} \tag{10.8}$$

where $\boldsymbol{\mu}$ is the mean value vector of the random variables.

The simplest and most widely used method starts by assuming that all the $(x_i - \mu_{x_i})$ terms are small; therefore, their squares, cubes, and higher powers will be even smaller, and can be ignored. Since only the first-order terms are included, the method based on this assumption is called the first-order reliability method. Then, the first order terms give

$$g(\mathbf{X}) = g(\boldsymbol{\mu}) + \sum_{i=1}^{n} \frac{\partial g}{\partial x_i}\left(x_i - \mu_{x_i}\right) \tag{10.9}$$

The first-order approximate mean and variance of $g(\mathbf{X})$ are given by

$$\mu_g = g\left(\mu_{x_1}, \mu_{x_2}, \ldots \mu_{x_n}\right) \tag{10.10}$$

$$Var(g) = \sigma_g^2 = \sum_{i=1}^{n}\sum_{j=1}^{n} \frac{\partial g}{\partial x_i}\frac{\partial g}{\partial x_j} Cov\left(x_i, x_j\right) \tag{10.11}$$

where $Cov(x_i, x_j)$ is the covariance of x_i and x_j.

One possible extension of the first-order approximation is to include terms in Equation (10.8) up to the second derivatives. Two principal justifications for this are (1) the terms $(x_i - \mu_{x_i})$ are not small enough to vanish when squared but do become small when cubed, and (2) the third and higher derivatives of g vanish or become small. When the second-order terms are included, the method is called the first-order second moment method (FOSM). Then, removing all terms with third or higher derivatives from Equation (10.8) gives

$$g(\mathbf{X}) = g(\boldsymbol{\mu}) + \sum_{i=1}^{n} \frac{\partial g}{\partial x_i}\left(x_i - \mu_{x_i}\right) + \frac{1}{2}\sum_{i=1}^{n}\sum_{j=1}^{n} \frac{\partial g}{\partial x_i}\frac{\partial g}{\partial x_j}\left(x_i - \mu_{x_i}\right)\left(x_i - \mu_{x_j}\right) \tag{10.12}$$

The first-order second-moment approximate mean of $g(\mathbf{X})$ are given by

$$\mu_g = g\left(\mu_{x_1}, \mu_{x_2}, \ldots \mu_{x_n}\right) + \frac{1}{2}\sum_{i=1}^{n}\sum_{j=1}^{n} \frac{\partial g}{\partial x_i}\frac{\partial g}{\partial x_j} Cov\left(x_i, x_j\right) \tag{10.13}$$

The expression for the variance of $g(\mathbf{X})$ when second-order terms are included is more complicated. Ang and Tang (2007) give the following equation for the variance when g is a function of one variable x:

$$\begin{aligned} Var(g) = \sigma_g^2 = \sigma_x^2\left(\frac{dg}{dx}\right)^2 - \frac{1}{4}\left(\sigma_x^2\right)^2\left(\frac{d^2 g}{dx^2}\right)^2 + E\left[\left(x - \mu_x\right)^3\right]\frac{dg}{dx}\frac{d^2 g}{dx^2} \\ + \frac{1}{4}E\left[\left(x - \mu_x\right)^4\right]\left(\frac{d^2 g}{dx^2}\right)^2 \end{aligned} \tag{10.14}$$

This expression involves the squares of the second derivatives and moments higher than second order. If these higher-order terms are removed, Equation (10.14) reduces to

$$Var(g) = \sigma_g^2 = \sigma_x^2 \left(\frac{dg}{dx} \right)^2 \tag{10.15}$$

The corresponding expressions for multiple variables are much more complicated; if the squares of the second derivatives and the moments higher than second order are removed, the result is the same as Equation (10.11). In other words, the simplified form of the first-order second-moment expression for the variance is the same as the first-order expression.

10.2.5 Monte Carlo Simulation

Another widely used method for reliability analysis is Monte Carlo simulation, which is used to obtain the probability distribution of an output variable given the probability distributions of a set of input random variables (Ang and Tang, 1990). This involves repeating a simulation process, using in each simulation a particular set of values of the random variables generated in accordance with the corresponding probability distributions. By repeating the process, a sample of outputs, each corresponding to a different set of values of the random variables, is obtained. Then the outputs can be analyzed to determine the probability of unsatisfactory performance of an engineering system. In the case of correlated random variables, the required random numbers can be generated using the marginal and conditional cumulative distribution function of random variables.

The technique has the advantage that it is relatively easy to implement on a computer and can deal with a wide range of functions, including those that cannot be expressed conveniently in explicit form (Baecher and Christian, 2003). It should be pointed out that Monte Carlo simulation solutions from finite samples are not exact unless the sample size is infinitely large. It is desirable to know the error underlying an estimated probability or, more importantly, how many simulations are required to obtain a certain accuracy (Ang and Tang, 1990). By approximating the binomial distribution with a normal distribution, Shooman (1968) developed the following expression for the following expression for the percent error:

$$\%error = 200\sqrt{\frac{1-p_f}{np_f}} \tag{10.16}$$

where p_f is the estimated probability of unsatisfactory performance and n is the sample size. There is a 95% chance that the percent error in the estimated probability will be less than that given by Equation (10.16). As an example, if 10,000 simulations are performed, obtaining an estimated probability of 0.01, Equation (10.16) will yield a 20% error. Thus, it is 95% likely that the actual probability will be within 0.01 ± 0.002. On the other hand, if an accuracy of 0.01 ± 0.001 is desired, Equation (10.16) can also be used to obtain the required number of simulations, which is n = 39,600.

10.2.6 Bayesian Networks

The technique of Bayesian networks was developed in the early 1980s by Judea Pearl at University of California at Los Angeles (UCLA) and Finn Jensen at Aalborg University, Denmark. Those initial studies aimed to integrate uncertainty in expert systems. A Bayesian network is a directed acyclic graph formed by variables (nodes) together with the directed edges, attached by tables of conditional probabilities of each variable on all its parents (Jensen, 1996).

A Bayesian network consists of a set of nodes and a set of directed arrows. A node represents either a discrete random variable with a finite number of states or a continuous random variable. The terms node and variable are often used interchangeably. The links between the nodes represent causal relationships between the nodes. The nodes without any arrows directed into them are called root nodes and they have unconditional probability tables. The nodes that have arrows directed into them are called child nodes, and the nodes that have arrows directed from them are called parent nodes. The nodes without children are called leaf nodes. Each child node has a conditional probability table, given the state of the parent nodes. Bayesian networks are also referred to as directed acyclic graphs, which indicate that cycles are not allowed.

Jensen (2001) summarized common components of a Bayesian network:

1. There is a set of variables and a set of directed edges between variables.
2. Each variable represents a probability distribution, which may be continuous or discrete.
3. The variables, together with the directed edges, form a directed acyclic graph.
4. There is a probability table attached to each variable with parents.

Figure 10.4 shows three typical simple Bayesian networks: a serial connection, a diverging connection, and a converging connection. In the serial connection in Figure 10.4(a), X_1 has an influence on X_2, which in turn influences X_3. In this case, evidence on X_1 will influence the certainty of X_3 through X_2, and vice versa. On the other hand, if the state of X_2 is known, then the channel is blocked, and X_1 and X_3 become independent, which can be described as saying that X_1 and X_3 are d-separated given X_2. A similar situation in a dam distress process is that incomplete sludge cleaning at foundation (X_1) may cause uneven-settlement cracks (X_2), and result in embankment leakage (X_3). The joint probability function of the three variables is

$$p(x_1, x_2, x_3) = p(x_3 \mid x_2)\, p(x_2 \mid x_1)\, p(x_1) \tag{10.17}$$

In the diverging connection in Figure 10.4(b), the influence can pass between the child nodes (X_2, X_3) unless the state of the parent node (X_1) is known. We say that X_2 and X_3 are d-separated given X_1. A similar situation in a dam distress process is that incomplete sludge cleaning at foundation (X_1) may cause foundation leakage (X_2) and uneven-settlement cracks (X_3). The joint probability function of the three variables is

$$p(x_1, x_2, x_3) = p(x_3 \mid x_1)\, p(x_2 \mid x_1)\, p(x_1) \tag{10.18}$$

In the converging connection in Figure 10.4(c), the parent nodes (X_2, X_3) are independent if nothing is known about the child node (X_1). In other words, the parent nodes (X_2, X_3) are dependent on the condition that there is knowledge on the child node (X_1). A similar situation in a dam distress process is that both incomplete sludge cleaning at foundation (X_2) and inappropriate or faulty embankment materials (X_3) may lead to uneven-settlement cracks (X_1). The joint probability function of the three variables is

$$p(x_1, x_2, x_3) = p(x_1 \mid x_2, x_3)\, p(x_2)\, p(x_3) \tag{10.19}$$

In practice, a Bayesian network often consists of all the above three types of connections. Figure 10.5 shows an example of a simple general Bayesian network.

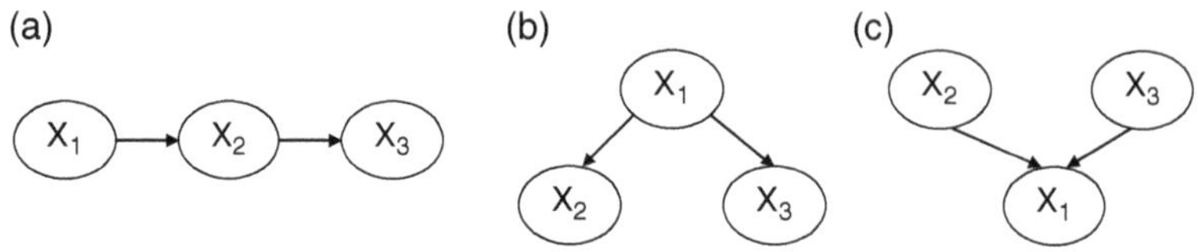

Figure 10.4 Illustration of three typical simple Bayesian networks: (a) serial connection, (b) diverging connection, and (c) converging connection. *Source:* Zhang et al. (2011b). Reproduced with permission from NRC Research Press

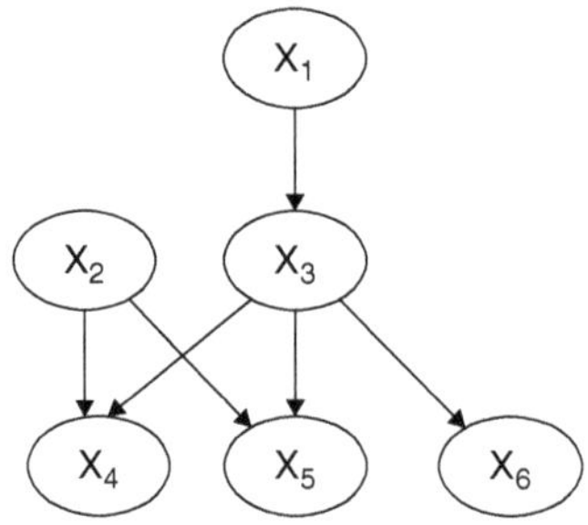

Figure 10.5 General Bayesian network

The joint probability function of random variables in a Bayesian network can be expressed in a general form as

$$p(x_1, \ldots, x_n) = \prod_{i=1}^{n} p(x_i \mid pa(X_i)) \tag{10.20}$$

in which $pa(X_i)$ is the parent set of X_i. Note that if X_i has no parents, then the function reduces to the unconditional probability of $p(x_i)$.

Two primary issues are involved in a Bayesian network: (1) establishing the cause and effect relationships between the variables, and (2) determining the conditional probabilities associated with these relationships (Smith, 2006). The underlying probabilistic representation allows the quantification of the strength of these relationships and the realization of inferences to aid decision-making. These calculations are based on Bayes' theorem, which allows inference based on observed evidence. For random variables A and B, the conditional probability of A given B is

$$P(A \mid B) = \frac{P(B \mid A)P(A)}{P(B)} \tag{10.21}$$

The probability construct has three fundamental components: (1) statistical estimates based on empirical data, (2) phenomenological models based on physical processes, and (3) expert judgment (Hartford and Baecher, 2004). In principle, probabilities can be derived directly from data if sufficient data is available or from phenomenological models if there is sufficient knowledge of the process that generates the parameter or state of interest. Expert judgment alone does not provide a reliable means of estimating probabilities without data and/or phenomenological models.

The matter of probability assignment can be approached from two perspectives, a top-down approach based fundamentally on available data and its interpretation, or a bottom-up approach of generating knowledge and then data based on physical processes (Moss and Eller, 2007). An ideal way to these issues is to collect as much data as possible, and at the same time to find as much as can be found about the phenomena that create the observations. Expert judgment plays an important role in either top-down approach or bottom-up approach, and the quality of judgment increases as both data and model completeness increase.

Bayesian networks are well suited for combining statistical data, physical analyses, and expert judgment in the same probabilistic model. They may replace fault trees and event trees, thus making the representation of the system more compact and intuitive (Li et al., 2007). In a Bayesian network, dependence is modeled by connecting the dependent nodes with an arrow and quantifying the dependence by specification of the conditional probability table. Therefore, Bayesian networks are preferable in cases where the causes of an event are mutually dependent. This is further emphasized by the fact that in a Bayesian network model, there will always be consistency between dependence/independence relations expressed at the local level and those expressed at the global level. In other words, a Bayesian network model can take into account local-level and global-level data in a systematic manner.

In a Bayesian network, sensitivity analysis can be conducted by simply altering the probability distributions for selected nodes and monitoring the changes in probability distribution of the target node. This may single out the most critical factors in a complex system. In addition, Bayesian networks serve as a dynamic tool in the sense that evidence may be inserted, probability updated, and the most probable cause-and-effect scenarios identified.

10.3 Examples of Probabilistic Analysis of Dam Failure

This section presents two examples of the use of Bayesian networks to conduct dam reliability analysis. In the first example, Bayesian networks are used to conduct a general reliability evaluation of Chinese on-service dams. In the second example, Bayesian networks are further applied to diagnose the distresses of a particular dam and to evaluate the effectiveness of rehabilitation measures for the dam. The use of the other methods presented in Section 10.2 in dam reliability and risk analysis has been illustrated in books such as Baecher and Christian (2003), Hartford and Baecher (2004), Ang and Tang (2007), and Fenton and Griffiths (2008).

10.3.1 Probabilistic Analysis of Chinese Dam Distresses

In this section, the technique of Bayesian networks (Jensen, 1996, 2001) is used to develop a robust probability-based tool for conducting probabilistic analysis of dam failure at the global level based on past performance of 993 cases of distressed dams. Based on prior experience of common characteristics extracted from the global-level data, Bayesian networks are further used to conduct probabilistic analysis of dam failure at the local level by combining global-level performance records and local-level performance records and/or analysis results. A case study on Chenbihe Dam is presented to illustrate the methodology.

10.3.1.1 Global-level Probabilistic Analysis of Dam Failure Using Bayesian Networks

In China, reservoirs are divided into three categories in terms of reservoir capacity: large-size reservoirs with capacities of over $100\times10^6\,m^3$, medium-size reservoirs with capacities of $10–100\times10^6\,m^3$, and small-size reservoirs with capacities of less than $10\times10^6\,m^3$. In this study,

Table 10.1 Subdivision of distressed embankment dams in terms of dam type

Dataset	Dam type	Number of cases	Sum
Dataset A	Homogeneous earthfill	591	610
	Composite earthfill	19	
Dataset B	Earthfill with clayey corewall	317	383
	Earthfill with inclined clayey corewall	59	
	Rockfill with clayey corewall	3	
	Rockfill with inclined clayey corewall	4	
Sum		993	993

1182 distressed dams associated with large- or medium-size reservoirs in China are collected and compiled into a database. The majority of the cases are embankment dams (993 cases), which are the focus of this study. Among these distressed embankment dams, 591 caes of homogeneous earthfill dams and 19 cases of composite earthfill dams are grouped into Dataset A due to their similar structure forms and distress characteristics. Similarly, 317 cases of earthfill dams with vertical clay cores, 59 cases of earthfill dams with inclined clay cores, three cases of rockfill dams with vertical clay-cores, and four cases of rockfill dams with inclined clay cores are grouped into Dataset B. Table 10.1 summarizes the types of embankment dam. In the probabilistic analysis that will be presented later, separate analyses will be conducted for homogeneous/composite and clay-core dams based on Dataset A and Dataset B, respectively.

Most of the past embankment dam failures have been caused by either overtopping or seepage erosion/piping (Li et al., 2006). Therefore this study focuses on dam failure by overtopping or seepage erosion/piping. Based on the details of embankment dam distresses in the database, characteristics of relevant distresses are studied, including the patterns, causes, and severities. A distress may be treated as the nonsatisfaction of a performance requirement for a component of a dam system as it was intended. Table 10.2 shows the common patterns and causes of the distresses associated with embankment dams, which are described as corresponding discrete variables with two to four states representing the changing severities. Owing to the limitation of the descriptive information provided by the database, most of the variables have only two states, satisfactory or unsatisfactory. A few variables have three or four states, from satisfactory to moderately unsatisfactory and even seriously unsatisfactory.

Based on the two datasets (Dataset A and Dataset B), four Bayesian networks are constructed for probabilistic analysis of failures of (1) homogeneous/composite dams by overtopping, (2) homogeneous/composite dams by seepage erosion/piping, (3) clay-core dams by overtopping, and (4) clay-core dams by seepage erosion/piping. Two primary issues are involved in a Bayesian network: (1) establishing the cause and effect relationships between the variables, and (2) determining the conditional probabilities associated with these relationships (Smith, 2006). The cause and effect relationships between the variables will be described in a causal network based on the database. The corresponding probability tables will be constructed by using a top-down approach based on available data and its interpretation (Hartford and Baecher, 2004).

10.3.1.2 Causal Networks for Analysis of Dam Failures

To develop a Bayesian network for probabilistic analysis of dam failure, the first issue is to identify all possible distress mechanisms and display them in a causal network. Based on the dam distress database, an inventory of possible dam distresses and corresponding causes is constructed.

Table 10.2 Common patterns and causes of the distresses associated with embankment dams. *Source:* Zhang et al. (2011b). Reproduced with permission from NRC Research Press

Code	Variable	State	
		Number	Description
ARS	Abutment rocks or soils	2	Satisfactory, Unsatisfactory
ASS	Abutment seepage situation	4	Normal, Unstable, Moderate leakage, Serious leakage
CF	Cut-off at foundation	2	Satisfactory, Unsatisfactory
CMG	Clay-core materials and geometries	2	Satisfactory, Unsatisfactory
DEH	Designed embankment height	2	Satisfactory, Unsatisfactory
DEW	Designed embankment width	2	Satisfactory, Unsatisfactory
EBI	Embankment-abutment interface	2	Satisfactory, Unsatisfactory
EC	Embankment cracking	2	Occurred, Non-occurred
ECE	Embankment crest elevation	2	Satisfactory, Unsatisfactory
EM	Embankment materials	2	Satisfactory, Unsatisfactory
ES	Embankment settlement	2	Normal, Excessive
ESP	Emergency spillway	3	Satisfactory, Moderately unsatisfactory, Seriously unsatisfactory
ESS	Embankment seepage situation	4	Normal, Unstable, Moderate leakage, Serious leakage
FD	Filtered drainage	2	Satisfactory, Unsatisfactory
FDC	Flood design criteria	2	Satisfactory, Unsatisfactory
FRC	Flood releasing capacity	2	Satisfactory, Unsatisfactory
FRT	Flood releasing tunnel	3	Satisfactory, Moderately unsatisfactory, Seriously unsatisfactory
FSS	Foundation seepage situation	4	Normal, Unstable, Moderate leakage, Serious leakage
GLD	Gate and lifting device	3	Satisfactory, Moderately unsatisfactory, Seriously unsatisfactory
OT	Overtopping	2	Yes, No
PS	Primary spillway	4	Satisfactory, Moderately unsatisfactory, Seriously unsatisfactory, no spillway
PW	Parapet wall	2	Satisfactory, Unsatisfactory
RBSS	Reservoir bank slope stability	2	Satisfactory, Unsatisfactory
RSS	Reservoir sedimentation situation	2	Satisfactory, Unsatisfactory
SCF	Sludge cleaning at foundation	2	Satisfactory, Unsatisfactory
SEP	Seepage erosion or piping	2	Yes, No
SSC	Seepage situation around embedded culverts	4	Normal, Unstable, Moderate leakage, Serious leakage
TB	Termite burrows	2	Occurred, Non-occurred
WRC	Water retaining capacity	2	Satisfactory, Unsatisfactory

Each distress mechanism is described as a causal chain consisting of several variables, which represents the distress process, from root causes to the final failure, in a transparent way. Table 10.2 shows the variables involved in probabilistic analysis of embankment dam failure. As mentioned before, each variable is assumed discrete with two to four states. A state indicates whether or not one variable can satisfy the performance requirement for which it was intended.

It is also interesting to note that some variables serve not only as distresses but also as causes for other distresses. For instance, embankment cracking caused by unsatisfactory sludge cleaning at the foundation may result in embankment leakage. This illustrates the need to group variables into different ranking categories. It starts with the leaf variable (i.e. overtopping or seepage erosion/piping) to model a set of causal chains, and progresses backwards toward detailed variables. The variables that have direct influence on the leaf variable are grouped into the first category. Some variables in the first category may have their specific causes, of which the corresponding variables are ranked in the second category. Such a process of dividing a category into subcategories continues until the root variables are reached.

By considering all the influencing factors in different ranking categories, the causal networks for probabilistic analysis of failures of homogeneous/composite and clay-core dams by overtopping and seepage erosion/piping are constructed in Figure 10.6. Figure 10.6(a) shows the causal network for probabilistic analysis of overtopping failures of both homogeneous/composite and clay-core dams. Figures 10.6(b) and (c) show the causal networks for probabilistic analysis of seepage erosion/piping failures of homogeneous/composite and clay-core dams, respectively.

10.3.1.3 Probability Tables for the Causal Network for Analysis of Dam Failures by Overtopping

The top-down approach based on available data is used to construct probability tables for the relationships between the causal networks in Figure 10.6. The prior probability tables for the root nodes and the conditional probability tables for the parent and child relationships are obtained based on the results of the historical frequencies from the dam distress database. The constructed probability tables reflect the statistics of the data currently available. Further case information may be gathered if necessary to update these probability tables following the same procedure.

Take the analysis of failure of homogeneous/composite dams by overtopping as an example. The nodes involved in the causal work in Figure 10.6(a) have only a few states. For instance, the root node of SCF has two states: "SCF = Satisfactory" indicating satisfactory sludge cleaning at foundation while "SCF = Unsatisfactory" indicates unsatisfactory sludge cleaning at foundation. Accordingly, P(SCF = Satisfactory) represents the prior probability of satisfactory sludge cleaning at foundation while P(SCF = Unsatisfactory) represents the prior probability of unsatisfactory sludge cleaning at foundation. In Dataset A, the number of cases with unsatisfactory sludge cleaning at foundation (SCF) is 90, as shown in Table 10.3. The prior probability of "SCF = Unsatisfactory" is the number of cases with unsatisfactory sludge cleaning at foundation divided by the total number of cases in Dataset A, which is P(SCF = Unsatisfactory) = 90/610 = 0.148, or P(SCF = Satisfactory) = 1 − 0.148 = 0.852. In other words, a distressed homogeneous/composite dam has a probability of 0.148 to involve unsatisfactory sludge cleaning at foundation. Similarly, the prior probabilities of the other root nodes in the causal network in Figure 10.6(a) can be calculated, and are summarized in Table 10.3. In the same way, the prior probabilities of the root nodes in the causal network for analysis of failure of clay-core dams by overtopping can be obtained and are also summarized in Table 10.3.

In addition to node SCF, node EM is also a parent of node ES, as shown in Figure 10.6(a). Hence, the conditional probability of each state of ES is estimated under different combinations of the SCF state and the EM state, as shown in Table 10.4. In Dataset A, there are 90 cases with unsatisfactory sludge cleaning at foundation (Table 10.3): 75 cases with satisfactory embankment materials and 15 cases with unsatisfactory embankment materials. There are 281 cases with unsatisfactory embankment materials (Table 10.3): 266 cases with satisfactory sludge cleaning at foundation and 15 cases with unsatisfactory sludge cleaning at foundation. In general, 75 cases exist with only unsatisfactory sludge cleaning at foundation, of which 4 cases involve excessive embankment

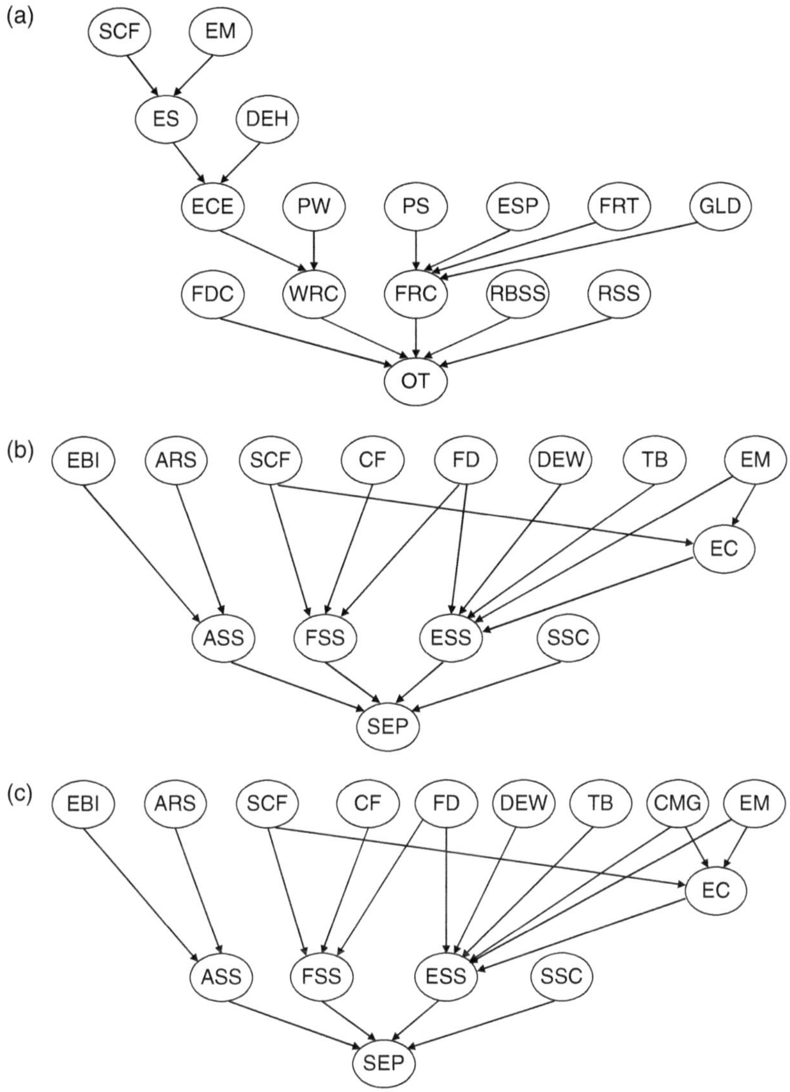

Figure 10.6 Summary of causal networks for probabilistic analysis of failures of (a) homogeneous/composite and clay-core dams by overtopping, (b) homogeneous/composite dams by seepage erosion/piping, and (c) clay-core dams by seepage erosion/piping (see full names of the variables in Table 10.6). *Source:* Zhang et al. (2011b). Reproduced with permission from NRC Research Press

settlement (SCF-ES); 266 cases involve unsatisfactory embankment materials, of which 6 cases involve excessive embankment settlement (EM-ES); 15 cases involve both unsatisfactory sludge cleaning at foundation and unsatisfactory embankment materials, of which 2 cases involve excessive embankment settlement (SCF&EM-ES). Hence, the conditional probability of excessive embankment settlement given unsatisfactory SCF is $P(ES=Excessive\,|\,SCF=Unsatisfactory)=4/75=0.053$; the conditional probability of excessive embankment settlement given unsatisfactory EM is $P(ES=Excessive\,|\,EM=Unsatisfactory)=6/266=0.023$; the conditional probability of excessive embankment settlement given both unsatisfactory SCF and unsatisfactory EM is

Table 10.3 Prior probability table for the root nodes involved in analysis of failures of embankment dams by overtopping. *Source:* Zhang et al. (2011b). Reproduced with permission from NRC Research Press

Code	State	Homogeneous/composite dams		Clay-core dams	
		Number of case	Probability	Number of case	Probability
DEH	Satisfactory	505	0.828	344	0.898
	Unsatisfactory	105	0.172	39	0.102
EM	Satisfactory	329	0.539	292	0.762
	Unsatisfactory	281	0.461	91	0.238
ESP	Satisfactory	595	0.975	379	0.989
	Moderately unsatisfactory	4	0.007	1	0.003
	Seriously unsatisfactory	11	0.018	3	0.008
FDC	Satisfactory	484	0.793	291	0.760
	Unsatisfactory	126	0.207	92	0.240
FRT	Satisfactory	548	0.898	366	0.955
	Moderately unsatisfactory	58	0.095	16	0.042
	Seriously unsatisfactory	4	0.007	1	0.003
GLD	Satisfactory	512	0.839	321	0.838
	Moderately unsatisfactory	79	0.130	55	0.144
	Seriously unsatisfactory	19	0.031	7	0.018
PS	Satisfactory	191	0.314	75	0.196
	Moderately unsatisfactory	301	0.493	224	0.585
	Seriously unsatisfactory	104	0.170	82	0.214
	No spillway	14	0.023	2	0.005
PW	Satisfactory	575	0.943	358	0.935
	Unsatisfactory	35	0.057	25	0.065
RBSS	Satisfactory	602	0.987	372	0.971
	Unsatisfactory	8	0.013	11	0.029
RSS	Satisfactory	606	0.993	380	0.992
	Unsatisfactory	4	0.007	3	0.008
SCF	Satisfactory	520	0.852	314	0.820
	Unsatisfactory	90	0.148	69	0.180

Table 10.4 Conditional probability table for the relationships among nodes SCF, EM, and ES involved in analysis of failure of homogeneous/composite dams by overtopping

	SCF = Satisfactory		SCF = Unsatisfactory	
	EM = Satisfactory	EM = Unsatisfactory	EM = Satisfactory	EM = Unsatisfactory
ES = Normal	1.000	0.977	0.947	0.867
ES = Excessive	0.000	0.023	0.053	0.133

P(ES = Excessive | SCF = Unsatisfactory & EM = Unsatisfactory) = 2/15 = 0.133. As a result, the conditional probability table for the relationships between the three nodes of SCF, EM, and ES can be obtained, as shown in Table 10.4.

If nodes SCF and EM are statistically independent, the conditional probability of excessive embankment settlement given both unsatisfactory SCF and unsatisfactory EM is P(ES = Excessive

| SCF = Unsatisfactory & EM = Unsatisfactory) = 1 − P(ES = Normal | SCF = SCF = Unsatisfactory & EM = Unsatisfactory) = 1 − P(ES = Normal | SCF = Unsatisfactory) × P(ES = Normal | EM = Unsatisfactory) = 1 − (1 − 0.053) × (1 − 0.023) = 0.075. In this case, the above estimated value of P(ES = Excessive | SCF = Unsatisfactory & EM = Unsatisfactory) = 0.133 based on the available data is larger than the calculated value of 0.075 based on the assumption of statistical independence. This is due to the negative combined effect of distresses. However, for some relationships between the nodes, when the corresponding available data is limited, the estimated value may be less than the calculated value based on the assumption of statistical independence. In such a case, the calculated value will be finally chosen.

Using the above methodology, the conditional probability tables for all the relationships involved in the causal network in Figure 10.6(a) for analysis of failure of homogeneous/composite dams by overtopping can be obtained, excluding the relationships of the leaf node of OT with the five parent nodes of FDC, WRC, FRC, RBSS, and RSS. There is no relevant data in the dataset on the relationships among node OT and its five parent nodes in Figure 10.6(a). Those distressed embankment dams may have great possibility of overtopping; but actually no overtopping failures have yet occurred. In this study, the conditional probability of OT given its five parents of FDC, WRC, FRC, RBSS, and RSS in Figure 10.6(a) is determined based on historical information and knowledge.

Before assigning the detailed probability values for the relationships of OT with its five parent nodes, the possibilities of these five parent nodes leading to overtopping have been ranked qualitatively, based on the past occurrence frequency. Li et al. (2006) reported 1737 cases of dam failures by overtopping in China, of which 1302 cases were caused by insufficient flood-releasing capacity and 435 cases were caused by extreme floods exceeding design criteria. Therefore, node FRC is the most important factor, which is followed by node FDC. Table 10.5 shows a typical transition from verbal descriptions of probability to numerical values for dam failures in China (Li et al., 2006). In practice, the two extreme conditions of virtually certain event and virtually unlikely event in Table 10.5 are very rare and hence not considered in this study. It can be said that the occurrence of overtopping is very likely given unsatisfactory flood releasing capacity, and hence the value of P(OT = Yes | FRC = Unsatisfactory) is within 0.1–0.5. In this study, the mean value is chosen, P(OT = Yes | FRC = Unsatisfactory) = 0.3. As mentioned before, the number of cases associated with overtopping caused by insufficient flood-releasing capacity is 1302, which is approximately three times of the number of cases caused by extreme floods exceeding design criteria (435 cases). For this reason, P(OT = Yes | FDC = Unsatisfactory) is assumed to be 0.1. There have been few overtopping failures caused by reservoir bank slope instability or excessive reservoir sedimentation in the past. Hence, the occurrence of overtopping given unsatisfactory RBSS and RSS is determined to be very unlikely. Correspondingly, both P(OT = Yes | RBSS = Unsatisfactory) and P(OT = Yes | RSS = Unsatisfactory) are assumed to be 0.001, the mean value of 0.0001 and 0.01 in log-scale. The occurrence of overtopping caused by unsatisfactory WRC is moderate, compared

Table 10.5 Verbal descriptions of likelihood and corresponding probability values for dam failures in China (Li et al. 2006)

Verbal descriptor	Probability
Event is virtually unlikely	0.000001–0.0001
Event is very unlikely	0.0001–0.01
Event is likely	0.01–0.1
Event is very likely	0.1–0.5
Event is virtually certain	0.5–1.0

with those caused by unsatisfactory FRC and FDC and those caused by unsatisfactory RBSS and RSS. Therefore, the occurrence of overtopping is determined to be likely, given unsatisfactory WRC. Correspondingly, P(OT=Yes | WRC=Unsatisfactory) is assumed to be 0.032, the mean value of 0.01 and 0.1 in log-scale. These five parent nodes of OT are assumed statistically independent, when calculating the probability of overtopping given the combined effect of unsatisfactory states of two parent nodes or more.

The prior probability tables for the root nodes and the conditional probability tables for the parent and child relationships involved in the causal network in Figure 10.6(a) for analysis of failure of homogeneous/composite dams by overtopping have been constructed. In the same way, the corresponding probability tables for the same casual network in Figure 10.6(a) for analysis of failure of clay-core dams by overtopping can also be constructed.

10.3.1.4 Probability Tables for the Causal Network for Analysis of Dam Failures by Seepage Erosion/Piping

Take the analysis of failure of homogeneous/composite dams by seepage erosion/piping as another example. Figure 10.6(b) shows the corresponding causal network for analysis of failure of homogeneous/composite dams by seepage erosion/piping. For instance, the root node of EBI has two states, "EBI=Satisfactory" indicating a satisfactory embankment–abutment interface while "EBI=Unsatisfactory" indicating an unsatisfactory embankment–abutment interface. The number of cases with unsatisfactory embankment–abutment interfaces in Dataset A is 53. Therefore, the prior probability of "EBI=Unsatisfactory" is the number of cases with unsatisfactory embankment–abutment interfaces divided by the total number of cases in Dataset A, P(EBI=Unsatisfactory)= 53/610=0.087, and then P(SCF=Satisfactory)=1−0.087=0.913. In other words, a distressed homogeneous/composite dam has a probability of 0.087 of involving unsatisfactory embankment–abutment interfaces. The prior probabilities of all the root nodes in the causal networks in Figure 10.6(b) can be calculated and summarized in Table 10.6. In the same way, the prior probabilities of the root nodes in the causal network for analysis of failure of clay-core dams by seepage erosion/piping in Figure 10.6(c) can be obtained and are also summarized in Table 10.6.

In addition to node EBI, node ARS is also a parent of node ASS, as shown in Figure 10.6(b). Hence, the conditional probability of ASS is estimated under different combinations of the SCF state and the EM state (Table 10.7). In Dataset A, there are 53 cases with unsatisfactory embankment–abutment interfaces (EBI) and 53 cases with unsatisfactory abutment rocks or soils (ARS), as shown in Table 10.6. Among the 53 cases with "EBI=Unsatisfactory", 35 cases involve moderate leakage at abutments while 18 cases involve serious leakage at abutments. Hence, the conditional probabilities of ASS given "EBI=Unsatisfactory" are P(ASS=Moderate leakage | EBI=Unsatisfactory)=35/53= 0.660, and P(ASS=Serious leakage | EBI=Unsatisfactory)=18/53=0.340. Similarly, the conditional probability of ASS given "ARS=Unsatisfactory" can also be calculated, as shown in Table 10.7. In Dataset A, there are no cases with both unsatisfactory embankment–abutment interfaces (EBI) and unsatisfactory abutment rocks or soils (ARS). Therefore, the conditional probability of ASS given both unsatisfactory EBI and unsatisfactory ARS, P(ASS | EBI=Unsatisfactory & ARS=Unsatisfactory) is calculated based on the assumption that EBI and ARS are statistically independent. As a result, the conditional probability table for the relationships among the three nodes of EBI, ARS, and ASS is obtained and summarized in Table 10.7.

Using the above methodology, the conditional probability tables for all the relationships involved in the causal network in Figure 10.6(b) for analysis of failure of homogeneous/composite dams by

Table 10.6 Prior probability table for the root nodes involved in analysis of failures of embankment dams by seepage erosion/piping. *Source:* Zhang et al. (2011b). Reproduced with permission from NRC Research Press

Code	State	Homogeneous/composite dams		Clay-core dams	
		Number of case	Probability	Number of case	Probability
ARS	Satisfactory	557	0.913	360	0.940
	Unsatisfactory	53	0.087	23	0.060
CF	Satisfactory	490	0.803	291	0.760
	Unsatisfactory	120	0.197	92	0.240
CMG	Satisfactory	—	—	281	0.734
	Unsatisfactory	—	—	102	0.266
DEW	Satisfactory	567	0.930	374	0.977
	Unsatisfactory	43	0.070	9	0.023
EBI	Satisfactory	557	0.913	354	0.924
	Unsatisfactory	53	0.087	29	0.076
EM	Satisfactory	329	0.539	292	0.762
	Unsatisfactory	281	0.461	91	0.238
FD	Satisfactory	490	0.803	315	0.822
	Unsatisfactory	120	0.197	68	0.178
SCF	Satisfactory	520	0.852	314	0.820
	Unsatisfactory	90	0.148	69	0.180
SSC	Normal	332	0.545	222	0.580
	Unstable	88	0.144	31	0.080
	Moderate leakage	113	0.185	62	0.162
	Serious leakage	77	0.126	68	0.178
TB	Occurred	86	0.141	44	0.115
	Non-occurred	524	0.859	339	0.885

Table 10.7 Conditional probability table for the relationships among nodes EBI, ARS, and ASS involved in analysis of failure of homogeneous/composite dams by seepage erosion/piping. *Source:* Zhang et al. (2011b). Reproduced with permission from NRC Research Press

	EBI = Satisfactory		EBI = Unsatisfactory	
	ARS = Satisfactory	ARS = Unsatisfactory	ARS = Satisfactory	ARS = Unsatisfactory
ASS = Normal	1.000	0.000	0.000	0.000
ASS = Unstable	0.000	0.000	0.000	0.000
ASS = Moderate leakage	0.000	0.717	0.660	0.473
ASS = Serious leakage	0.000	0.283	0.340	0.527

seepage erosion/piping can be obtained. Note that the conditional probability of SEP given its parents in Figure 10.6(b) is determined based on relevant data in the dataset. A few distressed dams experienced incidents of seepage erosion/piping that had been prevented from evolving into a final breach by taking immediate remedial actions, such as possibly drawing down the water level. These incidents related to seepage erosion/piping have been considered as the equivalent seepage erosion/piping failures.

The prior probability tables for the root nodes and the conditional probability tables for the parent and child relationships involved in the causal network in Figure 10.6(b) for analysis of failure of homogeneous/composite dams by seepage erosion/piping have been constructed. In the same way, the corresponding probability tables for the causal network in Figure 10.6(c) for analysis of failure of clay-core dams by seepage erosion/piping can also be constructed.

10.3.1.5 Calculated and Recorded Annual Probabilities of Dam Failure

The calculation for the annual probability of failure was conducted using the program Hugin Lite developed by Hugin Expert A/S (2004). Still using the analysis of failure of homogeneous/composite dams as an example, the probability calculations by using Hugin Lite for overtopping and seepage erosion/piping are shown in Figures 10.7(a) and 13(b), respectively. The calculated

(a)

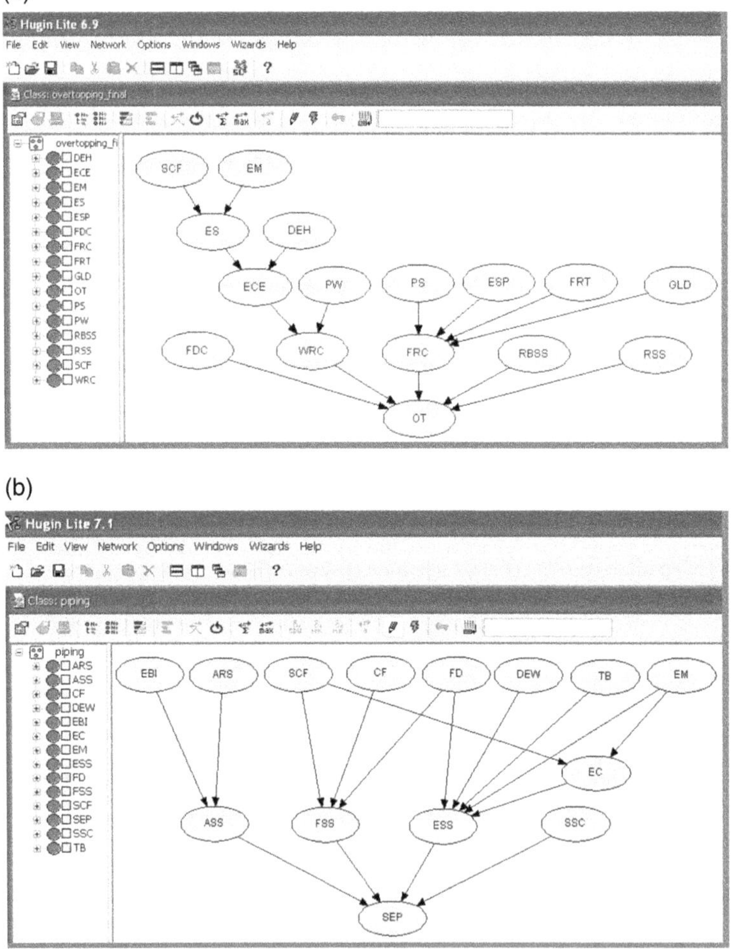

Figure 10.7 Calculations of annual probability of failure for (a) overtopping and (b) seepage erosion/piping of homogeneous/composite dams using program Hugin Lite. *Source:* Zhang et al. (2011b). Reproduced with permission from NRC Research Press

annual probability of the overtopping failure is P(OT=Yes)=0.0675. In other words, these distressed homogeneous/composite dams have an average frequency of 6.75% associated with the overtopping failure over their operation time. Similarly, the calculated annual probability of the seepage erosion/piping failure is P(SEP=Yes)=0.0377. That is to say, these distressed homogeneous/composite dams have an average frequency of 3.77% associated with the seepage erosion/piping failure over their operation time.

The average annual probability of a specific failure mode, P_f, can be calculated as

$$P_f = \left(\frac{N_f}{N_t}\right)\Big/T = \left(\frac{N_f}{N_d}\right)\left(\frac{N_d}{N_t}\right)\Big/T \tag{10.22}$$

in which N_f is the number of failed dams, N_d is the number of distressed dams, N_t is the total number of constructed dams, T is the dam operation time. Note that the ratio of N_f to N_t indicates the frequency of failure of constructed dams, the ratio of N_f to N_d indicates the frequency of failure of distressed dams, and the ratio of N_d to N_t indicates the frequency of distress of constructed dams.

As mentioned before, 1182 dams associated with large- or medium-size reservoirs in China are distressed. As of year 2006, the total number of constructed dams associated with large- or medium-size reservoirs in China is approximately 3770 (MWR, 2008). Accordingly, the historical frequency of distress of constructed dams is N_d/N_t=1182/3770=0.314. Most of these distressed dams were built during the 1950s to 1970s, hence the dam operation time is assumed to be T=50 years. As calculated by using Hugin Lite, the values of N_f/N_d for overtopping and seepage erosion/piping are 6.75% and 3.77%, respectively. Therefore, P_f can be calculated using Equation (10.22), P_f=0.0675×0.314/50=4.24×10^{-4} for overtopping while P_f=0.0377×0.314/50=2.37× 10^{-4} for seepage erosion/piping. Table 10.8 summarizes the calculated failure frequencies over the operation time and the calculated average annual failure probabilities for homogeneous/composite and clay-core dams in China.

The historical average annual failure probabilities for dams in China are also summarized in Table 10.8. Those values are particularly representative of embankment dams due to the fact that in China more than 90% of the constructed dams and 95% of the failed dams are embankment

Table 10.8 Calculated failure frequencies over the operation time and average annual failure probabilities (dam operation time=50 years). *Source:* Zhang et al. (2011b). Reproduced with permission from NRC Research Press

	Homogeneous/composite dams		Clay-core dams	
	Overtopping	Seepage erosion/piping	Overtopping	Seepage erosion/piping
Calculated failure frequency over the operation time	6.75%	3.77%	6.99%	2.72%
Calculated average annual failure probability	4.24×10^{-4}	2.37×10^{-4}	4.39×10^{-4}	1.71×10^{-4}
Historical average annual failure probability (Li et al., 2006)	[1]4.39×10^{-4}	[2]2.19×10^{-4}	[1]4.39×10^{-4}	[2]2.19×10^{-4}

[1]The probability is a general value for the overtopping failure of dams.

[2]The probability is a general value for the seepage erosion/piping failure of dams, including two parts in the original literature, 1.77×10^{-4} for failure at the dam body, abutment, and foundation, and 0.42×10^{-4} for failure at embedded culverts.

dams (Ru and Niu, 2001). It is found in Table 10.8 that the calculated average annual failure probabilities using Bayesian networks are close to the historical average annual failure probabilities, which shows that the outcome from the Bayesian networks is reasonable at the global level. It is not surprising that homogeneous/composite dams and clay-core dams have almost the same probabilities of overtopping failure but different probabilities of seepage erosion/piping failure. This is because whether overtopping occurs or not depends mainly on spillway capacities and flood design criteria but not on dam type. In contrast, a corewall and its filters have a good control of seepage through the dam body (Foster et al., 2000a), hence the seepage erosion/piping failure probability of clay-core dams is less than that of homogeneous/composite dams.

In Table 10.8, the historical average annual failure probability of dams due to seepage erosion/piping, 2.19×10^{-4}, includes two parts: 1.77×10^{-4} for failures at the dam body, abutment, and foundation, and 0.42×10^{-4} for failures at the embedded culverts (Li et al., 2006). Correspondingly, for homogeneous/composite dams the calculated average annual failure probability of 2.37×10^{-4} consists of 1.74×10^{-4} for failures at the embankment, abutment, and foundation and 0.63×10^{-4} for failures at the embedded culverts. For clay-core dams, the calculated average annual failure probability of 1.71×10^{-4} consists of 1.05×10^{-4} for failures at the embankment, abutment, and foundation and 0.66×10^{-4} for failures at the embedded culverts. For the seepage erosion/piping failure at the embankment, abutment, and foundation, the historical value of 1.77×10^{-4} is closer to the calculated value of 1.74×10^{-4} for homogeneous/composite dams than that of 1.05×10^{-4} for clay-core dams. This can be explained by the fact that most of the failed dams in China are homogeneous earthfill dams (e.g. Ru and Niu, 2001; Li et al., 2006). For the seepage erosion/piping failure at the embedded culverts, the calculated values for homogeneous/composite and clay-core dams are very similar, 0.63×10^{-4} and 0.66×10^{-4}, respectively, which are slightly larger than the historical value of 0.42×10^{-4}. As mentioned earlier, the historical average annual probability values are particularly representative of embankment dams, and hence can be directly compared with those calculated by using the Bayesian networks in this study. Note that the historical annual probabilities of failure are based on a separate dam failure dataset. Therefore, the predictions in Table 10.8 can be viewed as Class-A predictions.

10.3.2 Probabilistic Analysis of the Chenbihe Dam Distresses Using Bayesian Networks

In a probabilistic analysis, prior experience is often used as a first step for extracting common characteristics from global-level data, which provides a basis for allowing the incorporation of local-level data in a systematic way for a specific dam. Bayesian networks can deal with the combination of global-level performance records and local-level performance records and/or analysis results in order to make full use of available information for conducting probabilistic analysis of dam failure. In this section, a case study on the Chenbihe dam is presented to illustrate the methodology.

10.3.2.1 Introduction to the Chenbihe Dam (Niu, 2002)

The Chenbihe dam is located on Chenbihe River of Bai Se County in Guangxi Province, in south China. It is an earthfill dam with a height of 70.4 m, a crest length of 425.0 m, and a storage capacity of $11.3 \times 10^8\,m^3$. The crest elevation and the base elevation of the dam are 190.4 m and 120.0 m, respectively. Figure 10.8(a) shows a cross-section of the dam. The dam was constructed for hydropower generation, irrigation, and water supply.

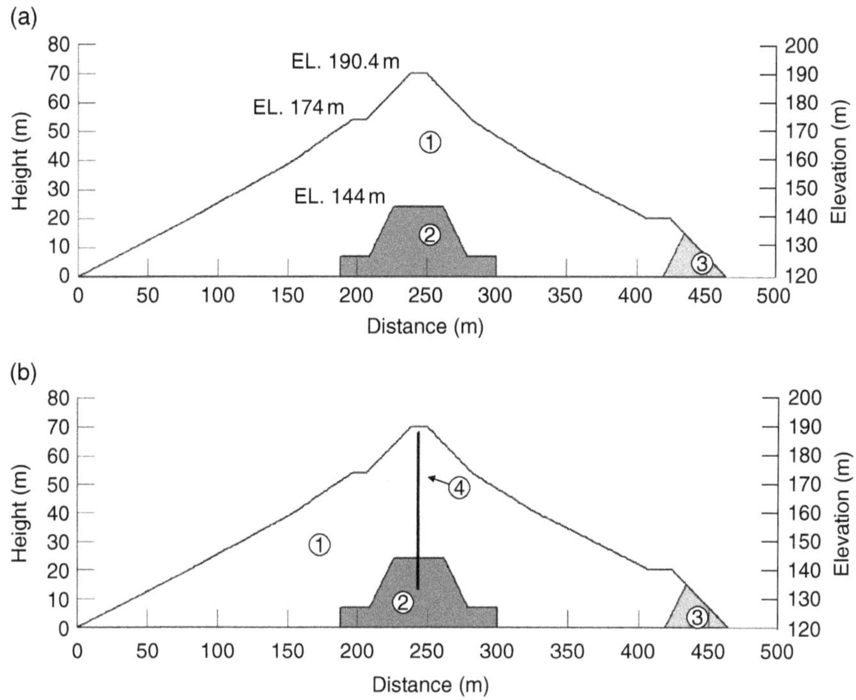

Figure 10.8 Sketch of the cross-section of Chenbihe Dam: (a) after construction in 1961 and (b) after rehabilitation in 1972 – (1) earthfill, (2) clay core, (3) drainage, and (4) concrete core. *Source:* Xu et al. (2011). Reproduced with permission from NRC Research Press

The dam construction was divided into three stages: (1) the dam on the left bank was raised from the base to elevation 150 m from December 1959 to May 1959, (2) the dam on the right bank was first raised from the base to elevation 150 m, and then the whole dam was raised to elevation 185 m from October 1959 to August 1960, (3) the dam was continually raised to the crest from June 1961 to October 1961.

The dam was originally designed as an earthfill dam with a clay core. In the first stage, both the quality of fill and clay materials and the construction quality were controlled rigorously and satisfied the design requirements. The thick clay core under elevation 144 m was constructed well, according to the required quality standard, as shown in Figure 10.8(a). In contrast, the quality control became loose in the second and third stages, due to the high-intensity construction required and the shortage of proper clayey soils for the core. The clay core constructed in the second and third stages became very thin and contained a large proportion of granular soils. Therefore, the dam above elevation 144 m can be treated as a homogeneous earthfill dam, as shown in Figure 10.8(a).

After the first filling of the reservoir in 1961 to elevation 176.7 m, leaking at the downstream slope at elevations of 173–174 m was found. The tail water level was at elevation 132 m. Then, the area and volume of the leakage increased as the water level rose. When the water level reached elevation 181.5 m in August 1971, there were five areas with serious leakage spreading around the downstream slope at elevations of 144–170 m, as shown in Figure 10.9. The total leakage area was 4315 m^2, approximately 10% of the downstream slope area. The seepage volume was measured at a rate of 200 L/min. The main dam distress was associated with seepage problems.

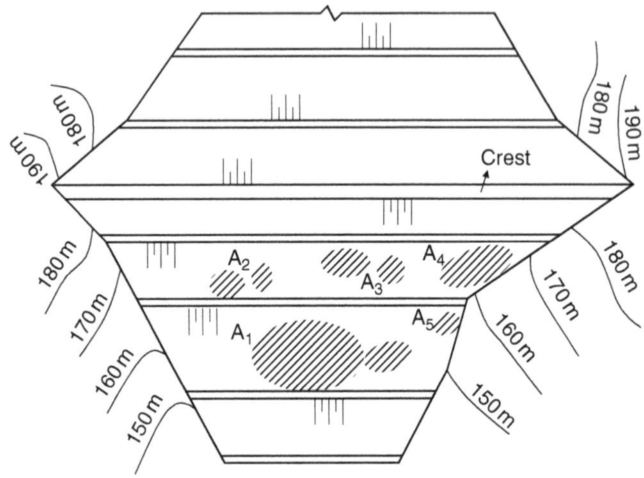

Figure 10.9 Sketch of the five leakage areas on the downstream of Chenbihe Dam ($A_1=2640\,m^2$, $A_2=225\,m^2$, $A_3=500\,m^2$, $A_4=850\,m^2$, and $A_5=100\,m^2$). *Source:* Xu et al. (2011). Reproduced with permission from NRC Research Press

10.3.2.2 Bayesian Network Analysis for the Chenbihe Dam before Rehabilitation

Bayesian networks can serve as a dynamic tool in the sense that monitoring evidence may be continuously utilized to update the knowledge. Therefore, based on the existing Bayesian network developed for probabilistic analysis of dam failure at the global level in the last section, a Bayesian network for probabilistic analysis of failure of a specific dam may be developed. The network can incorporate project-specific variables, which reflect local-level performance records and/or analysis results.

In this case, the seepage volume and the boundary conditions of the distressed embankment are known. Therefore, back analyses of the seepage through the embankment may be performed to improve the knowledge on embankment materials, such as coefficients of permeability of relevant materials. As indicated in Figure 10.8(a), the embankment involves three different types of material: (1) earthfill, (2) clay-core, and (3) drainage. The corresponding coefficients of permeability of these three materials are represented by k_1, k_2, and k_3, respectively. According to the typical ranges of k for different types of soil provided by British Standards Institution (BSI 1986), k_1 is within 10^{-4}–10^{-7} m/s, k_2 is within 10^{-7}–10^{-9} m/s, and k_3 is within 10^{-2}–10^{-4} m/s. The drainage was found to function normally, based on the historical performance records for the Chenbihe dam. Its coefficient of permeability is assumed to be the mean value of 10^{-2}–10^{-4} m/s in log-scale, $k_3=10^{-3}$ m/s. Therefore, two parameters of k_1 and k_2 need to be back analyzed.

To back analyze two or more parameters simultaneously under uncertainty, the back analysis can be implemented in a probabilistic way, in which uncertain parameters are modeled as random variables, and their distributions are updated based on the observed evidence. In this case, the prior distributions of k_1 and k_2 are assumed to be uniform within their respective ranges. Based on the seepage volume and the boundary conditions of the distressed embankment, k_1 and k_2 are back analyzed by using the program SEEP/W (Geo-Slope International Ltd. 2007). Then, the posterior distributions of k_1 and k_2 are obtained by combining the prior distributions and the probabilistic back analysis results of k_1 and k_2.

As mentioned before, the seepage problems exist mainly on the embankment above elevation 144 m, which can be treated as a homogeneous earthfill dam. Therefore, the causal network in Figure 10.6(b) for probabilistic analysis of failure of homogeneous/composite dams by seepage

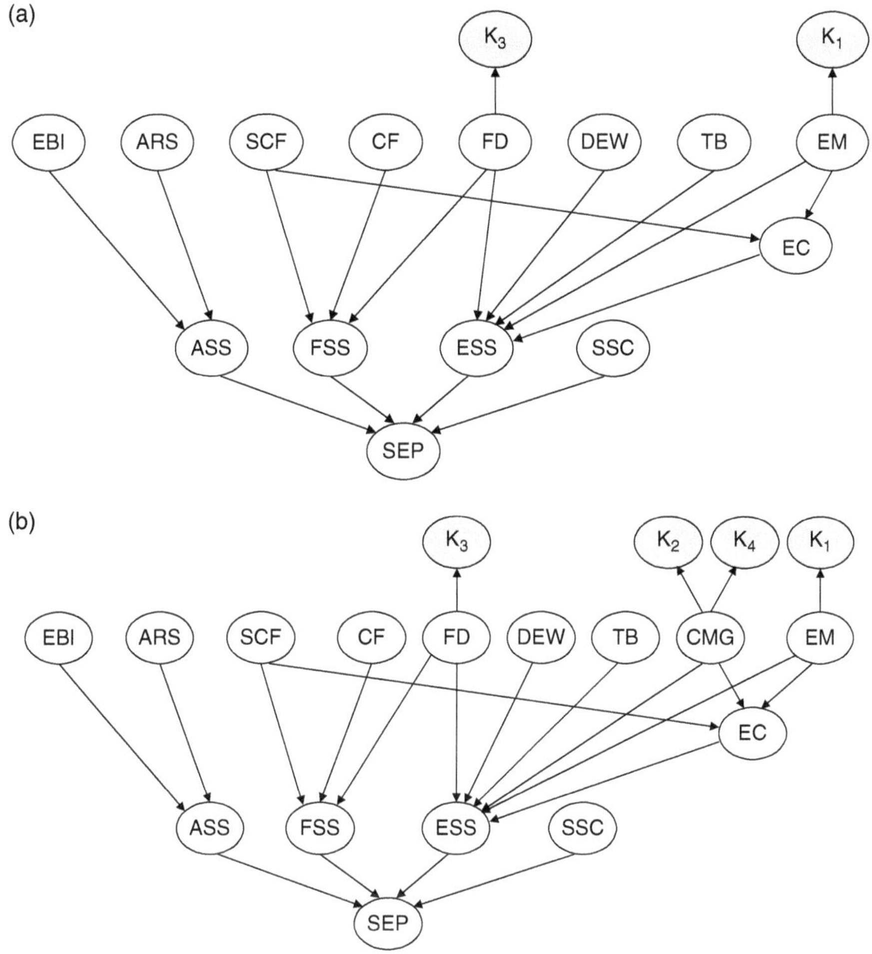

Figure 10.10 Causal network for probabilistic analysis of seepage erosion/piping failure of Chenbihe Dam: (a) after construction in 1961, and (b) after rehabilitation in 1972. *Source:* Xu et al. (2011). Reproduced with permission from NRC Research Press

erosion/piping is used as a basis. Further, two project-specific variables, the coefficients of permeability of the earthfill (K_1) and the drainage (K_3) are incorporated into the existing causal network, and the new causal network is shown in Figure 10.10(a). To be in consistent with the other variables, these two variables, K_1 and K_3, are also assumed as discrete variables with two states, "Satisfactory" and "Unsatisfactory". The conditional probability table for the relationships between nodes EM and K_1 and between nodes FD and K_3 is shown in Table 10.9. If the states of K_1 and K_3 are known, all the probability tables involved in the causal work in Figure 10.10(a) will be automatically updated, and hence the probability of seepage erosion/piping failure can be obtained.

In this case, the normally working drainage is associated with $k_3 = 10^{-3}$ m/s, which is larger than the design value, $k_{d3} = 10^{-5}$ m/s (Li et al., 2006). Hence, the state of K_3 is certain to be satisfactory, P(K_3 = Satisfactory) = 1 and P(K_3 = Unsatisfactory) = 0. It is known that k_1 is a continuous variable while K_1 is a discrete variable. Therefore, the continuous variable of k_1 needs to be transferred to

Table 10.9 Conditional probability table for the relationships between nodes EM and K_1, and nodes FD and K_3 involved in probabilistic analysis of seepage erosion/piping failure of Chenbihe dam

	EM = Satisfactory	EM = Unsatisfactory
K_1 = Satisfactory	1	0
K_1 = Unsatisfactory	0	1
	FD = Satisfactory	FD = Unsatisfactory
K_3 = Satisfactory	1	0
K_3 = Unsatisfactory	0	1

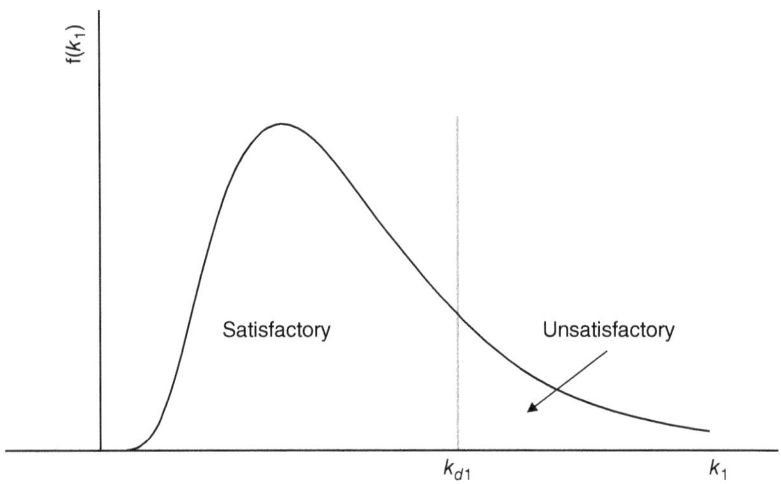

Figure 10.11 Comparison of k_l with a specified design requirement k_{dl}. *Source:* Xu et al. (2011). Reproduced with permission from NRC Research Press

the discrete variable of K_1. In this study, the distribution of k_l is compared with a specified design requirement k_{dl}, and hence divided into two regions in Figure 10.11, which correspond to the two states of K_1: "Satisfactory" and "Unsatisfactory". For either state of K_1, there is a probability value, which equals the area of the corresponding region in Figure 10.11. Also, the equation holds P(K_1 = Satisfactory) + P(K_1 = Unsatisfactory) = 1. The prior uniform distribution of k_1 is improved by a probabilistic back analysis using SEEP/W (Xu 2010) based on the observed evidence on the seepage volume, resulting in the posterior distribution of k_1 in Figure 10.12(a). Then, we obtain P(K_1 = Sati sfactory) = P($k_1 < 1.0 \times 10^{-6}$ m/s) = 2.4×10^{-9} while P(K_1 = Unsatisfactory) = P($k_1 > 1.0 \times 10^{-6}$ m/s) ≈ 1.0. Similarly, based on the posterior distribution of k_2 in Figure 10.12(b), we obtain P(K_2 = Satisfactory) = P($k_2 < 1.0 \times 10^{-7}$ m/s) = 1.0 while P(K_2 = Unsatisfactory) = P($k_2 > 1.0 \times 10^{-7}$ m/s) = 0.

The Bayesian network analysis is conducted using the program Hugin Lite. The analysis was firstly conducted without considering the results on K_1 and K_3 deduced from the project-specific evidence. Based on the dam performance records, there are no seepage problems at the abutment (ASS), the foundation (FSS), or along the embedded culverts (SSC). Hence, the states of nodes ASS, FSS, and SSC are set to be normal. The probability table for the nodes relevant to ESS in Figure 10.6(b) can be automatically updated in Hugin Lite, and the probability of seepage erosion/piping failure is summarized as the prior probability column in Table 10.10. Note that these prior probability values reflect the characteristics at the global level. In Table 10.10, the prior probability

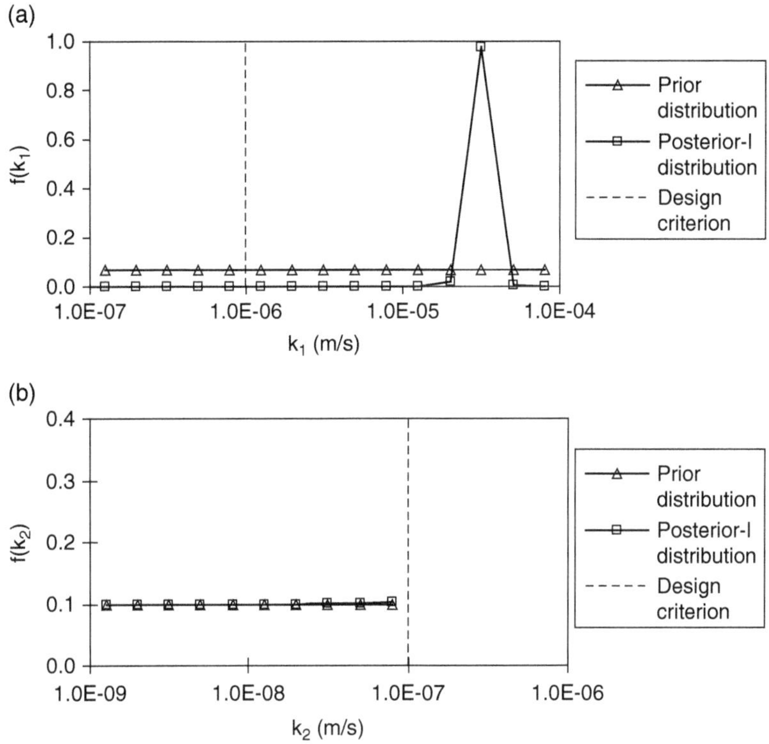

Figure 10.12 Prior and posterior distributions of (a) k_1 and (b) k_2. *Source:* Xu et al. (2011). Reproduced with permission from NRC Research Press

Table 10.10 Calculated failure frequencies over the design service life and average annual failure probabilities for seepage erosion/piping failure of the Chenbihe dam (Prior = prior values at the global level, Posterior-i = posterior values at the local level before rehabilitation, Posterior-ii = posterior values at the local level after rehabilitation)

	Prior	Posterior-I	Posterior-II
Calculated failure frequency over the design service life	1.41%	2.23%	0.11%
*Calculated average annual failure probability	1.41×10^{-4}	2.23×10^{-4}	1.10×10^{-5}
Criterion for average annual failure probability (Li et al. 2006)	$<1.0 \times 10^{-4}$ to 10^{-5} for dams associated with large- or medium-size reservoirs		

*Dam design service life = 100 years.

of SEP at the global level is P(SEP=Yes) = 0.0141. According to the classification criteria for hydraulic structures issued by MWR (1994), Chenbihe Dam belongs to the most important structure in the first grade, and the corresponding design service life is 100 years. Therefore, the prior average annual probability of seepage erosion/piping failure is $P_f = 0.0141/100 = 1.41 \times 10^{-4}$.

Considering the back analysis results on K_1 and K_3, P(K_1 = Unsatisfactory) ≈ 1.0 and P(K_3 = Satisfactory) = 1.0, the nodes of K_1 and K_3 are input into the existing Bayesian network in

Figure 10.6(b), resulting in a new Bayesian network shown in Figure 10.10(a). In addition to that, the states of nodes ASS, FSS, and SSC are set to be normal, the state of node K_1 is set to be unsatisfactory while the state of node K_3 is set to be satisfactory. The probability table for the nodes relevant to ESS in Figure 10.10(a) is automatically updated in Hugin Lite, and the probability of seepage erosion/piping failure is summarized as the posterior-I probability column in Table 10.10. Note that these posterior-I probability values reflect the characteristics at the local level considering the project-specific evidence on the seepage volume through the embankment. The posterior probability of SEP at the local level before rehabilitation is P(SEP = Yes | Posterior-I) = 0.0223. Therefore, the posterior-I average annual probability of seepage erosion/piping failure is $P_f = 0.0223/100 = 2.23 \times 10^{-4}$.

In Table 10.10, the prior average annual probability of seepage erosion/piping failure, 1.41×10^{-4}, is underestimated before considering the project-specific information. The posterior-I average annual probability of seepage erosion/piping failure at the local level is actually 2.23×10^{-4}. If we use the log-scale mean value, 3.16×10^{-5}, of the criterion shown in Table 10.10 as a basis, the prior global-level value of 1.41×10^{-4} is underestimated by 2.6 times, compared with the posterior-I local-level value. Dam failures are regarded as one of "low-probability, high-loss" events. From such a standpoint, the potential failure losses will be seriously underestimated if the prior global-level value is used in the dam risk management.

10.3.2.3 Bayesian Network Analysis for the Chenbihe Dam after Rehabilitation

In 1972, the distressed embankment was reinforced with a concrete-core, as shown in Figure 10.8(b). The crest elevation of the concrete-core is 188.2 m. The concrete-core has a thickness of 0.8–1.0 m, a maximum height of 55 m, and a length of 377.3 m. After the rehabilitation, the reservoir was filled gradually to the maximum water level at elevation 185 m. Meanwhile, the seepage volume continuously decreased and reached a very low and stable magnitude. The original leakage area on the downstream slope became dry, and back to the normal condition.

After the rehabilitation, Chenbihe Dam is an earthfill dam with a composite core of clay and concrete. In the former global-level analysis of embankment dam failure by seepage erosion/piping, the causal network in Figure 10.6(c) is constructed for clay-core embankment dams, while there is no causal network constructed particularly for concrete-core embankment dams. Therefore, the causal network in Figure 10.6(c) for clay-core dams is still used as a basis for the local-level analysis for reinforced Chenbihe Dam. The coefficients of permeability of the earthfill (K_1), the clay core (K_2), the drainage (K_3), and the concrete core (K_4) are incorporated into the existing causal network, and the new causal network is shown in Figure 10.10(b). The variables of K_2 and K_4 are also assumed discrete with two states: "Satisfactory" and "Unsatisfactory". The conditional probability table for the relationships between nodes EM and K_1, nodes CMG and K_2 and K_4, and nodes FD and K_3 is shown in Table 10.11.

It was reported in the literature (Niu, 2002) that the total water head behind the corewall decreased by a magnitude of 10–30 m compared with that before the rehabilitation, and the seepage rate decreased and reached a very low and stable magnitude. Based on these performance records on the seepage through the reinforced embankment and relevant seepage analyses, the probability distributions of the coefficients of permeability of the earthfill (k_1), the clay core (k_2), and the concrete core (k_4) can be updated by using the probabilistic back analysis, resulting the posterior-II distribution of k_1, k_2, and k_4, respectively, as shown in Figure 10.13. Based on the posterior-II distribution of k_1 in Figure 10.13(a), we obtain $P(K_1 = \text{Satisfactory}) = P(k_1 < 1.0 \times 10^{-4}\,\text{m/s}) = 1.0$

Table 10.11 Conditional probability table for the relationships between nodes EM and K_1, nodes CMG and K_2, and K_4, and nodes FD and K_3 involved in probabilistic analysis of seepage erosion/piping failure of the rehabilitated chenbihe dam

	EM = Satisfactory	EM = Unsatisfactory
K_1 = Satisfactory	1	0
K_1 = Unsatisfactory	0	1
	CMG = Satisfactory	CMG = Unsatisfactory
K_2 = Satisfactory	1	0.5
K_2 = Unsatisfactory	0	0.5
	FD = Satisfactory	FD = Unsatisfactory
K_3 = Satisfactory	1	0
K_3 = Unsatisfactory	0	1
	CMG = Satisfactory	CMG = Unsatisfactory
K_4 = Satisfactory	1	0.5
K_4 = Unsatisfactory	0	0.5

while P(K_1 = Unsatisfactory) = P($k_l > 1.0\times10^{-4}$ m/s) = 0. It is indicated that the value of k_l can satisfy the requirement of shell-fill for a clay-core earthfill dam, though it cannot satisfy the requirement of a seepage barrier for a homogeneous earthfill dam. Similarly, based on the posterior-II distribution of k_2 in Figure 10.13(b), we obtain P(K_2 = Satisfactory) = P($k_2 < 1.0\times10^{-7}$ m/s) = 1.0 while P(K_2 = Unsatisfactory) = P($k_2 > 1.0\times10^{-7}$ m/s) = 0. Based on the posterior-II distribution of k_4 in Figure 10.13(c), we obtain P(K_4 = Satisfactory) = P($k_4 < 1.0\times10^{-9}$m/s) ≈ 1.0 while P(K_4 = Unsatisfactory) = P($k_4 > 1.0\times10^{-9}$ m/s) = 1.7×10^{-21}. For the drainage, we know P(K_3 = Satisfactory) = 1.0 while P(K_3 = Unsatisfactory) = 0, as mentioned before.

In Figure 10.10(b), the states of nodes ASS, FSS, and SSC are still set to be normal, and the states of nodes K_1, K_2, K_3, and K_4 are set to be satisfactory. Accordingly, the probability table for the nodes relevant to ESS in Figure 10.10(b) is automatically updated in Hugin Lite, and the probability of seepage erosion/piping failure is summarized as the posterior-II probability column in Table 10.10. Note that these posterior-II probability values reflect the characteristics of the rehabilitated dam at the local level. The posterior probabilities of SEP at the local level after rehabilitation is P(SEP = Yes | Posterior-II) = 0.0011. Therefore, the posterior-II average annual probability of seepage erosion/piping failure is $P_f = 0.0011/100 = 1.10\times10^{-5}$.

In Table 10.10, the posterior-I average annual probability of seepage erosion/piping failure is 2.23×10^{-4}, which cannot satisfy the criterion for dams associated with large- or medium-size reservoirs (Li et al., 2006). After the rehabilitation, the probability of failure decreased to 1.10×10^{-5}, which satisfies the specified criterion. The above remedial action (i.e. concrete core) has successfully reduced the probability of failure of Chenbihe Dam.

10.3.2.4 Summary

Bayesian networks are used to conduct the global-level probabilistic analysis of embankment dam failure based on the database. Based on prior experience of common characteristics extracted from the global-level data, Bayesian networks are further used to conduct the local-level probabilistic

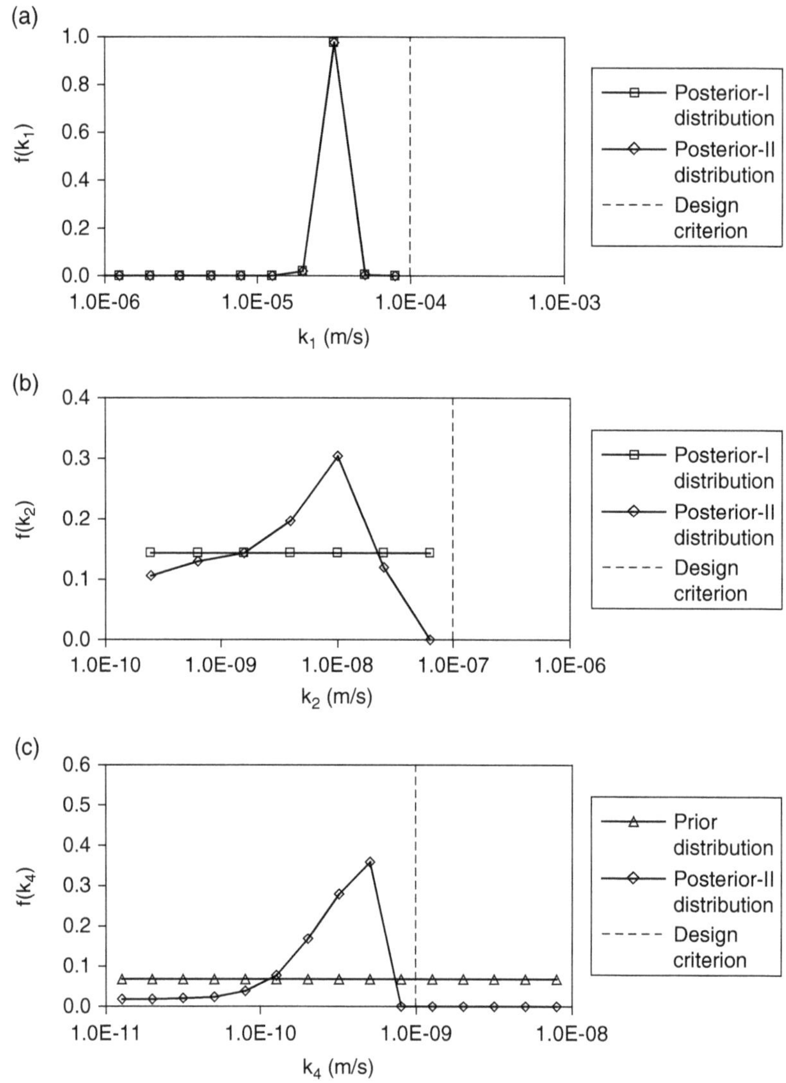

Figure 10.13 Posterior-I (or prior) and posterior-II distributions of (a) k_1, (b) k_2, and (c) k_4. *Source:* Xu et al. (2011). Reproduced with permission from NRC Research Press

analysis of Chenbihe dam failure by combining global-level performance records and local-level performance records and/or analysis results. The post-rehabilitation Bayesian network analysis proves the effectiveness of the concrete core, which has successfully reduced the probability of seepage erosion/piping failure of Chenbihe Dam to 1.10×10^{-5}, which satisfies the specified criterion ($<1.0\times10^{-4}$–10^{-5}). The original probability of seepage erosion/piping failure before the rehabilitation is 2.23×10^{-4}, which does not satisfy the specified criterion.

11

Vulnerability to Dam Breaching Floods

11.1 Concepts of Vulnerability

Vulnerability refers to the degree of loss to a given element or set of elements within the area affected by a hazard. It is expressed on a scale of 0 (no loss) to 1 (total loss). The product of elements at risk and vulnerability is defined as consequence, which is the one of the two important components of risk (the other is probability). Generally, there are four categories of vulnerability (Prina et al., 2004):

1. social vulnerability with respect to direct impact of life, physical and mental health, and living conditions
2. physical vulnerability with respect to direct danger to buildings, structures, infrastructures, networks, and objects
3. environmental vulnerability with respect to direct impact on natural environment (forests, animals, plants, water, etc.) considering their short- and long-term sensitivity
4. economic vulnerability with respect to indirect impact of the phenomenon on various activities that are affected by the hindrances induced (interruption of business, paralyzed traffic, reduced touristic demand, etc.).

The last two categories of vulnerabilities to dam breaching floods are seldom studied, as the environmental vulnerability is difficult to quantify and the economic vulnerability covers many aspects in both direct and indirect influences, such as agriculture, industry, business, transportation, and so on, in the flooded areas or in related areas. Therefore, in this chapter, only the first two categories of vulnerability due to dam breaching floods are discussed.

11.2 Human Vulnerability to Dam Breaching Floods

Blaikie et al. (1994) defined the social vulnerability as "the characteristics of a person or group in terms of their capacity to anticipate, cope with, resist and recover from the impacts of natural or man-made hazards." Dams serve the human beings with many benefits, such as flood control,

---

Dam Failure Mechanisms and Risk Assessment, First Edition. Limin Zhang, Ming Peng, Dongsheng Chang, and Yao Xu.

power generation and water supply. However, dam failure disasters, if occurring unfortunately, may cancel the benefits or even bring harm to people. Approximately 3462 dams failed during the fifty years from 1954 to 2004 in China, which accounted for 4% of all the dams built in the same period (Li el al., 2006). Many of the failures have caused loss of life. The failures of Banqiao Dam and Shimantan Dam in Henan Province in 1975 were evaluated as the world's most catastrophic dam failures (Graham, 1999). The number of deaths directly caused by the dam-break flood was 26,000, and 5.4 million people were impacted (Graham, 1999; Ru and Niu, 2001). In the past two centuries, at least 20 recorded dam failure events caused the death of 500 or more in each case, as shown in Table 11.1. The need for mitigation of human risks due to dam failure raises questions about the proper evaluation of potential loss of life.

11.2.1 Human Stability in Flood

Loss of human stability and subsequent drowning in flood flows is one of the important causes of loss of life. Several researchers have investigated the issue of human stability in flowing water.

Abt et al. (1989) conducted experiments on human stability in flowing water. Twenty human subjects and a monolith were tested on a flume slope (0.5% and 1.5% slope rates) with four kinds of surfaces (i.e. turf, smooth concrete, steel, and gravel). The tested people had weights of 41–91 kg, and heights of 1.52–1.91 m. The monolith was 53.4 kg in mass and 1.52 m in height. A critical product of water depth, *d*, and velocity, *v*, dv_{cr}, was used to describe human stability. In an attempt to provide a means to quantitatively predict the point of instability of the human subject, a linear regression equation was derived to relate the critical product to the weight and height of the subject, with a correlation coefficient of $R^2 = 0.48$:

$$dv_{cr} = \exp\left(0.222 W_g H_g / 1000 + 1.088\right) \tag{11.1}$$

where W_g is the weight (kg) and H_g is the height (m).

RESCDAM (2000) launched a program to define the limits of a safe rescue action in the case of a dam-break flood. One of the main contents of this program is human stability experiments in water flow. The experiments were conducted at a model basin of Helsinki University of Technology Ship Laboratory in Finland. The model basin was 130 m long, 11 m wide, and approximately 5.5 m deep, and the water temperature was about 16 °C. The model basin was equipped with a towing carriage with a maximum speed of 8 m/s. Seven human subjects were tested: five males and two females with heights of 1.60–1.95 m, masses of 48–100 kg, and ages of 17–60. The test flow velocities were 0.6–2.75 m/s and the water depths were 0.3–1.1 m. Due to differences in the individual properties, a wide range of values of *dv*, which describes a person's maneuverability in the flow, was obtained. The critical dv_{cr} causing loss of stability varies from 0.64 m²/s to 1.26 m²/s. Taller and heavier individuals managed better in flowing water.

Lind et al. (2004) derived theoretical relationships for the stability in water flows of three object shapes representing the human body: a circular cylindrical body, a square parallelepiped body, and composite cylinders (two small ones for the legs, and one for the torso). As shown in Figure 11.1, W_g is the weight, B_u is the buoyancy, M_a their common moment arm about the pivot of the toe, and A_w the wetted area's projection normal to the flow. The drag force (F_d) on the object can be calculated as:

$$F_d = C_d A_w \rho_w v^2 / 2 \tag{11.2}$$

where C_d is the drag coefficient, ρ_w is the density of water, and *v* is the water velocity.

Table 11.1 Dam failure cases that caused more than 500 deaths in each case

Dam name	Country	Failure time	Dam height (m)	Lake volume (million m^3)	Failure mode	Death	Reference
Baiguochong	China	1960	15.0	7	Overtopping	616	MWR 1981; Ru and Niu 2001
Banqiao, Shimantan	China	1975	24.5/25.0	492/92	Overtopping	26000	MWR 1981; Ru and Niu 2001
Dongchuankou	China	1963	31.0	27	Overtopping	510	MWR 1981; Ru and Niu 2001
Gleno	Italy	1923	35.0	5	Overtopping	600	Xu 2010
Hengjiang	China	1970	48.4	79	Piping	941	MWR 1981; Ru and Niu 2001
Hutai	China	1971	21.5	3	Overtopping	512	MWR 1981; Ru and Niu 2001
Irukaike	Japan	1868	28.0	19	Overtopping	1200	Xu 2010
Kuala Lumpur Dam	Malaya	1961	—	—	—	600	McClelland and Bowles 2002
Lijiazui	China	1973	25.0	1	Overtopping	580	MWR 1981; Ru and Niu 2001
Liugongshe	China	1973	—	300	Piping	702	MWR 1981
Liujiatai	China	1963	35.9	41	Overtopping	943	MWR 1981; Ru and Niu 2001
Longtun	China	1959	9.5	30	Overtopping	707	MWR 1981; Ru and Niu 2001
Machhu-II	India	1979	24.7	100	Overtopping	2500	Xu 2010
Mohne dam	Germany	1943	40.0	134	Piping	1200	McClelland and Bowles 2002
Panshet	India	1961	49.0	212	Overtopping	1000	Xu 2010
Puentes	Spain	1802	69.0	13	Piping	608	McClelland and Bowles 2002; Xu 2010
South Fork	US	1889	22.0	18	Overtopping	2209	McClelland and Bowles 2002; Xu 2010
Tiefusi	China	1960	14.5	20	Overtopping	898	MWR 1981; Ru and Niu 2001
Tigra	India	1917	25.9	124	Overtopping	1000	Xu 2010
Vaiont	Italy	1963	267.0	240	Reservoir landslide	2056	McClelland and Bowles 2002; Xu 2010

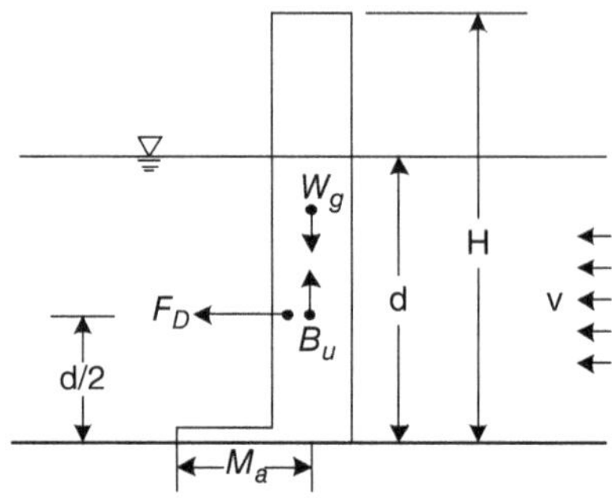

Figure 11.1 Human stability in water flow (modified from Lind et al. 2004)

Assuming that the resultant drag acts at one half of the water depth, *d/2*, as in uniform flow, the incipient instability condition is

$$\left(W_g - B_u\right)M_a = F_d d/2 \tag{11.3}$$

The human body is modelled by a rigid circular cylinder of diameter *a*. The critical depth–velocity product can be calculated as

$$dv_{cr} = K_D\left[W_g\left(1 - M_a / H_g\right)\right]^{1/2} \tag{11.4}$$

where K_D is a random variable that can be estimated from experimental observations. The equations can be simplified as follows:

$$dv_{cr} = K_E W_g^{\ 1/2} \tag{11.5}$$

$$dv_{cr} = K_F W_g \tag{11.6}$$

$$dv_{cr} = K_G \tag{11.7}$$

The coefficients, K_E, K_F, and K_G, can be obtained through regression analysis with experimental data. The results of the statistical analysis have been reported by Abt et al. (1989) and RESCDAM (2000). Both investigations show a major difference between females and males in their resistance to the pressure of floodwater, K_G. Men, as expected, have higher dv_{cr} values. Together, all 103 observations of K_G range from 0.65 to 2.13 m²/s, with a sample mean of 1.224 m²/s and a COV of 0.27. If the object's weight and height are taken into account, then the dispersion in the data is reduced significantly to a COV of 0.20 for K_D.

Jonkman and Penning-Rowsell (2008) conducted two standing tests and four walking tests in water flow to check the human instability. They considered two hydrodynamic mechanisms that can cause instability: moment instability (toppling) and friction instability (sliding). Moment instability occurs when the force of the oncoming flow exceeds the resultant moment due to the resultant weight of the body. Friction instability occurs if the drag force is larger than the frictional resistance between the person's feet and the substrate surface. Both mechanisms are schematically shown in Figure 11.2.

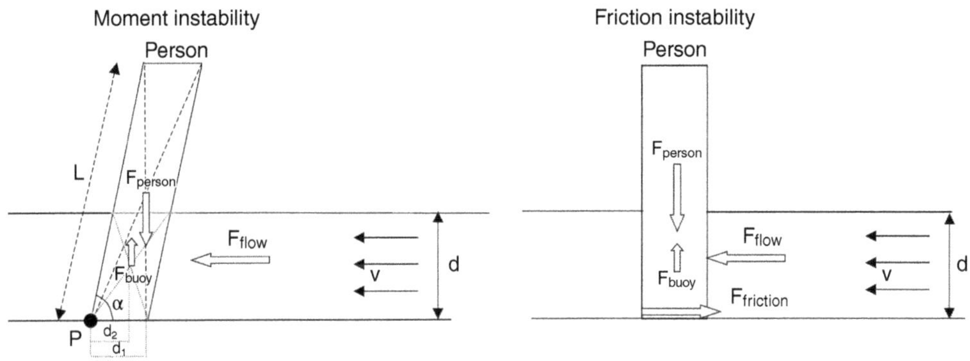

Figure 11.2 Models of human body for moment and for friction instability. *Source:* Jonkman, S. N. (2007). Reprinted with permission of Prof. Jonkman, S.N (Bas)

For moment stability, the critical condition $\Sigma\ Mp=0$ leads to

$$dv_{cr} = C_s W_g^{0.5} \tag{11.8}$$

$$C_s = \left(4gd_1 / C_D W_b \rho_w\right)^{0.5} \tag{11.9}$$

where W_b is the average body width exposed normal to the flow (m) and d_1 is the moment arm.

For friction stability, the critical condition $\Sigma\ Fp=0$ leads to

$$dv_{cr}^2 = C_F m^{0.5} \tag{11.10}$$

$$C_F = 2\mu g / C_D W_b \rho_w \tag{11.11}$$

where μ is the coefficient of static friction and m is the mass (kg).

When the water depth is relatively large and the velocity is small, the moment mechanism controls the human stability and dv_{cr} is a good measure. However, for the case of small depths but large flow velocities, the friction mechanism controls the human stability, and dv_{cr}^2 is a better measure.

11.2.2 Influence Factors

The mechanisms of casualties caused by dam-break floods are complex. McClelland and Bowles (1999) listed 41 means by which people may die in a flood. There are at least four categories of factors influencing fatality in a flood (McClelland and Bowles, 2002; Ramsbottom el al., 2003; Priest et al., 2007; Jonkman, 2007):

1. hydraulic factors: water depth, flow velocity, flood rise rate, wave, water temperature, flood duration, debris in water, flood severity, etc.

2. area and building factors: area topography, area usage, building type, number of building story, building quality, roads, traffic situation, traffic tools, etc.
3. time and weather factors: warning time, time of a day, time of week, time of year, holiday or festival, weather (e.g. foggy, rainy, and windy), etc.
4. population factors: age, gender, physical state, faith, experience/knowledge of flood, family relationship, financial state, population distribution, human decision, human behavior, etc.

Not all the factors directly affect the final fate of a person facing a flood. Some demonstrate their influences through other factors. For example, time of a day affects the population distribution and human behavior towards a flood; these intermediate factors then affect the fate of people.

11.2.3 Methods for Evaluating Human Vulnerability Factor in a Flood

Existing models for estimating the expected loss of life (*LOL*) caused by a flood can be divided into two categories: those that are empirically based and those that rely on parameters considered to be theoretically important (McClelland and Bowles, 2002). In order to have a thorough understanding of these models, the empirical models are separated into two subcategories: those based on experience or judgment and those based on statistical data. In addition, the models that involve physical analysis calibrated with statistical data (empirical analysis) are specially categorized and named as compromised models.

Figure 11.3 summarizes part of the existing models. The horizontal axial shows the change from empirical analysis to mechanistic analysis. The vertical axis shows the change from a macro scale, like the whole event of a flood, to a micro scale, like individual behavior. The empirical models are generally at the macro scale with empirical analysis; the physical models are generally at the micro scale with mechanistic analysis; and the compromised models are in between.

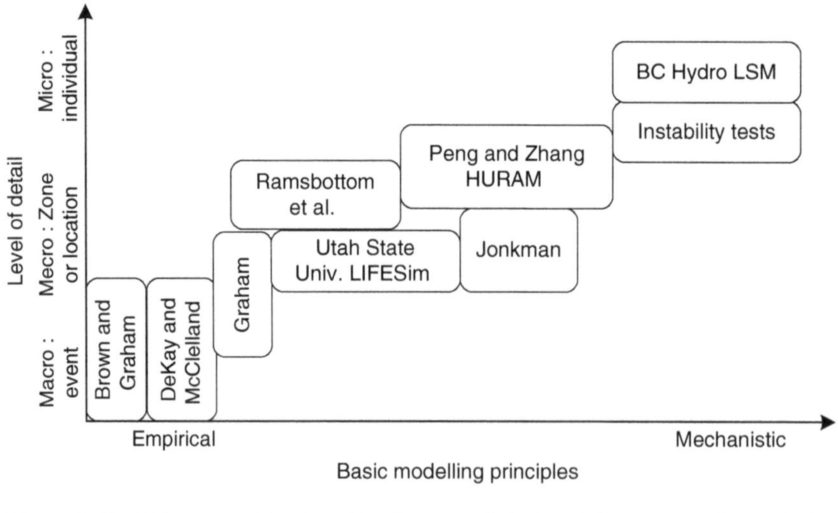

Figure 11.3 Existing models for estimating loss of life (Modified from Jonkman, 2007)

11.2.3.1 Empirical Models Based on Experience or Judgment (McClelland and Bowles, 2002)

Ayyaswamy et al. (1974) prepared a report for the US Atomic Energy Commission to evaluate the probabilities and potential consequences to ground-based population at risk from dam failures. This report focuses on earthquake-induced dam failures. The dams are assumed to suffer from a complete and instantaneous failure with total release of the impounded water. The *LOL* is provided as a function of flood depth and the evacuation rate. Both of these two parameters are assumed with subjective judgment.

Friedman (1975) developed a generic model that could be applied to many types of natural hazards: earthquakes, hurricanes, floods, tornadoes, wind, hail, and so on. A loss potential index is calculated based on four factors: (1) a natural hazard generator, which is used to determine the frequency of earthquakes or storms, (2) local conditions, which modify the severity pattern by the natural hazard generator, (3) population at risk, defined as the number of persons exposed to the hazard, and (4) the vulnerability of the population at risk (*PAR*), which is its susceptibility to life loss during an event of a given severity. The loss function is based on the estimated number of buildings expected to be damaged. Using the annual flood tabulations of the American Red Cross, it is assumed that one casualty would follow every 170 damaged dwellings, or every 85 dwellings in the case of flash floods.

Petak and Atkisson (1982) developed a natural disaster model that is implemented in three steps: (1) quantification of a hazard curve for a region (annual exceedance probability vs intensity of event), (2) quantification of a vulnerability probability distribution (expected structural damages vs intensity of event at the location of the structure), and (3) an exposure distribution (how many of each type of structure, parcels of property, people, etc., are exposed to each intensity level). These three components, hazard, vulnerability, and exposure, are then related sequentially in an event tree to generate values for annualized risk. The structural damage is treated as a fundamental factor, which means that fatalities are assumed to be directly linked to damage to structures.

McCann et al. (1985) proposed a model based on parameters of water depth (*d*), distance to dam site, *PAR*, and effectiveness of warning and evacuation. The flood area is divided into a number of subzones. For each subzone, the *LOL* can be estimated as follows:

$$LOL_i = \phi(d_i) r_i PAR_i \tag{11.12}$$

$$LOL_i = \phi(d_i) r_i p_i N_i \tag{11.13}$$

where i is the number of the subzone; *LOL* is the loss of life; $\phi(d)$ is the fraction of people losing their lives as a function of water depth; r is the fraction of people present when the flood wave arrives among the population at risk (*PAR*); p is the percent of time a given zone is occupied; and N is the number of people occupying a zone during business hours. Note that Equation (11.12) applies to a residential area and Equation (11.13) applies to a more transitory business district. The perish ratio, $\phi(d)$, and the exposed ratio, r, are set according to subjective judgment.

11.2.3.2 Empirical Models Based on Statistic Data

Lee at al. (1986) suggested a model with six variables that could be found in historical data. These six variables are loss of life (*LOL*), population at risk (*PAR*), Warning time (W_t), previous experience with flooding within the last 10 years, whether or not the area was urbanized, and the depth of flooding at peak stage. A dataset of 47 flood cases was assembled for regression analysis. Two equations are presented:

$$\frac{LOL}{PAR}=\frac{\exp\left[-6.2+3.1\left(\frac{1}{1+W_t}\right)-0.00034\left(W_tPAR\right)^{0.5}-0.0077PAR^{0.5}+1.4EX+0.0039d\right]}{1+\exp\left[-6.2+3.1\left(\frac{1}{1+W_t}\right)-0.00034\left(W_tPAR\right)^{0.5}-0.0077PAR^{0.5}+1.4EX+0.0039d\right]} \tag{11.14}$$

$$\frac{LOL}{PAR}=\frac{\exp\left[-0.18+1.7\left(\frac{1}{1+W_t}\right)-0.00044\left(W_tPAR\right)^{0.5}-0.0092PAR^{0.5}+0.26EX-0.18UR\right]}{1+\exp\left[-0.18+1.7\left(\frac{1}{1+W_t}\right)-0.00044\left(W_tPAR\right)^{0.5}-0.0092PAR^{0.5}+0.26EX-0.18UR\right]} \tag{11.15}$$

where W_t is the warning time in minutes, EX is the experience with floods in the last 10 years (1 = yes; 0 = no), d is the depth of flooding at peak stage (feet above flood stage), and UR is the urbanized area (1 = urban area with population ≥ 10,000; 0 = otherwise).

Brown and Graham (1988) published their empirical models for assessing the threat to life from a dam failure in 1988. The loss of life is estimated based on two variables: PAR and W_t. Two kinds of situation, adequate warning ($W_t \geq 1.5$ h) and insufficient warning ($W_t < 1.5$ h), are considered, and regression analysis is conducted with 24 cases of dam failures and major flash floods occurring since 1950. The following empirical equations are obtained:

$$LOL = 0.002PAR,\ W_t \geq 1.5\text{h} \tag{11.16}$$

$$LOL = PAR^{0.6},\ W_t < 1.5\text{h} \tag{11.17}$$

The above equations were later updated by USBR (1989) for cases with warning times less than 15 minutes and water depths greater than 0.9 m:

$$LOL = 0.5PAR,\ W_t < 0.25\text{h} \tag{11.18}$$

The USBR equations are considered as one of the most important attempts to estimate the loss of life based on regression analysis of historic failure events (Aboelata and Bowles, 2005). It should be noted that the formulas are discrete. For example, for PAR = 10,000 the loss of life jumps from 5000 at $W_t < 0.25$ h to 251 at $0.25 < W_t < 1.5$ h, and then jumps to 20 at $W_t > 1.5$ h.

DeKay and McClelland (1993) presented new formulas with three parameters: PAR, W_t and *Force* (Equation 11.19). For simplification, Force is assumed as 1.0 for high force situations when the PAR is located in a canyon, where flood waves are likely to be very deep and swift (Equation 11.20); and as zero for low force situations when the PAR is located in plain where flood waves are likely to be shallow and the flow velocity is slow (Equation 11.21).

$$LOL = PAR\left\{1+13.277\left(PAR^{0.440}\right)\exp\left[0.759\left(W_t\right)-3.790\left(Force\right)+2.223\left(W_t\right)\left(Force\right)\right]\right\}^{-1} \quad (11.19)$$

$$LOL = PAR\left\{1+13.277\left(PAR^{0.440}\right)\exp\left[2.982\left(W_t\right)-3.790\right]\right\}^{-1} \quad (11.20)$$

$$LOL = PAR\left\{1+13.277\left(PAR^{0.440}\right)\exp\left[0.759\left(W_t\right)\right]\right\}^{-1} \quad (11.21)$$

Another important model was provided by Graham (1999). A new dataset expanded from the work of Brown and Graham (1988) and DeKay and McClelland (1993) was compiled, including 26 dam failures, 40 floods, and 50 sub-populations. Graham divided the 50 sub-populations into 15 categories based on unique combinations of three dominant factors that influenced loss of life: the flood severity, the warning time, and the extent to which the population at risk understand the severity of the approaching flood. Graham suggested a table with the averages and ranges of fatality rates in 15 categories (Table 11.2). The flood severity is classified as low, medium, or high. A flood has (1) low severity when homes are flooded but not destroyed, (2) medium severity when some homes or businesses are destroyed or knocked off their foundations, but some homes and trees remain unsubmerged, and (3) high severity when the floodplain is swept clean with little trace of any prior existing structures or vegetation.

Table 11.2 Comparison of Graham's model (1999) and the HURAM model

Flood severity	Warning time (Minute)	Flood severity understanding	Fatality rate		
			Graham's model (1999)		HURAM
			Suggested value	Suggested range	
High	No warning	N/A	0.75	0.30–1.0	0.7707
	15–60	Vague	Use the values shown above and apply		0.4195
		Precise	to the number of people who remain in		
	More than 60	Vague	the dam failure floodplain after		0.0794
		Precise	warnings are issued.		
Medium	No warning	N/A	0.15	0.03–0.35	0.1071
	15–60	Vague	0.04	0.01–0.08	0.0593
		Precise	0.02	0.005–0.04	
	More than 60	Vague	0.03	0.005–0.06	0.0117
		Precise	0.01	0.0–0.02	
Low	No warning	N/A	0.01	0.0–0.02	0.0079
	15–60	Vague	0.007	0.0–0.015	0.0045
		Precise	0.002	0.0–0.004	
	More than 60	Vague	0.0003	0.0–0.0006	0.0009
		Precise	0.0002	0.0–0.0004	

11.2.3.3 Physical Models Based on Simulation

Physical models have been developed to simulate human behavior during floods. Human stability tests (Abt et al., 1989; RESCDAM, 2000; Jonkman and Penning-Rowsell, 2008) were conducted to find the human instability conditions in water flows with various combinations of water depth and velocity, which was introduced earlier.

Assaf et al. (1997) presented a one-dimensional model for estimating loss of life in dam failure flood. The flooded area is divided into several units. The expected loss of life in each unit is calculated as the product of number of *PAR* and the probability of loss of life. Water depth is a key parameter to define the people who are toppled or not. The probabilities of survival of both the toppled and un-toppled people are suggested based on subjective judgment or empirical data.

Watson et al. (2001), Assaf and Hartford (2001, 2002), and Johnstone et al. (2005) in BC Hydro presented a conceptual two-dimensional model considering hydraulic characteristics, the presence of people in the inundated area, the effectiveness of evacuation, and individual/group behavior of people. Decisions have to be made when a flood is likely coming with three choices: stay in buildings, evacuate or escape by walking, or evacuate or escape by vehicle. A time-related parameter, $PPC(t)$, represents the physical condition of people, which is influenced by the product of average water depth and velocity (*DV*). Critical values of *DV* are set to define the condition of toppling and drowning of people.

11.2.3.4 Compromised Models

Compromised models combine the knowledge of both empirical models and physical models. In compromised models, a flooded area is normally separated into some zones (sub-areas) based on physical concepts. The mechanisms of flood-caused fatality are considered to a certain degree. The loss of life in each zone is calibrated with expert judgment or historical data. As shown in Figure 11.3, the compromised model bridges between macro and micro scale, or the meso scale indicating the subzone or local location.

Ramsbottom et al. (2003) presented a method to estimate the number of deaths/injuries during flood:

$$N(I) = N \times X \times Y \tag{11.22}$$

where N = population within the floodplain, X = proportion of the population exposed to a given flood, and Y = proportion of those at risk who will suffer death/injury. X is determined with the combined effect of flood characteristics and location characteristics, which are assumed with scores (1–3) in each subzone. Y is scored based on population characteristics. The model provides a simple method for estimating injuries and deaths according to the characteristics of flood, area, and population. However, the influence of each factor is represented by a score based on subjective judgment or experience, which may not be accurate. Later on, the model was extended by Priest et al. (2007), who calibrated the model with more European cases and added a new variable of mitigating actions.

Utah State University (McClelland and Bowles, 1999, 2002; Aboelata and Bowles, 2005) has developed a model (LIFESim) for estimating loss of life induced by dam-break floods. Loss of shelters, warning, and evacuation are considered. The flooding areas are divided into three zones with unique life-loss distributions. The probabilities of proportional life loss for three flood zones are quantified with statistical data. The concepts of the three zones are:

1. safe zones that consist of safe havens that provide a high degree of safety and a consistently low fatality rate: Safe zones are typically dry, exposed to relatively quiescent floodwaters, or exposed to shallow flooding unlikely to sweep people off their feet. The fatality rate in safe zones is virtually zero.
2. compromised zones consisting of damaged buildings, trees or other objects that may provide shelters for people: In compromised zones, the available shelter has been severely damaged by the flood, increasing the exposure of flood victims to violent floodwaters. The historical fatality rate in compromised zones ranges from zero to about 50%, with an average rate near 12%.
3. chance zones: Inside these areas people are submerged or face the open flood and chance havens that might be reached while drifting. In the chance zones, flood victims are typically swept downstream or trapped underwater. The survival rate depends largely on chance. The historical fatality rate in chance zones ranges from about 38% to 100%, with an average rate over 91%.

A model for estimating the loss of life induced by dike failures has also been developed by Jonkman (2007) and Jonkman et al. (2008, 2009). In the model, three typical hazard zones are also distinguished for a breach of a flood defense protecting a low lying area: the breach zone, zones with rapidly rising waters, and the remaining zone.

1. breach zones: these areas are close to the breach of a flood defense, where high water flow velocity occurs. This leads to the collapse of buildings and instability of people standing in the flow.
2. zones with rapidly rising waters: Due to the rapid rising of water, people are not able to reach shelters on higher grounds or higher floors of buildings. This is particularly hazardous in combination with larger water depths.
3. remaining zone: In this zone the water level rises more slowly, offering better possibilities to find shelter. Fatalities may occur among those who did not find a shelter, or due to adverse health conditions associated with extended exposure of those in shelters.

Mortality in the breach zone, when $dv \geq 7\,m^2/s$ and $v \geq 2\,m/s$, is very high and assumed to be 1.0:

$$F_r = 1 \tag{11.23}$$

where F_r is the fatality. Mortality in the zone with rapidly rising water and the remaining zone are simulated with a best-fit trendline of lognormal distribution of water depth (*d*):

$$F_r(d) = \Phi_N\left(\frac{\ln(d) - \mu_N}{\sigma_N}\right) \tag{11.24}$$

where Φ_N is the probability function of a normal distribution; μ_N and σ_N are the mean and standard deviation of $ln(d)$. In the zone with rapidly rising water, μ_N equals 1.46 and σ_N equals 0.28 when $d \geq 2.1\,m$, the rise rate $w \geq 0.5\,m/h$, and $dv \leq 7\,m^2/s$ or $v < 2\,m/s$. While in the remaining zone, μ_N equals 1.46 and σ_N equals 0.28 when $w < 0.5\,m/h$ or $w \geq 0.5\,m/h$ and $d < 2.1\,m$), and $dv \leq 7\,m^2/s$ or $v < 2\,m/s$.

11.2.4 Database of Fatalities in Dam/Dike Breaching or Other Floods

Detailed data for analyzing human vulnerability to floods is rare but very important. Quantitative analysis of loss of life largely relies on historical fatality data, since the ways of human thinking and behavior are difficult to describe mathematically. A database of fatalities in 321 cases of dam/dike breaching or other floods has been compiled. Among these cases, 254 are directly or indirectly compiled by Jonkman (2007). In particular, 91 and 76 cases are associated with the floods caused by storm surges in the Netherlands in 1953 and the flooding of New Orleans caused by the dike failures during Hurricane Katrina in 2005, respectively. For each case, the database documents the location of flooding, occurrence time, number of fatalities, and factors influencing the fatalities such as warning, flood rise rate and building damage if such information is available.

11.3 Bayesian Network Analysis of Human Vulnerability to Floods

11.3.1 Bayesian Networks

Peng and Zhang (2012b,c) presented a new human risk assessment model (HURAM) using Bayesian networks. A Bayesian network combines the knowledge of graph theory and statistics theory. It consists of nodes and arcs/links with their (conditional) probabilities, which can solve uncertain problems by logical reasoning. The concepts of Bayesian networks were introduced in Section 10.2.6. Bayesian networks have been used in various fields (e.g. fire safety, air safety, and environmental issues). Applications in engineering risk analysis are also reported (Friis-Hansen, 2000; Faber et al., 2002; Grêt-Regamey and Straub, 2006; Straub and Der Kiureghian, 2010a,b). In dam engineering, Smith (2006) and Li et al. (2007) evaluated dam safety, and Zhang et al. (2011) and Xu et al. (2011b) diagnosed embankment dam distresses using Bayesian networks. Bayesian networks have been proven to be a robust method for risk analysis.

Figure 11.4 illustrates a simple Bayesian network for analyzing human vulnerability to dam breaching floods. The network has four nodes: two basic nodes (warning time, W_t, and flood severity, F_s, one medium node (evacuation, *Eva*), and one end node (loss of life, *LOL*). The nodes are connected by three arcs/links: *Wt–Eva*, *Eva–LOL*, and *Fs–LOL*. The prior probability of *LOL* can be calculated by

$$P(LOL = D) = \sum_{i=1}^{2}\sum_{j=1}^{2}\sum_{k=1}^{2} P(LOL = D, Eva = E_i, Wt = W_j, Fs = F_k) \tag{11.25}$$

where P=probability, D=died, E_1=evacuated, E_2=not-evacuated, W_1=sufficient warning time, W_2=little warning time, F_1=high flood severity, and F_2=low flood severity. According to the theorem of Bayesian network (Jensen, 2001), the joint probability can be calculated as

$$P(x_1, \ldots, x_n) = \prod_{i=1}^{n} P(x_i \mid Pa(X_i)) \tag{11.26}$$

where $Pa(X_i)$ is the parent set of X_i. If X_i has no parents, then the function reduces to unconditional probability of $P(x_i)$. In this case, as *Wt* is the parent of *Eva*, and both *Eva* and *Fs* are the parents of *LOL*, the following equation can be obtained:

$$\begin{aligned} &P(LOL = D, Eva = E_i, Wt = W_j, Fs = F_k) \\ &= P(Wt = W_j) \times P(Fs = F_k) \times P(Eva = E_i \mid Wt = W_j) \times P(LOL = D \mid Eva = E_i, Fs = F_k) \end{aligned} \tag{11.27}$$

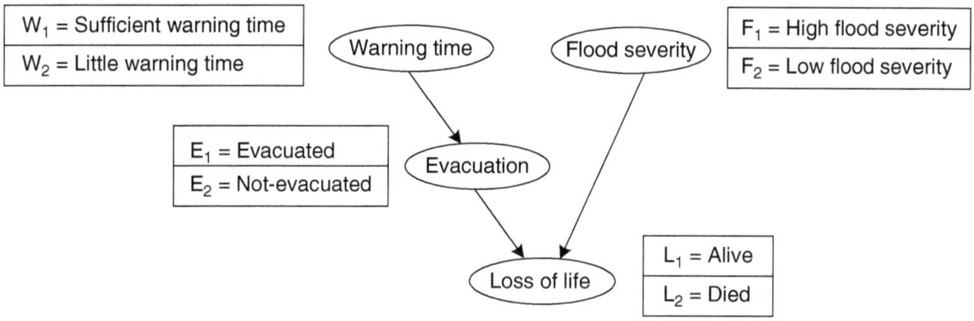

Figure 11.4 A simple Bayesian network for vulnerability to dam-breaking floods

where the (conditional) probabilities on the right hand side of the equation are quantified with available information (e.g. statistical data, judgment, and physical and empirical methods).

Consider a case with little warning time ($Wt = W_2$) and high flood severity ($Fs = F_1$), the posterior probability of *LOL*, $P(LOL = D)$, can be calculated based on the definition of conditional probability as:

$$P(LOL = D \mid Wt = W_2, Fs = F_1) = \frac{P(LOL = D, Wt = W_2, Fs = F_1)}{P(Wt = W_2, Fs = F_1)}$$
$$= \frac{\sum_{i=1}^{2} P(LOL = D, Eva = E_i, WT = W_2, Fs = F_1)}{\sum_{j=1}^{2}\sum_{i=1}^{2} P(LOL = L_j, Eva = E_i, WT = W_2, Fs = F_1)} \tag{11.28}$$

where the joint probabilities in Equation (11.28) are calculated with Equation (11.27).

The simple example shows the procedure of Bayesian networks for estimating of *LOL*:

1. building the a Bayesian network of *LOL*
2. quantifying the nodes (basic probability tables) and arcs (conditional probability tables) and obtaining the prior probabilities of *LOL*
3. finding the evidences of the applied case
4. calculating the posterior probabilities of *LOL* according to the evidences.

11.3.2 Building the Bayesian Network for Human Vulnerability

A logic structure is presented in Figure 11.5 to view the interactions between the key factors and loss of life. Theoretically, all the population downstream of a dam is within the scope of this study, until the water depth in each sub-area has been found. The population in the flooded area (water depth >0) before evacuation is regarded as the population at risk (*PAR*). The *PAR* may survive by evacuation before the arrival of a flood. The population that has not evacuated from the flooded area is called the exposed population (Jonkman, 2007). The exposed population is subject to one of four flood severity levels: high flood severity, medium flood severity, low flood severity, and safe.

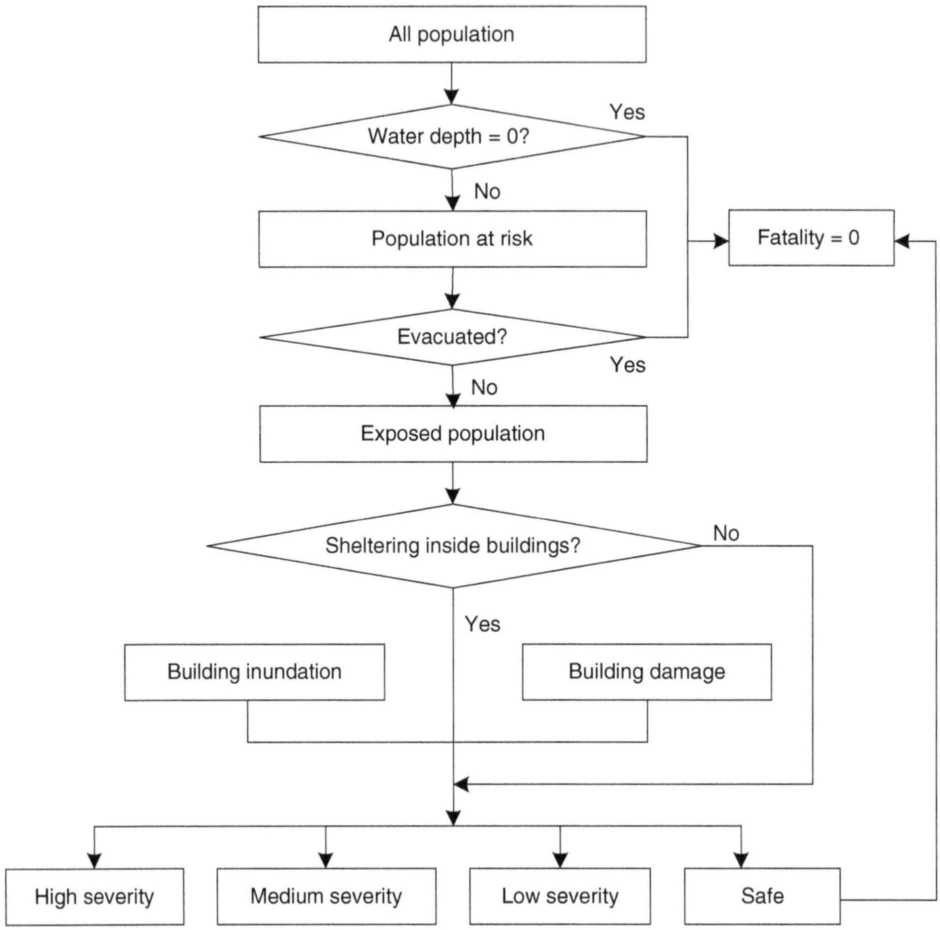

Figure 11.5 Logic structure of the HURAM model. *Source:* Peng and Zhang (2012b). Reproduced with permission from Springer

According to the logic structure in Figure 11.5, a Bayesian network is built with the program Hugin Lite (Hugin Expert A/S, 2004) as shown in Figure 11.6. There are 15 nodes and 23 arcs in the network. Each node is characterized by several discrete states, as shown in Table 11.3. The network consists of four main sub-networks (Figures 11.7a–d) for evacuation, sheltering inside buildings, flood severity, and LOL referring to the logic structure in Figure 11.5.

11.3.2.1 Evacuation Sub-network

Figure 11.7a shows the factors influencing evacuation and their interrelationships. People who manage to evacuate from a flooded area must have more available time than demanded. The available time is the sum of the warning time and the flood rise time. The warning time is defined as the elapsed time between the issuing of an evacuation warning and the arrival of a flood at the population at risk (Graham, 1999). The warning time consists of two parts: the time span between the issued warning and the dam breaching moment (i.e. warning initiation time), and the time of the flood routes from the dam site to the studied area (i.e. flood routing time). The flood rise time

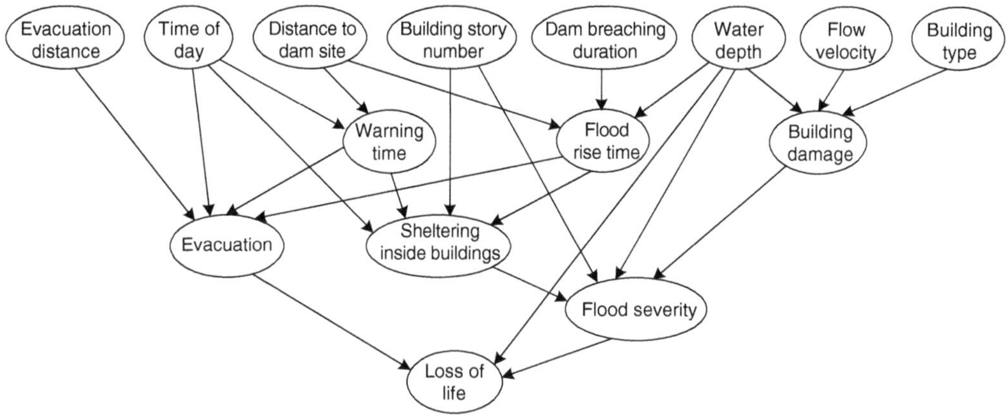

Figure 11.6 The Bayesian network for estimating loss of life. *Source:* Peng and Zhang (2012b). Reproduced with permission from Springer

Table 11.3 Nodes and states of the Bayesian network

Node	States	
	No.	States
Evacuation distance	4	0–100 m; 100–500 m; 500–2000 m; and >2000 m
Time of day	3	8:00–17:00; 17:00–22:00; and 22:00–08:00
Distance to dam site	4	0–4.8 km; 4.8–12 km; 12–36 km; and >36 km
Building story number	4	1 story; 2 stories; 3 stories; and >3 stories
Dam breaching duration	4	0–1 hour; 1–3 hours; 3–9 hours; and >9 hours
Water depth	7	0–1.5 m; 1.5–3 m; 3–4.5 m; 4.5–6 m; 6–7.5 m; 7.5–9 m; and >9 m
Water flow velocity	5	0–1 m/s; 1–2 m/s; 2–4 m/s; 4–6 m/s; and >6 m/s
Building type	3	Unanchored wood framed, anchored wood framed, and masonry, concrete and brick
Warning time	5	0–0.25 hour; 0.25–1 hours; 1–3 hours; 3–6 hours; and >6 hours
Flood rise time	4	0–0.25 hour; 0.25–1 hour; 1–3 hours; and >3 hours
Building damage	3	Slight damage; partial damage; and major damage
Evacuation	2	Evacuated; and not-evacuated
Sheltering in building	4	Unsheltered; the second story; the third story; higher than three stories
Flood severity	4	Safe; low flood severity; medium flood severity; and high flood severity
Loss of life	2	Alive; died

is defined as the time interval between the first arrival of the flood and the time the flood poses obvious threats to people. The end of the flood rise time in this study is defined as the time when the water depth reaches the peak depth if it is finally less than 1.5 m, or when the water depth reaches 1.5 m if it is finally larger than 1.5 m.

The demand time consists of the warning transmitting time, response time, and evacuation time (Frieser, 2004). The warning transmitting time is the span from the initiation of a warning to the receipt of the warning by the people at risk. The response time is the time for people to confirm and perceive the warning and prepare for evacuation. The evacuation time is the time span for people to travel to safe places. Quantitative analysis of the evacuation sub-network will be introduced later.

11.3.2.2 Sub-network for Sheltering Inside Buildings

The exposed population may shelter themselves inside buildings for safety. Organized sheltering inside safe buildings, such as inside the Superdome in New Orleans during Hurricane Katrina (Jonkman, 2007; Jonkman et al., 2009), is treated as part of evacuation. The sheltering of people is assumed as taking refuge at the top story of their buildings. This phenomenon has frequently been recorded in historical flood cases (Harrison, 1864; Sharpe, 1995; McClelland and Bowles, 2002). Figure 11.7b shows the main factors and their interrelationships influencing sheltering inside buildings. Sheltering is influenced by the population distribution, which varies with the time of day, the warning time and the flood rise time. However, the people at the shelter are not absolutely safe, depending on the possibility of building damage and/or building inundation.

11.3.2.3 Flood Severity Sub-network

The exposed people, whether sheltered or not, are subject to one of the four flood-severity levels: safe, low flood severity, medium flood severity, and high flood severity (Figure 11.5). The flood severity level depends on the building damage and inundation in the flood area. The levels of building damage can be referred to Clausen and Clark (1990):

1. slight (inundation) damage: Damage similar to that caused by a natural low-velocity river flood. No immediate structural damage (longer-term damage)
2. partial damage: Moderate structural damage, i.e. windows and doors knocked out. Little damage to key structural elements of the building
3. total (major) damage: Total structural collapse or major damage to the structure necessitating demolition and rebuilding.

Building damage is mainly influenced by the characteristics of both the flood and the buildings. RESCDAM (2000) presented criteria of building damage as shown in Table 11.4. The criteria are applied to calculate the probabilities of building damage in Figure 11.7c.

Building inundation includes building inundation and story inundation. The whole building is fully inundated if the water depth is higher than the height of the building. If the building is only partially inundated, some building stories may not be inundated, depending on the water depth and the number of building stories. The definitions of building-story inundation are shown in Table 11.5. Note that the cases with water depths greater than 12 m are not specially considered in this study, because these cases are rare.

Based on the concepts of building damage and building inundation, the flood severity in this study is defined as follows and shown in a matrix form in Table 11.6:

1. safe: where the buildings are slightly or partially damaged and the stories people stay are not inundated
2. low flood severity: where the buildings are slightly or partially damaged and the stories people stay are partially inundated
3. medium flood severity: where the buildings are slightly damaged and the buildings or stories people stay are fully inundated; or the buildings are partially damaged and the stories where people stay are fully inundated but the buildings are not fully inundated; or major damage to the buildings occurs and the stories where people stay are not or are partially inundated
4. high flood severity: where partial or major damage to the buildings occur in the zone and the whole buildings are fully inundated; or major damage to the buildings occur and the stories where people stay are fully inundated.

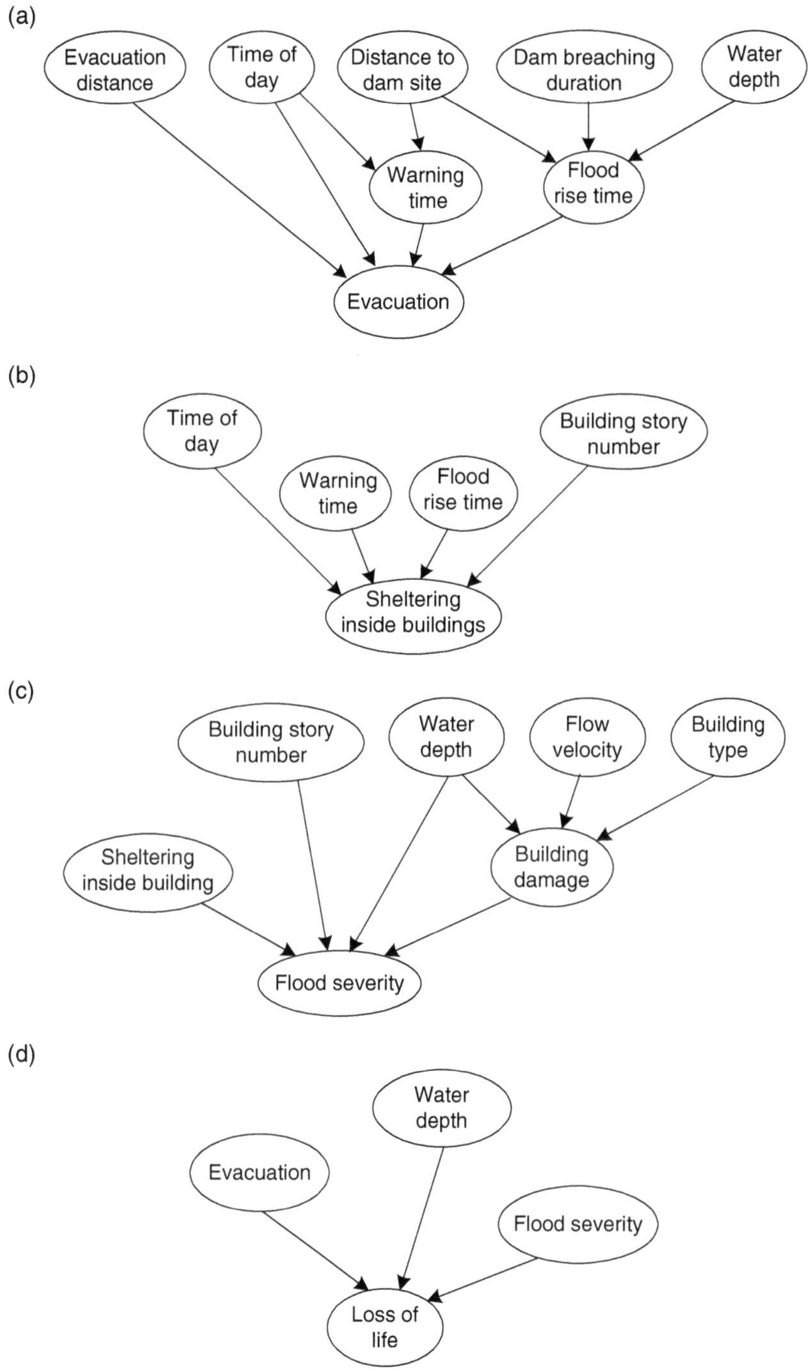

Figure 11.7 Sub-networks for (a) evacuation, (b) sheltering in buildings, (c) flood severity, and (d) loss of life. *Source:* Peng and Zhang (2012b). Reproduced with permission from Springer

Table 11.4 Recommended damage parameters for building structures (modified from RESCDAM, 2000)

Building type	Inundation damage	Partial damage	Major damage
Unanchored wood-framed	$dv < 2\,m^2/s$	$dv \geq 2\,m^2/s$	$dv \geq 3\,m^2/s$
Anchored wood-framed	$dv < 3\,m^2/s$	$dv \geq 3\,m^2/s$	$dv \geq 7\,m^2/s$
Masonry, concrete and brick	$dv < 3\,m^2/s$ or $v < 2\,m/s$	$dv \geq 3\,m^2/s$ and $v \geq 2\,m/s$	$dv \geq 7\,m^2/s$ and $v \geq 2\,m/s$

Note: d = water depth, v = water flow velocity

Table 11.5 Definitions of building-story inundation

Building story	Not inundated	Partially inundated	Fully inundated
1st story	0 m	0–3 m	>3 m
2nd story	0–3 m	3–6 m	>6 m
3rd story	0–6 m	6–9 m	>9 m
Higher than 3 stories	0–9 m	>9 m	N/A*

*Water depth larger than 12 m is not specially considered in this study.

Table 11.6 Flood severity matrix

Building inundation	Story inundation	Building damage		
		Slight damage	Partial damage	Major damage
Partially inundated	Not inundated	Safe	Safe	Medium severity
	Partially inundated	Low severity	Low severity	Medium severity
	Fully inundated	Medium severity	Medium severity	High severity
Fully inundated	Fully inundated	Medium severity	High severity	High severity

11.3.2.4 Loss of Life Sub-network

Generally, there are three ways to quantify *LOL* due to floods: expert judgment (Ramsbottom et al., 2003; Priest et al., 2007), suggested ranges based on historical data (Graham 1999; Aboelata and Bowles, 2005), and regression analysis with historical data (DeKay and McClelland, 1993; Jonkman, 2007; Jonkman et al., 2008). In this section, the third method is opted, since this method allows full utilization of the historical data and inclusion of the uncertainties of the influence parameters. Water depth is adopted as a key parameter to quantify the human vulnerability of the exposed people in each flood severity. The reasons for using water depth in this model are as follows:

1. Physically, water depth is one of the most important parameters influencing the stability and inundation of both human beings and buildings. Buildings are crucial for sheltering people. Statistically, there are strong relationships between *LOL* and water depth (Jonkman, 2007; Jonkman et al., 2008).
2. Comparing to other parameters, relatively more detailed and accurate records of water depth are available. Besides, water depth has frequently been used as a key parameter in previous models (Boyd et al., 2005; IPET, 2006; Jonkman, 2007).

As shown in Figure 11.7d, three parameters (evacuation, flood severity, and water depth) directly affect the *LOL*. The detailed quantification of *LOL* will be explained in the next section.

11.3.3 Quantifying the Networks

The quantification of a Bayesian network is to find the probabilities of the basic nodes and the conditional probabilities of the interrelationships between these nodes. Generally, there are four sources of data for quantifying a Bayesian network: statistical data, judgment based on experience, existing physical models or empirical models, and logical inference. Quantification based on statistical data, if sufficient, is a convenient and convincing way. A database is compiled for this purpose, with 343 dam failure cases with records of fatality. The detailed information on 321 cases is presented in Appendix D and described in Section 11.2.4. Of the 343 failure cases, 184 (53.9%) occurred in China, 115 (33.5%) in the USA, and 5 (1.5%) each in Italy and India, 4 (1.2%) in UK and 30 in other countries (Figure 11.8). The fatality numbers of each case range from 0 to 26,000 (Banqiao and Shimantan Dam failure in China, 1975). Ninety-six (28.0%) of these cases have no more than one fatality; 227 cases (66.2%) no more than 10; 293 cases (85.4%) no more than 100; and 337 (98.3%) no more than 1000, as shown in Figure 11.9. Jonkman (2007) also presented 244 fatality records in eight coast flood events. Even so, the available data alone may not be sufficient to quantify the Bayesian network for two main reasons. First, most of the available cases have few detailed records about the parameters influencing fatality. Second, the Bayesian network has a complex structure and a large number of combinations of states, which needs hundreds of detailed records for sufficient quantification. Owing to the limited statistical data, all the four sources of information mentioned above are used to quantify the model.

11.3.3.1 Quantification of Basic Factors

Records on four basic nodes (time of day, distance to dam site, dam breaching duration, and water depth) are available in the above dataset. The recorded number and probability of each state are shown in Table 11.7. Out of 47 cases with records of flood occurrence time, 18 occurred during 8:00–17:00, 8 during 17:00–22:00, and 21 during 22:00–08:00. Out of 62 cases with records of distance to dam site, 26 cases were located 0–4.8 km away from the dam sites, 12 cases 4.8–12 km away, 20 cases 12–36 km away and 4 cases >36 km away. There are 57 dam-break cases with

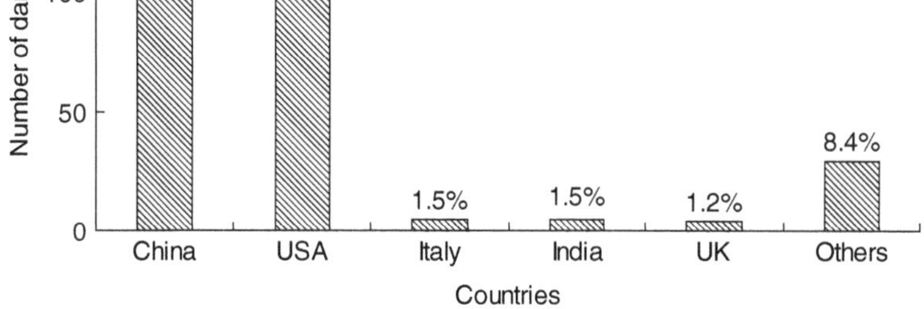

Figure 11.8 Locations of dam-failure cases with records of fatality. *Source:* Peng and Zhang (2012b). Reproduced with permission from Springer

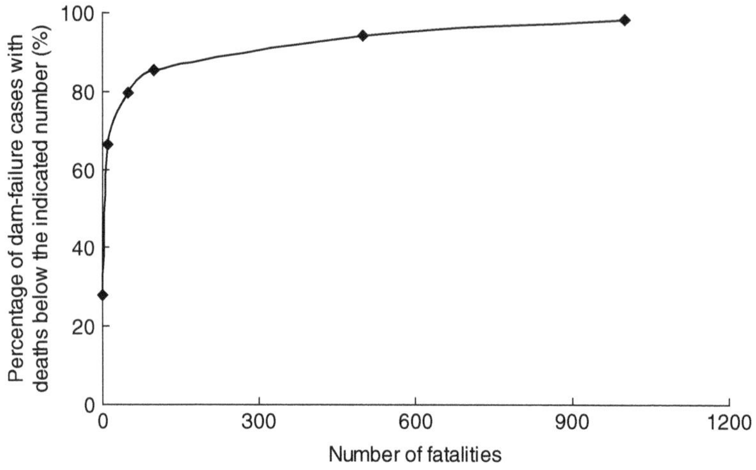

Figure 11.9 Number of fatalities in dam-failure cases

Table 11.7 Prior probabilities of basic nodes

Nodes	States	Cases	Probabilities
Time of day	8:00–17:00	18	0.3830
	17:00–22:00	8	0.1702
	22:00–08:00	21	0.4468
Distance to dam site	0–4.8 km	26	0.4194
	4.8–12 km	12	0.1935
	12–36 km	20	0.3226
	>36 km	4	0.0645
Dam breaching duration	0–1 hour	17	0.2982
	1–3 hours	30	0.5263
	3–9 hours	8	0.1404
	>9 hours	2	0.0351
Water depth	0–1.5 m	105	0.3777
	1.5–3 m	113	0.4065
	3–4.5 m	43	0.1547
	4.5–6 m	9	0.0324
	6–7.5 m	3	0.0108
	7.5–9 m	2	0.0072
	>9 m	3	0.0108

Note: The probabilities of the other four basic nodes (evacuation distance, evacuation distance, water flow velocity, and building type) are assumed to follow uniform distributions due to the lack of sufficient data.

information of breaching duration, 17, 30, 8, and 2 cases breached in 0–1 h, 1–3 h, 3–9 h, and >9 h, respectively. Information of water depth in 229 sub-areas was recorded as shown in Table 11.7. Insufficient records are available for the other four basic nodes: evacuation distance, building story, water flow velocity, and building type. They are assumed to be uniformly distributed.

11.3.3.2 Quantification of Evacuation

Evacuation is a process that consists of decision-making, warning, response, and evacuation (Frieser, 2004). Evacuation is defined as a success if the available time is larger than the demand time:

$$W_t + R_t > T_t + S_t + E_t \tag{11.29}$$

where W_t is the warning time, R_t is the flood rise time, T_t is the warning transmitting time, S_t is the response time, and E_t is the evacuation time.

Warning time consists of warning initiation time and flood routing time, as mentioned earlier. It is more likely to find signs of dam break during the daytime than during the night. Graham (1999) suggested values of warning initiation time (i.e. the time between issuing warning and the dam failure) for earthfill dam breaks at different times of a day, which are 0–1.5 h in the daytime and −1.0–0.5 h in the night-time. Negative values mean that the warning is issued after the dam breaks. Therefore, the warning initiation times in this study are assumed as uniform distributions U(0, 1.5), U(−0.5, 1) and U(−1, 0.5) for 8:00–17:00, 17:00–22:00 and 22:00–08:00, respectively (in hours), where U(a, b) means a uniform distribution with a low bound of "a" and an upper bound of "b". The flood routing time depends on the distance to the dam site and the flow velocity. The flood routing time is the ratio of the distance to the dam divided by the average flood routing speed (5 m/s is assumed). The conditional probabilities of warning time given the time of day and distance to the dam site are calculated using Monte Carlo simulation in Microsoft Excel. For example, when the distance to dam is 0–4.8 km and the time of day is 08:00–17:00, the routing time is U(0, 0.2667) and the warning initiation time is U(0, 1.5). The probabilities of warning time in 0–0.25 h, 0.25–1 h, 1–3 h, 3–6 h, and >6 h are 0.0772, 0.5009, 0.4219, 0, and 0, respectively from the Monte Carlo simulations.

As introduced above, the flood rise time is defined as the time from the flood initiation in a zone downstream of the dam to the time when water arrives the peak depth (D_p) if $D_p < 1.5$ m, and the time from flood initiation to the time when the water depth reaches 1.5 m if $D_p \geq 1.5$ m. The flood rise time is mainly influenced by breaching duration and flood severity. Also, the distance to the dam and the geometry of the flood route may affect the flood rise time, as the rise of water level may slow down with the consumption of energy. The flood rise time can be calculated by

$$R_t = B_t \alpha \beta \gamma \tag{11.30}$$

where R_t is the flood rise time; B_t is the breaching duration; α is the time to peak (P_t) divided by the breaching duration (B_t), where 0.2225 is suggested as the mean value of α; $\beta = 1$ when the peak water depth $D_p < 1.5$ m and $\beta = 1.5/D_p$ when $D_p > 1.5$ m; and γ is the coefficient of energy consumption. The flood rise time is prolonged along the river due to energy consumption. The values of γ are assumed as 1, 1.05, 1.1, and 1.2 when the distances to dam site are 0–4.8 km, 4.8–12 km, 12–36 km, and >36 km, respectively based on experiences of flood simulation for the breach of the Tangjiashan landslide dam (Peng and Zhang, 2011).

The effectiveness of warnings depends on many factors, e.g. disaster characteristics, government organization, warning methods, area characteristics, time, weather, and population characteristics (Frieser, 2004). Generally, empirical methods based on statistical data are applied to predict the warning transmitting time. Lindell and Perry (1992) presented a Weibull distribution for estimating the warning transmitting time:

$$P_t = 1 - \exp\left(-at^b\right) \tag{11.31}$$

where P_t is the proportion of the population that have been warned at time *t*, coefficients *a* and *b* are suggested as 1.30 and 0.70, 2.00 and 0.50, and 3.50 and 0.60 for moderate rapid warning, rapid warning, and very rapid warning, respectively (Lindell et al., 2002). In this study, the warning transmitting time is assumed as W(1.30, 0.70), W(2.00, 0.50), and W(3.50, 0.60) for times of a day of 22:00–08:00, 17:00–22:00, and 08:00–17:00, respectively. W(*a*, *b*) denotes a Weibull distribution (Equation 11.31) with coefficients *a* and *b*.

After receipt of warning, people need time to respond before making a decision. People may doubt about the warning message at first sight. They will try to confirm it from various sources. Besides, people need time to prepare for evacuation. The response time depends on the prediction lead time (i.e. the time span between the moment of forecasting and the predicted moment of occurrence). Lindell et al. (2002) assumed a Weibull distribution for estimating the response time for hurricanes, with mean values of 1.3–3.8 h in different situations. However, as the lead time of hurricanes is of the order of days, the response time may not be suitable for dam breaks that have a lead time of the order of hours (Frieser, 2004). Based on the records of the Dale Dyke dam failure (Harrison, 1864) and the Mill River dam failure (Sharpe, 1995), the response time before an emergent dam-break disaster is in the range of 0–0.5 h. In this section, W(4, 1) is assumed for the dam-break response time. The mean value is 15 minutes and the standard deviation is 15 minutes.

The evacuation time depends on the evacuation distance and speed. A dam break usually occurs in a short time and the influenced width (across river) is usually smaller than that of a hurricane or levee break flood. On-foot evacuation is assumed here. McClelland and Bowles (2002) estimated on-foot evacuation time for dam break disasters, assuming a speed of around 100 m in 1 minute in daytime and around 100 m in 1.3–1.5 minutes at night. Thus, the evacuation velocity is assumed as 100 m/min, 83 (=100/1.2) m/min, and 71 (=100/1.4) m/min during 08:00–17:00, 17:00–22:00, and 22:00–08:00, respectively.

In Equation (11.29), W_t, R_t, and E_t are assumed to follow uniform distributions, and T_t and S_t Weibull distributions. Conducting Monte Carlo simulations in Microsoft Excel, the conditional probabilities of evacuation under 240 combinations are obtained, with three states of time of day, four states of evacuation distance, four states of flood rise time, and five states of warning time.

11.3.3.3 Quantification of Sheltering Inside Buildings

The people who have not evacuated from the flooded area may shelter themselves inside buildings. The warned people may be assumed to find shelter at the top of the buildings if they have enough time to respond, while the unwarned people or the warned people without enough response time stay where they are. For example, in an area with three-story buildings, the sheltered people all stay in the third story, while the people unsheltered stay where they are, i.e. outdoors, in the first story, second story, or third story depending on their original locations.

Based on the study of Rogers and Sorensen (1989) and Aboelata and Bowles (2005), approximately 20%, 10%, and 2% of the people are outdoors (including those in transit) during 08:00–17:00, 17:00–22:00, and 22:00–08:00, respectively. The people indoors are assumed to be uniformly distributed in each floor. Take three-story buildings as an example. During 17:00–22:00, the people outdoors are 10%; the people in the first, second, and third story are each 30%.

The sheltering is successful if the available time is larger than the demand time:

$$W_t + R_t > T_t + S_t \tag{11.32}$$

Comparing to Equation (11.29), evacuation time is not needed for sheltering. As introduced above, W_t and R_t follow uniform distributions, and T_t and S_t follow Weibull distributions. Conducting Monte Carlo simulations in Microsoft Excel, the conditional probabilities of sheltering inside buildings have been obtained under altogether 240 combinations, with four states of building story, three states of time of day, five states of warning time, and four states of flood rise time.

11.3.3.4 Quantification of Flood Severity

The exposed people are subject to one of four flood severity levels: safe, low flood severity, medium flood severity, and high flood severity, depending on the status of building damage and inundation. Based on the definitions presented earlier, there are 336 combinations of conditional probabilities of flood severity, with three states of building damage, four states of building stories, four states of sheltering inside buildings, and seven states of water depth.

11.3.3.5 Quantification of Loss of Life

As shown in Figure 11.5, the population at risk is either evacuated or located in one of the four flood severities. The fatality rate of the evacuated people and the people in the safe zone is very low and assumed as zero (McClelland and Bowles, 2002; Aboelata and Bowles, 2005; Jonkman, 2007).

The fatality rates in the low flood severity and medium severity zones are quantified with water depth based on the historical data. A high water depth causes the inundation and instability of both people and buildings. Jonkman (2007) and Jonkman et al. (2008) suggested a lognormal distribution to simulate the fatality rate as a function of water depth:

$$F_D(h) = \Phi\left(\frac{\ln(h) - \mu_N}{\sigma_N}\right) \tag{11.33}$$

where $F_D(h)$ is the fatality rate as a function of water depth h, Φ is the standard normal distribution function, and μ_N and σ_N are the mean and standard deviation, respectively.

Pistrika and Jonkman (2010) reported the damage to residential buildings due to the flood in New Orleans during Hurricane Katrina in 2005. The buildings in the center of New Orleans were slightly damaged or partially damaged and partially inundated, which falls into 'low flood severity' in this study. Jonkman (2007) presented simulated water depths in the center of New Orleans and the corresponding fatality rates. A lognormal distribution with $\mu_N = 3.376$, $\sigma_N = 1.188$, and $R^2 = 0.637$ is obtained through a regression analysis with part of the data (excluding the partially flooded subareas). Figure 11.10 displays the simulated curve for the fatality in the low flood severity cases.

Table 11.8 shows the recorded flood data involving medium flood severity. Note the data of coastal floods in Netherland and Japan are selected from the dataset of Jonkman (2007) with building collapse ratios larger than 10% and large water rise rates (>0.5 m/h). Through regression analysis, a lognormal distribution with $\mu_N = 1.649$, $\sigma_N = 0.562$, and $R^2 = 0.643$ is obtained. Figure 11.11 displays the simulated curve for the fatality in the medium flood severity cases.

The fatality rate in the high flood severity zone is very large, as people lose their shelters in rapid and deep water flows. McClelland and Bowles (2002) presented 45 high flood severity cases with fatality rates ranging from 0.3830 to 1.0, with an average fatality rate of 0.918. There are also two other high flood severity cases. In the Dale Dyke dam break in 1864, 160 people suffered from

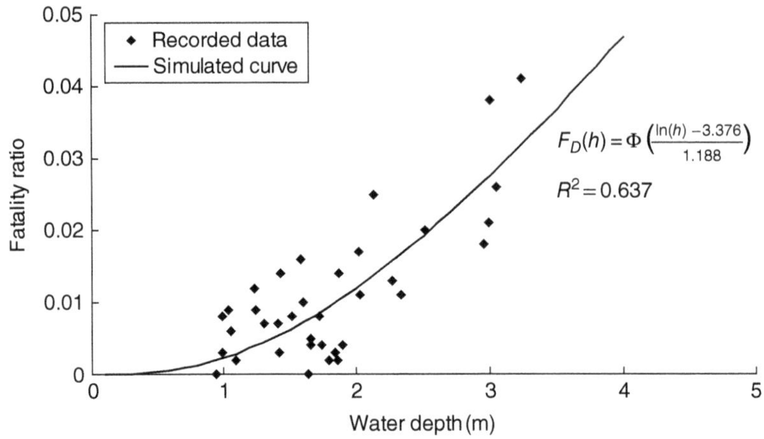

Figure 11.10 Quantification of loss of life in low flood severity cases (data selected from Jonkman, 2007). *Source:* Peng and Zhang (2012b). Reproduced with permission from Springer

Table 11.8 Recorded flood fatality data in medium flood severity cases

Flood hazard	Time	Location	Exposed people	Number of death	Fatality rate	Water depth (m)	References
Dale Dyke dam break in UK	1864	Hillsbrough	73	32	0.438	5.19	Harrison 1864; McClelland and Bowles 2002
Dale Dyke dam break in UK	1864	Neepsend	268	18	0.067	2.14	Harrison 1864; McClelland and Bowles 2002
Dale Dyke dam break in UK	1864	Harvest lane	132	8	0.061	2.75	Harrison 1864; McClelland and Bowles 2002
Coast flood in Netherland	1953	Burghsluis	50	10	0.200	3.85	Jonkman 2007
Coast flood in Netherland	1953	Nieuwerkerk	1871	288	0.154	3.30	Jonkman 2007
Coast flood in Netherland	1953	Ouwerkerk buiten	180	66	0.367	3.80	Jonkman 2007
Coast flood in Netherland	1953	Ouwerkerk kern	385	25	0.065	2.80	Jonkman 2007
Coast flood in Netherland	1953	Serooskerke	227	15	0.066	2.60	Jonkman 2007
Coast flood in Netherland	1953	Stavenisse	1645	156	0.095	3.20	Jonkman 2007
Coast flood in Netherland	1953	Oude Tonge	3011	305	0.101	3.00	Jonkman 2007
Coast flood in Netherland	1953	Nieuwe Tonge buiten	105	38	0.362	3.50	Jonkman 2007
Coast flood in Japan	1959	Nabeta reclamation	288	123	0.427	3.78	Tsuchiya and Yasuda 1980; Jonkman 2007
Coast flood in Japan	1959	Jyonan reclamation	279	55	0.197	3.58	Tsuchiya and Yasuda 1980; Jonkman 2007
Kelly Barnes Dam in USA	1977	Trailerville	64	15	0.234	3.05	McClelland and Bowles 2002

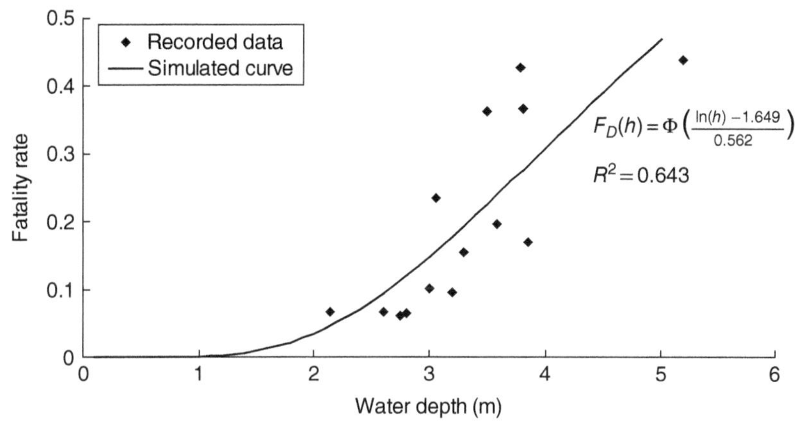

Figure 11.11 Quantification of loss of life in medium flood severity cases. *Source:* Peng and Zhang (2012b). Reproduced with permission from Springer

high severity floods (i.e. the buildings were swept away or fully inundated in rapid water flows); 154 out of those were killed, with a fatality rate of 0.9625 (Harrison, 1864). In the Bangladesh flood caused by a cyclone in 1991, 112 of the 285 people swept away died, with a fatality rate of 0.3930 (Bern, 1993). Combined with the 45 cases published by McClelland and Bowles (2002) and the above two cases, the average fatality in the high flood severity cases is 0.9078. Thus, the fatality rate in high flood severity zones is assumed as a constant value of 0.9078.

Conducting Monte Carlo simulations in Microsoft Excel, the conditional probabilities of sheltering inside buildings have been obtained with altogether 56 combinations, with two states of evacuation, four states of flood severity, and seven states of water depths.

11.3.3.6 The Integrated Model

With the prior probabilities of the basic nodes in Table 11.7 and the conditional probabilities quantified, the prior probabilities of the medium and final nodes are calculated with Hugin Lite, as shown in Table 11.9. The table shows the global human risks due to dam failure floods. The evacuation rate and fatality rate are 68.92% and 4.54%, respectively. To obtain the human risks in a dam failure case, more specific information on hydraulic parameters, population distribution and characteristics of terrains and buildings is needed.

11.3.4 Validation of the Model

Graham (1999) presented a model for predicting the *LOL* due to dam breaks or flash floods based on statistical data. Three discrete parameters are used in the model: flood severity (high, medium and low), warning time (0–15 minutes, 15–60 minutes, and >60 minutes) and flood severity understanding (vague and precise). The fatality rate in each combination of these states is quantified based on 50 historical records in 40 flood events. Table 11.2 shows the suggested fatality rates in Graham's model and those predicted using the HURAM model (with the same values of flood severity and warning time and keeping the rest of the parameters at their prior states). As the suggested fatality rates in Graham's model are based on historical data, they provide a good reference for other models. The parameter of flood severity understanding is not involved in HURAM.

Table 11.9 Prior probabilities for the medium and end nodes (the probabilities of the basic nodes are shown in Table 11.7)

Node	States	Probability
Warning time (hour)	0–0.25	0.2433
	0.25–1	0.3129
	1–3	0.3979
	3–6	0.0459
	>6	0
Flood rise time (hour)	0–0.25	0.4411
	0.25–1	0.4421
	1–3	0.0931
	>3	0.0107
Evacuation	Evacuated	0.6892
	Not-evacuated	0.3108
Sheltering inside building	Unsheltered	0.3267
	In the second story	0.2457
	In the third story	0.2139
	Higher than 3 stories	0.2137
Building damage	Slight damage	0.4001
	Partial damage	0.1510
	Major damage	0.4489
Flood severity*	Safe	0.3556
	Low severity	0.1718
	Medium severity	0.4083
	High severity	0.0643
Loss of life	Alive	0.0454
	Died	0.9546

*The probabilities of the flood severity are the conditional probabilities given the people not-evacuated.

The suggested fatality rates in Graham's model and the calculated fatality rates using the HURAM model are generally in good agreement. In the high flood severity case, the fatality rates of the two models for the no-warning cases are close (0.77 in HURAM and 0.75 in Graham's model). There is no suggested value for the high flood severity case with warning times of 15–60 minutes and more than 60 minutes in Graham's model due to lack of data. The results are available in HURAM as the warning, evacuation, and sheltering processes are simulated in the model. In the medium flood severity case, all the three predicted fatality rates from the new model are in the respective suggested ranges of Graham's model. In the low flood severity case, the values for the no-warning and some-warning (15–60 minutes) cases are quite similar, while the value for the case with warning times more than 60 minutes in HURAM are slightly larger than the suggested value in Graham's model (0.0009 and 0.0002, respectively). This is because, with sufficient warning time and low flood severity, a large number of people may take shelter in safe places (e.g. high rise buildings) instead of staying in the low flood severity zone.

Figure 11.12 shows the predicted fatality rates using the DeKay and McClelland model (1993) (D&M model), Graham's model and HURAM, as well as the recorded fatality rates in historical

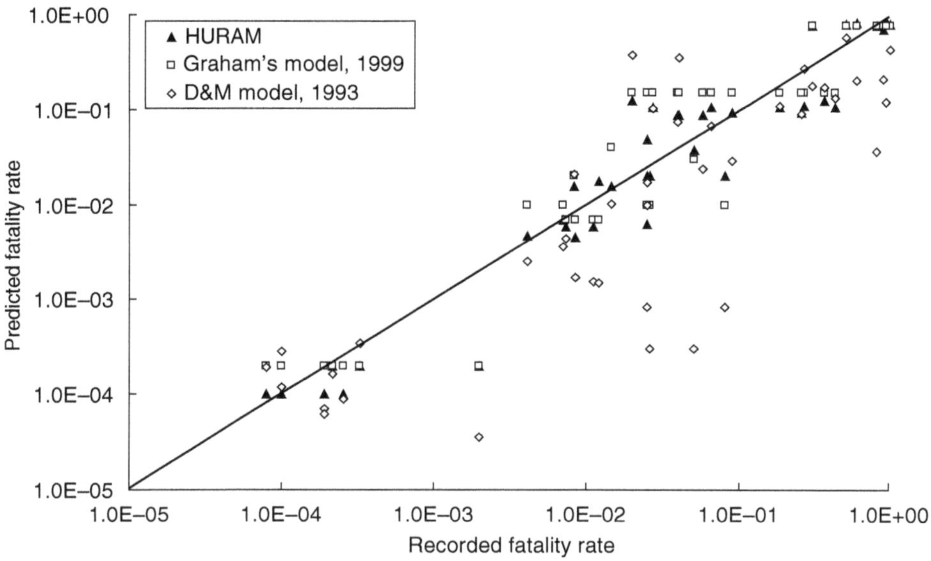

Figure 11.12 Comparison of recorded fatality rate and predicted fatality rate using the D&M model, Graham's model and the HURAM model. *Source:* Peng and Zhang (2012b). Reproduced with permission from Springer

cases. The D&M model is an empirical model with two continuous parameters (*PAR*, warning time) and one discrete parameter (flood force/severity). Most of the historic cases are compiled by Graham (1999) and used to quantify Graham's model as shown in Table 11.2. The diagonal line in Figure 11.12 means a perfect prediction as the predicted values are equal to the recorded values. Graham's model and HURAM give relatively good predictions. The results from the D&M model are more scattered, as the model is quantified with historic data at macro scale (i.e. the whole flood event).

Compared with the D&M model and Graham's model, the HURAM model using Bayesian Networks has several features:

1. Referring to the Bayesian network in Figure 11.5, a large number of parameters (14 in total) and their interrelationships are considered in a systematic structure in HURAM. The change in one parameter will influence the others in the network through their interrelationships.
2. This is a probabilistic model instead of a deterministic model. The uncertainties in the parameters and their interrelationships are represented with probabilities.
3. Besides the statistical data, HURAM incorporates knowledge from many existing studies in the Bayesian network, such as evacuation simulations (Lindell and Perry, 1992; Lindell et al., 2002), building damage criteria (Clausen and Clark, 1990; RESCDAM, 2000), flood severity concepts (Graham, 1999; McClelland and Bowles, 2002), human vulnerability functions (Jonkman, 2007; Jonkman et al., 2008) and so on.
4. The prior (conditional) probabilities in HURAM can be updated with information of specific cases by Bayes' theorem. As time goes on, more information can be collected, and the uncertainties in the parameters will become smaller. Thus, the updated results based on the new information will become more reliable.

11.4 Damage to Buildings and Infrastructures

Floods may cause direct damage to buildings and infrastructure, including lifelines (e.g. water supply, sewerage and drainage, gas and power supply, and telecommunication facilities), transportation facilities (e.g. roads, railways, and airports), essential facilities (e.g. hospitals, schools, and fire brigades). Besides, the indirect consequences due to the damage to infrastructure could be even greater. For example, the damage to hospitals may incur loss of life, and the damage to lifelines could paralyze the operation of a city. With regard to the damage to infrastructure by dam breaching floods, very little data and no well-established models exist (Merz et al., 2010). Occasionally, models for assessing earthquake risks are adopted to estimate indirect flood damage (Scawthorn et al., 2006a,b). In the *Multicoloured Manual* (Penning-Rowsell et al., 2005), the examination of damage to infrastructure is presented through case studies. Considering the variety of structures and limited studies on damage to infrastructure, the emphasis of this section is placed on the damage to buildings.

Flood-induced building damage is the negative effects on buildings which would not occur without floods, including wall failure, doors and windows being forced open, glass breaking, roofs collapsing and foundations being undermined. Clausen and Clark (1990) classified flood-caused building damage into three categories:

1. inundation damage: damage similar to that caused by a natural low-velocity river flood. No immediate structural damage is observable but long-term damage may occur as a result of saturation of foundations or structural timber.
2. partial (or minor) damage: moderate structural damage, i.e. windows and doors knocked out but with little damage to the major structural elements of the building.
3. total destruction (or major damage): total structural collapse or major damage to the structure necessitating demolition and rebuilding.

Kelman (2002) divided flood-induced building damage into six categories, from DS0 to DS5:

DS0: no water contact with structure
DS1: water contacts outside of structure but does not enter
DS2: water infiltrates (i.e. seeps or leaks in through small apertures), or external features are damaged or removed by water or debris
DS3: water or debris penetrates through a closed or covered opening (probably by breaking the opening), e.g. a window or a door
DS4: water or debris penetrates through a route not including an opening (structural integrity is attacked), e.g. a wall or roof
DS5: structure is damaged beyond repair, e.g. walls collapse, the structure moves, or the foundation is undermined

The two sets of definitions are equivalent to a certain degree: DS1 and DS2 are equivalent to inundation damage, DS3 belongs to partial damage, DS5 belongs to total destruction, and DS4 is a transition from partial damage to total destruction.

11.4.1 Flood Action on Buildings

Building damage occurs due to flood action, which includes hydrostatic action, hydrodynamic action, buoyancy action, erosion action, debris action, and no physical action (Roos, 2003; Kelman and Spence, 2004).

11.4.1.1 Hydrostatic Action

Hydrostatic action takes three forms: hydrostatic pressure, buoyancy action, and capillary rise. The hydrostatic pressure imparted by water against a building is given by:

$$P_s = \rho_w g d \tag{11.34}$$

where ρ_w is the density of water (kg/m^3), g is the gravitational acceleration (m/s^2), and d is the water depth (m) at the building. The difference of the hydrostatic water pressures inside and outside the building may cause some damage. The force with different water pressures is given by (Figure 11.13)

$$F = \frac{1}{2}\rho_w g\left(d_o^2 - d_i^2\right) \tag{11.35}$$

The moment about point O in Figure 11.13 is given by

$$M = \frac{1}{6}\rho_w g\left(d_o^3 - d_i^2\right) \tag{11.36}$$

Building damage occurs when the force or moment exceeds the resistance of any building component, such as a door, window, or wall.

The buoyance force is an uplift force which results in floating of the building or its components. The buoyancy force is given by

$$F_b = \rho_w g V = \rho_w g A d \tag{11.37}$$

where V is the inundated volume (m^3), A = inundated area (m^2), and d is the water depth (m). The buoyancy force is highly relevant for timber buildings, mobiles/caravans, furniture, vehicles, and so on. However, the buoyance force may be ignored for masonry, brick, or concrete buildings.

Capillary rise may cause damage beyond the water level. Huelman and Corrin (1997) suggested an upper limit of capillary rise of 0.45 m following a flood, depending on the building materials. Normally, capillary rise causes damage through chemical action. For example, Kelman and Spence (2004) presented special cases of cavity wall insulation damaged by capillary action along with "strawboard wall boards which can have a capillary effect of several meters".

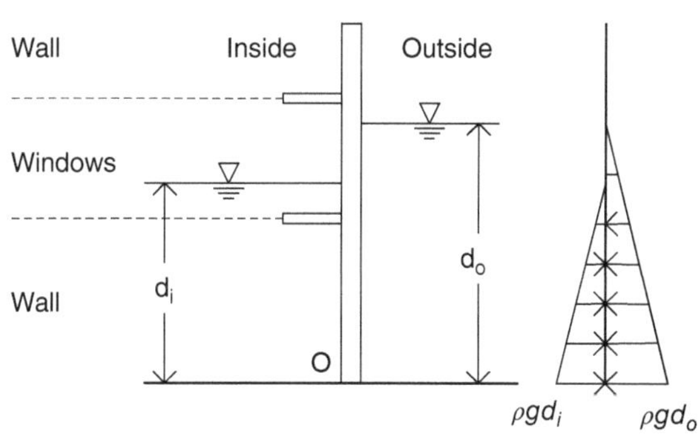

Figure 11.13 Lateral pressure distribution on a building

11.4.1.2 Hydrodynamic Action

Hydrodynamic action includes hydrodynamic pressure, wave action, and other action caused by turbulence, wave breaking, seepage, etc. The hydrodynamic pressure imparted by a water flow against a building is given by

$$P_d = \frac{1}{2} C_d \rho_w v^2 \tag{11.38}$$

where C_d is the coefficient (between 0 and 2) and v is the water flow velocity (m/s). The hydrodynamic force per unit length, F_d, can be calculated as

$$F_d = P_d d = \frac{1}{2} C_d \rho_w d v^2 \tag{11.39}$$

The peaks and troughs of non-breaking waves would increase and decrease the pressures and total force on the buildings, respectively. The exact change in total force depends on the ratio of wave height to water depth. Waves breaking can impart large pressures compared with other hydrodynamic action. The peak dynamic pressures of breaking waves can be as much as 15 to 18 times those calculated for non-breaking waves (USACE, 1984).

Turbulence is irregular fluctuations in velocity in both magnitude and direction. Turbulence can be highly variable over short spatial and temporal scales, making quantitative predictions difficult. Water seeping through soil may physically move the soil. The seepage force pre unit volume is given by

$$F_s = \rho_w g i \tag{11.40}$$

where i is the hydraulic gradient.

11.4.1.3 Debris Action

Moving debris can cause damage to buildings and bridges. Debris action can be divided into three categories: static action, erosion, and dynamic action. Static debris action occurs due to sediment accumulating externally or internally to a building. Debris could cause severe erosion, such as pebbles or household items being dragged along with the flow and gouging out soil from the sides or bed of the flow channel. Dynamic debris action occurs when moving debris impacts a building.

The impact force of the floating debris to building wall can be calculated as follows (Roos, 2003):

$$F_{de} = v\sqrt{mgk_t} \tag{11.41}$$

where m is the mass of the debris (kg), and k_t is the spring stiffness (N/m), which is given by

$$\frac{1}{k_t} = \frac{1}{k_w} + \frac{1}{k_d} \tag{11.42}$$

where k_w is the spring stiffness of wall (N/m) and k_d is the spring stiffness of debris (N/m).

11.4.1.4 Erosion Action

Moving water can cause erosion by scouring away soil from the sides or bed along which the water flows. Foundations with a shallow construction depth are sensitive to this mechanism. The buildings may tilt upon erosion of part of its foundation or wall. The erosion of soil depends on water flow velocity, and the soil properties such as particle size, cohesion, and friction angle. The mechanisms of soil erosion were introduced in Chapters 6 and 7.

11.4.1.5 Chemical Action

Flood damage can also be caused by chemical action which occurs when water contacts an object; it includes corrosion, rusting, and mustiness. Floodwater may be contaminated with sewage, petrol, oil, paint, household cleaners, or industrial chemicals. Any corrosiveness or flammability in the contaminants could result in chemical damage to buildings.

11.4.2 Models for Building Damage Evaluation

Three categories of methods for evaluating building damages caused by floods are available (Penning-Rowsell and Chatterton, 1977; Smith and Ward, 1998; Aglan et al., 2004; Sagala, 2006; Merz et al., 2010):

1. empirical methods, which evaluate building damage based on empirical information either from records collected after flood events or information collected via what-if-questions
2. experimental methods, in which experiments are conducted to measure the building damage when the building and its components are exposed to floodwater; two types of experiments may be included in this method: full-scale tests and reduced-scaled tests
3. numerical methods that rely on numerical analysis of flood-building interactions; a building damage is defined when the load from the flood exceeds the resistance of the building or its components.

11.4.2.1 Empirical Models

Stage damage curves are frequently used to estimate building damage (DS1–2) due to a slow-rising flood. In a slow-rising flood, the floodwater depth increases slowly, allowing for infiltration of water into the building through small openings of doors and windows. Slow-rising means that a large hydrostatic pressure differential between the inside and outside of a building does not occur. Building damage occurs through micro-physical action or chemical action during water–building contacts. Many stage damage curves have been presented, based on various data from different locations (Black, 1975; Penning-Rowsell and Chatterton, 1977; Parker et al., 1987; Penning-Rowsell et al., 1992; Scawthorn et al., 2006a,b; Apel et al., 2009).

For example, Black (1975) suggested criteria for critical depths to float different kinds of houses (Table 11.10). In the MURL Model (MURL, 2000), the damage ratio of buildings is given by a linear function

$$D_b = 0.02d \tag{11.43}$$

Table 11.10 Hydraulic condition for building float and building damage (modified from Black, 1975; USACE, 1985)

	Building type	1 story	1 and half story	2 story	3 story
Building float (Black, 1975)	Version I (lighter, drywall construction)	$d > 1.9$ m	$d > 2.7$ m	$d > 2.9$ m	—
	Version II (heavier, plaster wall construction)	$d > 2.8$ m	$d > 3.5$ m	$d > 4.7$ m	—
	Version I with brick veneer	—	$d < 5.2$ m	—	—
Building collapse (USACE, 1985)	Masonry or concrete	$v > 1.92$ m/s $dv^2 > 12.80$ m³/s	—	$v > 2.29$ m/s $dv^2 > 38.80$ m³/s	$v > 2.29$ m/s $dv^2 > 54.70$ m³/s
	Wood buildings	$d > 3.05$ m/s $dv^2 > 7.51$ m³/s	—	$d > 4.57$ m/s $dv^2 > 7.51$ m³/s	$d > 6.10$ m/s $dv^2 > 7.51$ m³/s
	Steel buildings	$v > 5.40$ m/s $dv^2 > 10.13$ m³/s	—	$v > 5.40$ m/s $dv^2 > 20.00$ m³/s	$v > 5.40$ m/s $dv^2 > 28.74$ m³/s

where D_b is damage ratio (maximum $D_b = 1$) and d = water depth (m). In the ICPR Model (ICPR, 2001), damage in residential buildings is estimated by

$$D_b = \left(2d^2 + 2d\right)/100 \tag{11.44}$$

HYDROTEC (2001) presented another equation:

$$D_b = \left(27\sqrt{d}\right)/100 \tag{11.45}$$

Penning-Rowsell et al. (1992) briefly discuss the impact of flow velocity on building damage in fast-rising floods (DS3–5). USACE (1985) uses water depth and/or flow velocity to set a criterion for identifying the collapse of buildings, as shown in Table 11.10. Clausen and Clark (1990) developed a criterion for predicting dam-break flood damage. They calculated flow velocities v (m/s) and depths d (m) in the Dale Dyke dam failure in the UK in 1864:

1. Inundation damage (no immediate structural damage): $dv < 2$ m/s or $dv < 3$ m²/s.
2. Partial damage (moderate structural damage, little damage to the major structural elements of the building): $v \geq 2$ m/s and $3 \leq dv < 7$ m²/s.
3. Total destruction (total structural collapse or major damage to the structure): $v \geq 2$ m/s and $dv \geq 7$ m²/s.

USACE (1995) suggests that only hydrostatic loads need to be considered in designing for $v < 1.5$ m/s, while flood-proofing measures would be economically ineffective for $v > 3.0$ m/s. For $v < 3.0$ m/s, dynamic effects of the moving water may be converted into equivalent hydrostatic loads with the excess depth proportional to v^2.

11.4.2.2 Experimental Models

Aglan et al. (2004) conducted an experiment by inundating some buildings to assess the damage of the building structure and its components. Vulnerability curves for the buildings between inundation duration and damage were developed. Normally, this method is time-consuming.

11.4.2.3 Numerical Methods

Two damage models are considered by Roos (2003) for studying the effects of flood on buildings: scour of the foundation and failure of walls. Scour of foundation is likely to occur when high velocity water flow moves the top soil of the base of buildings on shallow foundations. Failure of walls happens when large loads are exerted. The scour of a foundation depends on the type of the top soil layer and the velocity of water flow. Roos (2003) distinguished six types of soil for which critical water velocities of scour are 0.2 m/s for mold/sand, 0.6 m/s for clay/gravel, and 5.0 m/s for grass/paving. The wall failure mechanism depends on the resistance of the structures and the loading system. Generally two kinds of failure types can be distinguished: failure of shear force and failure of bending moment. When water is deep but the velocity is small, hydrostatic pressure may be the key factor. However, when the water flow is rapid, instead of depth, water velocity may control the failure.

Nadal et al. (2010) developed an analytical method to estimate the direct impact of flood actions on buildings and to determine the expected damage. Building vulnerability is modeled based on analytical representations of the failure mechanisms of individual building components. The flood actions generated during different flooding events are assessed and compared to the resistance of each building component. The assessed flood actions include hydrostatic and hydrodynamic forces, waves, turbulent bores, debris impacts, and time-dependent local soil scour. The results show how floodwater velocity increases the magnitude of the flood damage compared to those that solely consider water depth.

11.4.3 Relationship between Building Damage and Loss of Life

Building damage in floods not only implies loss of properties, but may also lead to loss of shelters for people in floods. High fatalities often occur in cases of catastrophic building damage. For example, the floods from the Banqiao and Shimantan dam failures in 1975 in Henan Province, China, swept most of the buildings downstream and killed more than 26,000 people. By contrast, the floods from the Changkai levee breach in 2010 in Jiangxi Province, China, did not cause any building collapses or fatality. Although the floods inundated four towns with an area of more than 100 km^2 and 100,000 people were at risk, most of the people managed to take shelter on the second or higher floors of buildings (Peng and Zhang, 2011).

Graham (1999) believes that the human vulnerability was highly related to the flood severity, which is in part measured by building damage. A dam-break flood was divided into three flood severities: the low severity occurring when no buildings are washed off their foundations; medium severity occurring when homes are destroyed but mangled homes and trees remain for people to seek refuge in or on; high severity occurring when the flood sweeps the area clean and nothing remains. The fatality rates in these three categories (Table 11.2) increase rapidly from low severity to high severity.

Similarly, Aboelata and Bowles (2005) divided the flooded areas into three zones based on loss of building shelters: safe zones, compromised zones, and chance zones. In this case, the loss of shelters was not limited to building damage, but also included buildings being submerged. The average fatality rates in these zones are 0, 0.12 and 0.91, respectively.

Zhai et al. (2006) emphasized the relationship between the number of inundated buildings and casualties. Based on statistical data from Japan, empirical equations were established to estimate the flood fatalities and injuries with the number of inundated buildings.

Peng and Zhang (2012b,c) presented a model for human risk analysis considering 14 parameters and their interrelationships using Bayesian networks, as shown in Figure 11.6. Both building damage and building inundation are used to define the flood severity with a matrix, as shown in Table 11.6. The influence path from building damage to loss of life was analyzed and clearly quantified. Also, building damage was evaluated by considering water depth, flow velocity, and building type.

11.5 Suggested Topics for Further Research

1. A complete database with detailed information about dam breaching parameters, floods, building damages, economic loss, and fatality is needed. Detailed case studies would also be helpful to understand the mechanisms of vulnerability due to dam breaching floods.
2. In social vulnerability analysis, a continuous-state Bayesian network could be developed without limiting it to Gaussian distributions. This could be realized by using approximation algorithms. In addition, the distributions of the continuous-state parameters should be quantified with expanded data. Dynamic Bayesian networks considering real-time effects may be helpful for conducting a dynamic analysis of human risk due to dam-break floods.
3. In building vulnerability analysis, model experiments on building actions in water flow are still necessary. Especially, collision by debris flowing in water is dangerous for building structures.
4. Studies on environmental and economic vulnerability are urgently needed, which may include, among other topics, actions of floods on ecology, recovery of the ecosystem after a severe flood, and evaluation of local macro-economic losses due to floods.

12

Dam Failure Risk Assessment

The breaching of a dam may have catastrophic consequences for the people and properties downstream of the dam through outburst of floods. Uncertainties prevail in the analysis of dam breaching consequences, particularly in the identification of failure modes, the estimation of dam material properties and breaching parameters, and the evaluation of flood routing and vulnerability. Hence it is important to consider various sources of uncertainties in dam risk analysis. This chapter aims to present the concepts and procedure of dam breaching risk assessment. First the concepts of risk, risk analysis, risk assessment, and risk management are described. Then procedures of risk analysis and risk assessment are illustrated. The concept of as-low-as-reasonably-practicable (ALARP) is introduced in particular, which is very important for risk assessment.

12.1 Risk and Risk Assessment

12.1.1 Definition of Risk

Risk is a measure of the probability and severity of an adverse effect on health, property, or the environment (Fell and Hartford, 1997; ISSMGE, 2004). Quantitatively, risk is the product of probability (P), elements at risk (E), and the vulnerability (V):

$$R = P \times E \times V \tag{12.1}$$

Probability is a measure of the degree of certainty between zero (impossibility) and 1.0 (certainty). It is an estimate of the likelihood of the magnitude of an uncertain quantity, or the likelihood of the occurrence of an uncertain future event. Elements at risk are the population, buildings and engineering works, infrastructure, environmental features, and economic activities in the area affected by a hazard. Vulnerability is defined as the degree of loss to a given element or set of elements within the area affected by a hazard. It is expressed on a scale of 0 (no loss) to 1 (total loss).

Dam Failure Mechanisms and Risk Assessment, First Edition. Limin Zhang, Ming Peng, Dongsheng Chang, and Yao Xu.

12.1.2 Risk Management

Risk management is the systematic application of management policies, procedures, and practices to the tasks of identifying, analyzing, assessing, mitigating, and monitoring risk (ISSMGE, 2004). Risk management comprises three components: risk analysis, risk assessment, and risk management (Fell and Hartford, 1997; Schanze et al., 2006), as shown in Figure 12.1. The glossary of risk and its components are summarized in Table 12.1 (Fell and Hartford, 1997; ISSMGE, 2004).

12.1.2.1 Risk Analysis

Risk analysis uses available information to estimate the risk to individuals or populations, property or the environment. The objective of risk analysis is to provide risk results to assist risk assessment (whether the risk is accepted or not and how will the risk mitigation measures work) and risk management (the decision and implementation of risk mitigation measures). Risk analysis includes hazard identification, probability analysis, and consequence analysis, as shown in Figure 12.1. There are two categories of risk analysis methods: qualitative risk analysis and quantitative risk analysis. Qualitative risk analysis uses word form, descriptive, or numeric rating scales to describe the magnitudes of potential consequences and the likelihood that those consequences will occur. Quantitative risk analysis is based on numerical values of the probability, vulnerability, and consequences, resulting in a numerical value of the risk (ISSMGE, 2004).

12.1.2.2 Risk Assessment

Risk assessment is the process of making a decision on whether existing risks are tolerable and whether present risk control measures are adequate, and if not, whether alternative risk control measures are justified or will be implemented (ISSMGE, 2004). Risk assessment incorporates the risk analysis and risk evaluation phases as shown in Figure 12.1. After obtaining the risk analysis results, such as the probability, consequence, and risk of the hazard, then risk acceptance and

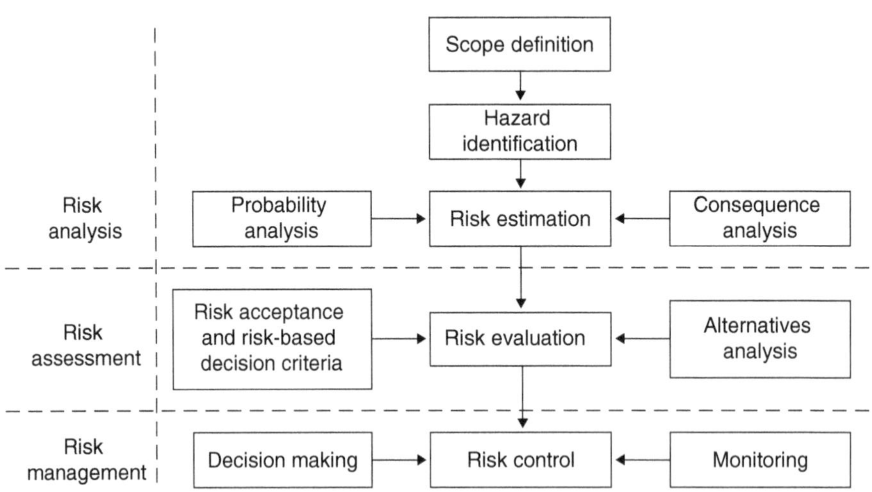

Figure 12.1 The process of risk management (modified from Fell and Hartford, 1997)

Table 12.1 Glossary of risk and its components (after ISSMGE, 2004)

Item	Definition
Acceptable risk	A risk which everyone impacted is prepared to accept. Action to further reduce such risk is usually not required unless reasonably practicable measures are available at low cost in terms of money, time and effort.
ALARP (as low as reasonably practicable) principle	That principle which states that risks, lower than the limit of tolerability, are tolerable only if risk reduction is impracticable or if its cost is grossly in disproportion (depending on the level of risk) to the improvement gained.
Conditional probability	The probability of an outcome, given the occurrence of some event.
Consequence	In relation to risk analysis, the outcome or result of a hazard being realized.
Danger (Threat)	The natural phenomenon that could lead to damage, described in terms of its geometry, mechanical, and other characteristics. The danger can be an existing one (such as a creeping slope) or a potential one (such as a rockfall). The characterization of a danger or threat does not include any forecasting.
Disaster	A serious disruption of the functioning of a community or a society causing widespread human, material, economic or environmental losses which exceed the ability of the affected community or society to cope using its own resources. A disaster results from the combination of hazards, vulnerability, and insufficient capacity or measures to reduce the negative consequences of risk.
Elements at risk	Population, buildings and engineering works, infrastructure, environmental features, and economic activities in the area affected by a hazard.
Frequency	A measure of likelihood expressed as the number of occurrences of an event in a given time or in a given number of trials.
F–N curves	Curves relating the probability per year of causing the number of lives (N) or more fatalities (F) to N. This is the complementary cumulative distribution function. Such curves may be used to express societal risk criteria and to describe the safety levels of particular facilities.
Hazard	Probability that a particular danger (threat) occurs within a given period of time.
Individual risk to life	The increment of risk imposed on a particular individual by the existence of a hazard. This increment of risk is an addition to the background risk to life, which the person would live with on a daily basis if the facility did not exist.
Likelihood	Conditional probability of an outcome given a set of data, assumptions, and information. Also used as a qualitative description of probability and frequency.
Loss	Any negative consequence, financial, or otherwise.
Population at risk (PAR)	All those persons who would be directly exposed to the consequences of failure of a structure or facility if they did not evacuate.
Probability	A measure of the degree of certainty. This measure has a value between zero (impossibility) and 1.0 (certainty). It is an estimate of the likelihood of the magnitude of the uncertain quantity, or the likelihood of the occurrence of the uncertain future event.
Reliability	Likelihood of successful performance of a given project element. Mathematically, Reliability = 1 – Probability of failure.
Risk	Measure of the probability and severity of an adverse effect to life, health, property, or the environment. Quantitatively, Risk = Hazard × Potential Worth of Loss. This can be also expressed as "Probability of an adverse event times the consequences if the event occurs".

(*Continued*)

Table 12.1 (*Continued*)

Item	Definition
Risk analysis	The use of available information to estimate the risk to individuals or populations, property or the environment, from hazards. Risk analyses generally contain the following steps: scope definition, hazard identification, and risk estimation.
Risk assessment	The process of making a decision recommendation on whether existing risks are tolerable and present risk control measures are adequate, and if not, whether alternative risk control measures are justified or will be implemented. Risk assessment incorporates the risk analysis and risk evaluation phases.
Risk-based decision-making	Decision-making, which has as a main input the results of risk assessment. It involves a balancing of social and other benefits and the residual risks.
Risk control	The implementation and enforcement of actions to control risk, and the periodic re-evaluation of the effectiveness of these actions.
Risk estimation	The process used to produce a measure of the level of health, property, or environmental risk being analyzed. Risk estimation contains the following steps: probability analysis, consequence analysis, and their integration.
Risk evaluation	The stage at which values and judgment enter the decision process, explicitly or implicitly, by including consideration of the importance of the estimated risks and the associated social, environmental, and economic consequences, in order to identify a range of alternatives for managing the risks.
Risk management	The systematic application of management policies, procedures, and practices to the tasks of identifying, analyzing, assessing, mitigating, and monitoring risk.
Risk mitigation	A selective application of appropriate techniques and management principles to reduce either likelihood of an occurrence or its adverse consequences, or both.
Societal risk	The risk of widespread or large scale detriment from the realization of a defined risk, the implication being that the consequence would be on such a scale as to provoke a socio/political response.
Tolerable risk	A risk within a range that society can live with so as to secure certain net benefits. It is a range of risk regarded as non-negligible and needing to be kept under review and reduced further if possible.
Uncertainty	Describes any situation without certainty, whether or not described by a probability distribution. Uncertainty is caused by natural variation and/or incomplete knowledge (lack of understanding or insufficient data). In the context of structural safety, uncertainty can be attributed to (i) aleatory uncertainty: inherent variability in natural properties and events, and (ii) epistemic uncertainty: incomplete knowledge of parameters and the relationships between input and output values.
Vulnerability	The degree of loss to a given element or set of elements within the area affected by a hazard. It is expressed on a scale of 0 (no loss) to 1 (total loss).

decision criteria are needed to evaluate whether or not the risk is tolerable, or accepted or not. Tolerable risk is the risk within a range that society can live with so as to secure certain net benefits. A tolerable risk needs to be kept under review and reduced further if possible. Acceptable risk is the risk that everyone impacted is prepared to accept. Action to further reduce such risk is usually not required unless reasonably practicable measures are available at low cost in terms of money, time, and effort. The ALARP principle is broadly selected as the risk assessment criterion that states that risks, lower than the limit of tolerability, are tolerable only if risk reduction is impracticable or if its cost is grossly in disproportion (depending on the level of risk) to the improvement gained (Fell and Hartford, 1997; ISSMGE, 2004). In risk assessment, the risks under some alternative control measures should be evaluated for decision-making.

12.1.2.3 Risk Management

Risk management is the systematic application of management policies, procedures, and practices to the tasks of identifying, analyzing, assessing, mitigating, and monitoring risk (ISSMGE, 2004). In risk management, risk control measures are applied to mitigate risk either by reducing the probability of the flood or reducing the flood consequences. Normally, there are two types of risk control measures towards dam breaching: engineering measures and non-engineering measures. Non-engineering measures (e.g. warning and evacuation) mitigate risks by reducing the elements at risk (i.e. people and both movable and unmovable properties); for instance, transferring people and movable properties out of the flooded areas. Engineering measures not only reduce the elements at risk by limiting flooded areas but also reduce dam failure probability, flood severity, and the vulnerability of the elements at risk. The residual risks after risk control measures need to be assessed based on monitoring information, which may still be large enough to bring threats to the people downstream. The details of risk management will be introduced in Chapter 13 and a case study will be presented in Chapter 14.

12.2 Dam Failure Risk Analysis

Analysis of risks posed by dam breaching generally consists of the following steps: scope definition, hazards identification, failure modes identification, probability estimation, prediction of elements at risk, vulnerability evaluation, and risk estimation (Fell and Hartford, 1997; ISSMGE 2004; Hartford and Baecher, 2004).

12.2.1 Scope Definition

Definition of analysis scope, which is the start of risk analysis, answers what should be studied. For dam breaching analysis, this step documents the risk analysis objectives and plans, areas to be studied, available data, suitable methods, and expected results (Hartford and Baecher, 2004). Data collection (e.g. historical hydrological data, rainfall data, and existing studies) and investigation (e.g. site studies, laboratory tests, and interviews) may be needed in this step.

12.2.2 Hazards Identification

Hazards identification answers the question "What can go wrong?" Normally, two aspects are to be investigated: "loads" and "resistances". The "loads" relate to the forces on a dam, its foundation and appurtenant facilities (e.g. spillway and outlet works) resulting from external factors such as flood conditions, earthquakes, other natural hazards (e.g. landslides, debris flow, and avalanche), and artificial hazards. The "resistances" denote the ability to withstand the loads, which relate to the structure of the dam and those from weakening of the dam structure (e.g. corrosion, metal fatigue, cracks leading to seepage and internal erosion), and operational errors (e.g. equipment errors and human errors).

12.2.3 Identification of Failure Modes

Identification of failure modes is an essential step in risk analysis, since the breaching parameters associated with different failure modes can be significantly different. Identification of failure modes requires a systematic investigation of the dam and the hydrological conditions. For example, a cracked dam tends to fail in piping; a heavy storm causing rapid rise of water level may lead to overtopping. Several methods are commonly used for identifying failure modes: failure modes and effects analysis (FMEA), event trees analysis (ETA), fault tree analysis (FTA), and Bayesian networks. The background of these methods was described in Chapter 10. Bayesian networks are less well known but have proven to be a robust tool. Xu et al. (2009) and Zhang et al. (2009) used Bayesian networks to diagnose dam distress modes.

12.2.4 Estimation of Failure Probability

The dam failure probability depends on a comparison of "load" and "resistance", and is closely related to dam type and failure mode. Most landslide dams have failed by overtopping (91.8%) because they were normally of massive volumes, large widths, and gentle slope angles both upstream and downstream (Peng and Zhang, 2012a). In comparison, 58% of man-made earth and rockfill dams failed by overtopping, 37% by piping failure, and 5% by slope failure.

There are two definitions of failure probability: classical definition (objective probability) and Bayesian definition (subjective probability). The classical definition of the probability of an event A can be formulated as

$$P(A) = \frac{N_A}{N_{tot}} \tag{12.2}$$

where N_A = number of equally likely ways by which an experiment may lead to A; N_{tot} = total number of equally likely ways in the experiment. In the Bayesian interpretation, the probability $P(A)$ of the event is formulated as

$$P(A) = \text{degree of belief that A will occur} \tag{12.3}$$

Based on the definitions of probability, three categories of methods are available to predict dam failure probability: statistical methods, physically based methods, and Bayesian methods.

Statistical methods, which use frequency based on historical data to replace probability, are straightforward and easy to apply. Physically based methods combine the classical probability concepts and physical analysis, which is commonly used in engineering practice. For example, a failure may be defined as the load being larger than the resistance (L > R) based on physical analysis; distributions are assigned for both the load and the resistance based on statistical analysis; finally, the failure probability is calculated using appropriate reliability methods, such as FORM or Monte Carlo simulation. Bayesian methods combine both objective probability and subjective probability. First, the prior probability is obtained with subjective information such as judgment, experience, or statistical data in similar projects. Then evidences of the concerned dam are investigated. Finally, the prior probability is updated with these site-specific evidences.

12.2.5 *Evaluation of Elements at Risk*

Dam breaching floods cause direct or indirect consequences to the people and properties downstream, as shown in Table 12.2 (Frieser, 2004). Some of the elements at risk can be monetized (e.g. properties, business etc.), the risks of which could be integrated for risk management. Some of them may not be monetized (e.g. human lives, environmental elements, societal welfare etc.), which need to be transferred or ignored before risk integration.

Evacuation is an effective measure to mitigate dam breaching risks. However, not all the risk elements can be reduced by evacuation. Some of them may be reduced or even prevented through evacuation (e.g. human lives, moveable properties such as vehicles and cash); a large part of them may not be reduced through evacuation (e.g. immoveable properties such as buildings, business, and environmental elements). Besides, evacuation itself would incur costs for resettling evacuated people and lead to interruption of business (Peng and Zhang, 2013a,b).

The frequently studied elements at risk are the population at risk (PAR) and impacted buildings. The analysis of risks to other infrastructure (e.g. bridges, tunnels, pipelines) may follow the same approach to the risk analysis for buildings.

PAR is the most important element at risk during a dam breaching event. PAR is defined as the number of people who would be exposed to a flood without evacuation or shelter (Jonkman, 2007). PAR depends on the flood inundation area, the population size, and the population distribution. The population distribution, however, may be difficult to evaluate since it varies spatially and temporally. For instance, the population density in the downtown is often higher than that in the suburbs; most of the people may not stay at home during work time; the population would increase dramatically at places of interest on holidays. For human risk analysis, the flooded areas may need to be divided into sub-areas with similar flood severities. People may move from one sub-area to another during a flood event, especially when an evacuation instruction has been issued (Peng and Zhang, 2012b,c).

Buildings are another important type of element at risk. The study of building damage aims not only to estimate economic consequences due to floods, but also to assist the human risk analysis. A large number of historical cases indicate that a flooded area with high human vulnerability is often associated with serious building damage. Major building damage often means loss of shelter for people in floodwater. Building damage depends on building type, building structure, story

Table 12.2 Elements at risk due to dam breaching (based on Frieser, 2004)

Category		Element at risk	Monetarily valuable?	Reduction of damage by evacuation?
Directly impacted by floods	Contact with water	People	No	Yes
		Moveable properties	Yes	Yes
		Immoveable properties	Yes	No
		Environmental elements	No	No
	Non-contact with water	Business (income, productions, and so on)	Yes	No
		Societal welfare	No	No
Indirectly impacted by floods		Traffics	No	No
		Business in surrounding areas	Yes	No
		Socictal disruption	No	No

number, and hydraulic conditions. Buildings can be classified according to their usage (e.g. residential, commercial, educational, government, and industrial). The safety factors for different types of building may be different. For example, the factor of safety for government buildings may be higher than that for residential buildings. Building structure is an essential factor of building damage evaluation during a flood. Normally, reinforced concrete buildings are more resistant than masonry or timber buildings. Building story influences not only the building damage but also the building inundation. The people may take shelter in the high stories above the water level. However, the buildings are not meaningful shelters if they are fully inundated (Peng and Zhang, 2012b,c).

12.2.6 *Vulnerability Evaluation*

The methods for estimating vulnerability can be divided into three categories: empirical methods based on judgments or statistical data, physical methods with physical analysis and tests, and compromise methods combining both empirical and physical information. The details of vulnerability analysis were described in Chapter 11.

12.2.7 *Risk Estimation*

Risk estimation is to quantify the risk by integrating the probability and consequence (the product of vulnerability and element at risk),

$$R(O) = P(H) \times P(S \mid H) \times P(T \mid S) \times V(O \mid T) \times E \tag{12.4}$$

where $R(O)$ = risk of a certain object (e.g. loss of life, property damage etc.); $P(H)$ = annual probability of the hazardous event (i.e. the dam breaching); $P(S|H)$ = probability of spatial impact (i.e. the breaching flood impacting an area) given the event; $P(T|S)$ = probability of temporal impact (i.e. the presence of the element at risk) given the spatial impact; $V(O|T)$ = vulnerability of the object (i.e. proportion of value lost or the probability of the loss); E = element at risk.

Sometimes people may only be concerned with the risk given a certain event instead of during a period; for instance, the risk of dam breaching given a hazard (e.g. an earthquake or a storm). The risk in this case is the conditional probability of the consequence and expressed as

$$R(O \mid H) = P(S \mid H) \times P(T \mid S) \times V(O \mid T) \times E \tag{12.5}$$

The flooded area sometimes needs to be divided into several sub-areas with similar severity parameters. The risk can be obtained by summing the risks in each sub-area (R_i):

$$R = \sum_{i=1}^{n} R_i \tag{12.6}$$

However, the risk in each sub-area may not be fully independent. For example, warning and evacuation may lead to the movement of people from one sub-area to another. In such cases, the correlations of the risks should be considered. Bayesian network is a robust tool to deal with the correlations of risk factors using conditional probabilities.

Human risk is always the most significant component of dam breaching risk management. Two concepts are frequently used to describe human risk: individual risk and societal risk. Individual risk gives the probability of individual death at a certain location regardless of the number of people. Individual risk is directly applicable for land use planning and corresponding zoning purposes. Societal risk takes into account the actual presence of people and gives a fatality number for a whole area, no matter precisely where the harm occurs within that area (Jonkman, 2007). Individual risk is inclined to describe the natural threat of a hazard to people, while societal risk is inclined to describe the concrete threats of a certain case. For example, suppose that in a certain dam breaching case, there are two sub-areas downstream of the dam. There are 100 people at risk in the first sub-area and 10,000 in the second sub-area. If other conditions are all the same, the individual risk in these two sub-areas is exactly the same; while the societal risk of the second is 100 times that of the first.

12.3 Risk Assessment

People have to live with risk, since it is impossible to reduce the risk to zero. The objective of risk assessment is to evaluate whether the evaluated risks are tolerable or not. There are several ways of risk evaluation: risk comparison, cost effectiveness of risk reduction, and cost–benefit analysis (Fell, 1994).

1. Risk comparison. Comparison of the calculated risk through risk analysis with socially "acceptable risk" (e.g. the annual probability of loss of life). The risk is accepted if it is lower than a certain value.
2. Cost effectiveness of risk reduction. This is commonly in terms of the cost of saving lives.
3. Cost–benefit analysis. Comparing the annual cost of risk reduction measures with the annual reduction in risk.

Fell (1994) argued that decisions based on risk assessment may involve consideration of legal, political, social, and financial issues. The decision will usually be made by owners, politicians, or others, not by the person who conducts the risk analysis. For example, more than 200,000 people were evacuated from their homes for two weeks in Mianyang City, Sichuan Province, China, in 2008 when exposed to the threat of the Tangjiashan landslide dam triggered by the Ms 8.0 Wenchuan earthquake. However, the peak discharge through Mianyang City was 7100 m^3/s, less than the defense discharge (12,000 m^3/s) (Peng and Zhang, 2013b). The evacuation decision was made considering political issues: no additional fatality was permitted after the earthquake claimed more than 80,000 lives. Risk comparison is one of the most popular ways of risk assessment and is further described in the following section.

12.3.1 Risk Tolerance Criteria

12.3.1.1 Qualitative Criteria

In qualitative criteria, the risks are ranked into several levels based on experience or judgment. The landslide dams caused by the Wenchuan earthquake were classified into four risk levels (i.e. extremely high, high, medium, and low, with symbols of I, II, III, and IV, respectively) according to their hazard levels and consequences to the people downstream, as shown in Tables 12.3–12.5 (MWR, 2009; Peng et al., 2014). The landslide dams at risk level I should be disposed with highest priority, for example, the Tangjiashan landslide dam with a dam height of 82 m, a lake volume of 3.16×10^8 m^3, and a potential population at risk of 1.3 million. The landslide dams at risk level II

Table 12.3 Classification of Landslide Dams according to Hazard (after MWR, 2009)

Hazard level	Influence factors		
	Dam height (m)	Lake capacity (million m^3)	Material and structure
Extremely high	≥70	≥100	Mostly soils
High	30–70	10–100	Soils with massive rubble
Medium	15–30	1–10	Massive rubble with soils
Low	<15	<1	Mostly massive rubble

Note: The hazard level is determined by the highest level of the three influence factors. In cases in which there is only one highest level factor and the lowest level is two or more levels lower than the highest level, then the hazard level can be lowered down by one level. For example, the hazard of a dam consisting of massive rubble with soils (medium level) with a height of 80 m (extremely high level) and a lake capacity of 80 million m^3 (high level) is rated as "high".

Table 12.4 Classification of Landslide Dams according to Consequences (after MWR, 2009)

Consequence level	Influence factors		
	Population at risk	Influence area	Facility
Extremely high	$\geq 10^6$	Center of a city above county-level	National or equivalent facilities
High	10^5–10^6	Center of a county-level city	Provincial or equivalent facilities
Relatively high	10^4–10^5	Center of a town	Municipal or equivalent facilities
Average	$<10^4$	Villages	Normal facilities

Note: The consequence level is determined by the highest level of any of the three factors

Table 12.5 Risk Classification of Landslide Dams (after MWR, 2009)

Risk level		Consequence level			
		Extremely high	High	Relatively high	Average
Hazard level	Extremely high	I	I	II	II
	High	I	II	II	III
	Medium	I	II	III	III
	Low	II	II	III	IV

Note: Risk levels: I = extremely high, II = high, III = medium and IV = low.

require engineering control measures as soon as possible. At least 16 landslide dams were classified into this risk level. The landslide dams at risk level III may also be mitigated with engineering measures, depending on available human resources and facilities. However, warning and evacuation should always be prepared. For the landslide dams at risk level IV, engineering measures may not be needed since such dams will not bring serious consequences to the people and the properties downstream. Warning and evacuation may still be needed during breaching period.

12.3.1.2 Quantitative Criteria

In quantitative criteria, dam risk falls into three regions from high to low: unacceptable region, region of tolerability, and broadly acceptable region (HSE, 2001; Munger et al., 2009), as shown in Figure 12.2. In unacceptable region, the risk is higher than the tolerable risk limit, which cannot be justified except in extraordinary circumstances (EEC). In EEC, risks, which would normally be unacceptable, can be tolerated on account of the special benefits that the dam brings to society (Munger et al., 2009). In this region, either engineering or non-engineering measures must be taken to mitigate the risks. In the region of tolerability, the risk is lower than the tolerable risk limit but higher than the broadly acceptable risk level. In this region, risk may be lowered to a tolerable level by meeting project-specific ALARP requirements. The concept of the use of ALARP considerations is that risks lower than the tolerable risk limit are tolerable only if further risk reduction is impracticable or if the cost is grossly disproportional to the risk reduction. ALARP will be discussed further in Section 12.3.2. In the broadly acceptable region, risk is regarded as insignificant and adequately controlled.

Theoretically, dam breaching risk should be assessed according to multi-tolerable criteria because of various categories of consequences, such as loss of human life, economic loss, environmental damage, and social impact. However, most of the studies of tolerable criteria focus on loss of human life for several reasons. First, human life is of the most significant concern in dam failure disasters. Second, other consequences, especially the environmental damage and social impact, may be difficult to evaluate. Finally, it is hard to integrate the human risk and economic risk for decision-making because of moral considerations. In this chapter, the discussion of risk tolerable criteria is limited to the human risk.

According to the interim tolerable risk guidelines for US Army Corps of Engineers (USACE) dams (Munger et al., 2009), there are three measures for human risk evaluation: annual probability of failure, individual risk, and societal risk. The societal risk could be expressed in two different ways: annual life loss and probability–fatality number chart (F–N chart).

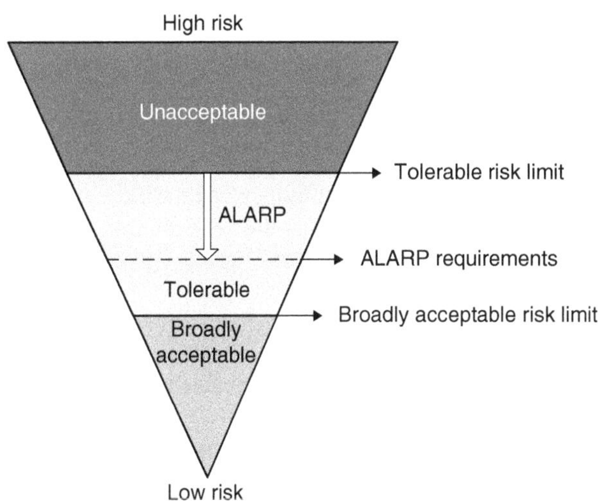

Figure 12.2 Tolerability of risk framework (Modified from HSE, 2001)

Annual Probability of Failure (APF)
The USACE policy for estimated APF is (Munger et al., 2009):

1. APF > 0.0001 per year: APF in this range is unacceptable except in exceptional circumstances.
2. APF < 0.0001 per year: APF in this range will be considered tolerable provided other tolerable risk guidelines are met.

Individual Risk
In 1989, the Health and Safety Executive (HSE) published a document entitled "Risk Criteria for Land Use Planning in the Vicinity of Major Industrial Hazards". Individual risk criteria proposed by the HSE are (Cornwell and Meyer, 1997):

1. Individual risk levels lower than 10^{-6} per year are defined as acceptable.
2. Individual risk levels greater than 10^{-5} per year are unacceptable for small developments.
3. Individual risk levels greater than 10^{-6} per year are unacceptable for large developments.

The Directorate General for Environmental Protection in the Netherlands published a document entitled "Premises for Risk Management, Dutch Environmental Policy Plan, 1989". For facility siting, the regulatory requirements are (Cornwell and Meyer, 1997):

1. Individual risk levels lower than 10^{-8} per year are defined as acceptable.
2. Individual risk levels greater than 10^{-5} per year are unacceptable for existing facilities.
3. Individual risk levels greater than 10^{-6} per year are unacceptable for new facilities.

In Hong Kong, an interim risk guideline (Geotechnical Engineering Office, 1998) was issued for landslides and builder falls from natural terrain. The maximum allowable individual risk was set at 10^{-5} for new developments and 10^{-4} for existing developments.

Australian National Committee on Large Dams (ANCOLD) (2003) presented risk tolerable criteria for existing dams and new dams respectively, which are also adopted by USACE. For existing dams, the tolerable individual risk should be less than a limit of 10^{-4} per year, except in exceptional circumstances. For new dams, the tolerable individual risk should be one order lower than that for the existing dams, namely, a limit value of 10^{-5} per year, except in exceptional circumstances.

Annual Life Loss (ALL)
The life loss may be expressed as the mean number of fatalities, which is also called annual life loss. The mathematical term for the mean is the "expected value" of the number of fatalities, *N*, which is estimated as the products of the probability–consequence pairs combined over all dam failure exposure scenarios. The USACE policy for the estimated ALL is (Munger et al., 2009):

1. ALL > 0.01 lives/year: Risk in this range is unacceptable except in exceptional circumstances and is a reason for urgent actions to reduce risk.
2. ALL between 0.01 and 0.001 lives/year: Risk in this range is unacceptable except in exceptional circumstances and is a reason for actions taken to reduce risk.
3. ALL < 0.001 lives/year: Risk in this range may be considered tolerable provided other guidelines are met.

Probability–fatality Number Chart (F–N Chart)

The societal risk is often presented as a probability–fatality number chart (F–N chart), which is a plot of the annual probability of exceedance of potential life loss versus potential life loss number on a log-log plot. The F–N chart displays the entire estimated probability distribution of life loss for a dam involving all failure modes and all population exposure scenarios for the particular dam (Munger et al., 2009).

USACE adopted the risk tolerable criteria of the ANCOLD (2003) guideline. The tolerable risk limit line of existing dams is shown in Figure 12.3(a), with the product of *F* and *N* of 0.001 when *N* is less than 1000, and with an *N* value of 1000 regardless of the value of *F*. The risks in the area on the right side of the risk limit line are unacceptable except in exceptional circumstances, while the risks in the area on the left of the line are tolerable only if they satisfy the ALARP requirements. The tolerable risk limit line for new dams is shown in Figure 12.3(a), which is similar to the one for existing dams except the product of *F* and *N* is 0.0001 instead of 0.001.

The Geotechnical Engineering Office of the Civil Engineering and Development Department of the Hong Kong SAR (HKGEO) provides risk tolerable criteria for landslides (e.g. Wong et al., 1997), as shown in Figure 12.3(b), which are frequently used in landslide risk assessment all over the world. The HKGEO criteria are similar to the ones for existing dams as shown in Figure 12.3(a). The risk limit line has the product of *F* and *N* of 0.001 when *N* is less than 1000. However, there are three zones in the HKGEO landslide risk criteria: unacceptable, tolerable with ALARP, and broadly acceptable. The broadly acceptable zone is not involved in the ANCOLD dam risk criteria. In addition, in the HKGEO criteria, the area with *N* larger than 1000 and the product of *F* and *N* smaller than 0.001 is specially defined as an "intense security region", which needs detailed studies during risk assessment. The same region in the ANCOLD criteria is unacceptable.

The *F*/*N* lines in Figure 12.3(c) were proposed by the Provincial Waterstaat, Groningen, Netherlands. The lines separate three categories of risk: acceptable, needing further assessment, and unacceptable. The *F*/*N* lines in Figure 12.3(d) were proposed by the Health and Safety Executive (HSE), United Kingdom. The lines separate three categories of risk: negligible, ALARP, and intolerable (Cornwell and Meyer, 1997).

12.3.2 ALARP Considerations

When the estimated life safety risk has been reduced to the tolerable risk limit or lower, ALARP requirements should be considered to decide whether the risk should be reduced further or not. In ALARP requirements, a tradeoff has to be made between the sacrifice in implementing the risk reduction measures (money, time, trouble, and effort) and the gain from these risk reduction measures. Normally, disproportionality is used as a justification for reducing the risk below the tolerable risk limit. The disproportionality is the ratio of "cost to save a statistical life" (CSSL) to "willingness to pay to prevent a statistical fatality" (WTP). With the disproportionality ratio and the WTP from guidelines, the CSSL can be estimated to reduce the risks.

The ANCOLD (2003) suggested disproportionality ratios in Table 12.6 for the risks just below the tolerable risk limit and far below the tolerable risk limit (or just above the broadly acceptable region). The WTP is set as USD 5 million/statistical fatality prevented. The WTP value for use in USACE dam safety risk assessments is the current value used by the US Department of Transportation (USDOT), which is USD 5.8 million with a standard deviation of USD 2.4 million (USDOT, 2008).

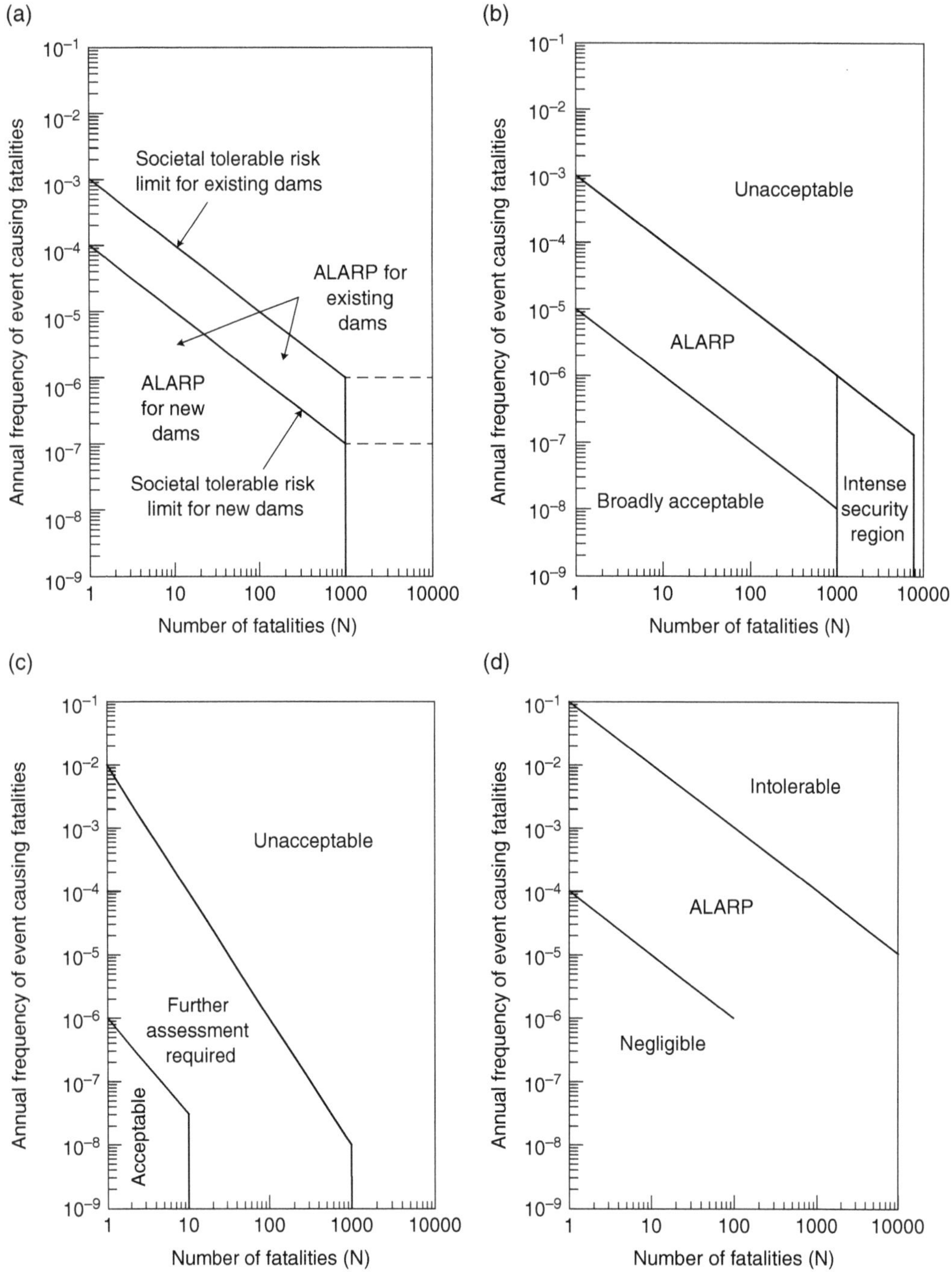

Figure 12.3 Societal risk guidelines: (a) ANCOLD (after ANCOLD, 2003); (b) HKGEO (after Wong et al., 1997); (c) Provincial Waterstaat, Groningen, Netherlands (after Cornwell and Meyer, 1997); (d) Health and Safety Executive (HSE), UK (after Cornwell and Meyer, 1997)

Table 12.6 ALARP Justifications (Modified from ANCOLD, 2003)

	ALARP justification rating	Range of disproportionality ratios	
		Greater than or equal to	Less than
For risks just below the tolerable risk limit	Very strong	0	1
	Strong	1	4
	Moderate	4	20
	Poor	20	–
For risks far below the tolerable risk limit (or just above the broadly acceptable region)	Very strong	0	0.3
	Strong	0.3	1
	Moderate	1	6
	Poor	6	–

12.4 Suggested Topics for Further Research

Further investigation on three topics is recommended:

1. There are four categories of consequences: physical, social, environmental, and economic. Most existing studies on dam breaching consequences are limited to social and economic consequences, while the quantitative assessment of physical and environmental risk has received little attention. Criteria for risk-based decision-making considering the four categories of consequences, which are of significance to decisions on planning, design, and operation of dams, are yet to be established. How to monetize loss of life considering economic developments and political concerns is also a practical issue.
2. The existing criteria for dam failure risk assessment concern the risks within a certain period (e.g. one year), yet conditional risk criteria are needed given the occurrence of a specific hazard. For instance, a serious storm may cause the rise of water level leading to possible overtopping failures. The existing criteria should be extended to assess the risks under the given storm and to decide the level of emergency plans to be implemented.
3. Methodologies for assessing the risks of a chain of dams need to be established. Cascading development of hydropower involves many dams along a river. A strong earthquake or a severe storm may trigger multiple landslide dams along a river. The breaching of an upstream dam may cause the failure of downstream dams one after another. The floods from cascading dam breaching may be very different from that from a single dam failure.

13

Dam Failure Contingency Risk Management

13.1 Process of Contingency Risk Management

Serious natural and engineering disasters have occurred worldwide at an alarming rate in the past few years, possibly due to climate change, active tectonic movements, human activities, and insufficient engineering design and management. Compared to such disasters as earthquakes, rapid landslides, and tsunamis that have little lead time (seconds to minutes), floods due to dam/levee breaks usually have longer lead time (hours to days), which provides an opportunity for implementing pre-disaster emergency management to save human life and property. Despite of the benefits of saving human life and property, pre-disaster emergency management should be treated as a very serious issue since it often incurs a large amount of economic expense at the same time (Frieser, 2004).

The process of pre-disaster emergency management can be divided into four phases: (1) observation, risk assessment, and decision-making, (2) warning, (3) response, and (4) evacuation and sheltering (Frieser, 2004) as shown in Figure 13.1.

Phase 1 starts from the identification of disaster signs. Professional bodies monitor indicators of the disaster (such as water depth and rainfall intensity) and report to the authorities. The authorities evaluate the probability and consequence of the disaster, and make decisions whether to issue evacuation warning or not. Phase 2 starts from disseminating warning messages. It takes time to transmit warning messages. An S-curve for the PAR warned and the proceeding of warning is shown in Figure 13.1. Phase 3 starts at the receipt of warning messages by the population at risk. People need time to confirm the warning messages, evaluate the situation, and make their own decisions, which may not follow those of the authorities. A part of the PAR may evacuate to safe places, as shown in Phase 4. The rest, either refusing to evacuate or having insufficient time, may try to shelter themselves in relatively safe places (e.g. high-rise buildings) in the flooded areas. After the occurrence of the disasters, the affected people may flee to safe heavens, but part of them might lose their lives.

Dam Failure Mechanisms and Risk Assessment, First Edition. Limin Zhang, Ming Peng, Dongsheng Chang, and Yao Xu.

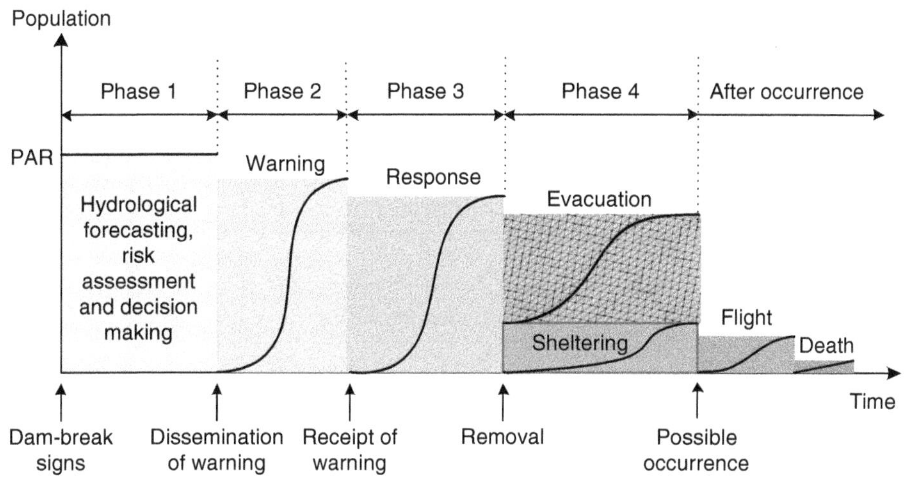

Figure 13.1 Emergency management of dam breaks in time scale (modified from Frieser, 2004)

13.1.1 Observation and Prediction

Before prediction, a large amount of information is needed, especially the observation of the hazard development. The potential to predict the occurrence and the prediction lead time (time span between the moment of forecasting and the predicted moment of the possible occurrence) depend on the type of hazard and the available prediction systems. Table 13.1 summarizes qualitative lead times in different kinds of hazard. Quick and efficient predictions based on intuition and experience are needed for emergency cases. Advanced equipment (e.g. remote sensing) and evaluation methods (e.g. probability methods) are needed to obtain a more accurate evaluation if time permits. The predicted lead time is also dependent on the development of professional technology in the research fields of disasters. For example, the forecast system of the National Hurricane Center (NHC) in USA is able to predict striking locations of a hurricane within a range of 160 km (100 miles) 24 hours before the expected landfall of the storm (Urbina and Wolshon, 2003). Thanks to data from geophysical, geochemical, and geological monitoring systems, the eruption of Volcán de Colima in Mexico could be predicted in adavnce. On 13 November 1998, the eruption was predicted to occur within three to five days (16–18 November). Volcán de Colima erupted on 20 November, a week after the prediction had been made (Zobin et al., 2002).

13.1.2 Decision-making

After the occurrence of a disaster has been predicted, authorities must decide whether to issue an evacuation to prevent or decrease loss of life and property. Despite the benefits of saving human life and property, an evacuation decision should be treated as a very serious issue since it often also incurs a large amount of economic expense (Frieser, 2004).

Generally, there are two categories of methods of decision-making for emergency management: deterministic methods and probabilistic methods. Deterministic methods are those based on deterministic analysis, experiences, and judgment without explicit consideration of uncertainties. In deterministic methods, some critical values (e.g. water level, period return flood) are often suggested as indexes for evacuation decision-making (Nielsen et al., 1994; Frieser, 2004). Some guidelines also

Table 13.1 Prediction lead time indications (modified from Frieser, 2004)

Prediction lead time	Type of hazard
Zero to seconds	Explosion, fire, airplane crash, car crash
Minutes	Earthquake, tsunami, landslides
Hours	Storm surge, man-made dam breaks
Days	Hurricane, volcanic eruption, river flood, landslide dam breaks

offer recommendations for decision-making based on judgments (Urbina and Wolshon, 2003; FEMA, 2004). Probabilistic decision methods under uncertainty are more realistic.

13.1.3 Warning

People at risk may be warned directly by the hazard itself or indirectly via an evacuation warning disseminated by the authorities. Warning should be treated as a very serious event, as it may cause unnecessary losses. For example, the evacuation in the 1995 Netherlands flood incurred an estimated loss of €140 million (Frieser 2004) but no flooding actually occurred.

13.1.3.1 Warning Messages

Rogers and Sorensen (1991) suggested that a warning message should contain two conceptually distinct aspects: alert and notice. Alert is to catch the attention of people and show them the potential danger. Notice denotes the content of a warning message. Mileti and Peek (2000) suggested that a reasonable warning should contain the following five pieces of information:

1. hazard: a description of the expected event and its threat to people's safety to minimize the likelihood of misperceiving the hazard and taking incorrect actions
2. location: a clear definition of the area at risk; it should be clear which inhabitants are at risk and which are not
3. guidance: information on how people should protect themselves
4. time: information on the amount of time until the possible occurrence; people will take protective actions according to the available time
5. source: information about the source disseminating the warning message; people are more likely to believe warning from official sources

13.1.3.2 Warning Influencing Factors

Based on the studies of Rogers et al. (2000), Wolshon et al. (2001), and Mileti (1995), the effectiveness of warning depends on at least six factors:

1. disaster characteristics (nature, severity, emergency, possible of occurrence, and influenced area)
2. government organization (government response, organizational effectiveness, and arrangement of labor and facilities)
3. warning methods (television, radio, internet, face-to-face notice, portable loudspeakers, sirens, telephones, mails, emails)

4. area characteristics (area usage, population density, communication conditions, and traffic conditions)
5. time and weather (time of a day, workday or holiday, rainy, foggy, or snowy day)
6. population characteristics (age, gender, physical state, disaster experience, family members, and societal relationship)

Warning messages can sometimes not disseminate well. Parker et al. (2009) gave five principal reasons why flood warnings may not be received: (1) patchiness in the geographical coverage of official flood warning systems, (2) failure to adequately forecast a flood, and therefore to issue warnings, (3) short flood warning lead time, (4) inefficiencies in flood warning dissemination leading to delays, access or other problems, and (5) the unavailability or inability of people to receive flood warning.

13.1.3.3 Estimation of Warning Time

The proportion of population warned is a function of warning time. Sorensen and Mileti (1988) suggested an exponential function to simulate the warned population in three ranges of prediction lead time. The empirical equations from regression analysis are expressed as follows:

$$P_t = 81.83t^{3.488}\left(R^2 = 0.98\right) \quad \text{when } t < 50-60\,\text{min} \tag{13.1}$$

$$P_t = 59.58t^{0.4753}\left(R^2 = 0.67\right) \quad \text{when } 60\,\text{min} < t < 3\,hours \tag{13.2}$$

$$P_t = 66.63t^{0.2089}\left(R^2 = 0.58\right) \quad \text{when } t > 3\,hours \tag{13.3}$$

where P_t is the cumulative percentage of population warned and t is time elapsed after warning issuance (hours).

Rogers and Sorensen (1989) published warning diffusion curves with different communication media (telephone, tone alert, siren, and media). Lindell and Perry (1992) suggested a Weibull distribution function to simulate the warning diffusion curve:

$$P_t = 1 - \exp\left(-at^b\right) \tag{13.4}$$

where a and b are two parameters of the Weibull distribution with suggested values of 1.3 and 0.7 respectively for moderately rapid disasters, 2.0 and 0.5 respectively for rapid floods, and 3.5 and 0.6 respectively for very rapid floods.

Rogers and Sorensen (1991) believed that transmission of warning could be characterized as a "birth process" whereby people first hear of the event and then sequentially tell others. A model based on mathematical inference was presented. Two components, the part directly warned by the government and the part informed indirectly from others, are taken into consideration in this model.

13.1.4 Response

After receipt of warning, people need time to respond before making a decision. The response process can be divided into three stages: receipt of warning, perception, and response (Frieser, 2004). People may doubt the warning message in the first instance. They will try to confirm it from various sources such as neighbors, relatives, and the authorities. After confirmation, they make their own decisions, which may not be exactly the same as the advice from the local government.

13.1.4.1 Compliance with Evacuation Warnings

Dow and Cutter (1998) suggested that the compliance of people strongly depends on perceived risk of people. When people do not feel safe, they tend to evacuate. Personal risk perception is related to the magnitude of the hazard, the accompanying expected damage, prior experience with the hazard and the type of housing. Drabek (2000) identified six individual characteristics that affect flood and other hazard warning responses: risk perception, age, gender, ethnic minority, socio-economic status, and fate control.

13.1.4.2 "Shadow Effect"

"Shadow effect" refers to the phenomenon that people beyond the evacuation area are evacuated unnecessarily. The shadow effect should be minimized as it not only increases the cost of evacuation, but also overloads the traffic. Sorensen and Mileti (1988) pointed out a few reasons for unnecessary evacuation:

1. unclear information from the authority
2. safety considerations
3. following the behavior of relatives or friends
4. misleading rumors or distorted messages
5. worrying about traffic jam if late

13.1.4.3 Estimation of Response Time

Response time refers to the duration from the receipt of warning to the start of mobilization. As human response mainly depends on societal and psychological factors, the response time is mainly estimated using empirical methods. Frieser (2004) presented records of compliance rates to evacuation warnings in various disasters (such as toxic release, nuclear accident, hurricane, volcanic eruption, and floods).

Lindell et al. (2002) suggested a Weibull distribution to simulate response time, as shown in Equation (13.4). The values of a and b in this equation are 0.085 and 2.55, respectively, for people at home, 0.015 and 2.90, respectively, for people at work (who need to return home first for collection of stuff), and 0.35 and 3.00, respectively, for transients.

13.1.5 Evacuation

Evacuation is the process between the start of leaving the risk area and the arrival at a safe place. Generally, people at risk may receive warning of recommended/voluntary or mandatory evacuation, depending on the magnitude of the disaster, the available time and the required time for evacuation (Frieser, 2004). Usually, evacuation involves housing arrangement, food supply, and traffic organization from the authorities. The concern of evacuation starts from wars. Systemic analysis of evacuation originated from the Three Mile Island nuclear releasing accident in 1979 (Mileti and Peek, 2000), which brought panic to the people around. During the past two decades, the USA has continually suffered from the strikes of hurricanes (especially Hurricane Katrina in 2005), which made the study of evacuation more meaningful.

13.1.5.1 Evacuation Time Factors

Jonkman (2007) summarized four categories of factors influencing the evacuation time:

1. disaster characteristics (nature, magnitude, rate of growth, location and time of possible occurrence)
2. area characteristics (area size, road network, escape routes, presence of shelters, distance to safety)
3. population characteristics (number of people, age distribution, spatial distribution, physical state, evacuation behavior)
4. procedures and organization (knowledge on the hazard, risk, protective action recommendation, transport, preparation, decision-making, skill of residents and emergency management personnel)

13.1.5.2 Evacuation Models

Generally, there are two categories of models for estimating evacuation time: empirical models and physical models. Empirical models are those mainly based on subjective judgment or statistical data. Tweedie et al. (1986) suggested a Rayleigh distribution to simulate the relationship between evacuation rate and time:

$$F(t)=1-\exp\left(-t^2/T\right) \tag{13.5}$$

where $F(t)$ is the evacuation rate, t is time in minutes, and T is assumed time for completing evacuation (1800 minutes suggested). Radwan et al. (1985) presented a logistic curve for estimating the traffic amount needed to be evacuated:

$$T_f(t)=\frac{1}{1+\exp\left(-u(t-v)\right)} \tag{13.6}$$

where $T_f(t)$ is the traffic amount to be evacuated at time t. Two parameters, u and v, have to be calibrated with statistic data. With the logistic equation, USACE (2000b) presented three curves for slow, medium, and rapid evacuation respectively calibrated with field investigation data.

Physical models mainly focus on simulating transportation processes at large scale, like evacuation during Katrina Hurricane in New Orleans. The process of these models can be generally divided into four steps: risk area zoning, traffic demand estimation, traffic capacity estimation, and traffic management (Urbanik II, 2000).

1. risk area zoning. A disaster-influenced area needs to be zoned based on risk category. The sequence and compliance of evacuation vary in each category. Normally, areas at high risk may be evacuated first with good compliance.
2. traffic demand estimation. Traffic demand depends on the number, density, and distribution of population, vehicle possession ratio, and the compliance ratio.

3. traffic capacity estimation. Traffic capacity strongly relies on characteristics of transportation networks, including traffic methods (e.g. trains, buses, and private vehicles), road characteristics (e.g. width, length, and quality), and traffic bottlenecks (e.g. bridges, tunnel, and segments under construction or repair).
4. traffic management. Evacuation needs to remove a large amount of traffic in a relatively short time, which would be much worse than rush-hour traffic. Thus, traffic management is needed to attain efficiency. The commonly used methods for traffic management is adjusting traffic flow based on information from a dynamic monitoring system, use of road shoulders, sidewalks and reverse lanes to increase the traffic capacity, and arranging people to guide the traffic at the bottlenecks.

Several physical models for simulating evacuation are available. Models developed in the 1980s mainly focus on nuclear releases, influenced by the Three Mile Island accident. Typical are those based on micro simulation (individual behavior) such as NETSIM (Urbanik and Desrosler, 1981), and micro simulation (traffic flow) such as NETVAC (Sheffi et al., 1982) and MASSVAC (Hobeika and Jamei, 1985). The latter is also used for evacuation during hurricanes. Models developed since 1990s focus on evacuation from natural disasters such as hurricanes and large-scale floods. Oak Ridge National Laboratory (1999) developed a program, OREMS, in Fortran. However, the traffic demand as a function of time needs to be input instead of being estimated. PBS&J (2000) developed a network and GIS based system – ETIS. The system can be used to estimate traffic demand and evacuation time for different hurricane categories.

Lindell et al. (2002) presented an EMBLEM model for predicting the evacuation time during the whole process (decision-making, warning, response, and evacuation). The time for decision-making, warning, and response is estimated using empirical methods. The evacuation time (t_e) is calculated as

$$t_e = t_c + t_q + t_h \tag{13.7}$$

where, t_c is the time from home to highway, t_q is the time waiting for access to enter the highway, and t_h is the time cost on the highway. Lindell et al. (2008) published an updated model, EMBLEM2, which considers time-varying traffic flow and traffic management.

13.2 Decision-making Under Uncertainty

Making technical decisions under uncertainties is a necessary part or even the major part of the work of an engineer. The objective of decision analysis is to maximize the benefit or minimize the loss. Risk-based decision-making should at least include the following components (Ang and Tang, 1990):

1. a list of all feasible alternatives, including the acquisition of additional information, if appropriate
2. a list of all possible outcomes associated with each alternative
3. an estimation of the probability associated with each possible outcome
4. an evaluation of the consequences associated with each combination of alternative and outcome
5. the criteria for decision
6. a systematic evaluation of all alternatives

BC Hydro (1993), USBR (1997), and ANCOLD (1998) published life-safety criteria for dam safety decision-making. Ranking of water-management options under uncertainty was reported by Vrijling and Van Gelder (2000) and many others. Two methods for decision-making are commonly used: decision tree and influence diagram.

13.2.1 Decision Tree

A decision tree integrates the relevant components of the decision analysis in a systematic manner suitable for an analytical evaluation of the optimal alternative. Decision trees are frequently used to conduct quantitative risk-based decision-making in various disasters (Frieser, 2004; Smith et al., 2006; Lindell et al., 2007; Woo, 2008; Liu, 2009). Figure 13.2 shows a generic example of a decision tree with two alternatives. The decision tree begins with a square node, referred to as a decision node, at which there are two alternatives a_1 and a_2. With each alternative, there may be several possible outcomes, shown as branches originating from a circular node called a chance node. In this example, θ_1 and θ_2 are the two possible outcomes associated with alternatives a_1 and a_2. The outcomes from a chance node are mutually exclusive and collectively exhaustive. The desirability of the consequence of each sequence or path in the tree is measured by its "utility value" recorded at the end of the sequence, such as $u(a_i, \theta_j)$. The decision is made by comparing the expected utility value in each alternative and choosing the one with minimum value. The expected utility value is calculated as the product of the probability and the utility value. For example, the expected utility value of a_1 is given by

$$u(a_1) = p(\theta_1 | a_1)u(a_1, \theta_1) + p(\theta_2 | a_1)u(a_1, \theta_2) \tag{13.8}$$

Decision trees are a useful decision support tool that uses a straightforward matter to show the logic relationships of the alternatives and outcomes. They are commonly used to help identify a strategy most likely to reach a goal. A decision tree appears to be similar to an event tree, since they are in tree-like graphs with an initiating node. Also, the methods for calculating the expected consequences in event trees and utility values in decision trees are almost the same. However, there are three major differences between these two methods:

1. Event trees are used to identify and evaluate the possible sequences of subsequent events following an initiating event, while decision trees are used to find out the expected consequences of each alternative and make the optimal decision.

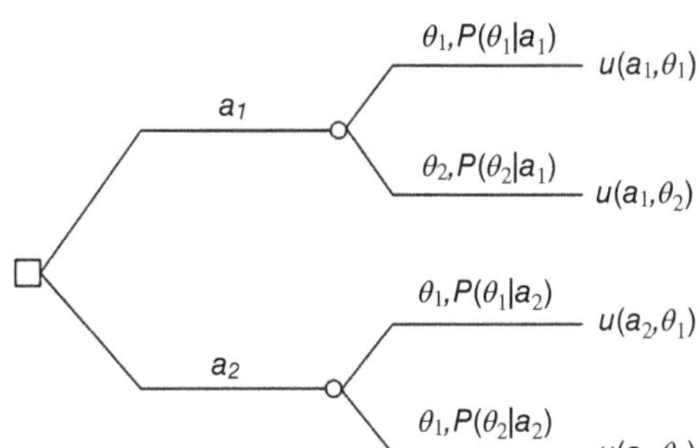

Figure 13.2 A decision tree

2. In event trees, there are only chance nodes without decision nodes. The branches of chance nodes are outcomes, which may occur with certain probabilities. The branches of decision nodes are alternatives, which can be chosen by the decision-makers. Thus, there are no probability values in each decision node.
3. The results in event trees are usually consequences, either negative or positive, while the results in a decision tree are usually utility values, which involve subjective feelings of the decision-maker.

13.2.2 Multi-phase Decision

Frieser (2004) and Smith et al. (2006) commented that the evacuation decision for floods can be delayed in cases with long prediction lead time (the period between the prediction and the arrival of the flood) and large uncertainties. Figure 13.3 shows a delayed decision tree for a flood disaster. The flood consequences include evacuation costs (*C*), flood damage (*D*), and loss of life (*L*). Evacuation costs include initial costs (e.g. costs of transport, accommodation, food supply, organization, and service), interruption of gross domestic product (GDP), and indirect influences (e.g. influence of the market in the surrounding areas). The evacuation costs increase with the warning time. Flood damage includes all the consequences caused by flooding except those to human life. Some moveable properties, such as cars and portable things, can be saved by evacuation. However, immoveable properties such as houses, GDP interruption during flooding, and environmental damage cannot be reduced by evacuation. Historical data shows that the fatality rate due to flooding could be largely reduced by allowing more warning time (DeKay and McClelland, 1993; Graham, 1999).

The evacuation may be delayed to obtain information with less uncertainty – variation $(P_1) <$ variation (P_2) – and to reduce the evacuation costs $(C_1 < C_2)$, as shown in Figure 13.3. However, it would run the risk of losing more lives $(L_1 > l_1 > l_2)$ and property $(D_1 > d_1 > d_2)$ given less available time for evacuation. A good decision should try to attain a minimum expected total loss.

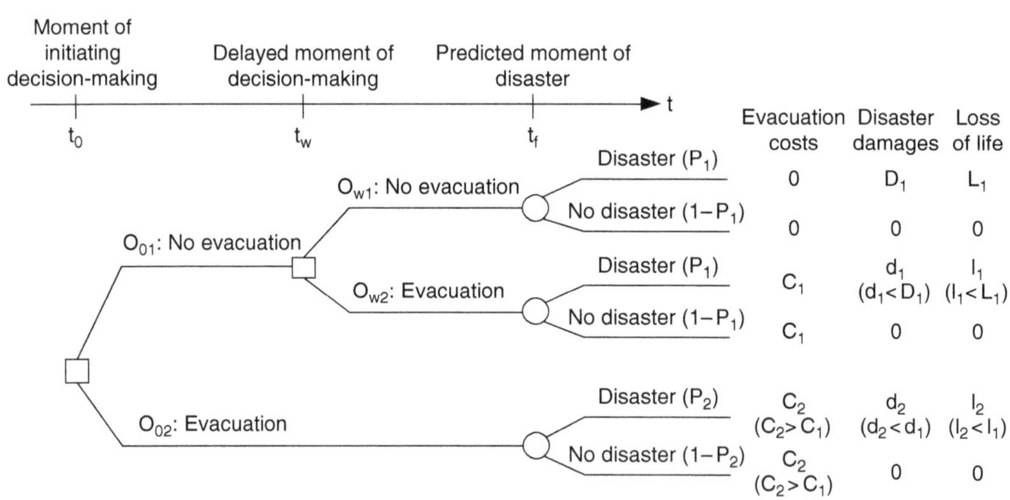

Figure 13.3 Decision tree for dam-break emergency management. *Source:* Jonkman, S. N. (2007). Modified with permission of Prof. Jonkman, S.N (Bas)

Assuming that we are now at time t_0, the predicted moment of disaster is at time t_f, and the decision may be delayed until t_w, then the optimal decision at t_0 is to find the minimum expected total loss at this moment:

$$OP(t_0) = Min\left[E(O_{01}), E(O_{02})\right] = Min\left\{E(O_{i1}), Min\left[E(O_{w1}), E(O_{w1})\right]\right\} \tag{13.9}$$

where the expected loss of decision O_{02} $[E(O_{02})]$ is given by

$$E(O_{02}) = P_2(C_2 + d_2 + l_2) + (1 - P_2)(C_2) = C_2 + P_2(d_2 + l_2) \tag{13.10}$$

Similarly, the expected loss of decision of O_{w1} and O_{w2} are given by

$$E(O_{w1}) = P_1(D_1 + L_1) \tag{13.11}$$

$$E(O_{w2}) = C_1 + P_1(d_1 + l_1) \tag{13.12}$$

Since the decision may be delayed to any time before the predicted moment, $E(O_{02})$ is actually a function of t_w. Usually, some discrete value of t_w is selected to obtain each $E(O_{02})$ under each t_w, then the minimum $E(O_{02})$ is selected as the final. Actually, the decision may be delayed a number of times. Thus, time-dependent evacuation decision can be analyzed using a multi-phase decision tree (Frieser 2004; Smith et al. 2006).

The probabilistic methods using decision trees are superior to deterministic methods owing to the inclusion of uncertainties. A premise of using a decision tree in the existing methods is to assume a predicted time of flooding, t_f, as shown in Figure 13.3. Normally, t_f is set as a target time (e.g. with enough lead time to evacuate the people) or the time of the worst predicted situation (e.g. highest water level or largest flood flow rate). This may not be reasonable due to the fact that a dam-break flood may occur at any future time. Therefore, the predicted flood probability should be a stochastic process instead. The loss of life and property would be largely underestimated if the flood occurs before the predicted time, and vice versa. A dynamic decision model will be presented in Section 3.3 to solve this problem.

Example

There was a serious flood in the region of Nijmegen, the Netherlands, in 1995. The highest water level of the Rhine River reached 16.68 m on 1 February 1995, which was just 0.1–0.2 m below the crest height of the dikes at some locations (Frieser, 2004). The population at risk was estimated as 300,000 if the dike failed. A model, Lobith, was applied to predict the water level, the uncertainties of which decreased with the prediction lead time. The model was able to make a prediction with the lead time of four days.

Table 13.2 shows the predicted and measured water level during the flood. There were four predictions for each of the coming four days. For instance, the predicted values with 1–4 days' lead time were 16.00, 16.20, 16.10, and 16.10 on 28 January 1995. The corresponding prediction errors were 0.07, 0.20, 0.30, and 0.50, respectively. The decision tree for multi-phase evacuation is shown in Figure 13.4. The evacuation costs and flood damage (economic loss and the monetized loss of life) in this figure refer to Frieser (2004). The corresponding predicted water level and the probability of flooding are shown in Table 13.3. On 28 January 1995, the expected loss of evacuation was €148.8 million, which was larger than that without evacuation. Thus, no evacuation should be

Table 13.2 Predicted and measured water levels during the flood. *Source:* Frieser (2004). Modified with permission from Prof. Bas Jonkman

Date (dd-mm-yy)	Predicted water level (m)						
	29-01-95	30-01-95	31-01-95	01-02-95	02-02-95	03-02-95	04-02-95
28-01-95	16.00	16.20	16.10	16.10	–	–	–
29-01-95	–	16.25	16.45	16.40	16.30	–	–
30-01-95	–	–	16.55	16.65	16.65	16.55	–
31-01-95	–	–	–	16.75	16.65	16.65	16.35
Measured	15.99	16.35	16.62	16.68	16.41	15.97	15.37

Note: All water levels in Tables 13.2 and 13.3 are given in meters relative to the standardized datum level in the Netherlands (NAP), which is approximately equal to the mean sea level. Since the area considered is located in an upstream part of the Dutch Rhine River, the values are well above the mean sea level.

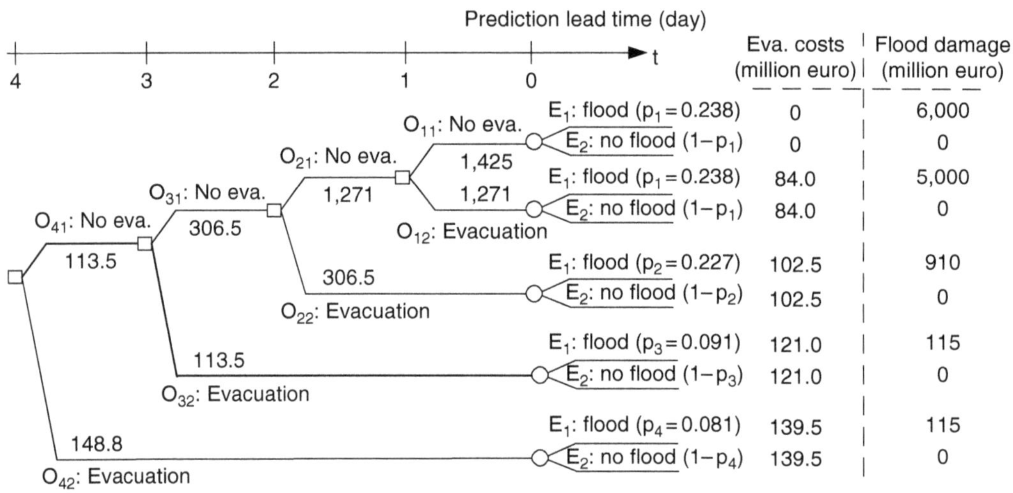

Figure 13.4 Decision tree for multi-phase decision-making. *Source:* Jonkman, S. N. (2007). Modified with permission of Prof. Jonkman, S.N (Bas)

Table 13.3 Predicted water level and flooding probability with different lead time. *Source:* Frieser (2004). Modified with permission from Bas Jonkman

Date (dd-mm-yy)	Prediction lead time (day)	Predicted value [m]	$\sigma\Delta_h$ [m]	P_f [1/year]
28-01-95	4	16.10	0.50	0.081
29-01-95	3	16.40	0.30	0.091
30-01-95	2	16.65	0.20	0.227
31-01-95	1	16.75	0.07	0.238

chosen with the lead time of 4 days. However, the decision on the next day was opposite, since the expected loss of evacuation was €113.5 million, which was smaller than that without evacuation, €306.5 million. The optimum decision is to evacuate the people on 29 January 1995 and the minimal total loss is €113.5 million based on the analysis with a multi-phase decision tree.

13.2.3 Influence Diagrams

Influence diagrams are an effective tool to solve complex decision problems under uncertainties. Influence diagrams were developed as a computer-aided modeling tool by Miller et al. (1976) and Howard and Matheson (1984). Olmsted (1983) explored the notion of solving a decision problem through influence diagram manipulations, and an algorithm to accomplish that was constructed by Shachter (1986).

Influence diagrams are an extension of Bayesian networks. Besides the chance nodes in Bayesian networks, two more types of nodes – decision nodes and value nodes – are necessary in influence diagrams. Decision nodes have several alternatives for the decision maker to choose. Value nodes are the consequences, either positive (losses) or negative (gains). As with Bayesian networks, the procedure of the application of influence diagrams can be divided into several parts: building the graphic structure by considering the interrelationships of nodes (qualitatively analysis), quantifying the structure with probabilities and conditional probabilities (prior quantitative analysis), finding evidences of a certain case, and updating the prior probabilities with the evidences (posteriorly quantitative analysis).

Example

An example of influence diagram for dam-break emergency decision is shown in Figure 13.5. Four parameters with discrete states are involved in the influence diagram: warning (no warning, 0–3 hours, 3–24 hours, and >24 hours), evacuation (evacuated and not-evacuated), water depth (0–1.5 m, 1.5–3 m, 3–6 m, and >6 m), and flood severity (safe, low severity, medium severity, and high severity). The flood severity is evaluated using HURAM in Chapter 11. The evacuation node is a decision node, which has four alternatives. Normally, no warning leads to no evacuation costs but may incur the largest flood damage and loss of life. Little warning leads to less evacuation costs and may incur more flood damage and loss of life. Therefore, a good warning strategy aims to balance the three types of consequences and obtain the minimum expected total loss. Water depth, evacuation,

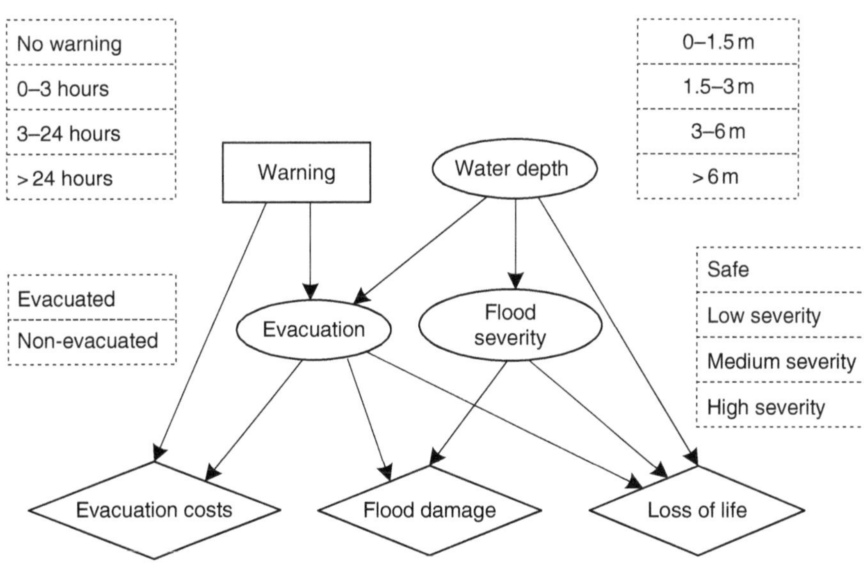

Figure 13.5 Influence diagram of the dam-break emergency decision model

Table 13.4 Prior probability table of water depth based on statistical data

Node	State	Number of cases	Frequency
Water depth	0–1.5 m	105	0.378
	1.5–3 m	113	0.406
	3–6 m	52	0.187
	>6 m	8	0.029

and flood severity are three chance nodes, the distributions of which need to be analyzed. Evacuation cost, flood damage, and loss of life are value nodes. Despite the moral consideration, loss of life may be monetized to achieve a rational decision. A commonly used method for quantifying the value of a human life is according to the average macro contribution of a person (Frieser, 2004). In that case, a human life is calculated as the product of average gross domestic product per person and the average human longevity. For example, according to Peng and Zhang (2013b), the monetized human life in 2008 in Mianyang in Sichuan Province, China is RMB 1.03 million.

Based on the statistical analysis, the frequencies of water depth in a flood are shown in Table 13.4. Out of all 278 cases of records with water depth, 105 cases (37.8%) have water depths of 0–1.5 m, 113 cases (40.6%) of 1.5–3 m, 52 cases (18.7%) of 3–6 m, and 8 cases (2.9%) of more than 6 m. Therefore, the probability of each state of water depth is set accordingly. The quantification of the conditional probabilities of evacuation and flood severity is implemented with HURAM. For example, the evacuation probability under a warning time of 0–3 hours and water depth of 0–1.5 m is obtained by setting the warning time and water depth in HURAM as 0–3 hours and 0–1.5 m, and keeping the other parameters as default. The conditional probabilities of evacuation and flood severity are shown in Tables 13.5 and 13.6, respectively. The flood consequences, including evacuation costs, flood damage and loss of life, are obtained with the information of evacuation rate, sheltering rate and fatality rate, the details of which will be introduced in a dynamic decision-making model (DYDEM) in Section 3.3. In a town with 1000 people that may be affected by a flood, the three types of consequences under certain conditions are shown in Tables 13.7–13.9.

The expected total losses – $E(L)$ – is the sum of the three expected consequences in each sub-area:

$$E(L)=\sum_{1}^{N}\left[C_i+P_i\left(D_i+LOL_i\right)\right] \tag{13.13}$$

where P_i means the probability of a flood in sub-area i. Thus, the optimal decision relies on the flood probability. Tables 13.10–13.13 show the expected total consequences and the optimum decision with flood probabilities from 1 to 0.001. Note that the prior means the expected total consequences without knowledge of water depth. Generally, the optimum warning strategy varies with the water depth and flood probability. When the water depth is low and the flood probability is small, no warning is needed.

13.3 Dynamic Decision-Making

Dynamic decision-making under emergency management is aimed at minimizing the possible dam-break consequences using primarily non-structure measures, such as warning, sheltering, and evacuation by taking advantage of sufficient time effect. The dynamic decision-making problems are first defined at both time and space scale in the next section.

Table 13.5 Prior conditional probability table for evacuation using HURAM

Water depth	0–1.5 m				1.5–3 m				3–6 m				>6 m			
Warning time (hour)	0	0–3	3–24	>24	0	0–3	3–24	>24	0	0–3	3–24	>24	0	0–3	3–24	>24
Evacuated	0.134	0.513	0.979	1.000	0.091	0.473	0.979	1.000	0.028	0.407	0.975	1.000	0.012	0.371	0.973	1.000
Non-evacuated	0.866	0.487	0.021	0.000	0.909	0.527	0.021	0.000	0.972	0.593	0.025	0.000	0.988	0.629	0.027	0.000

Table 13.6 Prior conditional probability table for flood severity using HURAM

Water depth (m)		0–1.5	1.5–3	3–6	>6
Flood severity	Safe	0.608	0.353	0.168	0.081
	Low severity	0.284	0.171	0.099	0.081
	Medium severity	0.108	0.476	0.529	0.481
	High severity	0.000	0.000	0.204	0.356

Table 13.7 Evacuation costs evaluated by DYDEM (PAR = 1000)

Evacuation Warning (hour)	Evacuated				Non-evacuated			
	No.	0–3	3–24	>24	No.	0–3	3–24	>24
Evacuation cost (RMB 10,000)	0	34.3	42.8	52.6	0	15.5	18.8	22.6

Table 13.8 Flood damage evaluated by DYDEM (PAR = 1000)

Evacuation Flood severity	Non-evacuated			
	Safe	Low severity	Medium severity	High severity
Evacuation cost (RMB 10,000)	0.0	134.5	313.7	448.2

13.3.1 Dam Failure Emergency Management

13.3.1.1 In Time Scale

The prediction lead time (the duration between the prediction moment and the predicted failure time) of dam-break floods is usually in the order of hours or days. During this period emergency management can be possibly implemented to save human lives and property. In time scale, the process of emergency management before the arrival of a dam-break flood can be divided into four phases: (1) hydrological forecasting, risk assessment, and decision-making, (2) warning, (3) response, and (4) evacuation and sheltering (Frieser, 2004), as presented in Section 13.1.

The dam-break emergency management in time scale displays the sequence of human activities and the population distributions with time before the flood occurrence. A good decision should consider not only the available time before the predicted arrival of the flood, but also the demand time for each phase and the corresponding population distribution.

13.3.1.2 In Space Scale

Dam break risk analysis covers a large area along the river from the catchment upstream of the dam to the potential flooding areas downstream of the dam. In space scale (Figure 13.6), the dam-break emergency management includes five steps: hydraulic parameter forecasting, dam-break probability evaluation, dam breaching simulation, flood routing simulation, and flood consequence evaluation.

Hydraulic parameters such as reservoir volume, inflow rate and water elevation directly influence the dam safety conditions. These parameters should be treated as stochastic processes instead of independent random variables due to the correlations in time scale.

Table 13.9 Monetized loss of life evaluated by DYDEM (PAR = 1000)

Evacuation [a]	Non-evacuated											
Flood severity	Low severity				Medium severity				High severity			
Water depth (m)	0–1.5	1.5–3	3–6	>6	0–1.5	1.5–3	3–6	>6	0–1.5[b]	1.5–3[b]	3–6	>6
Probability of LOL (%)	0.19	1.59	5.74	12.52	0.19	7.42	38.88	73.20	—	—	90.78	90.78
Monetized LOL (RMB 10,000)	196	1638	5907	12896	196	7643	40046	75396	—	—	93503	93503

Note:
[a] The probability of LOL is zero when the people are evacuated or the flood severity is Safe;
[b] The probability of high severity is zero when the water depth is less than 3 m.

Table 13.10 Total consequences (in RMB 10,000) and the optimal decision with a dam failure probability of 1.0

Water depth (m)		0–1.5	1.5–3	3–6	>6	Prior*
Warning time (hour)	No.	129	3717	39976	70138	11058
	0–3	98	2179	24400	44675	6774
	3–24	**45**	127	1062	1959	324
	>24	53	**53**	**53**	**53**	**53**

Table 13.11 Total consequences (in RMB 10,000) and the optimal decision with a dam failure probability of 0.1

Water depth (m)		0–1.5	1.5–3	3–6	>6	Prior
Warning (hour)	No.	**13**	372	3998	7014	1106
	0–3	32	240	2461	4488	699
	3–24	43	**51**	144	234	70
	>24	53	53	**53**	**53**	**53**

Table 13.12 Total consequences (in RMB 10,000) and the optimal decision with a dam failure probability of 0.01

Water depth (m)		0–1.5	1.5–3	3–6	>6	Prior
Warning (hour)	No.	**1.3**	**37**	400	701	111
	0–3	26	46	267	469	92
	3–24	42	43	**52**	61	**45**
	>24	53	53	53	**53**	53

Table 13.13 Total consequences (in RMB 10,000) and the optimal decision with a dam failure probability of 0.001

Water depth (m)		0–1.5	1.5–3	3–6	>6	Prior
Warning (hour)	No.	**0.13**	**3.7**	**40**	70	**11**
	0–3	25	27	48	67	31
	3–24	42	42	43	**44**	43
	>24	53	53	53	53	53

Failure mode is a key parameter for dam-break probability analysis. Most dam failures were caused by either overtopping or piping, with corresponding percentages of 56.0% and 35.7% for man-made earth and rockfill dams (Xu and Zhang, 2009; Zhang et al., 2009), and 91% and 8% for landslide dams (Peng and Zhang, 2012a). Since this study focuses on decision-making, only the overtopping failure mode is considered, which usually suffices for breached landslide dams. Overtopping occurs when the reservoir water elevation exceeds the elevation of the dam crest. The probability of dam failure is closely related to hydrological parameters.

The next two components are to simulate the dam breaching process and flood routing downstream of the dam. The hydraulic parameters significantly affect the flood consequence downstream. Models for both dam breaching and flood routing are needed to obtain the hydraulic parameters.

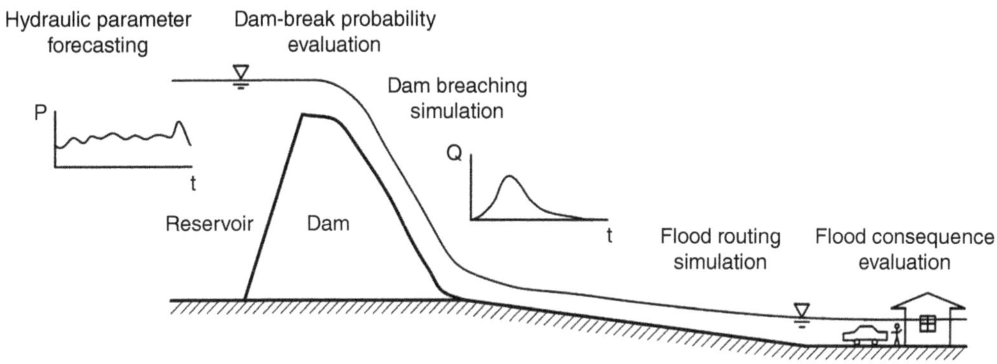

Figure 13.6 Emergency management of dam breaks in space scale

The flood consequences, including evacuation costs, flood damage and loss of life, are highly related to the warning time. Generally, evacuation costs increase while flood damage and loss of life decrease with more warning time. The dam-break emergency management should cover the evolution of the dam-break event in space scale, especially the dam-break probability and flood consequences. A proper decision should take the dam-break probability as a time series and the consequences as functions of warning time.

13.3.2 Dynamic Decision-making Framework

The framework of dynamic decision-making is intended to make a decision whether to evacuate the population at risk or to delay the decision; to predict the optimal time to evacuate the PAR with the minimum expected total loss; and to update the decision-making with new information when delayed decision is chosen.

Assume a continuous stochastic process of dam-break probability as shown in Figure 13.7. The studied period ranges from t_0 (i.e. the current time) to t_{end} (i.e. the moment after which the risk is not considered). The probability of dam failure in a short period, *dt*, is calculated as

$$P(t)=f(t)dt \tag{13.14}$$

where $f(t)$ is a continuous stochastic process of dam failure probability. Given the time of issuing evacuation warning ($t_0 \le t_w \le t_{end}$) as shown in Figure 13.7, the period between t_w and possible flood arrival time (t_f) is called the warning time (W_t), which is critical for the people at risk to evacuate to safe places. Warning time equals zero if the possible flood occurs before the time for evacuating the PAR. Thus, W_t is a non-negative parameter and given by:

$$W_t = 0, \text{ when } t_f < t_w \tag{13.15}$$

$$W_t = t_f - t_w, \text{ when } t_f \ge t_w \tag{13.16}$$

$$W_t = t_{end} - t_w, \text{ when } t_f \ge t_{end} \tag{13.17}$$

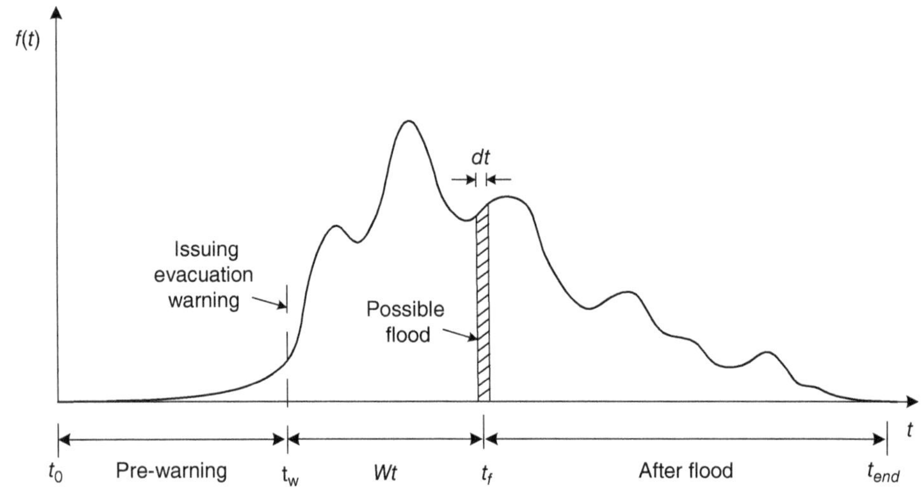

Figure 13.7 Probability of dam-break flood as a continuous stochastic process of time

The flood risk, or expected total loss, $E(L_t)$, is the sum of the expected evacuation costs, flood damage, and loss of life, which are obtained by integrating the product of these three categories of consequences and their corresponding probabilities along time:

$$\begin{aligned} E(L_t) &= \int L(t)f(t)dt = \int_{t_0}^{+\infty}\left[C(W_t)+D(W_t)+L(W_t)\right]f(t)dt \\ &= \int_{t_0}^{+\infty}C(W_t)f(t)dt + \int_{t_0}^{+\infty}\left[D(W_t)+L(W_t)\right]f(t)dt \end{aligned} \tag{13.18}$$

where $L(t)$ is the flood consequences or total loss; $C(W_t)$, $D(W_t)$ and $L(W_t)$ are the evacuation cost, flood damage, and loss of life as functions of warning time, respectively, which will be presented later in this chapter. Considering the definition of W_t in Equations (13.15)–(13.17), Equation (13.18) can be separated into two equations:

$$\int_{t_0}^{+\infty}C(W_t)f(t)dt = \int_{t_0}^{t_w}C(0)f(t)dt + \int_{t_w}^{t_{end}}C(t-t_w)f(t)dt + C(t_{end}-t_w)\left[1-\int_{t_0}^{t_{end}}f(t)dt\right] \tag{13.19}$$

$$\int_{t_0}^{+\infty}\left[D(W_t)+L(W_t)\right]f(t)dt = \int_{0}^{t_w}\left[D(0)+L(0)\right]f(t)dt + \int_{t_w}^{t_{end}}\left[D(t-t_w)+L(t-t_w)\right]f(t)dt + 0 \tag{13.20}$$

The last part of Equation (13.19) denotes that evacuation costs incurred even if there is no dam failure or if the failure time is beyond the studied period. The last part of Equation (13.20) denotes that the flood damage and loss of life are not considered if there is no dam failure in the studied period.

The optimal time to evacuate the PAR is the time to attain the minimum total loss or the time at which the derivative of the following function is zero:

$$Min\left[E\left(L_t\right)\right] \text{or} \frac{dE\left(L_t\right)}{dt_w}=0 \tag{13.21}$$

For a discrete time series as shown in Figure 13.8, the probability of flooding in the period from t_{j-1} to t_j is given by $P(t_j)$. The expected total loss is given by:

$$\begin{aligned} E\left(L_t\right) &= \sum_{j=1}^{+\infty} L\left(t_j\right)P\left(t_j\right) = \sum_{j=1}^{+\infty}\left[C\left(W_t\right)+D\left(W_t\right)+L\left(W_t\right)\right]P(t_j) \\ &= \sum_{j=1}^{+\infty} C\left(W_t\right)P\left(t_j\right)+\sum_{j=1}^{+\infty}\left[D\left(W_t\right)+L\left(W_t\right)\right]P(t_j) \end{aligned} \tag{13.22}$$

Similarly, considering the definition of W_t in Equations (13.15)–(13.17), Equation (13.22) can be separated into two equations:

$$\sum_{j=1}^{+\infty} C\left(W_t\right)P\left(t_j\right) = \sum_{j=1}^{w} C(0)P\left(t_j\right)+\sum_{j=w+1}^{N} C\left(t_j-t_w\right)P\left(t_j\right)+C\left(t_{end}-t_w\right)\left[1-\sum_{j=0}^{N}P\left(t_j\right)\right] \tag{13.23}$$

$$\sum_{j=1}^{+\infty}\left[D\left(W_t\right)+L\left(W_t\right)\right]P(t_j) = \sum_{j=1}^{w}\left[D(0)+L(0)\right]P(t_j)+\sum_{j=w+1}^{N}\left[D\left(t_j-t_w\right)+L\left(t_j-t_w\right)\right]P(t_j)+0 \tag{13.24}$$

where $t_N = t_{end}$. The optimal time to issue the evacuation warning is the time to attain the minimum total loss: Min($E(L_t)$).

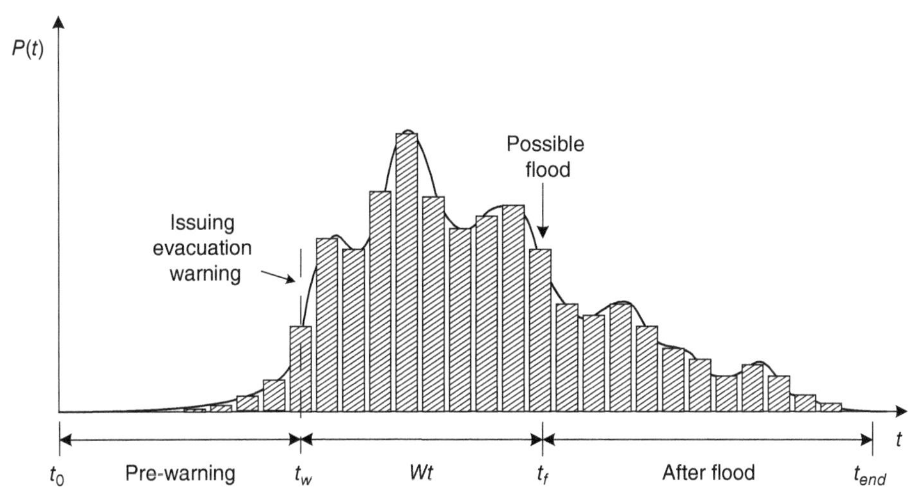

Figure 13.8 Probability of dam-break flood as a discrete stochastic process of time

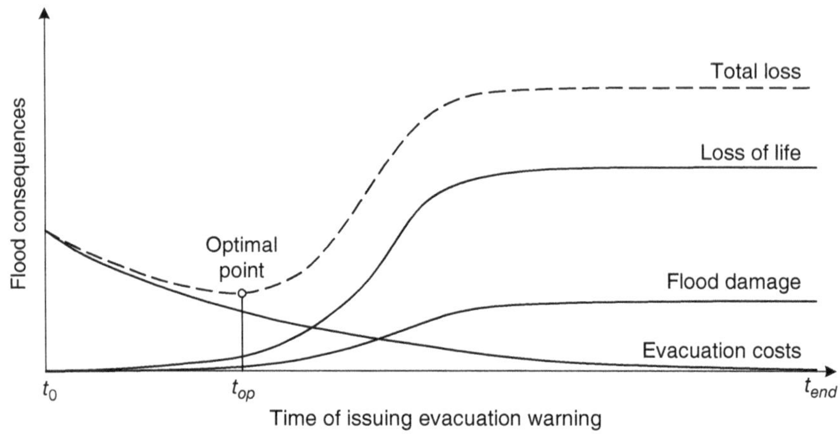

Figure 13.9 Flood consequences as functions of time for evacuating the PAR

Figure 13.9 shows the flood consequences as functions of the time for issuing a warning (t_w). The flood damage and loss of life increase and the evacuation costs decrease with t_w. Therefore, there is an optimal point (t_{op}) for evacuating the PAR to achieve the minimum expected total loss. The parameter, t_{op}, is very important in decision-making. If t_{op} is close to t_0, the PAR should be evacuated immediately; if $t_{op} = t_{end}$, no evacuation is decided; if $t_0 < t_{op} < t_{end}$, the PAR should be evacuated at t_{op}. If it is not in an urgent case – that is t_{op} is much larger than t_0 – we may delay the decision to gain more information to reduce the uncertainties.

As time goes on, more information for the decision will be available. The decision with a different initial time (t_0) is defined as a stage at which the available information and the evaluated results may be different. Thus, a dynamic decision should be a multi-stage decision process. In each stage, as the information gathered and its condition of uncertainty is different, the predicted dam-break probabilities and flood consequences could be different. It would be unnecessary to consider the decision-making as a continuous process of t_0 since the changes in information may be small in a short period. In this framework, multi-stage decisions are limited to stages with significant changes in information, such as a predicted storm, changes in knowledge on the properties of the dam materials, or implementing major flood control measures. In the multi-stage decision framework, both the predicted dam-break probabilities and consequences should be updated in each new stage.

From the overview of the dynamic decision framework, two important components of the framework are critical: prediction of dam-break probability as a stochastic process and evaluation of flood consequences as functions of warning time. These two components will be presented in the next two sections.

13.3.3 Time Series Models for Estimating Dam Failure Probability

Where overtopping failure is concerned, the probability of dam failure is the probability that the reservoir water volume reaches the reservoir capacity. The inflow rate into the reservoir is the only stochastic process in estimating the reservoir volume at time t, which can be forecast using time series analysis methods. A time series is a set of observations taken sequentially in time. An intrinsic feature of a time series is that adjacent observations are dependent. In this section, time series analysis is performed to forecast the future values of inflow rate based on available observations.

The analysis is divided into four steps: model identification, model estimation, model diagnostic checking, and forecasting following Brockwell and Davis (1996) and Box et al. (2008).

13.3.3.1 Time Series Models for Forecasting Inflow Rates

A stationary time series, which is one with its mean and variance independent of time, can usually be simulated as a mixed autoregressive-moving average (ARMA) model. In an ARMA model, a time-related variable x_t can be expressed as a finite, linear aggregate of previous values of the time series and random shocks, $a_t, a_{t-1},\ldots,a_{t-q}$:

$$x_t = \phi_1 x_{t-1} + \phi_2 x_{t-2} + \cdots + \phi_p x_{t-p} + a_t - \theta_1 a_{t-1} - \theta_2 a_{t-2} - \cdots - \theta_q a_{t-q} \tag{13.25}$$

in which ϕ_i and θ_i are the coefficients of the ARMA(p, q) model to be quantified. A random shock, a_t, is an independently and identically distributed (IID) stochastic process. Normally a_t can be assumed as a normal distribution with a mean value of zero, $N(0, \sigma_a^2)$.

If a time series is not a stationary model, it can often be transferred to a stationary one by differentiating it (Box et al., 2008). A difference equation Δx_t is defined as

$$\Delta x_t = x_t - x_{t-1} \tag{13.26}$$

and $\Delta^d x_t$ as

$$\Delta^d x_t = \Delta^{d-1} x_t - \Delta^{d-1} x_{t-1} \tag{13.27}$$

If x_t can be transferred to a stationary time series ω_t with difference equation, namely,

$$\omega_t = \Delta^d x_t = \phi_1 \omega_{t-1} + \phi_2 \omega_{t-2} + \cdots + \phi_p \omega_{t-p} + a_t - \theta_1 a_{t-1} - \theta_2 a_{t-2} - \cdots - \theta_q a_{t-q} \tag{13.28}$$

then, x_t is called an autoregressive integrated moving average time series, or ARIMA(p, d, q). Actually, ARMA(p, q) is a special case of ARIMA(p, d, q) with $d=0$.

For an ARMA(p, q) model as shown in Equation (13.25), if $q=0$, then

$$x_t = \phi_1 x_{t-1} + \phi_2 x_{t-2} + \cdots + \phi_p x_{t-p} + a_t \tag{13.29}$$

is called an *autoregressive* model of order p, or AR(p) for short. If the $p=0$ in an ARMA(p, q) model, then x_t can be expressed as a finite weighted sum of $a_t, a_{t-1},\ldots, a_{t-q}$,

$$x_t = a_t - \theta_1 a_{t-1} - \theta_2 a_{t-2} - \cdots - \theta_q a_{t-q} \tag{13.30}$$

is called a *moving average* model of order q, or MA(q) for short.

13.3.3.2 Model Identification

The objective of model identification is to find a suitable time series model with orders (p, d, q) to simulate the observations of a time series. For a given observed time series, the first step is to check whether it is stationary or not. If it is non-stationary, we need to transform it using Equation (13.26) until it becomes a stationary time series. The symptom of a non-stationary time series is that the autocorrelation function ρ_k at time lag k will not die out quickly and will fall off slowly and nearly linearly with the increase of k (Box et al., 2008). The autocorrelation function, ρ_k, is given by

$$\rho_k = \frac{\gamma_k}{\gamma_0} \tag{13.31}$$

where γ is called as *autocovariance* at time lag k and is given by

$$\gamma_k = \text{cov}\left[x_t, x_{t+k}\right] = E\left[(x_t - \mu)(x_{t+k} - \mu)\right] \tag{13.32}$$

For a given time series of inflow rate, $x_1, x_2, \dots x_N$, the estimate of γ_k is given by

$$\gamma_k{}^* = \frac{1}{N}\sum_{t=1}^{N-k}\left[(x_t - \bar{x})(x_{t+k} - \bar{x})\right] \quad k = 0, 1, \dots, N-1 \tag{13.33}$$

where $\bar{x}$ is the average value of the observations of x_t.

Another important parameter for a time series is its partial autocorrelation function ϕ_{kk}. The jth autocorrelation function ρ_j can be described as an autoregressive function as

$$\rho_j = \phi_{k1}\rho_{j-1} + \phi_{k2}\rho_{j-2} + \cdots \phi_{k(k-1)}\rho_{j-k+1} + \phi_{kk}\rho_{j-k} \qquad j = 1, 2, \dots, k \tag{13.34}$$

where the last coefficient ϕ_{kk} is called a partial autocorrelation function. ϕ_{kk} can be estimated using recursive formulas (Durbin, 1960; Box et al., 2008).

The model identification for a stationary time series is to find the model type [AR(p), MA(q) or ARMA(p, q)] and the corresponding order (i.e. p and q). According to Box et al. (2008), one feature of AR(p) is that the autocorrelation function ρ_k tails off, while its partial autocorrelation function ϕ_{kk} has a cut-off after lag p. Conversely, for MA(q), the autocorrelation function ρ_k cuts off after lag p, while its autocorrelation function ϕ_{kk} tails off. If both ρ_k and ϕ_{kk} tail off, then a mixed process is suggested, as shown in Table 13.14. For example, the autocorrelation function (ρ_k) and partial autocorrelation function (ϕ_{kk}) of an assumed time series are shown in Figure 13.10. As ρ_k has a cut-off after lag 1 and ϕ_{kk} tails off, the time series can be assumed as MA(1). The assumption needs to be tested, which will be introduced later.

13.3.3.3 Model Estimation and Diagnostic Checking

An ARMA(p, q) time series, shown in Equation (13.25), has $(p+q+1)$ parameters, namely, $\phi_1, \dots, \phi_p, \theta_1, \dots, \theta_q$, and σ_a. The objective of model estimation is to find proper parameters to fit the observations of the time series. The error of a ARMA(p, q) time series at time t is expressed as

$$a_t = x_t - \phi_1 x_{t-1} - \phi_2 x_{t-2} - \cdots - \phi_p x_{t-p} + \theta_1 a_{t-1} + \theta_2 a_{t-2} + \cdots + \theta_q a_{t-q} \tag{13.35}$$

The least squares method is used to find parameters ϕ_i and θ_i for achieving the least sum of the squares of a_t:

$$Min\left[\sum_{t=1}^{n} a_t^2(\phi_i, \theta_i)\right], \quad i = 1, 2, \dots, n \tag{13.36}$$

Table 13.14 Identification of time series

Model	Equation	Behavior of ρ_k	Behavior of ϕ_{kk}
AR(p)	$x_t = \phi_1 x_{t-1} + \cdots \phi_p x_{t-p} + a_t$	Tail off	Cut off at ϕ_{pp}
MA(q)	$x_t = a_t - \theta_1 a_{t-1} - \ldots \theta_q a_{t-q}$	Cut off at ρ_q	Tail off
ARMA(p, q)	$x_t = \phi_1 x_{t-1} + \cdots \phi_p x_{t-p} + a_t - \theta_1 a_{t-1} - \cdots \theta_q a_{t-q}$	Tail off	Tail off

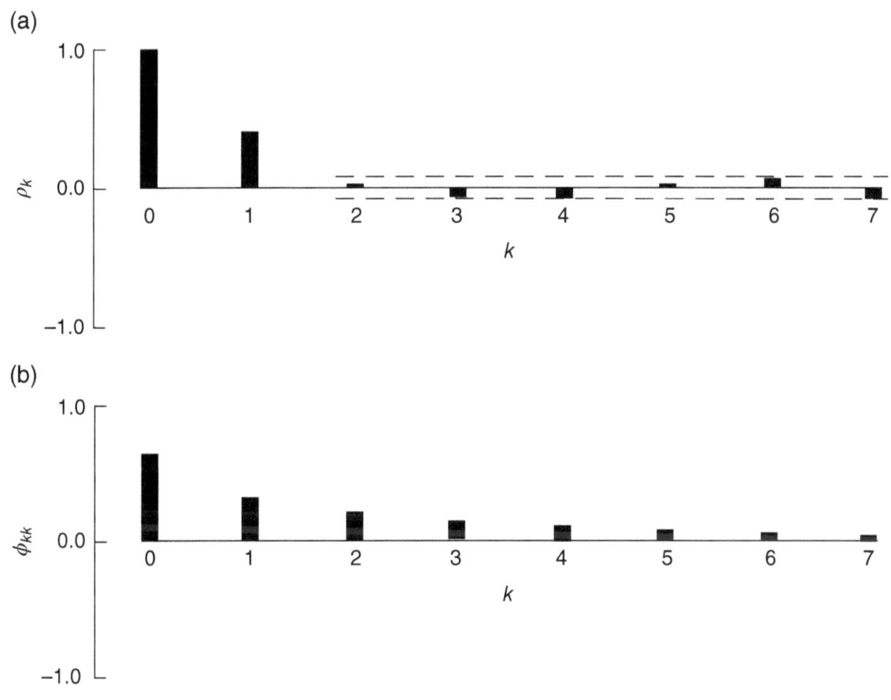

Figure 13.10 Parameters of an assumed time series: (a) autocorrelation function, (b) partial autocorrelation function

This can be implemented using a solver in Microsoft Excel.

After obtaining the parameters, the next step is to conduct model diagnostic checking to make sure the assumed model is suitable. Box et al. (2008) show that the equation

$$n\sum_{k=1}^{K}\left[\rho_k^*(a)\right]^2 \tag{13.37}$$

approximately follows a $\chi^2(K-p-q)$ distribution, where $\rho_k^*(a)$ is the estimated autocorrelation function of a_t, which is defined in Equations 13.31 and 13.32. The model can be checked through a χ^2 goodness-of-fit test at a confidence level (e.g. 5% or 10%).

13.3.3.4 Dam-break Probability Analysis

A dam is assumed to be overtopped once the reservoir volume exceeds its capacity (V_{cr}):

$$V_t > V_{cr} \tag{13.38}$$

where V_t is the reservoir volume at time *t*. If V_t is a normal variate with a mean of M_{Vt} and a standard deviation of σ_{Vt}, the probability of overtopping failure before time *t* (P_O) is given by

$$P_O(t) = P(V_t > V_{cr}) = 1 - P(V_t \le V_{cr}) = 1 - \Phi\left(\frac{V_{cr} - M_{Vt}}{\sigma_{Vt}}\right) \tag{13.39}$$

where $\Phi()$ is the probability function of a standard normal distribution.

For a discrete variable situation, the probability of overtopping in the period between $t - \Delta t$ and t, $P_D(t)$, is the probability that the predicted lake volume at time *t* is larger than V_{cr} and the predicted lake volumes before time *t* ($V_{t-k\Delta t}$, $k \ge 1$) are smaller than or equal to V_{cr}:

$$P_D(t) = P(V_t > V_{cr}, V_{t-\Delta t} \le V_{cr}, \ldots, V_{t-k\Delta t} \le V_{cr}, \ldots) \tag{13.40}$$

where the $V_{t-k\Delta t}$s denote all the predicted lake volumes before time *t*.

Based on mass conservation, the reservoir volume at time *t*, V_t, is given by:

$$V_t = V_{t-\Delta t} + (Q_t - Q_{ot} - Q_{et})\Delta t \tag{13.41}$$

where Δt is a time interval, and Q_t, Q_{ot}, and Q_{et} are the inflow rate, outflow rate, and evaporation rate of the reservoir at time *t*, respectively. For a specific dam before overtopping, the outflow rate Q_{ot} can be treated as a deterministic variable. The evaporation rate (Q_{et}) in a short time during the emergency management could be ignored. If the inflow rate is greater than the outflow rate, then V_t always increases monotonically with time; namely, $V_t > V_{t-\Delta t}$ and $V_{t-\Delta t} > V_t - V_{t-k\Delta t}$, $k > 1$. $P_D(t)$ can now be expressed as

$$\begin{aligned} P_D(t) &= P(V_t > V_{cr}, V_{t-\Delta t} \le V_{cr}, \ldots, V_{t-k\Delta t} \le V_{cr}, \ldots) \\ &= P(V_t > V_{cr}, V_{t-\Delta t} \le V_{cr}) \\ &= P(V_{t-\Delta t} \le V_{cr}) - P(V_t \le V_{cr}, V_{t-\Delta t} \le V_{cr}) \\ &= P(V_{t-\Delta t} \le V_{cr}) - P(V_t \le V_{cr}) \\ &= [1 - P_O(t-1)] - [1 - P_O(t)] \\ &= P_O(t) - P_O(t-1) \end{aligned} \tag{13.42}$$

13.3.3.5 Forecasting Dynamic Inflow Rates and Lake Volumes

After a time-series model and its model parameters have been identified as described in the appendices, forecasts of reservoir inflow rate can be obtained from the following difference equation:

$$x_t = \phi_1 x_{t-1} + \phi_2 x_{t-2} + \cdots + \phi_p x_{t-p} + a_t - \theta_1 a_{t-1} - \theta_2 a_{t-2} - \cdots - \theta_q a_{t-q} \tag{13.43}$$

For example, an AR(2) time series can be forecast as

$$\begin{aligned} x_t^*(1) &= \phi_1 x_t + \phi_2 x_{t-1} \\ x_t^*(2) &= \phi_1 x_t^*(1) + \phi_2 x_{t-1} \\ x_t^*(l) &= \phi_1 x_t^*(l-1) + \phi_2 x_t^*(l-2), \quad l = 3, 4, \ldots \end{aligned} \tag{13.44}$$

where x_t is the recorded value and $x_t^*(l)$ is the predicted value with lead time l. Note the expected value of a_t is zero as a_t follows a normal distribution of $N(0, \sigma_a^2)$.

A time series can also be expressed in a random-shock form of an infinite series (Box et al. 2008):

$$\begin{aligned} x_t &= a_t + \psi_1 a_{t-1} + \psi_2 a_{t-2} + \psi_3 a_{t-3} \cdots \\ &= a_t + \sum_{j=1}^{\infty} \psi_j a_j \end{aligned} \tag{13.45}$$

where the coefficients ψ_js can be obtained by substituting Equation (13.45) into (13.43) and comparing the coefficients of a_t in both sides.

As a_t is an identically distributed stochastic process, $N(0, \sigma_a^2)$, the standard deviation of x_t is calculated as

$$\sigma^2\left[x_t(l)\right] = \left(1 + \psi_1^2 + \psi_2^2 + \ldots + \psi_{l-1}^2\right)\sigma_a^2 \tag{13.46}$$

where σ_a is estimated as

$$\sigma_a^2 = \frac{1}{n-1}\sum_1^n a_t^2 \tag{13.47}$$

in which a_t can be obtained using Equation (13.35).

According to Equation (13.41), the reservoir volume V_t can be expressed as a function of the inflow rate, x_t:

$$V_t - V_{t-\Delta t} = \left(x_t + \mu_Q - Q_{ot}\right)\Delta t \tag{13.48}$$

where $Q_t = x_t + \mu_Q$, and μ_Q is the mean value of Q_t. Let us set

$$v_t = \frac{V_t}{\Delta t} \text{ and } v_{t-i} = \frac{V_{t-i\Delta t}}{\Delta t} \tag{13.49}$$

then v_t is given by

$$x_t = v_t - v_{t-1} - \left(\mu_Q - Q_{ot}\right) \tag{13.50}$$

Take x_t as an AP(2) model for example again:

$$x_t = \phi_1 x_{t-1} + \phi_2 x_{t-2} + a_t \tag{13.51}$$

Writing Equation (13.50) at times t, $t-1$ and $t-2$ and substituting these equations for x_t, x_{t-1}, and x_{t-2} into Equation (13.51), v_t can be expressed as

$$v_t = (1+\phi_1)v_{t-1} - (\phi_1 - \phi_2)v_{t-2} - \phi_2 v_{t-3} + (1-\phi_1-\phi_2)\mu_Q - (Q_{Ot} - \phi_1 Q_{Ot-1} - \phi_2 Q_{Ot-2}) + a_t \tag{13.52}$$

Set

$$C_Q = (1-\phi_1-\phi_2)\mu_Q - (Q_{Ot} - \phi_1 Q_{Ot-1} - \phi_2 Q_{Ot-2}) \tag{13.53}$$

then the means of v_t can be forecasted as

$$\begin{aligned} v_t^*(1) &= (1+\phi_1)v_t - (\phi_1-\phi_2)v_{t-1} - \phi_2 v_{t-2} + C_Q \\ v_t^*(2) &= (1+\phi_1)v_t^*(1) - (\phi_1-\phi_2)v_t - \phi_2 v_{t-1} + C_Q \\ v_t^*(3) &= (1+\phi_1)v_t^*(2)(\phi_1-\phi_2)v_t^*(1) - \phi_2 v_t + C_Q \\ v_t^*(l) &= (1+\phi_1)v_t^*(l-1)(\phi_1-\phi_2)v_t^*(l-2) - \phi_2 v_t^*(l-2) + C_Q \quad l = 4, 5, \ldots \end{aligned} \tag{13.54}$$

where v_t is the recorded value and $v_t^*(l)$ is the predicted value with lead time l.

The standard deviation of v_t and V_t can be obtained following the method as shown in Equations (13.40) and (13.41). With the means and standard deviations of V_t, the probabilities of dam failure as a time series can be predicted with Equations (13.40) and (13.42).

13.3.4 Evaluation of the Consequences of Dam Failures

The flood consequences, which include evacuation costs, flood damage, and loss of life, are closely related to evacuation, sheltering, and loss of life. Before evaluating the consequences, HURAM is used to simulate these three processes, as presented in Chapter 11.

13.3.4.1 Modifications to HURAM

HURAM is implemented in Hugin Lite (Hugin Expert A/S, 2009), which is a program for the analysis of Bayesian networks. Hugin Lite is powerful for the analysis of Bayesian networks involving discrete variables or continuous normal variates. However, in the dynamic decision-making framework (DYDEM), the parameters are not limited to these two types. Thus, the calculations for flood consequence in DYDEM are coded in Visual Basic in Microsoft Excel with Monte Carlo simulations instead of in Hugin Lite. The modifications are summarized as follows:

1. In HURAM, the parameter of time of day, with the states of 08:00–17:00, 17:00–22:00, and 22:00–08:00, is considered in the evacuation and sheltering components. In each state of time of day, the distributions of warning transmitting time (T_t), response time (S_t) and evacuation time (E_t) are different. These can be handled as the lead time in HURAM is of the order of minutes to hours (the lead time is often in one state of time of a day). While in DYDEM, the lead time for decision-making is often of the order of days. The distributions of T_t, S_t, and E_t would be complicated if the time of day is considered. Thus, the effect of time of day is not considered in DYDEM.

2. In HURAM, the warning transmitting distributions are W(3.5, 0.6), W(2.0, 0.5), W(1.3, 0.7) for times of a day of 08:00–17:00, 17:00–22:00, and 22:00–08:00, respectively. Here W(a, b) denotes a Weibull distribution with coefficients a and b:

$$P_t = 1 - \exp\left(-at^b\right) \tag{13.55}$$

In DYDEM, we use W(1.3, 0.7) only for convenience and safety. W(1.3, 0.7) is suggested for moderately rapid warning by Lindell et al. (2002).

3. The response time distribution is assumed as W(4, 1) for emergent dam break situations in HURAM, with a mean value of 15 minutes and a standard deviation of 15 minutes, while for decision-making in DYDEM, the response time should be much longer, as people need time to evacuate properties and prepare to live outside of their homes for several days. A distribution of W(0.085, 2.55) is used according to practices of hurricane evacuation (Lindell et al., 2002).

13.3.4.2 Estimation of the Flood Consequences

The evacuation cost consists of initial costs and GDP interruption. The initial costs (C_i) are the expenses for evacuating and arranging the people at risk and necessary services (e.g. security and medical care). The initial costs are generally proportional to the number of people to be evacuated and the time interrupted (warning time):

$$C_i = cP_{eva}\left(PAR\right)\left(W_t + 3\right) \tag{13.56}$$

where c is the expense per person per day (e.g. RMB 60 or US\$ 9.5 per person per day for the residents in Mianyang, Sichuan, Chine in 2008); P_{eva} is the proportion of the people evacuated, which is estimated using the modified HURAM; W_t is the warning time in days. The 3-day time is taken as the minimal period of time between the predicted moment of flooding and the return of the residents (Frieser, 2004). The GDP interruption (C_{GDP}) is calculated as

$$C_{GDP} = \frac{GDP_P}{365}\left(PAR\right)\left(W_t + 4\right) \tag{13.57}$$

where G_{GDP} is the average GDP per person in the flooded area. It is expected that economic sectors need time to restart their business (Frieser, 2004). Therefore, a duration of 4 days is added to the warning time. Thus the evacuation costs (C) are given by

$$C = C_i + C_{GDP} \tag{13.58}$$

The flood damage (D) is limited to the moveable properties in this study. The moveable properties are generally proportional to the number of people who have neither evacuated nor sheltered in safe zones:

$$D = \left(1 - P_{eva}\right)\left(1 - P_{safe}\right)\left(PAR\right)\alpha I_p \tag{13.59}$$

where P_{safe} is the ratio of the people taking sheltering in safe zones; α is the proportion of properties that can be transferred (0.1 is assumed); I_p is the property of each person, which is taken as the cumulative net income (i.e. income minus spending) per person:

$$I_p = (I - S)n \tag{13.60}$$

where I and S are the average income and spending per person, n is the average working period per person (e.g. 20 years).

Despite ethical considerations, a human life has to be measured for evacuation decision-making. Jonkman (2007) reviewed approaches of evaluating the human life. A method with macro economic considerations is chosen here (Van Manen and Vrijling, 1996, quoted by Jonkman, 2007). In this method, the value of a human life (V_L) is given as the product of GDP per person (GDP_p) and the average longevity (L):

$$V_L = (GDP_p)L \tag{13.61}$$

For example, the GDP_p and L in Mianyang City, China are RMB 13,745 and 75 years in 2008 (Mianyang Bureau of Statistics, 2009). Thus, the value of one person in 2010 in China is RMB 1.03 million. The monetized loss of life (M_L) is then calculated as

$$M_L = V_L(LOL) \tag{13.62}$$

where LOL is the loss of life predicted with HURAM as a function of warning time. As P_{eva}, P_{safe}, and LOL can be predicted as functions of warning time with HURAM, the three categories of flood consequences are expressed as functions of warning time.

13.3.5 Features of DYDEM

With the predicted dam-break probabilities as a time series and the flood consequences as functions of warning time, the expected total loss can be calculated with Equation (13.18) for continuous time series and Equation (13.22) for discrete time series. The optimal time to evacuate the PAR is obtained by obtaining the minimum expected total loss, as shown in Figure 13.9. The decision can be updated when additional information is available to reduce the uncertainties in the model parameters.

Frieser (2004) and Smith et al. (2006) published decision-making methods for floods with multi-phase decision trees, which are extended from the two-phase decision tree. The time of possible flood occurrence is set as a target time (e.g. with enough lead time to evacuate the people) or the time for achieving the worst predicted situation (e.g. highest water level or largest flood flow rate). The minimum expected total loss in each phase is obtained by comparing those in all alternatives. The optimal time to evacuate the PAR is the time achieving the minimum total expected loss.

Table 13.15 compares these two methods and DYDEM. The dynamic decision framework presented in this chapter has several features:

1. The framework takes the dam failure probability as a time series and the flood consequences as functions of warning time. The time effects on both dam break probability and consequence are sufficiently considered.

Table 13.15 Comparison of decision-making methods for dam break floods

Model	Frieser (2004)	Smith et al. (2006)	DYDEM (2012)
Decision method	Multi-phased decision tree	Multi-phased decision tree	Combine time series and Bayesian networks
Probability prediction	Statistic regression	Monte Carlo simulation with assumed distributions	Time series
Evacuation simulation	Based on statistical data	Assumed as linear function of warning time	Monte Carlo simulation with parameters from statistical data
Evaluation of evacuation cost	Based on statistic data	A constant value	HURAM based on Bayesian networks
Evaluation of flood damage	Based on statistic data	A constant value	HURAM based on Bayesian networks
Prediction of loss of life	A constant fatality rate	A constant fatality rate	HURAM based on Bayesian networks
Value of human life	Macro economic approach	A constant value	Macro economics approach

2. Decision trees have a premise of a fixed occurrence time, which may not be reasonable, as the probability of disaster occurrence is a stochastic process. The loss of life and property may be underestimated if the flood occurs before the predicted moment, and vice versa. The dam failure probability is simulated as a time series in DYDEM, in which the dam may break at any future time with a certain probability.
3. A successful evacuation can be attained when the available time (i.e. the sum of warning time and flood rise time) is more than the demanded time (i.e. the sum of warning transmitting time, response time, and evacuation time). The parameter distributions are based on excising models and statistical data. Monte Carlo simulation is used to simulate the evacuation process with these parameter distributions.
4. The human risk or loss of life, which is complex and often assumed as constant values in the existing methods, is simulated with HURAM based on Bayesian networks. Fourteen uncertain parameters and their interrelationships are considered in this model.
5. Flood consequences, including evacuation cost, flood damage and loss of life, are closely related to evacuation rate (the proportion of the people evacuated), sheltering rate (the proportion of the people sheltered), and fatality rate (the proportion of the people who die). The processes of evacuation, sheltering, and loss of life are simulated in HURAM with a Bayesian network.

13.4 Suggested Topics for Further Research

1. Predicting the inflow rate as a time series with information of rainfall or weather forecasting: the inflow rate is closely related to rainfall information. The uncertainty in inflow rate will be reduced if the information on rainfall or weather forecasting can be sufficiently integrated into the time-series analysis method.
2. Developing fast and efficient decision-making methods that cater for rapidly evolving situations in a specific dam case: the framework of dynamic decision-making in this book strongly depends on the hydraulic parameters from dam breaching simulation and flood routing analysis. The framework may not be very efficient as it often takes time to obtain such hydraulic parameters. However, for a specific dam, such information can be obtained before a potential

dam break event. Therefore it is possible to incorporate the information of a specific dam into the decision framework. In this way, we can efficiently obtain the hydraulic parameters of flood routing in any area downstream and hence estimate the populations at risk.

3. Studying characteristics of regional emergency management: dam-breach emergency management for a particular region ought to cater for the regional social-economic conditions (e.g. local administration and regulations), means of emergency management (e.g. mobilization of the armed forces to fight against floods, high sense of organization and discipline), and disaster relief.

14

Case Study: Risk-based Decision-making for the Tangjiashan Landslide Dam Failure

14.1 Timeline for Decision-making for the Tangjiashan Landslide Dam Failure

Affected by the Wenchuan earthquake on 12 May 2008 in Sichuan, China, a slope of Tangjiashan Hill failed, which blocked the Jianjiang River and formed a large landslide dam, as shown in Figures 3.12 and 3.13. The dam was located at 3.5 km upstream of Beichuan Town whose population was 30,000 before the earthquake, and 85 km upstream of Mianyang City with 1,127,000 residents (Figure 14.1). Refer to Section 3.3.4 in Chapter 3 for detailed information of the landslide dam.

Since the formation of the landslide dam, the landslide lake water level rose rapidly, posing an enormous risk to the residents downstream of the lake. A series of problems arose when making decisions on evacuating the people downstream. What was the dam failure probability? When would the dam fail? How large would the peak outflow rate be if the dam failed? How effective was a diversion channel for reducing the risks? How many people should be evacuated? When would the proper time be to evacuate the PAR? A late decision could lead to loss of life and property, whereas a very early evacuation would incur unnecessary expense. There was little information about the landslide dam when it was first identified. How could the best decision be progressively reached as relevant information on the hydrological, geological, and social-economic conditions became available gradually? A framework for decision-making based on dynamic risk analysis is needed to scientifically answer these questions.

A dynamic decision-making framework for dam break emergency management (DYDEM) under uncertainty was presented in Chapter 13. In this chapter, DYDEM is applied to the dynamic decision analysis for the Tangjiashan landslide dam failure. The optimal time to evacuate the PAR for achieving a minimum expected total risk is studied in different stages with increasing available hydrological, geological, and social-economic information.

Dam Failure Mechanisms and Risk Assessment, First Edition. Limin Zhang, Ming Peng, Dongsheng Chang, and Yao Xu.

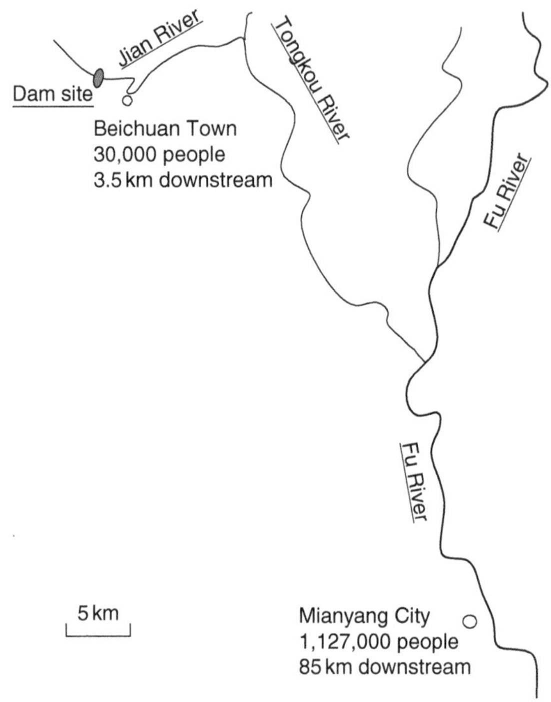

Figure 14.1 Locations of the dam, Beichuan Town and Mianyang City

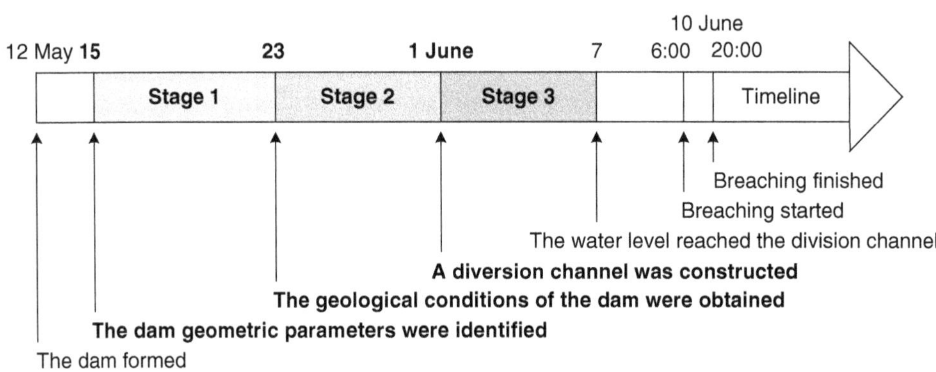

Figure 14.2 The timeline and three stages of decision-making

Figure 14.2 summarizes the main issues along the timeline of the Tangjiashan landslide dam failure event. Three stages can be distinguished for evacuation decision-making in Table 14.1:

1. Stage 1 started on 15 May, when the landslide dam was identified and the geometric parameters of the dam were estimated as shown in Table 3.5. As the soil conditions of the dam remained unknown, an empirical model (Peng and Zhang, 2012c) based on statistical data was used to predict the breaching parameters.

Table 14.1 Available information in the three stages of decision-making for Tangjiashan landslide dam

Stage	Start date	Geometric parameters	Soil conditions	Diversion channel	Dam crest elevation	Model for predicting breaching parameters
1	15 May 2008	Known	Unknown	No	752.2 m	Empirical model
2	23 May 2008	Known	Known	No	752.2 m	Physical model
3	1 June 2008	Known	Known	Constructed	740.4 m	Physical model

2. Stage 2 started on 23 May, when the soil conditions of the dam were investigated as shown in Figure 3.13. Both geometric parameters and soil conditions were available. Thus, a physical model for dam breaching analysis, DABA (Chang and Zhang, 2010), was applied to improve the prediction of the breaching parameters. In stages 1 and 2, the elevation of the dam crest was 752.2 m.
3. Stage 3 started on 1 June, when the diversion channel had been constructed. With the channel, the dam crest was lowered by 12 m (i.e. the elevation of the dam crest became 740.4 m) and the lake capacity was reduced by $69 \times 10^6\,m^3$. Moreover, a large part of the top soil layer with high erodibility was removed. Therefore, the peak outflow rate during breaching was significantly reduced in stage 3.

In the following sections, the analysis of the dam break probability and flood consequences and the dynamic decision-making in each of these three stages will be demonstrated.

14.2 Prediction of Dam Break Probability with Time Series Analysis

As 92% of past landslide dam failures were due to overtopping (Peng and Zhang, 2012c), only overtopping failure is considered in this study. A dam is overtopped at time t when the reservoir volume (V_t) exceeds its capacity (V_{cr}), or $V_t > V_{cr}$.

Based on conservation of mass as introduced in Chapter 13, the reservoir volume at time t, V_t, is given by:

$$V_t = V_{t-\Delta t} + \left(Q_t - Q_{ot} - Q_{et}\right)\Delta t \tag{14.1}$$

where Δt is the time interval, Q_t is the inflow rate at time t, Q_{et} is the evaporation rate which could be ignored for a short time during the emergency management, and Q_{ot} is the outflow rate at time t. In the Tangjiashan landslide dam case, the seepage flow rate was very small ($<1\,m^3/s$), which can be ignored ($Q_{ot}=0$). Therefore, the main task of predicting the dam break probability is to forecast the inflow rate at time t, Q_t.

14.2.1 Forecasting Inflow Rates

According to the records at a hydrological station 4 km upstream of the Tangjiashan landslide dam and the records at the dam site, the average daily inflow rates from 12 May to 13 June 2008 were estimated as shown in Table 14.2 (Zhang, 2009a). Only limited data is available to build a time

Table 14.2 Recorded time series of inflow rate into the Tangjiashan landslide lake (data from Zhang, 2009a)

Date	12 May	13	14	15	16	17	18	19	20
Inflow rate (m^3/s)	57.5	62.4	118.0	104.0	94.6	87.9	88.9	101.0	174.0
Date	21	22	23	24	25	26	27	28	29
Inflow rate (m^3/s)	128.0	105.0	98.3	93.2	91.1	98.9	90.0	88.0	121.0
Date	30	31	1 June	2	3	4	5	6	7
Inflow rate (m^3/s)	96.4	89.8	89.5	79.1	78.5	76.1	70.4	75.3	111.0
Date	8	9	10	11	12	13			
Inflow rate (m^3/s)	103.0	113.0	93.3	79.0	78.4	73.4			

Note: The average inflow rate is 94.18 m^3/s and the standard deviation is 21.64 m^3/s.

Table 14.3 Autocorrelation function and partial autocorrelation function for time series Q_t

k	1	2	3	4	5	6
ρ_k	0.42	0.01	−0.09	−0.12	0.00	0.13
ϕ_{kk}	0.42	−0.20	−0.02	−0.09	0.10	0.09

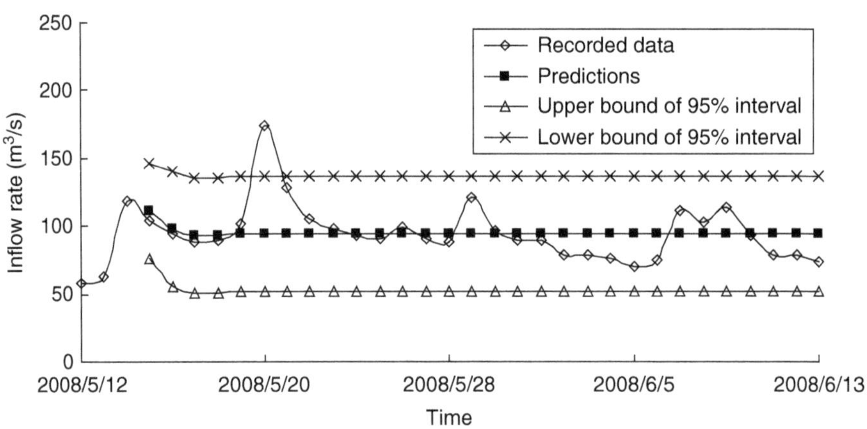

Figure 14.3 Predicted inflow rate in stage 1 based on the records before 15 May using the time series method

series model for inflow forecasting. Based on the method introduced in Chapter 13, autocorrelation function ρ_k and partial autocorrelation function ϕ_{kk} of time series x_t are shown in Table 14.3. Since ρ_k does not fall off slowly with the increase of k and the dataset in Figure 14.3 shows a stationary trend, a stationary model is suggested (Box et al., 2008). As the current daily inflow into the reservoir is closely related to the inflow values in the previous days according to the river confluence theory (Nash, 1957), an autoregressive model AR(p) or ARMA(p,q) is suggested, in which the current daily inflow is expressed as the weighted sum of the previous daily inflow values. Since the partial autocorrelation function, ϕ_{kk}, has a cut-off after lag 2 as shown in Table 14.3, the time series is identified to follow an autoregressive model AR(2) according to the rules in Chapter 13 (Peng and Zhang, 2013a,b):

$$x_t = \phi_1 x_{t-1} + \phi_2 x_{t-2} + a_t \tag{14.2}$$

where ϕ_1 and ϕ_2 are coefficients; the random shock a_t is an independent and identical time series with a mean of zero, which is often set as a normal distribution $N(0,\sigma^2)$. The assumption of AR(2) will be tested later.

The error of time series AR(2) is given by

$$a_t = x_t - \phi_1 x_{t-1} - \phi_2 x_{t-2} \tag{14.3}$$

The least squares method is used to find parameters ϕ_i and θ_i for achieving the least sum of the squares of a_t:

$$Min\left[\sum_{t=1}^{n} a_t^2(\phi_i,\theta_i)\right],\ i = 1, 2, \ldots, n \tag{14.4}$$

By using the solver in Microsoft Excel with the data in Table 14.2, we find ϕ_1=0.463 and ϕ_2=−0.181. Therefore, time series x_t is expressed as

$$x_t = 0.463x_{t-1} - 0.181x_{t-2} + a_t \tag{14.5}$$

According to Box et al. (2008), the assumption of AR(2) for inflow rate x_t is tested by checking whether $33\sum_{k=1}^{7}\left[\rho_k^*(a)\right]^2$ follows a $\chi^2(5)$ distribution. Here $\rho^*_k(a)$ is the estimated autocorrelation function of a_t. The 10% and 5% quantile points for $\chi^2(5)$ are 9.24 and 11.1, respectively. As 4.46 in Table 14.4 is smaller than either 9.24 or 11.1, the model in Equation (14.5) cannot be refused.

As the expected value of a_t is zero, the inflow rate with a lead time l at time t can be calculated by:

$$\begin{aligned} x^*_t(1) &= 0.463x_t - 0.181x_{t-1} \\ x_t^*(2) &= 0.463x_t^*(1) - 0.181x_{t-1} \\ x_t^*(l) &= 0.463x_t^*(l-1) - 0.181x_t^*(l-2), \quad l = 3, 4, \ldots \end{aligned} \tag{14.6}$$

where x_t is the recorded value, $x_t^*(l)$ is the predicted mean value of the inflow rate with a lead time of l, x_{t+l}.

In order to obtain the standard deviation of the predicted value $x_t^*(l)$, we express Equation (14.2) in a random shock form of an infinite series:

$$\begin{aligned} x_t &= a_t + \psi_1 a_{t-1} + \psi_2 a_{t-2} + \psi_3 a_{t-3} \cdots \\ &= a_t + \sum_{j=1}^{\infty} \psi_j a_j \end{aligned} \tag{14.7}$$

where the coefficients can be obtained by substituting Equation (14.7) into Equation (14.2) and comparing the coefficients of a_t (Box et al. 2008). As a_t is an identically distributed stochastic process of $N(0,\sigma_a^2)$, the standard deviation of x_t is calculated as (Box et al. 2008)

Table 14.4 Autocorrelation function for A_t in time series Q_t

k	1	2	3	4	5	6	7	$31\sum_{k=1}^{7}\rho_k^2(a)$
ρ_k	0.039	0.095	0.014	−0.121	−0.097	0.274	−0.040	4.462

Table 14.5 The ψ weights for time series Q_t in stage 1

l	1	2	3	4	5	6	7	8	9	...*
ψ_l	0.463	0.034	−0.068	−0.038	−0.005	0.004	0.003	0.001	−2.7E−04	...*

Note: *The ψ weights are neglected as they are too small.

$$\sigma^2\left[x_t(l)\right]=\left(1+\psi_1^2+\psi_2^2+\ldots+\psi_{l-1}^2\right)\sigma_a^2 \tag{14.8}$$

where the ψ weights (ψ_l) are shown in Table 14.5 according to the methods in Chapter 13, and σ_a is estimated as

$$\sigma_a^2=\frac{1}{33-1}\sum_1^{33}a_t^2=312.5 \tag{14.9}$$

in which a_t is calculated using Equation (14.3).

Taking stage 1 for example, Figure 14.3 shows the predicted inflow rates into the Tangjiashan landslide lake in stage 1 based on the records before 15 May. Generally, the records are within the 95% confidence interval in the stage. The first several predicted values are relatively close to the records. The predicted values with lead times larger than five days ($l>5$) gradually approach a constant value of 94.2 m^3/s and the corresponding standard deviations approach a constant value of 21.6 m^3/s, which are the estimated average value and standard deviation based on the statistical data, respectively, as shown in Table 14.2. One reason for this is that the dependence between two variables of the time series at a long time interval is very small. In other words, the variables at long time intervals appear to be independent. The predictions in stage 2 and stage 3 are updated by using the records during 12–22 May and during 12–31 May, respectively.

14.2.2 Forecasting Lake Volume

As the Δt in Equation (14.1) is one day, the reservoir volume at day t (V_t) is given by

$$V_t-V_{t-1}=3600\times24\times Q_t=3600\times24\times\left(x_t+\bar{Q}\right) \tag{14.10}$$

where $\bar{Q}=94.18$ m^3/s. Set $v_t=V_t/(3600\times24)$, then

$$x_t=v_t-v_{t-1}-\bar{Q} \tag{14.11}$$

Substituting the expressions for x_t, x_{t-1} and x_{t-2} into Equation (14.2) yields

$$v_t = 1.463v_{t-1} - 0.644v_{t-2} + 0.181v_{t-3} + 0.717\bar{Q} + a_t \tag{14.12}$$

Therefore, the forecasting equations for v_t are given by

$$\begin{aligned}
v^*_t(1) &= 1.463v_t - 0.644v_{t-1} + 0.181v_{t-2} + 67.53 \\
v_t^*(2) &= 1.463v_t^*(1) - 0.644v_t + 0.181v_{t-1} + 67.53 \\
v_t^*(3) &= 1.463v_t^*(2) - 0.644v_t^*(1) + 0.181v_t + 67.53 \\
v_t^*(l) &= 1.463v_t^*(l-1) - 0.644v_t^*(l-2) + 0.181v_t^*(l-3) + 67.53, l = 4,5,\ldots
\end{aligned} \tag{14.13}$$

Similar to x_t in the appendix, v_t can also be expressed in a random shock form of an infinite series:

$$\begin{aligned}
v_t &= a_t + \psi_1 a_{t-1} + \psi_2 a_{t-2} + \psi_3 a_{t-3} \cdots \\
&= a_t + \sum_{j=1}^{\infty} \psi_j a_j
\end{aligned} \tag{14.14}$$

where the coefficients of ψ_js can be obtained by substituting Equation (14.14) into (14.12) and comparing the coefficients of a_t on both sides (Box et al. 2008). Table 14.6 shows the ψ weights (ψ_l) for v_t. As a_t is an independently and identically distributed stochastic process with a normal distribution of $N(0, \sigma_a^2)$, the standard deviation of x_t is calculated as (Box et al. 2008)

$$\sigma^2\left[v_t(l)\right] = \left(1 + \psi_1^2 + \psi_2^2 + \ldots + \psi_{l-1}^2\right)\sigma_a^2 \tag{14.15}$$

Figure 14.4 shows the records, predictions, and the upper and lower 95% confidence intervals of the lake volume. The predictions are generally within the 95% confidence interval, closer to the records at small lead times. The variance of the predicted lake volume increases with time, as the prediction for the future is less supported by the existing data.

14.2.3 Prediction of Dam Failure Probability

Since v_t can be expressed as a weighted average of several normal variates as shown in Equation (14.14), v_t is also a normal distribution. With the predicted mean and variance of the lake volume, the probability of dam failure before time *t*, $P_O(t)$, is calculated as

$$P_O(t) = P(V_t > V_{cr}) = 1 - P(V_t \le V_{cr}) = 1 - \Phi\left(\frac{V_{cr} - M_{Vt}}{\sigma_{Vt}}\right) \tag{14.16}$$

Table 14.6 The ψ weights for time series V_t in stage 1

l	1	2	3	4	5	6	7	8	9	...*
ψ_l	1.463	1.497	1.429	1.392	1.386	1.391	1.394	1.394	1.394	...*

Note: *The ψ weights are close to a constant value of 1.394 when $l > 9$.

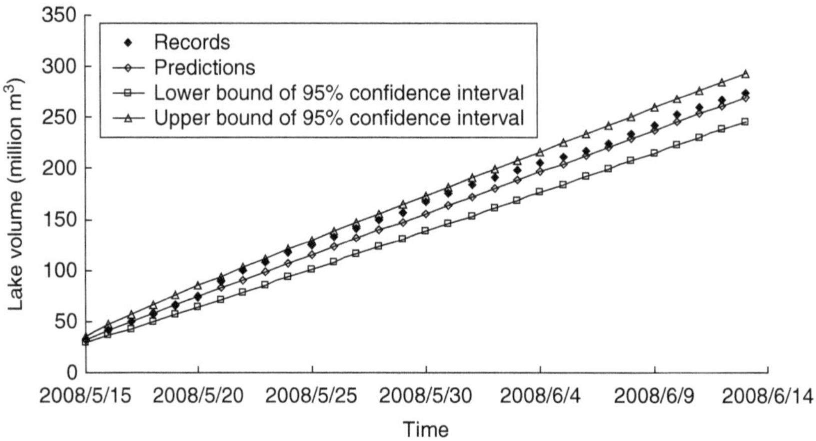

Figure 14.4 Predicted lake volume in stage 1 using the time series analysis method

where V_t is calculated using equation (14.13), M_{vt} and σ_{vt} are the mean and standard deviation of V_t, respectively. The probability of overtopping in the period between $t - \Delta t$ and t, $P_D(t)$, is calculated as the probability that the predicted V_t is larger than V_{cr} and the predicted lake volumes before time t are smaller than or equal to V_{cr}:

$$P_D(t) = P(V_t > V_{cr}, V_{t-1} \leq V_{cr}, \ldots, V_1 \leq V_{cr}) \tag{14.17}$$

As the inflow rate is greater than the outflow rate, the lake volume always increases. According to the analysis in Chapter 13, P_D is expressed as:

$$\begin{aligned} P_D(t) &= P(V_t > V_{cr}, V_{t-1} \leq V_{cr}, \ldots, V_1 \leq V_{cr}) \\ &= P_O(t) - P_O(t-1) \end{aligned} \tag{14.18}$$

For example, V_{cr} in stage 1 is $316 \times 10^6 \text{m}^3$; the predicted probabilities of P_O and P_D in stage 1 as two discrete time series are shown in Figure 14.5.

The time interval of the predicted hydraulic parameters is in days, because the inflow rates in Table 14.2 were recorded in days. However, more precise warning time is needed in DYDEM because the flood consequences are sensitive to the warning time. Therefore, in this study, interpolations are conducted for the predicted means and variances of the lake volume with a time interval of one hour. This is reasonable as the predicted mean values and variances of the lake volume increase nearly linearly at large lead times ($l \geq 5$).

Figure 14.6 shows the discrete probability of dam failure in an interval of one hour in all three stages. Note V_{cr} is $316 \times 10^6 \text{m}^3$ in stages 1 and 2, and $224 \times 10^6 \text{m}^3$ in stage 3 because the diversion channel was completed in stage 3. The dam break probabilities in discrete time series in the three stages will be used to conduct dynamic decision analysis. In stage 3, the dam may fail much earlier than in stages 1 and 2 because of the smaller lake capacity. Moreover, the estimated times of dam break in stage 3 become less scattered than in stages 1 and 2, since more information is available in stage 3 such as the geometric conditions, a constructed diversion channel, the decreased crest elevation, and the removal of a soil layer with high erodibility.

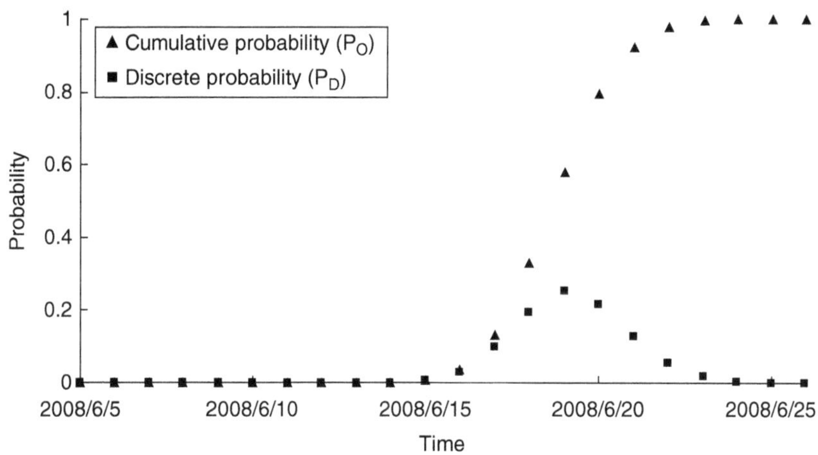

Figure 14.5 Dam break probability as a discrete time series at an interval of one day in stage 1

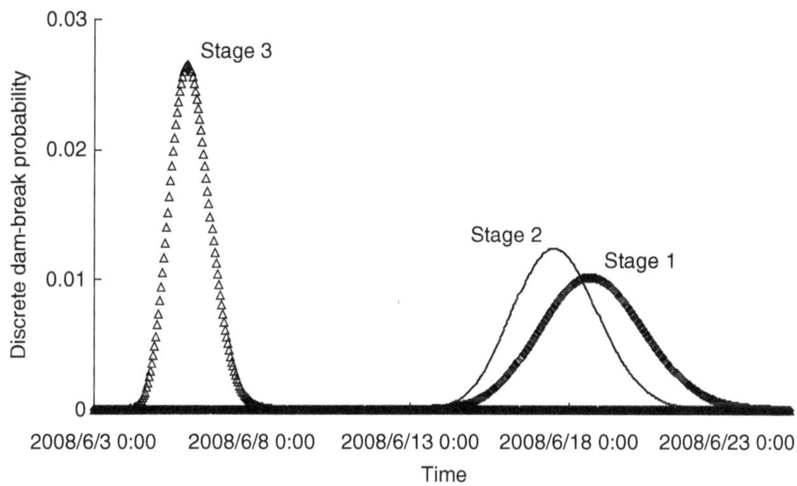

Figure 14.6 Discrete probability of dam failure in three stages with a time interval of one hour

14.3 Simulation of Dam Breaching and Flood Routing

The estimation of dam break probability answers the question of whether and when the dam would fail. In this section the questions of how it would fail and who would be affected by the dam break flood will be answered. The dam breaching process significantly influences the flood consequences downstream. The peak outflow rate affects the flood area and the corresponding PAR. The breach formation time, which is the duration from the lake water level reaching the dam crest to the start of breaching (i.e. the breach reaches the upstream bound of the dam crest), is part of the available warning time. The breach development time or breaching time, which is the duration from the start to the completion of the breaching process significantly influences the flood rise time.

14.3.1 Simulation of Dam Breaching and Flood Routing in Stage 1

In stage 1, when only the geometric parameters of the Tangjiashan landslide dam are available, an empirical model for landslide dams (Peng and Zhang, 2012c) is applied to predict the breaching parameters. The empirical model was developed based on 52 landslide dam failure cases from all over the world. These cases are divided into three groups with different dam erodibility (high, medium, or low) according to the dam type, rock and soil properties, and factors such as triggers (Knapen et al., 2007, Briaud et al., 2008). The equations with medium erodibility are used to simulate the Tangjiashan landslide dam since the dam is assumed to be of medium erodibility based on Peng and Zhang (2012c). The inputs of the model include dam height (82 m), dam width (802 m), dam volume ($20.4 \times 10^6 m^3$), lake volume ($316 \times 10^6 m^3$), and dam erodibility (medium). As no information about the dam erodibility is available at that time, the uncertainty of this parameter is estimated according to statistics. Out of the 52 cases with breaching-parameter records, 6 cases are of low erodibility, 31 cases of medium erodibility, and 15 cases of high erodibility. Thus, the probabilities of low, medium, and high erodibility are 0.116, 0.596, and 0.288, respectively (Peng and Zhang, 2012c). The outflow rates with high, medium, and low erodibility are shown in Figure 14.7. Table 14.7 shows the predicted peak outflow rates and breaching times. The peak outflow rate in the high erodibility case is 5 times that in the medium erodibility case and 16 times that in the low erodibility case. Note that the empirical model differs from the one for constructed embankment dams by Xu and Zhang (2009).

After obtaining the breaching parameters, a river analysis program, HEC-RAS 4.0 (Hydrologic Engineering Center, 2008), is used to simulate the flood routing process. This is a one-dimensional hydraulic analysis program developed by US Army Corps of Engineers. The main physical laws for the program are the conservation of energy for steady flows and the conservation of mass and momentum for unsteady flows (Hydrologic Engineering Center, 2008). Details of the simulations are reported by Peng (2012). The highest water depth, maximum flow velocity, and PAR in Beichuan Town and Mianyang City are shown in Table 14.7. The flood risks vary significantly from low to high erodibility. Mianyang City would not be flooded in the low and medium erodibility cases. However, the flood in the high erodibility case would put 210,000 people at risk. The highest water depth in Mianyang City would be 3.68 m.

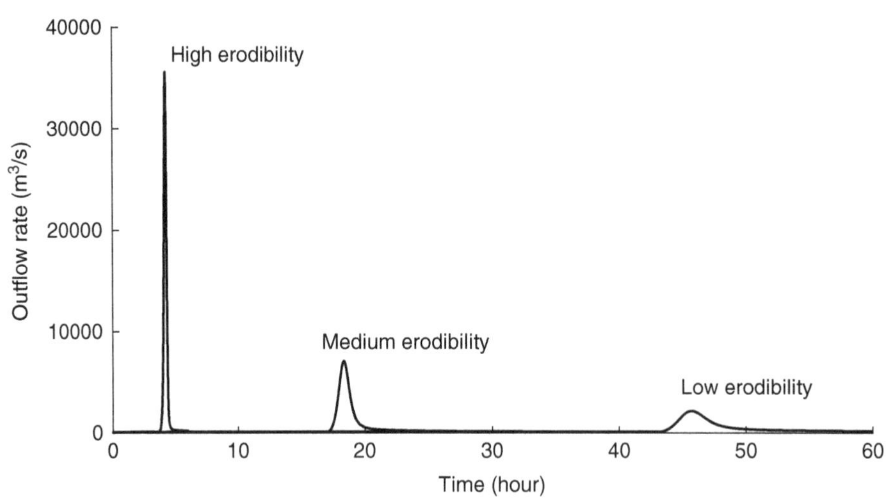

Figure 14.7 Outflow rates in the three scenarios of stage 1

Table 14.7 Dam break floods in Beichuan Town and Mianyang City in stage 1

Dam erodibility		Low	Medium	High
Probability		0.116	0.596	0.288
Peak discharge at dam site (m^3/s)		2160	7100	35,610
Breaching time (hour)		17.6	8.1	3.7
Beichuan Town	Peak discharge (m^3/s)	2160	7100	35,370
	Population at risk	780	20,100	30,000
	Highest water depth (m)	0.26	6.67	21.37
	Maximum flow velocity (m/s)	0.18	1.14	2.01
Mianyang City	Peak discharge (m^3/s)	2160*	6990*	28,720
	Population at risk	0	0	207,370
	Maximum water depth (m)	0	0	3.68
	Maximum flow velocity (m/s)	0	0	1.37

Note: *The design flood for the dike system in Mianyang City is 14,500 m^3/s with a 100-year return period.

14.3.2 Simulation of Dam Breaching and Flood Routing in Stage 2

In stage 2, after the geological conditions of the dam have been investigated, a physical model, DABA (Chang and Zhang, 2010), is used to improve the dam breaching simulation. The water–soil interactions in this model are simulated with theory of erosion and shallow water flow. The soil erodibility is described by the coefficient of erodibility (i.e. how fast the soil erodes) and the critical erosive shear stress (i.e. when the erosion starts and ends), which are obtained based on empirical relations with basic soil parameters. Chang et al. (2011) derived two empirical equations for estimating the coefficient of erodibility of landslide dams based on results of field erodibility tests. Chang and Zhang (2010) measured the basic properties of soil samples taken from the Tangjiashan landslide dam at different depths (10 m, 20 m, 21 m, and 50 m). The coefficients of erodibility of the soils at these depths are estimated as 120, 65, 62, and 10 $mm^3/N\text{-}s$, respectively using the two empirical equations.

The inputs for the model include both the geometric information and soil parameters of the dam. The outputs are the breaching parameters (e.g. instantaneous breach size, breaching time, water level, and outflow rate) during the breaching process. The peak outflow rate and the breaching time in stage 2 are predicted as 14,700 m^3/s and 14.5 hours, respectively. Peng and Zhang (2012c) showed that the predicted peak outflow rate (Q_p) and breaching time (T_b) can be simulated as log-normal distributions. The standard deviation of $ln(Q_p)$ and $ln(T_b)$ are 0.53 and 0.40, respectively. Therefore, in stage 2, $ln(Q_p)$ and $ln(T_b)$ are assumed to be normal distributions of $N[\ln(14700), 0.53)]$ and $N[\ln(14.5), 0.40]$. It is very tedious to consider $\ln(Q_p)$ and $\ln(T_b)$ at infinite continuous states. In this study, $\ln(Q_p)$ and $\ln(T_b)$ are discretized into five ranges, corresponding to five scenarios in Table 14.8. For example, scenario 1 is from $-\infty$ to $(\mu-1.5\sigma)$, where $\mu=\ln(14700)$ and $\sigma=0.53$. The probability of this range is 0.067 for a normal variate. The median $[P(Q_p<\text{median})=0.5]$ of this range is $e^{(\mu\ 1.83\sigma)}=5570\ m^3/s$, which is used as a representative value for this range. Similarly, the corresponding values of the other ranges are shown in Table 14.8. Figure 14.8 shows the outflow rate curves for the five scenarios.

HEC-RAS 4.0 is used to simulate the flood routing process in stage 2 and the results are shown in Table 14.8. Without the construction of the diversion channel in stage 2, Beichuan would be seriously flooded in all five scenarios. The predicted flood of 14,700 m^3/s in scenario 3 would

Table 14.8 Dam break floods in Beichuan Town and Mianyang City in stage 2

Scenario		1	2	3	4	5
Range		$-\infty$ – ($\mu-1.5\sigma$)	($\mu-1.5\sigma$) – ($\mu-0.5\sigma$)	($\mu-0.5\sigma$) – ($\mu+0.5\sigma$)	($\mu+0.5\sigma$) – ($\mu+1.5\sigma$)	($\mu+1.5\sigma$) – ∞
Probability		0.067	0.242	0.382	0.242	0.067
Median		$\mu-1.83\sigma$	$\mu-0.89\sigma$	μ	$\mu+0.89\sigma$	$\mu+1.83\sigma$
Peak discharge at dam site (m^3/s)		5570	9120	14,700	25,180	39,090
Breaching time (hour)		30.0	20.6	14.5	10.2	7.0
Beichuan Town	Peak discharge (m^3/s)	5550	9110	14,680	25,080	38,800
	Population at risk	18,090	25,080	30,000	30,000	30,000
	Maximum water depth (m)	6.03	8.36	12.10	17.30	22.43
	Maximum flow velocity (m/s)	1.03	1.24	1.48	1.80	2.07
Mianyang City	Peak discharge (m^3/s)	5170	8810	13,580	20,440	29,950
	Population at risk	0	0	0	101,990	227,090
	Maximum water depth (m)	0	0	0	1.81	4.03
	Maximum flow velocity (m/s)	0	0	0	0.86	1.46

Note: μ is the predicted mean; σ is the predicted standard deviation.

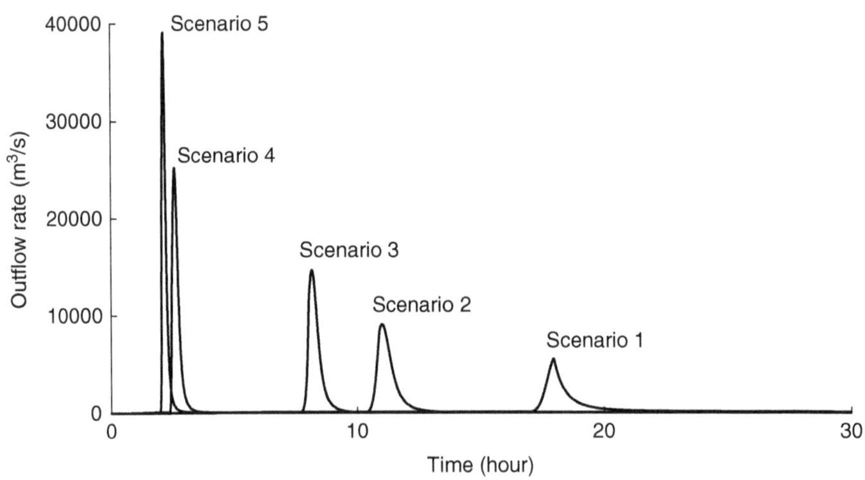

Figure 14.8 Outflow rates in the five scenarios of stage 2

inundate the entire Beichuan Town with a maximum water depth of 12.1 m. Mianyang would not be flooded in scenarios 1–3. However, the floods in scenarios 4 and 5, especially scenario 5, would cause a large number of PAR. The highest water depth in these two scenarios would be 1.8 m and 4.0 m, respectively. The PAR in scenario 5 in stage 2 in Mianyang would be 230,000 million.

14.3.3 Simulation of Dam Breaching and Flood Routing in Stage 3

Stage 3 starts when the diversion channel has been completed. The diversion channel largely reduced the peak outflow rate during the breaching of Tangjiashan Landslide Dam. Similar to stage 2, stage 3 is also divided into five scenarios. Table 14.9 shows the peak outflow rate and breaching time in each scenario. The predicted mean peak outflow rate and mean breaching time are 6540 m^3/s and 16.3 hours, respectively, which are close to the recorded values of 6500 m^3/s and 14 hours (Cui et al., 2009; Liu et al., 2010). Figure 14.9 shows the dam break outflow rate curves for the five scenarios.

The flood parameters and populations at risk in both Beichuan and Mianyang are shown in Table 14.9. The PAR in Beichuan increases significantly from scenario 1 to scenario 5. With the predicted mean peak outflow rate, Beichuan would be flooded with a maximum water depth of 6.1 m, which is in good agreement with the observation that the river water level rose to the third story of many buildings according to the wet marks observed by the authors. Mianyang City would not be flooded in any of the scenarios in stage 3, as the maximum flow rate, 13,690 m^3/s, is smaller than the design flow rate of 14,500 m^3/s for the city (Liu, 2008).

14.4 Evaluation of Flood Consequences

The flood consequences include evacuation costs, flood damage, and loss of life. The flood consequences are closely related to the evacuation rate, sheltering rate and loss of life, which are estimated using a human risk analysis model (HURAM) based on Bayesian networks (Peng and Zhang, 2012a,b). The methodology of evaluating flood consequences, which is coded in Microsoft Excel, was introduced in Chapter 13. A human life is monetized according to macro-economic contributions.

Table 14.9 Dam break floods in Beichuan Town and Mianyang City in stage 3

Scenario		1	2	3	4	5
Range		$-\infty$ – $(\mu-1.5\sigma)$	$(\mu-1.5\sigma)$ – $(\mu-0.5\sigma)$	$(\mu-0.5\sigma)$ – $(\mu+0.5\sigma)$	$(\mu+0.5\sigma)$ – $(\mu+1.5\sigma)$	$(\mu+1.5\sigma)$ – ∞
Probability		0.067	0.242	0.382	0.242	0.067
Median		$\mu-1.83\sigma$	$\mu-0.89\sigma$	μ	$\mu+0.89\sigma$	$\mu+1.83\sigma$
Peak discharge at dam site (m^3/s)		2450	4050	6540	10,420	17,260
Breaching time (hour)		33.7	23.2	16.3	11.5	7.9
Beichuan Town	Peak discharge (m^3/s)	2210	4010	6520	10,390	17,120
	Population at risk	2190	9600	18,300	27,000	30,000
	Maximum water depth (m)	0.35	3.20	6.10	9.31	13.85
	Maximum flow velocity (m/s)	0.22	0.83	1.10	1.29	1.57
Mianyang City	Peak discharge (m^3/s)	2190	3940	6280	9520	13,690
	Population at risk	0	0	0	0	0
	Maximum water depth (m)	0	0	0	0	0
	Maximum flow velocity (m/s)	0	0	0	0	0

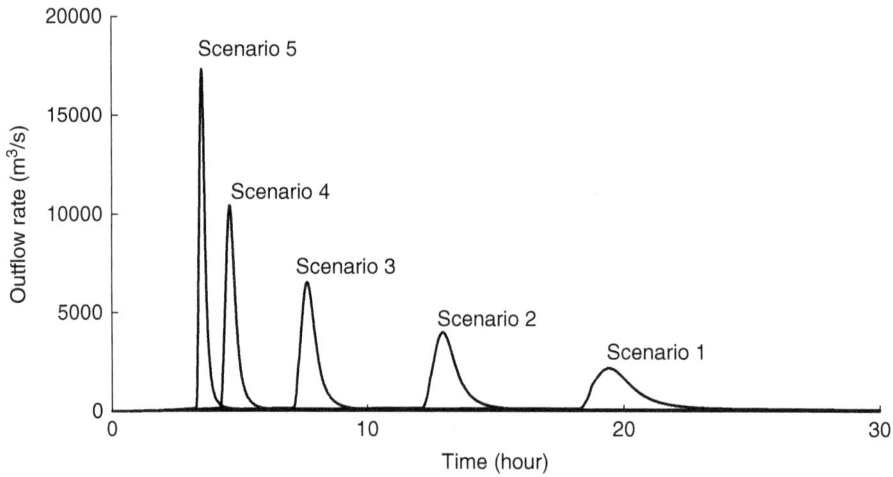

Figure 14.9 Outflow rates in the five scenarios of stage 3

14.4.1 Methodology

The evaluation of flood consequences in Beichuan in stage 1 is taken as an example to illustrate the analysis methodology. In the medium erodibility case with a peak outflow rate of 7100 m^3/s, the inputs for HURAM (Peng and Zhang, 2012a, b) are listed in Figure 14.10 for five sub-areas (S1–S5). In this case, the maximum water depth of 6.13 m occurs in sub-area 5. The probabilities of evacuation, sheltering, and loss of life at a warning time of zero, calculated with HURAM, are shown in Figure 14.10. For example, the evacuation rate is 0.208 in sub-area 1. As the buildings in this sub-area are mostly three-story brick structures, they are just slightly damaged and can serve as shelters. The ratios of the people unsheltered, on the 2nd floor and on the 3rd floor are 19.3%, 11.1%, and 69.6%, respectively. Since the water depth is 0–1.5 m in this sub-area, the people on the 2nd and 3rd floors are in the safe zone. The fatality ratio in this sub-area is 0.02% and the expected loss of life is 1.12 if without any warning.

Given the sheltering rate, evacuation rate and fatality rate, the flood consequences can be estimated based on the methods presented in Chapter 13. The evacuation costs include arrangement fees (C_i) and GDP interruption (C_{GDP}). The flood damage (D) is limited to the moveable properties in this study. The moveable properties are generally proportional to the number of people who have neither evacuated nor sheltered in safe zones. The monetized loss of life is calculated as:

$$V_L = (LOL)(GDP_p)L \tag{14.19}$$

where LOL is the predicted loss of life, GDP_p is the annual gross domestic product per person, and L is the average longevity (75 years). LOL is 1.12 in sub-area 1 in Figure 14.10; hence V_L for sub-area 1 is calculated as $V_L = 1.12 \times 13{,}745 \times 75 =$ RMB 1,154,580.

In additional to the medium erodibility scenario, two other scenarios (low erodibility and high erodibility) are also involved in stage 1. The probabilities of low, medium, and high erodibility are 0.116, 0.596, and 0.288, respectively as shown in Table 14.7. Therefore, the flood consequence (FC) in stage 1 is the weighted sum of those of the three scenarios:

$$FC_1 = 0.116FC_L + 0.596FC_M + 0.288FC_H \tag{14.20}$$

	A	B	C	D	E	F	G
1			**Medium erodibility case in stage 1, Beichuan**				
2							
3	Sub-area		S1	S2	S3	S4	S5
4	Evacuation distance	Lower bound (m)	0.00	0.00	100.00	100.00	100.00
5		Upper bound (m)	100.00	100.00	500.00	500.00	500.00
6	Distance to dam	(m)	3500	3500	3500	3500	3500
7	Building story number	1, 2, 3, 4	3	3	3	3	3
8	Dam breaching time	(Hours)	8.10	8.10	8.10	8.10	8.10
9	Water depth	Lower bound (m)	0.00	1.50	3.00	4.50	6.00
10		Upper bound (m)	1.50	3.00	4.50	6.00	6.67
11	Flow velocity	(m/s)	1.14	1.14	1.14	1.14	1.14
12	Building type	UW = 1, AW = 2, MCB = 3	3	3	3	3	3
13	Population at risk	Number	4500	4500	4500	4500	2010
14	Warning initiation time	Hours	0.00				
15			**Calculation**				
26	Evacuation	Evacuation Rate	0.2077	0.1103	0.0211	0.0078	0.0046
29		Unsheltered	0.1933	0.2981	0.4241	0.4497	0.4562
30	Sheltering inside buildings	2nd floor	0.1105	0.1704	0.2423	0.2570	0.2607
31		3rd floor	0.6962	0.5315	0.3336	0.2934	0.2831
32		>3rd floor	0.0000	0.0000	0.0000	0.0000	0.0000
40	P(LOL)		0.0002	0.0042	0.1267	0.2441	0.5141
41	LOL		1.12	19.08	569.95	1,098.35	1,033.29
42							
43	GDP	Per person	1.37E+04	1.37E+04	1.37E+04	1.37E+04	1.37E+04
44	Value of life	Per person	1.03E+06	1.03E+06	1.03E+06	1.03E+06	1.03E+06
45	L(LOL)	Total LOL	**1.15E+06**	**1.97E+07**	**5.88E+08**	**1.13E+09**	**1.07E+09**
46		GDP Interruption/person	1.51E+02	1.51E+02	1.51E+02	1.51E+02	1.51E+02
47		GDP Interruption	6.79E+05	4.52E+06	4.52E+06	4.52E+06	4.52E+06
48	Evacuation costs	Arrangement fee/person	1.80E+02	1.80E+02	1.80E+02	1.80E+02	1.80E+02
49		Arrangement fee	1.69E+05	8.93E+04	1.71E+04	6.33E+03	1.66E+03
50		Total costs	**8.48E+05**	**4.61E+06**	**4.54E+06**	**4.53E+06**	**4.52E+06**
51		Per person	4.48E+03	4.48E+03	4.48E+03	4.48E+03	4.48E+03
52	Flood damage	Person number	6.89E+02	1.19E+03	2.94E+03	3.16E+03	2.00E+03
53		Total damage	**3.09E+06**	**5.35E+06**	**1.32E+07**	**1.41E+07**	**8.97E+06**
54							
55	Note: UW = Unanchored wood-framed; AW = Anchored wood-framed, MCB = Masonry, concrete and brick						

Figure 14.10 Evaluation of flood consequences in Beichuan Town in the medium erodibility case in stage 1

where FC_L, FC_M, and FC_H are the flood consequences in the low, medium, and high erodibility cases, respectively.

Similarly, the flood consequences in stages 2 and 3 are also the weighted sum of those of the five scenarios and given by

$$FC_i = 0.067FC_{i1} + 0.242FC_{i2} + 0.382FC_{i3} + 0.242FC_{i4} + 0.067FC_{i5}, i = 2,3 \qquad (14.21)$$

in which the five coefficients are the probabilities of the five scenarios in Tables 14.8 and 14.9.

14.4.2 *Calculated Dam Break Flood Consequences*

With the methods introduced above, the flood consequences as functions of warning time in Beichuan Town and Mianyang City are presented in Figures 14.11 and 14.12, respectively. In Beichuan Town (Figure 14.11), the flood severities in the three stages are very high. The monetized loss of life dominates when the warning time is short. This is reasonable as human life is the first concern upon a serious dam failure. However, if the warning time is relatively long, the evacuation

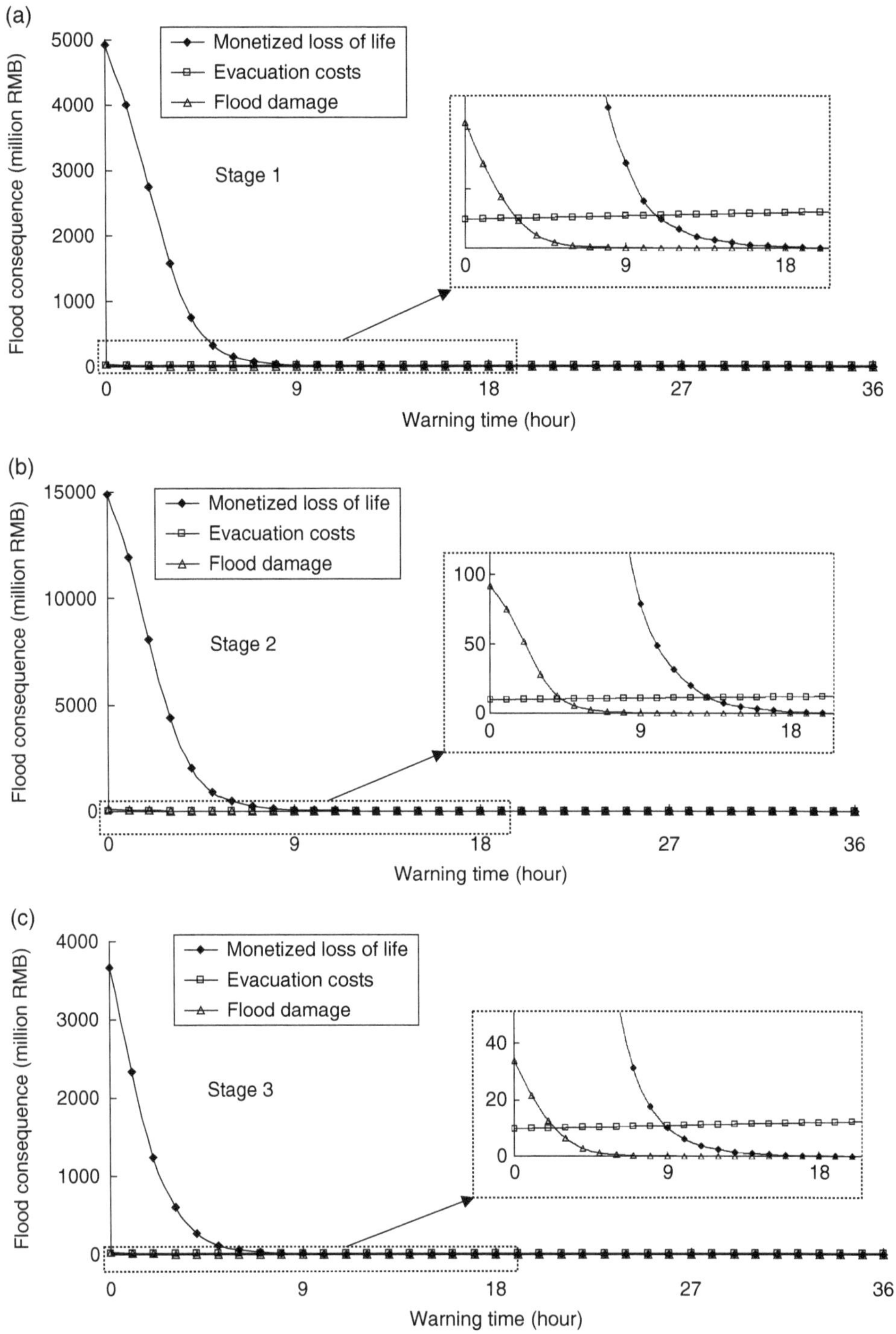

Figure 14.11 Flood consequences in Beichuan Town: (a) stage 1, (b) stage 2, (c) stage 3

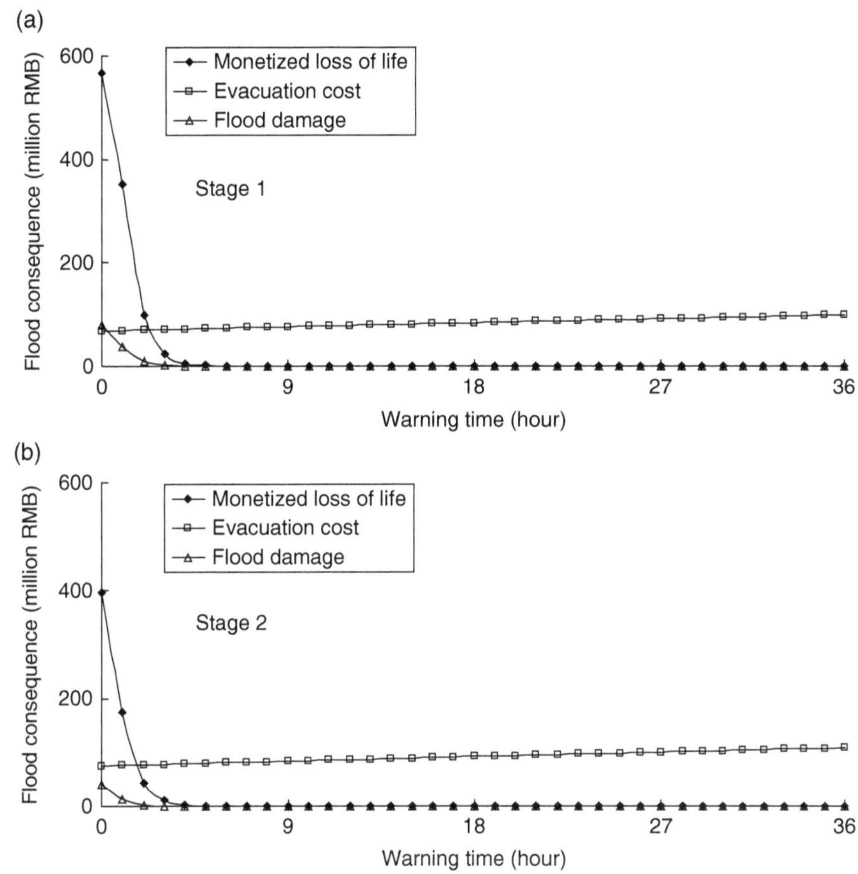

Figure 14.12 Flood consequences in Mianyang City: (a) stage 1, (b) stage 2. No flooding is predicted in stage 3

costs would be the largest expenditure. Compared to stage 1, the estimated flood consequences in stage 2 are much larger. In stage 2, the erodibility of the top soil layer of the dam is very high (120 mm^3/N-s). The rapid erosion of this soil layer would cause a larger peak outflow rate and a shorter breaching time. With the diversion channel built in stage 3, the flood consequences would be smaller. The monetized potential loss of life with zero warning time is RMB 3674 million (3564 fatalities) in stage 3, compared to RMB 4928 million (4780 fatalities) in stage 1 and RMB 14,904 million (14,458 fatalities) in stage 2. The larger number of fatalities in stage 2 is due to two reasons: the higher flood severity (water depth and flow velocity) and the shorter flood rise time. The flood rise time is closely related to the breaching time and floodwater depth as shown in Tables 14.7–14.9.

In Mianyang City (Figure 14.12), no flooding is predicted in stage 3 and no evacuation is needed. In stages 1 and 2, the flood damage and evacuation costs are much larger, but the monetized loss of life is much smaller than those of Beichuan Town with the same warning time. The reason is that the number of PAR in Mianyang City is much larger but the flood severity is much lower than those in Beichuan Town. The larger number of PAR would cause more evacuation costs and flood damage. However, because the values of the maximum water depth and flow velocity in Mianyang City are much smaller than those in Beichuan, as shown in Tables 7 and 8, the fatality rates in

Mianyang City are much lower. In fact, the maximum water depth is much lower than the brick structures with three and more stories, so the people in Mianyang can take refuge in the buildings, which is simulated as sheltering in HURAM.

14.5 Dynamic Decision-making

14.5.1 Methodology

In this example, the decision for the Tangjiashan landslide dam break is divided into three stages with different levels of information. Figure 14.6 shows the probabilities of dam failure in the three stages. There is still quite some time from the moment of V_t reaching V_{cr} to the start of breaching, which is called the breach initiation time. As little data is available to predict the breaching initiation time and the corresponding uncertainties, a constant value of 72 hours is selected, based on the records of the breaching process (Chang and Zhang, 2010; Peng and Zhang, 2012a,b). This means that the dam starts to breach three days after the water level reaches the dam crest.

In order to demonstrate the method of decision-making, the flood in Beichuan Town in stage 1 is taken as an example. Figure 14.13 shows the probability of dam break in Beichuan as a discrete time series with an interval of one hour. The studied time range is set between t_0 of 0:00 on 14 May and t_{end} or t_{360} of 0:00 on 29 May. The predicted dam failure probability outside of this range is very small (3.2×10^{-5}). Theoretically, a dam would fail at any time in this range. Let us assume that the dam break flood arrives at t_f and the warning is issued at t_w. The flood consequences with any t_w in this range can be obtained with the method presented in Chapter 13. For example, assume t_w is at 0:00 on 17 May, then $t_w=t_{72}$. The total flood consequence is given by

$$\begin{aligned} E(L_t) &= \sum_{j=0}^{360} L(t_j)P_D(t_j) = \sum_{j=0}^{360}\left[C(W_t)+D(W_t)+L(W_t)\right]P_D(t_j) \\ &= \sum_{j=0}^{360} C(W_t)P_D(t_j) + \sum_{j=0}^{360}\left[D(W_t)+L(W_t)\right]P_D(t_j) \end{aligned} \tag{14.22}$$

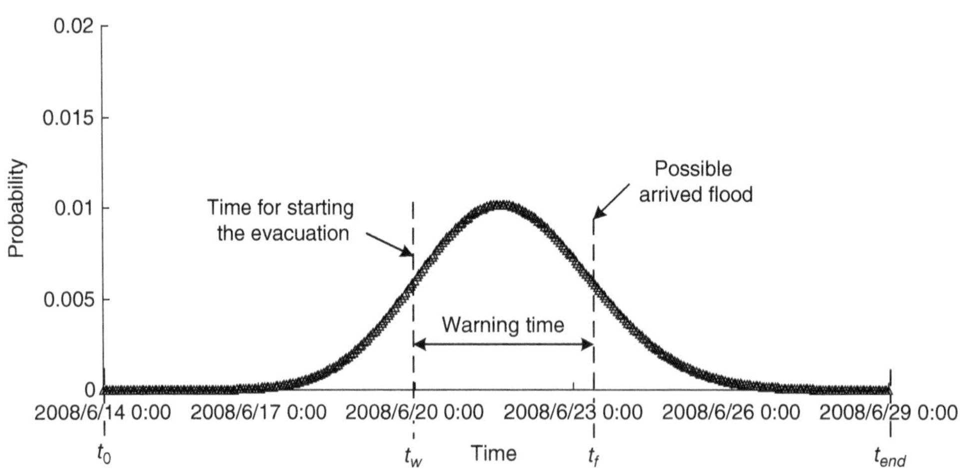

Figure 14.13 The probability of dam break flood in Beichuan as a discrete time series at an interval of one hour in stage 1 (with a breach formation time of three days)

where $C(W_t)$, $D(W_t)$, and $L(W_t)$ are the evacuation costs, flood damage, and monetized loss of life as functions of warning time (W_t), which can be found in Figure 14.11; $P_D(t_j)$ is the flood probability in the duration of t_{j-1} to t_j, which is expressed in Figure 14.13. Therefore, the optimal time to evacuate the PAR is the time to achieve the minimum total loss [Min($E(L_t)$)] in Equation (14.22).

14.5.2 *Dynamic Decision-making in Three Stages*

Figures 14.11 and 14.12 show the expected flood consequences as functions of the time to evacuate the PAR in Beichuan Town and Mianyang City, respectively. The evacuation costs decrease but the flood damage and monetized loss of life increase with delayed time for evacuation. The expected total consequence decreases at first but then increases. Therefore, an optimum decision time can be obtained corresponding to the minimum expected total loss (MTC).

In Beichuan Town, the optimal decision and flood consequences are shown in Table 14.10. The optimum evacuation time is at 08:00 on 16 June in stage 1, when the MTC is RMB 41.93 million. The optimum time is brought forward 14 hours in stage 2 with an MTC of RMB 41.60 million. In stage 3, as the diversion channel has been constructed, the dam crest elevation is reduced by 12 m. The optimum time to evacuate the PAR is at 0:00 on 7 June, when the MTC is RMB 36.20 million. The 8–9 days' difference between stage 3 and stages 1 and 2 is due to the reduced reservoir capacity ($69 \times 10^6 m^3$) in stage 3. It takes 8–9 days to fill this part of the lake. As shown in Figure 14.14, the expected total loss increases rapidly about one half day after the optimum time because of the increased monetized loss of life with later warning. Generally, an evacuation is not avoidable in all three stages for Beichuan. The optimal times for evacuating the PAR, as shown in Figure 14.14, may be brought forward by a certain period for convenience (e.g. at daytime) since the sensitivity is not high.

In Mianyang City, the optimal decision and flood consequences are shown in Table 14.11. Since there is no predicted flooding in stage 3, which is a fact, no evacuation is needed. In stage 1 as shown in Figure 14.15, the optimal time for evacuating the PAR is set at 10:00 on 19 June with an expected minimum total consequence of RMB 370 million. As in Beichuan, the optimal time is brought forward 16 hours in stage 2 with an MTC of RMB 381 million. The MTC is more sensitive to the time for evacuating the PAR than that in Beichuan. The optimum times for evacuating the PAR in Mianyang are 2–3 days later than those in Beichuan. There are three reasons for this: (1) the probabilities of the floods in both stages 1 and 2 in Beichuan are very large, but the probabilities of the floods in Mianyang are much smaller (i.e. 0.12 in stage 1 and 0.31 in stage 2), (2) the higher flood severity in Beichuan needs more warning time for

Table 14.10 Optimal decision and flood consequences in Beichuan Town

Stage	1	2	3
Maximum PAR	30,000	30,000	30,000
Optimal time*	08:00 on 16 June	18:00 on 15 June	00:00 on 7 June
Evacuation cost (RMB)	40.7 million	40.8 million	35.8 million
Flood damage (RMB)	9.64 thousand	5.23 thousand	3.46 thousand
Monetized loss of life (RMB)	1.2 million	0.8 million	0.4 million
Total loss (RMB)	41.9 million	41.6 million	36.2 million

Note: *The crest elevation in stages 1 and 2 was the natural crest elevation before the excavation of the diversion channel.

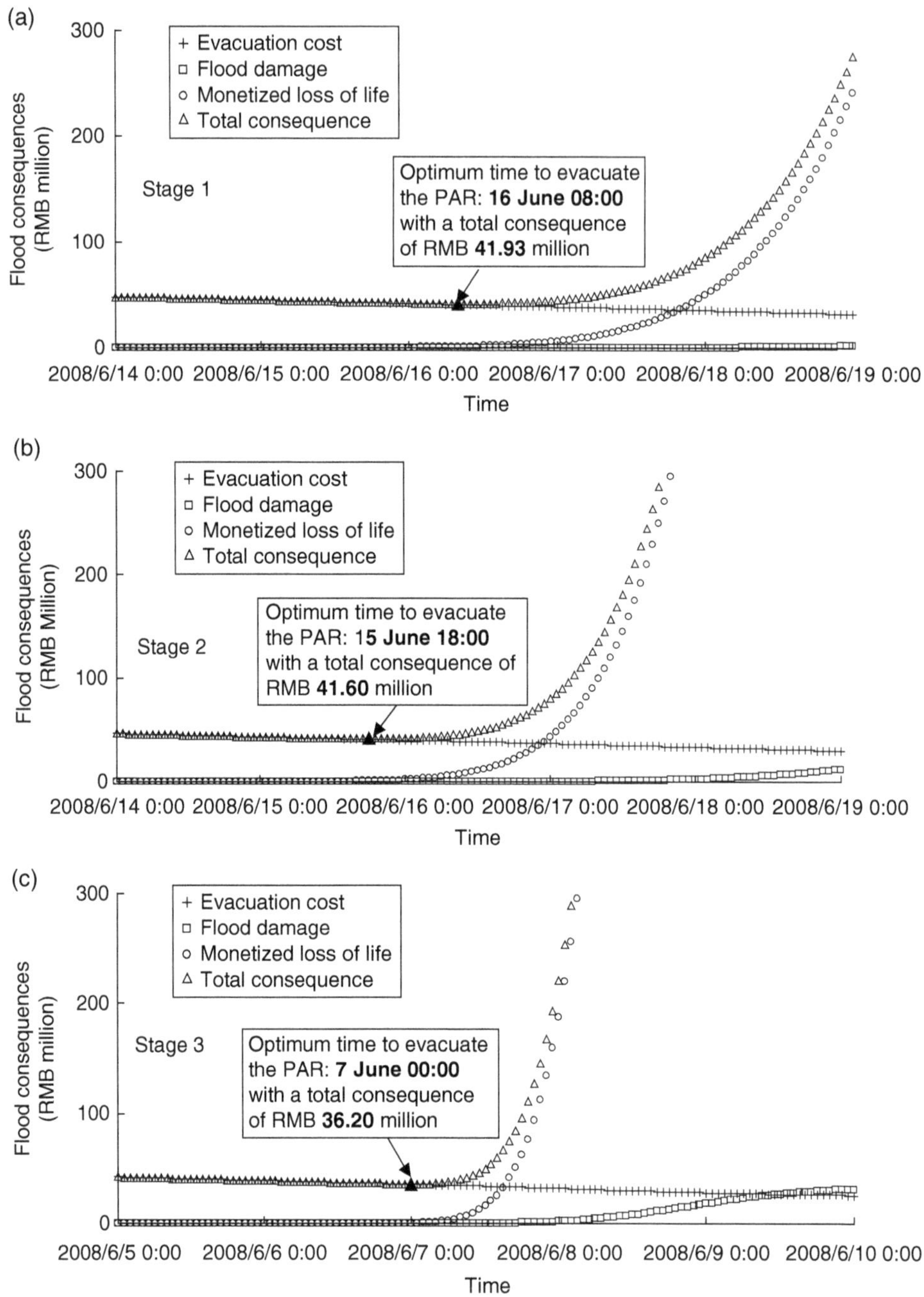

Figure 14.14 Optimal decision-making for Beichuan Town: (a) stage 1, (b) stage 2, (c) stage 3

evacuation, and (3) the flood rise time in Mianyang is much longer and a certain time is needed for the flood to travel 85 km from Beichuan to Mianyang. There is a trend of convergence of the monetized loss of life, flood damage, and total consequences in both stages in Mianyang

Table 14.11 Optimal decision and flood consequences in Mianyang City

Stage	1	2	3
Maximum PAR	207,370	227,090	0
Optimal time*	10:00 on 19 June	18:00 on 18 June	–
Evacuation cost (RMB)	340.0 million	353.2 million	–
Flood damage (RMB)	3.7 million	2.6 million	–
Monetized loss of life (RMB)	26.3 million	25.1 million	–
Total loss (RMB)	370.1 million	380.9 million	–

Note: *The crest elevation in stages 1 and 2 was the natural crest elevation before the excavation of the diversion channel.

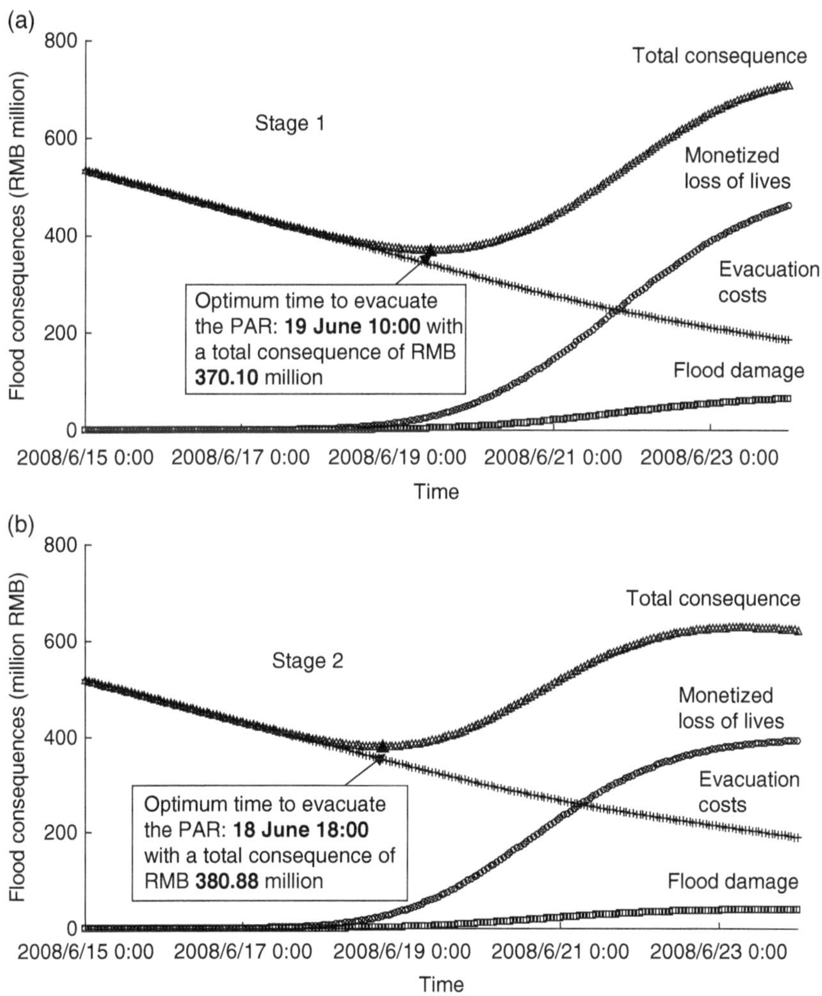

Figure 14.15 Optimal decision-making for Mianyang City: (a) stage 1, (b) stage 2 (No flooding is predicted in stage 3)

City. This is because there are limits in the maximum loss of life, flood damage, and total consequence even if no evacuation is implemented.

According to Sina News (2008a,b) and Gov.cn (2008), the government organized evacuation exercises in the areas downstream of the dam on 27 May 2008. An evacuation instruction was issued at 16:00 on 30 May 2008. More than 200,000 people were evacuated from Mianyang City, Beichuan and other towns by 20:00 on 31 May, 10 days before the start of the breaching. The evacuated people were allowed to go back home on 11 June when the flow rate at the dam site was smaller than 300 m^3/s. The people in both Beichuan and Mianyang were evacuated earlier than the optimal times suggested in this study. In Beichuan, the earlier evacuation would not incur much expense since a serious flood was not avoidable and the population at risk was relatively small. In Mianyang, nearly 200,000 people had been evacuated for 12 days but no extraordinary flood actually occurred in this city. An early evacuation decision was made for two main reasons: (1) possible societal impacts of the flood, (2) A high flood severity associated with a predicted peak discharge of 33,770 m^3/s based on experiences of constructed embankment dams (Huang et al., 2008). The study of Peng and Zhang (2012c) shows that direct application of empirical models for man-made earth and rockfill dams to landslide dams would, on average, overestimate the breach size by more than 60% and the peak outflow rate by approximately 200%, and underestimate the breaching duration by approximately 50%. DYDEM helps on a rational decision in the Tangjiashan landslide case. However, DYDEM can offer little help on the decision with a short-lived landslide dam (Dong et al., 2011). In such a case, an immediate decision is needed on evacuation of the PAR.

14.6 Discussions

The framework of dynamic decision-making takes the dam failure probability as a time series and flood consequences as functions of warning time. It sets a criterion for making an evacuation decision and updates the decision with new information when delayed decision is chosen. In this section, the influences of a few key factors – value of human life, failure mode, and sensitivity of consequences – are discussed.

14.6.1 Influence of the Value of Human Life

In this study, a value of the macro-economic contribution of a person is used to monetize a human life. One may argue that the value is too low or too high for decision-making. Therefore, the influence of value of life on the final decision-making is discussed here. If the value of life becomes higher, the minimum expected total consequences (MTC) in Figures 14.14 and 14.15 would be larger than the present values. The curve of the monetized loss of life would increase more rapidly and meet the evacuation costs earlier. Therefore, an earlier evacuation with a higher MTC is needed if the value of life is higher, and vice versa. If political or societal influences are involved, which puts saving a life as the highest priority, new criteria should be used. In this case, the human risk should be first reduced to a certain degree before considering economic issues.

14.6.2 Influence of Failure Mode

Only overtopping failure is considered in this study as most failed landslide dams (92%) were overtopped. Other failure modes such as slope failure and piping may influence the final decision. Slope failures are mass soil movements on either the upstream or the downstream side of a dam.

A slope failure reduces the soil thickness and, in some cases, the crest height of the dam. A slope failure by itself may not result in flooding, but it can affect the future ability of a dam to perform as intended. An earlier dam breaching with a higher peak outflow rate would happen if a slope failure occurs and hence an earlier evacuation warning should be issued with more MTC. However, as landslide dams are caused by rapid deposition of landslide debris materials, the slopes on both upstream and downstream sides are relatively gentle and the secondary slope failure is less likely to happen. Statistical data shows that only two landslide dams (1%) are recorded with slope failures. As with the slope failure, piping failure can lead to an earlier failure and hence an earlier evacuation warning as it may occur before the water level reaching the dam crest. Few landslide dams failed in piping (12 cases, 7% of all failures) as a large ratio of dam width and dam height often leads to a small hydraulic gradient.

14.6.3 Sensitivity of the Minimum Expected Total Consequence

In this study, the MTC is not very sensitive to the warning time before the optimal time in Beichuan and the evacuation may be brought forward for safety, while in Mianyang, the MTC is much more sensitive to the warning time than that in Beichuan and should be considered carefully. Several factors influence the sensitivity. The number of the PAR is a main factor, since the evacuation cost, which is related to the PAR, dominates the MTC before the optimal time of evacuation. The flood severity is another factor, as it directly influences the flood damage and loss of life. If the flood severity in Beichuan were not so high, the MTC would not increase so fast over time. Then the time of evacuation could be delayed from the optimal time since it would not cost much more. The probability of flood is also a very important factor as it is also highly related to the expected flood damage and loss of life. Other factors (e.g. dam parameters, hydraulic parameters, city size, and building characteristics) that affect the PAR, the flood severity, or the flood probability, also indirectly influence the sensitivity of MTC.

Part IV

Appendixes: Dam Failure Databases

Appendix A

Database of 1443 Cases of Failures of Constructed Dams

Dam Failure Mechanisms and Risk Assessment, First Edition. Limin Zhang, Ming Peng, Dongsheng Chang, and Yao Xu.

Table A.1 Database of 1443 Cases of Failures of Constructed Dams

No.	Name	Country[1]	Type[2]	Type divided[3]	Height (m)	Crest width (m)	Crest length (m)	Year of cons.	Year of failure	Failure cause[4]	Failure cause divided[5]	Reservoir capacity (10^6 m^3)	Water volume stored above breach invert (10^6 m^3)	Water depth above breach (m)	Breach geometry: Height (m)	Breach geometry: Top width (m)	Breach geometry: Bottom width (m)	Breach geometry: Aver. width (m)	Breach geometry: Aver. side slope Z:1 (h:v)	Peak outflow rate (m^3/s)	Fail. time	Life loss
1	Abbott	USA	MAS	—	2.1	—	—	—	1982	OT	—	—	—	—	—	—	—	—	—	—	—	—
2	Abenaki	USA	—	—	—	—	—	—	1929	QP	SEP	—	0.6164	—	—	—	—	—	—	—	—	—
3	Abeniki	USA	—	GR	11.3	—	—	—	1960	OT	—	—	—	—	—	—	—	—	—	—	—	—
4	Able Acres	USA	EFL	—	5.8	—	—	—	1994	OT	—	—	—	—	—	—	—	—	—	—	—	—
5	Acton	USA	EFL	—	—	—	—	—	1946	—	—	0.1	—	—	—	—	—	—	—	—	—	—
6	Aero Plantation	USA	EFL	—	8.5	—	—	1967	1976	OT	—	—	—	—	—	—	—	—	—	—	—	—
7	Ahraura	IN	EFL	CF	26	—	—	1953	1953	QP	—	61	—	—	—	—	—	—	—	—	—	—
8	Alamo Arroyo	USA	EFL	—	21	—	—	1960	1960	QP	SEP	6.6	—	—	—	—	—	—	—	—	—	—
9	Albrethsen	USA	EFL	—	6.1	—	—	1932	1982	OT	—	—	—	—	—	—	—	—	—	—	—	—
10	Alcova	USA	EFL	HD	81	—	233	1938	1968	QP	—	227	—	—	—	—	—	—	—	—	—	—
11	Alcyon	USA	EFL	—	3.7	—	—	—	1896	—	—	—	—	—	—	—	—	—	—	—	—	—
12	Alexander	USA	EFL	—	38.1	—	189.1	1930	1930	QP	SF-SL	2.8	—	—	—	—	—	—	—	—	—	6
13	Alexander	USA	EFL	—	—	—	—	—	1982	OT	ISC	—	—	—	—	—	—	—	—	—	—	—
14	Allegan City	USA	EFL	—	4.2	—	—	—	1978	OT	—	—	—	—	—	—	—	—	—	—	—	—
15	Allen	USA	CON	GR	7.3	—	—	1934	1981	QP	SEP	—	—	—	—	—	—	—	—	—	—	—
16	Allen Subdivision	USA	EFL	—	4.9	—	—	—	2005	DT	ABA	—	—	—	—	—	—	—	—	—	—	—
17	Allens Mill	USA	EFL	—	4.3	—	—	—	1999	OT	—	—	—	—	—	—	—	—	—	—	—	—
18	Allentown	USA	CON	GR	8.2	—	171.9	1937	1937	—	—	—	—	—	—	—	—	—	—	—	—	—
19	Alto Pass	USA	EFL	—	13.7	—	—	—	1965	QP	SEP	—	—	—	—	—	—	—	—	—	—	—
20	Amanda Lowe	USA	—	—	—	—	—	—	1959	QP	SF-SL	—	—	—	—	—	—	—	—	—	—	—
21	American River	USA	TIM	—	8.2	—	53.3	—	1950	OT	—	—	—	—	—	—	—	—	—	—	—	—
22	Anaconda	USA	EFL	DC	22	—	—	1898	1938	QP	SEP	0.2	0.2466	—	—	—	—	—	—	—	—	—
23	Anderson	USA	MAS	GR	12.8	—	—	1897	1901	OT	—	—	—	—	—	—	—	—	—	—	—	—
24	Anderson Powell	USA	EFL	—	7.6	—	—	—	2004	OT	—	—	—	—	—	—	—	—	—	—	—	—
25	Andrews	USA	—	—	—	—	—	—	1994	OT	—	—	—	—	—	—	—	—	—	—	—	—
26	Angels	USA	MAS	GR	15.6	—	120	—	1895	QP	SEP	—	—	—	—	—	—	—	—	—	—	1
27	Anita Dam	USA	EFL	—	—	—	—	—	1997	QP	SEP	—	—	—	—	—	—	—	—	—	—	—
28	Annapolis Mall	USA	EFL	—	9.1	—	—	—	1993	QP	SEP	—	—	—	—	—	—	—	—	—	—	—
29	Ansonia Brass	USA	CON	—	—	—	—	—	1912	QP	SF-OV	—	—	—	—	—	—	—	—	—	—	—
30	Ansonia-I	USA	EFL	—	6	—	—	—	1884	—	—	0.2	—	—	—	—	—	—	—	—	—	—
31	Ansonia-II	USA	EFL	—	6	—	—	—	1894	QP	SEP	0.2	—	—	—	—	—	—	—	—	—	—
32	Anzalduas	USA	EFL	—	—	—	—	—	1972	PM	—	12.3	—	—	—	—	—	—	—	—	—	4
33	Apishapa	USA	EFL	HD	34.14	4.88	178.6	1920	1923	QP	SEP	22.5	22.2	28	31.1	91.5	81.5	86.5	0.44	6850	2.5	—
34	Appert	USA	—	—	3	—	—	—	1995	OT	—	—	—	—	—	—	—	5.3	—	—	—	—
35	Apple Valley	USA	EFL	—	6.1	—	—	—	1997	QP	SEP	—	—	—	—	—	—	6.1	—	—	—	—
36	Archusa	USA	EFL	—	9.1	—	—	—	1998	QP	QIS	—	—	—	—	—	—	—	—	—	—	—

37	Armando	BR	EFL	—	41	—	660	1958	1977	OT	—	27	—	—	—	—	—	—	—	—	—	—
38	Aromos	CL	EFL	—	42.5	—	220	1979	1984	QP	SEP	60	—	—	—	—	—	—	—	—	—	—
39	Arrowhead	USA	EFL	—	5.5	—	—	—	1994	QP	SEP	—	—	—	—	—	—	—	—	—	—	—
40	Ascalmore	USA	EFL	—	10.4	—	—	1959	2000	DT	ABA	—	—	—	—	—	—	—	—	—	—	—
41	Ashburnham	USA	—	—	—	—	—	—	1850	OT	—	—	—	—	—	—	—	—	—	—	—	2
42	Asherville	USA	EFL	—	15.2	—	—	—	1976	QP	SF-SL	—	—	—	4.6	—	—	15.3	—	—	—	4
43	Ashizawa	JP	EFL	—	15	—	—	1912	1956	OT	—	—	—	—	—	—	—	—	—	—	—	—
44	Ashland	USA	EFL	—	23.2	—	—	—	1901	DT	BUA	—	—	—	—	—	—	—	—	—	—	—
45	Ashland Little	USA	MAS	GR	4.3	—	21.3	1874	1902	QP	SF-SL	—	—	—	—	—	—	—	—	—	—	—
46	Ashley Brook	USA	CON	BU	18	—	133	1908	1909	QP	SEP	0.09	—	—	—	—	—	—	—	—	—	—
47	Au Sable River	USA	—	—	—	—	—	—	1912	—	—	—	—	—	—	—	—	—	—	—	—	—
48	Aurora West	USA	—	GR	2.1	—	—	—	1996	OT	—	—	—	—	—	—	—	—	—	—	—	—
49	Austin PA I	USA	CON	—	—	—	—	—	1911	QP	SF-SL	1.05	—	—	—	—	—	—	—	—	—	80
50	Austin PA II	USA	TIM	—	9.8	—	—	—	1942	OT	IF	0.3	—	—	—	—	—	—	—	—	—	—
51	Austin TX I	USA	MAS	GR	18.3	—	332.5	1893	1893	QP	SEP	—	—	—	—	—	—	—	—	—	—	—
52	Austin TX II	USA	MAS	GR	20.7	—	388.6	1899	1900	QP	SEP	21	—	—	—	—	—	—	—	—	0.25	8
53	Austin TX III	USA	MAS	BU	20.7	—	388.6	1915	1915	PM	—	21	—	—	—	—	—	—	—	—	—	—
54	Avalon I	USA	EFL	CF	17.5	—	320.9	1893	1893	OT	—	6.9	—	—	—	—	—	—	—	—	—	—
55	Avalon II	USA	EFL	CF	17.5	—	420.6	1893	1904	QP	SEP	7.8	—	—	—	—	—	—	—	—	—	—
56	Avion	USA	EFL	—	4.3	—	—	—	1989	QP	—	—	—	—	—	—	—	—	—	—	—	—
57	Avoca	USA	EFL	—	—	—	—	—	1894	OT	ISC	0.01	—	—	—	—	—	—	—	—	—	—
58	Bad Axe	USA	EFL	CF	15.8	—	226	1965	1967	QP	SEP	1.23	—	—	—	—	—	—	—	—	—	—
59	Bad Axe No. 12	USA	EFL	—	22.3	—	—	—	1978	QP	SEP	5	—	—	—	—	—	—	—	—	—	—
60	Baker	USA	EFL	—	5.5	—	—	—	1999	QP	SEP	—	—	—	—	—	—	—	—	—	—	—
61	Baker Austin	USA	—	—	—	—	—	—	1994	OT	—	—	—	—	—	—	—	—	—	—	—	—
62	Baldhill	USA	EFL	—	19	—	540	1951	1970	QP	SF-SO	0.0872	—	—	—	—	—	—	—	—	—	—
63	Baldwin Hills	USA	EFL	HD	71	19.2	198	1951	1963	QP	SEP	11	0.91	12.2	21.3	—	—	25	0.31	1130	1.3	5
64	Ballard Mill	USA	EFL	CF	3	—	—	—	1984	OT	—	—	—	—	—	—	—	—	—	—	—	—
65	Balsams	USA	EFL	—	18	—	91.4	1927	1929	QP	SF	—	—	—	18.3	91.4	—	—	—	—	—	—
66	Bancroft I	USA	—	—	—	—	—	—	1848	QP	—	—	—	—	—	—	—	—	—	—	—	—
67	Bancroft II	USA	—	—	—	—	—	—	1884	—	—	—	—	—	—	—	—	—	—	—	—	—
68	Bankhead Lock	USA	CON	GR	34.1	—	—	1975	—	PM	—	—	—	—	—	—	—	—	—	—	—	—
69	Barbours	USA	EFL	—	6.1	—	—	—	1945	OT	—	—	—	—	6.1	—	—	9.2	—	—	—	—
70	Barcroft	USA	EFL	—	21	—	—	—	1972	OT	—	—	—	—	—	—	—	—	—	—	—	—
71	Barnes	USA	EFL	—	4	—	—	—	1995	OT	—	—	—	—	—	—	—	—	—	—	—	—
72	Barnesville	USA	EFL	—	2.4	—	—	—	1994	OT	—	—	—	—	—	—	—	—	—	—	—	—
73	Barryton	USA	EFL	—	4.6	—	—	—	1986	OT	—	—	—	—	—	—	—	—	—	—	—	—
74	Bartlett	USA	CON	GR	87.5	—	—	1939	—	QP	QIS	—	—	—	—	—	—	—	—	—	—	—
75	Barton	USA	EFL	—	14.6	—	—	1910	1922	QP	SF-SL	—	—	—	—	—	—	—	—	—	—	—
76	Basin	USA	EFL	—	5.5	—	—	—	—	DT	ABA	—	—	—	—	—	—	—	—	—	—	—
77	Bass Haven	USA	EFL	—	9.1	—	—	—	1984	OS	—	—	—	—	—	—	—	—	—	—	—	—
78	Battle River	CA	EFL	—	14	—	548	1956	1956	QP	SEP	15	—	—	—	—	—	—	—	—	—	—

(Continued)

Table A.1 (*Continued*)

No.	Name	Country[(1)]	Type[(2)]	Type divided[(3)]	Height (m)	Crest width (m)	Crest length (m)	Year of cons.	Year of failure	Failure cause[(4)]	Failure cause divided[(5)]	Reservoir capacity ($10^6 m^3$)	Water volume stored above breach invert ($10^6 m^3$)	Water depth above breach (m)	Breach geometry: Height (m)	Breach geometry: Top width (m)	Breach geometry: Bottom width (m)	Breach geometry: Aver. width (m)	Breach geometry: Aver. side slope Z:1 (h:v)	Peak outflow rate (m^3/s)	Fail. time	Life loss
79	Battlement	USA	EFL	—	7.3	—	—	1909	—	QP	SEP	—	—	—	—	—	—	—	—	—	—	—
80	Bay Meadows	USA	—	—	—	—	—	—	1998	QP	SEP	—	—	—	—	—	—	—	—	—	—	—
81	Bayless Austin	USA	CON	GR	15.8	—	166	1909	1911	QP	SF-SL	1.3	—	—	—	—	—	—	—	—	—	100
82	Bear Gulch	USA	EFL	—	19.2	—	222.5	1896	1914	QP	SF-SL	—	—	—	—	—	—	—	—	—	—	—
83	Bearwallow	USA	EFL	HD	10	3.1	152	1963	1976	QP	SF-SL	—	0.0493	5.79	6.4	—	—	12.2	1.43	—	—	4
84	Beaver Creek	USA	EFL	—	—	—	—	—	1931	—	—	—	—	—	—	—	—	—	—	—	—	—
85	Beaver Lake	USA	EFL	—	3	—	—	—	1983	QP	SEP	—	—	—	—	—	—	—	—	—	—	—
86	Beaver Meadow	USA	EFL	—	—	—	—	—	1939	OT	—	—	—	—	—	—	—	—	—	—	—	—
87	Beaver Pond	USA	EFL	—	3	—	—	—	1990	OT	—	—	—	—	—	—	—	—	—	—	—	—
88	Becket	USA	EFL	—	6.1	—	—	—	1927	OT	—	—	—	—	—	—	—	—	—	—	—	—
89	Belci	RO	EFL	—	16	—	428	1963	1991	OT	ISC	12	—	—	—	—	—	—	—	—	—	116
90	Belden	USA	RFL	—	50	—	152	1958	1967	OT	ISC	2.96	—	—	—	—	—	—	—	—	—	—
91	Belding	USA	CON	GR	4.6	—	—	—	1987	OT	—	—	—	—	—	—	—	—	—	—	—	—
92	Beldon Pond	USA	—	—	—	—	—	—	1999	QP	SEP	—	—	—	—	—	—	—	—	—	—	—
93	Belle Fourche	USA	EFL	—	35	—	1981.2	1909	1933	QP	SF-SL	—	—	—	—	—	—	—	—	—	—	—
94	Bennett	USA	EFL	—	9.1	—	—	—	1957	OT	ISC	—	—	—	—	—	—	—	—	—	—	—
95	Bennett Lake I	USA	EFL	—	—	—	—	—	1953	—	—	—	—	—	—	—	—	—	—	—	—	—
96	Bennett Lake II	USA	EFL	—	9.1	—	—	—	1998	OT	—	—	—	—	9.1	20	9.1	—	—	—	—	—
97	Bennett York	USA	EFL	—	8.2	—	—	—	2004	OS	—	—	—	—	—	—	—	—	—	—	—	—
98	Bent Tree	USA	—	—	—	—	—	—	1999	QP	SEP	0.004	—	—	—	—	—	—	—	—	—	—
99	Bergeron	USA	EFL	—	11	—	—	—	1996	QP	SEP	—	0.4186	—	—	—	—	—	—	—	—	1
100	Bibbins	USA	RFL	—	4.6	—	—	—	1884	—	—	—	—	—	—	—	—	—	—	—	—	—
101	Big Bay	USA	EFL	—	17.4	—	—	—	2004	QP	SEP	—	—	—	—	—	—	—	—	—	—	—
102	Big Millpond	USA	MAS	GR	3	—	—	1900	1979	OT	—	—	—	—	—	—	—	—	—	—	—	—
103	Big Sand I	USA	EFL	—	11	—	—	1962	1965	OT	—	—	—	—	—	—	—	—	—	—	—	—
104	Big Sand II	USA	EFL	—	7.6	—	—	—	2002	DT	ABA	—	—	—	—	—	—	—	—	—	—	—
105	Big Sandy	USA	EFL	—	3	—	—	—	1998	OT	—	—	—	—	—	—	—	—	—	—	—	—
106	Bila Desna	CZ	EFL	—	18	—	244	1915	1916	—	—	0.4	—	—	—	—	—	—	—	—	—	—
107	Bilberry	GB	EFL	—	20	—	90	1845	1852	QP	SEP	0.31	—	—	—	—	—	—	—	—	—	81
108	Bilk Creek	USA	EFL	—	12.2	—	—	—	1992	QP	QIS	—	—	—	—	—	—	—	—	—	—	—
109	Birch Creek	USA	—	—	—	—	—	—	1994	OT	—	—	—	—	—	—	—	—	—	—	—	—
110	Birchwood	USA	EFL	—	2.7	—	—	1955	2004	OT	—	—	—	—	—	—	—	—	—	—	—	—
111	Bischel	USA	EFL	—	3.7	—	—	—	1988	QP	SEP	—	—	—	—	—	—	—	—	—	—	—
112	Bishop	USA	EFL	—	6	—	18	1904	1982	OT	—	—	—	—	—	—	—	—	—	—	—	1
113	Bishop Creek	USA	EFL	—	10.4	—	135	1908	1909	OT	—	0.15	—	—	—	—	—	—	—	—	—	—
114	Bison Park	USA	EFL	—	8.2	—	—	—	1901	OT	ISC	—	0.3185	—	—	—	—	—	—	—	—	—

115	Black Creek	USA	EFL	—	13.1	—	—	1964	1983	QP	QIS	—	—	—	—	—	—	—	—	—	—	—
116	Black River	USA	TIM	—	6.7	—	—	—	1947	OT	—	—	—	—	—	—	—	—	—	—	—	—
117	Black Rock	USA	—	—	—	—	—	—	1992	OT	—	—	—	—	—	—	—	—	—	—	—	—
118	Black Rock I	USA	EFL	CF	21.3	—	208	1907	1909	QP	SEP	18	—	—	—	—	—	—	—	—	—	—
119	Black Rock II	USA	EFL	CF	—	—	—	—	1936	QP	SEP	—	—	—	—	—	—	—	—	—	—	—
120	Blackbrook-I	GB	EFL	—	28	—	160	1797	1799	QP	SEP	0.23	—	—	—	—	—	—	—	—	—	—
121	Blackbrook-II	GB	MAS	GR	—	—	—	1801	1804	QP	SEP	—	—	—	—	—	—	—	—	—	—	—
122	Blackstone	USA	EFL	—	6.1	—	—	—	1976	OT	—	—	—	—	—	—	—	—	—	—	—	—
123	Blackwood	USA	EFL	—	—	—	—	—	1940	OT	—	—	—	—	—	—	—	—	—	—	—	—
124	Blairtown	USA	EFL	—	—	—	—	—	1888	QP	SEP	—	—	—	—	—	—	—	—	—	—	—
125	Blandin	USA	TIM	—	6.4	—	—	1902	1948	QP	SEP	—	—	—	—	—	—	—	—	—	—	—
126	Blue Spgs Pwr	USA	CON	—	9.4	—	—	—	2004	OT	—	—	—	—	—	—	—	—	—	—	—	—
127	Bluewater	USA	RFL	—	10.5	—	99.1	1908	1909	OT	—	3.2	—	—	—	—	—	—	—	—	—	—
128	Blyderivier	ZA	EFL	—	21.6	—	341	1922	1922	OT	—	2.3	—	—	—	—	—	—	—	—	—	—
129	Boardman	USA	CON	GR	—	—	—	1904	1907	—	—	—	—	—	—	—	—	—	—	—	—	—
130	Bockelman	USA	EFL	—	7.6	—	—	—	1993	OT	—	—	—	—	—	—	—	—	—	—	—	—
131	Boeing Creek	USA	EFL	—	6.1	—	—	—	1996	OT	ISC	—	—	—	—	—	—	—	—	—	—	—
132	Bog Brook	USA	—	—	—	—	—	—	1988	DT	ABA	—	—	—	—	—	—	—	—	—	—	—
133	Bolan	PK	EFL	CF	22.9	—	533	1958	1976	OT	ISC	90	—	—	—	—	—	—	—	—	—	—
134	Bolton-I	USA	EFL	—	—	—	—	1856	1938	DT	BUA	—	—	—	—	—	—	—	—	—	—	—
135	Bolton-II	USA	EFL	—	5.5	—	—	—	1941	QP	—	—	—	—	—	—	—	—	—	—	—	—
136	Bon Esperanca	BR	EFL	—	17	—	450	1976	1977	—	—	37	—	—	—	—	—	—	—	—	—	—
137	Bookhamer	USA	EFL	—	—	—	—	—	1998	QP	QIS	—	—	—	—	—	—	—	—	—	—	—
138	Boston Felt	USA	TIM	—	3.7	—	—	—	2006	OT	IF	—	—	—	2.4	—	—	6.1	—	—	—	—
139	Bostwick Pond I	USA	EFL	—	4	—	—	—	1934	OT	—	—	—	—	—	—	—	7.6	—	—	—	—
140	Bostwick Pond II	USA	EFL	—	4	—	—	—	1999	QP	QIS	—	—	—	—	—	—	—	—	—	—	—
141	Bouzey	FR	MAS	GR	22.9	—	525	1880	1895	QP	SF-OV	7	—	—	—	—	—	—	—	—	—	90
142	Box Lake	USA	EFL	—	6.1	—	—	1890	1973	OT	ISC	—	—	—	—	—	—	—	—	—	—	—
143	Boy Scout Camp	USA	—	—	—	—	—	—	1998	OT	—	—	—	—	—	—	—	—	—	—	—	—
144	Boyd	USA	EFL	—	9.8	—	—	—	1995	QP	SEP	—	—	—	—	—	—	—	—	—	—	—
145	Boyd Lake	USA	EFL	—	9.1	—	—	—	1993	QP	SF-SL	—	—	—	—	—	—	—	—	—	—	—
146	Boydstown I	USA	EFL	—	7.4	—	93	1896	1897	OT	—	—	—	—	—	—	—	30.5	—	—	—	—
147	Boydstown II	USA	EFL	—	8.5	—	93	1897	1903	OT	ISC	—	—	—	—	—	—	—	—	—	—	—
148	Braddock	USA	EFL	—	4.3	—	—	—	1973	OT	—	—	—	—	—	—	—	—	—	—	—	—
149	Bradfield	GB	EFL	CF	29	—	382	1863	1864	QP	SEP	3.2	—	—	—	—	—	—	—	1150	<0.5	—
150	Brainerd	USA	—	—	—	—	—	—	1917	QP	SEP	—	—	—	—	—	—	—	—	—	—	—
151	Break Neck Run	USA	—	—	7	—	30.5	1877	1902	OT	—	0.049	—	—	7	—	—	30.5	—	9.2	3	—
152	Breakneck	USA	EFL	—	4.3	—	—	1870	1940	—	—	—	—	—	—	—	—	—	—	—	—	—
153	Briar Pipe	USA	MAS	—	5.2	—	—	—	1989	—	—	—	—	—	—	—	—	—	—	—	—	—
154	Bridgeport	USA	EFL	—	—	—	—	1855	1905	OT	ISC	—	—	—	—	—	—	—	—	—	—	—
155	Bridgewater	USA	TIM	—	5.8	—	—	1936	1973	QP	QIS	—	—	—	—	—	—	—	—	—	—	—
156	Bridgeway	USA	EFL	—	5.8	—	—	1968	1968	OT	—	—	—	—	—	—	—	—	—	—	—	—

(Continued)

Table A.1 *(Continued)*

No.	Name	Country(1)	Type(2)	Type divided(3)	Height (m)	Crest width (m)	Crest length (m)	Year of cons.	Year of failure	Failure cause(4)	Failure cause divided(5)	Reservoir capacity ($10^6 m^3$)	Water volume stored above breach invert ($10^6 m^3$)	Water depth above breach (m)	Breach geometry: Height (m)	Breach geometry: Top width (m)	Breach geometry: Bottom width (m)	Breach geometry: Aver. width (m)	Breach geometry: Aver. side slope Z:1 (h:v)	Peak outflow rate (m^3/s)	Fail. time	Life loss
157	Briggsville	USA	—	—	—	—	—	—	1993	QP	—	—	—	—	—	—	—	—	—	—	—	—
158	Brindley	USA	EFL	—	—	—	—	—	2003	DT	ABA	—	—	—	—	—	—	—	—	—	—	—
159	Briseis	AU	RFL	—	17	—	137	1926	1929	OT	ISC	1	—	—	—	—	—	—	—	—	—	14
160	Broad Brook	USA	CON	—	15	—	—	—	1890	—	—	—	—	—	—	—	—	—	—	—	—	—
161	Brokaw I	USA	TIM	—	6.1	—	—	1909	1912	OT	IF	—	—	—	—	—	—	—	—	—	—	—
162	Brokaw II	USA	EFL	—	8.5	—	243.8	1923	1938	OT	—	—	—	—	—	—	—	—	—	—	—	—
163	Bronson	USA	MAS	—	3.7	—	—	—	1982	OT	—	—	—	—	—	—	—	—	—	—	—	—
164	Brooktrails No. 3	USA	EFL	—	15	—	117	1970	1971	QP	SF-SL	0.34	—	—	—	—	—	—	—	—	—	—
165	Brookville I	USA	EFL	—	4.9	—	109.7	1912	1912	OT	ISC	—	—	—	—	—	—	—	—	—	—	—
166	Brookville II	USA	EFL	—	6.1	—	—	—	1996	OT	—	—	—	—	—	—	—	—	—	—	—	—
167	Browder	USA	EFL	—	8.2	—	—	—	1972	QP	SEP	—	—	—	—	—	—	—	—	—	—	—
168	Brownes Lake	USA	EFL	—	3.7	—	—	1894	1984	OT	BOS	0.62	—	—	—	—	—	46	—	—	3	—
169	Browns Millpond	USA	EFL	—	4	—	—	—	1994	OT	—	—	—	—	—	—	—	—	—	—	—	—
170	Brownville	USA	TIM	—	6.1	—	—	1903	1978	OT	—	—	—	—	6.1	—	—	27.4	—	—	—	—
171	Bruce Lake	USA	CON	GR	—	—	—	—	1911	DT	WTA	—	—	—	—	—	—	—	—	—	—	—
172	Bruce Nordland	USA	—	—	—	—	—	—	1986	OT	—	—	—	—	—	—	—	—	—	—	—	—
173	Bruceton Mills	USA	MAS	—	6.1	—	—	—	1996	OT	—	—	—	—	—	—	—	—	—	—	—	—
174	Brush Hollow	USA	EFL	—	30.5	—	266.7	1910	1928	QP	SF-SL	5.2	—	—	—	—	—	—	—	—	—	—
175	Buckhaven No.2	USA	—	—	—	—	—	—	—	OT	—	—	0.0247	6.1	6.1	—	—	4.72	0.73	—	—	—
176	Buckhorn	USA	EFL	—	—	—	—	—	1923	OT	ISC	—	—	—	—	—	—	—	—	—	—	—
177	Bucks Lake	USA	—	—	—	—	—	—	1970	OT	BOS	—	—	—	—	—	—	—	—	—	—	—
178	Buffalo Creek I	USA	EFL	—	32	—	—	1970	1973	OT	—	0.61	—	—	14	153	125	—	—	1420	0.5	—
179	Buffalo Creek II	USA	EFL	HD	14	—	366	1967	1972	QP	SF-SO	0.5	—	—	—	—	—	—	—	—	4	125
180	Bullock Draw	USA	EFL	HD	5.8	4.3	—	1971	1971	QP	SEP	1.13	0.74	3.05	5.79	13.6	11	12.5	0.21	—	—	—
181	Bully Creek	USA	RFL	—	38.1	—	61	1925	1925	OT	BOP	20	—	—	—	—	—	—	—	—	—	—
182	Burgess Falls	USA	—	—	—	—	—	—	1928	—	—	—	—	—	—	—	—	—	—	—	—	5
183	Burney Creek	USA	EFL	—	3.4	—	—	—	1937	OT	—	—	—	—	—	—	—	46	—	—	—	—
184	Burton Pond	USA	—	—	—	—	—	—	1991	QP	QIS	—	—	—	—	—	—	—	—	—	—	—
185	Bushy Hill	USA	EFL	—	8.8	—	—	—	1982	OT	—	0.6	—	—	—	—	—	—	—	—	—	—
186	Bushy Pond	USA	MAS	—	3.7	—	—	—	1982	OT	—	—	—	—	—	—	—	—	—	—	—	—
187	Butler	USA	EFL	HD	—	—	—	—	1982	OT	—	—	2.38	7.16	7.16	—	—	62.5	0.85	810	—	—
188	C D Batchelor	USA	EFL	—	1.2	—	—	—	1982	OT	—	—	—	—	—	—	—	—	—	—	—	—
189	Cade Lake No. 3	USA	EFL	—	6.1	—	—	—	1994	OT	ISC	—	—	—	—	—	—	—	—	—	—	—
190	Calaveras	USA	EFL	—	73.2	—	384.1	1914	1918	QP	SF-SL	201	—	—	—	—	—	—	—	—	—	—
191	Calaveras C.	USA	CON	BU	4.3	—	—	—	1986	OT	—	—	—	—	—	—	—	—	—	—	—	—
192	Calhoun Lake	USA	EFL	—	7.3	—	—	—	1950	QP	QIS	—	—	—	—	—	—	—	—	—	—	—

193	Callon Dam	USA	RFL	—	—	—	—	—	1912	OT	—	—	—	—	—	—	—	—	—	—	—	—
194	Caloosa Sand	USA	—	—	—	—	—	—	1999	OT	—	—	—	—	—	—	—	—	—	—	—	—
195	Cambria	USA	—	—	—	—	—	—	1993	PM	—	—	—	—	—	—	—	—	—	—	—	—
196	Camden	USA	EFL	—	6.1	—	—	1853	1902	OT	—	—	—	—	—	—	—	—	—	—	—	—
197	Cameron	USA	EFL	—	6	—	—	—	1982	OT	—	0.5	—	—	—	—	—	—	—	—	—	—
198	Camp 5	USA	—	GR	7.6	—	—	—	—	DT	WTA	—	—	—	—	—	—	—	—	—	—	—
199	Camp Ernst	USA	EFL	—	18.6	—	—	—	1978	QP	SF-SL	—	—	—	—	—	—	—	—	—	—	—
200	Camp La Junta	USA	EFL	—	3.7	—	—	—	2000	OT	—	—	—	—	3.7	—	—	—	—	—	—	—
201	Camp Ockinicken	USA	EFL	—	2.1	—	—	—	2004	—	—	—	—	—	—	—	—	—	—	—	—	—
202	Camp Ritchie	USA	EFL	—	7.9	—	125	—	1929	QP	—	0.12	—	—	—	—	—	—	—	—	—	—
203	Camp Weona	USA	EFL	—	3.4	—	—	—	1998	OT	—	—	—	—	3.4	9.1	9.1	9.1	—	—	—	—
204	Campbell	USA	—	—	—	—	—	—	1990	—	—	—	—	—	—	—	—	—	—	—	—	—
205	Caney Coon Creek Site 2	USA	EFL	—	15.3	—	—	1964	1964	QP	—	—	—	—	—	—	—	—	—	—	—	—
206	Canterbury	USA	—	—	—	—	—	—	1914	—	—	—	—	—	—	—	—	—	—	—	—	—
207	Canyon Ferry	USA	TIM	—	8.8	—	147.8	—	1898	QP	—	—	—	—	—	—	—	—	—	—	—	—
208	Canyon MT	USA	EFL	—	6.4	—	—	—	1996	OT	—	—	—	—	—	—	—	—	—	—	—	—
209	Canyon SD	USA	EFL	—	6.1	—	152	1938	1972	OT	BOS	0.985	—	—	—	—	—	—	—	850	0.1	237
210	Cardinal Lake	USA	—	—	—	—	—	—	1994	QP	SEP	—	—	—	—	—	—	—	—	—	—	—
211	Carl Smith	USA	EFL	—	17.4	—	—	—	1998	QP	SF-SL	—	—	—	—	—	—	—	—	—	—	—
212	Carlanna Lake	USA	RFL	—	9.4	—	—	—	1973	OT	—	—	—	—	—	—	—	—	—	—	—	—
213	Carlton Wood	USA	EFL	—	7.6	—	—	1960	1977	OT	BOS	—	—	—	—	—	—	—	—	—	—	—
214	Carson City	USA	—	—	—	—	—	—	1986	—	—	—	—	—	—	—	—	—	—	—	—	—
215	Carter Pond	USA	EFL	—	5.5	—	—	—	2004	—	—	—	—	—	—	—	—	—	—	—	—	—
216	Casa Monte	USA	CON	—	4.6	—	—	—	1996	OT	—	0.41	—	—	—	—	—	—	—	—	—	—
217	Cascade	USA	RFL	—	7.6	—	—	—	1981	DT	EQ	0.03	—	—	—	—	—	—	—	—	—	—
218	Cascade Lake	USA	CON	GR	3	—	—	—	1982	DT	BUA	0.57	—	—	—	—	—	—	—	—	—	—
219	Cash	USA	—	—	14.9	—	—	—	1985	OT	—	—	—	—	—	—	—	—	—	—	—	—
220	Castlewood I-1	USA	RFL	—	21.3	—	182.9	1890	1897	QP	SEP	4.3	—	—	—	—	—	—	—	—	—	—
221	Castlewood I-2	USA	RFL	—	21.3	—	182.9	1890	1899	QP	SEP	4.3	—	—	—	—	—	—	—	—	—	—
222	Castlewood II	USA	RFL	DC	28	4.9	182.9	1899	1933	OT	—	4.3	6.17	21.6	21.3	54.9	33.5	44.2	0.5	3570	—	2
223	Cat Creek Dam	USA	—	—	—	—	—	—	1986	—	—	—	—	—	—	—	—	—	—	—	—	—
224	Caulk Lake	USA	EFL	—	20	—	134	1950	1973	QP	SF	0.7	0.698	11.1	9.1	36.6	—	—	2	—	—	—
225	Cazedero	USA	RFL	—	21.3	—	55	1906	1965	OT	ISC	16	—	—	—	—	—	—	—	—	—	—
226	Cedar Falls	USA	CON	GR	66.1	—	—	—	1918	QP	SF-SL	—	—	—	—	—	—	—	—	—	—	—
227	Cedar Hill Lake	USA	EFL	—	6.7	—	—	—	1977	QP	SEP	—	—	—	—	—	—	—	—	—	—	—
228	Cedar Lake	USA	EFL	—	13.7	—	—	—	1986	QP	SEP	—	—	—	—	—	—	—	—	—	—	—
229	Centennial Narrows	USA	—	—	—	—	—	—	1997	QP	SEP	—	—	—	—	—	—	—	—	—	—	—
230	Center Creek #1	USA	EFL	—	14	—	182.9	1869	1973	OT	BOS	0.4	—	—	—	—	—	—	—	—	—	—
231	Central Maine	USA	EFL	—	9.1	—	—	1850	—	—	—	—	—	—	—	—	—	—	—	—	—	—
232	Chambers	USA	EFL	—	7.6	—	—	—	1977	OT	—	—	—	—	—	—	—	—	—	—	—	—
233	Chambers Lake I	USA	EFL	—	15	—	213	1885	1891	QP	QIS	—	—	—	—	—	—	—	—	—	—	—
234	Chambers Lake II	USA	EFL	—	15	—	244	1892	1907	QP	QIS	10	—	—	—	—	—	—	—	—	—	—

(Continued)

Table A.1 *(Continued)*

No.	Name	Country[(1)]	Type[(2)]	Type divided[(3)]	Height (m)	Crest width (m)	Crest length (m)	Year of cons.	Year of failure	Failure cause[(4)]	Failure cause divided[(5)]	Reservoir capacity (10^6 m^3)	Water volume stored above breach invert (10^6 m^3)	Water depth above breach (m)	Breach geometry					Peak outflow rate (m^3/s)	Fail. time	Life loss
															Height (m)	Top width (m)	Bottom width (m)	Aver. width (m)	Aver. side slope Z:1 (h:v)			
235	Chambersburg	USA	—	—	—	—	—	—	1912	QP	—	—	—	—	—	—	—	—	—	—	—	—
236	Channahon I	USA	EFL	—	7.6	—	—	—	1948	OT	—	—	—	—	—	—	—	—	—	—	—	—
237	Channahon II	USA	EFL	—	7.6	—	—	—	1996	OT	—	—	—	—	—	—	—	—	—	—	—	—
238	Chapman	USA	EFL	—	20.1	—	—	1980	1983	QP	QIS	—	—	—	—	—	—	—	—	—	—	—
239	Cheaha Creek	USA	EFL	CF	7	4.3	—	1970	1970	OT	—	0.069	—	—	—	—	—	—	—	—	5.5	—
240	Cheney Dam	USA	EFL	—	38	—	7455	1965	1971	QP	SEP	306	—	—	—	—	—	—	—	—	—	—
241	Chenowith Dam	USA	—	—	—	—	—	—	2000	DT	ABA	—	—	—	—	—	—	—	—	—	—	—
242	Cherokee Sandy Creek Site 8A	USA	EFL	—	8.2	—	—	1963	1964	QP	SEP	—	—	—	—	—	—	—	—	—	—	—
243	Childsdale Dam	USA	—	—	—	—	—	—	1986	—	—	—	—	—	—	—	—	—	—	—	—	—
244	Chinook Water District Dam	USA	EFL	—	7.9	—	—	—	1990	OT	ISC	—	—	—	—	—	—	—	—	—	—	—
245	Chitauni	IN	RFL	—	—	—	—	—	1968	—	—	—	—	—	—	—	—	—	—	—	—	—
246	Chopper'S Lake	USA	EFL	—	8.1	—	—	—	1994	OT	—	—	—	—	2.4	—	—	—	—	—	—	—
247	Christiansen	USA	EFL	—	5.8	—	—	—	1990	OS	—	—	—	—	—	—	—	—	—	—	—	—
248	Church Rock Mine #1	USA	EFL	—	5.2	—	—	1976	1977	QP	SEP	—	—	—	—	—	—	—	—	—	—	—
249	City Of Kake	USA	TIM	—	7	—	—	—	2000	OS	—	—	—	—	—	—	—	—	—	—	—	—
250	City Reservoir	USA	MAS	—	10	—	—	—	1912	QP	SEP	—	—	—	—	—	—	45.7	—	—	—	—
251	Clarke Apple	USA	—	—	—	—	—	—	2002	QP	SEP	—	—	—	—	—	—	—	—	—	—	—
252	Clay Brook	USA	EFL	—	1.8	—	—	1973	1998	OT	—	—	—	—	—	—	—	—	—	—	—	—
253	Clear Creek #2	USA	EFL	—	3.7	—	—	—	1980	QP	SEP	—	—	—	—	—	—	—	—	—	—	—
254	Clear Fork	USA	—	—	—	—	—	—	1894	—	—	—	—	—	—	—	—	—	—	—	—	—
255	Clearwater Lake	USA	EFL	HD	—	—	—	1965	1994	OT	—	—	0.466	4.05	3.78	—	—	22.8	1.03	—	—	—
256	Clendening	USA	EFL	—	21.3	—	—	1934	1934	QP	SEP	—	—	—	—	—	—	—	—	—	—	—
257	Cleveland	USA	EFL	—	—	—	—	—	1940	OT	—	—	—	—	—	—	—	—	—	—	—	—
258	Cloud Lake	USA	EFL	—	4.8	—	—	—	1994	OT	—	—	—	—	—	—	—	—	—	—	—	—
259	Coal	USA	EFL	—	61	—	—	—	1973	QP	SF-SL	—	—	—	—	—	—	—	—	—	—	—
260	Cobden	USA	—	—	—	—	—	—	1894	—	—	—	—	—	—	—	—	—	—	—	—	—
261	Cockram Mill	USA	CON	GR	6.1	—	—	—	1990	OT	—	—	—	—	—	—	—	—	—	—	—	—
262	Coedty	GB	EFL	DC	11	3.1	262	1924	1925	DT	BUA	0.31	0.311	>11	11	67	18.2	42.7	2.22	—	—	60
263	Coffin Lake	USA	EFL	—	4.7	—	—	—	1994	OT	—	—	—	—	—	—	—	—	—	—	—	—
264	Cold Brook	USA	—	—	—	—	—	—	1999	—	—	—	—	—	—	—	—	—	—	—	—	—
265	Cold Springs	USA	EFL	—	15.2	—	—	—	1912	QP	SEP	—	—	—	—	—	—	—	—	—	—	—
266	Colee Naylor	USA	—	—	—	—	—	—	1999	OT	—	—	—	—	3.7	—	—	12.2	—	—	—	—
267	Coleman	USA	EFL	—	—	—	—	—	1954	DT	EQ	—	—	—	—	—	—	—	—	—	—	—
268	Colley Lake	USA	EFL	—	19	—	—	—	1963	—	—	—	—	—	—	—	—	—	—	—	—	—

269	Colonial	USA	CON	GR	9.5	—	62.8	1906	1912	QP	—	—	—	—	—	—	—	—	—	—	—	—
270	Colt Dam	USA	CON	—	5.6	—	—	—	2005	QP	SEP	—	—	—	4.6	—	—	30.5	—	—	—	—
271	Columbian	USA	—	—	—	—	—	—	1903	—	—	—	—	—	—	—	—	—	—	—	—	—
272	Columbus	USA	MAS	GR	11.9	—	167.6	1901	1901	OT	—	—	—	—	—	—	—	—	—	—	—	—
273	Comstock Pond	USA	MAS	—	2.7	—	—	—	1982	OT	—	—	—	—	—	—	—	—	—	—	—	—
274	Concow	USA	—	—	—	—	—	—	1903	—	—	—	—	—	—	—	—	—	—	—	—	—
275	Condit	USA	CON	GR	38.1	—	—	—	1983	QP	SEP	—	—	—	—	—	—	—	—	—	—	—
276	Connedogwinet- Creek-I	USA	CON	GR	3.4	—	883.9	1911	1911	—	—	—	—	—	—	—	—	—	—	—	—	—
277	Connedogwinet- Creek-II	USA	CON	GR	3.4	—	883.9	1911	1912	—	—	—	—	—	—	—	—	—	—	—	—	—
278	Connelsville	USA	EFL	—	7	—	63.1	1887	1902	—	—	0.04	—	—	—	—	—	—	—	—	—	—
279	Conodoguinet	USA	MAS	—	2.7	—	—	—	1912	QP	SF-OV	—	—	—	—	—	—	—	—	—	—	—
280	Conshohocken Hill-I	USA	EFL	—	—	—	—	—	1873	QP	—	3.8	—	—	—	—	—	—	—	—	—	—
281	Conshohocken Hill-II	USA	EFL	—	—	—	—	—	1876	QP	—	3.8	—	—	—	—	—	—	—	—	—	—
282	Conshohocken Hill-III	USA	EFL	—	—	—	—	—	1879	QP	—	3.8	—	—	—	—	—	—	—	—	—	—
283	Conshohocken Hill-IV	USA	EFL	—	—	—	—	—	1886	QP	—	3.8	—	—	—	—	—	—	—	—	—	—
284	Coon Creek 41	USA	EFL	—	11.9	—	—	—	1978	QP	SEP	—	—	—	—	—	—	—	—	—	—	—
285	Coon Rapids	USA	CON	GR	6.4	—	39.6	1914	1917	QP	SEP	—	—	—	—	—	—	—	—	—	—	—
286	Cordrays Pond	USA	—	—	—	—	—	—	1994	OT	—	—	—	—	—	—	—	—	—	—	—	—
287	Corpus Christi	USA	EFL	—	31.7	—	1244	1930	1930	QP	SEP	79	—	—	—	—	—	61	—	—	—	—
288	Costilia-I	USA	EFL	—	36.9	—	182.9	1920	1924	QP	SEP	19.4	—	—	—	—	—	—	—	—	—	—
289	Costilia-II	USA	EFL	—	36.9	—	182.9	1920	1941	QP	SF-SL	19.4	—	—	—	—	—	—	—	—	—	—
290	Cottonwood Flandrau Park-I	USA	EFL	—	11.3	—	—	1936	1947	OT	—	—	—	—	—	—	—	—	—	—	—	—
291	Cottonwood Flandrau Park-II	USA	EFL	—	11.3	—	—	1936	1965	OT	—	—	—	—	—	—	—	—	—	—	—	—
292	Cottonwood Flandrau Park-III	USA	EFL	—	11.3	—	—	1936	1969	OT	—	—	—	—	—	—	—	—	—	—	—	—
293	Cougar Creek	CA	—	—	—	—	—	—	—	OT	—	—	0.0298	11.1	10.4	—	—	—	—	—	—	—
294	County Dam	USA	EFL	—	5.2	—	—	—	1890	QP	SF-SL	—	—	—	—	—	—	—	—	—	—	—
295	County Road 15	USA	—	—	—	—	—	—	1998	OT	—	—	—	—	—	—	—	23	—	—	—	—
296	Covey Dam	USA	EFL	—	6.4	—	—	—	1999	OT	—	—	—	—	—	—	—	—	—	—	—	—
297	Cow Creek	USA	EFL	—	4.3	—	—	1935	1999	OT	—	—	—	—	—	—	—	—	—	—	—	—
298	Cranberry	USA	—	—	—	—	—	—	1996	QP	SEP	—	—	—	—	—	—	—	—	—	—	—
299	Crater Lake	USA	EFL	—	11	—	—	—	1928	OT	BOS	—	—	—	—	—	—	—	—	—	—	—
300	Crisp County	USA	EFL	—	12.8	—	—	—	1994	OT	—	—	—	—	—	—	—	152.4	—	—	—	—
301	Crow Creek	USA	EFL	—	—	—	—	—	1955	QP	QIS	—	—	—	—	—	—	—	—	—	—	—
302	Crown City Mining Pond No. 24 Dam	USA	—	—	—	—	—	—	1999	OT	BOS	—	—	—	—	—	—	—	—	—	—	—
303	Crump	USA	EFL	—	4.6	—	—	—	1980	QP	SEP	—	—	—	3.7	9.1	9.1	9.1	0	—	—	—
304	Crystal Lake	USA	EFL	—	15.2	—	—	—	1961	QP	SEP	—	—	—	—	—	—	—	—	—	—	—
305	CSC Orchards Frost Protection	USA	—	—	—	—	—	—	1995	QP	SEP	—	—	—	—	—	—	—	—	—	—	—

(Continued)

Table A.1 (*Continued*)

No.	Name	Country[(1)]	Type[(2)]	Type divided[(3)]	Height (m)	Crest width (m)	Crest length (m)	Year of cons.	Year of failure	Failure cause[(4)]	Failure cause divided[(5)]	Reservoir capacity ($10^6 m^3$)	Water volume stored above breach invert ($10^6 m^3$)	Water depth above breach (m)	Breach geometry: Height (m)	Breach geometry: Top width (m)	Breach geometry: Bottom width (m)	Breach geometry: Aver. width (m)	Breach geometry: Aver. side slope Z:1 (h:v)	Peak outflow rate (m^3/s)	Fail. time	Life loss
306	Cuba Reservoir	USA	EFL	—	15.7	—	—	1851	1868	QP	—	0.49	—	—	—	—	—	—	—	—	—	—
307	Currant Creek	USA	EFL	—	25	—	—	—	1959	QP	—	—	—	—	—	—	—	—	—	—	—	—
308	Current River Port Arthur	USA	—	—	—	—	—	—	1908	—	—	—	—	—	—	—	—	—	—	—	—	—
309	Cypress Shores	USA	EFL	—	4.6	—	—	—	1999	OT	—	—	—	—	—	—	—	—	—	—	—	—
310	D.M.A.D.	USA	EFL	—	10.4	—	823	1960	1983	QP	—	12.6	—	—	—	—	—	—	—	—	—	1
311	Daggs Dam	USA	EFL	CF	19.2	—	—	1919	1973	QP	SEP	4.9	—	—	—	—	—	—	—	—	—	—
312	Daguerre Point	USA	—	—	—	—	—	1906	1963	OT	—	—	—	—	—	—	—	—	—	—	—	—
313	Dale Dike	GB	EFL	—	29	—	380	1863	1864	QP	SEP	3.2	—	—	—	—	—	—	—	—	—	245
314	Dallas	USA	EFL	—	8.8	—	—	—	1891	QP	—	0.5	—	—	—	—	—	—	—	—	—	—
315	Dalton	USA	EFL	—	8.8	—	55.2	1910	1912	QP	—	0.1	—	—	—	—	—	—	—	—	—	—
316	Dam No. 2	USA	EFL	—	12.8	—	219.5	1884	1912	—	—	—	—	—	—	—	—	—	—	—	—	—
317	Dam Mahanoy	USA	EFL	—	7.5	—	—	—	1892	—	—	0.01	—	—	—	—	—	—	—	—	—	1
318	Dam No. 26	USA	—	—	6.1	—	182.9	1912	1912	—	—	—	—	—	—	—	—	—	—	—	—	—
319	Dam on Pierce Park Gulch	USA	—	—	—	—	—	—	1979	OT	—	—	—	—	—	—	—	—	—	—	—	—
320	Danaher Lake	USA	EFL	—	6.4	—	—	—	1986	OT	—	—	—	—	—	—	—	—	—	—	—	—
321	Dansville	USA	CON	BU	4.6	—	112.2	—	1909	QP	—	—	—	—	—	—	—	—	—	—	—	—
322	Dansville Reservoir	USA	EFL	—	16.5	—	—	—	1909	QP	SEP	—	—	—	—	—	—	—	—	—	—	—
323	Dantiwada	IN	EFL& CON	—	61	—	137.5	1969	1973	OT	—	464	—	—	—	—	—	—	—	7500	—	—
324	Danville	USA	CON	GR	3	—	53	1903	1930	QP	SF-SL	—	—	—	—	—	—	—	—	—	—	—
325	Davis Lake Dam Okanogan	USA	EFL	CF	2.1	—	—	—	1974	OT	ISC	0.62	—	—	—	—	—	—	—	—	—	—
326	Davis Lake No. 1	USA	—	—	—	—	—	—	1974	OT	—	—	—	—	—	—	—	—	—	—	—	—
327	Davis Reservior	USA	EFL	FD	11.9	6.1	292.6	1914	1914	QP	SEP	58	58	11.58	11.9	21.3	—	—	0.25	510	—	—
328	Decker Dam	USA	—	—	—	—	—	—	1996	QP	QIS	—	—	—	—	—	—	—	—	—	—	—
329	Deckers	USA	EFL	—	5.5	—	—	1953	1983	QP	—	0.1	—	—	—	—	—	—	—	—	—	—
330	Deer Creek	USA	EFL	—	6.1	2.4	61	—	1999	OT	—	—	—	—	—	—	—	—	—	—	—	—
331	Deer Lake	USA	MAS	—	3.7	—	—	—	1982	OT	—	—	—	—	—	—	—	—	—	—	—	—
332	Deer Lake CA	USA	EFL	—	7.6	—	—	1952	1977	OT	—	—	—	—	—	—	—	—	—	—	—	—
333	Deering	USA	EFL	—	6.7	—	—	1867	1938	QP	SEP	—	—	—	—	—	—	—	—	—	—	—
334	Del Monte	CO	—	—	—	—	—	1975	1976	—	—	—	—	—	—	—	—	—	—	—	—	80
335	Dells	USA	EFL	—	18.3	—	292.6	1908	1911	OT	—	13	—	—	—	112.8	—	—	—	5440	0.67	—
336	Demery's Lake	USA	—	—	—	—	—	—	1986	OT	BOS	—	—	—	—	—	—	—	—	—	—	—
337	Dennery Lake	USA	EFL	—	6.7	—	—	—	2005	QP	SEP	—	—	—	—	—	—	—	—	—	—	—
338	Derby-I	USA	MAS	GR	11	—	—	1867	1869	—	—	—	—	—	—	—	—	—	—	—	—	—

339	Derby-II	USA	MAS	GR	11	—	—	1867	1891	—	—	—	—	—	—	—	—	—	—	—	—	—
340	Des Moines	USA	CON	GR	—	—	—	—	1893	QP	—	—	—	—	—	—	—	—	—	—	—	—
341	Desabia	USA	EFL	—	16.2	—	315.5	1903	1932	QP	SEP	0.4	—	—	—	—	—	—	—	—	—	—
342	Dhanibara	IN	EFL	—	20.7	—	—	1975	1976	OT	—	61	—	—	—	—	—	—	—	—	—	—
343	Diamond Mills Paper Company	USA	EFL	—	10.7	—	—	—	1978	PM	—	—	—	—	—	—	—	—	—	—	—	—
344	Dillard Dam	USA	—	—	—	—	—	—	1996	OT	—	—	—	—	—	—	—	—	—	—	—	—
345	Dimock	USA	EFL	—	9.1	—	—	1937	1984	OT	—	0.71	—	—	—	—	—	—	—	—	—	—
346	Dixie Springs Refuge Lk	USA	EFL	—	5.5	—	—	—	2002	OT	—	—	—	—	—	—	—	—	—	—	—	—
347	Dixon Dam	USA	EFL	—	2.7	—	—	—	1938	OT	—	—	—	—	2.7	—	—	67	—	—	—	—
348	Dmad	USA	EFL	HD	10.4	—	—	—	1983	QP	—	19.7	19.7	—	—	—	—	—	—	793	—	1
349	Donelson Lake	USA	—	—	—	—	—	—	1967	—	—	—	—	—	—	—	—	—	—	—	—	—
350	Doty-Tex	USA	—	—	—	—	—	—	1984	OT	—	—	—	—	—	—	—	—	—	—	—	—
351	Dozier Lake	USA	—	—	—	—	—	—	1994	—	—	—	—	—	—	—	—	—	—	—	—	3
352	Dresser #4	USA	EFL	CF	32	—	—	—	1975	QP	SEP	—	—	—	—	—	—	91.4	—	—	—	—
353	Drews	USA	—	—	8.8	—	—	—	1978	OT	IF	—	—	—	—	—	—	—	—	—	—	—
354	Dry Creek	USA	EFL	—	13.8	—	1114	1938	1939	QP	SEP	—	—	—	—	—	—	—	—	—	—	—
355	Dubose Lake	USA	EFL	—	6.1	—	—	—	1999	OT	—	—	—	—	1.8	—	—	9.1	—	—	—	—
356	Durham	USA	EFL	—	6.7	—	—	—	1978	OT	—	—	—	—	—	—	—	—	—	—	—	—
357	Dyer Dam	USA	TIM	—	3.7	—	63.5	1880	1901	QP	—	—	—	—	—	—	—	—	—	—	—	—
358	Dykstra	USA	RFL	—	15.2	—	—	—	1926	OT	—	—	—	—	—	—	—	—	—	—	—	—
359	Eagle Lake	USA	EFL	—	—	—	—	—	2001	QP	SEP	—	—	—	—	—	—	—	—	—	—	—
360	East Bank 1	USA	EFL	—	—	—	—	—	2005	QP	SEP	—	—	—	—	24.4	24.4	24.4	—	—	—	—
361	East Bank II	USA	EFL	—	—	—	—	—	2005	QP	QIC	—	—	—	—	—	—	—	—	—	—	—
362	East Burke	USA	TIM	—	—	—	—	—	1927	—	—	—	—	—	—	—	—	—	—	—	—	—
363	East Ellijay	USA	—	—	—	—	—	—	2003	—	—	—	—	—	—	—	—	—	—	—	—	—
364	East Fork Pond	USA	EFL	—	13.4	—	—	1978	1978	QP	SEP	—	1.87	9.8	11.4	—	—	17.2	0.44	—	—	—
365	East Head Pond	USA	EFL	—	4.6	—	—	—	1997	QP	SEP	—	—	—	—	—	—	—	—	—	—	—
366	East Lee	USA	EFL	—	7.6	—	144.8	1965	1968	QP	SEP	0.45	—	—	—	—	—	—	—	—	—	2
367	East Lemmon	USA	EFL	—	9.1	—	—	—	1980	OT	—	—	—	—	7.6	85.3	30.5	—	—	—	—	—
368	East Liverpool	USA	EFL	—	—	—	—	1901	1901	QP	SEP	0.04	—	—	—	—	—	—	—	—	—	—
369	East Peoria	USA	EFL	—	—	—	—	—	1994	QP	SEP	—	—	—	—	—	—	—	—	—	—	—
370	East Pittsford	USA	EFL	—	12.2	—	33.5	—	1947	OT	—	4.3	—	—	—	—	—	—	—	—	—	—
371	East Troy	USA	EFL	—	2.1	—	—	1840	1973	OT	ISC	—	—	—	—	—	—	—	—	—	—	—
372	Eastover Mining Company Dam	USA	—	—	—	—	—	—	1981	DT	WTA	—	—	—	—	—	—	—	—	—	—	1
373	Eastwick	USA	EFL	—	—	—	—	—	1932	QP	SF-SL	—	—	—	—	—	—	—	—	—	—	7
374	Eblen No. 2	USA	EFL	—	4.3	—	—	—	1977	QP	SEP	—	—	—	—	—	—	—	—	—	—	—
375	Echo Lake	USA	RFL	—	5.5	—	—	—	1958	QP	SEP	—	—	—	—	—	—	—	—	—	—	—
376	Edgemon	USA	—	—	—	—	—	—	1994	OT	—	—	—	—	—	—	—	—	—	—	—	—
377	Edison Pond	USA	EFL	—	—	—	—	—	2000	QP	SEP	—	—	—	—	—	—	—	—	—	—	—
378	Edwards	USA	TIM	—	13	—	—	—	1974	—	—	—	—	—	—	—	—	—	—	—	—	—

(*Continued*)

Table A.1 *(Continued)*

No.	Name	Country[(1)]	Type[(2)]	Type divided[(3)]	Height (m)	Crest width (m)	Crest length (m)	Year of cons.	Year of failure	Failure cause[(4)]	Failure cause divided[(5)]	Reservoir capacity (10^6 m^3)	Water volume stored above breach invert (10^6 m^3)	Water depth above breach (m)	Breach geometry Height (m)	Top width (m)	Bottom width (m)	Aver. width (m)	Aver. side slope Z:1 (h:v)	Peak outflow rate (m^3/s)	Fail. time	Life loss
379	El Dorado Canyon	USA	EFL	—	16.8	—	—	1973	1986	QP	QIS	—	—	—	—	—	—	—	—	—	—	—
380	Eight Trout Club	USA	EFL	—	6.4	—	—	1935	1990	QP	SEP	—	—	—	—	—	—	—	—	—	—	—
381	Eigiau	GB	CON	GR	10.5	—	991.5	1911	1925	QP	SEP	4.52	—	—	—	—	—	—	—	396.5	—	16
382	Eildon	AU	EFL	—	40	—	213.4	1927	1929	—	—	3390	—	—	—	—	—	—	—	—	—	—
383	Eklutna	USA	EFL	—	7.9	—	169.2	1929	1964	DT	EQ	—	—	—	—	—	—	—	—	—	—	—
384	El Salto	BO	EFL	—	15	—	30	—	1976	QP	SEP	0.45	—	—	—	—	—	—	—	—	—	—
385	Elandsdrift	ZA	EFL	—	28	—	600	1974	1974	QP	SF	3.3	—	—	—	—	—	—	—	—	—	—
386	Eleva Roller Mill	USA	EFL	—	4.3	—	—	—	1994	DT	ABA	—	—	—	—	—	—	—	—	—	—	—
387	Elk City	USA	EFL	DC	9.1	4.9	564	1925	1936	OT	—	0.74	1.18	9.44	9.14	45.5	27.7	36.6	1	—	0.83	—
388	Ellington	USA	—	—	—	—	—	—	1890	—	—	—	—	—	—	—	—	—	—	—	—	—
389	Elwha River	USA	CON	GR	33	—	135	1912	1912	QP	SEP	9.2	—	—	—	—	—	—	—	—	—	—
390	Ema	BR	EFL	—	18.5	—	370	1932	1940	QP	SEP	10	—	—	—	—	—	—	—	—	—	—
391	Emery	USA	EFL	—	16	—	130	1850	1966	QP	SEP	0.5	0.425	6.55	8.23	—	—	10.8	0.35	—	—	—
392	Empire	USA	EFL	—	12	—	—	1906	1909	QP	SEP	47	—	—	—	—	—	—	—	—	—	—
393	Enderlin Park	USA	MAS	—	3	—	—	—	2004	OT	—	—	—	—	—	—	—	—	—	—	—	—
394	English	USA	RFL	—	30.5	—	100.9	1878	1883	QP	—	18	—	—	—	—	—	—	—	—	—	—
395	English Lake	USA	—	—	—	—	—	—	1994	OT	—	—	—	—	—	—	—	—	—	—	—	—
396	English Water	USA	EFL	—	15.8	—	148	1965	1965	QP	SEP	0.61	—	—	—	—	—	—	—	—	—	—
397	Erindale-I	CA	EFL	—	10.7	—	213.4	1910	1910	OT	—	—	—	—	—	—	—	—	—	—	—	—
398	Erindale-II	CA	EFL	DC	10.7	—	213.4	1910	1912	OT	—	—	—	—	4.6	39.5	—	—	—	—	<0.5	—
399	Esperanza Farms Lake	USA	EFL	—	8.6	—	—	—	1994	OT	—	—	—	—	—	—	—	—	—	—	—	—
400	Essex Mill Dam	USA	EFL	—	4.3	—	—	—	1999	OT	—	—	—	—	—	—	—	—	—	—	—	—
401	Euclides da Cunha	BR	EFL	HD	60	—	312	1960	1977	PM	—	13	—	58.22	53	131	—	—	—	1020	7.3	—
402	Eureka Holding Pond Dike	USA	EFL	—	12.2	—	—	—	1995	QP	SEP	—	—	—	—	—	—	9.1	—	—	—	—
403	Evans	USA	EFL	—	5.5	—	—	—	1989	—	—	—	—	—	—	—	—	—	—	—	—	2
404	Fairchild	USA	EFL	—	—	—	—	—	1993	OT	—	—	—	—	—	—	—	—	—	—	—	—
405	Fairfield Swamp	USA	EFL	DC	4.6	—	—	1967	1980	QP	SEP	—	—	—	—	—	—	—	—	—	—	—
406	Fairhaven Dam	USA	EFL	—	7.6	—	—	—	1967	OT	—	—	—	—	—	—	—	—	—	—	—	—
407	Falls River	USA	MAS	—	2.1	—	—	—	1982	OT	—	—	—	—	—	—	—	—	—	—	—	—
408	Farda	USA	EFL	—	4.6	—	—	—	—	OT	—	—	—	—	—	—	—	—	—	—	—	—
409	Fergoug I	DZ	MAS	GR	43	—	316	1871	1881	OT	—	30	—	—	—	—	—	—	—	—	—	—
410	Fergoug II	DZ	MAS	GR	43	—	—	1885	1927	OT	—	30	—	—	—	—	—	—	—	—	—	—
411	Fergus Falls	USA	CON	GR	7.6	—	94.5	1908	1909	QP	—	—	—	—	—	—	—	—	—	—	—	—
412	Ferguson Lake	USA	—	—	—	—	—	—	1994	OT	—	—	—	—	—	—	—	—	—	—	—	—
413	Fertile	USA	CON	GR	—	—	—	—	1910	OT	—	—	—	—	—	—	—	—	—	—	—	—
414	Fertile Mill Dam	USA	EFL	—	3.4	—	—	—	1979	QP	SEP	—	—	—	—	—	—	—	—	—	—	—

415	Fishpond	USA	EFL	—	11.9	—	—	—	1994	QP	QIS	—	—	—	—	—	—	—	—	—	—	—
416	Flagstaff	AU	RFL	—	15.7	—	—	1963	1963	QP	SEP	—	—	—	—	—	—	—	—	—	—	—
417	Flat Creek Lake	USA	EFL	—	4.9	—	—	—	1994	PM	—	—	—	—	—	—	—	—	—	—	—	—
418	Flint River	USA	EFL	—	18.3	—	—	—	1994	OT	—	—	—	—	—	—	—	—	—	—	—	15
419	Florida Power	USA	EFL	—	12.6	—	—	1977	1979	—	—	—	—	—	—	—	—	—	—	—	—	—
420	Flowage Lake	USA	EFL	—	7.3	—	—	1916	1985	QP	SEP	—	—	—	—	—	—	—	—	—	—	—
421	Fogelman	USA	—	—	—	—	—	—	—	QP	SEP	—	0.493	11.1	12.6	—	—	7.62	0.36	—	—	—
422	Fontenelle	USA	EFL	—	47.9	—	1524	—	1965	QP	SF-SL	440	—	—	—	—	—	—	—	—	—	—
423	Fool Creek No 1	USA	EFL	—	7.3	—	—	—	1974	QP	SEP	—	—	—	—	—	—	—	—	—	—	—
424	Forbes Lake	USA	—	—	—	—	—	—	1994	OT	—	—	—	—	—	—	—	—	—	—	—	—
425	Foreman Branch	USA	EFL	—	2.1	—	—	—	1999	OT	ISC	—	—	—	—	—	—	—	—	—	—	—
426	Forman Pond	USA	EFL	—	2.7	—	—	—	1982	OT	—	—	—	—	—	—	—	—	—	—	—	—
427	Forsyth	USA	EFL	—	6.1	—	—	—	1997	QP	SEP	—	—	—	—	—	—	—	—	—	—	—
428	Forsythe	USA	EFL	—	19.8	—	—	1920	1921	QP	SEP	—	—	—	—	—	—	—	—	—	—	—
429	Fort Peck	USA	EFL	—	76	—	—	1938	1938	QP	SF-SL	24000	—	—	—	—	—	—	—	—	—	—
430	Fort Pitt-I	USA	EFL	—	1.8	—	53.3	1896	1896	OT	—	—	—	—	—	—	—	—	—	—	—	—
431	Fort Pitt-II	USA	EFL	—	3	—	53.3	1896	1903	OT	ISC	—	—	—	3	—	—	27.4	—	—	—	2
432	Fort Ritchie I	USA	EFL	—	5.8	—	—	1898	1929	—	—	—	—	—	—	—	—	12.2	—	—	—	—
433	Fort Ritchie II	USA	EFL	—	—	—	—	—	1999	QP	QIS	—	—	—	—	—	—	—	—	—	—	—
434	Fort Wayne	USA	EFL	—	—	—	—	—	1905	—	—	0.01	—	—	—	—	—	—	—	—	—	—
435	Fountain Lake	USA	—	—	—	—	—	—	1994	OT	—	—	—	—	—	—	—	—	—	—	—	—
436	Fox Trail Lake	USA	EFL	—	—	—	—	—	2000	OT	—	—	—	—	—	—	—	—	—	—	—	—
437	FP&L Cooling-Martin	USA	EFL	—	9.4	—	—	1977	1979	—	—	—	—	—	—	—	—	—	—	—	—	—
438	Frackville	USA	—	—	—	—	—	—	1901	OT	—	—	—	—	—	—	—	—	—	—	—	—
439	Francis Galloway Lake	USA	EFL	—	4.3	—	—	—	2001	QP	SEP	—	—	—	—	—	—	—	—	—	—	—
440	Frankfurt	DE	EFL	HD	9.75	—	120	1975	1977	QP	SEP	0.35	0.352	8.23	9.75	9.2	4.6	6.9	0.4	79	2.5	—
441	Franktown Parker Fpm-1	USA	EFL	—	13.4	—	—	—	1965	OT	—	—	—	—	—	—	—	—	—	—	—	—
442	Fraser Valley	USA	EFL	—	3.7	—	30.5	1946	1947	OT	—	—	—	—	—	—	—	—	—	—	—	—
443	Frazers Dam	USA	EFL	—	5.8	—	—	—	1999	OT	—	—	—	—	—	—	—	—	—	—	—	—
444	Frazier	USA	EFL	—	7.6	—	—	1915	1935	QP	SEP	—	—	—	—	—	—	—	—	—	—	—
445	Fred Burr	USA	EFL	HD	16	—	99	1947	1948	QP	SEP	0.63	0.75	10.2	10.4	—	—	—	—	654	—	—
446	Fred Trick #1357	USA	EFL	DC	4.6	—	—	1916	1974	QP	SEP	—	—	—	—	—	—	—	—	—	—	—
447	Freddies Lake	USA	EFL	—	9.4	—	—	—	1993	OT	—	—	—	—	—	—	—	10.7	—	—	—	—
448	Fredonia	USA	EFL	—	8.2	—	—	—	1938	OT	ISC	—	—	—	—	—	—	—	—	—	—	—
449	Free Lake Dam	USA	—	—	—	—	—	—	1994	OT	—	—	—	—	—	—	—	—	—	—	—	—
450	Freeman Run	USA	TIM	—	9.8	—	177	—	1942	OT	ISC	0.25	—	—	—	—	—	—	—	—	—	—
451	French Landing	USA	EFL	HD	12.2	2.4	—	1924	1925	QP	SEP	—	3.87	8.53	14.2	41	13.8	27.4	0.97	929	0.58	—
452	Frenchman Creek	USA	EFL	—	7.3	—	—	—	1995	QP	SEP	—	—	—	—	—	—	—	—	—	—	—
453	Frenchman Dam	USA	EFL	HD	12.5	6.1	884	1951	1952	QP	SEP	21	16	10.8	12.5	67	54.4	54.6	0.5	1420	3	—
454	Frias	AR	RFL	FD	17.7	—	60	1939	1970	OT	—	0.14	—	—	15	62	—	—	—	40	0.25	—
455	Friendship Hill Lower	USA	EFL	—	5.1	—	—	—	1979	OT	—	—	—	—	—	—	—	—	—	—	—	—
456	Fruit Growers	USA	EFL	—	11	—	277.4	1898	1937	QP	SF-SL	41.2	—	—	—	—	—	—	—	—	—	—

(Continued)

Table A.1 *(Continued)*

No.	Name	Country[1]	Type[2]	Type divided[3]	Height (m)	Crest width (m)	Crest length (m)	Year of cons.	Year of failure	Failure cause[4]	Failure cause divided[5]	Reservoir capacity ($10^6 m^3$)	Water volume stored above breach invert ($10^6 m^3$)	Water depth above breach (m)	Breach geometry: Height (m)	Breach geometry: Top width (m)	Breach geometry: Bottom width (m)	Breach geometry: Aver. width (m)	Breach geometry: Aver. side slope Z:1 (h:v)	Peak outflow rate (m^3/s)	Fail. time	Life loss
457	Funks	USA	EFL	—	2.1	—	—	—	1973	OT	—	—	—	—	—	—	—	—	—	—	—	—
458	Furnace Pond	USA	EFL	—	—	—	—	—	2000	QP	SEP	—	—	—	—	—	—	—	—	—	—	—
459	Galbreath Sediment	USA	—	—	—	—	—	—	1997	QP	SEP	—	—	—	—	—	—	—	—	—	—	—
460	Gallinas	USA	MAS	GR	29	—	210	1910	1957	OT	—	0.23	—	—	—	—	—	—	—	—	—	—
461	Gallipolis Locks	USA	EFL	—	7	—	—	—	1912	QP	SEP	—	—	—	—	—	—	—	—	—	—	—
462	Garant Lake	USA	EFL	—	5.5	—	—	—	1994	OT	—	—	—	—	—	—	—	—	—	—	—	—
463	Garfield	USA	EFL	—	6.1	—	—	1924	1924	—	—	0.02	—	—	—	—	—	—	—	—	—	—
464	Garza	USA	EFL	—	37	—	3352.8	1926	1926	QP	SF-SL	239	—	—	—	—	—	—	—	—	—	—
465	Geneva Dam	USA	TIM	—	4	—	—	—	1952	PM	—	—	—	—	—	—	—	—	—	—	—	—
466	George Dam	USA	—	—	—	—	—	—	1914	—	—	—	—	—	—	—	—	—	—	—	—	—
467	George Lake	USA	RFL	DC	9.1	—	—	1909	1927	—	—	—	—	—	—	—	—	—	—	—	—	—
468	Ghattara	LY	EFL	—	38.5	—	217	1972	1977	QP	SEP	5.5	—	—	—	—	—	—	—	—	—	—
469	Giles Lake Dam	USA	—	—	—	—	—	—	1994	OT	—	—	—	—	—	—	—	—	—	—	—	—
470	Gleno	IT	CON	MAS	35	—	224	1923	1923	QP	SF-OV	5	—	—	—	—	—	—	—	—	—	600
471	Goffs Mill Lake	USA	—	—	—	—	—	—	1994	OT	—	—	—	—	—	—	—	—	—	—	—	—
472	Goldsboro	USA	—	—	—	—	—	—	1890	—	—	55.5	—	—	—	—	—	—	—	—	—	—
473	Golf Course	USA	EFL	—	3	—	—	1948	1998	OT	—	—	—	—	3	—	—	—	—	—	—	—
474	Gomulak	USA	EFL	—	5.2	—	—	—	1994	OT	BOP	—	—	—	—	—	—	—	—	—	—	—
475	Goodenough	USA	EFL	—	13.1	—	—	—	1941	DT	ABA	—	—	—	—	—	—	—	—	—	—	—
476	Goodrich Creek I	USA	EFL	—	15.2	—	—	1860	1896	OT	—	—	—	—	—	—	—	—	—	—	—	7
477	Goodrich Creek II	USA	EFL	—	19.8	—	—	—	1956	QP	SEP	—	—	—	—	—	—	—	—	—	—	—
478	Goodroe Lake	USA	—	—	—	—	—	—	1994	OT	—	—	—	—	—	—	—	—	—	—	—	—
479	Goose Creek	USA	EFL	HD	6.1	3	701	1903	1916	OT	—	10.6	10.6	1.37	4.1	30.5	22.3	26.4	0.5	565	<0.5	—
480	Goose Lake	USA	EFL	—	10.4	—	—	—	1994	OT	—	—	—	—	—	—	—	—	—	—	—	—
481	Goose Neck	USA	RFL	—	21.3	—	—	1900	1900	OT	—	—	—	—	—	—	—	—	—	—	—	—
482	Gorton Pond	USA	RFL	—	2.4	—	—	—	1982	OT	—	—	—	—	—	—	—	—	—	—	—	—
483	Goshen	USA	EFL	—	2.4	—	—	1856	1979	OT	—	—	—	—	—	—	—	—	—	—	—	—
484	Gossett Lake	USA	EFL	—	8.5	—	—	—	1977	QP	—	—	—	—	—	—	—	—	—	—	—	—
485	Gouldtown - Mill 5 West Channel	USA	EFL	—	4.6	—	—	—	1998	OT	—	—	—	—	4.6	—	—	61	—	—	—	—
486	Graham Lake	USA	EFL	—	34.1	—	335.3	1922	1923	QP	SEP	200	—	—	—	—	—	—	—	—	—	—
487	Gran Cheurfas	DZ	MAS	GR	42	—	155	1884	1885	QP	SEP	17	—	—	—	—	—	—	—	—	—	10
488	Grand Forks Co. Com. #1	USA	EFL	—	8.8	—	—	—	2000	QP	QIS	—	—	—	—	—	—	—	—	—	—	2
489	Grand Mesa #8	USA	EFL	—	5.5	—	—	—	1983	OT	BOS	—	—	—	—	—	—	—	—	—	—	—
490	Grand Rapids	USA	EFL	DC	7.62	3.7	441	1874	1900	OT	—	0.22	0.0255	6.4	6.4	12.2	6	9.1	2.26	—	0.5	1
491	Granite Creek	USA	—	—	—	—	—	—	1971	—	—	—	—	—	—	—	—	—	—	1841	—	—

492	Grass Valley	USA	EFL	—	14.7	—	159	1891	1895	—	—	—	—	—	—	—	—	—	—	—	—	—
493	Great Falls	USA	CON	GR	28	—	—	—	1926	OT	—	—	—	—	—	—	—	—	—	—	—	—
494	Great Works	USA	TIM	—	6.1	—	—	—	1981	—	—	—	—	—	—	—	—	—	—	—	—	—
495	Greeley	USA	—	—	2.7	—	—	—	1993	DT	ABA	—	—	—	—	—	—	—	—	—	—	—
496	Green Acres	USA	—	—	—	—	—	—	1997	OT	—	—	—	—	—	—	—	—	—	—	—	—
497	Greenlick	USA	EFL	CF	19	—	259	1901	1904	QP	SEP	0.6	—	—	—	—	—	—	—	—	—	—
498	Greenview Dam	USA	EFL	—	7.3	—	—	—	2004	QP	SEP	—	—	—	—	—	—	—	—	—	—	—
499	Grenloch	USA	EFL	—	—	—	—	—	1940	OT	—	—	—	—	—	—	—	—	—	—	—	—
500	Grindstone	USA	EFL	—	7	—	—	—	1944	OT	—	—	—	—	—	—	—	—	—	—	—	—
501	Gropp's Lake	USA	EFL	—	7.6	—	—	1900	1979	QP	QIS	—	—	—	6.1	—	—	15.2	—	—	—	—
502	Gros Ventre	USA	EFL	—	54.9	—	—	1925	1927	QP	SEP	80.2	—	—	—	—	—	—	—	—	—	—
503	Guddah	IN	EFL	—	28	—	—	1956	1956	QP	—	—	—	—	—	—	—	—	—	—	—	—
504	Gunnison	USA	EFL	—	6.5	—	—	—	1890	QP	SEP	—	—	—	—	—	—	—	—	—	—	—
505	Haas Pond	USA	EFL	—	4	—	—	—	1984	QP	SEP	—	0.0234	2.99	4	12.2	9.1	10.7	0.38	—	—	—
506	Haeker Dam	USA	EFL	—	7	—	—	—	1986	OT	—	—	—	—	2	—	—	4.6	—	—	—	—
507	Hagmaeir North	USA	EFL	—	9.1	—	—	—	1982	OT	—	—	—	—	—	—	—	—	—	—	—	—
508	Haledon	USA	EFL	—	7.6	—	—	—	1903	—	—	—	—	—	—	—	—	—	—	—	—	—
509	Halfmoon Bay	USA	EFL	—	24.4	—	—	—	1926	OT	—	—	—	—	—	—	—	—	—	—	—	—
510	Hall Lake Dam	USA	EFL	—	3.7	—	—	—	1999	OT	—	—	—	—	1.8	—	—	3	—	—	—	—
511	Halls Lake	USA	EFL	—	3.4	—	—	1919	1984	QP	SEP	—	—	—	—	—	—	—	—	—	—	—
512	Hamilton	USA	TIM	—	2.4	—	—	—	1978	—	—	—	—	—	—	—	—	—	—	—	—	—
513	Hamilton Dam	USA	EFL	—	4	—	—	—	1997	OT	—	—	—	—	—	—	—	—	—	—	—	—
514	Hamilton Mill	USA	—	—	—	—	—	—	1996	QP	SEP	—	—	—	—	—	—	—	—	—	—	—
515	Hamlin Lake I	USA	TIM	—	3.6	—	—	1888	1912	QP	—	—	—	—	—	—	—	—	—	—	—	—
516	Hamlin Lake II	USA	EFL	—	4	—	—	1888	1912	—	—	—	—	—	—	—	—	—	—	—	—	—
517	Hannawa Falls I	USA	EFL	—	10.7	—	99.1	—	1900	OT	—	—	—	—	—	—	—	—	—	—	—	—
518	Hannawa Falls II	USA	MAS	GR	9.1	—	—	1900	1914	QP	—	—	—	—	—	—	—	—	—	—	—	—
519	Hans Strijdom	ZA	RFL	—	18.5	—	400	1977	1977	OT	—	—	—	—	—	—	—	—	—	—	—	—
520	Harell Lake	USA	—	—	—	—	—	—	1994	OT	—	—	—	—	—	—	—	—	—	—	—	—
521	Harper Lake	USA	EFL	—	7	—	—	—	1994	OT	—	—	—	—	—	—	—	—	—	—	—	—
522	Harrison	USA	EFL	—	3	—	—	—	1931	DT	BUA	—	—	—	—	—	—	61	—	—	—	—
523	Harrison County	USA	EFL	—	17.3	—	—	—	1993	OT	—	—	—	—	—	—	—	—	—	—	—	—
524	Hart Lake	USA	EFL	HD	—	4.3	—	1920	1986	QP	SEP	—	6.35	10.7	10.8	—	—	73.9	3.03	—	—	—
525	Hatchery Lake	USA	EFL	—	7.3	—	—	1931	1985	QP	QIS	—	—	—	—	—	—	—	—	—	—	—
526	Hatchtown	USA	EFL	CF	18.9	6.1	237.7	1908	1914	QP	SEP	15	14.8	16.8	18.3	180	140	151	2.42	3080	3	—
527	Hatfield Headrace	USA	—	—	—	—	—	—	1993	—	—	—	—	—	—	—	—	—	—	—	—	—
528	Hatfield I	USA	EFL	—	6.8	—	1524	1908	1911	OT	ISC	12.3	—	—	6.8	—	6.1	91.5	—	3400	2	—
529	Hatfield II	USA	EFL	—	29.6	—	—	—	1993	QP	SEP	—	—	—	—	—	—	—	—	—	—	—
530	Hauser Lake I	USA	STL	—	21	—	192	1906	1908	QP	SEP	66	—	—	—	—	—	—	—	—	—	—
531	Hauser Lake II	USA	CON	GR	40	—	219	1911	1969	—	—	84	—	—	—	—	—	—	—	—	—	—
532	Hazel Lake	USA	EFL	—	2.6	—	—	—	1995	QP	SEP	—	—	—	—	—	—	—	—	—	—	—
533	Hebron I	USA	EFL	HD	17.1	3.7	1127.8	1913	1914	QP	SEP	—	—	12.2	15.3	61	30.4	45.7	0.5	—	2.25	—

(Continued)

Table A.1 *(Continued)*

No.	Name	Country[(1)]	Type[(2)]	Type divided[(3)]	Height (m)	Crest width (m)	Crest length (m)	Year of cons.	Year of failure	Failure cause[(4)]	Failure cause divided[(5)]	Reservoir capacity ($10^6 m^3$)	Water volume stored above breach invert ($10^6 m^3$)	Water depth above breach (m)	Breach geometry					Peak outflow rate (m^3/s)	Fail. time	Life loss
															Height (m)	Top width (m)	Bottom width (m)	Aver. width (m)	Aver. side slope Z:1 (h:v)			
534	Hebron II	USA	EFL	—	17.1	—	1127.8	1918	1942	OT	—	—	—	—	—	—	—	—	—	—	—	—
535	Hein Coulee	USA	EFL	—	6.7	—	—	—	1988	QP	SEP	—	—	—	—	—	—	—	—	—	—	—
536	Heiwaike	JP	EFL	—	22.3	—	82.5	1949	1951	OT	ISC	0.22	—	—	—	—	—	—	—	—	—	—
537	Hell Hole I	USA	RFL	DC	30	—	—	1964	1964	OT	—	2.6	—	—	—	—	—	—	—	—	—	—
538	Hell Hole II	USA	RFL	—	67.1	21.3	—	1964	1964	QP	SEP	30.6	30.6	35.1	56.4	—	—	121	0.96	7360	—	—
539	Hematite	USA	EFL	—	4	—	—	—	1998	QP	SEP	—	—	—	3.4	—	—	4.1	—	—	—	—
540	Hemet	USA	—	—	6.1	—	83.2	1923	1927	OT	—	8.63	—	—	6.1	83.2	—	—	—	1600	3	—
541	Hempstead	USA	EFL	—	2.1	—	—	—	1982	OT	—	—	—	—	—	—	—	—	—	—	—	—
542	Henry	USA	—	—	—	—	—	—	1996	QP	SEP	—	—	—	—	—	—	—	—	—	—	—
543	Henry Kaufman Pond Dam	USA	EFL	—	—	—	—	—	1997	OT	—	—	—	—	1.8	—	—	—	—	—	—	—
544	Herrin	USA	EFL	CF	8.2	4.6	—	—	1935	OT	—	—	—	>10.7	10.7	—	—	47.2	1.14	—	—	—
545	Hesperia Dam	USA	EFL	—	3.4	—	—	—	1986	OT	—	—	—	—	—	—	—	—	—	—	—	—
546	Hester Lake	USA	EFL	—	—	—	—	—	1990	QP	SEP	—	—	—	—	—	—	—	—	—	—	—
547	Higgins Millpond	USA	EFL	—	2.4	—	—	—	1979	OT	—	—	—	—	—	—	—	—	—	—	—	—
548	High Falls	USA	EFL	—	9.1	—	—	—	1999	OT	ISC	—	—	—	—	—	—	—	—	—	—	—
549	Highland Lake	USA	CON	—	3.7	—	—	—	1996	—	—	—	—	—	—	—	—	—	—	—	—	—
550	Highline	USA	EFL	—	10.4	—	—	1930	1984	OT	—	—	—	—	—	—	—	—	—	—	—	—
551	Hi-Land Lake	USA	EFL	—	4.3	—	—	1882	1982	QP	SEP	—	—	—	—	—	—	—	—	—	—	—
552	Hill #1	USA	EFL	—	7.6	—	—	—	2001	OT	—	—	—	—	—	—	—	—	—	—	—	—
553	Hill Woodward	USA	CON	GR	10.7	—	30.5	1913	1918	QP	SEP	—	—	—	—	—	—	—	—	—	—	1
554	HilltopAka Fairmont	USA	EFL	—	8.5	—	—	—	1940	PM	—	—	—	—	—	—	—	—	—	—	—	—
555	Hinds Lake	CA	EFL	CF	12	—	5180	1980	1982	QP	SEP	7500	—	—	—	—	—	—	—	—	—	—
556	Hog Waste Lagoon Dike	USA	—	—	—	—	—	—	1999	DT	WTA	—	—	—	—	—	—	—	—	—	—	—
557	Holbrook Pond	USA	CON	—	4.6	—	—	—	1982	OT	—	—	—	—	—	—	—	—	—	—	—	—
558	Holland Dam Site A	USA	MAS	BU	4	—	—	—	1997	QP	SEP	—	—	—	—	—	—	—	—	—	—	—
559	Hollenbeck Dam	USA	—	—	—	—	—	—	1996	OT	—	—	—	—	—	—	—	—	—	—	—	—
560	Holliday Creek	USA	EFL	—	—	—	—	—	1901	—	—	—	—	—	—	—	—	—	—	—	—	—
561	Holmdel Park	USA	EFL	—	—	—	—	—	1989	QP	SEP	—	—	—	—	—	—	—	—	—	—	—
562	Holmes Creek	USA	EFL	—	19.8	—	—	1903	1924	QP	SF-SL	—	—	—	—	—	—	—	—	—	—	—
563	Holoka Lake	USA	EFL	—	5.8	—	—	—	1994	OT	—	—	—	—	—	—	—	—	—	—	—	—
564	Holyoke-I	USA	TIM	—	9	—	—	1847	1848	—	—	—	—	—	—	—	—	—	—	—	—	—
565	Holyoke-II	USA	TIM	—	9	—	—	1847	1879	—	—	—	—	—	—	—	—	—	—	—	—	—
566	Holyoke-III	USA	TIM	—	9	—	—	1847	1881	—	—	—	—	—	—	—	—	—	—	—	—	—
567	Holyoke-IV	USA	TIM	—	9	—	—	1847	1882	—	—	—	—	—	—	—	—	—	—	—	—	—
568	Holyoke-V	USA	TIM	—	9	—	—	1847	1884	—	—	—	—	—	—	—	—	—	—	—	—	—

569	Home Ranch	USA	EFL	—	6.1	—	—	—	1982	OT	—	—	—	—	—	—	—	—	—	—	—	—
570	Homestake	USA	EFL	—	27.7	—	—	—	1977	QP	SEP	—	—	—	27.7	76	30.5	—	—	—	—	—
571	Hope Mills Dam	USA	EFL	—	10	—	—	—	2003	—	—	—	—	—	—	—	—	—	—	—	—	—
572	Hope Ranch	USA	—	—	—	—	—	—	1902	—	—	0.01	—	—	—	—	—	—	—	—	—	—
573	Hope Reservoir	USA	EFL	—	7	—	—	1882	1907	QP	SEP	—	—	—	—	—	—	—	—	—	—	—
574	Horn Rapids	USA	TIM	—	1.5	—	—	1893	1997	OT	—	—	—	—	—	—	—	—	—	—	—	—
575	Hornell	USA	EFL	—	—	—	—	1912	1912	QP	SEP	—	—	—	—	—	—	—	—	—	—	—
576	Horse Creek I	USA	EFL	FD	16.9	4.9	5150	1912	1914	QP	SEP	21	12.8	7.01	12.8	76.2	70	73.1	0.83	3890	3	—
577	Horse Creek II	USA	EFL	—	12.2	—	701	1914	1935	OT	—	3.1	—	—	—	—	—	—	—	—	—	—
578	Horse Lake I	USA	EFL	—	3.7	—	—	—	1938	OT	ISC	—	—	—	—	—	—	—	—	—	—	—
579	Horse Lake II	USA	EFL	—	3.7	—	—	—	1952	DT	ABA	—	—	—	—	—	—	—	—	—	—	—
580	Horsehead Creek Lake	USA	EFL	—	7.6	—	—	—	1994	OT	—	—	—	—	—	—	—	—	—	—	—	—
581	Horseshoe CO	USA	EFL	—	10.1	—	—	—	1973	QP	SEP	—	—	—	—	—	—	—	—	—	—	—
582	Horseshoe WA	USA	EFL	—	—	—	—	—	1974	QP	SF-SO	—	—	—	—	—	—	—	—	—	—	—
583	Hortmans Pond	USA	EFL	—	4.8	—	—	—	1994	OT	—	—	—	—	—	—	—	—	—	—	—	—
584	Horton	USA	EFL	—	10.4	—	356.6	1924	1925	OT	ISC	0.05	—	—	—	—	—	—	—	—	—	—
585	Housatonic	USA	MAS	ARC	12.2	—	193.9	1869	1891	QP	—	—	—	—	—	—	—	—	—	—	—	—
586	House Autry Mill	USA	—	—	—	—	—	—	1999	OT	—	—	—	—	3	—	—	14.3	—	—	—	—
587	Housers Millpond	USA	—	—	—	—	—	—	1994	OT	—	—	—	—	—	—	—	—	—	—	—	—
588	Houston Lake	USA	EFL	—	7.9	—	—	—	1994	OT	—	—	—	—	—	—	—	—	—	—	—	—
589	Howard	USA	EFL	—	7.6	—	—	1929	1982	DT	BUA	0.4	—	—	—	—	—	—	—	—	—	—
590	Huaccoto	PE	—	—	—	—	—	—	1974	—	—	—	—	—	—	—	—	—	—	—	—	—
591	Hubacov	CZ	EFL	—	5.7	—	112	1760	1974	—	—	0.15	—	—	—	—	—	—	—	—	—	—
592	Huffman Antelope	USA	EFL	—	4.6	—	—	—	1952	QP	SEP	—	—	—	—	—	—	—	—	—	—	—
593	Humboldt Milldam	USA	CON	GR	4	—	—	—	1969	OT	—	—	—	—	—	—	—	—	—	—	—	—
594	Hunter's Creek	USA	EFL	—	12.2	—	152.4	—	1932	PM	—	—	—	—	—	—	—	—	—	—	—	—
595	Huntington	USA	EFL	—	10.1	—	—	1954	1984	OT	—	—	—	—	—	—	—	—	—	—	—	—
596	Hunts Brook	USA	EFL	—	3	—	—	—	1982	OT	—	—	—	—	—	—	—	—	—	—	—	—
597	Hutchinson	USA	EFL	HD	—	—	—	1960	1994	OT	—	—	1.17	4.42	3.75	—	—	33.4	1.14	—	—	—
598	Huttos Lake	USA	EFL	—	3.7	—	—	—	1979	OT	—	—	—	—	—	—	—	45.7	—	—	—	—
599	Hyogiri	KR	EFL	—	15.6	—	109.5	1940	1961	OT	—	0.22	—	—	—	—	—	—	—	—	—	250
600	Idbar	YU	CON	ARC	38	—	108	1959	1960	QP	SEP	1.9	—	—	—	—	—	—	—	—	—	—
601	Idylwilde	USA	CON	GR	17.4	—	—	1913	1976	OT	—	—	—	—	—	—	—	—	—	—	—	—
602	IHNC East Bank North	USA	EFL	—	—	—	—	—	2005	QP	QIC	—	—	—	—	76.3	76.3	76.3	—	—	—	—
603	IHNC East Bank South	USA	EFL	—	—	—	—	—	2005	QP	QIC	—	—	—	—	274.5	274.5	274.5	—	—	—	—
604	IMC-AGRICO	USA	—	—	—	—	—	—	1994	QP	SEP	—	—	—	—	—	—	—	—	—	—	—
605	Indian Creek	USA	MAS	GR	—	—	—	—	1894	OT	—	—	—	—	—	—	—	—	—	—	—	—
606	Indian Lake	USA	EFL	—	7.9	—	—	1940	1983	QP	SEP	—	—	—	—	—	—	—	—	—	—	—
607	Ingles	USA	—	—	—	—	—	—	2001	OT	ISC	—	—	—	—	—	—	—	—	—	—	—
608	Institute Pond	USA	EFL	—	7.9	—	—	—	1911	—	—	—	—	—	—	—	—	—	—	—	—	—
609	Iowa Beef Processors	USA	EFL	HD	4.6	—	305	1971	1993	DT	ABA	0.333	0.333	4.42	4.57	—	—	16.8	0.33	—	—	—
610	Ireland #5 I	USA	EFL	—	—	—	—	—	1973	OT	—	—	—	—	—	—	—	—	—	—	—	—

(Continued)

Table A.1 (*Continued*)

No.	Name	Country[1]	Type[2]	Type divided[3]	Height (m)	Crest width (m)	Crest length (m)	Year of cons.	Year of failure	Failure cause[4]	Failure cause divided[5]	Reservoir capacity ($10^6 m^3$)	Water volume stored above breach invert ($10^6 m^3$)	Water depth above breach (m)	Breach geometry Height (m)	Top width (m)	Bottom width (m)	Aver. width (m)	Aver. side slope Z:1 (h:v)	Peak outflow rate (m^3/s)	Fail. time	Life loss
611	Ireland #5 II	USA	EFL	HD	5.2	2.4	—	1946	1984	QP	SEP	—	0.16	3.81	5.18	—	—	13.5	0.38	110	—	—
612	Iron River	USA	EFL	—	—	—	—	—	1922	OT	—	—	—	—	—	—	—	—	—	—	—	—
613	Irukaike	JP	EFL	—	28	—	724	1633	1868	OT	ISC	19	—	—	—	—	—	—	—	—	—	1200
614	Isleton Earth	USA	EFL	—	2.4	—	—	—	1972	—	—	—	—	—	—	—	—	—	—	—	—	—
615	ISP Minerals	USA	—	—	—	—	—	—	1992	OT	BOP	—	—	—	16.8	15.3	6.1	—	—	—	—	—
616	Ithaca	USA	—	—	4.6	—	—	—	1905	—	—	—	—	—	—	—	—	—	—	—	—	—
617	Ivex of Ohio Upper Lake	USA	EFL	—	—	—	—	—	1994	OT	—	—	—	—	—	15.2	12.2	—	—	—	—	—
618	Ivoryton Pond	USA	EFL	—	1.8	—	—	—	1982	OT	—	—	—	—	—	—	—	—	—	—	—	—
619	J. A. Phillips	USA	EFL	—	7.6	—	—	—	1977	OT	—	—	—	—	—	—	—	—	—	—	—	—
620	J.B. Dunnell	USA	EFL	—	8.8	—	—	—	1999	OT	—	—	—	—	8.8	—	—	—	—	—	—	—
621	Jackson Bluff	USA	EFL	—	18.3	—	—	—	1957	—	—	—	—	—	—	—	—	366	—	—	—	—
622	Jackson Creek	USA	EFL	—	—	—	—	—	1977	QP	SEP	—	—	—	—	—	—	9.1	—	—	—	—
623	Jackson Reif	USA	—	—	—	—	—	—	1977	QP	SEP	—	—	—	—	—	—	—	—	—	—	—
624	Jacobs Creek	USA	—	—	—	—	—	—	—	QP	SEP	—	0.423	20.1	21.3	—	—	17.5	0.61	—	—	—
625	Jan Land Company Lake	USA	EFL	—	5.8	—	—	—	1998	OT	—	—	—	—	4.9	7.6	4.6	—	—	—	—	—
626	Janesville	USA	CON	BU	2.7	—	30.5	—	1912	QP	—	—	—	—	—	—	—	—	—	—	—	—
627	Jay	USA	CON	GR	5.2	—	—	—	1987	OT	—	—	—	—	—	—	—	30.5	—	—	—	—
628	Jaycees Pond	USA	EFL	—	5.2	—	—	—	1995	QP	SEP	—	—	—	—	12.2	7.6	—	—	—	—	—
629	Jefferson County	USA	—	—	—	—	—	—	1897	—	—	—	—	—	—	—	—	—	—	—	—	2
630	Jennings	USA	EFL	—	6.4	—	—	—	1977	QP	SEP	—	—	—	—	—	—	—	—	—	—	—
631	Jennings Creek 16	USA	RFL	—	16.8	—	113.4	1960	1964	QP	SEP	0.31	—	—	—	—	—	—	—	—	—	—
632	Jennings Creek 3	USA	RFL	—	21	—	92	1962	1963	QP	SEP	0.43	—	—	—	—	—	—	—	—	—	—
633	Jim Falls	USA	EFL	—	10.7	—	—	—	1941	OT	—	—	—	—	—	—	—	—	—	—	—	—
634	Johnson Ck # 4	USA	EFL	—	8.8	—	—	—	1997	QP	QIS	—	—	—	—	—	—	—	—	—	—	—
635	Johnson Dam	USA	EFL	—	14	—	—	—	1945	QP	SEP	—	—	—	—	—	—	—	—	—	—	—
636	Johnson Lake I	USA	EFL	—	7.6	3	—	—	1998	OT	ISC	—	—	—	7.6	30.5	10.7	—	—	—	—	—
637	Johnson Lake II	USA	EFL	—	7.6	—	—	—	2004	OT	—	—	—	—	4.6	—	12.2	—	—	—	—	—
638	Johnson Pond	USA	RFL	—	3.7	—	—	—	1982	OT	—	—	—	—	—	—	—	12.2	—	—	—	—
639	Johnston City	USA	EFL	HD	4.3	1.8	—	1921	1981	QP	SEP	0.575	0.575	3.05	5.18	13.4	2	8.23	1	—	—	—
640	Johnstown 1 South Fork 1	USA	EFL	CF	38.1	3.05	284	1842	1889	OT	—	18.9	18.9	24.6	24.4	128	61	94.5	1.38	8500	—	2209
641	Jones Lake	USA	EFL	—	4	—	—	—	1999	OT	—	—	—	—	—	—	—	—	—	—	—	—
642	Juba Old Debris Barrier 1	USA	EFL	—	—	—	—	—	1907	OT	ISC	—	—	—	—	—	—	—	—	—	—	—
643	Julesburg	USA	EFL	—	16.8	—	—	1905	1910	QP	SEP	34.7	—	—	—	—	—	—	—	—	—	—
644	Jund	USA	—	—	—	—	—	—	1993	DT	ABA	—	—	—	—	—	—	27.5	—	—	—	—
645	K. U. - Graham	USA	EFL	—	8.5	—	—	1935	—	DT	ABA	—	—	—	—	—	—	—	—	—	—	—
646	Kaddam	IN	EFL	HD	41	—	—	1957	1958	OT	ISC	215	—	—	15.2	30	—	—	—	11000	1	—

647	Kaila I	IN	EFL	—	26.3	—	213.4	1955	1959	QP	SF-SL	14	—	—	—	—	—	—	—	—	—	—
648	Kaila II	IN	EFL	—	24.5	—	—	—	1965	OT	ISC	—	—	—	—	—	—	—	—	—	—	—
649	Kantale	LK	EFL	—	26.8	—	2524	1869	1986	QP	SEP	136	—	—	—	—	—	—	—	—	—	135
650	Karval	USA	EFL	—	18.3	—	—	—	1975	QP	SEP	—	—	—	—	—	—	—	—	—	—	—
651	Kedar Nala	IN	EFL	—	20	—	—	1964	1964	QP	SEP	17	—	—	—	—	—	—	—	—	—	—
652	Keene	USA	RFL	—	4.5	—	—	—	1895	OT	—	—	—	—	—	—	—	—	—	—	—	—
653	Keith Kindschi	USA	—	—	—	—	—	—	1978	OT	ISC	—	—	—	—	—	—	—	—	—	—	—
654	Kelly Barnes	USA	EFL	HD	11.6	6.1	122	1948	1977	OT	—	0.505	0.777	11.3	12.8	35	18	27.3	0.85	680	0.5	39
655	Kelly Dam	USA	RFL	—	—	—	—	1862	1912	OT	—	—	—	—	—	—	—	—	—	—	—	—
656	Kellys Pond	USA	EFL	—	4.6	—	—	—	1999	OT	—	—	—	—	—	—	—	—	—	—	—	—
657	Kendall Lake	USA	EFL	HD	6.7	—	128	1990	1990	OT	—	0.728	—	—	—	—	—	—	—	—	—	4
658	Kenilworth Lake	USA	EFL	—	5.2	—	—	—	1995	—	—	—	—	—	—	—	—	—	—	—	—	—
659	Kennedy Lake	USA	EFL	—	5.6	—	—	—	1994	OT	—	—	—	—	—	—	—	—	—	—	—	—
660	Kern Brothers	USA	EFL	CF	16.5	—	—	—	1949	—	—	—	—	—	—	—	—	—	—	—	—	—
661	Kerrville Ponding	USA	—	GR	10.7	—	—	1980	1984	QP	—	—	—	—	—	—	—	—	—	—	—	—
662	Kersey Lake	USA	EFL	—	2.8	—	—	—	1994	OT	—	—	—	—	—	—	—	—	—	—	—	—
663	Ketner	USA	EFL	—	13.7	—	179.1	1911	1911	OT	—	—	—	—	—	—	—	—	—	—	—	—
664	Khadakwvasla	IN	MAS	GR	33	—	1471	1879	1961	OT	—	137	—	—	—	—	—	—	—	2780	<4	250
665	Kharagpur	IN	EFL	—	24	—	—	1961	1961	OT	—	55	—	—	—	—	—	—	—	—	—	—
666	Kilbourn City	USA	—	—	—	—	—	—	1897	—	—	—	—	—	—	—	—	—	—	—	—	—
667	Killington	GB	EFL	—	18	—	250	1820	1836	OS	—	3.5	—	—	—	—	—	—	—	—	—	—
668	Killingworth I	USA	MAS	GR	3.7	—	—	—	1938	OT	—	0.1	—	—	—	—	—	—	—	—	—	—
669	Killingworth II	USA	EFL	—	5.5	—	—	—	1938	OT	—	0.5	—	—	—	—	—	—	—	—	—	—
670	King Meadow	USA	—	—	—	—	—	—	1965	OT	—	—	—	—	—	—	—	—	—	—	—	—
671	Kirby Dam	USA	EFL	—	—	—	—	1905	1990	QP	SEP	—	—	—	—	—	—	—	—	—	—	—
672	Kirbys Mill Dam	USA	TIM	—	4.3	—	—	—	1999	OT	—	—	—	—	—	—	—	—	—	—	—	—
673	Kisatchie Lake	USA	EFL	—	7.6	—	—	—	1991	OT	—	—	—	—	—	—	—	—	—	—	—	—
674	Klickitat Mill Pond	USA	TIM	—	—	—	—	—	1998	PM	—	—	—	—	—	—	—	—	—	—	—	—
675	Knife Lake	USA	EFL	—	6.1	—	61	—	1972	—	—	9.86	—	—	6.1	12.2	—	—	—	1099	5	4
676	Knodle	USA	—	—	—	—	—	—	1993	QP	SEP	—	—	—	—	—	—	—	—	—	—	—
677	Knolbrook	USA	—	—	—	—	—	—	1894	—	—	—	—	—	—	—	—	—	—	—	—	—
678	Knoxville	USA	EFL	—	—	—	—	—	1883	QP	—	—	—	—	—	—	—	—	—	—	—	—
679	Kodaganar	IN	EFL	—	11.5	—	2425	1977	1977	OT	—	12.3	—	—	—	—	—	—	—	1280	—	—
680	Kohlsville Dam	USA	—	—	—	—	—	—	1980	QP	SEP	—	—	—	—	—	—	—	—	—	—	—
681	Kohodiar	IN	EFL& CON	—	36	—	459	1963	1983	—	—	41	—	—	—	—	—	—	—	—	—	—
682	Komoro	JP	CON	BU	16	—	96	1927	1928	QP	SEP	0.13	—	—	—	—	—	—	—	—	—	—
683	Kosterville Mill	USA	TIM	—	—	—	—	—	1984	—	—	—	—	—	—	—	—	—	—	—	—	—
684	Kraftsmen	USA	EFL	HD	—	—	—	—	1994	OT	—	—	0.177	3.66	3.2	—	—	14.5	1.48	—	—	—
685	Krone Dam	USA	—	—	—	—	—	—	1993	OT	—	—	—	—	—	—	—	—	—	—	—	—
686	Kundli	IN	MAS	GR	45	—	—	1924	1925	QP	—	—	—	—	—	—	—	—	—	—	—	—
687	La Blonde	USA	EFL	—	5.2	—	—	—	1992	OT	BOS	—	—	—	—	—	—	—	—	—	—	—

(Continued)

Table A.1 *(Continued)*

No.	Name	Country[(1)]	Type[(2)]	Type divided[(3)]	Height (m)	Crest width (m)	Crest length (m)	Year of cons.	Year of failure	Failure cause[(4)]	Failure cause divided[(5)]	Reservoir capacity (10^6 m^3)	Water volume stored above breach invert (10^6 m^3)	Water depth above breach (m)	Breach geometry: Height (m)	Breach geometry: Top width (m)	Breach geometry: Bottom width (m)	Breach geometry: Aver. width (m)	Breach geometry: Aver. side slope Z:1 (h:v)	Peak outflow rate (m^3/s)	Fail. time	Life loss
688	La Fruta	USA	EFL	HD	—	4.9	—	1930	1930	QP	SEP	—	78.9	7.9	14	—	—	58.8	0.3	—	—	—
689	La Laguna	MX	EFL	—	16	—	675	1912	1969	QP	SEP	4.3	—	—	—	—	—	—	—	—	—	—
690	La Regadera	CO	EFL	—	36.9	—	—	—	1937	QP	—	—	—	—	—	—	—	—	—	—	—	—
691	Lac Vieux Desert	USA	EFL	—	2.6	—	—	—	1941	OT	—	—	—	—	—	—	—	—	—	—	—	—
692	Lackawaxen	USA	—	—	—	—	—	—	1942	OT	—	—	—	—	—	—	—	—	—	—	—	8
693	Lacomb Diversion	USA	EFL	—	—	—	—	—	1998	—	—	—	—	—	—	—	—	—	—	—	—	—
694	Ladysmith	USA	TIM	—	8.5	—	—	—	1994	OT	—	—	—	—	—	—	—	—	—	—	—	—
695	Lafayette	USA	EFL	—	36.6	—	—	1928	1928	QP	SF-SL	—	—	—	—	—	—	—	—	—	—	—
696	Lake	USA	EFL	CF	15.9	—	—	—	1893	OT	—	—	—	—	—	—	—	—	—	—	—	—
697	Lake Jimmy Carter	USA	EFL	—	4.6	—	—	—	1999	OT	—	—	—	—	—	—	—	—	—	—	18	—
698	Lake O Hills	USA	EFL	—	4	—	—	—	1972	QP	SEP	—	—	—	—	—	—	—	—	—	—	1
699	Lake Adelpha	USA	—	—	—	—	—	—	1974	—	—	—	—	—	—	—	—	—	—	—	—	—
700	Lake Algonquin	USA	CON	GR	5.5	—	—	—	1949	—	—	—	—	—	—	—	—	—	—	—	—	—
701	Lake Avalon	USA	EFL	HD	14.5	—	420	1894	1904	QP	SEP	7.75	31.5	13.7	14.6	—	—	130	0.52	2320	2	—
702	Lake Barcroft	USA	EFL	HD	21	—	62.5	1913	1972	OT	—	3.1	—	—	11	23	—	—	—	—	>1	—
703	Lake Bray Dam	USA	EFL	—	4.6	—	—	—	1999	OT	—	—	—	—	—	—	—	—	—	—	—	—
704	Lake Carroll Pond 2	USA	EFL	—	4.3	—	—	—	1990	OT	—	—	—	—	—	—	—	—	—	—	—	—
705	Lake Cawndilla	AU	EFL	—	12.5	—	65.8	1961	1962	QP	SEP	730	—	—	—	—	—	—	—	—	—	—
706	Lake Center	USA	EFL	—	8.2	—	—	—	1991	—	—	—	—	—	—	—	—	—	—	—	—	—
707	Lake Charles	USA	EFL	—	7.3	—	—	1964	1979	QP	QIS	—	—	—	—	—	—	—	—	—	—	—
708	Lake Clopine	USA	EFL	—	4.1	—	—	—	1994	OT	—	—	—	—	—	—	—	—	—	—	—	—
709	Lake Como	USA	EFL	—	4.6	—	—	—	1980	OT	—	—	—	—	—	—	—	—	—	—	—	—
710	Lake Dawn	USA	EFL	—	—	—	—	—	1950	OT	—	—	—	—	—	—	—	—	—	—	—	—
711	Lake Dixie	USA	EFL	—	—	—	—	—	1940	OT	—	—	—	—	—	—	—	—	—	—	—	—
712	Lake Dockery	USA	EFL	—	7.3	—	—	—	2004	DT	ABA	—	—	—	—	—	—	—	—	—	—	—
713	Lake Emily	USA	EFL	—	—	—	—	—	1993	OT	—	—	—	—	—	—	—	—	—	—	—	—
714	Lake Forest	USA	EFL	—	8.2	—	—	—	2003	QP	SF-SO	—	—	—	—	—	—	—	—	—	—	—
715	Lake Francis I	USA	EFL	HD	15.2	4.9	302.4	1899	1899	QP	SEP	0.865	0.789	14	17.1	30	10.4	18.9	0.65	—	1	—
716	Lake Francis II	USA	EFL	—	24	—	396	1901	1935	QP	SEP	2.3	—	—	—	—	—	—	—	—	—	—
717	Lake Gary Dam	USA	EFL	—	12.2	—	—	—	1995	QP	SEP	—	—	—	—	—	—	—	—	—	—	—
718	Lake Genevieve	USA	EFL	HD	7.6	—	—	1930	1985	DT	ABA	—	0.68	6.71	7.92	—	—	16.8	1.54	—	—	—
719	Lake George I	USA	EFL	—	5.8	—	335.3	1907	1909	OT	—	2.5	—	—	—	—	—	—	—	—	—	—
720	Lake George II	USA	EFL	—	5.8	—	335.3	1909	1914	OT	ISC	2.5	—	—	—	—	—	—	—	—	—	—
721	Lake Hemet	USA	EFL	—	45	—	64	1923	1927	OT	—	17	—	—	—	—	—	—	—	—	—	1
722	Lake Housatonic I	USA	MAS	GR	7.6	—	—	1869	1891	QP	SF-OV	—	—	—	—	—	—	—	—	—	—	—
723	Lake Housatonic II	USA	MAS	GR	7.6	—	—	—	1902	DT	ABA	—	—	—	—	—	—	—	—	—	—	—

724	Lake Hyenga	USA	EFL	—	4.3	—	—	—	1999	OT	—	0.02	—	—	—	—	—	—	—	—	—	—
725	Lake Jennifer Lower	USA	—	—	—	—	—	—	1994	OT	—	—	—	—	—	—	—	—	—	—	—	—
726	Lake Jennifer Upper	USA	—	—	—	—	—	—	1994	OT	—	—	—	—	—	—	—	—	—	—	—	—
727	Lake Keowee	USA	STL	—	—	—	—	1978	1978	—	—	—	—	—	—	—	—	—	—	—	—	7
728	Lake Lanahan	USA	EFL	—	7.9	—	—	—	1999	OT	—	—	—	—	7.9	—	—	—	—	—	—	—
729	Lake Lanier	USA	CON	ARC	18.9	—	71.9	—	1926	QP	SEP	—	—	—	—	—	—	—	—	—	—	—
730	Lake Latonka	USA	EFL	HD	13.1	6.1	701	1965	1966	QP	SEP	1.59	4.09	6.25	8.69	—	—	39.2	1.18	290	3	—
731	Lake Leanna	USA	—	—	—	—	—	—	1979	—	—	—	—	—	—	—	—	—	—	—	—	—
732	Lake Leigh	USA	CON	BU	9.8	—	97.5	1906	1911	QP	—	—	—	—	—	—	—	—	—	—	—	—
733	Lake Litchfield	USA	EFL	—	19	—	133	1975	1975	QP	SEP	956	—	—	—	—	—	—	—	—	—	—
734	Lake Ludlow Club	USA	EFL	CF	7.3	—	—	—	1935	OT	—	—	—	—	—	—	—	—	—	—	—	3
735	Lake Lynn Dam	USA	EFL	—	2.1	—	—	—	1995	QP	SEP	—	—	—	—	—	—	—	—	—	—	—
736	Lake Marian	USA	—	—	14	—	227	—	—	—	—	3.28	—	—	—	—	—	—	—	864	—	—
737	Lake Metamora	USA	EFL	—	4.9	—	—	1966	1974	QP	QIS	—	—	—	—	—	—	—	—	—	—	—
738	Lake Nancy	USA	EFL	—	4.9	—	—	1924	1976	QP	—	—	—	—	—	—	—	—	—	—	—	—
739	Lake North I	USA	EFL	—	8.2	—	—	—	1947	—	—	—	—	—	—	—	—	—	—	—	—	—
740	Lake North II	USA	EFL	—	8.2	—	—	1961	1966	—	—	—	—	—	—	—	—	—	—	—	—	—
741	Lake Paran	USA	EFL	—	7.6	—	—	—	1852	QP	SEP	—	—	—	—	—	—	—	—	—	—	1
742	Lake Park	USA	EFL	—	3.7	—	—	—	2000	DT	ABA	—	—	—	—	—	—	—	—	—	—	—
743	Lake Pauline	USA	EFL	—	3.7	—	—	—	1994	DT	BUA	—	—	—	—	—	—	—	—	—	—	—
744	Lake Philema	USA	EFL	HD	—	—	—	1965	1994	OT	—	—	4.78	9	8.53	—	—	47.2	0.33	—	—	—
745	Lake Powell	USA	EFL	—	5.8	—	—	—	1999	DT	ABA	—	—	—	5.5	—	—	35.1	—	—	—	—
746	Lake Roland-1	USA	EFL	—	9.4	—	—	1861	1972	OT	—	—	—	—	—	—	—	—	—	—	—	—
747	Lake Roland-2	USA	EFL	—	9.4	—	—	1861	1979	OT	—	—	—	—	—	—	—	—	—	—	—	—
748	Lake Runnemede	USA	EFL	—	4.6	—	—	—	1998	QP	SEP	—	—	—	—	—	—	—	—	—	—	—
749	Lake Seneca	USA	EFL	—	9.8	—	—	1965	1974	QP	SEP	—	—	—	4.6	—	—	12.2	—	—	—	—
750	Lake Shaftsbury	USA	EFL	—	3.4	—	—	1906	1936	OT	—	—	—	—	—	—	—	—	—	—	—	—
751	Lake Shannon	USA	EFL	—	4.6	—	—	1972	1975	OT	—	—	—	—	—	—	—	—	—	—	—	—
752	Lake Sherburne	USA	EFL	—	33.2	—	—	—	1916	QP	QIS	—	—	—	—	—	—	—	—	—	—	—
753	Lake Springfield	USA	—	—	14.6	—	68.6	—	—	—	—	66	—	—	—	—	—	—	—	5100	—	—
754	Lake Stockwell	USA	EFL	—	3.7	—	—	—	2004	OT	—	—	—	—	—	—	—	—	—	—	—	—
755	Lake Success	USA	EFL	—	5.5	—	—	—	—	—	—	—	—	—	—	—	—	—	—	—	—	—
756	Lake Tanglewood	USA	EFL	—	25.9	—	—	1960	1960	OT	—	—	—	—	—	—	—	30.5	—	—	—	—
757	Lake Thompson	USA	EFL	—	—	—	—	—	1960	OT	—	—	—	—	—	—	—	—	—	—	—	—
758	Lake Tinkle	USA	EFL	—	6.1	—	—	—	1994	QP	—	—	—	—	—	—	—	—	—	—	—	—
759	Lake Toxaway	USA	EFL	—	18.9	—	117.4	1902	1916	QP	SEP	13	—	—	—	—	—	82.4	—	—	—	—
760	Lake Venita	USA	EFL	—	9.2	—	—	—	1997	QP	SEP	—	—	—	—	—	—	—	—	—	—	—
761	Lake Vera	USA	RFL	—	15	—	51	1880	1905	OT	ISC	—	—	—	4.3	—	—	—	—	—	—	—
762	Lake Vermilion	USA	EFL	—	13.7	—	—	—	1930	QP	SF-SL	—	—	—	—	—	—	—	—	—	—	—
763	Lake Waterford	USA	EFL	—	4.6	—	—	1925	1979	OT	—	—	—	—	—	—	—	—	—	—	—	—
764	Lake Waxahachie	USA	EFL	—	17.7	—	—	1956	—	QP	—	—	—	—	—	—	—	—	—	—	—	—
765	Lake Yohola	USA	EFL	—	4.2	—	—	—	1994	OT	—	—	—	—	—	—	—	—	—	—	—	—

(Continued)

Table A.1 *(Continued)*

No.	Name	Country[(1)]	Type[(2)]	Type divided[(3)]	Height (m)	Crest width (m)	Crest length (m)	Year of cons.	Year of failure	Failure cause[(4)]	Failure cause divided[(5)]	Reservoir capacity (10^6 m³)	Water volume stored above breach invert (10^6 m³)	Water depth above breach (m)	Breach geometry: Height (m)	Breach geometry: Top width (m)	Breach geometry: Bottom width (m)	Breach geometry: Aver. width (m)	Breach geometry: Aver. side slope Z:1 (h:v)	Peak outflow rate (m³/s)	Fail. time	Life loss
766	Lake Yosemite	USA	EFL	—	16.2	—	1425	1884	1943	—	—	9	—	—	—	—	—	—	—	—	—	—
767	Lakeside	USA	EFL	—	15.2	—	—	—	1975	QP	SF	—	—	—	—	—	—	—	—	—	—	1
768	Lakeview	USA	EFL	—	3.4	—	—	—	1983	QP	SEP	—	—	—	—	—	—	—	—	—	—	—
769	Lakewood Villages	USA	EFL	—	7.9	—	—	—	1995	QP	SEP	—	—	—	—	—	—	—	—	—	—	—
770	Lambert Lake	USA	EFL	HD	16.5	—	—	—	1963	QP	SEP	—	0.296	12.8	14.3	—	—	7.62	0.21	—	—	—
771	Lancaster	USA	EFL	—	7.6	—	—	1894	1894	QP	SEP	0.06	—	—	—	—	—	—	—	—	—	—
772	Lancaster Dam	USA	EFL	—	4.6	—	—	—	1995	OT	—	0.006	—	—	—	—	—	—	—	—	—	—
773	Landrum Lake	USA	EFL	—	13.4	—	—	—	1990	QP	SF	—	—	—	—	—	—	—	—	—	—	—
774	Lang	USA	EFL	—	2.4	—	—	—	1973	OT	—	—	—	—	—	—	—	—	—	—	—	—
775	Laraway	USA	TIM	—	4.6	—	—	1909	1984	—	—	—	—	—	—	—	—	—	—	—	—	—
776	Laural Run	USA	EFL	HD	12.8	6.1	189.9	—	1977	OT	—	0.385	0.555	14.1	13.7	—	—	35.1	2.4	1050	—	40
777	Lawn Lake	USA	EFL	HD	7.9	2.4	115.5	1903	1982	QP	SEP	0.9	0.798	6.71	7.62	—	—	22.2	0.96	510	—	4
778	Layman	USA	—	—	20	3.7	—	1913	1915	—	—	49.5	—	—	20	107	87	97	—	—	—	—
779	Leader Middle Clear Boggy Creek Site 15-I	USA	EFL	—	7.6	—	—	—	1958	—	—	—	—	—	—	—	—	—	—	—	—	—
780	Leader Middle Clear Boggy Creek Site 15-II	USA	EFL	—	7.6	—	—	1965	1965	—	—	—	—	—	—	—	—	—	—	—	—	—
781	Leader Middle Creek Boggy Creek Site 33	USA	EFL	—	7	—	—	1963	1969	—	—	—	—	—	—	—	—	—	—	—	—	—
782	Lebanon	USA	EFL	—	9.1	—	—	—	1882	OT	—	—	—	—	—	—	—	—	—	—	—	—
783	Lebanon Forest	USA	EFL	—	3	—	—	—	2004	OT	—	—	—	—	—	—	—	—	—	—	—	—
784	Lee Lake CA	USA	EFL	—	14	—	—	1923	1938	QP	QIS	—	—	—	—	—	—	—	—	—	—	—
785	Lee Lake MA	USA	—	—	—	—	—	—	1968	—	—	—	—	—	—	—	—	—	—	—	—	2
786	Leech Lake	USA	EFL	—	4	—	—	1884	1957	QP	—	—	—	—	—	—	—	—	—	—	—	—
787	Leesville Dam	USA	EFL	—	7.6	—	—	—	1982	—	—	—	—	—	—	—	—	—	—	—	—	—
788	Leeuw Gamka	ZA	EFL	—	15	—	548	1920	1928	QP	SEP	10	—	—	—	—	—	—	—	—	—	—
789	Leguaseca	ES	CON	MAR	20	—	70	1958	1987	QP	SF	0.02	—	—	—	—	—	—	—	—	—	—
790	Leland	USA	—	—	—	—	—	—	1990	OT	—	—	—	—	—	—	—	—	—	—	—	—
791	Lemont	USA	—	—	—	—	—	—	1903	—	—	—	—	—	—	—	—	—	—	—	—	—
792	Leroux Creek	USA	EFL	—	7.6	—	—	—	1905	OT	ISC	—	—	—	—	—	—	—	—	—	—	—
793	Lewis Creek	USA	—	—	—	—	—	—	1896	OT	—	—	—	—	—	—	—	—	—	—	—	5
794	Lewiston	USA	EFL	—	25	—	—	—	1893	—	—	—	—	—	—	—	—	—	—	—	—	—
795	Lies Lake Dam	USA	—	—	—	—	—	—	1994	OT	—	—	—	—	—	—	—	—	—	—	—	—
796	Lily Lake	USA	EFL	HD	—	—	—	1913	1951	QP	SEP	—	0.0925	3.35	3.66	—	—	10.8	0.13	71	—	—
797	Lima	USA	EFL	—	—	—	—	—	1894	OT	ISC	—	—	—	—	—	—	—	—	—	—	—
798	Linchester	USA	EFL	—	4	—	—	1930	1979	OT	—	—	—	—	—	—	—	—	—	—	—	—

799	Lincoln Pond-I	USA	CON	GR	7.6	—	76.5	—	1909	OT	—	—	—	—	—	—	—	—	—	—	—	—
800	Lincoln Pond-II	USA	CON	GR	7.6	—	76.5	—	1912	QP	—	—	—	—	—	—	—	—	—	—	—	—
801	Lindauer	USA	—	—	—	—	—	—	1912	OT	—	—	—	—	—	—	—	—	—	—	—	—
802	Linnie Lac Dam	USA	EFL	—	—	—	—	—	1973	OT	—	—	—	—	—	—	—	—	—	—	—	—
803	Linville	USA	EFL	—	48.8	—	—	1919	1919	—	—	—	—	—	—	—	—	—	—	—	—	—
804	Little Deer Creek	USA	EFL	HD	26	6.1	109.7	1962	1963	QP	SEP	1.8	1.36	22.9	27.1	23	—	29.6	0.75	1330	0.33	1
805	Little Falls	USA	EFL	—	9.2	—	—	—	1979	QP	QIC	—	—	—	—	—	—	—	—	—	—	—
806	Little Indian Creek	USA	—	—	—	—	—	—	1928	—	—	—	—	—	—	—	—	—	—	—	—	3
807	Little Juniper I	USA	EFL	—	6.7	—	—	—	1932	OS	—	—	—	—	—	—	—	—	—	—	—	—
808	Little Juniper II	USA	EFL	—	6.7	—	—	—	1949	OT	ISC	—	—	—	—	—	—	—	—	—	—	—
809	Little Muskego	USA	EFL	—	4.6	—	—	—	1973	OT	—	—	—	—	—	—	—	—	—	—	—	—
810	Little Ocmulgee	USA	CON	—	2	—	—	—	1998	OT	—	—	—	—	—	—	—	—	—	—	—	—
811	Little Rock	USA	MAS	GR	11	—	—	—	1887	QP	—	0.03	—	—	—	—	—	—	—	—	—	—
812	Little Washita River Site-13	USA	EFL	—	10.7	—	—	—	1987	QP	SEP	—	—	—	—	—	—	—	—	—	—	—
813	Little Wewoka Creek Site 17	USA	EFL	—	7.3	—	—	1960	1960	QP	SEP	—	—	—	—	—	—	—	—	—	—	—
814	Little Wolf Creek Herrin	USA	EFL	—	12.2	—	213.4	—	1936	—	—	—	—	—	—	—	—	—	—	—	—	—
815	Littlefield	USA	RFL	—	37	—	91	1929	1929	QP	SEP	—	—	—	—	—	—	—	—	—	—	—
816	Lliu Lliu	CL	EFL	—	20	—	550	1934	1985	DT	EQ	3	—	—	—	—	—	—	—	—	—	—
817	Lo Ovalle	CL	EFL	—	12.5	—	1520	1932	1985	DT	EQ	13	—	—	—	—	—	—	—	—	—	—
818	Loch Alpine	USA	EFL	—	7.5	—	39.6	1926	1926	—	—	—	—	—	—	—	—	—	—	—	—	—
819	Loch Highland Lake Lower	USA	EFL	—	8.6	—	—	—	1994	—	—	—	—	—	—	—	—	—	—	—	—	—
820	Lockwood Dam	USA	EFL	—	4.3	—	—	—	1989	—	—	—	—	—	—	—	—	—	—	—	—	2
821	Loden Dam	USA	—	—	—	—	—	—	1999	OT	—	—	—	—	—	—	—	—	—	—	—	—
822	Log Falls	CA	TIM	—	11	—	—	1921	1923	QP	—	30	—	—	—	—	—	—	—	—	—	—
823	Long Tom	USA	EFL	HD	15.2	—	—	1906	1916	QP	SEP	—	—	—	—	—	—	—	—	—	—	—
824	Long Valley	USA	EFL	—	46.3	—	—	—	1892	—	—	—	—	—	—	—	—	—	—	—	—	—
825	Longo Pond	USA	EFL	—	3.7	—	—	—	1999	OT	—	—	—	—	—	—	—	—	—	—	—	—
826	Longwald Pond	USA	EFL	—	9.1	—	91.4	1910	1922	QP	SEP	—	—	—	—	—	—	—	—	—	—	—
827	Lookout Shoals	USA	EFL	—	24.9	—	832	1915	1916	OT	ISC	49	—	—	—	—	—	—	—	—	—	—
828	Lookover Lake	USA	EFL	DC	3.7	—	—	—	1999	OT	ISC	—	—	—	—	—	—	—	—	—	—	—
829	Lott Dam	USA	—	—	—	—	—	—	2000	PM	—	—	—	—	—	—	—	—	—	—	—	—
830	Loup Loup Dam	USA	EFL	—	15.3	—	—	—	1938	QP	QIS	—	—	—	—	—	—	—	—	—	—	—
831	Lower Aetna Lake	USA	EFL	—	3.7	—	—	—	2004	OT	—	—	—	—	—	—	—	—	—	—	—	—
832	Lower Friendship	USA	EFL	—	9.1	—	—	—	—	DT	BUA	—	—	—	—	—	—	—	—	—	—	—
833	Lower Howell	USA	EFL	—	11	—	—	1877	1906	—	—	—	—	—	—	—	—	—	—	—	—	—
834	Lower Idaho Falls	USA	EFL& CON	—	15.2	—	275	1914	1976	DT	BUA	—	—	—	—	—	—	—	—	—	—	—
835	Lower Khajuri	IN	EFL	CF	16	—	—	1949	1949	QP	—	43	—	—	—	—	—	—	—	—	—	—
836	Lower Latham	USA	EFL	HD	8.2	4.6	—	—	1973	QP	SEP	7.08	7.08	5.79	7.01	—	—	79.2	6.3	340	—	—

(Continued)

Table A.1 *(Continued)*

No.	Name	Country[(1)]	Type[(2)]	Type divided[(3)]	Height (m)	Crest width (m)	Crest length (m)	Year of cons.	Year of failure	Failure cause[(4)]	Failure cause divided[(5)]	Reservoir capacity ($10^6 m^3$)	Water volume stored above breach invert ($10^6 m^3$)	Water depth above breach (m)	Breach geometry: Height (m)	Breach geometry: Top width (m)	Breach geometry: Bottom width (m)	Breach geometry: Aver. width (m)	Breach geometry: Aver. side slope Z:1 (h:v)	Peak outflow rate (m^3/s)	Fail. time	Life loss
837	Lower Melzingah	USA	EFL	—	10.4	—	120	1887	1897	OT	—	0.04	—	—	—	—	—	—	—	—	—	7
838	Lower Mill Pond	USA	MAS	—	3.1	—	—	—	1982	OT	—	—	—	—	3.1	—	—	—	—	—	—	—
839	Lower Monumental	USA	EFL	—	69	—	—	—	1964	OT	—	—	—	—	—	—	—	—	—	—	—	—
840	Lower North Eden Reservoir	USA	EFL	—	4.3	—	—	—	1980	OT	—	—	—	—	3.1	—	—	10.7	—	—	—	—
841	Lower Otay	USA	RFL	DC	46.6	3.7	173	1901	1916	OT	ISC	52	49.3	>39.6	39.6	172	93.8	133	1	—	0.25	30
842	Lower Pond	USA	EFL	—	1.5	—	—	—	1982	OT	—	—	—	—	—	—	—	—	—	—	—	—
843	Lower Providence	USA	EFL	—	5.2	—	182.9	1816	1901	OT	—	—	—	—	—	—	—	—	—	—	—	1
844	Lower Robertson	USA	TIM	—	5.5	—	—	—	2005	OT	—	—	—	—	—	—	—	—	—	—	—	—
845	Lower Rosegill	USA	EFL	—	2.1	—	—	—	1999	OT	—	—	—	—	—	—	—	—	—	—	—	—
846	Lower Shelton	USA	MAS	GR	7.6	—	33.5	1881	1903	OT	—	—	—	—	—	—	—	—	—	—	—	—
847	Lower Stichcomb	USA	EFL	—	4.2	—	—	—	1978	DT	ABA	—	—	—	—	—	—	—	—	—	—	—
848	Lower Tallassee	USA	MAS	GR	9.1	—	396.2	1901	1901	OT	—	—	—	—	—	—	—	—	—	—	—	—
849	Lower Turney	USA	EFL	—	4.6	—	—	—	1982	OT	—	—	—	—	—	—	—	—	—	—	—	—
850	Lower Twin Lakes	USA	EFL	—	3.7	—	52	—	1991	DT	BUA	—	—	—	—	—	—	—	—	—	—	—
851	Lower Two Medicine	USA	EFL	HD	11.3	3.7	—	1913	1964	OT	ISC	19.6	29.6	11.3	11.3	—	—	67	1.5	1800	—	9
852	Luther Pond	USA	EFL	—	4.6	—	—	—	1986	OT	—	—	—	—	—	—	—	—	—	—	—	—
853	Luxton Lake	USA	EFL	—	4.6	—	—	—	—	OT	—	—	—	—	—	—	—	—	—	—	—	—
854	Lyman	USA	EFL	CF	19.8	3.7	256	1913	1915	QP	SEP	43	35.8	16.2	19.8	107	87	97	1	—	—	8
855	Lynde Brook	USA	EFL	DC	12.5	15.2	87.5	1871	1876	QP	SEP	2.52	2.88	11.6	12.5	45.7	15.3	30.5	1.22	—	—	—
856	Lynx Creek	USA	MAS	ARC	8.5	—	45.7	1891	1891	QP	—	—	—	—	8.5	—	—	10.7	—	—	—	—
857	Mac Donalton	USA	MAS	GR	4.9	—	127.7	1911	1911	—	—	0.1	—	—	—	—	—	—	—	—	—	—
858	Machhu-II	IN	EFL& CON	—	24.7	6	3905	1972	1979	OT	—	100	—	—	60	540	—	—	—	7690	2	2500
859	Mackville Pond	USA	MAS	GR	5.5	—	—	1900	1927	OT	—	—	—	—	—	—	—	—	—	—	—	—
860	Madison	USA	CON	GR	4.9	—	—	1906	1916	QP	—	—	—	—	—	—	—	—	—	—	—	—
861	Mafeteng	LS	EFL	CF	23	—	500	1988	1988	QP	—	—	—	—	—	—	—	—	—	—	—	—
862	Magic Valley	USA	EFL	—	39.6	—	884	1909	1911	QP	SEP	236.8	—	—	—	—	—	—	—	—	—	—
863	Mahanoy Township	USA	EFL	—	10.4	—	—	—	1892	—	—	—	—	—	—	—	—	—	—	—	—	1
864	Malcolm B. Rawls	USA	EFL	—	6.4	—	—	—	1997	DT	BUA	—	—	—	—	—	—	—	—	—	—	—
865	Mallard Lake	USA	EFL	—	6.4	—	—	—	1996	DT	ABA	—	—	—	—	—	—	—	—	—	—	—
866	Mallison Falls	USA	MAS	—	4.3	—	—	—	1983	QP	QIS	—	—	—	—	—	—	—	—	—	—	—
867	Malpasset	FR	CON	ARC	66	1.5	222	1954	1959	QP	SF-SL	47	—	—	—	—	—	—	—	—	—	421
868	Mammoth	USA	EFL	DC	23	—	—	1916	1917	OT	ISC	14	—	—	21.3	—	—	9.2	—	2520	3	1
869	Manchester	USA	MAS	GR	9	—	52.5	—	1902	QP	SF-SL	—	—	—	—	—	—	—	—	—	—	—
870	Manitou	USA	CON	ARC	15.2	—	91.4	—	1924	QP	SEP	—	—	—	—	—	—	—	—	—	—	—

871	Manivali	IN	EFL	—	18.4	—	—	1975	1976	QP	SEP	4.8	—	—	—	—	—	—	—	—	—	—
872	Manley	USA	EFL	—	5.5	—	—	—	1955	OT	—	—	—	—	—	—	—	—	—	—	—	—
873	Mann Creek	USA	—	—	—	—	—	—	1982	QP	SEP	—	—	—	—	—	—	—	—	—	—	—
874	Mansure Pond	USA	MAS	—	3.7	—	—	—	1982	OT	—	—	—	—	—	—	—	—	—	—	—	—
875	Maple Grove	USA	EFL	—	18	—	—	—	1979	DT	WTA	—	—	—	—	—	—	—	—	—	—	—
876	Maquoketa Milldam	USA	EFL	—	7.9	—	—	—	1927	QP	SEP	—	—	—	—	—	—	—	—	—	—	—
877	Marathon Oil Company	USA	EFL	—	—	—	—	—	1979	QP	SF-SO	—	—	—	—	—	—	—	—	—	—	—
878	Maready Chicken	USA	EFL	—	2.9	—	—	—	1995	—	—	—	—	—	—	—	—	—	—	—	—	—
879	Marquette No. 3	USA	—	—	10.1	—	—	—	2003	DT	BUA	—	—	—	—	—	—	—	—	—	—	—
880	Marschke Lake	USA	EFL	—	5.5	—	—	—	1988	—	—	—	—	—	—	—	—	—	—	—	—	—
881	Marsh Lake	USA	EFL	—	8.5	—	—	—	2001	OT	—	—	—	—	—	—	—	—	—	—	—	—
882	Marshall Creek	USA	EFL	—	25.6	—	472.4	—	1937	QP	—	—	—	—	—	—	—	—	—	—	—	—
883	Marshall Lake	USA	EFL	—	26.2	—	661.4	1908	1909	QP	SEP	12.6	—	—	—	—	—	—	—	—	—	—
884	Martin Cooling	USA	EFL	—	10.4	—	—	—	1979	QP	SEP	136	136	8.53	12.8	—	—	186	—	3115	—	—
885	Martin Davey	USA	EFL	—	—	—	—	—	1940	OT	—	—	—	—	—	—	—	—	—	—	—	—
886	Mascoma River	USA	TIM	—	4.9	—	—	—	1981	QP	SF	0.21	—	—	4.9	—	—	10.7	—	—	—	—
887	Masters Pond	USA	EFL	—	4.6	—	—	—	1982	QP	—	—	—	—	—	—	—	—	—	—	—	—
888	Masterson	USA	EFL	—	18.3	—	—	1950	1951	QP	SEP	—	—	—	—	—	—	—	—	—	—	—
889	Mauch Chunk	USA	EFL	—	16.6	—	—	—	1942	—	—	—	—	—	—	—	—	—	—	—	—	—
890	Maur Hill	USA	EFL	—	10.7	—	—	—	1978	QP	SF-SL	—	—	—	—	—	—	—	—	—	—	—
891	Max Wilcox	USA	—	—	7.6	—	—	—	1973	QP	SEP	—	—	—	—	—	—	—	—	—	—	—
892	Mayfield	USA	—	—	—	—	—	1965	1965	QP	SF-SL	—	—	—	—	—	—	—	—	—	—	—
893	McCarty	USA	EFL	—	15.3	—	—	1942	1978	OT	—	—	—	—	—	—	—	—	—	—	—	—
894	Mcgill Lake	USA	EFL	—	4.3	—	—	—	1994	OT	—	—	—	—	—	—	—	—	—	—	—	—
895	Mcguire Lake	USA	EFL	—	4.9	—	—	—	2004	OT	—	—	—	—	—	—	—	—	—	—	—	—
896	Mckemie Lake	USA	EFL	—	6.1	—	—	—	1994	OT	—	—	—	—	—	—	—	—	—	—	—	—
897	Mckemie Lake North	USA	EFL	—	8.2	—	—	—	1994	OT	—	—	—	—	—	—	—	—	—	—	—	—
898	McKinley	USA	—	—	—	—	—	—	1912	—	—	—	—	—	—	—	—	—	—	—	—	—
899	Mcknight Lake	USA	EFL	—	10.1	—	—	—	1994	DT	BUA	—	—	—	—	—	—	—	—	—	—	—
900	McMahon Gulch	USA	EFL	—	17	—	137	1924	1925	OT	—	0.03	—	—	—	—	—	—	—	—	—	—
901	McNeil Lake	USA	—	—	—	—	—	—	1994	OT	—	—	—	—	—	—	—	—	—	—	—	—
902	Meadow Lake	USA	EFL	CF	9.1	—	—	—	1962	OT	—	0.75	—	—	—	—	—	—	—	—	—	—
903	Meadow Pond	USA	—	—	—	—	—	—	1996	—	—	—	—	—	—	—	—	—	—	—	—	1
904	Medicine Creek	USA	EFL	—	10.8	—	—	—	1964	OT	—	—	—	—	—	—	—	—	—	—	—	9
905	Melville	USA	EFL	CF	11	3	243.8	1907	1909	QP	SEP	—	24.7	7.92	9.75	40	25.6	32.8	0.7	—	—	—
906	Melzingah Dam	USA	EFL	—	15.2	—	—	—	1897	OT	—	—	—	—	—	—	—	—	—	—	—	7
907	Melzingah Lower	USA	EFL	DC	9.1	—	—	—	1897	OT	—	—	—	—	—	—	—	—	—	—	—	—
908	Mena Valparaiso	CL	EFL	—	17	—	200	1885	1888	QP	SEP	0.06	—	—	—	—	—	—	—	—	—	—
909	Mendham	USA	EFL	—	—	—	—	—	1996	QP	SEP	—	—	—	1.5	—	—	6.1	—	—	—	—
910	Mendota	USA	TIM	—	4.9	—	106.7	1898	1916	QP	—	—	—	—	—	—	—	—	—	—	—	—
911	Menno	USA	EFL	—	16.2	—	—	1935	1984	OT	IF	—	—	—	16.2	—	—	—	—	—	—	—
912	Merced Falls	USA	EFL	—	11.6	—	—	—	1950	OT	ISC	—	—	—	11.6	—	—	—	—	—	—	—

(Continued)

Table A.1 *(Continued)*

No.	Name	Country[1]	Type[2]	Type divided[3]	Height (m)	Crest width (m)	Crest length (m)	Year of cons.	Year of failure	Failure cause[4]	Failure cause divided[5]	Reservoir capacity (10^6 m^3)	Water volume stored above breach invert (10^6 m^3)	Water depth above breach (m)	Breach geometry: Height (m)	Breach geometry: Top width (m)	Breach geometry: Bottom width (m)	Breach geometry: Aver. width (m)	Breach geometry: Aver. side slope Z:1 (h:v)	Peak outflow rate (m^3/s)	Fail. time	Life loss
913	Mercedes	USA	CON	GR	7.6	—	91.4	—	1967	—	—	—	—	—	—	—	—	—	—	—	—	—
914	Merimac Upper Lake	USA	EFL	HD	—	—	—	1939	1994	OT	—	—	0.0696	3.44	3.05	—	—	14.2	0.41	—	—	—
915	Merlyn Schrunk	USA	—	—	—	—	—	1983	1984	OT	—	—	—	—	2.4	—	—	12.8	—	—	—	—
916	Mermentau-I	USA	TIM	—	—	—	—	1904	1905	DT	WTA	—	—	—	—	—	—	—	—	—	—	—
917	Mermentau-II	USA	TIM	—	—	—	—	1904	1907	DT	WTA	—	—	—	—	—	—	—	—	—	—	—
918	Mermentau-III	USA	TIM	—	—	—	—	1904	1911	DT	WTA	—	—	—	—	—	—	—	—	—	—	—
919	Merrill	USA	EFL	—	7	—	—	—	1912	OT	—	—	—	—	—	—	—	—	—	—	—	—
920	Merritt Lake	USA	EFL	—	4.9	—	—	—	1994	OT	—	—	—	—	—	—	—	—	—	—	—	—
921	Middle Pond	USA	—	—	—	—	—	—	2000	DT	ABA	—	—	—	1.7	—	—	4.3	—	—	—	—
922	Middlebrook	USA	EFL	—	5.2	—	—	—	1977	OT	—	0.03	—	—	—	—	—	—	—	—	—	—
923	Middlefield I	USA	EFL	—	9.1	—	—	—	1874	—	—	—	—	—	—	—	—	—	—	—	—	—
924	Middlefield II	USA	EFL	—	6.1	—	152.4	1874	1901	OT	—	—	—	—	—	—	—	—	—	—	—	—
925	Middlesex	USA	TIM	—	14	—	—	1896	1927	OT	—	—	—	—	—	—	—	—	—	—	—	—
926	Middletown	USA	EFL	—	10.7	—	45.7	1860	1961	QP	SEP	—	—	—	—	—	—	—	—	—	—	—
927	Middletown Hydraulic Dam I	USA	EFL	—	—	—	—	—	1993	—	—	—	—	—	—	—	—	—	—	—	—	—
928	Middletown Hydraulic Dam II	USA	EFL	—	—	—	—	—	1997	OT	BOS	—	—	—	4.1	6.9	3.4	—	—	—	—	—
929	Mid-South Youth Camp	USA	—	—	7.6	—	—	—	1975	OT	—	—	—	—	—	—	—	—	—	—	—	—
930	Milburn	USA	EFL	—	—	—	—	1893	1893	QP	—	0.2	—	—	—	—	—	—	—	—	—	—
931	Military Park	USA	EFL	—	6.4	—	—	—	1911	QP	SEP	—	—	—	—	—	—	—	—	—	—	—
932	Milk Pond Dam	USA	EFL	—	—	—	—	—	1982	OT	—	—	—	—	—	—	—	—	—	—	—	—
933	Milk Ranch	USA	EFL	—	11.6	—	—	—	1995	PM	—	—	0.6164	—	—	—	—	—	—	—	—	—
934	Mill Brook	USA	CON	BU	12.2	—	—	1909	1966	QP	SEP	0.03	—	—	—	—	—	—	—	—	—	—
935	Mill Creek	USA	EFL	—	20	—	84	1889	1957	QP	SEP	0.3	—	—	—	—	—	—	—	—	—	—
936	Mill Pond CT	USA	EFL	—	4.6	—	—	—	1982	OT	—	—	—	—	—	—	—	—	—	—	—	—
937	Mill Pond RI	USA	EFL	—	—	—	—	—	2000	—	—	—	—	—	2.1	—	1.4	—	—	—	—	—
938	Mill Pond MA	USA	EFL	—	—	—	—	—	2001	QP	SEP	—	—	—	—	—	—	—	—	—	—	—
939	Mill River	USA	EFL	DC	13	—	183	1865	1874	QP	SF-SL	2.5	2.5	—	—	—	—	—	—	1645	—	144
940	Millard Pond	USA	MAS	—	3	—	—	1890	1955	OT	—	—	—	—	—	—	—	—	—	—	—	—
941	Miller Pond	USA	EFL	—	3.7	—	—	1901	1936	DT	BUA	—	—	—	—	—	—	—	—	—	—	—
942	Millsboro Pond	USA	EFL	—	3.7	—	—	—	1979	QP	SEP	—	—	—	—	—	—	—	—	—	—	—
943	Minatare	USA	EFL	CF	34.8	—	1127.8	1915	1920	QP	SF-SL	—	—	—	—	—	—	—	—	—	—	—
944	Mine #20 Slurry	USA	EFL	—	19.8	—	—	—	1974	OT	—	—	—	—	—	—	—	—	—	—	—	—
945	Minneapolis	USA	CON	GR	28.3	—	161.5	—	1899	QP	SF-SL	—	—	—	—	—	—	—	—	—	—	—
946	Minor's Millpond	USA	—	—	—	—	—	—	1994	OT	—	—	—	—	—	—	—	—	—	—	—	—

947	Mirror Lake	USA	—	—	—	—	—	—	1976	—	—	—	—	—	—	—	—	—	—	—	—	—
948	Mockingbird	USA	EFL	—	7.6	—	—	1900	1979	OT	—	—	—	—	—	—	—	—	—	—	—	—
949	Model Reservoir	USA	EFL	—	14	—	—	1900	1920	—	—	25	—	—	—	—	—	—	—	—	—	—
950	Modest Creek	USA	EFL	—	7	—	—	—	1977	QP	—	—	—	—	—	—	—	—	—	—	—	—
951	Moffittsville	USA	—	—	—	—	—	—	1912	—	—	—	—	—	—	—	—	—	—	—	—	—
952	Mohamed V	MA	CON	GR	62	—	—	—	1963	QP	SF	—	—	—	—	—	—	—	—	—	—	—
953	Mohawk	USA	EFL	—	5.5	—	121.9	1913	1915	OT	—	—	—	—	—	—	—	—	—	—	—	—
954	Mohegan Park	USA	EFL	CF	6.1	—	—	1893	1963	QP	SEP	0.17	—	—	—	—	—	—	—	—	—	6
955	Molteno	ZA	EFL	—	15	—	800	1881	1882	QP	—	0.22	—	—	—	—	—	—	—	—	—	—
956	Monashka Creek	USA	RFL	—	14.3	—	—	—	1978	OT	—	0.62	—	—	6.1	—	—	30.5	—	—	—	—
957	Montreal	USA	MAS	GR	—	—	—	—	1896	QP	—	—	—	—	—	—	—	—	—	—	—	—
958	Moose Lodge	USA	EFL	—	3.7	—	—	—	1995	OT	—	—	—	—	—	—	—	—	—	—	—	—
959	Morgan County	USA	—	—	—	—	—	—	1950	—	—	—	—	—	—	—	—	—	—	—	—	—
960	Morgan Dam	USA	EFL	—	7.3	—	—	—	1994	QP	QIS	—	—	—	—	—	—	—	—	—	—	—
961	Mosinee	USA	TIM	—	6.7	—	—	—	1965	—	—	—	—	—	—	—	—	—	—	—	—	—
962	Moss Creek Lake	USA	EFL	—	20.4	—	—	—	2000	QP	SEP	—	—	—	—	—	—	—	—	—	—	—
963	Moss Mill Lake	USA	EFL	—	2.1	—	—	—	1997	OT	—	—	—	—	—	—	—	4.6	—	—	—	—
964	Mossy Lake	USA	EFL	HD	2.8	—	—	1963	1994	OT	—	—	4.13	4.41	3.44	—	—	41.5	1.24	—	—	—
965	Mount Lake	USA	EFL	—	—	—	—	1937	1938	OT	—	—	—	—	—	—	—	—	—	—	—	—
966	Mount Morris	USA	—	—	—	—	—	—	1995	QP	SEP	—	—	—	—	—	—	—	—	—	—	—
967	Mountain Lake	USA	EFL	—	4	—	45.8	—	2000	—	—	0.18	—	—	—	—	—	—	—	—	—	—
968	Mountjoy Hill	USA	EFL	—	13.7	—	—	—	1893	QP	SEP	0.075	—	—	—	—	—	—	—	—	—	4
969	Moyie River	USA	CON	ARC	16	—	47	—	1926	OT	—	0.34	—	—	—	—	—	—	—	—	—	—
970	Moyle Dam	USA	EFL	—	—	—	—	—	1978	QP	SF-SL	—	—	—	—	—	—	—	—	—	—	—
971	Mud Creek Dam	USA	EFL	—	3.1	—	—	—	1979	OT	—	—	0.0012	—	—	—	—	—	—	—	—	—
972	Mud Lake	USA	TIM	—	2	—	—	1870	1922	—	—	—	—	—	—	—	—	—	—	—	—	—
973	Mud Lake CA I	USA	EFL	—	3	—	—	—	1932	QP	SEP	—	—	—	—	—	—	—	—	—	—	—
974	Mud Lake CA II	USA	EFL	—	3	—	—	—	—	QP	SF-SO	—	—	—	—	—	—	—	—	—	—	—
975	Mud Pond I	USA	EFL	—	4.6	—	99.1	—	1886	QP	—	—	—	—	—	—	—	—	—	—	—	7
976	Mud Pond II	USA	EFL	—	3.4	—	—	1890	1890	—	—	—	—	—	—	—	—	—	—	—	—	—
977	Muddy Bayou	USA	EFL	—	—	—	—	—	1973	OT	—	—	—	—	—	—	—	—	—	—	—	—
978	Mulkey Lake	USA	EFL	—	4.6	—	—	—	1994	OT	—	—	—	—	—	—	—	—	—	—	—	—
979	Mullens Farm	USA	EFL	—	—	—	—	—	1997	QP	SF-SO	—	—	—	—	—	—	—	—	—	—	—
980	Murphy	USA	EFL	—	6.7	—	—	—	1970	OT	BOS	—	—	—	—	—	—	—	—	—	—	—
981	Murtha Pond	USA	EFL	—	2.1	—	—	—	2000	OT	—	—	—	—	1.5	—	—	5.3	—	—	—	—
982	Muscotatuck	USA	MAS	GR	2.4	—	39.6	1909	1910	—	—	—	—	—	—	—	—	—	—	—	—	—
983	Mush Creek #2	USA	EFL	—	12.2	—	—	1978	1985	QP	SF-SL	—	—	—	—	—	—	—	—	—	—	—
984	Nacimiento	USA	EFL	—	22.9	—	—	—	1973	QP	SF	—	—	—	—	—	—	—	—	—	—	—
985	Nagels Mill Pond	USA	EFL	—	4.9	—	—	—	1999	QP	SEP	—	—	—	4.9	—	—	18.3	—	—	—	—
986	Nahzille	USA	EFL	HD	5.5	—	130	—	1996	OT	—	—	—	—	5.03	6.71	6.71	6.71	0	—	—	—
987	Nanaksagar	IN	EFL	—	16.5	—	19300	1962	1967	QP	SEP	210	—	—	16	—	—	46	—	9700	12	100
988	Narraguinnep-I	USA	EFL	—	29.6	—	1280.2	1910	1928	QP	SF-SL	23	—	—	—	—	—	—	—	—	—	—

(Continued)

Table A.1 *(Continued)*

No.	Name	Country[(1)]	Type[(2)]	Type divided[(3)]	Height (m)	Crest width (m)	Crest length (m)	Year of cons.	Year of failure	Failure cause[(4)]	Failure cause divided[(5)]	Reservoir capacity (10^6 m^3)	Water volume stored above breach invert (10^6 m^3)	Water depth above breach (m)	Breach geometry: Height (m)	Breach geometry: Top width (m)	Breach geometry: Bottom width (m)	Breach geometry: Aver. width (m)	Breach geometry: Aver. side slope Z:1 (h:v)	Peak outflow rate (m^3/s)	Fail. time	Life loss
989	Narraguinnep-II	USA	EFL	—	29.6	—	1280.2	1910	1951	QP	SF-SL	23	—	—	—	—	—	—	—	—	—	—
990	Nash Bog Pond	USA	TIM	—	10.1	—	274.3	—	1969	QP	SEP	6.2	—	—	13.7	—	—	22.9	—	—	—	—
991	Nashville	USA	MAS	GR	10.7	—	—	1889	1912	QP	—	0.2	—	—	—	—	—	—	—	—	—	—
992	Natural Dam	USA	—	—	—	—	—	—	1998	OT	—	—	—	—	—	—	—	—	—	—	—	—
993	Navajo Lake	USA	EFL	—	3.7	—	45.8	—	1940	QP	SEP	—	—	—	—	—	—	—	—	—	—	—
994	Nebraska City	USA	EFL	—	6.1	—	—	1890	1890	QP	—	—	—	—	—	—	—	—	—	—	—	—
995	Necaxa	MX	EFL	—	59	—	297.2	1909	1909	QP	SF-SL	—	—	—	—	—	—	—	—	—	—	—
996	New Bedford	USA	EFL	—	7.6	—	182.9	1866	1868	QP	SEP	—	—	—	—	—	—	—	—	—	—	—
997	New Shoal Creek	USA	EFL	—	13.7	—	—	—	—	OT	—	—	—	—	10.7	61	22.9	—	—	—	—	—
998	Newell	USA	EFL	CF	65.5	—	—	1961	1961	QP	SF-SL	—	—	—	—	—	—	—	—	—	—	4
999	Newport	USA	—	—	—	—	—	—	1884	—	—	—	—	—	—	—	—	—	—	—	—	—
1000	Newport No. 11 Diversion Dam	USA	—	—	—	—	—	—	1994	QP	SEP	—	—	—	—	—	—	—	—	—	—	—
1001	Newton Falls Paper	USA	EFL	—	9.1	—	—	1965	—	—	—	—	—	—	—	—	—	—	—	—	—	—
1002	Newton Gulch	USA	EFL	—	12.8	—	—	—	1973	QP	SEP	—	—	—	—	—	—	—	—	—	—	—
1003	Nine Mile	USA	CON	GR	25.7	—	—	—	1996	QP	QIS	—	—	—	—	—	—	—	—	—	—	—
1004	Niobrara	USA	EFL	—	5.5	—	548.6	1926	1936	QP	—	—	—	—	—	—	—	—	—	—	—	—
1005	Nix Club Lake	USA	EFL	—	7	—	—	—	1989	—	—	—	—	—	—	—	—	—	—	—	—	1
1006	Nizhne Svirskaya	USSR	EFL	—	28	—	1860	1934	1935	DT	EQ	1190	—	—	—	—	—	—	—	—	—	—
1007	No Name Stock	USA	EFL	—	—	—	—	—	1972	—	—	—	—	—	—	—	—	—	—	—	—	—
1008	Noonan	USA	EFL	—	2.7	—	—	—	1986	QP	SEP	—	—	—	—	—	—	—	—	—	—	—
1009	Noppikoski	SE	EFL	—	18.5	—	175	1966	1985	OT	—	0.7	—	—	—	—	—	—	—	—	—	—
1010	Norman Swinney	USA	EFL	—	—	—	—	—	1993	OT	IF	—	—	—	—	—	—	—	—	—	—	—
1011	North Branch	USA	EFL	HD	5.5	—	—	—	1977	—	—	—	0.0222	5.49	—	—	—	—	—	29.4	—	—
1012	North Dike	USA	EFL	—	25	—	—	1905	1907	QP	SF-SL	—	—	—	—	—	—	—	—	—	—	—
1013	North Lake	USA	EFL	—	—	—	—	1957	1974	QP	—	—	—	—	—	—	—	—	—	—	—	—
1014	North Montpelier Pond I	USA	CON	GR	4.6	—	—	1920	1984	OT	—	—	—	—	—	—	—	—	—	—	—	—
1015	North Montpelier Pond II	USA	CON	GR	4.6	—	—	1920	1990	—	—	—	—	—	—	—	—	—	—	—	—	—
1016	North Scituate	USA	EFL	—	1.8	—	—	—	1926	OT	—	0.4	—	—	—	—	—	—	—	—	—	—
1017	Northfield	USA	TIM	—	7.6	—	30.5	—	1890	QP	—	—	—	—	—	—	—	—	—	—	—	—
1018	Norton Brook	USA	EFL	—	11	—	—	1935	1942	QP	SEP	—	—	—	11	—	—	—	—	—	—	—
1019	Nubble Pond	USA	EFL	—	6.4	—	—	—	1999	—	—	—	—	—	—	—	—	—	—	—	—	—
1020	Nuclear Lake	USA	—	—	4.9	—	—	—	—	OT	—	—	—	—	—	—	—	—	—	—	—	—
1021	Oak Orchard	USA	—	—	—	—	—	—	1894	DT	WTA	—	—	—	—	—	—	—	—	—	—	—
1022	Oakford Park	USA	EFL	DC	6.1	2.6	106.7	1895	1903	OT	ISC	—	—	—	4.6	23	—	—	—	—	1	23
1023	Oakland	USA	MAS	GR	2.4	—	—	—	1896	QP	—	—	—	—	—	—	—	—	—	—	—	—

1024 Oberon Lake #1	USA	EFL	—	8.4	—	—	—	1974	OT	—	—	—	—	—	—	—	—	—	—	—	—
1025 Odiel	ES	RFL	—	35	—	154	1968	1968	OT	—	3.3	—	—	—	—	—	—	—	—	—	—
1026 Ogayarindo Tameike	JP	EFL	—	24.5	—	99	1944	1953	OT	ISC	0.16	—	—	—	—	—	—	—	—	—	—
1027 Oklahoma City	USA	EFL	—	16.5	—	365.8	—	1923	OT	—	26.5	—	—	—	—	—	—	—	—	—	—
1028 Old Forge Pond	USA	EFL	—	3.7	—	—	—	1999	—	—	—	—	—	—	—	—	—	—	—	—	—
1029 Old Town #37	USA	EFL	—	6.7	—	—	1974	1983	QP	SF-SL	—	—	—	—	—	—	—	—	—	—	—
1030 Olothe	USA	—	—	—	—	—	—	1903	—	—	—	—	—	—	—	—	—	—	—	—	—
1031 Orleans Village	USA	TIM	—	4.6	—	—	1904	1979	QP	QIS	—	—	—	—	—	—	—	—	—	—	—
1032 Oros	BR	EFL	CF	35	5	620	1960	1960	OT	—	730	660	35.8	35.5	200	130	165	1	9630	—	—
1033 Oswego	USA	MAS	GR	4.3	—	109.7	1870	1912	QP	—	—	—	—	—	—	—	—	—	—	—	—
1034 Otter Lake	USA	EFL	HD	6.1	—	—	—	1978	QP	SEP	0.15	0.109	5	6.1	—	—	9.3	1.28	—	—	—
1035 Otto Run	USA	EFL	HD	5.8	—	—	—	1977	—	—	—	0.0074	5.79	—	—	—	—	—	60	—	—
1036 Ovcar Banja	YU	EFL	—	—	—	—	1952	1965	PM	—	—	—	—	—	—	—	—	—	—	—	—
1037 Overholser	USA	RFL	—	16.5	—	247	1920	1923	OT	ISC	18	—	—	18.6	—	—	91.5	—	—	—	—
1038 Owasco	USA	MAS	GR	4.3	—	—	1860	1912	QP	SEP	—	—	—	—	—	—	—	—	—	—	—
1039 Owen Reservoir	USA	EFL	—	17.1	—	247	1914	1914	QP	SEP	60	—	—	—	—	—	—	—	—	—	—
1040 Owens Lake	USA	—	—	—	—	—	—	1994	OT	—	—	—	—	—	—	—	—	—	—	—	—
1041 Owl Creek Site 13	USA	EFL	—	8.5	—	—	1957	1957	QP	SEP	—	—	—	—	—	—	—	—	—	—	—
1042 Owl Creek Site 7	USA	EFL	—	6.4	—	—	1957	1957	QP	SEP	—	—	—	—	—	—	—	—	—	—	—
1043 Oxford	USA	EFL	—	—	—	—	—	1896	OT	—	—	—	—	—	—	—	—	—	—	—	—
1044 Pace Lake	USA	EFL	—	4	—	—	—	1994	OT	—	—	—	—	—	—	—	—	—	—	—	—
1045 Pace Lake South	USA	EFL	—	4.9	—	—	—	1994	OT	—	—	—	—	—	—	—	—	—	—	—	—
1046 Pagara I	IN	MAS	GR	30.5	—	—	—	1923	OT	ISC	166	—	—	—	—	—	—	—	—	—	—
1047 Pagara II	IN	CON	GR	27.3	—	1439	1927	1943	OT	ISC	100	—	—	—	—	—	—	—	—	—	—
1048 Page Mill	USA	—	—	—	—	—	—	1893	—	—	—	—	—	—	—	—	—	—	—	—	—
1049 Palestine	USA	—	—	—	—	—	—	1890	—	—	—	—	—	—	—	—	—	—	—	—	—
1050 Pampulha	BR	EFL	—	18	—	350	1941	1954	QP	SEP	18	—	—	—	—	—	—	—	—	—	—
1051 Panguitch	USA	EFL	—	8.5	—	—	1903	1903	QP	—	—	—	—	—	—	—	—	—	—	—	—
1052 Panshet	IN	EFL	DC	49	—	740	1961	1961	OT	—	212	—	—	—	—	—	—	—	4870	—	1000
1053 Parish Lake South	USA	—	—	—	—	—	—	1994	OT	—	—	—	—	—	—	—	—	—	—	—	—
1054 Park Reservoir	USA	EFL	CF	4.6	—	—	1914	1914	—	—	0.2	—	—	—	—	—	—	—	—	—	—
1055 Partridge Lake	USA	—	—	—	—	—	—	1993	QP	SEP	—	—	—	—	—	—	—	—	—	—	—
1056 Patton	USA	—	—	—	—	—	—	1997	OT	ISC	—	—	—	—	—	—	—	—	—	—	—
1057 Peace Dale	USA	EFL	CF	4.9	—	—	—	1998	—	—	—	—	—	—	—	—	—	—	—	—	—
1058 Peapack Brook	USA	EFL	—	9.8	—	109.7	1927	1927	OT	—	—	—	—	—	—	—	—	—	—	—	—
1059 Penn Forest	USA	EFL	CF	46.1	—	—	—	1960	QP	SEP	—	—	—	—	—	—	—	—	—	—	—
1060 Perry Knitting	USA	EFL	—	—	—	—	—	2003	—	—	—	—	—	—	—	—	—	—	—	—	—
1061 Peru	USA	EFL	—	1.8	—	—	—	1998	OT	ISC	—	—	—	3.7	—	—	12.2	—	—	—	—
1062 Peters Reservoir No. 1 I	USA	EFL	—	—	—	—	1915	1974	—	—	—	—	—	—	—	—	—	—	—	—	—
1063 Peters Reservoir No. 1 II	USA	EFL	—	—	—	61	1974	1983	OT	—	—	—	—	4.1	10.7	—	—	—	—	—	—
1064 Peters Reservoir No. 2	USA	EFL	—	—	—	—	1959	1983	OT	—	—	—	—	6.1	—	—	—	—	—	—	—
1065 Phillips Pond	USA	EFL	—	10.4	—	—	—	1994	OT	—	—	—	—	—	—	—	—	—	—	—	—

(Continued)

Table A.1 (*Continued*)

No.	Name	Country[(1)]	Type[(2)]	Type divided[(3)]	Height (m)	Crest width (m)	Crest length (m)	Year of cons.	Year of failure	Failure cause[(4)]	Failure cause divided[(5)]	Reservoir capacity ($10^6 m^3$)	Water volume stored above breach invert ($10^6 m^3$)	Water depth above breach (m)	Breach geometry Height (m)	Top width (m)	Bottom width (m)	Aver. width (m)	Aver. side slope Z:1 (h:v)	Peak outflow rate (m^3/s)	Fail. time	Life loss
1066	Pickwick I	USA	EFL	—	—	—	—	—	1910	QP	SEP	—	—	—	—	—	—	—	—	—	—	—
1067	Pickwick II	USA	EFL	—	11	—	—	—	1980	OT	—	—	—	—	3.1	—	—	3.1	—	—	—	—
1068	Pierce Reservoir	USA	EFL	HD	—	3.1	—	—	1986	QP	SEP	—	4.07	8.08	8.69	—	—	30.5	0.77	—	—	—
1069	Pillar Rock	USA	CON	GR	7.6	—	—	—	1970	QP	SF	—	—	—	—	—	—	—	—	—	—	—
1070	Pine Cove Pond	USA	—	—	6.1	—	—	—	1998	—	—	—	—	—	—	—	—	—	—	—	—	—
1071	Pinhook	USA	EFL	—	6.1	—	137.2	1924	1927	—	—	—	—	—	—	—	—	—	—	—	—	—
1072	Pinkston	USA	EFL	—	21.4	—	—	—	1968	QP	SEP	—	—	—	—	—	—	—	—	—	—	—
1073	Pioneer Monument	USA	—	—	—	—	—	—	1978	QP	SEP	—	—	—	—	—	—	—	—	—	—	—
1074	Pittsfield	USA	EFL	—	10.7	—	—	—	1999	QP	SEP	—	—	—	—	—	—	—	—	—	—	—
1075	Pittston	USA	EFL	—	4.9	—	91.4	—	1901	—	—	0.01	—	—	—	—	—	—	—	—	—	125
1076	Plattsburg West Brook Dam 3	USA	CON	BU	10.7	—	100.6	1915	1916	QP	—	—	—	—	—	—	—	—	—	—	—	—
1077	Pleasant	USA	EFL	—	19.2	—	—	1928	1928	QP	SEP	76.5	—	—	—	—	—	—	—	—	—	—
1078	Pleasant CO	USA	EFL	—	5.8	—	—	—	1935	DT	ABA	—	—	—	—	—	—	—	—	—	—	—
1079	Point Of Rocks	USA	EFL	—	26.2	—	—	—	1915	QP	SF-SO	—	—	—	—	—	—	—	—	—	—	—
1080	Ponca Dam	USA	EFL	—	—	—	—	—	2000	QP	SEP	—	—	—	—	—	—	—	—	—	—	—
1081	Port Mayaca	USA	EFL	—	—	—	—	—	1979	—	—	—	—	—	—	—	—	—	—	—	—	—
1082	Port Republic I	USA	EFL	—	—	—	152.5	—	1992	—	—	—	—	—	—	—	—	—	—	—	—	—
1083	Port Republic II	USA	EFL	—	—	—	152.5	—	1997	OT	—	—	—	—	—	—	—	15.3	—	—	—	—
1084	Port Wing	USA	EFL	—	20.1	—	—	—	1985	OT	—	—	—	—	—	—	—	—	—	—	—	—
1085	Portland	USA	EFL	—	13.7	—	—	1889	1893	QP	SEP	0.6	—	—	—	—	—	—	—	—	—	4
1086	Portsville	USA	—	—	—	—	—	—	1903	—	—	—	—	—	—	—	—	—	—	—	—	—
1087	Potato Hill Lake	USA	EFL	HD	—	7.3	—	1947	1977	OT	—	—	0.105	>7.77	7.77	—	—	16.5	1.25	—	—	—
1088	Potsdam	USA	—	—	—	—	—	—	1914	—	—	—	—	—	—	—	—	—	—	—	—	—
1089	Potsdam - West	USA	TIM	—	3.4	—	—	—	1988	—	—	—	—	—	—	—	—	—	—	—	—	—
1090	Powell Lake	USA	EFL	—	10.7	—	—	—	2000	OT	—	—	—	—	—	—	—	—	—	—	—	—
1091	Pratt Fork Creek	USA	EFL	—	6.3	—	—	1934	1938	OT	—	—	—	—	—	—	—	—	—	—	—	—
1092	Price Lake	USA	EFL	—	10.1	—	—	—	—	QP	SEP	—	—	—	—	—	—	—	—	—	—	—
1093	Pritchard Lake	USA	EFL	—	11	—	—	—	2001	QP	SF-SL	—	—	—	—	—	—	—	—	—	—	—
1094	Prospect Reservoir	USA	EFL	HD	13.7	4.3	1842.5	1910	1980	DT	ABA	7.5	3.54	1.68	4.42	—	—	88.4	0.69	116	—	—
1095	Prospect Valley	USA	EFL	—	13.4	—	—	1910	—	—	—	—	—	—	—	—	—	—	—	—	—	—
1096	Prosser	USA	CON	GR	10.7	—	30.5	—	1928	OT	—	—	—	—	—	—	—	—	—	—	—	—
1097	Puddingstone	USA	EFL	HD	15.2	—	91.4	1926	1926	OT	—	0.617	0.617	>15.2	15.2	91.4	—	—	—	480	3	—
1098	Puddle Pond	USA	EFL	—	5.2	—	—	—	1996	OT	BOP	—	—	—	5.2	—	—	9.2	—	—	—	—
1099	Puentes	ES	MAS	GR	69	—	291	1791	1802	QP	SEP	13	—	—	—	—	—	—	—	—	—	680
1100	Quail Creek	USA	EFL	HD	24	—	610	1984	1988	QP	SEP	50	30.8	16.7	21.3	—	—	70	0.1	3110	—	—

1101	Rail Road Lake	USA	EFL	—	5.6	—	—	—	1980	QP	QIS	0.05	—	—	6.1	—	—	14.6	—	—	2	—
1102	Rail Yard Behind New Orleans Port	USA	EFL	—	—	—	—	—	2005	QP	QIC	—	—	—	—	—	—	—	—	—	—	—
1103	Rainbow Lake	USA	EFL	HD	14	11.3	—	—	1986	OT	—	—	6.78	10	9.54	—	—	38.9	2.52	—	—	—
1104	Rainbow Pond	USA	EFL	—	3	—	—	1974	1977	—	—	—	—	—	—	—	—	—	—	—	—	—
1105	Ramseur Lake	USA	MAS	—	4.3	—	—	—	1998	OT	—	—	—	—	2.1	—	—	18.3	—	—	—	—
1106	Randall's Pond	USA	EFL	—	—	—	—	—	1901	OT	—	—	—	—	—	—	—	—	—	—	—	8
1107	Readsboro	USA	TIM	—	12.2	—	—	—	1915	—	—	—	—	—	—	—	—	—	—	—	—	—
1108	Red Cliff	USA	EFL	—	—	—	—	—	1902	—	—	—	—	—	—	—	—	—	—	—	—	—
1109	Red Rock #3	USA	EFL	—	10.4	—	—	1910	1984	QP	SEP	—	—	—	—	—	—	—	—	—	—	—
1110	Redlands	USA	—	—	—	—	—	—	1914	—	—	—	—	—	—	—	—	—	—	—	—	—
1111	Reeder	USA	EFL	—	9.1	—	—	—	1905	OT	—	—	—	—	—	—	—	—	—	—	—	—
1112	Reedy Swine Farm Lagoon Dike	USA	EFL	—	—	—	—	—	1995	QP	—	—	—	—	—	—	—	3	—	—	—	—
1113	Reeves Lake	USA	EFL	—	6.1	—	—	—	1994	OT	—	—	—	—	—	—	—	—	—	—	—	—
1114	Renegade Resort Lake	USA	—	—	—	—	—	—	1973	OT	—	—	0.0139	3.66	3.66	—	—	2.29	0.63	—	—	—
1115	Reservoir No. 1	USA	EFL	—	9	—	—	1959	1961	—	—	—	—	—	—	—	—	—	—	—	—	—
1116	Reservoir No. 4	USA	EFL	—	15.2	—	304.8	—	1912	QP	SF-SL	—	—	—	—	—	—	—	—	—	—	—
1117	Reynolds No. 1	USA	EFL	—	14	—	—	—	1978	OT	—	—	—	—	—	—	—	61	—	—	—	—
1118	RGC Minerals	USA	—	—	—	—	—	—	1997	—	—	—	—	—	—	—	—	—	—	—	—	—
1119	Rhodesworth	GB	EFL	—	21	—	194	1852	1852	—	—	2.3	—	—	—	—	—	—	—	—	—	—
1120	Richardet Dam	USA	EFL	—	12.2	—	—	—	1985	QP	SF-SL	—	—	—	5.5	—	—	—	—	—	—	—
1121	Riddel Pond	USA	EFL	—	1.8	—	—	1980	1990	QP	SEP	—	—	—	—	—	—	—	—	—	—	—
1122	Ridgetop	USA	EFL	—	8.5	—	—	—	1974	OT	—	—	—	—	—	—	—	—	—	—	—	—
1123	Riley Mill Pond	USA	EFL	—	4.3	—	—	—	1999	OT	—	—	—	—	—	—	—	—	—	—	—	—
1124	Rincon del Bonete	UY	EFL	—	41.3	—	1172.3	1945	1959	OT	—	9000	—	—	—	—	—	—	—	—	—	—
1125	Rinse	USA	EFL	—	2.4	—	—	—	1996	QP	SEP	—	—	—	—	—	—	—	—	—	—	—
1126	Rito Manzanares	USA	EFL	HD	7.3	3.7	—	—	1975	OT	—	0.0247	0.0247	4.57	7.32	19	7.6	13.3	0.77	—	—	—
1127	Riverside	USA	EFL	—	12.5	—	—	—	1910	QP	SF-SO	—	—	—	—	—	—	—	—	—	—	—
1128	Riverview	USA	EFL	—	9.8	—	—	1947	1985	QP	SEP	—	—	—	—	—	—	—	—	—	—	—
1129	Roanoke	USA	—	—	—	—	—	—	1888	QP	—	—	—	—	—	—	—	—	—	—	—	—
1130	Roberts Pond	USA	—	—	—	—	—	—	1922	OT	—	—	—	—	—	—	—	—	—	—	—	—
1131	Roberts Tank	USA	EFL	—	6.1	—	—	—	1996	QP	SEP	—	—	—	6.1	—	—	9.2	—	—	—	—
1132	Rochester	USA	—	—	—	—	—	—	1914	—	—	—	—	—	—	—	—	—	—	—	—	—
1133	Rock	USA	—	—	—	—	—	—	1993	OT	—	—	—	—	—	—	—	—	—	—	—	—
1134	Rockport	USA	RFL	—	6.4	—	45.7	—	1912	QP	SF-OV	—	—	—	—	—	—	—	—	—	—	—
1135	Rogers	USA	EFL	—	—	—	—	—	1954	DT	EQ	—	—	—	—	—	—	—	—	—	—	—
1136	Rolling Green Community Lake	USA	—	—	—	—	—	—	1999	QP	SEP	—	—	—	—	—	—	—	—	—	—	—
1137	Rosegill Upper	USA	EFL	—	2.7	—	—	—	1999	OT	—	—	—	—	—	—	—	—	—	—	—	—
1138	Rothchild	USA	CON	GR	12.8	—	—	1912	1912	OT	—	—	—	—	—	—	—	—	—	—	—	—
1139	Roundy I	USA	EFL	—	3	—	—	—	1958	—	—	—	—	—	—	—	—	—	—	—	—	—

(Continued)

Table A.1 *(Continued)*

No.	Name	Country[1]	Type[2]	Type divided[3]	Height (m)	Crest width (m)	Crest length (m)	Year of cons.	Year of failure	Failure cause[4]	Failure cause divided[5]	Reservoir capacity (10^6 m^3)	Water volume stored above breach invert (10^6 m^3)	Water depth above breach (m)	Breach geometry: Height (m)	Breach geometry: Top width (m)	Breach geometry: Bottom width (m)	Breach geometry: Aver. width (m)	Breach geometry: Aver. side slope Z:1 (h:v)	Peak outflow rate (m^3/s)	Fail. time	Life loss
1140	Roundy II	USA	EFL	—	—	—	—	1965	1973	QP	SEP	—	—	—	—	—	—	—	—	—	—	—
1141	Route 649 Dam	USA	EFL	—	—	—	61	—	1997	OT	—	—	—	—	—	—	—	—	—	—	—	—
1142	Rowe´s Pond	USA	—	—	—	—	—	—	1912	—	—	—	—	—	—	—	—	—	—	—	—	—
1143	Roxboro Municipal	USA	EFL	—	10	—	—	1955	1984	QP	SEP	—	—	—	—	—	—	—	—	—	—	—
1144	Roxborough	USA	EFL	—	—	—	—	1893	1894	QP	SEP	—	—	—	—	—	—	—	—	—	—	—
1145	Roxbury	USA	MAS	GR	7.2	—	—	1870	1903	QP	—	—	—	—	—	—	—	—	—	—	—	—
1146	Royal Oaks	USA	EFL	—	7.6	—	—	—	2002	QP	SEP	—	—	—	—	—	—	—	—	—	—	—
1147	Ruahihi	NZ	RFL	—	32	—	67000	1981	1981	QP	SEP	31	—	—	—	—	—	—	—	—	—	—
1148	Rye Grass Swale I	USA	EFL	—	6.1	—	—	—	1932	DT	ABA	—	—	—	—	—	—	—	—	—	—	—
1149	Rye Grass Swale II	USA	EFL	—	6.1	—	—	—	1936	OT	ISC	—	—	—	—	—	—	—	—	—	—	—
1150	Rye Grass Swale III	USA	EFL	—	6.1	—	—	—	1938	OT	ISC	—	—	—	—	—	—	—	—	—	—	—
1151	Saddle Lake Camp	USA	EFL	—	6.4	—	—	1947	1974	QP	SEP	—	—	—	—	—	—	—	—	—	—	—
1152	Saddler	USA	EFL	—	2.4	—	—	—	1999	OT	—	—	—	—	—	—	—	—	—	—	—	—
1153	Sage Creek	USA	EFL	—	12.2	—	—	1911	1985	OT	—	—	—	—	—	—	—	—	—	—	—	—
1154	Saint John	USA	EFL	—	11.9	—	—	—	1980	QP	SEP	—	—	—	—	—	—	—	—	—	—	—
1155	Saint Lucien	DZ	EFL	—	27	—	—	1861	1862	QP	SEP	2	—	—	—	—	—	—	—	—	—	—
1156	Salles Oliveira	BR	EFL	HD	35.1	—	167	1966	1977	OT	—	25.9	71.5	38.4	35	—	—	167	—	7200	2	—
1157	Saluda	USA	EFL	—	63.4	—	2389	1903	1903	QP	SEP	—	—	—	—	—	—	—	—	—	—	—
1158	San Luis	USA	EFL	—	116.5	—	5630	1976	1981	QP	SF-SL	3083.7	—	—	—	—	—	—	—	—	—	—
1159	Sand Creek	USA	EFL	—	7.3	—	—	—	1915	QP	SEP	—	—	—	—	—	—	—	—	—	—	—
1160	Sandy Run	USA	EFL	HD	8.5	—	—	—	1977	OT	—	0.0568	0.0567	8.53	—	—	—	—	—	435	—	5
1161	Santa Helena	BR	RFL	—	25	—	250	1979	1985	—	—	250	—	—	—	—	—	—	—	—	—	—
1162	Sanville	USA	TIM	—	2.1	—	—	1973	1998	OT	—	—	—	—	—	—	—	—	—	—	—	—
1163	Saranac	USA	MAS	GR	13.7	—	46.3	1895	1912	—	—	—	—	—	—	—	—	—	—	—	—	—
1164	Sarnia	USA	EFL	—	4.6	—	—	1936	1978	QP	SEP	—	—	—	—	—	—	—	—	—	—	—
1165	Sassafras Mill	USA	EFL	—	3	—	—	—	1999	OT	—	—	—	—	—	—	—	—	—	—	—	—
1166	Saxe Gotha Millpond	USA	—	—	—	—	—	—	1994	PM	—	—	—	—	—	—	—	—	—	—	—	—
1167	Schaeffer	USA	EFL	DC	30	4.6	335	1911	1921	OT	—	3.92	4.44	>30.5	30	210	64	137	2.25	4500	0.5	—
1168	Schenectady	USA	EFL	—	9.1	—	45.7	—	1916	OT	—	—	—	—	9.1	24.4	—	—	—	—	—	—
1169	Schofield	USA	TIM	—	2.5	—	106.7	—	1912	OT	—	—	—	—	—	—	—	—	—	—	—	—
1170	Schweiger	USA	EFL	—	4.6	—	—	1982	—	—	—	—	—	—	—	—	—	—	—	—	—	—
1171	Scofield	USA	EFL	—	23.8	—	—	—	1928	QP	SEP	—	—	—	—	—	—	—	—	—	—	—
1172	Scott Dam	USA	EFL	—	4	—	—	—	1997	OT	—	—	—	—	6.9	—	—	9.9	—	—	—	—
1173	Scott Falls	CA	EFL& CON	—	15	—	610	1921	1923	OT	—	11	—	—	—	—	—	—	—	—	—	—
1174	Scott Farm No. 2	CA	—	—	—	—	—	—	—	QP	SEP	—	0.086	10.4	11.9	15	15	15	0	—	—	—

1175	Scottdale Greenlick	USA	EFL	—	18.3	—	—	1901	1904	QP	SEP	—	—	—	—	—	—	—	—	—	—	—
1176	Scout Lake	USA	—	—	—	—	—	—	1994	OT	—	—	—	—	—	—	—	—	—	—	—	—
1177	Scranton	USA	MAS	GR	7.6	—	—	—	1895	QP	—	—	—	—	—	—	—	—	—	—	—	—
1178	Sedro Wooley	USA	—	—	—	—	—	—	1983	—	—	—	—	—	—	—	—	—	—	—	—	—
1179	Seefield	USA	EFL	—	39.6	—	—	—	1925	OT	—	—	—	—	—	—	—	—	—	—	—	—
1180	Sella Zerbino Molare	IT	CON	GR	16.5	—	72	1925	1935	OT	—	10	—	—	—	—	—	—	—	—	—	130
1181	Selsfors	SE	MAS	BU	21	—	—	1943	1943	QP	SEP	—	—	—	—	—	—	—	—	—	—	—
1182	Seminary Hill	USA	—	—	—	—	—	—	1991	QP	SF-SL	—	—	—	—	—	—	—	—	—	—	—
1183	Senaca	USA	EFL	—	19	—	168	1975	1975	OT	—	3.6	—	—	—	—	—	—	—	—	—	—
1184	Seneca Falls	USA	CON	GR	—	—	144.8	1915	1915	—	—	—	—	—	—	—	—	—	—	—	—	—
1185	Seneca Lake	USA	EFL	—	3.4	—	—	—	2000	QP	SEP	—	—	—	—	—	—	15.3	—	—	—	—
1186	Sepulveda	USA	EFL	—	19.8	—	—	—	1914	OT	ISC	—	—	—	—	—	—	—	—	—	—	—
1187	Sequoyah	USA	EFL	—	19.2	—	17.1	1956	1956	OT	—	2.1	—	—	—	—	—	—	—	—	—	—
1188	Sevier Bridge	USA	EFL	—	15.9	—	—	1914	1914	—	—	—	—	—	—	—	—	—	—	—	—	—
1189	Sewalls Island	USA	TIM	—	4.7	—	—	—	1978	—	—	—	—	—	—	—	—	—	—	—	—	—
1190	Seymour	USA	EFL	—	7.3	—	—	—	1976	QP	SEP	—	—	—	—	—	—	—	—	—	—	—
1191	Shady Brook	USA	EFL	—	4.3	—	—	—	1982	OT	—	—	—	—	—	—	—	—	—	—	—	—
1192	Shale Creek	USA	EFL	—	4.6	—	—	—	2004	QP	SEP	—	—	—	—	—	—	—	—	—	—	—
1193	Shannon Lake	USA	EFL	—	6.7	—	—	—	1975	DT	BUA	—	—	—	—	—	—	—	—	—	—	—
1194	Sheep Creek	USA	EFL	HD	18.6	6.1	330	1969	1970	QP	SEP	1.43	2.91	14.02	17.1	30.5	13.5	22	0.5	—	—	—
1195	Sheffield	USA	EFL	—	8.5	—	219.5	1917	1925	DT	EQ	—	—	—	—	—	—	—	—	—	—	—
1196	Shellhouse Lake	USA	—	—	—	—	—	—	1994	OT	—	—	—	—	—	—	—	—	—	—	—	—
1197	Sherburne I	USA	EFL	—	—	—	—	1883	1905	OT	—	0.3	—	—	—	—	—	—	—	—	—	—
1198	Sherburne II	USA	EFL	DC	10.3	—	91.4	1892	1905	OT	—	0.042	—	—	—	46	—	—	—	960	2	—
1199	Sherman Dam	USA	EFL	—	41	—	1356	—	1979	OT	—	85.2	—	—	—	—	—	—	—	—	—	—
1200	Sherry Lake	USA	TIM	—	3.7	—	—	—	1994	OT	—	—	—	—	—	—	—	—	—	—	—	—
1201	Shipp Lake	USA	EFL	—	4.9	—	—	—	1994	OT	—	—	—	—	—	—	—	—	—	—	—	—
1202	Shofill Lake	USA	—	—	—	—	—	—	1994	OT	—	—	—	—	—	—	—	—	—	—	—	—
1203	Shopiere	USA	EFL	—	1.8	—	—	1840	1973	OT	—	—	—	—	—	—	—	—	—	—	—	—
1204	Short Mountain Creek	USA	EFL	—	13.1	—	225.6	1939	1939	OT	—	—	—	—	—	—	—	—	—	—	—	—
1205	Sibley	USA	TIM	—	3.7	—	—	1897	1997	QP	QIS	—	—	—	—	—	—	—	—	—	—	—
1206	Sid White	USA	EFL	—	4.3	—	—	—	1971	OT	ISC	—	—	—	—	—	—	—	—	—	—	—
1207	Sig	DZ	CON	GR	21	—	97.4	1858	1885	OT	—	3.3	—	—	—	—	—	—	—	—	—	—
1208	Silberman Lake	USA	EFL	—	4	—	—	—	1994	OT	—	—	—	—	—	—	—	—	—	—	—	—
1209	Silver Lake	USA	EFL	—	9.2	—	—	—	2003	—	—	—	—	—	—	—	—	—	—	—	—	—
1210	Simpson Dam	USA	EFL	—	5	—	—	—	1986	QP	SEP	—	—	—	—	—	—	—	—	—	—	—
1211	Sinker Creek	USA	EFL	HD	21.5	—	335	1910	1943	QP	SF-SL	3.33	3.33	21.34	21.3	92	49.2	70.6	0.5	—	2	—
1212	Six-Acre lake on Forrest Road	USA	—	—	—	—	—	—	1994	OT	—	—	—	—	—	—	—	—	—	—	—	—
1213	Skagway	USA	RFL	—	24.1	—	—	—	1965	—	—	—	—	—	—	—	—	—	—	—	—	2
1214	Sky Lake No. 1	USA	—	—	—	—	—	—	1987	QP	SEP	0.39	—	—	—	—	—	—	—	—	—	—
1215	Slate Creek	USA	CON	ARC	60.4	—	—	—	1938	OT	ISC	—	—	—	—	—	—	—	—	—	—	—

(*Continued*)

Table A.1 *(Continued)*

No.	Name	Country[1]	Type[2]	Type divided[3]	Height (m)	Crest width (m)	Crest length (m)	Year of cons.	Year of failure	Failure cause[4]	Failure cause divided[5]	Reservoir capacity ($10^6 m^3$)	Water volume stored above breach invert ($10^6 m^3$)	Water depth above breach (m)	Breach geometry: Height (m)	Breach geometry: Top width (m)	Breach geometry: Bottom width (m)	Breach geometry: Aver. width (m)	Breach geometry: Aver. side slope Z:1 (h:v)	Peak outflow rate (m^3/s)	Fail. time	Life loss
1216	Smart Sindicate	ZA	EFL	—	28	—	2802	1912	1961	QP	SEP	98	—	—	—	—	—	—	—	—	—	—
1217	Smith River	USA	EFL	—	8.2	—	—	—	2002	QP	SEP	—	—	—	—	—	—	—	—	—	—	—
1218	Snake Ravine I	USA	EFL	—	19.2	—	88.2	1893	1893	QP	—	—	—	—	—	—	—	—	—	—	—	—
1219	Snake Ravine II	USA	EFL	—	19.2	—	88.2	1898	1898	QP	—	—	—	—	—	—	—	—	—	—	—	—
1220	Snow Bird Lake	USA	EFL	—	5.5	—	—	—	1980	QP	SEP	—	—	—	—	—	—	—	—	—	—	—
1221	South Eden	USA	EFL	—	3.7	—	—	—	1980	OT	—	—	—	—	2.4	—	—	22.9	—	—	—	—
1222	South Fork	USA	EFL	CF	22	6	256	1852	1889	OT	ISC	18	—	—	—	128.1	38.1	—	—	122	0.75	2209
1223	South Fork Tributar	USA	EFL	HD	1.8	—	—	—	1977	—	—	—	0.0037	1.83	—	—	—	—	—	122	—	—
1224	Southern Clay Company Dam 2	USA	—	—	—	—	—	—	1989	QP	SEP	—	—	—	—	—	—	—	—	—	—	—
1225	Southern Pines Country Club Golf Course Dam 1	USA	—	—	—	—	—	—	1995	—	—	—	—	—	—	—	—	—	—	—	—	—
1226	Southern Pines Country Club Golf Course Dam 2	USA	—	—	—	—	—	—	1995	—	—	—	—	—	—	—	—	—	—	—	—	—
1227	Southern Pines Country Club Golf Course Dam 3	USA	—	—	—	—	—	—	1995	—	—	—	—	—	—	—	—	—	—	—	—	—
1228	Southern States Lake Dam	USA	EFL	—	8.3	—	—	—	1998	QP	SEP	—	—	—	—	—	—	—	—	—	—	—
1229	Spartansburg	USA	EFL	CF	3	—	54.9	—	1892	OT	—	—	—	—	—	—	—	—	—	—	—	—
1230	Speer Dam	USA	—	—	—	—	—	—	1996	OT	—	—	—	—	—	—	—	—	—	—	—	—
1231	Spencer	USA	—	—	—	—	—	—	1999	QP	SEP	—	—	—	—	—	—	—	—	—	—	—
1232	Spitskop	ZA	EFL	—	18	—	683	1974	1988	—	—	61	—	—	—	—	—	—	—	—	—	—
1233	Spring Lake	USA	EFL	CF	5.5	2.4	251.5	1887	1889	QP	SEP	0.135	0.136	5.49	5.49	20	9	14.5	1	—	—	3
1234	Spruce Lake	USA	EFL	—	6.1	—	—	—	1969	QP	SEP	—	—	—	—	—	—	—	—	—	—	—
1235	St. Cloud	USA	—	—	—	—	—	1970	1970	QP	SF	—	—	—	—	—	—	—	—	—	—	—
1236	St. Francis	USA	CON	GR	62.5	4.88	213	1926	1928	QP	SEP	47	47	—	—	—	—	—	—	14160	1.17	426
1237	St. Joe State Park Sediment Impoundment	USA	RFL	—	3	—	—	—	1990	OT	ISC	—	—	—	—	—	—	—	—	—	—	—
1238	St. Kinsman	USA	—	—	—	—	—	—	1886	QP	—	—	—	—	—	—	—	—	—	—	—	—
1239	Staffordville	USA	EFL	—	7.9	—	45.7	—	1877	QP	SEP	—	—	—	—	—	—	—	—	—	—	—
1240	Standley Lake-I	USA	EFL	—	34.4	—	2133.6	1911	1914	QP	SF-SL	—	—	—	—	—	—	—	—	—	—	—
1241	Standley Lake-II	USA	EFL	—	34.3	—	2133.6	1911	1916	QP	SF-SL	—	—	—	—	—	—	—	—	—	—	—
1242	Stanton	USA	EFL	—	—	—	—	1902	1902	—	—	—	—	—	—	—	—	—	—	—	—	—
1243	Starnes	USA	EFL	—	4.9	—	—	—	1997	DT	BUA	—	—	—	—	—	—	—	—	—	—	—
1244	State	USA	—	—	—	—	—	—	1893	—	—	—	—	—	—	—	—	—	—	—	—	—
1245	State Hospital Lake	USA	EFL	—	6.1	—	—	—	2002	DT	ABA	—	—	—	—	—	—	—	—	—	—	—

1246	Statham Lake	USA	EFL	HD	5.5	—	—	1955	1994	OT	—	—	0.564	5.55	5.12	—	—	21	0.54	—	—	—
1247	Staunton	USA	EFL	—	14	—	—	—	1896	—	—	—	—	—	—	—	—	—	—	—	—	5
1248	Stephenson Creek	USA	MAS	GR	—	—	—	—	1892	—	—	—	—	—	—	—	—	—	—	—	—	—
1249	Stevens Lake	USA	EFL	—	11	—	—	—	1993	OT	—	—	—	—	11	12.2	2.4	—	—	—	—	—
1250	Stewart	USA	EFL	—	2.4	—	—	—	1980	QP	SEP	—	—	—	—	—	—	—	—	—	—	—
1251	Stockton College	USA	EFL	—	2.3	3	213.5	—	1997	OT	—	—	—	—	—	—	—	—	—	—	—	—
1252	Stockton Creek	USA	EFL	HD	28	—	100	1949	1950	QP	SEP	0.5	—	—	—	—	—	—	—	—	—	—
1253	Stoneridge	USA	EFL	—	11.6	—	—	1969	1974	OT	ISC	—	—	—	—	—	—	—	—	—	—	—
1254	Stony	USA	CON	BU	15.5	—	320	1913	1914	QP	SEP	11.4	—	—	—	—	—	—	—	—	—	—
1255	Stouffer Pond	USA	EFL	—	2.4	—	—	—	1977	DT	BUA	—	—	—	—	—	—	—	—	—	—	—
1256	Stubbs Farm	USA	EFL	—	—	—	—	—	1999	QP	SEP	—	—	—	—	—	—	—	—	—	—	—
1257	Sudduth	USA	EFL	—	4	—	—	1907	1982	OT	—	—	—	—	—	—	—	—	—	—	—	—
1258	Summit Landfill	USA	—	—	4.9	—	—	—	1989	OT	ISC	—	—	—	—	—	—	—	—	—	—	—
1259	Sunset Lake	USA	EFL	CF	3.7	—	—	—	1998	QP	QIS	—	—	—	—	—	—	—	—	—	—	—
1260	Swanson	USA	EFL	—	4.9	—	—	1920	1991	QP	SEP	—	—	—	—	—	—	—	—	—	—	—
1261	Swearingen Lake	USA	—	—	—	—	—	—	1994	OT	—	—	—	—	—	—	—	—	—	—	—	—
1262	Sweet's Mill	USA	EFL	CF	—	—	—	—	2002	DT	ABA	—	—	—	—	—	—	—	—	—	—	—
1263	Sweetwater I	USA	EFL	DC	36	—	213	1911	1916	OT	—	34	—	—	9.2	—	—	22.9	—	—	—	—
1264	Sweetwater II	USA	EFL	DC	34.2	—	—	—	1927	OT	—	—	—	—	—	—	—	—	—	—	—	—
1265	Swift Dam No. 2	USA	RFL	FD	57.6	—	225.6	1914	1964	OT	ISC	37	37	47.85	57.6	225	225	225	0	24947	—	28
1266	Swift No. 2	USA	EFL	—	25.3	—	—	—	2002	QP	SEP	—	—	—	—	—	—	—	—	—	—	—
1267	Swimming Pool	USA	EFL	—	—	—	—	—	1979	OT	—	—	—	—	—	—	—	—	—	—	—	4
1268	Sydnors Millpond	USA	EFL	—	3.4	—	—	—	1999	OT	—	—	—	—	—	—	—	—	—	—	—	—
1269	Tabia	DZ	EFL	—	25	—	—	1876	1876	OT	—	—	—	—	—	—	—	—	—	—	—	—
1270	Table Rock Cove	USA	EFL	—	42.7	—	228.6	1927	1928	QP	SEP	34.5	—	—	—	—	—	—	—	—	—	—
1271	Tacoma	USA	TIM	—	6.1	—	36.6	—	1892	QP	—	—	—	—	—	—	—	—	—	—	—	—
1272	Talcville	USA	CON	GR	4	—	—	—	1998	OT	—	—	—	—	—	—	—	—	—	—	—	—
1273	Tampa	USA	MAS	GR	6.6	—	180	—	1898	DT	WTA	—	—	—	—	—	—	—	—	—	—	—
1274	Tannersville Reservoir #1	USA	EFL	—	6.1	—	—	—	1980	QP	SEP	0.005	—	—	—	—	—	—	—	—	—	—
1275	Tappan	USA	EFL	—	15.9	—	—	1934	1934	QP	—	—	—	—	—	—	—	—	—	—	—	—
1276	Tarkiln Pond	USA	EFL	—	—	—	—	—	1997	OT	—	—	—	—	—	—	—	15.3	—	—	—	—
1277	Tecumseh	USA	—	—	—	—	—	—	1894	—	—	—	—	—	—	—	—	—	—	—	—	—
1278	Telluride	USA	EFL	CF	9.1	—	182.9	1894	1909	OT	ISC	—	—	—	—	—	—	—	—	—	—	—
1279	Tennessee Valley	USA	EFL	—	6.1	—	—	—	—	OT	—	—	—	—	—	—	—	—	—	—	—	—
1280	Terrace	USA	EFL	—	51.2	—	182.9	—	1957	QP	SF-SL	22.2	—	—	—	—	—	—	—	—	—	—
1281	Terry Sanford	USA	EFL	—	12.2	—	—	1960	1976	QP	—	—	—	—	—	—	—	—	—	—	—	—
1282	Teton	USA	EFL	CF	93	10.7	945	1976	1976	QP	SEP	356	310	77.4	86.9	—	—	151	1	65120	4	11
1283	Tharpe Lake	USA	EFL	—	7.9	—	—	—	1994	OT	—	—	—	—	—	—	—	—	—	—	—	—
1284	Thomas	USA	EFL	—	8.8	—	—	—	1973	OT	—	—	—	—	—	—	—	—	—	—	—	—
1285	Thomas Pond	USA	EFL	—	8.5	—	—	—	1997	OT	BOS	—	—	—	5.8	—	—	—	—	—	—	—
1286	Thompsonville	USA	—	—	—	—	—	—	1901	—	—	—	—	—	—	—	—	—	—	—	—	—
1287	Thorton Place Pond	USA	EFL	—	—	—	—	—	1994	OT	—	—	—	—	—	—	—	—	—	—	—	—

(Continued)

Table A.1 *(Continued)*

No.	Name	Country[1]	Type[2]	Type divided[3]	Height (m)	Crest width (m)	Crest length (m)	Year of cons.	Year of failure	Failure cause[4]	Failure cause divided[5]	Reservoir capacity (10^6 m^3)	Water volume stored above breach invert (10^6 m^3)	Water depth above breach (m)	Breach geometry Height (m)	Top width (m)	Bottom width (m)	Aver. width (m)	Aver. side slope Z:1 (h:v)	Peak outflow rate (m^3/s)	Fail. time	Life loss
1288	Three Mile Flat Reservoir	USA	EFL	—	8.2	—	—	—	1955	DT	ABA	—	—	—	—	—	—	—	—	—	—	—
1289	Tigra	IN	MAS	GR	25.9	—	1341	1917	1917	OT	—	124	—	—	—	—	—	—	—	—	—	1000
1290	Timber Creek Watershed Dam 1	USA	EFL	—	9.5	—	—	—	1967	QP	SEP	—	—	—	—	—	—	—	—	—	—	—
1291	Timber Lake NJ	USA	EFL	—	5.8	—	—	—	2004	—	—	—	—	—	—	—	—	—	—	—	—	—
1292	Timber Lake VA	USA	EFL	—	9.3	—	—	—	1995	OT	—	—	—	—	—	—	—	—	—	—	—	2
1293	Timperley Wildlife Res I	USA	EFL	—	3.4	—	—	—	1990	DT	ABA	—	—	—	—	—	—	—	—	—	—	—
1294	Timperley Wildlife Res II	USA	EFL	—	3.4	—	—	—	1999	QP	SEP	—	—	—	—	—	—	—	—	—	—	—
1295	Tolna #2	USA	EFL	—	5.5	—	—	1935	1948	QP	SEP	—	—	—	—	—	—	—	—	—	—	—
1296	Tomahawk Lake	USA	EFL	—	4.3	—	—	—	2000	—	—	—	—	—	—	—	—	9.2	—	—	—	—
1297	Tomkins Lake I	USA	EFL	—	24.4	3.7	91.5	—	1987	OT	—	—	0.0185	—	—	—	—	—	—	—	—	—
1298	Tomkins Lake II	USA	EFL	—	24.4	3.7	91.5	—	1988	OT	—	—	—	—	—	—	—	—	—	—	—	—
1299	Tomkins Lake III	USA	EFL	—	24.4	3.7	91.5	—	2002	OT	ISC	—	—	—	24.4	21.4	18.3	—	—	—	—	—
1300	Toreson	USA	EFL	—	15	—	96	1898	1953	QP	SEP	1.4	—	—	—	—	—	—	—	—	—	—
1301	Torside	GB	EFL	—	31	—	270	1854	1855	QP	QIS	6.7	—	—	—	—	—	—	—	—	—	—
1302	Tous	ES	RFL	—	69.5	—	785	1982	1982	OT	ISC	53	—	—	—	—	—	—	—	—	—	40
1303	Towanda	USA	EFL	—	—	—	—	1882	1939	QP	SEP	9.5	—	—	—	—	—	—	—	—	—	—
1304	Town Bridge Pond	USA	EFL	—	4	—	—	—	1999	OT	—	—	—	—	—	—	—	—	—	—	—	—
1305	Trial Lake	USA	EFL	—	9.5	—	—	—	1986	QP	SEP	—	1.48	5.18	5.18	—	—	21	0.82	—	—	—
1306	Tropic	USA	EFL	—	8.8	—	—	—	1935	QP	QIS	—	—	—	—	—	—	—	—	—	—	—
1307	Trout Lake	USA	EFL	—	7.6	—	—	1894	1909	OT	—	—	0.493	8.53	8.53	—	—	26.2	1.79	—	—	—
1308	Troy	USA	EFL	—	13.1	—	—	—	1995	QP	SEP	0.03	—	—	—	—	—	—	—	—	—	—
1309	Tuckahoe State Park	USA	EFL	—	4.3	—	—	—	1999	OT	—	—	—	—	—	—	—	—	—	—	—	—
1310	Tupelo Bayou Site 1	USA	EFL	—	15.2	—	442	1973	1973	QP	SEP	4	—	—	—	—	—	—	—	—	—	—
1311	Tupper Lake	USA	EFL	—	5.5	—	222.5	—	1906	QP	SEP	0.02	—	—	—	—	—	—	—	—	—	—
1312	Turkey Creek	USA	EFL	—	6.7	—	—	—	1910	QP	—	—	—	—	—	—	—	—	—	—	—	—
1313	Turney Lower	USA	EFL	—	4.6	—	—	—	1982	OT	—	—	—	—	—	—	—	—	—	—	—	—
1314	Turtle Lake	USA	—	—	—	—	—	—	2001	QP	SEP	—	—	—	—	—	—	—	—	—	—	—
1315	Tutens Mill Pond	USA	EFL	—	—	—	—	—	1980	OS	—	—	—	—	—	30.5	—	—	—	—	—	—
1316	Udall	USA	EFL	—	7.3	—	—	—	1997	QP	SEP	—	—	—	—	—	—	—	—	—	—	—
1317	Underwood Millpond Dam	USA	—	—	—	—	—	—	1994	OT	—	—	—	—	—	—	—	—	—	—	—	—
1318	United Nuclear Corp	USA	EFL	—	12.2	—	—	1976	1979	QP	SEP	—	2.4655	—	—	—	—	—	—	—	—	—
1319	United Nuclear Home	USA	EFL	—	25.9	—	—	1976	1977	—	—	—	—	—	—	—	—	—	—	—	—	—
1320	Unknown, Aiken, SC	USA	—	—	—	—	—	—	1990	QP	SEP	—	—	—	—	—	—	—	—	—	—	—

1321	Unknown, Ashuelot River	USA	—	—	—	—	—	—	1991	QP	SEP	—	—	—	—	—	—	—	—	—	—	—
1322	Unknown, Bradford, VT	USA	EFL	—	—	—	—	—	1998	OT	—	—	—	—	—	—	—	—	—	—	—	—
1323	Unknown, Del Rio Creek, TN	USA	—	—	—	—	—	—	1984	QP	SEP	—	—	—	—	—	—	—	—	—	—	—
1324	Unknown, VT	USA	EFL	—	4.6	—	—	—	1998	QP	SEP	—	—	—	—	—	—	—	—	—	—	—
1325	Unknown, KY	USA	—	—	—	—	—	—	1989	OT	—	—	—	—	—	—	—	—	—	—	—	—
1326	Unknown, McMinnville	USA	—	—	—	—	—	—	1902	—	—	—	—	—	—	—	—	—	—	—	—	5
1327	Unknown, MD	USA	EFL	—	—	—	—	—	1999	OT	—	—	—	—	—	—	—	—	—	—	—	—
1328	Unknown, MO	USA	—	—	16.8	—	—	—	1992	OT	BOP	—	—	—	—	—	—	—	—	—	—	—
1329	Unknown, Portland	USA	EFL	—	—	—	—	—	1893	—	—	—	—	—	—	—	—	—	—	—	—	4
1330	Unknown, WA	USA	—	—	—	—	—	—	1990	OT	BOS	—	—	—	—	—	—	—	—	—	—	—
1331	Unknown1, GA	USA	—	—	—	—	—	—	1998	—	—	—	—	—	—	—	—	—	—	—	—	—
1332	Unknown2, GA	USA	—	—	—	—	—	—	2003	DT	ABA	—	—	—	—	—	—	3.1	—	—	—	—
1333	Upper & Lower Marimac	USA	—	—	—	—	—	—	1994	OT	—	—	—	—	—	—	—	—	—	—	—	—
1334	Upper Aetna Lake	USA	EFL	—	4.3	—	—	—	2004	OT	—	—	—	—	—	—	—	—	—	—	—	—
1335	Upper Bolton Lake	USA	EFL	—	4.6	—	—	—	1938	OT	—	—	—	—	—	—	—	—	—	—	—	—
1336	Upper Bond	USA	EFL	—	5.2	—	119	—	1982	OT	—	0.2	—	—	—	—	—	—	—	—	—	—
1337	UpperClear Boggy Creek Site 50	USA	EFL	—	8.8	—	—	1970	1970	—	—	—	—	—	—	—	—	—	—	—	—	—
1338	Upper Clear Boggy Creek Site 53	USA	EFL	—	7.6	—	—	1964	1964	QP	SEP	—	—	—	—	—	—	—	—	—	—	—
1339	Upper Jackson north of County	USA	—	—	—	—	—	—	1994	OT	—	—	—	—	—	—	—	—	—	—	—	—
1340	Upper Lebanon #1	USA	EFL	—	14.6	—	—	1929	—	QP	SEP	—	—	—	—	—	—	—	—	—	—	—
1341	Upper Melzingah	USA	EFL	—	8.4	—	93	1889	1897	OT	—	0.05	—	—	—	—	—	—	—	—	—	7
1342	Upper Millpond	USA	RFL	—	3.7	—	—	—	1982	OT	—	—	—	—	—	—	—	—	—	—	—	—
1343	Upper Moore	USA	EFL	—	5.8	—	—	1900	1973	QP	SEP	—	—	—	—	—	—	—	—	—	—	—
1344	Upper North Eden	USA	—	—	—	—	—	—	1981	QP	—	0.58	—	—	12.2	—	—	53.4	—	—	—	—
1345	Upper Pond	USA	EFL	—	5.2	—	—	—	1982	OT	—	0.22	0.222	5.18	5.18	—	—	16.5	1.71	—	—	—
1346	Upper Providence	USA	EFL	—	5.4	—	—	1816	1901	QP	—	—	—	—	—	—	—	—	—	—	—	—
1347	Upper Red Rock Creek Site 42	USA	EFL	—	7	—	—	1966	1967	QP	SEP	—	—	—	—	—	—	—	—	—	—	—
1348	Upper Red Rock Creek Site 48	USA	EFL	—	7	—	—	—	1964	QP	SEP	—	—	—	—	—	—	—	—	—	—	—
1349	Upper Red Rock Creek Site-20	USA	EFL	—	9.5	—	—	—	1986	QP	SEP	—	—	—	—	—	—	—	—	—	—	—
1350	Upper Shelton	USA	EFL	—	—	—	—	1881	1903	QP	SEP	0.15	—	—	—	—	—	—	—	—	—	—
1351	Upper Twin Lakes	USA	EFL	—	3.7	—	91.5	—	1991	OT	—	—	—	—	—	—	—	—	—	—	—	—
1352	Upriver Dam	USA	CON	GR	11	—	—	—	1986	PM	—	—	—	—	—	—	—	—	—	—	—	—
1353	Urban Pond	USA	EFL	—	3.7	—	—	—	1982	OT	—	—	—	—	—	—	—	—	—	—	—	—
1354	Utica Reservoir	USA	EFL	—	21.3	—	—	1873	1902	QP	SF-SL	0.7	—	—	21.3	—	—	38.1	—	—	—	—

(Continued)

Table A.1 *(Continued)*

No.	Name	Country[(1)]	Type[(2)]	Type divided[(3)]	Height (m)	Crest width (m)	Crest length (m)	Year of cons.	Year of failure	Failure cause[(4)]	Failure cause divided[(5)]	Reservoir capacity ($10^6 m^3$)	Water volume stored above breach invert ($10^6 m^3$)	Water depth above breach (m)	Breach geometry: Height (m)	Breach geometry: Top width (m)	Breach geometry: Bottom width (m)	Breach geometry: Aver. width (m)	Breach geometry: Aver. side slope Z:1 (h:v)	Peak outflow rate (m^3/s)	Fail. time	Life loss
1355	Vaiont	IT	CON	ARC	267	3.4	11.2	1961	1963	DT	RLR	240	—	—	—	—	—	—	—	—	0.12	2117
1356	Val Marie	CA	EFL	—	—	—	—	—	1952	OT	—	—	—	—	—	—	—	—	—	—	—	—
1357	Valentine Fish Hatchery Dam	USA	EFL	—	13.1	—	—	—	1911	OT	—	—	—	—	—	—	—	—	—	—	—	—
1358	Van Norman	USA	RFL	—	43	—	664.5	1921	1971	DT	EQ	25	—	—	—	—	—	—	—	—	—	—
1359	Vance Lake	USA	EFL	—	7.3	—	—	—	1979	QP	SEP	—	—	—	7.3	15.3	7.6	—	—	—	—	—
1360	Vaughn Creek	USA	CON	ARC	19	—	91.5	1926	1926	QP	SEP	—	—	—	—	—	—	—	—	—	—	—
1361	Vega de Tera	ES	MAS	BU	33.5	—	270	1956	1959	QP	SF-SL	7.3	—	—	—	—	—	—	—	—	—	144
1362	Vernon Marsh-Ref. Flowage	USA	EFL	—	2.1	—	—	—	1996	QP	SEP	—	—	—	—	—	—	—	—	—	—	—
1363	Vertrees	USA	EFL	—	8.2	—	—	—	1998	QP	SEP	—	—	—	—	—	—	—	—	—	—	—
1364	Veteran's Memorial State Park	USA	EFL	—	—	—	—	—	1999	OT	—	—	—	—	—	—	—	—	—	—	—	—
1365	Viktor	USA	EFL	—	7.6	—	131.1	—	1901	OT	—	0.2	—	—	—	—	—	—	—	—	—	—
1366	Village of Granville, Dam No. 1101	USA	EFL	—	2.4	—	—	1885	1976	—	—	—	—	—	2.4	—	—	7.6	—	—	—	—
1367	Vincent No. 2	USA	EFL	—	—	—	—	—	1995	OT	BOS	—	—	—	—	—	—	—	—	—	—	—
1368	Virden Creek	USA	EFL	—	—	—	—	—	1968	OT	—	—	—	—	—	—	—	—	—	—	—	1
1369	Virgin River	USA	RFL	—	19.8	—	—	1929	1929	QP	—	—	—	—	—	—	—	—	—	—	—	—
1370	Virginia Kendall	USA	EFL	—	7.6	—	—	—	—	QP	SEP	—	—	—	—	—	—	—	—	—	—	—
1371	W. Day	USA	EFL	—	6.4	—	—	—	1999	OT	—	—	—	—	—	—	—	—	—	—	—	—
1372	Wachi	JP	CON	GR	25	—	141	1967	1967	OT	ISC	0.63	—	—	—	—	—	—	—	—	—	—
1373	Waco Lake I	USA	EFL	—	44.2	—	—	—	1947	QP	SEP	—	—	—	—	—	—	—	—	—	—	—
1374	Waco Lake II	USA	EFL	—	42.7	—	5468.4	1961	1961	QP	SF-SL	895.5	—	—	—	—	—	—	—	—	—	—
1375	Waghad	IN	EFL	—	32	—	—	1883	1883	OT	—	17	—	—	—	—	—	—	—	—	—	—
1376	Wagner Creek	USA	EFL	—	15.2	—	98	1918	1938	QP	QIS	0.74	—	—	—	—	—	—	—	—	—	1
1377	Wainwright Lake	USA	—	—	—	—	—	—	1994	OT	—	—	—	—	—	—	—	—	—	—	—	—
1378	Wallace Lake	USA	EFL	—	4.3	—	—	—	1988	QP	SEP	—	—	—	—	—	—	—	—	—	—	—
1379	Walnut Canyon	USA	EFL	—	48.8	—	—	1968	1975	QP	—	—	—	—	—	—	—	—	—	—	—	—
1380	Walnut Grove	USA	RFL	—	33	—	122	1888	1890	OT	ISC	11	—	—	—	—	—	—	—	—	—	30–150
1381	Walter Bouldin	USA	EFL	DC	51.8	—	2268	1967	1975	QP	SEP	—	—	—	33.5	—	—	122	—	—	14	—
1382	Ward	USA	—	—	—	—	—	—	1897	OT	—	—	—	—	—	—	—	—	—	—	—	2
1383	Wardens Pond North	USA	EFL	—	4	—	—	—	2001	—	—	—	—	—	—	—	—	—	—	—	—	—
1384	Wards Pond	USA	MAS	—	3.7	—	—	1909	1984	QP	SEP	—	—	—	—	—	—	—	—	—	—	—
1385	Warrensburg Board	USA	—	GR	7.3	—	—	1909	1976	—	—	—	—	—	—	—	—	—	—	—	—	—
1386	Washington County	USA	EFL	—	7.9	—	—	1962	1962	QP	SEP	—	—	—	—	—	—	—	—	—	—	—

1387	Washington Lake	USA	—	—	—	—	—	—	1978	OT	—	0.37	—	—	3	—	—	6.1	—	—	—	—
1388	Washita	USA	EFL	—	3.7	—	—	—	1914	QP	SEP	—	—	—	—	—	—	—	—	—	—	—
1389	Wateree	USA	EFL	—	39.3	—	—	—	1916	—	—	—	—	—	—	—	—	—	—	—	—	—
1390	Waterman Dam	USA	EFL	—	10	—	—	—	1994	OT	—	—	—	—	—	—	—	—	—	—	—	—
1391	Wausau	USA	CON	GR	8.5	—	—	—	1912	OT	—	—	—	—	8.5	—	—	38.1	—	—	—	—
1392	Weare Reservoir	USA	EFL	—	10.4	—	—	—	1938	OT	ISC	—	—	—	—	—	—	—	—	—	—	4
1393	Wells Millpond Lake Dam	USA	—	—	—	—	—	—	1994	OT	—	—	—	—	—	—	—	—	—	—	—	—
1394	Weslake Schoenberger Creek	USA	—	—	14.6	—	100	—	—	—	—	0.276	—	—	—	—	—	—	—	—	—	—
1395	Wesley E. Seale	USA	EFL	—	25	—	1804	1958	1965	QP	QIS	374	—	—	—	—	—	—	—	—	—	—
1396	West Bank	USA	EFL	—	—	—	—	—	2005	QP	QIC	—	—	—	—	—	—	—	—	—	—	—
1397	West Brook	USA	CON	—	10.7	—	—	—	1916	QP	SEP	—	—	—	—	—	—	—	—	—	—	—
1398	West Creek	USA	EFL	—	6.7	—	—	1963	1983	OT	ISC	—	—	—	—	—	—	—	—	—	—	—
1399	Weston Mill	USA	TIM	—	4	—	—	1937	1977	QP	QIS	—	—	—	—	—	—	—	—	—	—	—
1400	Westport	USA	—	—	—	—	—	—	1914	—	—	—	—	—	—	—	—	—	—	—	—	—
1401	Whalebone Creek Pond	USA	MAS	—	4.6	—	—	—	1982	OT	—	—	—	—	—	—	—	—	—	—	—	—
1402	Whatley Lake	USA	EFL	—	7	—	—	—	1994	OT	—	—	—	—	—	—	—	—	—	—	—	—
1403	Wheatland	USA	EFL	HD	13.5	6.1	—	1893	1969	OT	—	11.5	11.6	12.2	13.7	46	41	43.5	0.75	—	1.5	—
1404	Wheeler Lock	USA	CON	GR	22	—	—	1936	1961	QP	—	—	—	—	—	—	—	—	—	—	—	—
1405	White Cloud	USA	EFL	—	6.1	—	—	—	1986	—	—	—	—	—	—	—	—	—	—	—	—	—
1406	White Miller	USA	—	—	—	—	—	—	1978	QP	SEP	—	—	—	—	—	—	—	—	—	—	—
1407	White Rapids-I	USA	CON	GR	14.9	—	—	1927	1956	QP	QIC	2	—	—	—	—	—	—	—	—	—	—
1408	White Rapids-II	USA	CON	GR	14.9	—	—	1927	1963	QP	QIC	2	—	—	—	—	—	—	—	—	—	—
1409	White River	USA	EFL	—	9.1	—	—	—	1910	OT	—	—	—	—	—	—	—	—	—	—	—	—
1410	Whitewater Brook Upper	USA	EFL	—	19	—	137	1949	1972	OT	ISC	0.52	—	—	6.1	6.4	—	—	—	70.8	3	—
1411	Whitewater Creek	USA	EFL	—	4.9	—	—	—	1994	OT	—	—	—	—	—	—	—	—	—	—	—	—
1412	Wiggins Mill	USA	EFL	—	4.6	—	—	1960	1979	QP	SEP	0.1	—	—	—	—	—	—	—	—	—	—
1413	Wild Rice	USA	—	—	7.6	—	—	—	2002	DT	BUA	—	—	—	—	—	—	—	—	—	—	—
1414	Wilkinson	USA	EFL	—	13.4	—	—	1956	1986	QP	SF-SO	—	—	—	—	—	—	—	—	—	—	—
1415	Wilkinson Lake	USA	EFL	DC	3.2	—	—	1956	1994	QP	SEP	—	0.533	3.57	3.72	—	—	29	1.74	—	—	—
1416	Williamette Falls	USA	—	—	—	—	—	—	1944	—	—	—	—	—	—	—	—	—	—	—	—	—
1417	Willow Creek	USA	EFL	—	7.6	—	—	—	1942	QP	SEP	—	—	—	—	—	—	—	—	—	—	—
1418	Wilmington I	USA	EFL	—	3	—	—	—	1900	QP	SEP	—	—	—	—	—	—	—	—	—	—	—
1419	Wilmington II	USA	—	—	—	—	—	—	1990	—	—	—	—	—	—	—	—	30.5	—	—	—	—
1420	Wilmot	USA	—	—	1.8	—	—	—	1992	QP	SEP	—	—	—	—	—	—	—	—	—	—	—
1421	Wilson No. 1	USA	—	—	—	—	—	—	1912	—	—	—	—	—	—	—	—	—	—	—	—	—
1422	Windom	USA	—	GR	3	—	—	—	1984	OT	—	—	—	—	—	—	—	—	—	—	—	—
1423	Windy Hill Lake	USA	EFL	—	5.5	—	—	—	2002	QP	—	—	—	—	—	—	—	—	—	—	—	—
1424	Winfield	USA	CON	GR	4.5	—	—	1876	1916	QP	—	0.02	—	—	—	—	—	—	—	—	—	—
1425	Winkler Lake	USA	EFL	—	7.6	—	—	—	1999	QP	SEP	—	—	—	—	—	—	—	—	—	—	—
1426	Winston I	USA	MAS	GR	12.2	—	—	1882	1904	QP	SF-OV	—	—	—	—	—	—	—	—	—	—	9

(Continued)

Table A.1 *(Continued)*

No.	Name	Country[(1)]	Type[(2)]	Type divided[(3)]	Height (m)	Crest width (m)	Crest length (m)	Year of cons.	Year of failure	Failure cause[(4)]	Failure cause divided[(5)]	Reservoir capacity ($10^6 m^3$)	Water volume stored above breach invert ($10^6 m^3$)	Water depth above breach (m)	Breach geometry: Height (m)	Breach geometry: Top width (m)	Breach geometry: Bottom width (m)	Breach geometry: Aver. width (m)	Breach geometry: Aver. side slope Z:1 (h:v)	Peak outflow rate (m^3/s)	Fail. time	Life loss
1427	Winston II	USA	EFL	DC	7.2	2.1	133	1904	1912	OT	—	0.664	0.662	6.4	6.1	21.3	18.3	19.8	0.2	—	5	—
1428	Winter	USA	—	—	7.6	—	—	—	1977	QP	SEP	—	—	—	—	—	—	—	—	—	—	—
1429	Wisconsin Dells	USA	RFL	—	18	—	79	1909	1911	OT	—	25	—	—	—	—	—	—	—	—	—	—
1430	Wise River	USA	EFL	CF	—	—	—	—	1927	OT	—	—	—	—	—	—	—	—	—	—	—	4
1431	Wishkah	USA	TIM	—	14.3	52	—	—	1997	QP	SEP	—	—	—	—	—	—	—	—	—	—	—
1432	Wolcott Pond	USA	EFL	—	6.4	—	—	—	1995	—	—	—	—	—	—	—	—	—	—	—	—	—
1433	Wolhwender	USA	—	—	—	—	—	—	1994	OT	—	—	—	—	—	—	—	—	—	—	—	—
1434	Woodbridge Div	USA	—	—	10.7	—	—	—	1965	QP	SEP	—	—	—	—	—	—	—	—	—	—	—
1435	Woodfin	USA	EFL	—	24.4	—	—	1965	1977	QP	QIS	—	—	—	—	—	—	—	—	—	—	—
1436	Woodruff	USA	EFL	—	9.8	—	—	—	1997	OT	—	—	—	—	—	—	—	—	—	—	—	—
1437	Wyoming	USA	EFL	—	14.9	—	—	—	1969	QP	SEP	—	—	—	—	—	—	—	—	—	—	—
1438	Wyoming Hereford Ranch	USA	EFL	—	9.5	—	—	—	1992	QP	QIS	—	—	—	—	—	—	—	—	—	—	—
1439	Xonxa	ZA	EFL	CF	24	—	300	1972	1972	OT	—	—	—	—	—	—	—	—	—	—	—	—
1440	Xuriguera	ES	CON	GR	42	—	165	1902	1944	QP	SF-SL	1.1	—	—	—	—	—	—	—	—	—	—
1441	Yara	USA	EFL	—	7.8	—	—	—	1994	OT	—	—	—	—	—	—	—	—	—	—	—	—
1442	Zaugg	USA	—	—	3.7	—	213.5	1885	1937	QP	SEP	0.025	0.0197	—	—	—	—	—	—	—	—	—
1443	Zuni Black Rock	USA	EFL	CF	21.3	—	219.5	1907	1909	QP	SEP	19.5	—	—	—	—	—	—	—	—	—	—

Notes:

[(1)] Country: AR = Argentina; AU = Australia; BO = Bolivia; BR = Brazil; CA = Canada; CL = Chile; CO = Colombia; CZ = Czech; DE = Germany; DZ = Algeria; ES = Spain; FR = France; GB = United Kingdom; IN = India; IT = Italy; JP = Japan; KR = Korea; LK = Sri Lanka; LS = Lesotho; LY = Libya; MA = Morocco; MX = Mexico; NZ = New Zealand; PE = Peru; PK = Pakistan; RO = Romania; SE = Sweden; USA = United States of America; USSR = The former Union of Soviet Socialist Republics; UY = Uruguay; YU = Yugoslavia; ZA = South Africa.

[(2)] Type: CON = concrete dam; EFL = earthfill dam; MAS = masonry dam; RFL = rockfill dam; STL = steel dam; TIM = timber dam.

[(3)] Type divided: ARC = arch dam; BU = buttress dam; CF = composite-fill embankment dam; DC = embankment dam with corewall; FD = concrete-faced embankment dam; GR = gravity dam; HD = homogeneous-fill embankment dam; MAR = multi-arch dam.

[(4)] Failure cause: DT = disasters; OS = others; OT = overtopping; PM = poor management; QP = quality problems.

[(5)] Failure cause divided: ABA = animal or biological attacks; BOP = blockage of pipes; BOS = blockage of spillways; BUA = breaching of upstream dams; EQ = earthquakes; IF = insufficient freeboards due to settlement or poor design; ISC = insufficient spillway capacities; RLR = reservoir landslides; QIC = quality issues in culverts and other embedded structures; QIS = quality issues in spillways; SEP = seepage erosion or piping; SF = structural failures of dam body-foundation unit; SF-OV = structural failures of dam body-foundation unit by overturning; SF-SL = structural failures of dam body-foundation unit by sliding; SF-SO = structural failures of dam body-foundation unit by sloughing; WTA = wars or terrorist attacks.

Appendix B

Database of 1044 Cases of Failures of Landslide Dams

Table B.1 Database of 1044 Cases of Landslide Dams

No	Country[(1)]	Name	Date of formation[(2)]	Landslide Type[(3)]	Trigger[(4)]	Landslide volume (×10⁶ m³)	Dam Type[(5)]	Dam volume (×10⁶ m³)	Dam height (m)	Dam length (m)	Dam width (m)	Lake length (m)	Lake volume (×10⁶ m³)	Time to failure (day)[(6)]	Failure mech.[(7)]	Breach depth (m)	Breach top width (m)	Breach bottom width (m)	Breaching time (hour)	Peak discharge (m³/s)	Loss of life	Reference (Below the table)
1	AF	Ajar river	1960s	RF	EQ	—	—	—	—	—	—	—	—	NF	NF	—	—	—	—	—	—	[1]
2	AT	Brixen Torrent	1946	DF	RF	—	—	—	—	—	—	—	—	—	—	—	—	—	—	—	5	[1]
3	AT	Gail River	328	DF	RF	—	—	—	—	—	—	—	—	—	—	—	—	—	—	—	—	[1]
4	AT	Gail River	1348	RA	EQ	30	—	—	—	—	—	—	—	<10	—	—	—	—	—	—	—	[1]
5	AT	Ill River	1894	DF	RF	2	—	—	—	—	—	—	—	—	—	—	—	—	—	—	—	[1]
6	AT	Lavant Valley	1660	DF	RF	—	—	—	—	—	—	—	—	—	—	—	—	—	—	—	29	[1]
7	AT	Moll River	1827	DF	RF, SM	—	—	—	—	—	—	—	—	—	—	—	—	—	—	—	—	[1]
8	AT	Muhlbach Torrent	1798	DF	RF	—	—	—	20–30	—	—	—	—	—	—	—	—	—	—	—	—	[1]
9	AT	Mur River	1958	DF	RF	—	—	—	30	—	—	—	—	—	—	—	—	—	—	—	12	[1]
10	AT	Palten River	1768	DF	RF	—	—	—	—	—	—	—	—	—	—	—	—	—	—	—	—	[1]
11	AT	Salzach River	1947	DF	RF	0.1–0.2	—	—	15	—	—	—	—	<1	—	—	—	—	—	—	2	[1]
12	AT	Salzach River	1794	EF	RF	—	—	—	—	—	—	3000	—	SY	—	—	—	—	—	—	—	[1]
13	AT	Velber Brook	1495	—	EQ	—	—	—	—	—	—	—	—	NF	NF	—	—	—	—	—	—	[1]
14	AT	Ziller River	1908	DF	RF	0.3	—	—	—	—	—	—	—	—	—	—	—	—	—	—	—	[1]
15	AU	Lake Elizabeth	1952	RS	RF	6	II	—	36	—	—	1600	—	414	OT	26	—	—	—	—	—	[1]
16	BO	Allpacoma landslide dam	2005	—	—	—	—	—	—	—	—	—	—	—	PP	—	—	—	—	—	—	[24]
17	BT	Tsatichuu River	2003	RA	EQ	10–15	II	5	110	580	700	1000	1.5	300	SF, PP	—	—	—	—	6900	—	[11]
18	CA	Attachie	1973	—	—	—	—	—	—	—	—	—	—	—	—	—	—	—	—	—	—	[3]
19	CA	Blanche River	1898	LS	SM	2.7	III	—	8	460	3200	—	—	—	—	—	—	—	—	—	—	[1]
20	CA	Britannia Creek	1921	—	RF	—	—	—	—	—	—	—	—	—	—	—	—	—	—	—	37	[1][12]
21	CA	Cheakamus River	1958	DF	RF	—	—	—	5	—	—	—	—	—	—	—	—	—	—	—	—	[1][12]
22	CA	Cheakamus River	1855-1856	RA	—	25	II	—	—	200–350	3500	—	—	—	OT	15	—	—	—	—	—	[1]
23	CA	Chilcotin River	1964	Slump	—	—	II	—	—	—	—	—	—	7	OT	—	—	—	—	—	—	[1]
24	CA	Clinton Creek	1976	—	—	—	—	3.43	26	—	—	—	—	—	—	—	—	—	—	—	—	[2]
25	CA	Crowsnest River	1903	RS	—	36.5	II	—	10	—	—	—	—	1.5	OT	—	—	—	—	—	—	[1]
26	CA	Dunvegan Creek	1959	—	—	—	—	—	—	—	—	—	—	—	—	—	—	—	—	—	—	[12]
27	CA	Dusty Creek	1963	DA	PW	5	—	—	—	—	—	—	—	—	OT	—	—	—	—	—	—	[1]
28	CA	Eureka River	1990	—	—	—	—	—	—	—	—	—	—	—	—	—	—	—	—	—	—	[3]
29	CA	Eureka River	1990	—	—	50	—	—	—	—	—	—	—	—	—	—	—	—	—	—	—	[3]
30	CA	Fraser River	1921	—	—	—	—	—	—	—	—	—	—	—	—	—	—	—	—	—	—	[13]
31	CA	Grand River	1943	SF	—	—	I	—	—	150	137	—	—	<10	—	—	—	—	—	—	—	[1]
32	CA	Halden Creek	1996	RS	—	—	II	1	—	—	900	1100	—	—	—	—	—	—	—	—	—	[4]
33	CA	Hines Creek	1990	—	PW	—	—	—	—	—	—	—	—	—	—	—	—	—	—	—	—	[3]
34	CA	Homathko River	1971-1973	DF	DF	—	III	—	20	—	—	1000	—	—	—	—	—	—	—	—	—	[1]

35	CA	Inklin River	1978	DA	HC	1	II	—	30	—	100	11000	—	—	—	—	—	—	—	—	—	[1]
36	CA	Kennedy River	1970	—	—	—	—	—	—	—	—	—	—	—	—	—	—	—	—	—	—	[12]
37	CA	Lievre River	1903	LS	RF	—	III	—	6.0–9.0	120	600	—	—	—	—	—	—	—	—	—	—	[1]
38	CA	Lievre River	1908	LS	—	—	—	—	2.7	—	—	—	—	—	—	—	—	—	—	—	—	[1]
39	CA	Maskinonge River	1840	LS	SM	—	III	—	23	—	—	14400	—	2	OT	—	—	—	—	—	—	[1]
40	CA	Meager Creek	1975	DS, DA	SM	29	—	—	—	—	—	—	—	—	OT	—	—	—	—	—	—	[1][13]
41	CA	Meager Creek	1998	DF	SM	—	—	1.2	—	—	—	800	—	400	OT	—	—	—	—	—	—	[14]
42	CA	Meager Creek tributary	1931	—	—	—	—	—	—	—	—	—	—	—	—	—	—	—	—	—	—	[1]
43	CA	Mess Creek	1947	—	—	—	—	—	—	—	—	—	—	—	—	—	—	—	—	—	—	[13]
44	CA	Montagneuse River	1939	—	SM	76	—	—	15.2	1500	—	4000	—	14	—	—	—	—	—	—	—	[15][16]
45	CA	Ryan River	1984	DF	RF	—	—	—	2.0–3.0	—	—	—	—	—	—	—	—	—	—	—	—	[1]
46	CA	Saddle River	1990	ES	RF	15	VI	—	20–30	—	800	—	—	—	—	—	—	—	—	—	—	[1]
47	CA	South Nation River	1971	LS	SM	—	III	—	11	200	2450	—	—	—	—	—	—	—	—	—	—	[1]
48	CA	Spirit River	1995	—	—	—	—	—	—	—	—	—	—	—	—	—	—	—	—	—	—	[3]
49	CA	Squamish River	1984	DF	—	—	—	—	—	—	—	—	—	—	—	—	—	—	—	—	—	[1]
50	CA	St. Anne River	1894	LS	—	19.6	III	—	—	—	—	—	—	—	—	—	—	—	—	—	4	[1]
51	CA	Tahltan River	1964	—	—	—	—	—	—	—	—	—	—	—	—	—	—	—	—	—	—	[12]
52	CA	Thompson River	1880	Slump	HC	15	II	—	18–25	274	880	14000	65	1.8	OT	—	—	—	—	—	—	[1][12][13]
53	CA	Thompson River	1899	Slump	—	—	—	—	—	—	—	—	—	—	—	—	—	—	—	—	—	[1]
54	CA	Thompson River	1905	Slump	HC	—	II	—	6	—	—	—	—	0.2	OT	—	—	—	—	—	—	[1]
55	CA	Thompson River	1921	Slump	HC	—	II	—	—	—	—	—	—	<1	OT	—	—	—	—	—	—	[1][13]
56	CA	Turbid Creek	1963	DA	PW	5	—	—	—	—	—	—	—	—	OT	—	—	—	—	—	—	[1]
57	CA	Unnamed Creek	1891	—	RF	—	—	—	—	—	—	—	—	0.3	—	—	—	—	—	—	13–40	[1][12]
58	CA	Wolverine Creek	1974	—	—	—	—	0.07	10	—	—	—	—	0.01	OT	—	—	—	—	—	—	[2][13]
59	CA	Yamaska River	1945	LS	RF	0.118	III	—	3.4	67	330	—	—	—	OT	—	—	—	—	—	—	[1]
60	CH	Biasca	1513	—	—	—	—	20	50	—	—	—	—	—	—	—	—	—	—	—	—	[2]
61	CH	Birse River	1937	DS	RF	2	V	—	9	30	110	100	—	NF	NF	—	—	—	—	—	—	[1]
62	CH	Birse River(upper dam)	1937	DS	RF	2	V	—	6	25	80	60	—	NF	NF	—	—	—	—	—	—	[1]
63	CH	Brenno River	1513	RA	RF	10–20.	—	—	50	—	—	5000	100	597	OT	—	—	—	—	—	600	[1]
64	CH	Brenno River	1868	DF	RF	—	—	—	—	—	—	—	—	—	—	—	—	—	—	—	—	[1]
65	CH	Derborence Torrent	1749	RA	—	30	—	—	—	—	—	—	—	NF	NF	—	—	—	—	—	—	[1]
66	CH	Grosse Schliere Torrent	1565	RS	—	—	—	—	—	—	—	—	—	350	—	—	—	—	—	—	—	[1]
67	CH	Grosse Schliere Torrent	1910	RS	RF, SM	—	—	—	—	—	—	—	—	—	—	—	—	—	—	—	—	[1]
68	CH	Illgraben Torrent	1961	DF	—	3.5	—	—	—	—	—	—	—	—	OT	—	—	—	—	—	—	[1]
69	CH	Linth River tributary	1594	RA	EQ	—	—	—	—	—	—	—	—	9	—	—	—	—	—	—	—	[1]
70	CH	Navisence Torrent	1200-1300	RA	—	2	—	—	—	—	—	500	—	—	—	—	—	—	—	—	—	[1]

(*Continued*)

Table B.1 *(Continued)*

No	Country[(1)]	Name	Date of formation[(2)]	Landslide Type[(3)]	Trigger[(4)]	Landslide volume ($\times10^6$ m^3)	Dam Type[(5)]	Dam volume ($\times10^6$ m^3)	Dam height (m)	Dam length (m)	Dam width (m)	Lake length (m)	Lake volume ($\times10^6$ m^3)	Time to failure (day)[(6)]	Failure mech.[(7)]	Breach depth (m)	Breach top width (m)	Breach bottom width (m)	Breaching time (hour)	Peak discharge (m^3/s)	Loss of life	Reference (Below the table)
71	CH	Poschiavino River	1987	DF	RF	0.35	—	—	—	—	—	—	—	—	—	—	—	—	—	—	—	[1]
72	CH	Rhine River	1585	DF	RF	—	—	—	—	—	—	—	—	—	—	—	—	—	—	—	—	[1]
73	CH	Rhine River(upper)	1807	DF	RF	—	—	—	12	—	—	—	—	—	—	—	—	—	—	—	—	[1]
74	CH	Rhine River(upper)	1868	DF	RF	—	—	—	10.-12	—	—	—	—	—	—	—	—	—	—	—	—	[1]
75	CH	Rhone River	563	DF	RF, SM	—	—	—	—	—	—	—	—	—	—	—	—	—	—	—	Many	[1]
76	CH	Rhone River	1636	DF	RF,SM	—	—	—	—	—	—	—	—	0.02	OT	—	—	—	—	—	—	[1]
77	CH	Rhone River	1926	DF	RF	0.55–1	—	—	—	—	—	—	—	—	—	—	—	—	—	—	—	[1]
78	CH	Schachen Torrent	1887	DA	SM	—	—	—	—	—	—	—	—	—	—	—	—	—	—	—	—	[1]
79	CH	Traversagna Torrent	1928	RS, RA	RF	30–40	—	—	—	—	1500	—	—	—	—	—	—	—	—	—	—	[1]
80	CH	Valais Canton	1749	RA	—	30	—	—	—	—	—	—	—	NF	NF	—	—	—	—	—	—	[1]
81	CH	Vispa	1991	—	—	—	—	20	25	—	—	—	—	—	—	—	—	—	—	—	—	[2]
82	CH	Vorderrhein River	1683	RA	—	10.0–20	—	—	—	—	—	—	—	0.1	OT	—	—	—	—	—	—	[1]
83	CL	Lake Pellaifa	1960	RA	EQ	—	—	—	8	—	—	—	—	NF	NF	—	—	—	—	—	—	[1]
84	CL	San Pedro River	1575	—	EQ	100	II	—	—	—	2000	—	—	134	—	—	—	—	—	—	—	[1]
85	CL	San Pedro River	1960	LS	EQ	30	II	—	26	1100	—	—	2.5	63	OT	—	—	—	—	—	—	[1]
86	CN	Amandi	1966	—	RF	0.5	—	—	—	—	—	—	—	—	—	—	—	—	—	—	—	[6]
87	CN	Anjiagou	1951	—	RF	2.8	—	—	—	—	—	—	—	—	—	—	—	—	—	—	—	[5]
88	CN	Ansai	1951	—	RF	—	—	—	—	—	—	—	—	—	—	—	—	—	—	—	—	[5]
89	CN	Ashigong	Q4	Slide	EQ	300	—	—	200	—	—	—	—	—	—	—	—	—	—	—	—	[5]
90	CN	Atushi	1902	Ava.	EQ	0.2–0.3	—	—	—	—	—	—	—	—	—	—	—	—	—	—	—	[5]
91	CN	Bagacun	1959	—	RF	24	—	—	150	—	—	—	—	7	—	—	—	—	—	—	—	[6]
92	CN	Baiguo village	2008	RS	EQ	—	II	0.4	10–20.	200	100	—	0.8	—	OT	—	—	—	—	—	0	[7][8]
93	CN	Bailong River	1879	—	EQ	—	—	—	—	—	—	—	—	—	—	—	—	—	—	—	—	[1]
94	CN	Bailong River	1879	—	EQ	—	—	—	—	—	—	—	—	—	—	—	—	—	—	—	—	[1]
95	CN	Bailong River	1963	ES, EF	RF	25	—	—	17	—	—	—	7	—	—	—	—	—	—	—	—	[1]
96	CN	Bailong River	1981	EF	—	40	—	—	25	—	—	—	19	NF	NF	—	—	—	—	—	—	[1]
97	CN	Baimeiya	1974	—	RF	7	—	—	—	—	—	—	—	LT	—	—	—	—	—	—	159	[6]
98	CN	Baisha industrial park	1988	—	MC	1.64	—	—	30	—	—	3500	216	—	—	—	—	—	—	—	—	[5]
99	CN	Baishagou	1919	—	RF	—	—	—	—	—	—	—	—	LT	—	—	—	—	—	—	—	[6]
100	CN	Baishagou	1953	—	RF	—	—	—	—	—	—	—	—	—	—	—	—	—	—	—	—	[5]
101	CN	Baishiyunshan	1480	—	RF	—	—	—	—	—	—	—	—	—	—	—	—	—	—	—	—	[5]
102	CN	Bai-Shui River	1984	DF	RF	0.218	III	—	—	100	420	—	—	—	—	—	—	—	—	—	—	[1]
103	CN	Baishuihegou	1983	DF	RF	0.1	—	—	—	—	—	—	—	—	—	—	—	—	—	—	—	[5]
104	CN	Baiyi'an	Q4	—	RF	44.3	—	—	—	—	—	—	—	LT	—	—	—	—	—	—	—	[6]
105	CN	Baizhidashitou	Q4	—	RF	—	—	—	—	—	—	—	—	LT	—	—	—	—	—	—	—	[6]
106	CN	Baota	3500BP	—	RF	105	—	—	—	—	—	—	—	LT	—	—	—	—	—	—	—	[6]

107	CN	Basu county	2000	DF	RF	—	—	—	—	—	—	—	—	—	—	—	—	—	—	—	—	[5]
108	CN	Batang	1870	RA	EQ	—	II	—	—	—	—	—	—	—	—	—	—	—	—	—	2000	[1][5]
109	CN	Batang	1997	Slide	EQ	0.08	—	—	15	—	—	—	0.05	—	—	—	—	—	—	—	—	[5]
110	CN	Bawafeng	1969	—	RF	—	—	—	—	—	—	—	1	—	—	—	—	—	—	—	—	[5]
111	CN	Bawangshan	Q4	—	RF	3	—	—	—	—	—	—	—	LT	—	—	—	—	—	—	—	[6]
112	CN	Bayi Reservoir	1974	DF	RF	—	—	—	—	—	—	—	—	—	—	—	—	—	—	—	—	[5]
113	CN	Beiguanhe	1917	—	EQ	—	—	—	—	—	—	—	—	ST	—	—	—	—	—	—	1800	[6]
114	CN	Beiyu River	1979	DF	RF	—	—	—	—	—	—	—	—	—	—	—	—	—	—	—	—	[5]
115	CN	Beiyu River	1980	DF	RF	—	—	—	—	—	—	—	—	—	—	—	—	—	—	—	—	[5]
116	CN	Beiyu River	1987	DF	RF	1.44	III	—	—	—	—	—	—	0.04	OT	—	—	—	—	—	—	[1]
117	CN	Bitang Creek	1961	Slump	—	80	—	—	65	—	—	1000	4.2	NF	NF	—	—	—	—	—	—	[1]
118	CN	Bitanggou	1963	—	—	>100	—	—	—	—	—	—	1.5–2	LT	—	—	—	—	—	—	—	[6]
119	CN	Bogetang	—	Slide	RF, EQ	—	—	—	—	—	—	—	—	—	—	—	—	—	—	—	—	[5]
120	CN	Bogexi	—	—	EQ,RF	54	—	—	—	—	—	—	—	LT	—	—	—	—	—	—	—	[6]
121	CN	Bomiguxiang	1953	DF	RF	17	—	—	—	—	—	—	—	ST	—	—	—	—	—	—	—	[5][6]
122	CN	Boqu River	1981	DF	DF	1	—	—	—	—	—	—	—	<1	—	—	—	—	—	—	—	[1]
123	CN	Boyangxiang	1978	DF	RF	2.5	—	—	—	—	—	—	—	<1	OT	—	—	—	—	—	—	[5]
124	CN	Caopingzi	Q4	—	—	150	—	—	—	—	—	—	—	LT	—	—	—	—	—	—	—	[6]
125	CN	Chahandasi	1961	—	RF	100	—	—	—	—	—	—	—	—	—	—	—	—	—	—	—	[5]
126	CN	Chaijiapo	1982	—	RF	2.07	—	—	—	—	—	—	—	730	—	—	—	—	—	—	—	[6]
127	CN	Changde	1631	Ava.	EQ	—	—	—	—	—	—	—	—	—	—	—	—	—	—	—	—	[5]
128	CN	Changhualin	1623	—	RF	—	—	—	—	—	—	—	—	—	—	—	—	—	—	—	—	[5]
129	CN	Changma	1932	Ava.	EQ	—	—	—	—	—	—	—	—	SD	—	—	—	—	—	—	—	[5]
130	CN	Changtian	1985	—	RF	6	—	—	—	—	—	—	—	730	—	—	—	—	—	—	—	[6]
131	CN	Chayu	1950	Ava.	EQ	—	—	—	—	—	—	—	—	7	—	—	—	—	—	—	—	[5]
132	CN	Chayuhe	1950	—	EQ	—	—	—	—	—	—	—	—	1	—	—	—	—	—	—	—	[5]
133	CN	Chengmenjiangou	1981	—	RF	1.8	—	—	—	—	—	—		0.17	—	—	—	—	—	—	—	[5]
134	CN	Chengyu highway	Q4	—	RF	1	—	—	—	—	—	—	—	LT	—	—	—	—	—	—	—	[6]
135	CN	Chenxian	1545	—	RF	—	—	—	—	—	—	—	—	365	—	—	—	—	—	—	—	[5]
136	CN	Cheshuiba	—	—	RF	0.16	—	—	—	—	—	—	—	—	—	—	—	—	—	—	—	[5]
137	CN	Chingshakiang River	1935	—	EQ	—	—	—	—	—	—	—	—	—	—	—	—	—	—	—	—	[1]
138	CN	Chinlungpei River	1927	—	EQ	—	—	—	—	—	—	—	—	—	—	—	—	—	—	—	—	[1]
139	CN	Chixigou	1969	DF	RF	—	—	—	—	—	—	—	—	—	—	—	—	—	—	—	—	[5]
140	CN	Chouni River	1654	Ava.	EQ	—	—	—	—	—	—	—	—	—	—	—	—	—	—	—	—	[1][5]
141	CN	Chulusongjie	2004	RA	—	—	—	—	—	—	—	—	—	—	—	—	—	—	—	—	—	[5]
142	CN	Ci county	1830	Ava.	EQ	—	—	—	—	—	—	—	—	—	—	—	—	—	—	—	—	[5]
143	CN	Cuihuashan	731	Slide	EQ	35	—	—	—	—	—	—	—	LT	—	—	—	—	—	—	—	[6]
144	CN	Cuiping-Guanin	1933	—	—	91	—	—	130	—	—	—	—	—	—	—	—	—	—	—	—	[5]
145	CN	Dabainigou	1968	DF	RF	—	—	—	—	—	—	—	—	0.5	—	—	—	—	—	—	—	[6]
146	CN	Dabainigou	1980	DF	RF	—	—	—	—	—	—	—	—	0.08	—	—	—	—	—	—	—	[5]
147	CN	Dabainigou	1985	DF	RF	—	—	—	—	—	—	—	—	ST	—	—	—	—	—	—	—	[5]

(Continued)

Table B.1 *(Continued)*

No	Country[(1)]	Name	Date of formation [(2)]	Landslide Type [(3)]	Trigger [(4)]	Landslide volume ($\times10^6 m^3$)	Dam Type [(5)]	Dam volume ($\times10^6 m^3$)	Dam height (m)	Dam length (m)	Dam width (m)	Lake length (m)	Lake volume ($\times10^6 m^3$)	Time to failure (day) [(6)]	Failure mech.[(7)]	Breach depth (m)	Breach top width (m)	Breach bottom width (m)	Breaching time (hour)	Peak discharge (m^3/s)	Loss of life	Reference (Below the table)
148	CN	Dabanqiao	28000BP	Slide	EQ	31	—	—	300	—	700	—	—	—	—	—	—	—	—	—	—	[5]
149	CN	Dadegou	1984	DF	RF	—	—	—	—	—	—	—	—	—	—	—	—	—	—	—	—	[5]
150	CN	Dadibian	1984	—	RF	0.16	—	—	—	—	—	—	—	LT	—	—	—	—	—	—	—	[6]
151	CN	Dadihe	1987	—	RF	3	—	—	—	—	—	—	—	—	—	—	—	—	—	—	—	[5]
152	CN	Dadu River	1786	RS,RA	EQ	—	—	—	—	—	—	—	—	10	OT	—	—	—	—	—	—	[1]
153	CN	Dadu River Tributary	1971	Slump	RF	—	II	—	15	—	200	—	—	—	—	—	—	—	—	—	—	[1]
154	CN	Daguan River	1917	—	EQ	—	—	—	—	—	—	—	—	—	—	—	—	—	—	—	—	[1]
155	CN	Daguanbei	1974	Slide	EQ	0.1	—	—	—	—	—	—	—	LT	—	—	—	—	—	—	—	[6]
156	CN	Dajiawadi	1952	—	RF	—	—	—	—	—	—	—	—	—	—	—	—	—	—	—	—	[5]
157	CN	Daluba	1856	Slide	EQ	72	—	—	70	—	—	—	67	—	—	—	—	—	—	—	—	[5]
158	CN	Dangyuanchi	1951	—	RF	—	—	—	—	—	—	—	—	—	—	—	—	—	—	—	—	[5]
159	CN	Dapingshan	1985	—	RF	0.4	—	—	—	—	—	—	—	0.03	—	—	—	—	—	—	22	[6]
160	CN	Dasuzizhen	1991	—	RF	300	—	—	—	—	—	—	—	—	—	—	—	—	—	—	—	[5]
161	CN	Dawanzi	1991	—	RF	2.7	—	—	40	—	—	—	0.2	LT	—	—	—	—	—	—	—	[6]
162	CN	Dawuxiang	2004	Ava.	SM	—	—	—	—	2200	1500	—	—	—	—	—	—	—	—	—	—	[5]
163	CN	Daxi	Q4	DF	RF	22.38	—	—	—	—	—	—	—	LT	—	—	—	—	—	—	—	[6]
164	CN	Dayingpan	1957	—	RF	—	—	—	—	—	—	—	—	0.08	—	—	—	—	—	—	—	[5]
165	CN	Denglongshan	1991	—	HC	1	—	—	—	—	—	—	—	—	—	—	—	—	—	—	7	[6]
166	CN	Diaobanya	1988	—	HC	0.6	—	—	—	—	—	—	—	—	—	—	—	—	—	—	—	[6]
167	CN	Diexi Maoxian	Q3–4	—	RF	100	—	—	—	—	—	—	—	73000	—	—	—	—	—	—	—	[6]
168	CN	Diexi(Dahaizi)	1933	Slide	EQ	12.76	—	—	—	—	—	—	—	—	—	—	—	—	—	—	Many	[6]
169	CN	Diexi(Maoxian)	1713	Slide	EQ	20	—	—	—	—	—	—	—	ST	—	—	—	—	—	—	—	[1][5]
170	CN	Diexi(Xiaohaizi)	1933	Slide	EQ	75	—	—	100	—	—	—	0.5	LT	—	—	—	—	—	—	100	[6]
171	CN	Diexi(xiaoqiao)	1933	Slide	EQ	46.5	—	—	160	—	—	—	80	45	—	—	—	—	—	—	2500	[6]
172	CN	Dong River	1965	RS	RF	29	II	—	51	—	650	1000	2.7	210	OT	—	—	—	—	—	—	[1]
173	CN	Dongge	1951	—	RF	—	—	—	—	—	—	—	—	0.005	—	—	—	—	—	—	—	[6]
174	CN	Donghekou	2008	RS	EQ	—	—	12	20	500	750	3000	6	10	OT	10	25	15	—	800–1000	0	[7][8]
175	CN	Dongyashanpo	1958	—	EQ	—	—	—	—	—	—	—	—	—	—	—	—	—	—	—	—	[5]
176	CN	Dongyemiao	Q4	—	—	211.2	—	—	—	—	—	—	—	LT	—	—	—	—	—	—	—	[6]
177	CN	Duigaqiaba	1838	—	RF	—	—	—	—	—	—	—	—	2	—	—	—	—	—	—	—	[5]
178	CN	Duozuo, Gonghe	Q4	—	RF	120	—	—	120	—	—	—	—	—	—	—	—	—	—	—	—	[5]
179	CN	Enshi	1490	—	RF	—	—	—	—	—	—	—	—	—	—	—	—	—	—	—	—	[5]
180	CN	Erli	420BC	—	RF	—	—	—	—	—	—	—	—	—	—	—	—	—	—	—	—	[5]
181	CN	Erpu	1979	DF	RF	0.1	—	—	—	—	—	—	—	—	—	—	—	—	—	—	—	[5]
182	CN	Fang Couty	788	—	EQ	—	—	—	—	—	—	—	—	—	—	—	—	—	—	—	—	[5]

183	CN	Fanjiaping	Q4	—	RF	125	—	—	—	—	—	—	—	LT	—	—	—	—	—	—	—	[6]
184	CN	Fenghuangzui	1951	—	RF	—	—	—	—	—	—	—	—	—	—	—	—	—	—	—	—	[5]
185	CN	Fengming bridge	2008	—	EQ	—	—	0.14	10	100	300	—	1.8	35	OT	—	—	—	—	500	0	[7][8]
186	CN	Fengshan	1983	RF	RF	—	—	—	—	—	—	—	—	—	—	—	—	—	—	—	—	[5]
187	CN	Fengyi	1925	Slide	EQ	—	—	—	—	—	—	—	—	—	—	—	—	—	—	—	—	[5]
188	CN	Fulin	1824	—	RF	—	—	—	—	—	—	—	—	—	—	—	—	—	—	—	—	[5]
189	CN	Futan	Q4	—	RF	6.3	—	—	—	—	—	—	—	LT	—	—	—	—	—	—	—	[6]
190	CN	Fuwei	1931	Ava.	EQ	—	—	—	—	—	—	—	—	—	—	—	—	—	—	—	—	[5]
191	CN	Ganhaizi	1920	—	RF	50.98	—	—	—	—	—	—	—	LT	—	—	—	—	—	—	—	[6]
192	CN	Ganhekou	2008	RS	EQ	—	II	0.01	10	—	—	—	0.5	—	OT	—	—	—	—	—	0	[7][8]
193	CN	Ganjiagou	1984	DF	RF	—	—	—	—	—	—	—	—	—	—	—	—	—	—	—	—	[5]
194	CN	Gaosongshu	1981	—	—	2.1	—	—	—	—	—	—	—	LT	—	—	—	—	—	—	—	[6]
195	CN	Gaozhiwan	1998	—	—	15.6	—	—	—	—	—	—	0.095	—	—	—	—	—	—	—	—	[5]
196	CN	Gengdiyanjiao	—	—	RF	—	—	—	—	—	—	—	—	—	—	—	—	—	—	—	—	[5]
197	CN	Gongpeng	1933	Slide	EQ	30	—	—	120	—	—	—	10	LT	—	—	—	—	—	—	—	[6]
198	CN	Goumaohe	1978	—	RF	0.8	—	—	—	—	—	—	—	—	—	—	—	—	—	—	—	[5]
199	CN	Guan County	952	—	RF	—	—	—	—	—	—	—	—	—	—	—	—	—	—	—	5000	[5]
200	CN	Guanjiagou	1982	DF	RF	—	—	—	—	—	—	—	—	0.001	—	—	—	—	—	—	—	[5]
201	CN	Guanjiayuanzi	1981	—	HC	5	—	—	—	—	—	—	—	LT	—	—	—	—	—	—	30	[6]
202	CN	Guanjiuping	1968	—	RF	—	—	—	10	—	—	—	0.4	—	—	—	—	—	—	—	—	[5]
203	CN	Guanmiaogou	1984	DF	RF	—	—	—	—	—	—	—	—	—	—	—	—	—	—	—	—	[5]
204	CN	Guantan	2008	ES	EQ	—	III	1.2	60	200	120	2000	10	—	OT	—	—	—	—	—	0	[7][8]
205	CN	Guanxian	1930	—	RF	—	—	—	—	—	—	—	—	—	—	—	—	—	—	—	—	[5]
206	CN	Guanxian	10BC	—	RF	—	—	—	—	—	—	—	—	3	—	—	—	—	—	—	—	[5]
207	CN	Guanyingou	1989	DF	EQ, RF	—	—	—	—	—	—	—	—	ST	—	—	—	—	—	—	—	[9]
208	CN	Guanzipu	2008	RS	EQ	—	II	—	60	390	—	—	5.85	—	OT	—	—	—	—	—	0	[7][8]
209	CN	Gulang	1927	Ava.	EQ	—	—	—	—	—	—	—	—	—	—	—	—	—	—	—	—	[5]
210	CN	Gulin	Q4	—	RF	131.2	—	—	—	—	—	—	—	LT	—	—	—	—	—	—	—	[6]
211	CN	Gushiqun	Q4	—	RF	1.5	—	—	—	—	—	—	—	—	—	—	—	—	—	—	—	[5]
212	CN	Guyuan	1921	Ava.	EQ	—	—	—	—	—	—	—	—	—	—	—	—	—	—	—	—	[5]
213	CN	Haiyuan (60 dams)	1921	Ava.	EQ	—	—	—	—	—	—	—	—	—	—	—	—	—	—	—	—	[5]
214	CN	Haiziping	2008	—	EQ	—	—	0.67	8	50	500	—	3	—	OT	—	—	—	—	—	0	[7][8]
215	CN	Hanyuan	1880	—	RF	—	—	—	—	—	—	—	—	—	—	—	—	—	—	—	—	[5]
216	CN	Hanyuan	1917	DF	RF	—	—	—	—	—	—	—	—	—	—	—	—	—	—	—	Many	[5]
217	CN	Heidongya	2008	—	EQ	—	—	0.4	50–80	120	700	400	1.8	—	OT	—	—	—	—	—	0	[7][8]
218	CN	Heishe	1983	—	RF	8.1	—	—	—	—	—	—	—	—	—	—	—	—	—	—	—	[6]
219	CN	Heishigou	1,925	—	RF	—	—	—	—	—	—	—	—	LT	—	—	—	—	—	—	—	[6]
220	CN	Henan	611	—	RF	—	—	—	—	—	—	—	—	—	—	—	—	—	—	—	—	[5]
221	CN	Henan	413BC	—	RF	—	—	—	—	—	—	—	—	—	—	—	—	—	—	—	—	[5]
222	CN	Hetian	1975	Ava.	EQ	—	—	—	—	—	—	—	—	0.04	—	—	—	—	—	—	—	[5]
223	CN	Heyangxiandong	774	—	RF	—	—	—	—	—	—	—	—	—	—	—	—	—	—	—	—	[5]
224	CN	Hongcun	2008	RS	EQ	—	III	0.4	40–50	80	—	—	1–1.5	—	OT	—	—	—	—	—	0	[7][8]

(*Continued*)

Table B.1 (*Continued*)

No	Country[(1)]	Name	Date of formation [(2)]	Landslide Type [(3)]	Trigger [(4)]	Landslide volume ($\times10^6$ m^3)	Dam Type [(5)]	Dam volume ($\times10^6$ m^3)	Dam height (m)	Dam length (m)	Dam width (m)	Lake length (m)	Lake volume ($\times10^6$ m^3)	Time to failure (day) [(6)]	Failure mech.[(7)]	Breach depth (m)	Breach top width (m)	Breach bottom width (m)	Breaching time (hour)	Peak discharge (m^3/s)	Loss of life	Reference (Below the table)
225	CN	Hongshancun	1990	—	RF	5.09	—	—	—	—	—	—	—	LT	—	—	—	—	—	—	—	[6]
226	CN	Hongshihe	2008	RS	EQ	—	III	18	50	400	500	—	4	10	OT	10	—	8.0–10	—	400–600	0	[7][8]
227	CN	Hongshuigou	1964	DF	RF	0.4	—	—	—	—	—	—	—	—	—	—	—	—	—	—	—	[5]
228	CN	Hongtupo	1990	—	RF	4.33	—	—	—	—	—	—	—	LT	—	—	—	—	—	—	—	[6]
229	CN	Hongyanbao	1609	Ava.	EQ	—	—	—	—	—	—	—	—	—	—	—	—	—	—	—	—	[5]
230	CN	Houshigou	2008	—	EQ	—	—	2.4	120	40	500	—	1.5	—	OT	—	—	—	—	—	0	[7][8]
231	CN	Hua County	1556	Ava.	EQ	—	—	—	—	—	—	—	—	—	—	—	—	—	—	—	—	[5]
232	CN	Huahongyuan	1713	—	EQ	20	—	—	—	—	—	—	—	ST	—	—	—	—	—	—	—	[5][6]
233	CN	Hualian Taiwan	1999	—	EQ	—	—	—	—	—	—	—	—	—	—	—	—	—	—	—	—	[5]
234	CN	Huamaxiang	1976	DF	RF	7	—	—	—	—	—	—	—	40	—	—	—	—	—	—	—	[5]
235	CN	Huangguancao	1986	—	RF	10	—	—	—	—	—	—	—	LT	—	—	—	—	—	—	—	[6]
236	CN	Huangjinkeng	1998	DF	RF	—	—	—	3	—	3000	—	—	—	—	—	—	—	—	—	—	[5]
237	CN	Huangliang 1#	—	—	RF	—	—	—	—	—	—	—	—	—	—	—	—	—	—	—	—	[5]
238	CN	Huangliang 2#	—	—	RF	20	—	—	—	—	—	—	—	—	—	—	—	—	—	—	—	[5]
239	CN	Huangliang 3#	—	—	RF	—	—	—	—	—	—	—	—	—	—	—	—	—	—	—	—	[5]
240	CN	Huangliang 5#	—	—	RF	200	—	—	—	—	—	7600	—	NF	NF	—	—	—	—	—	—	[5]
241	CN	Huanglianxia	1982	—	RF	—	—	—	40	—	—	—	20	ST	—	—	—	—	—	—	—	[6]
242	CN	Huangtupo	1995	—	RF	1.28	—	—	—	—	—	—	—	—	—	—	—	—	—	—	—	[5]
243	CN	Huaning	1789	Ava.	EQ	—	—	—	—	—	—	—	—	—	—	—	—	—	—	—	—	[5]
244	CN	Huashikuai	1996	—	RF	1.35	—	—	—	—	—	—	—	0.02	—	—	—	—	—	—	—	[5]
245	CN	Huilijian	1830	Ava.	EQ	—	—	—	—	—	—	—	—	1	—	—	—	—	—	—	—	[5]
246	CN	Hunshuigou	1982	DF	RF	0.6	—	—	—	—	—	—	—	—	—	—	—	—	—	—	—	[5]
247	CN	Huobuxun Lake, Qinghai Province	1962	—	EQ	—	—	—	—	—	—	—	—	—	—	—	—	—	—	—	—	[9]
248	CN	Huoshan	1969	—	RF	15.67	—	—	—	—	—	—	—	—	—	—	—	—	—	—	—	[5]
249	CN	Huoyanshi	Q4	—	RF	1.8	—	—	—	—	—	—	—	LT	—	—	—	—	—	—	—	[6]
250	CN	Jianchuan	1948	Ava.	EQ	—	—	—	—	—	—	—	—	—	—	—	—	—	—	—	—	[5]
251	CN	Jiangchuan, Yunnan Province	1951	—	EQ	—	—	—	—	—	—	1000-1500	—	—	—	—	—	—	—	—	—	[9]
252	CN	Jiangjiagou (Xiao river)	1919	DF	RF		III	—	10–11	—	—	10000	—	48	OT	—	—	—	—	—	—	[1][5]
253	CN	Jiangjiagou (Xiao river)	1919	DF	RF	—		—	—	—	—	—	—	48	—	—	—	—	—	—	—	[5]
254	CN	Jiangjiagou (Xiao river)	1937	DF	RF	—	—	—	—	—	—	10000	—	—	—	—	—	—	—	—	—	[5]
255	CN	Jiangjiagou (Xiao river)	1937	DF	RF		III	—	—	—	—	10000	—	40	OT	—	—	—	—	—	—	[1][5]

256	CN	Jiangjiagou (Xiao river)	1954	DF	RF		III	—	10	—	—	9000	—	20	OT	—	—	—	—	—	—	[1][5]
257	CN	Jiangjiagou (Xiao river)	1961	DF	RF		III	—	9.5	—	—	—	—	10	OT	—	—	—	—	—	—	[1][5]
258	CN	Jiangjiagou (Xiao river)	1964	DF	RF		III	—	—	—	—	—	—	10	OT	—	—	—	—	—	—	[1][5]
259	CN	Jiangjiagou (Xiao river)	1966	RA	EQ	—	—	—	—	—	—	106	—	—	—	—	—	—	—	—	—	[9]
260	CN	Jiangjiagou (Xiao river)	1968	DF	RF		III	—	10	—	—	10000	—	180	OT	—	—	—	—	—	—	[1][5]
261	CN	Jiangjiatuo	Q4	DF	RF	11.75	—	—	—	—	—	—	—	LT	—	—	—	—	—	—	—	[6]
262	CN	Jiangjin	1786	—	RF	—	—	—	—	—	—	—	—	—	—	—	—	—	—	—	—	[5]
263	CN	Jiangwei	152BC	Ava.	EQ	—	—	—	—	—	—	—	—	—	—	—	—	—	—	—	—	[5]
264	CN	Jiangwei	35 BC	Ava.	EQ	—	—	—	—	—	—	—	—	—	—	—	—	—	—	—	—	[1][5]
265	CN	Jianshui	1588	—	EQ	—	—	—	—	—	—	—	—	—	—	—	—	—	—	—	—	[5]
266	CN	Jiaojiayatou	1989	—	RF	0.06	—	—	—	—	—	—	—	—	—	—	—	—	—	—	—	[5]
267	CN	Jiawucun	1942	—	RF	—	—	—	—	—	—	—	—	ST	—	—	—	—	—	—	120	[6]
268	CN	Jiaxi	1952	—	—	8	—	—	—	—	—	—	—	3	—	—	—	—	—	—	—	[6]
269	CN	Jiege	1950	—	EQ	—	—	—	—	—	—	—	—	—	—	—	—	—	—	—	—	[5]
270	CN	Jiguanling	1994	—	HC	5.3	II	—	—	110		20000	—	LT	—	—	—	—	—	—	—	[6]
271	CN	Jinan	1966	—	RF	18.7	—	—	—	—	—	—	—	—	—	—	—	—	—	—	—	[5]
272	CN	Jinchuandianchang	1981	—	—	10	—	—	—	—	—	—	—	LT	—	—	—	—	—	—	—	[6]
273	CN	Jinghe	1973	—	EQ	—	—	—	—	—	—	—	—	—	—	—	—	—	—	—	—	[5]
274	CN	Jingyugou	1953	—	RF	4	—	—	—	—	—	—	—	—	—	—	—	—	—	—	—	[5]
275	CN	Jinniuhe	1955	—	RF	—	—	—	—	—	—	—	—	—	—	—	—	—	—	—	—	[5]
276	CN	Jinpin power station	2004	—	—	—	—	—	—	—	—	—	—	0.17	—	—	—	—	—	—	—	[5]
277	CN	Jipazi	1982	—	RF	15	—	—	—	—	—	—	—	LT	—	—	—	—	—	—	—	[6]
278	CN	Jishixia	8000BP	—	RF	200	—	—	200	800–1000	1500	—	—	365000	—	—	—	—	—	—	—	[5]
279	CN	Jisiamo	1963	—	RF	3	—	—	—	—	—	—	—	—	—	—	—	—	—	—	—	[5]
280	CN	Jiuxianping	Q4	—	RF	52.2	—	—	—	—	—	—	—	LT	—	—	—	—	—	—	—	[6]
281	CN	Jujinzhou	1383	—	RF	—	—	—	—	—	—	—	—	—	—	—	—	—	—	—	—	[5]
282	CN	Kangding, Sichuan	1999	RA	EQ	—	—	—	—	—	—	—	—	—	—	—	—	—	—	—	—	[9]
283	CN	Kangle	1936	RF	EQ	—	—	—	—	—	—	—	—	—	—	—	—	—	—	—	—	[5]
284	CN	Kenkouwuwan	1991	—	HC	0.4	—	—	—	—	—	—	—	—	—	—	—	—	—	—	—	[5]
285	CN	Koushan	Q4	—	—	150	—	—	265	—	—	—	—	LT	—	—	—	—	—	—	—	[6]
286	CN	Kuitun	1987	—	RF	0.45	—	—	—	—	—	—	—	<1	—	—	—	—	—	—	A few	[6]
287	CN	Kuzhuba	2008	RS	EQ	1.65	II	1.65	60	300	200	800	2	—	OT	—	—	—	—	—	0	[7][8]
288	CN	Langjiaogou	1956	—	RF	—	—	—	—	—	—	—	—	—	—	—	—	—	—	—	92	[5]
289	CN	Lanqiangbei	1988	Slide	EQ	—	—	—	—	—	—	—	—	—	—	—	—	—	—	—	—	[5]
290	CN	Lantian	879	Ava.	EQ	—	—	—	—	—	—	—	—	—	—	—	—	—	—	—	—	[5]
291	CN	Lantian	35 BC	Ava.	EQ	—	—	—	—	—	—	—	—	—	—	—	—	—	—	—	—	[1][5]

(Continued)

Table B.1 (*Continued*)

No	Country[(1)]	Name	Date of formation [(2)]	Landslide Type [(3)]	Trigger [(4)]	Landslide volume ($\times 10^6$ m^3)	Dam Type [(5)]	Dam volume ($\times 10^6$ m^3)	Dam height (m)	Dam length (m)	Dam width (m)	Lake length (m)	Lake volume ($\times 10^6$ m^3)	Time to failure (day) [(6)]	Failure mech.[(7)]	Breach depth (m)	Breach top width (m)	Breach bottom width (m)	Breaching time (hour)	Peak discharge (m^3/s)	Loss of life	Reference (Below the table)
292	CN	Laoguankou	1000	—	RF	3	—	—	—	—	—	—	—	LT	—	—	—	—	—	—	—	[6]
293	CN	Laojinshan	1996	—	RF	0.56	—	—	9–14	—	—	—	—	0.29	—	—	—	—	—	—	—	[5]
294	CN	Laoyingyan	2008	RS	EQ	—	III	4.7	106–140	300	—	—	10.1	—	OT	—	—	—	—	—	0	[7][8]
295	CN	Laozuochang	1989	—	RF	10	—	—	50	100	—	—	—	LT	—	—	—	—	—	—	—	[6]
296	CN	Layuequ	1967	—	RF	25	—	—	—	—	—	—	—	—	—	—	—	—	—	—	—	[5]
297	CN	Ledu	847	Ava.	EQ	—	—	—	—	—	—	—	—	3	—	—	—	—	—	—	—	[5]
298	CN	Leguluoduogou	1987	—	—	0.1	—	—	—	—	—	—	—	ST	—	—	—	—	—	—	—	[6]
299	CN	Leigongtang	—	—	RF	—	—	—	—	—	—	—	—	—	—	—	—	—	—	—	—	[5]
300	CN	Leijiadashan	1998	—	—	0.488	—	—	—	—	—	—	—	—	—	—	—	—	—	—	—	[5]
301	CN	Liangjiazhuang	1983	—	RF	4.12	—	—	68	—	—	—	1.5	LT	—	—	—	—	—	—	—	[6]
302	CN	Liangmudi	1986	—	—	0.011	—	—	—	—	—	—	—	ST	—	—	—	—	—	—	—	[6]
303	CN	Libaluo	1984	—	RF	—	—	—	—	—	—	—	—	ST	—	—	—	—	—	—	—	[6]
304	CN	Lijiachi	1951	—	RF	—	—	—	—	—	—	—	—	—	—	—	—	—	—	—	—	[5]
305	CN	Lijie	1983	DF	RF	—	—	—	—	—	—	—	—	—	—	—	—	—	—	—	—	[5]
306	CN	Lingzikou	20000BP	—	RF	—	—	—	—	—	—	—	—	LT	—	—	—	—	—	—	—	[6]
307	CN	Lintao	1461	—	RF	—	—	—	—	—	—	—	—	SD	—	—	—	—	—	—	—	[5]
308	CN	Litang	1948	Ava.	EQ	—	—	—	—	—	—	—	—	5	—	—	—	—	—	—	—	[5]
309	CN	Liujiagou	1952	—	EQ	—	—	—	—	—	—	—	—	—	—	—	—	—	—	—	—	[5]
310	CN	Liujiaping	1989	—	RF	10	—	—	—	—	—	—	—	LT	—	—	—	—	—	—	—	[6]
311	CN	Liujiawuchang	Q4	—	RF	17.19	—	—	—	—	—	—	—	LT	—	—	—	—	—	—	—	[6]
312	CN	Liulaiguan	Q4	—	RF	33	—	—	—	—	—	—	—	LT	—	—	—	—	—	—	—	[6]
313	CN	Liushi	1950	—	EQ	—	—	—	—	—	—	—	—	—	—	—	—	—	—	—	—	[5]
314	CN	Liuxianggou	2008	—	EQ	—	—	1.5	60	50	500	—	3	—	OT	—	—	—	—	—	0	[7][8]
315	CN	Liuzuizhen	1998	—	RF	30–40	—	1.5	—	—	—	—	—	—	—	—	—	—	—	—	—	[5]
316	CN	Lixian	1708	—	RF	—	—	—	—	—	—	—	—	—	—	—	—	—	—	—	—	[5]
317	CN	Lixian	1858	—	RF	—	—	—	—	—	—	—	—	—	—	—	—	—	—	—	—	[5]
318	CN	Lixian	1984	DF	RF	—	—	—	—	—	—	—	—	—	—	—	—	—	—	—	—	[5]
319	CN	Liyuping	1955	—	RF	—	—	—	—	—	—	—	—	730	—	—	—	—	—	—	—	[5]
320	CN	Liziyidagou	1981	—	RF	3	—	—	—	—	—	—	—	0.125	—	—	—	—	—	—	—	[6]
321	CN	Longmengou	1983	DF	RF	0.1	—	—	—	—	—	—	—	0.125	—	—	—	—	—	—	—	[5]
322	CN	Longtanxiang	1991	DF	RF	—	—	—	—	—	—	—	—	—	—	—	—	—	—	—	27	[5]
323	CN	Longwan	—	—	RF	20.5	—	—	—	—	—	—	—	—	—	—	—	—	—	—	—	[5]
324	CN	Longwan - Baozi	1954	—	RF	4.5	—	—	—	—	—	—	—	—	—	—	—	—	—	—	—	[5]
325	CN	Longyang	1300BP	—	RF	43	—	—	—	—	—	—	—	LT	—	—	—	—	—	—	—	[6]
326	CN	Luchedu	1935	—	RF	100	—	—	—	—	—	—	—	3	—	—	—	—	—	—	286	[6]
327	CN	Lugu iron mine	1970	DF	RF	—	—	—	—	—	—	—	—	—	—	—	—	—	—	—	—	[5]

328	CN	Luntai	1949	—	EQ	—	—	—	—	—	—	—	—	—	—	—	—	—	—	—	—	[5]
329	CN	Luoyugou	1965	—	RF	5.6	—	—	—	—	—	—	—	—	—	—	—	—	—	—	—	[5]
330	CN	Luquan	1965	—	RF	309	—	—	—	—	—	—	—	240	—	—	—	—	—	—	—	[6]
331	CN	Luquan	1966	DF	RF	—	—	—	—	—	—	—	—	—	—	—	—	—	—	—	—	[5]
332	CN	Lushan	1970	RA	EQ	—	—	—	10	40	150	—	—	0.42	—	—	—	—	—	—	—	[9]
333	CN	Maanmeikuang	1983	—	RF	—	—	—	—	—	—	—	—	—	—	—	—	—	—	—	—	[5]
334	CN	Ma'anshi	2008	—	EQ	—	—	5.8	67.6	950	—	—	1.15	—	—	—	—	—	—	—	—	[8]
335	CN	Mabian	1794	ES	RF	—	—	—	—	—	—	—	—	—	—	—	—	—	—	—	54	[5]
336	CN	Mabian	1935	Ava.	EQ	—	—	—	—	—	—	—	—	—	—	—	—	—	—	—	—	[5]
337	CN	Mabian	1936	Ava.	EQ	—	—	—	—	—	—	—	—	—	—	—	—	—	—	—	—	[5]
338	CN	Macaotan downstream	2008	RS	EQ	—	II	0.14	30	100	—	—	0.1	—	OT	—	—	—	—	—	0	[7][8]
339	CN	Macaotan site	2008	RS	EQ	—	II	0.2	40–50	90	—	—	0.25	—	OT	—	—	—	—	—	0	[7][8]
340	CN	Macaotan upstream	2008	RS	EQ	—	II	1	40–50	160	—	—	0.6	—	OT	—	—	—	—	—	0	[7][8]
341	CN	Machigai	208	—	EQ	—	—	—	—	60	—	—	1	NF	NF	—	—	—	—	—	—	By visit
342	CN	Mahu	1216	—	EQ	—	—	—	—	—	—	—	—	LT	—	—	—	—	—	—	—	[1][6]
343	CN	Mahuangxiang	1981	—	RF	—	—	—	—	—	—	—	—	1460	—	—	—	—	—	—	13	[6]
344	CN	Maidi	1971	—	HC	1	—	—	—	—	—	—	—	LT	—	—	—	—	—	—	40	[6]
345	CN	Maipengzi	1988	—	RF	0.26	—	—	—	—	—	—	—	LT	—	—	—	—	—	—	—	[6]
346	CN	Majiaba	1986	—	RF	28.8	—	—	—	—	—	—	—	LT	—	—	—	—	—	—	—	[6]
347	CN	Maoping	11490BP	—	RF	—	—	—	—	—	—	—	—	LT	—	—	—	—	—	—	—	[6]
348	CN	Menshishan	1971	—	—	7	—	—	—	—	—	—	—	730	—	—	—	—	—	—	—	[6]
349	CN	Mianyuanhe	1934	—	RF	—	—	—	—	—	—	—	—	—	—	—	—	—	—	—	200	[5]
350	CN	Min Rive (Deixi)	1933	RS, ES	EQ	150	II	—	255	400	1300	17000	400	7	OT	—	—	—	—	—	2423	[1][5]
351	CN	Min River (Deixi)	1933	RS	EQ	—	II	—	125	—	—	—	—	16	OT	—	—	—	—	—	—	[1][5]
352	CN	Min River (Deixi)	1933	RS, RA	EQ	—	IV	—	156	800	1700	12500	73	19	OT	—	—	—	—	—	—	[1][5]
353	CN	Minfeng	1924	Ava.	EQ	—	—	—	—	—	—	—	—	—	—	—	—	—	—	—	—	[5]
354	CN	Mingdonggou	1986	DF	RF	—	—	—	—	—	—	—	—	0.04	—	—	—	—	—	—	—	[5]
355	CN	Mingshanxian	991	—	RF	—	—	—	—	—	—	—	—	—	—	—	—	—	—	—	—	[5]
356	CN	Mogangling	1786	RS, RA	EQ	25	—	—	—	—	—	—	—	10	—	—	—	—	—	—	—	[6]
357	CN	Motuo	1950	Ava.	EQ	—	—	—	—	—	—	—	—	0.65	—	—	—	—	—	—	—	[5]
358	CN	Moximian	1786	RS, RA	EQ	—	—	—	—	—	—	—	—	9	—	—	—	—	—	—	—	[6]
359	CN	Muguaping	2008	ES	EQ	—	IV	0.2	15	100	—	—	0.04	—	OT	—	—	—	—	—	0	[7][8]
360	CN	Nanao	1600	Ava.	EQ	—	—	—	—	—	—	—	—	—	—	—	—	—	—	—	—	[5]
361	CN	Nanba	2008	RS	EQ	—	II	5.32	25–50	625	200	6000	6.86	—	OT	—	—	—	—	—	0	[7][8]
362	CN	Nanbujiongshan	1950	—	EQ	—	—	—	—	—	—	—	—	<1	—	—	—	—	—	—	—	[5]
363	CN	Nangou	—	—	RF	8.1	—	—	—	—	—	—	—	LT	—	—	—	—	—	—	—	[6]
364	CN	Nanguangou	1989	DF	RF	—	—	—	—	—	—	—	—	—	—	—	—	—	—	—	—	[5]
365	CN	Nanjiabawashan	1950	—	EQ	—	—	—	—	—	—	—	—	<1	—	—	—	—	—	—	—	[5]
366	CN	Nanjiang	—	—	RF	15	—	—	—	—	—	—	—	LT	—	—	—	—	—	—	—	[6]
367	CN	Nanjiang County	1975	RS, DF	—	7	III	—	—	200	450	450	—	—	—	—	—	—	—	—	—	[1]
368	CN	Nanmenwan	1987	RA	RF	0.007	—	—	—	—	—	—	—	—	—	—	—	—	—	—	—	[5]

(Continued)

Table B.1 (*Continued*)

No	Country[(1)]	Name	Date of formation[(2)]	Landslide Type[(3)]	Trigger[(4)]	Landslide volume ($\times 10^6 m^3$)	Dam Type[(5)]	Dam volume ($\times 10^6 m^3$)	Dam height (m)	Dam length (m)	Dam width (m)	Lake length (m)	Lake volume ($\times 10^6 m^3$)	Time to failure (day)[(6)]	Failure mech.[(7)]	Breach depth (m)	Breach top width (m)	Breach bottom width (m)	Breaching time (hour)	Peak discharge (m^3/s)	Loss of life	Reference (Below the table)
369	CN	Nanxucun	1981	—	RF	2.83	—	—	—	—	—	—	—	—	—	—	—	—	—	—	—	[5]
370	CN	Nileke	1812	Ava.	EQ	10	—	—	—	—	—	—	—	—	—	—	—	—	—	—	—	[5]
371	CN	Ninglang	1976	Ava.	EQ	—	—	—	—	—	—	—	—	—	—	—	—	—	—	—	—	[5]
372	CN	Ninglang	1998	Slide	EQ	—	—	—	—	—	—	—	1	—	—	—	—	—	—	—	—	[9]
373	CN	Niqiu	10700BP	—	RF	3	—	—	240	—	—	—	—	14600	—	—	—	—	—	—	—	[5]
374	CN	Niugunqiu	1968	—	RF	40	—	—	—	—	—	—	—	LT	—	—	—	—	—	—	—	[6]
375	CN	Niujiaodong	1982	—	RF	6	—	—	40	—	—	—	—	LT	—	—	—	—	—	—	—	[6]
376	CN	Niurihe	1982	—	RF	18	—	—	—	—	—	—	—	—	—	—	—	—	—	—	—	[5]
377	CN	Peilonggou	1985	DF	RF	39	—	—	—	—	—	—	—	—	—	—	—	—	—	—	—	[5]
378	CN	Pingan Village	1955	Ava.	EQ	—	—	—	—	—	—	—	—	—	—	—	—	—	—	—	—	[5]
379	CN	Pingwu	1976	Ava.	EQ	—	—	—	—	—	—	2000	—	—	—	—	—	—	—	—	—	[5]
380	CN	Pingwu	1976	Ava.	EQ	—	—	—	—	—	—	—	—	—	—	—	—	—	—	—	—	[5]
381	CN	Pinshan	1876	—	RF	—	—	—	—	—	—	—	—	—	—	—	—	—	—	—	—	[5]
382	CN	Piyang	1961	—	EQ	—	—	—	—	—	—	—	—	—	—	—	—	—	—	—	—	[5]
383	CN	Pubugou	2010	DF	RF	—	—	0.91	9	421	239	153	0.11	315	OT	—	—	—	—	—	0	By visit
384	CN	Pubugou	10000BP	—	—	4.75	—	—	—	—	—	—	—	LT	—	—	—	—	—	—	—	[6]
385	CN	Pucheng	1366	—	RF	—	—	—	—	—	—	—	—	3	—	—	—	—	—	—	—	[5]
386	CN	Pu'er	1970	RA	EQ	0.001	—	—	—	—	—	—	—	—	—	—	—	—	—	—	—	[9]
387	CN	Pufu	1965	—	—	309	—	—	140	—	—	—	—	—	—	—	—	—	—	—	—	[5]
388	CN	Pufugou	1921	—	RF	—	—	—	—	—	—	—	—	ST	—	—	—	—	—	—	7	[6]
389	CN	Pugejian	1850	Ava.	EQ	—	—	—	—	—	—	—	—	—	—	—	—	—	—	—	—	[5]
390	CN	Pujiagou	1974	—	RF	—	—	—	—	—	—	—	—	—	—	—	—	—	—	—	31	[5]
391	CN	Puqin	1960	—	RF	—	—	—	—	—	—	—	—	—	—	—	—	—	—	—	—	[5]
392	CN	Qiancaotuo	Q4	—	RF	3.6	—	—	—	—	—	—	—	LT	—	—	—	—	—	—	—	[6]
393	CN	Qianjiangpingcun	2003	Slide	—	—	—	—	20	1000	1200	—	—	—	—	—	—	—	—	—	10	[5]
394	CN	Qiaojia	1918	Ava.	EQ	—	—	—	—	—	—	—	—	—	—	—	—	—	—	—	—	[5]
395	CN	Qiaojia	1983	DF	RF	—	—	—	—	—	—	—	—	0.002	—	—	—	—	—	—	—	[5]
396	CN	Qina	1966	—	—	—	—	—	—	—	—	—	—	LT	—	—	—	—	—	—	—	[6]
397	CN	Qingyijiang	1979	DF	RF	0.05	—	—	—	40	110	—	—	—	—	—	—	—	—	—	—	[5]
398	CN	Qiongshangou	2003	DF	RF	—	—	—	6	—	—	—	2.88	0.003	—	—	—	—	—	—	—	[5]
399	CN	Qishan	780BC	Ava.	EQ	300	—	—	—	—	—	—	—	—	—	—	—	—	—	—	—	[1][5]
400	CN	Quxi	1970	—	EQ	—	—	—	—	—	—	—	—	ST	—	—	—	—	—	—	—	[6]
401	CN	Ranwuhu	200BP	—	RF	—	—	—	—	—	—	—	—	LT	—	—	—	—	—	—	—	[6]
402	CN	Saimihe	1985	—	RF	0.1	—	—	—	—	—	—	—	0.42	—	—	—	—	—	—	—	[6]
403	CN	Saleshan	1983	—	—	31	—	—	—	—	—	—	—	LT	—	—	—	—	—	—	—	[6]
404	CN	Sanhuisi	1964	—	RF	—	—	—	—	—	—	—	—	—	—	—	—	—	—	—	—	[5]
405	CN	Shabagou	1935	—	—	—	—	—	—	—	—	—	—	7	—	—	—	—	—	—	280	[6]

406	CN	Shadixiang	1976	—	—	—	—	—	—	—	—	—	—	—	—	—	—	—	—	—	—	[6]
407	CN	Shaling	1982	—	RF	10–15.	—	—	20	—	—	—	2	—	—	—	—	—	—	—	—	[6]
408	CN	Shangbailazhai	1933	Slide	EQ	0.2	—	—	20	—	—	—	10	—	—	—	—	—	—	—	—	[6]
409	CN	Shangshuimogou	1933	Slide	EQ	7.5	—	—	50	—	—	—	0.8	LT	—	—	—	—	—	—	—	[6]
410	CN	Shankou	1982	—	RF	18	—	—	—	—	—	—	—	LT	—	—	—	—	—	—	—	[6]
411	CN	Shanxi	586BC	—	RF	—	—	—	—	—	—	—	—	3	—	—	—	—	—	—	—	[5]
412	CN	Shanxi 1#	25BC	—	RF	—	—	—	—	—	—	—	—	—	—	—	—	—	—	—	—	[5]
413	CN	Shanxi 2#	16	—	RF	—	—	—	—	—	—	—	—	—	—	—	—	—	—	—	—	[5]
414	CN	Shanyang	1990	—	RF	0.07	—	—	—	—	—	—	—	—	—	—	—	—	—	—	—	[5]
415	CN	Shawan	1906	Ava.	EQ	—	—	—	—	—	—	—	—	—	—	—	—	—	—	—	—	[5]
416	CN	She'er	1978	DF	RF	—	—	—	—	—	—	—	—	—	—	—	—	—	—	—	—	[5]
417	CN	Shibangou	2008	RS	EQ	—	II	15	30–75	450	800	4000	11	—	OT	—	—	—	—	—	0	[7][8]
418	CN	Shigaodi	1881	—	RF	536	—	—	—	—	—	—	—	3	—	—	—	—	—	—	10	[6]
419	CN	Shijiapo	1981	Slide	EQ	0.48	II	30	100	1060	160	—	—	ST	—	—	—	—	—	—	—	[1][6]
420	CN	Shimengou	1978	DF	RF	—	—	—	—	—	—	—	—	—	—	—	—	—	—	—	—	[5]
421	CN	Shimenkan	Q4	—	—	100	—	—	405	—	—	—	—	365000	—	—	—	—	—	—	—	[6]
422	CN	Shiping	1887	Ava.	EQ	—	—	—	—	—	—	—	—	—	—	—	—	—	—	—	—	[5]
423	CN	Shiqian county	1995	DF	RF	0.13	—	—	—	—	—	—	—	—	—	—	—	—	—	—	—	[5]
424	CN	Shiquluoxu	1896	Ava.	EQ	—	—	—	—	—	—	—	—	—	—	—	—	—	—	—	—	[5]
425	CN	Shuangniancun	1981	—	—	—	—	—	—	—	—	—	—	ST	—	—	—	—	—	—	—	[6]
426	CN	Shuizhuyuan	Q4	DF	RF	21.1	—	—	—	—	—	—	—	LT	—	—	—	—	—	—	—	[6]
427	CN	Shunhe	—	—	RF	78.2	—	—	—	—	—	—	—	—	—	—	—	—	—	—	—	[5]
428	CN	Shuyexiang	1983	DF	RF	—	—	—	—	—	—	—	—	0.14	—	—	—	—	—	—	—	[5]
429	CN	Sijigou	1989	—	—	12	—	—	—	—	—	—	—	LT	—	—	—	—	—	—	—	[6]
430	CN	Sina	1966	—	RF	—	—	—	—	—	—	—	—	—	—	—	—	—	—	—	—	[5]
431	CN	Songping River	1933	RS	EQ	—	II	—	50	200	100	—	—	—	—	—	—	—	—	—	—	[1]
432	CN	Sunjiayuanzi	2008	RS	EQ	—	II	1.6	50	180	—	—	5.6	—	OT	—	—	—	—	—	0	[7][8]
433	CN	Suwalong	1997	—	SM	0.25	—	—	—	—	—	—	—	0.003	—	—	—	—	—	—	—	[5]
434	CN	Suzhouya	1957	—	HC	0.85	—	—	—	—	—	—	—	LT	—	—	—	—	—	—	—	[6]
435	CN	Taipingya	20000BP	—	RF	180	—	—	—	—	—	—	—	LT	—	—	—	—	—	—	—	[6]
436	CN	Tangbulanggou	1964	DF	SM	5	—	—	—	—	—	—	—	0.42	—	—	—	—	—	—	—	[6]
437	CN	Tangfanggou	Q4	DF	RF	7.79	—	—	—	—	—	—	—	LT	—	—	—	—	—	—	—	[6]
438	CN	Tanggudong	1967	DA,DS	RF	68	III	—	175	—	3000	53000	680	9	OT	88	—	55	6	53000	—	[1][5]
439	CN	Tanggudong	1969	—	RF	110	—	—	—	—	—	—	—	9	—	—	—	—	—	—	—	[6]
440	CN	Tangjiashan	2008	RS	EQ	20.37	II	20.37	82	611.8	802	20000	316	29	OT	42	145–235	80–100	14	6500	0	[7][8]
441	CN	Tangjiawan	2008	RS	EQ	—	III	4	30	300	—	—	15.2	—	OT	—	—	—	—	—	0	[7][8]
442	CN	Tangling Taiwan	1977	Slide	EF, EQ	48	—	—	—	—	—	—	—	—	—	—	—	—	—	—	—	[5]
443	CN	Tangyanguang	1961	—	RF	1.65	—	—	—	—	—	—	—	LT	—	—	—	—	—	—	40	[6]
444	CN	Tanlao	Q4	DF	—	0.78	—	—	—	—	—	—	—	LT	—	—	—	—	—	—	—	[6]
445	CN	Taoyuan	Q4	—	RF	15.75	—	—	—	—	—	—	—	LT	—	—	—	—	—	—	—	[6]
446	CN	Tianbao	1982	—	RF	7	—	—	—	—	—	—	—	LT	—	—	—	—	—	—	—	[6]

(Continued)

Table B.1 (*Continued*)

No	Country[(1)]	Name	Date of formation [(2)]	Landslide Type [(3)]	Trigger [(4)]	Landslide volume ($\times 10^6\,m^3$)	Dam Type [(5)]	Dam volume ($\times 10^6\,m^3$)	Dam height (m)	Dam length (m)	Dam width (m)	Lake length (m)	Lake volume ($\times 10^6\,m^3$)	Time to failure (day) [(6)]	Failure mech.[(7)]	Breach depth (m)	Breach top width (m)	Breach bottom width (m)	Breaching time (hour)	Peak discharge (m^3/s)	Loss of life	Reference (Below the table)
447	CN	Tianjiawa	1962	—	RF	6	—	—	—	—	—	—	—	—	—	—	—	—	—	—	—	[5]
448	CN	Tianquan	181bc	Ava.	EQ	—	—	—	—	—	—	—	—	—	—	—	—	—	—	—	—	[5]
449	CN	Tiantaixiang	2004	—	—	65	—	—	—	—	—	20000	200	—	—	—	—	—	—	—	—	[5]
450	CN	Tianzhuxian	1665	—	RF	—	—	—	—	—	—	—	—	—	—	—	—	—	—	—	—	[5]
451	CN	Tiaoshi	Q4	—	RF	2.5	—	—	—	—	—	—	—	LT	—	—	—	—	—	—	—	[6]
452	CN	Tietan	Q4	DF	RF	9.48	—	—	—	—	—	—	—	LT	—	—	—	—	—	—	—	[6]
453	CN	Tongwei	1921	Ava.	EQ	—	—	—	—	—	—	—	—	—	—	—	—	—	—	—	—	[5]
454	CN	Touzhaigou	1991	—	RF	18	—	—	—	—	—	—	—	LT	—	—	—	—	—	—	216	[6]
455	CN	Tuanjianxiang	1988	RA	EQ	1	—	—	—	—	—	—	—	—	—	—	—	—	—	—	—	[5][9]
456	CN	Tubagou	1984	—	—	1.2	—	—	25	—	—	—	—	LT	—	—	—	—	—	—	—	[6]
457	CN	Tumen	1992	—	RF	—	—	—	—	—	—	—	—	—	—	—	—	—	—	—	—	[5]
458	CN	Tupatang	1961	—	EQ	—	—	—	—	—	—	—	—	—	—	—	—	—	—	—	—	[5]
459	CN	Unkown, Hunan Province	1786	—	EQ	—	—	—	—	—	—	—	—	—	—	—	—	—	—	—	—	[1]
460	CN	Wangxicun	1991	—	RF	—	—	—	—	—	—	—	—	—	—	—	—	—	—	—	—	[5]
461	CN	Wanjiagou	1981	—	RF	1.68	—	—	—	—	—	—	—	—	—	—	—	—	—	—	—	[5]
462	CN	Weining	1948	Ava.	EQ	—	—	—	—	—	—	—	—	—	—	—	—	—	—	—	—	[5]
463	CN	Wenchuan	1657	Ava.	EQ	—	—	—	—	—	—	—	—	—	—	—	—	—	—	—	—	[5]
464	CN	Wenjiagou	1973	DF	RF	—	—	—	—	—	—	—	—	—	—	—	—	—	—	—	—	[5]
465	CN	Wuchiba	—	—	RF	1.8	—	—	—	—	—	—	—	LT	—	—	—	—	—	—	—	[6]
466	CN	Wudangqu	1963	—	RF	—	—	—	—	—	—	—	—	—	—	—	—	—	—	—	7	[5]
467	CN	Wudu	986	—	RF	—	—	—	—	—	—	—	—	—	—	—	—	—	—	—	—	[5]
468	CN	Wudu	1879	Ava.	EQ	—	—	—	—	—	—	—	—	—	—	—	—	—	—	—	—	[5]
469	CN	Wujiawan	1887	—	RF	33.75	—	—	—	—	—	6500	—	7	—	—	—	—	—	—	—	[6]
470	CN	Wujin River	1982	RA	EQ	0.27	—	—	—	—	—	—	0.1	ST	—	—	—	—	—	—	—	[9]
471	CN	Wulian River	1948	—	EQ	—	—	—	—	—	—	—	—	5	—	—	—	—	—	—	—	[1]
472	CN	Xiabailazhai	1933	Slide	EQ	13.2	—	—	60	—	—	—	0.7	45	—	—	—	—	—	—	—	[6]
473	CN	Xianfeng, Hubei	1856	Slide	EQ	4	—	4	100	—	600	—	—	—	—	—	—	—	—	—	—	[9]
474	CN	Xiao River	1949	DF	RF	—	III	—	—	—	—	—	—	30	OT	—	—	—	—	—	—	[1]
475	CN	Xiao River	1985	DF	RF	—	III	—	—	—	—	—	—	0.01	OT	—	—	—	—	—	—	[1]
476	CN	Xiaogangjian downstream	2008	RS	EQ	—	II	0.45	30	120	400	400	7	—	OT	—	—	—	—	—	0	[7][8]
477	CN	Xiaogangjian upstream	2008	RS	EQ	—	III	2	70–120	120	300	400	12	—	OT	30	80	—	—	3950	0	[7][8]
478	CN	Xiaojiagou	2010	DF	RF	—	—	0.75	15	500	100	150	0.23	315	OT	—	—	—	—	—	—	By visit
479	CN	Xiaojiaqiao	2008	RS	EQ	2.42	II	2.42	57–67	200	200	7000	20	25	OT	37.3	131.6	8	6.5	1000	0	[7][8]
480	CN	Xiaojiaya	1935	—	RF	0.1	—	—	—	—	—	—	—	LT	—	—	—	—	—	—	—	[6]

481	CN	Xiaonanhai	1856	Ava.	EQ	72	—	—	30	100	1500	12000	420	LT	—	—	—	—	—	—	—	[1][6][9]
482	CN	Xiashuimogou	1933	Slide	EQ	0.5	—	—	20	—	—	—	0.2	LT	—	—	—	—	—	—	—	[6]
483	CN	Xichang	624	Ava.	EQ	—	—	—	—	—	—	—	—	—	—	—	—	—	—	—	—	[5]
484	CN	Xiejiadianzi	2008	—	EQ	—	—	0.12	10	70	250	1000	1	—	OT	—	—	—	—	—	0	[7][8]
485	CN	Xieliupo	1981	—	—	25	—	—	22	—	—	—	13	—	—	—	—	—	—	—	—	[5]
486	CN	Xieliupo	1991	—	HC	25	—	—	—	—	—	—	—	LT	—	—	—	—	—	—	—	[6]
487	CN	Xietangou	Q4	DF	RF	5.32	—	—	—	—	—	—	—	LT	—	—	—	—	—	—	—	[6]
488	CN	Xihecun	1998	—	RF	2.5	—	—	—	—	—	—	—	—	—	—	—	—	—	—	—	[5]
489	CN	Xiji (30 lakes)	1920	ES, EF	EQ	—	—	—	—	—	—	5000	—	—	—	—	—	—	—	—	—	[1]
490	CN	Ximula	1819	—	RF	—	—	—	—	—	—	24000	—	—	—	—	—	—	—	—	—	[5]
491	CN	Xingou	1972	DF	RF	—	—	—	—	—	—	—	—	—	—	—	—	—	—	—	—	[5]
492	CN	Xinhuaxiangqun	Q4	—	RF	30	—	—	—	—	—	—	—	LT	—	—	—	—	—	—	—	[6]
493	CN	Xinjie Village	2008	ES	EQ	—	II	0.7	20	350	200	—	2	—	OT	—	—	—	—	—	0	[7][8]
494	CN	Xinpu	Q4	—	RF	57	—	—	—	—	—	—	—	LT	—	—	—	—	—	—	—	[6]
495	CN	Xintan	1030	Ava.	EQ	—	—	—	—	—	—	—	—	7300	—	—	—	—	—	—	—	[5]
496	CN	Xintan 1#	101	—	RF	—	—	—	—	—	—	—	—	—	—	—	—	—	—	—	—	[5]
497	CN	Xintan 2#	378	—	RF	—	—	—	—	—	—	—	—	—	—	—	—	—	—	—	—	[5]
498	CN	Xintan, Hubei	1985	—	RF	30	—	—	—	—	—	—	—	LT	—	—	—	—	—	—	—	[6]
499	CN	Xintanbei	1543	—	RF	—	—	—	—	—	—	—	—	29930	—	—	—	—	—	—	—	[5]
500	CN	Xintang	1970	—	RF	0.4	—	—	—	—	—	—	—	—	—	—	—	—	—	—	—	[5]
501	CN	Xixi River	1988	RS	—	—	—	—	12	—	—	—	—	—	—	—	—	—	—	—	—	[1]
502	CN	Xixiangkou	1950	—	RF	5.62	—	—	—	—	—	—	—	ST	—	—	—	—	—	—	—	[6]
503	CN	Xuecheng	888	—	RF	—	—	—	—	—	—	—	—	—	—	—	—	—	—	—	—	[5]
504	CN	Xundiao	1713	—	EQ	—	—	—	—	—	—	—	—	—	—	—	—	—	—	—	—	[5]
505	CN	Yading	1991	—	RF	7.5	—	—	—	—	—	—	7.55	—	—	—	—	—	—	—	—	[5]
506	CN	Yanchihe	1980	—	HC	1	—	—	—	—	—	—	—	LT	—	—	—	—	—	—	281	[6]
507	CN	Yanggao	1989	Ava.	EQ	—	—	—	4	15	40	—	—	—	—	—	—	—	—	—	—	[5]
508	CN	Yanglin	1833	Ava.	EQ	—	—	—	—	—	—	—	—	—	—	—	—	—	—	—	—	[5]
509	CN	Yangpingguan	1981	—	RF	—	—	—	—	—	—	—	—	—	—	—	—	—	—	—	—	[5]
510	CN	Yangtze River	377	—	—	26	—	—	—	—	—	50000	—	—	—	—	—	—	—	—	—	[1]
511	CN	Yangtze River	1026	—	—	15	—	—	—	—	—	—	—	—	—	—	—	—	—	—	—	[1]
512	CN	Yangtze River	1880	—	—	—	—	—	—	—	—	50000	—	3	—	—	—	—	—	—	—	[1]
513	CN	Yanjiayinpo	1988	—	RF	4	—	—	—	—	—	—	—	—	—	—	—	—	—	—	—	[5]
514	CN	Yankou	1996	—	RF	2.6	—	—	—	—	—	—	—	—	—	—	—	—	—	—	—	[5]
515	CN	Yankuang	1978	—	RF	2	—	—	—	—	—	—	—	LT	—	—	—	—	—	—	—	[6]
516	CN	Yanyangtan	2008	RS	EQ	—	II	1.6	20–30	150	—	3000	4	—	OT	—	—	—	—	—	0	[7][8]
517	CN	Yanzigou	1989	DF	RF	6.27	—	—	—	—	—	—	—	ST	—	—	—	—	—	—	—	[6]
518	CN	Yanziwo	1986	Slide	—	0.6	—	—	—	—	—	—	—	0.01	—	—	—	—	—	—	—	[5][6]
519	CN	Yanziyan	1988	—	HC	0.6	—	—	—	—	—	—	—	0.01	—	—	—	—	—	—	—	[6]
520	CN	Yanziyan	2008	RS	EQ	—	III	0.006	10	40	—	—	0.03	—	OT	—	—	—	—	—	0	[7][8]
521	CN	Yecheng, Xinjiang Province	1980	—	EQ	—	V	—	—	—	—	—	—	—	—	—	—	—	—	—	—	[9]

(*Continued*)

Table B.1 *(Continued)*

No	Country[(1)]	Name	Date of formation [(2)]	Landslide Type [(3)]	Trigger [(4)]	Landslide volume ($\times10^6 m^3$)	Dam Type [(5)]	Dam volume ($\times10^6 m^3$)	Dam height (m)	Dam length (m)	Dam width (m)	Lake length (m)	Lake volume ($\times10^6 m^3$)	Time to failure (day) [(6)]	Failure mech.[(7)]	Breach depth (m)	Breach top width (m)	Breach bottom width (m)	Breaching time (hour)	Peak discharge (m^3/s)	Loss of life	Reference (Below the table)
522	CN	Yellow River	1943	DA, ES	—	100–150	II	—	—	—	—	1500–2000	—	—	—	—	—	—	—	—	—	[1]
523	CN	Yi and Luo Rivers	1800BC	—	EQ	—	—	—	—	—	—	—	—	—	—	—	—	—	—	—	—	[1]
524	CN	Yibadao	2008	RS	EQ	—	II	0.15	25	120	140-180	—	3.79	—	OT	8	—	—	—	—	0	[7][8]
525	CN	Yidai	26BC	Ava.	EQ	—	—	—	—	—	—	—	—	—	—	—	—	—	—	—	13	[5]
526	CN	Yigong	2000	Slide	SM	300	—	300	60–100	—	3000	—	3000	—	OT	58.39	—	128	9.25	124000	—	[5]
527	CN	Yimen mine	1990	DF	RF	0.15	—	—	—	—	—	4000	—	—	—	—	—	—	—	—	—	[6]
528	CN	Yinjiang	—	—	RF	—	—	—	—	—	—	—	—	—	—	—	—	—	—	—	—	[5]
529	CN	Yongning	1960	—	RF	—	—	—	—	—	—	—	—	—	—	—	—	—	—	—	—	[5]
530	CN	Youzhegou	Q4	DF	RF	1.31	—	—	—	—	—	—	—	LT	—	—	—	—	—	—	—	[6]
531	CN	Yuanguoliang	1978	—	RF	0.8	—	—	—	—	—	—	—	—	—	—	—	—	—	—	—	[5]
532	CN	Yuanping, Shanxi Province	1952	Ava.	EQ	—	—	—	—	—	—	—	—	—	—	—	—	—	—	—	—	[9]
533	CN	Yuerzhai	1933	—	—	6.75	—	—	135	—	—	—	—	—	—	—	—	—	—	—	—	[5]
534	CN	Yuerzhai	1933	Slide	EQ	6.75	—	—	—	—	—	—	—	LT	—	—	—	—	—	—	—	[6]
535	CN	Yuexi	623	—	RF	—	—	—	—	—	—	—	—	—	—	—	—	—	—	—	—	[5]
536	CN	Yuexi	1902	Ava.	RF	—	—	—	—	—	—	—	—	—	—	—	—	—	—	—	700	[5]
537	CN	Yulong county	1991	DF	RF	—	—	—	—	—	—	—	—	0.08	—	—	—	—	—	—	—	[5]
538	CN	Yunyangxian	1896	—	RF	—	—	—	—	—	—	—	—	—	—	—	—	—	—	—	—	[5]
539	CN	Zaduo, Qinghai Province	1971	RA	EQ	—	—	—	—	—	—	—	—	—	—	—	—	—	—	—	—	[9]
540	CN	Zaoyang	1984	—	—	0.1	—	—	—	—	—	—	—	LT	—	—	—	—	—	—	—	[6]
541	CN	Zhachehe	1900	RF	DF	—	—	—	—	—	—	—	—	LT	—	—	—	—	—	—	—	[6]
542	CN	Zhalimuhe River	1927	—	EQ	—	—	—	—	—	—	—	—	24	—	—	—	—	—	—	—	[1]
543	CN	Zhalonggou	1995	DF	DF	—	III	—	—	—	—	—	—	—	—	—	40	>10	>10	—	1080	[5]
544	CN	Zhamu valley	1902	—	—	—	—	—	—	—	—	—	—	—	—	—	—	—	—	—	—	[5]
545	CN	Zhamulongba	1902	DF	RF	—	—	—	—	—	—	—	—	15	—	—	—	—	—	—	—	[5]
546	CN	Zhana	1943	—	—	150 ~ 160	—	—	—	—	—	—	—	1	—	—	—	—	—	—	100	[6]
547	CN	Zhangjiadagou	Q4	DF	RF	3.29	—	—	—	—	—	—	—	LT	—	—	—	—	—	—	—	[6]
548	CN	Zhangye	1548	—	EQ	—	—	—	—	—	—	—	—	—	—	—	—	—	—	—	—	[5]
549	CN	Zhaoan	1600	Ava.	EQ	—	—	—	—	—	—	—	—	—	—	—	—	—	—	—	—	[5]
550	CN	Zhaojiatang	1973	—	RF	2	—	—	—	—	—	—	—	LT	—	—	—	—	—	—	—	[6]
551	CN	Zhaotong	1935	—	EQ	—	—	—	—	—	—	—	—	—	—	—	—	—	—	—	—	[5]
552	CN	Zepozhu	1965	RS	RF	29	II	29	51	—	650	1000	2.7	210	OT	—	—	—	4	560	—	[1][6]
553	CN	Zhicheng	2008	—	EQ	—	—	5.8	67.6	950	—	—	1.15	—	—	—	—	—	—	—	—	[8]
554	CN	Zhinaxiang	1983	DF	RF	0.5	—	—	—	—	—	—	—	—	—	—	—	—	—	—	—	[5]

555	CN	Zhipingchuan	1718	Ava.	EQ	—	—	—	—	—	—	—	—	—	—	—	—	—	—	—	—	[1][5]
556	CN	Zhongning	1561	Ava.	EQ	—	—	—	—	—	—	—	—	—	—	—	—	—	—	—	—	[5]
557	CN	Zhongtingxiang	1983	—	—	1.65	—	—	—	—	—	—	—	—	—	—	—	—	—	—	7	[6]
558	CN	Zhongyangcun	1988	—	RF	7.65	—	—	30	150	600	—	—	—	—	—	—	—	—	—	26	[5]
559	CN	Zhongyangcun	1988	—	RF	10	—	—	40	—	200	10000	—	LT	—	—	—	—	—	—	26	[6]
560	CN	Zhoucanping	1981	—	—	10	—	—	10	—	—	—	0.25	—	—	—	—	—	—	—	—	[5]
561	CN	Zhouchangping	1982	—	RF	10	—	—	—	—	—	—	—	LT	—	—	—	—	—	—	—	[6]
562	CN	Zhouqu	1981	DF	RF	—	—	—	—	—	—	—	—	—	—	—	—	—	—	—	—	[5]
563	CN	Zhugen bridge	2008	—	EQ	—	—	3	90	68	500	—	4.5	—	OT	—	—	—	—	—	0	[7][8]
564	CN	Zhujuhe	1970	—	EQ	0.032	—	—	—	—	—	—	1	—	—	—	—	—	—	—	—	[5]
565	CN	Ziniupo	1733	Ava.	EQ	—	—	—	—	—	—	—	—	2	—	—	—	—	—	—	—	[6]
566	CN	Zuishanya	1959	—	RF	0.3	—	—	—	—	—	—	—	—	—	—	—	—	—	—	—	[5]
567	CN	Zuoyituo	Q4	—	RF	20.62	—	—	—	—	—	—	—	LT	—	—	—	—	—	—	—	[6]
568	CO	Chicamocha River	1979	Slump	RF	14.2	VI	—	15	100	600	3000	—	NF	NF	—	—	—	—	—	—	[1]
569	CO	Lagunilla River	1984	RS	RF	—	II	—	25	—	—	1500	1.3	365	OT	—	—	—	—	—	—	[1]
570	CR	Rio Toro River	1992	RS	RF	3	II	3	70–100	75	600	1200	0.5	30	PP	30–50	40–80	—	—	400	—	[17]
571	CZ	Handlovka River	1961	EF	RF	20	I	—	5–8.0	200	300	—	—	—	—	—	—	—	—	—	—	[1]
572	CZ	Unknown	1872	RS	RF	6	II	—	25	—	—	700	—	—	—	—	—	—	—	—	—	[1]
573	EC	Coca River	1987	DF	EQ	—	III	—	15	—	—	—	—	0.1	OT	—	—	—	—	—	—	[1]
574	EC	La Josefina	1993	RA	PW	20–44	II	20–44	100	300	1100	10000	200	33	OT	43	—	—	6	10000	0	[18]
575	EC	Pisque River	1990	ES	HC	1.8	II	1	58	60	450	2600	2.5	24	OT	30	50	—	3.5	700	—	[1][19]
576	EC	Rio Jadan	1993	ES	PW	25	II	—	—	—	—	—	—	—	—	—	—	—	—	—	—	[17]
577	EC	Rio Paute	1993	ES	PW	25	II	25	112	—	800	—	210	33	OT	—	—	—	4.0–6	8250	—	[17]
578	FR	Arc River	1740	DF	RF	—	—	—	—	—	—	—	—	—	—	—	—	—	—	—	—	[1]
579	FR	Arve River	1471	RA	—	—	—	—	—	—	—	—	—	NF	NF	—	—	—	—	—	—	[1]
580	FR	Arve River	'Early Christian era'	RS	—	—	—	—	—	—	—	—	—	—	—	—	—	—	—	—	—	[1]
581	FR	Brevon Torrent	1943	EF, ES	SM	2	—	—	—	—	—	1000	—	—	OT	—	—	—	—	—	—	[1]
582	FR	Cheran River	1000–1100	RS	—	3	—	—	20	—	—	—	—	—	—	—	—	—	—	—	—	[1]
583	FR	Doron River	1450	—	—	0.07	—	—	—	—	—	—	—	—	—	—	—	—	—	—	—	[1]
584	FR	Drome River	1442	RA	EQ	2	V	—	—	600	—	8000	—	NF	NF	—	—	—	—	—	—	[1]
585	FR	Drome River	100–200	RA	—	—	—	—	—	—	—	—	—	—	—	—	—	—	—	—	—	[1]
586	FR	Giffre River	1602	DA	RF, SM	—	—	—	—	—	—	—	—	—	—	—	—	—	—	—	—	[1]
587	FR	Grand-Creux stream	1635	DF	DF	—	—	—	—	—	—	—	—	—	—	—	—	—	—	—	—	[1]
588	FR	Isere River	163	DF	RF	—	—	—	—	—	—	—	—	—	—	—	—	—	—	—	—	[1]
589	FR	Isere River	1219	DF	DF	—	—	—	10	—	—	8000	—	10220	—	—	—	—	—	—	—	[1]
590	FR	Isere River	1732	DF	RF, SM	—	—	—	—	—	—	—	—	—	—	—	—	—	—	—	—	[1]
591	FR	Romanche River	1191	RA, DF	—	—	II	—	10–15	—	—	15000	—	10220	—	—	—	—	—	—	—	[1]
592	FR	Romanche River	1465	RA	RF	—	—	—	15–18	—	—	—	—	—	—	—	—	—	—	—	—	[1]
593	FR	Romanche River	1612	—	—	—	—	—	—	—	—	—	—	4	—	—	—	—	—	—	—	[1]

(*Continued*)

Table B.1 (*Continued*)

No	Country[1]	Name	Date of formation [2]	Landslide Type [3]	Trigger [4]	Landslide volume (×10⁶ m³)	Dam Type [5]	Dam volume (×10⁶ m³)	Dam height (m)	Dam length (m)	Dam width (m)	Lake length (m)	Lake volume (×10⁶ m³)	Time to failure (day) [6]	Failure mech.[7]	Breach depth (m)	Breach top width (m)	Breach bottom width (m)	Breaching time (hour)	Peak discharge (m³/s)	Loss of life	Reference (Below the table)
594	FR	St. Claude torrent	1877	RA	SM	3	—	—	—	—	500	—	—	1825	—	—	—	—	—	—	—	[1]
595	GT	Los Chocoyos River	1976	DA	EQ	0.75–1	III	—	20–50	300–400	800	300	—	NF	NF	—	—	—	—	—	—	[1]
596	GT	Pixcaya River	1976	DA	EQ	6	II	—	20	200	600	800	—	SD	—	—	—	—	—	—	—	[1]
597	GT	Quemaya River	1976	Slump	EQ	2	II	—	50	—	400	250	—	—	OT	—	—	—	—	—	A few	[1]
598	GT	Rio La Lima	1998	—	—	—	—	500	20	—	—	—	—	—	—	—	—	—	—	—	—	[2]
599	GT	Teculcheya River	1976	DA, DS	EQ	3.5	III	—	15	—	1200	1000	—	143	—	—	—	—	—	—	—	[1]
600	HU	Szohony Stream	1863–1910–1929	Slide	HC	5	II	—	90	100	400	650	0.15	NF	NF	—	—	—	—	—	—	[1]
601	ID	Baliem River	1989	—	EQ	—	—	—	—	—	—	—	—	—	—	—	—	—	—	—	—	[1]
602	ID	Solo River	1981	DF	EQ	0.1	III	—	10	30	200	800	0.1	16	—	—	—	—	—	—	—	[1]
603	ID	Unknown	1006	RS	—	—	—	—	—	—	—	—	—	—	—	—	—	—	—	—	—	[1]
604	IE	Clare River	1745	Slide	RF	—	—	—	—	—	—	—	—	—	—	—	—	—	—	—	—	[1]
605	IL	Jordan River	1267	—	—	—	—	—	—	—	—	—	—	0.2	—	—	—	—	—	—	—	[1]
606	IL	Jordan River	1546	—	EQ	—	—	—	—	—	—	—	—	2	—	—	—	—	—	—	—	[1]
607	IL	Jordan River	1906	—	EQ	—	—	—	—	—	—	—	—	1	—	—	—	—	—	—	—	[1]
608	IL	Jordan River	1927	—	EQ	—	—	—	—	—	—	—	—	<1	—	—	—	—	—	—	—	[1]
609	IL	Jordan River	1250BC	—	EQ	—	—	—	—	—	—	—	—	—	—	—	—	—	—	—	—	[1]
610	IN	Alaknanda River	1970	RS, DF	RF	—	—	—	60	—	—	—	—	—	—	—	—	—	—	—	—	[1]
611	IN	Alaknanda River		RF	—	—	—	—	—	—	—	—	—	—	—	—	—	—	—	—	—	[1]
612	IN	Bhagirathi River	1978	DF	—	—	—	—	30	—	—	—	—	—	—	—	—	—	—	—	—	[1]
613	IN	Birehi Ganga River	1893	RS	RF	—	III	286	274	760	2750	3930	460	338	OT	97.5	—	—	1	56650	—	[1]
614	IN	Delei River	1948	—	RF	—	—	—	46	—	—	—	—	2.5	OT	—	—	—	—	—	—	[1]
615	IN	Dihang River	1950	—	EQ	—	—	—	—	—	—	—	—	0.08	—	—	—	—	—	—	—	[1]
616	IN	Luhit River tributary	1950	RA, DA	EQ	—	—	—	—	—	—	—	—	7	—	—	—	—	—	—	—	[1]
617	IN	Para Chu River	1975	—	EQ	—	—	—	20	—	—	—	—	9	—	—	—	—	—	—	—	[1]
618	IN	Pauri Ganga River	1890's	RS	HC	—	—	—	350	—	—	—	—	NF	NF	—	—	—	—	—	—	[1]
619	IN	Rishiganga River	1968	RS,RA	—	—	II	0.43	40	117	183	1000	—	900	—	—	—	—	—	—	—	[1][2]
620	IN	Satluj river	2000	—	—	—	—	—	—	—	—	—	—	—	OT	—	—	—	—	1800	—	[20]
621	IN	Satluj river	2005	—	—	—	—	—	—	—	—	2100	64	—	—	—	—	—	—	2000	—	[20]
622	IN	Scob River	1897	—	EQ	—	—	—	—	—	—	—	—	87	—	—	—	—	—	—	—	[1]
623	IN	Subansiri River	1950	—	EQ	—	—	—	—	—	—	—	—	4	—	—	—	—	—	—	—	[1]
624	IN	Teesta River	—	—	RF	—	—	—	—	—	—	—	—	—	—	—	—	—	—	—	—	[1]
625	IN	Tiding River	1950	—	EQ	—	—	—	25	—	—	—	—	—	—	—	—	—	—	—	—	[1]
626	IN	Tirthan River	1905	RA	EQ	—	—	—	20–25	—	—	1200	—	450	OT	—	—	—	—	—	—	[1]
627	IT	Adda River	1807	RS	RF	—	—	—	—	—	—	—	—	180	OT	—	—	—	—	—	—	[1]

628	IT	Adda river	1987	DA	RF, HC	35	III	—	33	2700	550	2900	22	NF	NF	—	—	—	—	—	—	[1][2]
629	IT	Alba River	1896	RS	—	—	—	—	—	—	—	—	—	1460	—	—	—	—	—	—	—	[1]
630	IT	Ambria Stream	1888	RS	RF	—	—	—	15	—	—	400	—	—	—	—	—	—	—	—	—	[1]
631	IT	Arno River	1898	—	—	—	—	0.4	15	—	—	—	—	—	—	—	—	—	—	—	—	[2]
632	IT	Aurina Torrent	1867	DF	RF	—	—	—	—	—	—	—	—	—	—	—	—	—	—	—	—	[1]
633	IT	Boesimo Creek	1690	Slide	—	—	—	—	—	—	—	—	—	—	—	—	—	—	—	—	—	[21]
634	IT	Boite Torrent	1814	RA	—	20	—	—	30	750	750	—	—	—	—	—	—	—	—	—	—	[1]
635	IT	Buonamico River	1973	RS, RA	RF	21	II	21	90	400	700	1200	7.5	31	PP	50	—	—	—	—	—	[1][2]
636	IT	Buonamico River	1973	RA	RF	—	—	—	20	—	—	—	—	0.5	OT	—	—	—	—	—	—	[1]
637	IT	Buthier Torrent	1952	DF	SM	—	—	—	—	—	—	—	—	4	—	—	—	—	—	—	—	[1][2]
638	IT	Campigno Creek	1899	RS	—	—	—	0.93	15	—	—	—	—	—	—	—	—	—	—	—	—	[2]
639	IT	Cava S. Calogero	—	RS	—	1.5	II	1.5	<40	400	600	—	—	—	—	—	—	—	—	—	—	[22]
640	IT	Cava S. Giuseppe	—	RS	—	1.8	IV	1.8	30–40	250	500	—	—	—	—	—	—	—	—	—	—	[22]
641	IT	Codera stream	1988	DF	RF	—	—	—	3	—	—	200	—	NF	NF	—	—	—	—	—	—	[1]
642	IT	Cordevole Torrent	1771	RA, RS	—	20	II	—	80	500	1750	4000	—	NF	NF	—	—	—	—	—	—	[1][2]
643	IT	Corsanico creek	1717	DS	—	—	II	—	—	—	—	—	—	<1	OT	—	—	—	—	—	10	[21]
644	IT	Diveria River	1958	RA, DA	RF	—	—	—	—	—	—	—	—	1	—	—	—	—	—	—	—	[1][2]
645	IT	Dora di Veny	1920	RF, RA	—	4.5	—	—	—	—	—	—	—	—	—	—	—	—	—	—	—	[1][2]
646	IT	Dragone River	1954	—	—	—	—	7	25	—	—	—	—	—	—	—	—	—	—	—	—	[2]
647	IT	Duverso Torrent	1783	ES	EQ	—	II	—	26	170	—	830	—	NF	NF	—	—	—	—	—	—	[1]
648	IT	Duverso Torrent	1783	ES	EQ	—	IV	—	12	100	—	470	—	NF	NF	—	—	—	—	—	—	[1]
649	IT	Fiume Anapo	1693	RF	—	0.05	II	—	—	150	200	—	—	—	OT	—	—	—	—	—	—	[22]
650	IT	Gader Torrent	1821	EF	SM, RF	—	—	—	—	—	—	1000	—	2200	OT	—	—	—	—	—	—	[1]
651	IT	Isarco River	1891	DF	RF	0.5	—	—	—	—	—	1200	—	—	—	—	—	—	—	—	—	[1]
652	IT	Isarco River	1921	DF	RF	—	—	1	7	—	—	—	—	SD	—	—	—	—	—	—	—	[1][2]
653	IT	Lago Stream	1783	ES	EQ	—	IV	—	41	—	—	1300	—	—	—	—	—	—	—	—	—	[1]
654	IT	Lago Stream	1783	ES	EQ	—	II	—	52	—	—	1250	—	—	OT	—	—	—	—	—	—	[1]
655	IT	Lake Passo	1783	ES	EQ	—	II	—	14	—	—	380	—	—	—	—	—	—	—	—	—	[1]
656	IT	Lake Tofilo	1783	ES	EQ	—	II	—	18	—	—	620	—	—	—	—	—	—	—	—	—	[1]
657	IT	Leo River	1590	—	—	—	—	12.6	40	—	—	—	—	—	—	—	—	—	—	—	—	[2]
658	IT	Lima River	1814	—	—	—	—	20	15	—	—	—	—	—	—	—	—	—	—	—	—	[2]
659	IT	Lindo River	1783	ES	EQ	—	II	—	4	—	—	60	—	—	OT	—	—	—	—	—	—	[1]
660	IT	Lindo River	1783	ES	EQ	—	IV	—	32	80	—	1380	—	NF	NF	—	—	—	—	—	—	[1]
661	IT	Marecchia River	1945	—	—	—	—	1.5	20	—	—	—	—	—	—	—	—	—	—	—	—	[2]
662	IT	Marro River	1783	ES	EQ	—	II	—	34	—	—	1000	—	—	—	—	—	—	—	—	—	[1]
663	IT	Mera River	1618	RA, DA	RF	3.0–4	—	—	—	—	—	—	—	NF	NF	—	—	—	—	—	—	[1]
664	IT	Nera River	1906	DF	RF	—	—	—	—	—	—	—	—	SD	OT	—	—	—	—	—	—	[1]
665	IT	Ospitale Creek	1952	—	—	—	—	0.08	8	—	—	—	—	—	—	—	—	—	—	—	—	[2]
666	IT	Ovesca Torrent	1642	RA	—	12	—	12	45	650	1500	800	—	NF	NF	—	—	—	—	—	—	[1][2]
667	IT	Parma River	1902	—	—	—	—	8.37	30	—	—	—	—	—	—	—	—	—	—	—	—	[2]
668	IT	Passirio River	1404	RS	—	—	—	4.5	50	—	—	1000	—	5475	—	—	—	—	—	—	400+	[1][2]
669	IT	Pieve Santo Stefano	1855	RS	—	—	—	—	—	—	—	—	—	—	—	—	—	—	—	—	—	[21]

(*Continued*)

Table B.1 *(Continued)*

No	Country[1]	Name	Date of formation[2]	Landslide Type[3]	Trigger[4]	Landslide volume ($\times10^6 m^3$)	Dam Type[5]	Dam volume ($\times10^6 m^3$)	Dam height (m)	Dam length (m)	Dam width (m)	Lake length (m)	Lake volume ($\times10^6 m^3$)	Time to failure (day)[6]	Failure mech.[7]	Breach depth (m)	Breach top width (m)	Breach bottom width (m)	Breaching time (hour)	Peak discharge (m^3/s)	Loss of life	Reference (Below the table)
670	IT	Poschiavo landslide	1987	—	—	—	—	81.56	100	—	—	—	—	—	—	—	—	—	—	—	—	[2]
671	IT	Quarto di Savio	1812	RS, DS	—	16	—	—	—	—	—	—	—	—	—	—	—	—	—	—	—	[2]
672	IT	Reno River	1996	—	—	0.7	—	—	—	—	—	—	—	—	—	—	—	—	—	—	—	[21]
673	IT	Reno River	1700s	—	—	—	—	4.3	25	—	—	—	—	—	—	—	—	—	—	—	—	[2]
674	IT	Riganati Stream	1783	ES	EQ	—	II	—	31	—	—	820	—	—	—	—	—	—	—	—	—	[1]
675	IT	Rio Amerillo	—	RS	—	34	II	70	—	1000	1200	—	4.4	—	—	—	—	—	—	—	—	[22]
676	IT	Rio Boesimo Creek	1693	—	—	—	—	1.5	20	—	—	—	—	—	—	—	—	—	—	—	—	[2]
677	IT	Rotbach Torrent	1878	RS, DA	RF	14	—	—	—	—	—	—	—	1	OT	—	—	—	—	—	—	[1]
678	IT	Sambro River	1762	—	—	—	—	3.5	35	—	—	—	—	—	—	—	—	—	—	—	—	[2]
679	IT	Sambro River	1994	—	—	5	2	1.13	25	—	—	—	0.33	—	—	—	—	—	—	—	—	[2]
680	IT	Sarca River tributary	200–1000	RA	—	375	—	—	130	1500	3250	4500	—	NF	NF	—	—	—	—	—	—	[1]
681	IT	Savena River	1951	—	—	—	—	4	45	—	—	—	—	—	—	—	—	—	—	—	—	[2]
682	IT	Savena River	End'800	—	—	—	—	0.4	15	—	—	—	—	—	—	—	—	—	—	—	—	[2]
683	IT	Savio River	1812	—	—	—	—	16	70	—	—	—	—	—	—	—	—	—	—	—	—	[2]
684	IT	Savio River	1855	—	—	—	—	1.8	15	—	—	—	—	—	—	—	—	—	—	—	—	[2]
685	IT	Scanno lake	100s	—	—	—	—	112	33.1	—	—	—	—	—	—	—	—	—	—	—	—	[2]
686	IT	Scapriano landslide	1927	—	—	—	—	0.65	15	—	—	—	—	—	—	—	—	—	—	—	—	[2]
687	IT	Scoltenna River	1879	—	—	—	—	8	30	—	—	—	—	—	—	—	—	—	—	—	—	[2]
688	IT	Secchia River	1960	—	—	—	—	4.38	33	—	—	—	26	—	—	—	—	—	—	—	—	[2]
689	IT	Sernio landslide	1807	—	—	—	—	2.5	43	—	—	—	—	—	—	—	—	—	—	—	—	[2]
690	IT	Serra Torrent	1783	ES	EQ	—	II	—	17	100	—	535	—	—	—	—	—	—	—	—	—	[1]
691	IT	Tagliamento River	1692	RA	RF	30	—	—	80	625	1100	7000	—	50	OT	—	—	—	—	—	—	[1]
692	IT	Tevere River	1855	—	—	—	—	4.5	25	—	—	—	—	—	—	—	—	—	—	—	—	[2]
693	IT	Torrente Pisciarello	1693	RS	EQ	2	II	—	60	370	250	—	—	—	—	—	—	—	—	—	—	[22]
694	IT	Tramazzo Creek	1895	—	—	—	—	30	15	—	—	—	—	—	—	—	—	—	—	—	—	[2]
695	IT	Tributary of Fiume Irminio	3000BC	RS	—	14	II	—	60	550	1100	—	2.2	—	—	—	—	—	—	—	—	[22]
696	IT	Tricucio River	1783	ES	EQ	—	II	—	19	—	—	1065	—	—	OT	—	—	—	—	—	—	[1]
697	IT	Unnamed canyon	—	RS	—	1	IV	1	30	300	400	—	—	—	—	—	—	—	—	—	—	[22]
698	IT	Vaiont Torrent	1963	RS	RF	250	II	—	275	1000	1800	5000	—	—	OT	—	—	—	—	—	—	[1][2]
699	IT	Vallucciole Creek	1992	—	—	—	—	0.02	12	—	—	—	—	—	PP	—	—	—	—	—	—	[2][21]
700	IT	Vanoi Creek	1923	—	—	—	—	10	40	—	—	—	—	—	—	—	—	—	—	—	—	[2]
701	IT	Vanoi Torrent	1823	DF	RF	—	—	—	—	—	—	—	—	0.5	—	—	—	—	—	—	—	[1]
702	IT	Vanoi Torrent	1825	DF	—	—	—	—	—	—	—	—	—	150	—	—	—	—	—	—	52	[1]
703	IT	Venola Creek	1996	—	—	—	—	0.12	15	—	—	—	—	—	—	—	—	—	—	—	—	[2]
704	JM	Yallahs River	1692	RS	RF	66	II	—	—	—	—	—	—	—	OT	—	—	—	—	—	—	[1]

705	JP	Abe River	1702	—	EQ	120	III	4.88	30	500	650	—	4.7	NF	NF	—	—	—	—	—	— [1][10]
706	JP	Agatsuma River	1783	DA	VE	—	—	—	60	—	—	—	—	—	—	—	—	—	—	—	Many [1]
707	JP	Arida River	1953	RS	RF	0.64	II	0.18	10	80	150	300	0.047	0.01	OT	—	—	—	—	890*	— [1][10]
708	JP	Arida River	1953	—	RF	1.4	—	0.3	20	170	250	—	—	0.01	—	—	—	—	—	—	— [1][10]
709	JP	Arida River	1953	RS, ES	RF	5.2	II	2.6	60	300	500	5000	17	67	OT	—	—	—	—	750*	— [1][10]
710	JP	Asahi River	1889	RS, ES	RF	8.8	II	0.45	25	160	300	2700	0.92	0.2	—	—	—	—	—	—	— [1][10]
711	JP	Azusa River	1915	DF	VE	1.7	—	0.9	4.5	300	600	2000	0.53	NF	NF	—	—	—	—	—	— [1][10]
712	JP	Azusa River	1926	DF	VE	—	—	2	10	600	330	2300	1.2	NF	NF	—	—	—	—	—	— [1][10]
713	JP	Banjo River	1943	RS, ES	RF	1.5	II	1.5	80	400	250	2500	14	—	—	—	—	—	—	—	— [1][10]
714	JP	Chubetsu River	1980	DA	RF	0.02	II	—	3	—	—	—	—	—	—	—	—	—	—	—	— [1]
715	JP	Haya River	1910	DF	RF	—	—	—	—	—	—	—	—	—	—	—	—	—	—	—	— [1]
716	JP	Hibara River	1888	DA	VE	1500	III	—	25	800	—	9000	150	—	—	—	—	—	—	—	— [1][10]
717	JP	Higashi Takezawa	2004	Slide	EQ	1	II	1	24	350	260	—	—	—	—	—	—	—	—	—	— [23]
718	JP	Hime River	1911	DF	RF	20	—	1.9	60	250	500	4000	16	4	—	—	—	—	—	1800*	— [1][10]
719	JP	Hime River	1971	ES	RF	0.3	I	—	—	80	80	—	—	NF	NF	—	—	—	—	—	— [1]
720	JP	Hiramaru River	1962	Mud F	SM	—	—	—	—	—	—	—	—	—	—	—	—	—	—	—	— [1]
721	JP	Hiramaru River	1970	DF	SM	0.27	—	—	7	—	—	200	0.0028	—	—	—	—	—	—	—	— [1]
722	JP	Iketsu River	1889	—	RF	5.4	II	3.4	140	400	180	4000	26	6	—	—	—	—	—	480*	— [1][10]
723	JP	Imanishi River	1889	RS, ES	RF	1.5	II	1.1	60	250	250	2500	6.4	1	—	—	—	—	—	—	— [1][10]
724	JP	Imanishi River	1889	RS, ES	RF	1.4	II	1.1	75	350	125	1500	9	1.5	—	—	—	—	—	—	— [1][10]
725	JP	Ishikari River tributary	1969	ES	SM	2	II	—	15	—	500	—	—	—	—	—	—	—	—	—	— [1]
726	JP	Kaifu River	1892	RS, ES	—	2	II	2	45	250	350	5000	14	1157	—	—	—	—	—	—	— [1][10]
727	JP	Kaminirau River	1788	—	RF	2	II	2	50	250	500	1500	2.2	579	—	—	—	—	—	440*	— [1][10]
728	JP	Kano River	1889	—	RF	3.6	II	0.094	15	130	130	2500	1.3	0.02	—	—	—	—	—	1600*	— [1][10]
729	JP	Kano River	1889	DF	RF	5.2	II	0.1	20	200	120	1600	0.6	1	—	—	—	—	—	1300*	— [1][10]
730	JP	Kano River	1889	—	RF	1.3	II	0.15	20	100	100	1800	1	—	—	—	—	—	—	—	— [1][10]
731	JP	Kano River	1889	RS, ES	RF	1.5	II	0.44	25	180	200	2000	1.8	—	—	—	—	—	—	—	— [1][10]
732	JP	Kashiwa River	1889	RS, ES	RF	4.9	II	2.6	70	200	450	600	1.7	22	—	—	—	—	—	—	— [1][10]
733	JP	Kawarabitsu River	1889	RS, ES	RF	26	II	13	80	300	700	9000	40	17	OT	—	—	—	—	2000*	— [1][10]
734	JP	Kawarada river	2007	DF	EQ	—	—	—	40	30	—	—	—	—	—	—	—	—	—	—	— [25]
735	JP	Kose River	1984	Slump	SM	5.6	II	—	20	—	90	500	—	NF	NF	—	—	—	—	—	— [1]
736	JP	Koshibu River	1961	RS, ES, DA	RF	3	II	2.4	6	500	800	—	0.4	<10	OT	—	—	—	—	—	— [1][10]
737	JP	Ma River	1858	DA	EQ	150	III	12	110	600	200	2000	2.6	59	—	—	—	—	—	—	140 [1][10]
738	JP	Matsu River	1891	RS, EA	RF	3.2	II	3.2	55	500	230	500	3.1	NF	NF	—	—	—	—	—	— [1][10]
739	JP	Nagasawa River	1978	RS, ES	SM	0.3	II	—	10	50	80	300	0.03	—	—	—	—	—	—	—	— [1]
740	JP	Naka River	1893	RS, ES	RF	4	II	3.3	80	250	330	10000	75	3	—	—	—	—	—	5600*	— [1][10]
741	JP	Nakatsu River	1888	DA	VE	1500	III	—	34	550	—	4000	44	84	—	—	—	—	—	—	— [1][10]
742	JP	Nakaya River	1953	RS	RF	0.46	II	0.4	40	100	200	400	0.27	68	—	—	—	—	—	—	— [1][10]
743	JP	Naruse River	1984	RS	RF	35	II	—	—	—	950	—	—	—	—	—	—	—	—	—	— [1]
744	JP	Nishi River	1889	—	RF	4.4	—	0.6	20	200	250	2500	1.3	0.4	—	—	—	—	—	980*	— [1][10]

(Continued)

Table B.1 (*Continued*)

No	Country[(1)]	Name	Date of formation [(2)]	Landslide Type [(3)]	Trigger [(4)]	Landslide volume (×10^6 m^3)	Dam Type [(5)]	Dam volume (×10^6 m^3)	Dam height (m)	Dam length (m)	Dam width (m)	Lake length (m)	Lake volume (×10^6 m^3)	Time to failure (day) [(6)]	Failure mech.[(7)]	Breach depth (m)	Breach top width (m)	Breach bottom width (m)	Breaching time (hour)	Peak discharge (m^3/s)	Loss of life	Reference (Below the table)
745	JP	Nishi River	1889	—	RF	0.63	II	0.63	20	120	120	1000	0.4	0.4	—	—	—	—	—	1100*	—	[1][10]
746	JP	Nishi River	1889	RS, ES	RF	2.7	II	0.63	25	200	250	3500	1.8	0.4	—	—	—	—	—	1200*	—	[1][10]
747	JP	Nishi River	1889	RS, ES	RF	4.4	II	0.93	25	130	250	250	0.11	NF	NF	—	—	—	—	—	—	[1][10]
748	JP	Niu River	1982	RS, DS	RF	0.61	II	0.18	15	50	180	3000	1.3	NF	NF	—	—	—	—	—	—	[1][10]
749	JP	Odokoro River	1967	DF	—	3.6	—	—	30	150	200	900	0.9	—	—	—	—	—	—	—	—	[1]
750	JP	Oi River	1889	RS, ES	RF	3.4	II	2.6	100	400	150	300	2.3	10	—	—	—	—	—	—	—	[1][10]
751	JP	Ojika River	1683	RS	EQ	3.8	II	3.8	70	400	400	—	64	14600	OT	—	—	—	—	620*	1005	[1][10]
752	JP	Ono River	1888	DA	VE	1500	III	—	18	500	—	4000	14	272	—	—	—	—	—	—	—	[1][10]
753	JP	Oshiro River	1586	DF	EQ	3	II	1.2	60	300	300	—	6.4	—	—	—	—	—	—	—	—	[1][10]
754	JP	Oshiro River	1586	DF	EQ	2.4	II	1	60	250	250	—	6	—	—	—	—	—	—	—	—	[1][10]
755	JP	Osusawa River	1888	DA	VE	1500	III	—	18	—	—	5000	—	NF	NF	—	—	—	—	—	—	[1]
756	JP	Otaki River	1984	DA	EQ, RF	36	III	12.5	40	250	2500	1000	—	NF	NF	—	—	—	—	—	—	[1]
757	JP	Oya River	1949	RS, ES	EQ	0.009	II	0.0045	10	50	100	—	0.0033	NF	NF	—	—	—	—	—	—	[1][10]
758	JP	Sai River	1847	RS, ES	EQ	84	II	21	65–100	1000	650	23000	350	19	OT	—	—	—	—	3700*	100+	[1][10]
759	JP	Sakauchi River	1895	RF	RF	1.5	II	0.96	38	110	350	2000	2	6	OT	—	—	—	—	—	—	[1][10]
760	JP	Shinano River	1984	Slump	RF	0.5	II	—	15	—	230	—	—	—	—	—	—	—	—	—	—	[1]
761	JP	Shinsei Lake	1923	ES	EQ	0.23	II	0.18	10	100	200	200	0.037	NF	NF	—	—	—	—	—	—	[1][10]
762	JP	Shiratani River	1889	RS, ES	RF	20	II	10	190	600	500	—	38	5	—	—	—	—	—	580*	—	[1][10]
763	JP	Shiratani River	1953	RS	RF	0.25	II	0.09	25	100	100	200	0.06	NF	NF	—	—	—	—	—	—	[1][10]
764	JP	Shiratani River	1965	RS	RF	1.8	II	1.4	25	200	250	650	1.3	NF	NF	—	—	—	—	—	—	[1]
765	JP	Sho River	1586	RS, AS	EQ	25	II	19	100	900	600	12000	150	20	OT	—	—	—	—	1900*	—	[1][10]
766	JP	Susobana River	1847	RS, ES	EQ	1.2	II	1.2	54	250	300	2100	16	110	—	—	—	—	—	510*	—	[1][10]
767	JP	Tajiri River	1978	DS, DA	EQ	0.05	II	—	10	20	125	—	—	NF	NF	—	—	—	—	—	—	[1]
768	JP	Terano landslide dam	2004	ES	EQ	1	II	1	25	60–80	200-300	—	—	—	—	—	—	—	—	—	—	[23]
769	JP	Tokonami River	1961	RS, ES	RF	0.3	II	—	20	50	170	700	0.35	30	OT	—	—	—	—	—	—	[1]
770	JP	Totsu River	1889	DA	RF	23	II	0.036	18	100	450	2000	0.78	0.04	—	—	—	—	—	3400*	—	[1][10]
771	JP	Totsu River	1889	RS, ES	RF	0.11	II	0.073	7	100	250	3000	0.65	0.08	—	—	—	—	—	6900*	—	[1][10]
772	JP	Totsu River	1889	RS, ES	RF	36	II	0.15	10	150	150	1000	0.56	0.1	—	—	—	—	—	—	—	[1][10]
773	JP	Totsu River	1889	RS, ES	RF	1.6	II	0.23	10	130	380	2800	0.93	0.2	OT	—	—	—	—	3500*	—	[1][10]
774	JP	Totsu River	1889	RS, ES	RF	50	II	2.5	80	100	350	6000	17	0.3	—	—	—	—	—	2400*	—	[1][10]
775	JP	Totsu River	1889	RS, ES	RF	3.7	II	3.1	110	200	690	5000	42	0.7	—	—	—	—	—	4800*	—	[1][10]
776	JP	Totsu River	1889	RS, ES	RF	2.5	II	0.85	50	180	300	2000	1.6	0.7	OT	—	—	—	—	—	—	[1][10]
777	JP	Totsu River	1889	—	RF	—	—	—	6	—	70	1700	0.26	4	—	—	—	—	—	—	—	[1][10]
778	JP	Totsu River	1889	RS, RA	RF	20	II	1.7	28	250	500	3000	3.2	—	—	—	—	—	—	—	—	[1][10]
779	JP	Totsu River	1889	RS, ES	RF	1.7	II	0.28	10	160	220	1600	0.52	—	—	—	—	—	—	—	—	[1][10]
780	JP	Totsu River	1889	DS, DA	RF	5.6	II	—	12	200	250	2300	0.72	—	—	—	—	—	—	—	—	[1][10]

781	JP	Unknown river	1662	—	—	—	—	—	—	—	—	—	—	—	—	—	—	—	—	—	—	[1]
782	JP	Yamato River	1889	RS, ES	RF	6.6	II	4.2	80	300	350	—	12	22	OT	—	—	—	—	—	—	[1][10]
783	JP	Yamato River	1931-1932	RS, DS	RF	22	VI	0.11	20–36	50	170	—	10	NF	NF	—	—	—	—	—	—	[1]
784	JP	Yanagikubo River	1847	RS, ES	EQ	1.5	II	0.65	35	150	250	500	1.4	NF	NF	—	—	—	—	—	—	[1][10]
785	JP	Yu River	1858	DA	EQ	150	—	0.4	125	600	700	1000	27	12	—	—	—	—	—	—	140	[1][10]
786	JP	宝永.大谷崩れ(西日影沢)	1707	DF	EQ, RF	—	—	—	—		200	—	0.081	—	—	—	—	—	—	—	—	[10]
787	JP	宝永.大谷崩れ(タチ沢)	1707	DF	EQ, RF	—	—	—	—		120	—	0.044	—	—	—	—	—	—	—	—	[10]
788	JP	長野縣鬼無里村	1997	DF	RF	0.93	—	0.08	40	415	50	—	0.21	NF	NF	—	—	—	—	—	—	[10]
789	JP	長野縣西部地震.御嶽崩れ	1984	DF	EQ	34	—	26	40	3300	280	—	3.7	NF	NF	—	—	—	—	—	—	[10]
790	JP	姫川.小土山	1971	Ava.,	—	0.2	—	—	—	150	60	—	—	NF	NF	—	—	—	—	—	—	[10]
791	JP	姫川.真那板山	1502	Ava.,	—	50	—	50	150	500	200	—	120	—	—	—	—	—	—	—	—	[10]
792	JP	濃尾地震.根尾西谷川	1891	Ava.,	EQ	1.5	—	1.8	60	235	250	—	8.1	—	—	—	—	—	—	—	—	[10]
793	JP	濃尾地震.水鳥	1891	—	EQ	0.088	—	—	—	—	—	—	1.4	—	—	—	—	—	—	—	—	[10]
794	JP	濃尾地震後.德山白谷	1965	Ava.,	RF	1.8	—	0.98	65	150	260	—	2	—	—	—	—	—	—	—	—	[10]
795	JP	濃尾地震後.越山谷	1965	DF	RF	0.98	—	0.63	10	280	450	—	0.29	—	—	—	—	—	—	—	—	[10]
796	JP	琵琶湖西岸.(町居崩れ)	1662	Ava.,	EQ	24	—	24	110	350	362	—	5.9	14	—	—	—	—	—	—	—	[10]
797	JP	善光寺.當信川	1847	Ava.,	EQ	6	—	4	60	250	400	—	8.6	—	—	—	—	—	—	—	—	[10]
798	JP	天正.(庄川下流)	1586	Ava.,	EQ	—	—	30	100	400	750	—	140	20	—	—	—	—	—	—	—	[10]
799	JP	新潟縣上川村	2000	Ava.,	RF	0.036	—	0.022	20	24	90	—	0.076	NF	NF	—	—	—	—	—	—	[10]
800	JP	有田川.高野谷	1953	—	RF	—	—	0.017	10	80	30	—	0.03	NF	NF	—	—	—	—	—	—	[10]
801	JP	有田川.箕谷	1953	—	RF	—	—	—	3		50	—	0.04	—	—	—	—	—	—	—	—	[10]
802	JP	有田川.梁瀬一ノ谷	1953	Ava.,	RF	0.045	—	0.024	5	120	40	—	0.015	—	—	—	—	—	—	—	—	[10]
803	JP	神戸市.清水	1985	Ava.,	EQ	0.018	—	0.0012	6	30	7	—	0.0008	—	—	—	—	—	—	—	—	[10]
804	JP	福島.半田新沼	1901	Ava.,	RF	13	—	—	—	—	—	—	—	3241	—	—	—	—	—	—	—	[10]
805	NO	Gaula River	1345	LS	—	55	III	—	—	7000	1200	5000	—	—	—	—	—	—	—	—	—	[1]
806	NO	Ulvadal river	1960	DS	RF	—	II	—	2–3.0	—	—	—	—	NF	NF	—	—	—	—	—	—	[1]
807	NO	Vaerdalselven River	1863	LS	—	—	—	—	—	—	—	—	—	0.03	—	—	—	—	—	—	—	[1]
808	NO	Vaerdalselven River	1893	LS	—	55	III	—	—	7000	1200	5000	—	—	—	—	—	—	—	—	111	[1]
809	NP	Bhairab Kunda Stream	1996	DF	RF	0.18	II	—	—	—	300	—	—	—	—	—	—	—	—	—	—	[26]
810	NP	Chirling Khola River	1978	RS	RF	0.02	II	—	6	10	60	100	0.004	350	—	—	—	—	—	—	—	[1]
811	NP	Dhankuta Khola River	1974	RS	RF	—	II	—	—	—	—	—	—	—	—	—	—	—	—	—	—	[1]

(Continued)

Table B.1 *(Continued)*

No	Country[(1)]	Name	Date of formation [(2)]	Landslide Type [(3)]	Trigger [(4)]	Landslide volume ($\times 10^6$ m^3)	Dam Type [(5)]	Dam volume ($\times 10^6$ m^3)	Dam height (m)	Dam length (m)	Dam width (m)	Lake length (m)	Lake volume ($\times 10^6$ m^3)	Time to failure (day) [(6)]	Failure mech.[(7)]	Breach depth (m)	Breach top width (m)	Breach bottom width (m)	Breaching time (hour)	Peak discharge (m^3/s)	Loss of life	Reference (Below the table)
812	NP	Kaligandaki River	1936	RS	RF	—	I	—	—	—	—	—	—	1.25	—	—	—	—	—	—	3	[1]
813	NP	Labu Khola	1968	RS	RF	>1	—	—	60	90	150	—	—	3.5	PP,OT	—	—	—	—	—	—	[1]
814	NP	Saptagandaki River	1930	RS	RF	—	—	—	—	—	—	—	—	—	—	—	—	—	—	—	—	[1]
815	NP	Sunkoshi River	1984	RS	RF	12	I	—	8	300	300	1000	2.4	0.08	OT	—	—	—	—	—	—	[1]
816	NP	Sunkoshi River	1984	RS	RF	3	I	—	8	300	200	1000	1.5	0.08	—	—	—	—	—	—	—	[1]
817	NP	Tadi Khola River	1927	RS	RF	—	II	—	—	—	—	250	—	0.2	—	—	—	—	—	—	3	[1]
818	NP	Trisuli River	1985	—	—	—	—	—	—	—	—	2500	—	0.25–0.5	OT	—	—	—	5	2010	—	[1]
819	NP	Yangma Khola river	1980	RS, DS	RF	—	II	—	—	—	—	—	—	<1	—	—	—	—	—	—	—	[1]
820	NZ	Buller River	1908	—	—	—	—	—	—	—	—	—	—	—	—	—	—	—	—	—	—	[1]
821	NZ	Buller River	1968	DA	EQ	4.3	II	—	12	100	250	6000	—	0.9	—	—	—	—	—	—	—	[1]
822	NZ	Buller River	1971	DA	RF	—	III	—	—	100	350	—	—	—	—	—	—	—	—	—	—	[1]
823	NZ	Coppermine Creek	1976	Slump	RF	0.035	III	—	10	—	—	—	—	—	—	—	—	—	—	—	—	[1]
824	NZ	Drysdale Creek	1913	Slide	EQ	—	—	—	—	—	—	—	—	—	—	—	—	—	—	—	—	[1]
825	NZ	Falls Creek	1929	—	EQ	—	—	—	—	—	—	—	—	—	—	—	—	—	—	—	—	[1]
826	NZ	Glasseye Creek	1929	—	EQ	—	—	—	—	—	—	—	—	—	—	—	—	—	—	—	—	[1]
827	NZ	Hangaroa River	1988	RS	RF	—	II	—	—	—	—	6000	—	—	—	—	—	—	—	—	—	[1]
828	NZ	Lake Marina	1929	—	EQ	—	—	—	—	—	—	—	—	—	—	—	—	—	—	—	—	[1]
829	NZ	Lower Lindsay Lake	1929	—	EQ	—	—	—	—	—	—	—	—	—	—	—	—	—	—	—	—	[1]
830	NZ	Maruia River	1929	RS	EQ	—	II	—	—	—	—	—	—	2	—	—	—	—	—	—	—	[1]
831	NZ	Matakitaki River	1929	RS	EQ	—	II	—	25	—	—	5000	—	—	—	—	—	—	—	—	—	[1]
832	NZ	Matiri River	1929	RS	EQ	—	—	—	—	—	—	—	—	NF	NF	—	—	—	—	—	—	[1]
833	NZ	Matiri River	1929	RS	EQ	—	—	—	—	—	—	—	—	NF	NF	—	—	—	—	—	—	[1]
834	NZ	Mokihinui River	1929	RS	EQ	—	II	—	23	100	—	11000	—	17	—	7.6	—	—	—	—	—	[1]
835	NZ	Moonstone Lake	1929	—	EQ	—	—	—	—	—	—	—	—	—	—	—	—	—	—	—	—	[1]
836	NZ	Mt Adams	1999	RA	RF	10.0–15	II	10.0–15	80–100	—	700	—	5.0–7.0	6	OT	40–50	100	30	5.5	2000-3000	—	[27][28]
837	NZ	Mt Ruapehu Tephra	2007	—	—	—	—	—	7	—	—	—	—	—	PP	—	—	—	—	—	—	[29]
838	NZ	Ponui Stream	1976	RS, DS	RF	2.5	II	2.5	50–60	—	570	—	2.89	NF	NF	—	—	—	—	—	—	[1][2]
839	NZ	Poulter River	2200 BP	—	—	—	—	2200	248	—	—	—	—	—	—	—	—	—	—	—	—	[30]
840	NZ	Ram Creek	1968	RA	EQ	5	III	2.8	40	150	1200	325	1.1	5475	OT	30	100	30	—	1000	—	[1][33]
841	NZ	Ruamahanga River	1855	RS	EQ	—	II	—	—	—	—	—	—	NF	NF	—	—	—	—	—	—	[1]
842	NZ	Sandstone Lake	1929	—	EQ	—	—	—	—	—	—	—	—	—	—	—	—	—	—	—	—	[1]
843	NZ	Stanley River	1929	DA	EQ	—	—	—	40	—	—	2000	—	—	—	—	—	—	—	—	—	[1]
844	NZ	Te Hoe River	1931	—	EQ	—	—	—	25	—	—	2000	5.0–6	2555	—	—	—	—	—	—	—	[1]

845	NZ	Thompson Stream	1905	—	—	—	—	16	30	—	—	—	—	—	—	—	—	—	—	—	—	[2]
846	NZ	Thompson Stream	1929	RS, RA	EQ	—	II	—	30	—	400	—	—	—	—	—	—	—	—	—	—	[1]
847	NZ	Tuki Tuki River	1968	RS, DA	HC	0.765	II	—	15	—	—	—	—	NF	NF	—	—	9	—	—	—	[1]
848	NZ	Tunawaea Stream Valley	1991	Slide	—	—	—	4	70	170–370	550	—	0.9	332	OT	15–20	—	—	1	250	—	[2][31][44]
849	NZ	Waikeraimoana Lake	2200 BP	—	—	—	—	2200	248	—	—	—	—	—	—	—	—	—	—	—	—	[2]
850	PE	Huancapara River	—	—	—	—	—	—	20	500	200	750	—	—	OT	—	—	—	—	—	—	[1]
851	PE	Mantaro River	1930	DA	—	—	II	—	—	300	450	—	—	60	—	—	—	—	—	—	—	[1]
852	PE	Mantaro River	1945	RS	HC	3.5	II	3.5	133	250	580	21000	301	73	SF, PP	56	—	—	—	35400	—	[1]
853	PE	Mantaro River	1974	DA	RF	1600	III	1300	160	1000	3800	31000	670	42	OT	107	243	30	12	10000	—	[1][32]
854	PE	Maranon River	1946	—	EQ	—	—	—	—	—	—	—	—	41	—	—	—	—	—	—	—	[1]
855	PE	Nepena River	1970	RS	EQ	—	—	—	20	—	—	480	—	NF	NF	—	—	—	—	—	—	[1]
856	PE	Pelagatos River	1946	RA	EQ	25	—	—	—	—	—	—	—	—	—	—	—	—	—	—	—	[1]
857	PE	Santa River	1941	DF	DF	10	—	—	—	—	—	—	—	2	—	—	—	—	—	—	—	[1]
858	PE	Santa River	1962	DA	SM	13	II	—	—	300	1300	—	—	0.002	—	—	—	—	—	3000	—	[1]
859	PE	Santa River	1970	Slump	EQ	25	VI	—	—	150	300	700	—	NF	NF	—	—	—	—	—	—	[1]
860	PE	Santa River	1970	DA	EQ	50–100	III	—	—	300	3500	1000–2000	—	0.02	—	—	—	—	—	—	—	[1]
861	PE	Shacsha River	1970	RF, DA	EQ	0.5	II	—	—	—	—	—	—	NF	NF	—	—	—	—	—	—	[1]
862	PE	Tincog River	1967	—	RF	—	—	—	—	—	—	—	—	—	—	—	—	—	—	—	—	[1]
863	PE	Yuracyacu River	1968	RS	—	0.3	—	—	—	—	—	—	—	—	—	—	—	—	—	—	—	[1]
864	PG	Bairaman River	1985	RS, DA	EQ	200	III	200	200	1000	3000	3000	50	489	OT	70	—	—	3	8000	—	[1][34]
865	PG	Bairaman River	1985	RS, DA	EQ	150	III	—	50	500	1000	3500	2.1	NF	NF	—	—	—	—	—	—	[1]
866	PG	Clearwater Creek	1935	RS, ES	EQ	2	III	—	45	100	200	500	1	NF	NF	—	—	—	—	—	—	[1]
867	PG	Ok Ma River	1984	Slide	HC	15	—	—	3	100	300	800	0.1	731	—	—	—	—	—	—	—	[1]
868	PG	Tiaru River	1985	DF	EQ	0.6	—	—	30	100	300	700	0.28	19	—	—	—	—	—	—	—	[1]
869	PG	Tiaru River	1985	RS, DS	EQ	5	—	—	50	200	500	1000	2	NF	NF	—	—	—	—	—	—	[1]
870	PG	Undal River	1941	—	EQ	—	—	—	11	137	411	—	—	—	—	—	—	—	—	—	—	[1]
871	PH	Bued River	1968	DF	RF	1	—	—	10	—	300	—	—	—	—	—	—	—	—	—	—	[1]
872	PH	Jalaur River	1938	Slump	RF	—	—	—	55	—	—	8000	—	33	—	—	—	—	—	—	—	[1]
873	PH	Naporoc River	1628	DA	—	1500	III	—	90	—	—	5000	—	NF	NF	—	—	—	—	—	—	[1]
874	PK	Ghizar River	1980	DF	DF	—	III	—	30	200	300	5000	—	1.3	OT	—	—	—	—	—	—	[35]
875	PK	Gilgit River	1981	DF	—	—	—	—	5	500	—	—	—	—	—	—	—	—	—	—	—	[1]
876	PK	Gilgit River	1984	DF	RF	—	—	—	—	—	—	—	—	0.1	—	—	—	—	—	—	—	[1]
877	PK	Hattian Bala	2005	—	EQ	65	—	—	—	—	—	—	—	—	—	—	—	—	—	—	—	—
878	PK	Hunza River	1858	RS	RF, SM	—	—	—	100	—	—	45000	—	180	OT	—	—	—	—	—	—	[1]
879	PK	Hunza River	1974	DF	RF	—	—	—	4	40	—	—	—	0.04	—	—	—	—	—	—	—	[1]
880	PK	Hunza River	1976	DF	—	—	—	—	20	—	—	12000	—	NF	NF	—	—	—	—	—	—	[1]
881	PK	Hunza River	2010	Slide	PW	45	—	30	120–200	1300	300	21000	450	NF	NF	—	—	—	—	—	—	[48][49]
882	PK	Indus River	1840	RS, DA	EQ	—	—	—	224	1600	—	64000	—	150	OT	—	—	—	—	—	—	[1]

(Continued)

Table B.1 *(Continued)*

No	Country[1]	Name	Date of formation[2]	Landslide Type[3]	Trigger[4]	Landslide volume ($\times10^6 m^3$)	Dam Type[5]	Dam volume ($\times10^6 m^3$)	Dam height (m)	Dam length (m)	Dam width (m)	Lake length (m)	Lake volume ($\times10^6 m^3$)	Time to failure (day)[6]	Failure mech.[7]	Breach depth (m)	Breach top width (m)	Breach bottom width (m)	Breaching time (hour)	Peak discharge (m^3/s)	Loss of life	Reference (Below the table)
883	PK	Indus River	1841	RS	—	—	—	—	200	—	—	65000	—	—	—	—	—	—	—	540000	—	[37]
884	PK	Karli stream	2005	DA	EQ	85	II	—	130	—	450	800	86	1580	OT	—	—	—	—	5500	—	[38][47]
885	PK	Tang stream	2005	DA	EQ	85	II	—	22	—	130	400	5	1735	OT	—	—	—	—	—	—	[38][47]
886	PL	Wetlina River	1980	RF, DF	RF	0.15	—	—	6.0–10	25–30	60	200	—	2920	OT	3.0–5	10.0–15	—	—	—	—	[1]
887	RO	Lake Rosu	1828	RS	—	—	—	—	—	—	—	—	—	NF	NF	—	—	—	—	—	—	[1]
888	SE	Gota Alv River	1950	ES	HC	3	VI	—	5	110	110	—	—	—	—	—	—	—	—	—	—	[1]
889	SI	Mongga River tributary	1986	RA	RF	5	III	—	30	400	1500	3000	12	108	OT	—	—	—	—	—	—	[1]
890	SL	Unknown river	1982	DS,DF	RF	—	—	—	—	—	—	—	—	<1	—	—	—	—	—	—	Some	[1]
891	SP	Velillos River	1986	ES,EF	HC	0.5	II	—	3	15	35	—	—	NF	NF	—	—	—	—	—	—	[1]
892	TH	Haui Sao River	1988	RS	RF	—	II	—	12	50–100	35	—	—	1	OT	—	—	—	—	—	—	[1]
893	TH	Tha Di River	1988	RS	RF	—	I	—	6	80	25	—	—	4	OT	—	—	—	—	—	—	[1]
894	TH	Tha Di River	1988	DF	RF	—	I	—	3	40	5	—	—	2	OT	—	—	—	—	—	—	[1]
895	TR	Degirmen River	1988	RA	RF	0.25	—	—	6	50	130	300	—	NF	NF	—	—	—	—	—	—	[1]
896	TR	Gorge from Cehennem-Dere Glacier	1840	RA	EQ	—	—	—	—	—	—	—	—	3	—	—	—	—	—	—	—	[1]
897	TR	Kıratlı	1987	—	RF	—	—	—	—	—	—	—	—	—	—	—	—	—	—	—	—	[46]
898	TR	Sera River	1950	—	RF	5	—	—	—	—	—	—	1.8	—	—	—	—	—	—	—	—	[46]
899	TR	Solaklı River	1929	—	RF	—	—	10	—	—	—	—	10	NF	NF	—	—	—	—	—	146	[46]
900	TR	Tortum River	1700s	RS	—	223	—	180	270	1350	790–1380	8500	538	—	OT	—	—	—	—	—	—	[46]
901	TW	Hsiaolin village	2009	Slide	RF	25.2	III	15.4	44–68	370	1500	—	9.9	0.04	OT	40–46	—	—	8.4 Min	70649	—	[39][40]
902	TW	Tsao-Ling Lake	1862	DS	EQ	—	II	—	—	—	—	—	—	13140	OT	—	—	—	—	—	—	[1]
903	TW	Tsao-Ling Lake	1941	DS	EQ	48	II	—	140	—	—	—	12.8	3669	OT	—	—	—	—	—	—	[1]
904	TW	Tsao-Ling Lake	1942	DS	RF	—	II	282.1	217	1100	1600	7200	157	3435	OT	—	—	—	4.5	—	154	[1]
905	TW	Tsao-Ling Lake	1979	DS	RF	5.00	II	5	90	—	—	—	40	9	OT	—	—	—	3	—	—	[1]
906	TW	Tsao-Ling Lake	1999	—	EQ	120.00	—	25	50	5000	—	—	46	—	—	—	—	—	—	—	—	[42]
907	TW	寶來溪(寶山村)	2009	—	RF	—	III	—	—	—	—	180	—	—	OT	—	—	—	—	—	—	[41]
908	TW	大安溪內灣段	1999	—	EQ	—	—	—	10	—	—	—	—	—	—	—	—	—	—	—	—	[42]
909	TW	大社溪(大社村)	2009	—	RF	—	—	—	—	—	—	—	—	—	—	—	—	—	—	—	—	[41]
910	TW	東埔蚋溪	2000	—	RF	—	—	0.036	15	—	—	—	—	—	—	—	—	—	—	—	—	[42]
911	TW	旱溪	1999	—	EQ	—	—	0.018	6	—	—	—	—	—	—	—	—	—	—	—	—	[42]
912	TW	郝馬戛班溪（神木村）	2009	—	RF	—	—	—	—	—	—	—	—	—	—	—	—	—	—	—	—	[41]

913	TW	和社溪(神木村)	2009	Ava.	RF	—	III	—	—	—	—	—	—	—	OT	—	—	—	—	—	—	[41]
914	TW	侯硐大粗坑溪	2000	—	RF	—	—	—	4	—	—	—	—	—	—	—	—	—	—	—	—	[42]
915	TW	九份二山韭菜湖溪	1999	—	EQ	—	—	—	29	—	—	—	0.68	—	—	—	—	—	—	—	—	[42]
916	TW	九份二山瀧仔坑溪	1999	—	EQ	—	—	—	37.5	—	—	—	1.1	—	—	—	—	—	—	—	—	[42]
917	TW	拉克斯溪(梅蘭村)	2009	RA	RF	—	II	—	—	—	—	—	—	—	NF	—	—	—	—	—	—	[41]
918	TW	荖濃溪(寶來村)	2009	Ava.	RF	—	—	—	—	—	—	—	—	—	OT	—	—	—	—	—	—	[41]
919	TW	荖濃溪(復興村)	2009	—	RF	—	—	—	—	—	—	—	—	—	—	—	—	—	—	—	—	[41]
920	TW	荖濃溪(梅山村)	2009	—	RF	—	III	—	—	—	—	—	—	—	OT	—	—	—	—	—	—	[41]
921	TW	荖濃溪(梅山口)	2009	—	RF	—	—	—	—	—	—	—	—	—	—	—	—	—	—	—	—	[41]
922	TW	立霧溪	2002	—	EQ	—	—	0.005	—	—	—	—	—	—	—	—	—	—	—	—	—	[42]
923	TW	平坑(關山村)	2009	—	RF	—	—	—	—	—	—	—	—	—	—	—	—	—	—	—	—	[41]
924	TW	旗山溪(那瑪夏鄉)	2009	DF	RF	—	III	—	9	—	—	—	—	—	OT	—	—	—	—	—	—	[41]
925	TW	汝仍溪(牡丹村)	2009	—	RF	—	—	—	—	—	—	—	—	—	—	—	—	—	—	—	—	[41]
926	TW	沙里仙溪(同富村)	2009	—	RF	—	—	—	—	—	—	—	—	—	—	—	—	—	—	—	—	[41]
927	TW	沙漣河	1999	—	EQ	—	—	—	—	—	—	—	—	—	—	—	—	—	—	—	—	[42]
928	TW	生毛樹溪	1999	—	EQ	—	—	—	—	—	—	—	—	—	—	—	—	—	—	—	—	[42]
929	TW	石盤溪	1999	—	EQ	—	—	—	—	—	—	—	—	—	—	—	—	—	—	—	—	[42]
930	TW	士文溪(春日鄉)	2009	S	RF	—	II	—	25	—	—	700	1.85	—	OT	—	—	—	—	—	—	[41]
931	TW	太麻里溪(包盛社)	2009	DA	RF	—	III	—	10	—	1200	—	5.33	—	OT	—	—	—	—	—	—	[41]
932	TW	頭汴坑溪龍寶橋	1999	—	EQ	—	—	0.006	5	—	—	—	—	—	—	—	—	—	—	—	—	[42]
933	TW	頭汴坑溪一江橋	1999	—	EQ	—	—	1.87	—	—	—	—	—	—	—	—	—	—	—	—	—	[42]
934	TW	新武呂溪	2002	—	RF	—	—	0.45	30	—	—	—	—	—	—	—	—	—	—	—	—	[42]
935	TW	雪山坑溪	1999	—	EQ	—	—	0.16	15	—	—	—	—	—	—	—	—	—	—	—	—	[42]
936	USA	Box Canyon Creek	1964	RS, DS	EQ	—	II	—	3–5.0	—	—	—	—	NF	NF	—	—	—	—	—	—	[1]
937	USA	Cache Creek	1906	DS	EQ	2.40	II	0.55	32	120	240	9000	14.9	5	PP,OT	—	—	—	—	—	—	[1][2]
938	USA	Cascade Creek	1941	DF	RF	—	IV	—	—	—	—	120	—	—	—	—	—	—	—	—	—	[1]
939	USA	Cedar Creek	1988	ES	—	1.70	I	1.7	3	30	150	1300	0.053	105	OT	2	10	8	—	—	—	[1]
940	USA	Chakachatna river	1953	DF	RF, VE	—	—	—	—	—	—	8000	—	2	—	4.6	—	—	—	—	—	[1]
941	USA	Chicken Creek	1984	ES	SM	—	II	—	6	50	—	170	0.02	0.13	OT	—	—	—	—	—	—	[1]
942	USA	Claverack Creek	1915	Slump	RF	—	VI	—	6	—	—	—	—	—	—	—	—	—	—	—	—	[1]
943	USA	Coldwater Creek	1980	DA	VE	2800	III	—	71	—	—	3500	83	NF	NF	—	—	—	—	—	—	[1]
944	USA	Colorado River	1923	MF	RF	—	III	—	—	—	—	—	—	1	OT	—	—	—	—	—	—	[1]
945	USA	Colorado River	1966	DF	RF	—	II	—	—	—	—	—	—	—	—	—	—	—	—	—	—	[1]
946	USA	Columbia River	1450	—	—	—	—	—	—	—	—	—	—	—	—	—	—	—	—	220000	—	[37]
947	USA	Corralitos Creek	1989	RF	EQ	0.0035	II	—	7	10–15.0	20	200	0.006	NF	NF	—	—	—	—	—	—	[1]
948	USA	Corralitos Creek	1989	RS	EQ	0.0035	II	—	3	25	15	30	0.0012	NF	NF	—	—	—	—	—	—	[1]
949	USA	Cottonwood Creek	1983	DS, DA	SM	—	II	—	5	—	—	—	—	0.6	OT	—	—	—	—	—	—	[1]
950	USA	Cottonwood Creek	1984	DF	SM	—	III	—	2.5	20	50	50	—	NF	NF	—	—	—	—	—	—	[1]
951	USA	Cowlitz River	1949	EF, ES	EQ	0.015–0.02	II	—	5	40–50	50–80	—	—	—	OT	—	—	—	—	—	—	[1]
952	USA	Dead Horse Creek	1984	DF	SM	0.15	III	—	4	100	200	170	0.009	—	—	—	—	—	—	—	—	[1]
953	USA	Dragon Canyon	1931	DF	RF	—	—	—	—	—	—	—	—	1	OT	—	—	—	—	—	—	[1]

(Continued)

Table B.1 (*Continued*)

No	Country[1]	Name	Date of formation[2]	Landslide Type[3]	Trigger[4]	Landslide volume (×10⁶ m³)	Dam Type[5]	Dam volume (×10⁶ m³)	Dam height (m)	Dam length (m)	Dam width (m)	Lake length (m)	Lake volume (×10⁶ m³)	Time to failure (day)[6]	Failure mech.[7]	Breach depth (m)	Breach top width (m)	Breach bottom width (m)	Breaching time (hour)	Peak discharge (m³/s)	Loss of life	Reference (Below the table)
954	USA	East Fork Hood River	1980	DF	RF	0.07–0.1	III	0.07–0.1	10.7	100	225	—	0.105	0.01	OT	—	—	—	—	850	—	[1][35]
955	USA	Elk Rock Lake	1980	DA	VE	2800	III	—	9	65.6	—	—	0.31	63	OT	15.8	—	—	0.67	450	0	[1][36][44]
956	USA	Fraser River	1985	DF	SM	0.003	III	—	1.0–2	—	50–70	300	—	NF	NF	—	—	—	—	—	—	[1]
957	USA	Garfield Creek	1867	DF, DA	RF	—	III	—	13	100	350	200	—	1	—	—	—	—	—	—	—	[1]
958	USA	Ginseng Hollow creek	1969	MF	RF	0.0016	—	—	2	5	100	—	0.00014	0.01	OT	—	—	—	—	—	—	[1]
959	USA	Gros Ventre River	1909	—	—	—	—	37.5	25	—	—	—	—	—	—	—	—	—	—	—	—	[2]
960	USA	Gros Ventre River	1923	—	—	—	—	67.5	75	—	—	—	—	—	—	—	—	—	—	—	—	[2]
961	USA	Gros Ventre River, Wyoming	1925	DA, RS	SM	38	III	—	70–75	600	3000	6500	80	696	PP, OT	15	—	—	—	—	6	[1][35]
962	USA	Gros Ventre River	1909–1910	EF	SM	—	III	—	—	1000	3000–4000	2000	—	NF	NF	—	—	—	—	—	—	[1]
963	USA	Grouse Creek	1955	EF	RF	—	—	—	—	—	—	—	—	250	OT	—	—	—	—	—	—	[1]
964	USA	Grouse Creek	1964	EF	RF	—	—	—	—	—	—	—	—	—	—	—	—	—	—	—	—	[1]
965	USA	Gualala River	1906	—	EQ	—	IV	—	—	—	—	—	—	—	—	—	—	—	—	—	—	[1]
966	USA	Hackberry Creek	1988	RF	HC	—	II	—	10	30	—	75	—	NF	NF	—	—	—	—	—	—	[1]
967	USA	Hebgen Lake	1959	—	—	—	—	—	—	—	—	10000	—	—	—	—	—	—	—	—	—	[36]
968	USA	Hinckley Creek	1906	—	EQ	—	—	—	20	—	—	—	—	—	—	—	—	—	—	—	—	[1]
969	USA	Hinckley Creek	1989	RF	EQ	0.0016	II	—	5	15	15	130	—	NF	NF	—	—	—	—	—	—	[1]
970	USA	Hurdygurdy Creek	1964	—	RF	—	—	—	—	—	—	5000	—	—	—	—	—	—	—	—	—	[1]
971	USA	Jackson Creek Lake	1980	DA	VE	2800	III	0.77	4.5	975	180–425	820	2.47	644	OT	—	—	—	—	477	—	[1][36]
972	USA	Jap Creek	1964	RS, DS	EQ	—	II	—	5–8.0	—	125	—	—	NF	NF	—	—	—	—	—	—	[1]
973	USA	Kaweah River	1867	—	—	—	—	—	20	—	—	—	—	41	—	—	—	—	—	—	—	[45]
974	USA	Kolob Creek	1990	RA, DA	—	—	II	—	6	9	40	400	0.006	15	—	—	—	—	—	—	—	[1]
975	USA	Lake Fork	700BP	—	—	50–100	—	—	40	500	1700	3000	—	NF	NF	—	—	—	—	—	—	[1][35]
976	USA	Lake Whatcom tributary	1983	RS	RF	—	II	—	10	—	250	—	—	NF	NF	—	—	—	—	—	—	[1]
977	USA	Leigh Creek	1941	DF	RF	—	III	—	—	100	—	300	—	NF	NF	—	—	—	—	—	—	[1]
978	USA	Los Gatos Creek	1906	—	EQ	—	—	—	—	—	—	—	—	—	—	—	—	—	—	—	—	[1]
979	USA	Love Creek	1982	DF	RF	0.46	III	—	6	50	200	300	—	—	—	—	—	—	—	—	—	[1]
980	USA	Maacama Creek	1906	RS	EQ	—	II	—	—	—	—	—	—	—	—	—	—	—	—	—	—	[1]
981	USA	Madison River	1959	RS	EQ	21	II	26	60–70	500	1600	10000	—	NF	NF	—	—	—	—	—	—	[1][36]
982	USA	Mattole River	1983	DF	RF	9.5	III	—	12	90	400	2200	—	NF	NF	—	—	—	—	—	—	[1]
983	USA	Monumental Creek	1909	—	—	—	—	0.3	10	—	—	—	—	—	—	—	—	—	—	—	—	[2]

984	USA	Mosquito Creek	1979–1980	DF	RF	0.4	II	0.023	9	—	60	—	—	—	—	—	—	—	—	—	—	[1]
985	USA	Monument Creek	1984	DA	RF	—	—	—	6	—	—	—	—	—	—	—	—	—	—	—	—	[1]
986	USA	Muddy Creek	1986	ES	SM	180	VI	—	1	30	1000	—	—	NF	NF	—	—	—	—	—	—	[1]
987	USA	Nisqually River	1990	Slump	HC	0.2	I	—	7	240	170	1000	—	NF	NF	—	—	—	—	—	—	[1]
988	USA	North Fork of South Branch Potomac River	1949	DA	RF	—	—	—	—	—	—	—	—	—	—	—	—	—	—	—	—	[1]
989	USA	North Fork of Toutle River	1880	—	—	—	—	0.02	4.5	—	—	—	—	—	—	—	—	—	—	—	—	[2]
990	USA	Otatso Creek	1910	RA	RF	7	V	—	5	100	120	1000	—	NF	NF	—	—	—	—	—	—	[1]
991	USA	Otatso Creek	1910	RA	RF	7	V	—	5	100	120	400	—	NF	NF	—	—	—	—	—	—	[1]
992	USA	Ottauqueche River tributary	1984	EF	RF, SM	0.021	—	—	2.5	—	—	—	—	—	OT	—	—	—	—	—	—	[1]
993	USA	Powder River	1984	RS, DA	RF	6	II	—	9	60	200	—	0.29	NF	NF	—	—	—	—	—	—	[1]
994	USA	Presumpscot River	1868	ES	—	0.6	II	—	5	60	500	1000	—	—	—	—	—	—	—	—	—	[1]
995	USA	Provo River	1930	DF	RF	—	—	—	2.5	—	—	—	—	14	OT	—	—	—	—	—	—	[1]
996	USA	Provo River	1931	DF	RF	—	—	—	6	—	120	—	—	—	OT	—	—	—	—	—	—	[1]
997	USA	Provo River	1938	DF	RF	—	—	—	—	—	100	—	—	NF	NF	—	—	—	—	—	—	[1]
998	USA	Provo River	1938	DF	RF	0.038	—	—	4	—	120	2500	—	—	—	—	—	—	—	—	—	[1]
999	USA	Purisma Creek	1906	—	EQ	—	II	—	8.0–10	—	—	—	—	—	—	—	—	—	—	—	—	[1]
1000	USA	Quartz Creek	1995	—	—	—	—	0.05	20	—	—	—	—	—	—	—	—	—	—	—	—	[2]
1001	USA	San Cristobal Lake	700s	—	—	—	—	29.4	70	—	—	—	—	—	—	—	—	—	—	—	—	[2]
1002	USA	San Gregorio Creek	1906	—	EQ	—	—	—	2	—	—	—	—	—	—	—	—	—	—	—	—	[1]
1003	USA	Santa Maria Lake	Preh.	—	—	—	—	389.12	152	—	—	—	—	—	—	—	—	—	—	—	—	[2]
1004	USA	Silver Creek	1988	RF	—	0.017	II	0.02	12	30	92	610	0.0574	365	—	—	—	—	—	—	—	[1]
1005	USA	South Fork American River	1983	RS	RF, HC	3	II	0.02	15	20	130	—	0.395	0.35	OT	—	—	—	—	—	—	[1][2]
1006	USA	South Fork Castle Creek	1980	DA	VE	2800	III	—	37	600	425	2000	24	NF	NF	—	—	—	—	—	—	[1][43]
1007	USA	South Fork Kaweah River	1867	DF, DA	RF	—	III	—	20	100	200	300	—	1	—	—	—	—	—	—	—	[1]
1008	USA	South Fork of Smith River	1970	—	—	—	—	0.28	15	—	—	—	—	—	—	—	—	—	—	—	—	[2]
1009	USA	South Fork Smith River	1965	DF	RF	1.9	III	—	15	75	500	—	2.7	0.25	OT	—	—	—	—	—	—	[1]
1010	USA	South Fork Smith River	1970	DF	RF	1.1	III	—	13.5	75	400	—	1.7	0.05	OT	—	—	—	—	—	—	[1]
1011	USA	Spanish Fork River	1983	DS	SM	22	II	—	63	200	450	6000	78	NF	NF	—	—	—	—	—	—	[1]
1012	USA	Spitit Lake	1980	DA	VE	2800	III	—	69	—	—	5000	330	NF	NF	—	—	—	—	—	—	[1][36]
1013	USA	Spruce Creek	1986	DA	RF	0.09	II	—	—	50	200	—	—	—	—	—	—	—	—	—	—	[1]
1014	USA	Sweetwater Creek	1976	DF	RF	—	—	—	—	—	—	—	—	—	—	—	—	—	—	—	—	[1]

(Continued)

Table B.1 *(Continued)*

No	Country[(1)]	Name	Date of formation [(2)]	Landslide Type [(3)]	Trigger [(4)]	Landslide volume ($\times 10^6\,m^3$)	Dam Type [(5)]	Dam volume ($\times 10^6\,m^3$)	Dam height (m)	Dam length (m)	Dam width (m)	Lake length (m)	Lake volume ($\times 10^6\,m^3$)	Time to failure (day) [(6)]	Failure mech.[(7)]	Breach depth (m)	Breach top width (m)	Breach bottom width (m)	Breaching time (hour)	Peak discharge (m^3/s)	Loss of life	Reference (Below the table)
1015	USA	Swift Creek	1984	DA	SM	0.01	III	—	5	25	100	300	—	SD	OT	—	—	—	—	—	—	[1]
1016	USA	Tolt River	1967	Slump	HC	—	II	—	—	—	—	—	—	1	—	—	—	—	—	—	—	[1]
1017	USA	Toutle River	2500BP	—	VE	—	—	—	—	—	—	—	—	—	—	—	—	—	—	260000	—	[37]
1018	USA	Trinity River	1890	RS	RF, SM	—	II	—	—	—	—	22000	—	0.6	OT, PP	—	—	—	—	—	—	[1]
1019	USA	Uncompahgre River	1971	DF	RF	—	—	—	—	—	—	—	—	—	—	—	—	—	—	—	—	[1]
1020	USA	Van Duzen River	1964	DA	RF	0.45	—	0.43	10	—	—	3000	2	1	—	—	—	—	—	—	—	[1][2]
1021	USA	Virgin River	1923	RS	—	—	—	—	—	—	—	—	—	—	—	—	—	—	—	—	—	[1]
1022	USA	Virgin River	1941	RS	—	—	—	—	—	—	—	—	—	—	—	—	—	—	—	—	—	[1]
1023	USA	Weat Branch Soquel Creek	1989	Slump	EQ	—	II	—	4	12	25	75	0.00185	61	OT	—	—	—	—	—	—	[1]
1024	USA	Wilson River	1964	DF	RF	0.45	—	—	20	—	135	—	—	NF	NF	—	—	—	—	—	—	[1]
1025	USA	Wilson River	1991	DS	RF	0.76	II	—	6	50	100	490	—	NF	NF	—	—	—	—	—	—	[1]
1026	USA	Yakima River	1947	Slump	HC	3.0–4	II	—	5.0–10	200	400	—	—	5.0	—	—	—	—	—	—	—	[1]
1027	USSR	Amtkel Lake	1891	—	EQ	—	—	—	200	—	—	—	—	—	—	—	—	—	—	—	—	[1]
1028	USSR	Angren River	1973	RS	HC	600–700	II	—	50–60	—	—	—	—	—	—	—	—	—	—	—	—	[1]
1029	USSR	Charukha River	1989	—	EQ	—	—	—	90	—	—	—	—	—	—	—	—	—	—	—	—	[1]
1030	USSR	Chavkhun-bak River	1970	RS	EQ	10	—	—	—	—	—	—	—	—	—	—	—	—	—	—	—	[1]
1031	USSR	Dubursa River Tributary	1949	DA	EQ	—	—	—	—	—	—	—	—	—	—	—	—	—	—	—	—	[1]
1032	USSR	Isfayram River	1977	DF	RF, SM	—	—	—	—	—	—	—	—	—	OT	—	—	—	—	—	—	[1]
1033	USSR	Medeo Dam	1966–1967	—	HC	2.5	II	—	110	—	—	—	6.20	NF	NF	—	—	—	—	—	—	[1]
1034	USSR	Murgab River; Lake Sarez	1911	RS	EQ	2000	II	—	800	1,000	1,000	53,000	16,000	NF	NF	—	—	—	—	—	—	[1]
1035	USSR	Naryn River	1946	—	EQ	—	—	—	50–60	—	—	—	—	—	—	—	—	—	—	—	—	[1]
1036	USSR	Obi-Kabut River	1949	DA	EQ	—	II	—	—	—	—	—	—	—	—	—	—	—	—	—	—	[1]
1037	USSR	Ritseuli River	1972	—	—	30.	—	—	20	—	—	—	—	NF	NF	—	—	—	—	—	—	[1]
1038	USSR	Tegermach River (Lake Yashinkul)	1835	RF, DF	EQ	20	III	20	120	—	60	—	6.60	48,000	PP	90	280–340	50–60	—	4960	—	[1]
1039	USSR	Utch-Terek River	1989	—	HC	3	II	—	42	—	295	—	—	NF	NF	—	—	—	—	—	—	[1]
1040	USSR	Yasman River	1949	DA	EQ	—	III	—	—	1,500	—	—	—	—	—	—	—	—	—	—	—	[1]
1041	USSR	Zeravshan River	1964	—	EQ	15–20	—	—	200	400	1,800	—	—	NF	NF	—	—	—	—	—	—	[1]
1042	YU	Jovatz River	1977	DS	RF	200	II	—	15–20	150	700	2500	0.5	NF	NF	—	—	—	—	—	—	[1]
1043	YU	Vatasha River	1956	RS, ES	HC	20	II	—	70	400	800	—	—	NF	NF	—	—	—	—	—	—	[1]
1044	YU	Visotchiza River	1963	DS	SM	4	II	—	35	150	500	5000	14	NF	NF	—	—	—	—	—	—	[1]

Note:

* The peak outflow rates of these 22 cases in Japan were estimated values.

(1) Country: AF = Afghanistan; AT = Austria; BO = Bolivia; BT = Bhutan; AU = Australia; CA = Canada; CH = Switzerland; CL = Chile; CN = China; CO = Colombia; CR = Costa Rica; CZ = Czech; DE = Germany; FR = France; GT = Guatemala; ID = Indonesia; IE = Ireland; IL = Israel; IN = India; IT = Italy; JM = Jamaica; JP = Japan; KG = Republic of Kirgizstan; KR = Korea; NO = Norway; NP = Nepal; NZ = New Zealand; PE = Peru; PG = Papua New Guinea; PH = Philippines; PK = Pakistan; PL = Poland; RO = Romania; SE = Sweden; SI = Solomon Islands; SL = Sri Lanka; SP = Spain; TH = Thailand; TR = Turkey; TW = Chinese Taiwan; USA = United States of America; USSR = The former Union of Soviet Socialist Republics; YU = Yugoslavia.

(2) Data of formation: BC = Before Christ; BP = Before present; Preh. = Prehistoric; Q3 = Tertiary Period; Q4 = Quaternary Period.

(3) Landslide type: Ava. = Avalanche; DA = Debris avalanche; DF = Debris flow; DS = Debris slide; EF = Earth flow; ES = Earth slide; LS = Lateral spread; MF = Mud flow; RA = Rock avalanche; RF = Rock fall; RS = Rock slide.

(4) Trigger: EQ = Earthquake; HC = Human-caused; PW = Progressive weakening; RF = Rainfall; SM = Snowmelt.

(5) Dam type: Definitions refer to Costa and Schuster (1988).

(6) Time to failure: NF = did not fail by the reported date; SD = several days; LT = long term; ST = short term (LT and ST refer to Chai et al., 1998).

(7) Failure mechanism: OT = Overtopping; PP = Piping; SF = Slope failure; NF = did not fail by the reported date.

References for Appendix B

[1] Costa, J. E., and Schuster, R. L. (1991). *Documented historical landslide dams from around the world. US* Geological Survey Open-File Report, 91–239, 486 pp.

[2] Ermini, L., and Casagli, N. (2003). "Prediction of the behaviour of landslide dams using a geomorphological dimensionless index." *Earth Surface Processes and Landforms*, **28**(1), 31–47.

[3] Miller, B. G. N., and Cruden, D. M. (2002). "The Eureka River landslide and dam, Peace River Lowlands, Alberta." *Canadian Geotechnical Journal*, **39**(4), 863–878.

[4] Geertsema, M., and Clague, J. J. (2006). "1000-year record of landslide dams at Halden Creek, north-eastern British Columbia." *Landslides*, **3**(3), 217–227.

[5] Yan, R. (2006). *Secondary disaster and environmental effect of landslides and collapsed dams in the upper reaches of Minjiang River*. Master Thesis, Sichuan University, Chengdu, China, 143 pp.

[6] Chai, H. J., Liu, H. C., and Zhang, Z. Y. (1998). "Study on the categories of landslide-damming of rivers and their characteristics." *Journal of the Chengdu Institute of Technology*, **25**(3), 411–416.

[7] Xu, Q., Fan, X. M., Huang, R. Q., and Westen, C. V. (2009). "Landslide dams triggered by the Wenchuan Earthquake, Sichuan Province, south-west China." *Bulletin of Engineering Geology and the Environment*, **68**(3), 373–386.

[8] Cui, P., Han, Y. S., Chao, D., and Chen X. Q. (2011). *Formation and treatment of landslide dams emplaced during the 2008 Wenchuan earthquake, Sichuan, China*. Natural And Artificial Rockslide Dams, Edited by Evans, S. G., Hermanns, R. L., Storm, A., and Scarascia-Mugnozza, G., *Lecture Notes in Earth Sciences*, **133**, 295–321.

[9] Nie, G. Z., Gao, J. G., and Deng, Y. (2004). Preliminary study on earthquake-induced dammed lake. *Quaternary Sciences*, **24**(3), 293–301.

[10] Tung, Y. X. (2008). *Quantitative Analysis for Stability of Landslide Dams*. Master thesis, National Central University, Taiwan.

[11] Dunning, S. A., Rosser, N. J., Petley, D. N., and Massey, C. R. (2006). "Formation and failure of the Tsatichhu landslide dam, Bhutan." *Landslides*, **3**(2),107–113.

[12] Evans, S. G. (1986). Landslide damming in the Cordillera of western Canada, in R. L. Schuster (ed.), *Landslide Dams: Processes, Risk and Mitigation*. Geotechnical Special Publication No. 3, ASCE, New York, 111–130.

[13] Clague, J. J., and Evans, S. G. (1994). "Formation and failure of natural dams in the Canadian Cordillera." *Geological Survey of Canada*, Bulletin 464.

[14] Bovis, M. J., and Jakob, M. (2000). "The July 29, 1998, debris flow and landslide dam at Capricorn Creek, Mount Meager Volcanic Complex, southern Coast Mountains, British Columbia." *Canadian Journal of Earth Sciences*, **37**(10), 1321–1334.

[15] Cruden, D. M., Lu, Z. Y., and Thomson, S. (1997). "The 1939 Montagneuse River landslide, Alberta." *Canadian Geotechnical Journal*, **34**(5), 799–810.

[16] Miller, B. G. N., Cruden, D. M. (2002). "The Eureka River landslide and dam, Peace River Lowlands, Alberta." *Canadian Geotechnical Journal*, **39** (4): 863–878.

[17] Mora, S., Madrigal, C., Estrada, J. and Schuster, R. L. (1993). "The 1992 Rio Toro landslide dam, Costa Rica." *Landslide News*, **7**, 19–22.

[18] Plaza-Nieto, G., and Zevallos, O. (1994). "The 1993 La Josefina rockslide and Rio Paute landslide dam, Ecuador." *Landslide News*, **8**: 4–6.

[19] Plaza-Nieto, G., Yepes, H., and Schuster, R. L. (1990). "Landslide dam on the Pisque river, northern Ecuador." *Landslide News*, **4**: 2–4.

[20] Gupta, V., and Sah, M. P. (2008). "Impact of the Trans-Himalayan Landslide Lake Outburst Flood (LLOF) in the Satluj catchment, Himachal Pradesh, India." *Natural Hazards*, **45**(3), 379–390.

[21] Casagli, N. and Ermini., L. (1999). "Geomorphic analysis of landslide dams in the northern Apennines." *Transaction Japanese Geomorphological Union*, **20,** 219–249.

[22] Nicoletti, P. G., and Parise, M. (2001). "Seven landslide dams of old seismic origin in south-eastern Sicily (Italy)." *Geomorphology*, **46**(3–4), 203–222.

[23] Sassa, K., Fukuoka, H., Wang, F., Wang, G. H., Okada, K., and Marui, H. (2006). "Landslide Disasters Triggered by the 2004 Mid-Niigata Prefecture Earthquake in Japan." *Annuals of Disas. Prev. Res.* Inst., Kyoto Univ., No. 49C, 119–136.

[24] Hermanns, R. (2005). "All Clear for Allpacoma Landslide Dam." *MAP:GAC Newsletter*, **5**(1).

[25] Sabo Dept., MLIT (2007). Landslide dam formed in Kawaradariver in Ishikawa Pref. online: http://www.sabo-int.org/case/2007japan_wajima.pdf.

[26] Adhikari, D. P., and Koshimizu, S. (2005). "Debris flow disaster at Larcha, upper Bhotekoshi Valley, central Nepal." *Island Arc*, **14**(4), 410–423.

[27] Hancox, G. T., McSaveney, M. J., Manville, V. R., and Davies, T. R. (2005). "The October 1999 Mt Adams rock avalanche and subsequent landslide dam-break flood and effects in Poerua River, Westland, New Zealand." *New Zealand Journal of Geology and Geophysics*, **48**(4), 683–705.

[28] Becker, J. S., Johnston, D. M., Paton, D., Hancox, G. T., Davies, T. R., McSaceney, M. J., and Manville, V. R. (2007). "Response to landslide dam failure emergencies: issues resulting from the October 1999 Mount Adams Landslide and dam-break flood in the Poerua River, Westland, New Zealand." *Natural Hazards Review*, **8**(2), 35–42.

[29] Massey, C. I., Manville, V., Hancox, G. H., Keys, H. J., Lawrence, C., and McSaveney, M. (2010). "Out-burst flood (lahar) triggered by retrogressive landsliding, 18 March 2007 at Mt Ruapehu, New Zealand-a successful early warning." *Landslides*, **7**(3): 303–315.

[30] Adams, J. (1981). "Earthquake triggered landslides from lakes in New Zealand." *Earthquake Information Bulletin (USGS)*, **13**(6), 204–215.

[31] Jennings, D. N., Webby, M. G., and Parkin, D. T. (1993). "Tunawaea landslide dam, King Country, New Zealand." *Landslide News*, **7**, 25–27.

[32] Lee, K. L., and Duncan, J. M. (1975). *Landslide of April 25, 1974, on the Mantaro River, Peru. Washington, D. C.*, Committee on Natural Disasters, Commission on Sociotechnical Systems, National Research Council, 71 pp.

[33] Nash, T., Bell, D., Davies, T., and Nathan, S. (2008). "Analysis of the formation and failure of Ram Creek landslide dam, South Island, New Zealand." *New Zealand Journal of Geology & Geophysics*, **51**(3), 187–193.

[34] King, J., Loveday, I., and Schuster, R. L. (1989). "The 1985 Bairaman landslide dam and resulting debris flow, Papua New Guinea." *Quarterly Journal of Engineering Geology*, **22**(4), 257–270.

[35] Schuster, R. L., and Costa, J. E. (1986). "A perspective on landslide dams." In: Landslide Dams: Processes, Risk, and Mitigation, Geotechnical Special Publication No. 3, ASCE, New York, 1–20.

[36] Schuster, R. L. (1985). "Landslide dam in the Western United States." *US Geological Survey*, pp. 411–418.

[37] O'Connor, J. E., and Costa, J. E. (2004). "The world's largest floods, past and present: Their causes and magnitudes." *US Geological Survey Circular*, **1254**, 1–13.

[38] Edwin, L. H., and Anthony, J. C. (2006). *Landslides Triggered by the October 8, 2005, Pakistan Earthquake and Associated Landslide Dammed Reservoirs*. US Geological Survey, Reston, Virginia, pp. 1–10.

[39] Dong, J. J., Li, Y. S., Kuo, C. Y., Sung, R. T., Li, M. H., Lee, C. T., Chen, C. C., and Lee, W. R. (2011). "The formation and breach of a short-lived landslide dam at Hsiaolin village, Taiwan – Part I: Post-event reconstruction of dam geometry." *Engineering Geology*, **123**(1–2), 40–59.

[40] Li, M. H., Sung, R. T., Dong, J. J., Lee, C. T., and Chen, C. C. (2011). "The formation and breaching of a short-lived landslide dam at Hsiaolin Village, Taiwan – Part II: Simulation of debris flow with landslide dam breach." *Engineering Geology*, **123**(1–2), 60–71.

[41] Chen, K. T., Tsang, Y. C., Kuo, Y. S, Lu, M. H., and Shieh, C. L. (2010). "Case Analysis of Landslide Dam Formation by Typhoon Morakot." *Journal of the Taiwan Disaster Prevention Society*, **2**(1), 43–50.

[42] Liao, Z. Z., Dong, J. J., Shi, T. Y., Ye, K. J., Pan, Y. W., Wang, C. D., and Qian, X. Z. (2002). *Study on the disaster mitigation strategies for landslide dams (1/3)*. Report for the Water Resource Planning Institute, Water Resources Agency, Ministry of Economic Affairs, Taiwan, China.

[43] Meyer, W., Schuster, R. L., and Sabol, M. A. (1994). "Potential for seepage erosion of landslide dam." *Journal of Geotechnical Engineering, ASCE*, **120**(7), 1211–1229.

[44] Walder, J. S., O'Connor, J. E. (1997). "Methods for predicting peak discharge of floods caused by failure of natural and constructed earthen dams." *Water Resources Research*, **33**(10), 2337–2348.

[45] Schuster, R. L., Wieczorek, G. F., Hope D. G., II. (1998). "Landslide dams in Santa Cruz County, California, resulting from the earthquake." *US Geological Survey Professional Paper*, **1551-C**, 51–70.

[46] Duman, T. Y. (2009). "The largest landslide dam in Turkey: Tortum landslide." *Engineering Geology*, **104**(1–2), 66–79.

[47] Konagai, K., and Sattar, A. (2012). "Partial breaching of Hattian Bala Landslide Dam formed in the 8th October 2005 Kashmir Earthquake, Pakistan." *Landslides*, **9**(1), 1–11.

[48] Petley, D. (2010). The landslide at Attabad in Hunza, Gilgit/Baltistan: current situation and hazard management needs. Report prepared for Focus Humanitarian Assistance, Pakistan, based upon a rapid field assessment on 26th February – 4th March. Online: http://blog.travel-culture.com/wp-content/uploads/2010/03/10_02_Hunza_landslide_short_report_final.pdf.

[49] Petley, D. (2010). The Attabad landslide crisis in Hunza, Pakistan – lessons for the management of valley blocking landslides. Report prepared for Focus Humanitarian Assistance, Pakistan. Online: http://blogs.agu.org/landslideblog/files/2011/05/11_01-Geohazards-5-Attabad-final.pdf.

References

163 News (2014). "National Administration of Surveying, Mapping and Geoinformation released aerial images of seismic area," *163 News*. Retrieved from http://news.163.com/photoview/00AN0001/71294.html.

Aberg, B. (1993). "Washout of grains from filtered sand and gravel materials," *Journal of Geotechnical Engineering*, **119**(1), 36–53.

Aboelata, M. A. and Bowles, D. S. (2005). "LIFESim: a model for estimating dam failure life loss," Report to Institute for Water Resources, US Army Corps of Engineers and Australian National Committee on Large Dams, Institute for Dam Safety Risk Management, Utah State university, Logan, Utah, USA.

Abt, S. R., Wittler, R. J., Taylor, A., and Love, D. J. (1989). "Human stability in a high flood hazard zone." *Water Resources Bulletin*, **25**(4), 881–890.

Ackers, P. and White, W. R. (1973). "Sediment transport: A new approach and analysis," *Journal of Hydraulic Division*, **99**(11), 2041–2061.

Adel, H. D., Bakker, K. J., and Breteler, M. K. (1988). "Internal stability of minestone," in *Proceedings of International Symposium on Modelling Soil-Water-Structure Interactions*, Balkema, Rotterdam, 225–231.

Aglan, H., Wendt, R., and Livengood, S. (2004). "Field testing of energy-efficient flood-damage-resistance residential envelope systems," Summary Report, Oak Ridge National Laboratory, Oak Ridge, Tennessee, USA.

Ammerlaan, P. R. M. (2007). "Levees and levee evaluation: the Dutch and US practice compared," MSc thesis, Delft University of Technology, The Netherland.

ANCOLD (Australian National Committee on Large Dams) (2003). *Guidelines on Risk Assessment. Australian National Committee on Large Dams Inc.*, Tasmania, Australia.

ANCOLD (1998). *Guidelines on Risk Assessment*. Working Group on Risk Assessment, Australian National Committee on Large Dams, Sydney, New South Wales, Australia.

Ang, A. H.-S. and Tang, W. H. (1990). *Probability Concepts in Engineering Planning and Design, Vol. 2: Decision, Risk, and Reliability*, Wiley, New York.

Ang, A. H.-S. and Tang, W. H. (2007). *Probability Concepts in Engineering*, 2nd ed., Wiley, New York.

Annandale, G. W. (2006). *Scour Technology-Mechanics and Engineering Practice*, McGraw-Hill, New York.

Apel, H., Aronica, G. T., Kreibich, H., and Thieken, A. H. (2009). "Flood risk analyses – how detailed do we need to be?" *Natural Hazards*, **49**(1), 79–98.

ASCE/EWRI Task Committee on Dam/Levee Breaching (2011). "Earthen embankment breaching," *ASCE Journal of Hydraulic Engineering*, **137**(12), 1549–1564.

Assaf, H. and Hartford, D. N. D. (2002). "A virtual reality approach to public protection and emergency preparedness planning in dam safety analysis," in *Proceedings of the Canadian Dam Association Conference*, Victoria, Canada.

Assaf, H. and Hartford, D. N. D. (2001). "Physically-based modelling of life safety considerations in water resource decision-making," World Water Congress 2001, ASCE, Reston, 1–10.

Dam Failure Mechanisms and Risk Assessment, First Edition. Limin Zhang, Ming Peng, Dongsheng Chang, and Yao Xu.

Assaf, H., Hartford, D. N. D., and Cattanach, J. D. (1997). "Estimating dam breach flood survival probabilities," *ANCOLD Bulletin*, **107**, 23–42.

ASTM (2009). *Standard Practice for Description and Identification of Soils (Visual-Manual Procedure)*, ASTM D2488-09a, ASTM International, West Conshohocken, Pa.

ASTM (2007). *Standard Test Method for Dispersive Characteristics of Clay Soil by Double Hydrometer*, ASTM D4221-06, ASTM International, West Conshohocken, Pa.

ASTM (2006). *Standard Test Method for Identification and Classification of Dispersive Clay Soils by the Pinhole Test*, ASTM D4647-06, ASTM International, West Conshohocken, Pa.

ASTM (2000a). Standard Practice for Classification of Soils for Engineering Purposes (Unified Soil Classification System), ASTM D2487, ASTM International, West Conshohocken, Pa.

ASTM (2000b). Standard Test Method for Soil Compaction Determination at Shallow Depths Using 5 lb (2.3 kg) Dynamic Cone Penetrometer, ASTM D7380-08, ASTM International, West Conshohocken, Pa.

ASTM (2000c). Standard Test Method for Erodibility Determination of Soil in the Field or in the Laboratory by the Jet Index Method, D5852-00, ASTM International, West Conshohocken, Pa.

Ayyaswamy, P., Hauss, B., Hseih, T., Moscati, A., Hicks, T. E., and Okrent, D. (1974). "Estimates of the risks associated with dam failure," School of Engineering and Applied Science, University of California, Los Angeles, California. 375 pp.

Baecher, G. B. and Christian, J. T. (2003). *Reliability and Statistics in Geotechnical Engineering*, Wiley, England.

Bagnold, R. A. (1980). "An empirical correlation of bed-load transport rates in flumes and natural rivers," in *Proceedings of Royal Society of London*, Series A, Vol. 372, pp. 453–473.

Bagnold, R. A. (1956). "The flow of cohesionless grains in fluids," *Philosophical Transactions of the Royal Society of London*, Series A, **249**, 291–293.

Baidu Baike, (2014). Deze Reservoir. baidu.com. online: http://baike.baidu.com/view/9805592.htm.

Bates, P. D. and De Roo, A. P. J. (2000). "A simple raster-based model for flood inundation simulation," *Journal of Hydrology*, **236**(1–2), 54–77.

Bayoumi, A. and Meguid, M. A. (2011). "Wildlife and safety of earthen structures: a review," *Journal of Failure Analysis and Prevention*, **11**(4), 295–319.

BC Hydro (1993). Guidelines for Consequence-based Dam Safety Evaluations and Improvements. Hydroelectric Engineering Division Report No. H2528, BC Hydro, Burnaby, BC, Canada.

Bear, J. (1972). *Dynamics of Fluids in Porous Media*, American Elsevier, New York.

Becker, J. S., Johnston, D. M., Paton, D., Hancox, G. T., Davies, T. R., McSaceney, M. J., and Manville, V. R. (2007). "Response to landslide dam failure emergencies: issues resulting from the October 1999 Mount Adams Landslide and dam-break flood in the Poerua River, Westland, New Zealand." Natural Hazards Review, **8**(2), 35–42.

Béguin, R. (2011). *Etude Multi-échelle de L'érosion de Contact au Sein des Ouvrages Hydrauliques en Terre*, Ph.D. thesis, University of Grenoble.

Bendahmane, F., Marot, D., Rosquoët, F., and Alexis, A. (2008). "Experimental parametric study of suffusion and backward erosion," *Journal of Geotechnical and Geoenvironmental Engineering*, **134**(1), 57–67.

Bezuijen, A., Klein-Breteler, M., and Bakker, K. J. (1987). "Design criteria for placed block revetments and granular filters," *Coastal and Port Engineering*, China Ocean Press, Beijing, 1852–1866.

Black, R. D. (1975). "Flood proofing rural residences," Project Agnes report, Pennsylvania, New York State College of Agriculture and Life Sciences, Ithaca, Prepared for Economic Development Administration, Washington, DC Office of Technical Assistance.

Blaikie, P., Cannon, T., Davis, I., and Wisner, B. (1994). At Risk: Natural Hazards, People's Vulnerability, and Disasters. Routledge, New York.

Bligh, W. G. (1910). "Dams barrages and weirs on porous foundations," *Engineering News*, **64**(26).

Bolle, A., et al. (2006). "Hydraulic modeling of the two-directional interaction between sewer and river systems," in *Proceeding of the Urban Drainage Modeling and Water Sensitive Urban Design*, Monash Univ., Melbourne, Australia.

Bonelli, S. (2013). *Erosion in Geomechanics Applied to Dams and Levees*, ISTE Ltd and John Wiley & Sons, Inc.

Bonelli, S. and Benahmed, N. (2011). "Piping flow erosion in water retaining structures," *International Journal on Hydropower and Dams*, **18**(3), 94–99.

Bonelli, S. and Marot, D. (2008). "On the modelling of internal soil erosion" in *Proceedings of the 12th International Conference of the International Association for Computer Methods and Advances in Geomechanics*, Goa, India.

Bonelli, S., Brivois, O., Borghi, R., and Benahmed, N. (2006). "On the modelling of piping erosion," *C. R. Mecanique*, **334**(8–9), 555–559.

Bonnard, C. (2004). "Technical and human aspects of historic rockslide dammed lakes and landslide dam breaches," in *Security of Natural and Artificial Rockslide Dams*, K. Abdrakhmatov, S. G. Evans, R. Hermanns, G. Scarascia-Mugnozza, and A. L. Strom, Eds., Extended Abstracts Volume, NATO Advanced Research Workshop, Bishkek, Kyrgyzstan, 13–19.

BOSS (1999). *BOSS DAMBRK: User's Manual*, BOSS International, Madison, Wisconsin.

Box, G. E. P., Jenkins, G. M., and Reinsel, G. C. (2008). "Time series analysis: forecasting and control," *Wiley Series in Probability and Statistics*, Hoboken, New Jersey, USA.

Boyd, E., Levitan, M., and van Heerden, I. (2005). "Further specification of the dose-Response relationship for flood fatality estimation," *US-Bangladesh Workshop on Innovation in Windstorm/Storm Surge Mitigation Construction*, National Science Foundation and Ministry of Disaster & Relief, Government of Bangladesh, Dhaka, Bangladesh.

Brauns, J. (1985). *Erosionsverhalten geschichteten Bodens bei horizontaler Durchstromung. Wasserwirtschaft*, **75**, 448–453.

Briaud, J. L. (2008). "Case histories in soil and rock erosion: Woodrow Wilson Bridge, Brazos River Meander, Normandy Cliffs, and New Orleans Levees," *Journal of Geotechnical and Geoenvironmental Engineering*, **134**(10), 1425–1447.

Briaud, J. L., Chen, H. C., Govindasamy, A. V., and Storesund, R. (2008). "Levee erosion by overtopping in the New Orleans during the Katrina Hurricane," *Journal of Geotechnical and Geoenvironmental Engineering*, **134**(5), 618–632.

Briaud, J. L., Ting, F. C. K., Chen, H. C., Cao, Y., Han, S. W., and Kwak, K. W. (2001). "Erosion function apparatus for scour rate predictions," *Journal of Geotechnical and Geoenvironmental Engineering*, **127**(2), 105–113.

British Standards Institution (BSI) (1991). *Reliability of Systems Equipment and Components – Guide to Failure Modes, Effects, and Criticality Analysis, BS-5760*, British Standards Institution, London.

British Standards Institution (BSI) (1986). *Code of Practice for Foundations, BS-8004*, British Standards Institution, London.

Brockwell, P. J. and Davis, R. A. (1996). "Introduction to time series and forecasting" in *Springer Texts in Statistics*, New York, USA.

Brown, C. B. (1950). *Sediment Transport*, H. Rouse (Ed.), Engineering Hydraulics, Wiley, New York.

Brown, C. A. and Graham, W. J. (1988). "Assessing the threat to life from dam failure," *Water Resources Bulletin*, **24**(6), 1303–1309.

Brown, R. J. and Rogers, D. C. (1981). *Users Manuals for Program BRDAM*, US Bureau of Reclamation, Denver, Colorado.

Brown, R. J. and Rogers, D. C. (1977). "A simulation of the hydraulic events during and following the Teton Dam failure," in *Proceedings of the Dam-Break Flood Routing Workshop*, Water Resources Council, Bethesda, Maryland, pp. 131–163.

Brownlie, W. R. (1981). "Compilation of fluvial channel data: laboratory and field," Report No. KH-R-43B, W. M. Keck Laboratory of Hydraulic and Water Resources, California Institute of Technology, Pasadena, California.

Burenkova, V. V. (1993). "Assessment of suffusion in non-cohesive and graded soils" in *Filters in Geotechnical and Hydraulic Engineering*, Balkema, Rotterdam, 357–360.

Cai, W. (2014). "30 soldiers from Armed Police Hydropower Troops took risk mitigation measures for landslide dams in Yunnan," Online: http://js.xhby.net/system/2014/08/08/021583443.shtml.

Canadian Standards Association. (1993). *Guidelines and Requirements for Reliability Analysis Methods, Q636–93*, Canadian Standards Association, Ontario.

Canuti, P., Casagli, N., and Ermini, L. (1998). "Inventory of landslide dams in the Northern Apennine as a model for induced flood hazard forecasting," in *Managing Hydro-geological Disasters in a Vulnerate Environment*, Andah K (Ed.), CNR-GNDCI Publication 1900. CNR-GNDCI-UNESCO (IHP): Perugia, pp.189–202.

Canuti, A., Frassoni, A., and Natale, L. (1993). "Failure of the Rio Paute Landslide dam," *Landslide News*, **8**, 6–7.

Cao, D. (1994). "Countermeasures for seepage erosion of Yangtze River main dikes," *Yangtze River*, **25**(1), 25–29.

Cao, Z., Pender, G., Wallis, S., and Carling, P. (2004). "Computational dam-break hydraulics over erodible sediment bed," *Journal of Hydraulic Engineering*, **130**(7), 689–703.

Cao, Z., Yue, Z., and Pender, G. (2011). "Landslide dam failure and flood hydraulics, Part I: Experimental investigation. Part II: Coupled mathematical modeling," *Natural Hazards*, **59**(2), 1003–1045.

Carr, R. S. and Smith, G. P. (2006). "Linking of 2D and pipe hydraulic models at fine spatial scales," *Urban Drainage Modeling and Water Sensitive Urban Design*, Monash Univ., Melbourne, Australia.

Casagli, N. and Ermini, L. (1999). "Geomorphic analysis of landslide dams in the Northern Apennine," *Transactions of the Japanese Geomorphological Union*, **20**(3), 219–249.

CCTV News (2008, July 5). "Shocking 29 days of Tangjiashan Landslide Dam: Three breaching scenarios were predicted," Online http://scitech.people.com.cn/GB/7473434 .html.

Cencetti, C., Fredduzzi, A., Marchesini, I., Naccini, M., and Tacconi, P. (2006). "Some considerations about the simulation of breach channel erosion on landslide dams." *Computational Geosciences*, **10**, 201–219.

Chai, H. J., Liu, H. C., and Zhang, Z. Y. (1998). "Study on the categories of landslide-damming of rivers and their characteristics." *Journal of the Chengdu Institute of Technology*, **25**(3):411–416 (in Chinese).

Chang, D. S. (2012). *Internal Erosion and Overtopping Erosion of Embankment Dams and Landslide Dams*, Ph.D. thesis, The Hong Kong University of Science and Technology, Hong Kong.

Chang, D. S. and Zhang, L. M. (2010). "Simulation of the erosion process of landslide dams due to overtopping considering variations in soil erodibility along depth," *Natural Hazards and Earth System Sciences*, **10**(4), 933–946.

Chang, D. S. and Zhang, L. M. (2011). "A stress-controlled erosion apparatus for studying internal erosion in soils," *Geotechnical Testing Journal*, **34**(No), 579–589.

Chang, D. S. and Zhang, L. M. (2013a). "Critical hydraulic gradients of internal erosion under complex stress states," *Journal of Geotechnical and Geoenvironmental Engineering*, **139**(9), 1454–1467.

Chang, D. S. and Zhang, L. M. (2013b). "Extended internal stability criteria for soils under seepage," *Soils and Foundations*, **53**(4), 569–583.

Chang, D. S., Zhang, L. M., and Xu, T. H. (2012). "Laboratory investigation of initiation and development of internal erosion in soils under complex stress states," in *Proceedings of the Sixth International Conference on Scour and Erosion*, Paris, ISSMGE, pp. 895–902.

Chang, D. S., Zhang, L. M., Xu, Y. and Huang, R. Q. (2009). "Analysis of overtopping failure of Hongshihe Landslide dam after Wenchuan earthquake." *Chinese Journal of Engineering Geology*, **17**(1), 50–55.

Chang, D. S., Zhang, L. M., Xu, Y., and Huang, R. Q. (2011). "Field testing of erodibility of two landslide dams triggered by the 12 May Wenchuan earthquake," *Landslides*, **8**(3), 321–332.

Chapuis, R. P. (1992). "Similarity of internal stability criteria for granular soils," *Canadian Geotechnical Journal*, **29**(4), 711–713.

Chapuis, R. P. and Gatien, T. (1986). "An improved rotating cylinder technique for quantitative measurements of the scour resistance of clays," *Canadian Geotechnical Journal*, **23**(1), 83–87.

Chapuis, R. P., Contant, A., and Baass, K. A. (1996). "Migration of fines in 0–20 mm crushed base during placement, compaction, and seepage under laboratory conditions," *Canadian Geotechnical Journal*, **33**(8), 168–176.

Chen, A. S., Djordjevic, S., Leandro, J., and Savic, D. (2007). "The urban inundation model with bidirectional flow interaction between 2D overland surface and 1D sewer networks," in *Proceeding of the 6th NOVATECH Int. Conf.*, Workshop I, Graie, Lyon, France, pp. 465–472.

Chen, X. Q. (2000). "The types and failure modes of landslide dams induced by the Chi-chi Earthquake," in *Symposium on Hill-slope Hazards and Strategies after the 9–21 Earthquake*, Taichung, Taiwan.

Chen, X. Q., Zhao, W. Y., Gao, Q., Jia, S. T., and Zhu, X. H. (2011). "Experimental research on effect of man-made structure controlling dam-break flood," *Chinese Journal of Mountain Science*, **29**(2), 217–225.

Chen, Z. Y. (2003). "Reflection on some problems in Yangtze River embankment reinforcement project," *Advances in Science and Technology of Water Resources*, **23**(4), 1–7.

Chiew, Y. M. and Paker, G. (1994). "Incipient sediment motion on non-horizontal slopes," *Journal of Hydraulic Research*, **32**(5), 649–660.

China Earthquake Administration, (2014a). Ms 6.5 earthquake occurred in Ludian County, Zhaotong City, Yunnan Province at 16:30 on 3 August. cea.gov.cn. online: http://www.cea.gov.cn/publish/dizhenj/587/597/20140803164417654530904/index.html (in Chinese).

Chinnarasri, C., Jirakidlert, S., and Wongwises, S. (2004). "Embankment dam breach and its outflow characteristics," *Civil Engineering and Environmental System*, **21**(4), 247–264.

Chou, J. W., Lu, J. K., Li, N., and Chen, H. (1999). "The development of flood simulation model in flood risk information management system in Beijing area," *Journal of Catastrophology*, **14**(4), 17–21.

Chow, V. T. (1959). *Open Channel Hydraulics*, McGraw-Hill Book Company, New York.

CIRIA (Construction Industry Research and Information Association), US Ministry of Ecology and USACE (United States Army Corps of Engineers) (2013). International Levee Handbook, CIRIA, eBook.

Cividini, A., Bonomi, S., Vignati, G. C., and Gioda, G. (2009). "Seepage-induced erosion in granular soil and consequent settlements," *International Journal of Geomechanics*, **9**(4), 187–194.

Clague, J. J. and Evans, S. G. (1994). "Formation and failure of natural dams in the Canadian Cordillera." Geological Survey of Canada Bulletin **464**, 35 pp.

Clague, J. J. and Evans, S. G. (2000). "A review of catastrophic drainage of moraine-dammed lakes in British Columbia," *Quaternary Science Reviews*, **19**, 1763–1783.

Clausen, L. and Clark, P. B. (1990). "The development of criteria for predicting dam-break food damages using modelling of historical dam failures," in *International Conference on River Flood Hydraulics*, White, W. R. (Ed.). John Wiley & Sons Ltd. Hydraulics Research Limited, Wallingford, England, pp. 369–380.

Coleman, S. E., Andrews, D. P., and Webby, M. G. (2002). "Overtopping breaching of non-cohesive homogeneous embankment," *Journal of Hydraulic Engineering*, **128**(9), 829–838.

Cook, A. C. (2008). "Comparison of one-dimensional HEC-RAS with two-dimensional FESWMS model in flood inundation mapping," Master thesis, Purdue University, West Lafayette, Indiana, USA.

Cornell, C. A. and Newmark, N. M. (1978). "On the seismic reliability of nuclear power plants" in *ANS Topical Meeting on Probabilistic Reactor Safety*, Newport Beach, California.

Cornwell, J. B. and Meyer, M. M. (1997). "Risk acceptance criteria or 'how safe is safe enough?'" in *Presented at II Risk Control Seminar*, Petróleos de Venezuela, Puerto La Cruz, Venezuela.

Costa, J. E. (1985). "Floods from dam failures," US Geological Survey, Open-File Report 85–560, 54 pp.

Costa, J. E. and Schuster, R. L. (1991). "Documented historical landslide dams from around the world," US Geological Survey Open-File Report, 91–239, 486 pp.

Costa, J. E. and Schuster, R. L. (1988). "The formation and failure of natural dams." *Geological Society of America Bulletin*, **100**(7), 1054–1068.

Cui, P., Dang, C., Zhuang, J. Q., You, Y., Chen, X. Q., and Scott, K. M. (2012). "Landslide-dammed lake at Tangjiashan, Sichuan Province, China (triggered by the Wenchuan Earthquake, May 12, 2008): risk assessment, mitigation strategy, and lessons learned," *Environmental Earth Sciences*, **65**(4): 1055–1065.

Cui, P., Zhu, Y. Y., Han, Y. S., Chen, X. Q., and Zhuang, J. Q. (2009). "The 12 May Wenchuan earthquake-induced landslide lakes: distribution and preliminary risk evaluation," *Landslides*, **6**(3), 209–223.

Cunge, J. A., Holly, F. M., and Verwey, A. (1980). *Practical Aspects of Computational River Hydraulics*, Pitman Publishing Limited, London.

Dai, R. Q., Yan, K. A., and Zhou, F., (2008). Construction of emergency risk removing works of Wenjiaba barrier lake in Pingwu County of Sichuan Province. *Yangtze River* **39** (22), 7–8.

Damgaard, J. S., Whitehouse, R. J. S., and Soulsby, R. L. (1997). "Bed-load sediment transport on steep longitudinal slopes," *Journal of Hydraulic Engineering*, **123**(12), 1130–1139.

Danka, J. (2015). *Mathematical Model for the Assessment of Overtopping Breaching Erosion of River Dikes*, Ph.D. thesis, The Hong Kong University of Science and Technology, Hong Kong.

Danka, J. and Zhang, L. M. (2015). "Dike failure mechanisms and breaching parameters," *Journal of Geotechnical and Geoenvironmental Engineering*, **141**(9), 04015039.

DeKay, M. L. and McClelland, G. H. (1993). "Predicting loss of life in cases of dam failure and flash flood," *Risk Analysis*, **13**(2), 193–205.

D'Eliso, C. (2007). "Breaching of sea dikes initiated by wave overtopping, A tier and modular modeling approach," Ph.D. thesis, University of Florence, Florence, Italy.

Den Adel, H., Koenders, M. A., and Bakker, K. J. (1994). "The analysis of relaxed criteria for erosion-control filters," *Canadian Geotechnical Journal*, **31**(6), 829–840.

Dewey, R. L. and Gillette, D. R. (1993). "Prediction of embankment dam breaching for hazard assessment," in *Proceedings of the ASCE Conference on Geotechnical Practice in Dam Rehabilitation*, Rayleigh, North Carolina, pp. 131–144.

DHI (Danish Hydraulic Institute) (2003). *MIKE 11–A Modeling System for Rivers and Channels*, Short Introduction Tutorial, Danish Hydraulic Institute, Denmark.

DHI (Danish Hydraulic Institute) (2007). *MIKE 21 Flow Model. Hydrodynamic Module User Guide*, Danish Hydraulic Institute, Denmark.

Dong, J. J., Li, Y. S., Kuo, C. Y., Sung, R. T., Li, M. H., Lee, C. T., Chen, C. C., and Lee, W. R. (2011). "The formation and breach of a short-lived landslide dam at Hsiaolin village, Taiwan – part I: Post-event reconstruction of dam geometry." *Engineering Geology*, **123**(1–2), 40–59.

Dong, J. J., Tung, Y. H., Chen, C. C., Liao, J. J., and Pan, Y. W. (2009). Discriminant analysis of the geomorphic characteristics and stability of landslide dams. *Geomorphology*, **110**, 162–171.

Dow, K. and Cutter, S. L. (1998). "Crying wolf: repeat responses to hurricane evacuation orders." *Coastal Management*, **26**(4), 237–252.

Drabek, T. E. (2000). *The social factors that constrain human responses to flood warnings*, Vol. 1, Parker, D. J., Ed., Routledge Hazards and Disasters Series.

Du Boys, M. P. 1879. "Le Rhône et les rivièrs a lit affouillable." Ann. Ponts Chaussées, **5**(18), 141–195.

Dunning, S. A., Rosser, N. J., Petley, D. N., and Massey, C. R. (2006). "Formation and failure of the Tsatichhu landslide dam, Bhutan," *Landslides*, **3**(2), 107–113.

Durbin, J. (1960). "The fitting of time-series models." *Review of the International Statistical Institute*, **28**(3), 233–244.

Egiazaroff, I. V. (1965). "Calculation of non-uniform sediment concentration," *Journal of the Hydraulics Division*, **91**(4), 225–247.

El Shamy, U., and Aydin, F. (2008). "Multi-scale modelling of flood-induced piping in river levees," *Journal of Geotechnical and Geoenvironmental Engineering*, **134**(9), 1385–1398.

Einstein, H. A. (1950). "The bed-load function for sediment transportation in open channel flows," Technical Bulletin No. 1026, US Department of Agriculture, Soil Conservation Service, Washington, DC, USA.

Einstein, H. A. (1942). "Formulas for the transport of bed-load," *Transaction of the American Society of Civil Engineers*, **107**, 561–573.

Engelund, F. and Hansen, E. (1967). *A Monograph on Sediment Transport in Alluvial Streams*, Teknisk Vorlag, Copenhagen, Denmark.
Ergun, S. (1952). "Fluid flow through packed columns," *Chemical Engineering Progress*, **48**(6), 89–94.
Ermini, L. and Casagli, N. (2003). "Prediction of the behavior of landslide dams using a geomorphological dimensionless index," *Earth Surface Processes and Landforms*, **28**(1), 31–47.
Faber, M. H., Kroon, I. B., Kragh, E., Bayly, D., and Decosemaeker, P. (2002). "Risk assessment of decommissioning options using Bayesian networks," *Journal of Offshore Mechanics and Arctic Engineering*, **124**(4), 231–238.
Faeh, R. (2007). "Numerical modeling of breach erosion of river embankments," *Journal of Hydraulic Engineering*, **133**(9), 1000–1009.
Fand, R. M., Kim, B. Y. K., Lam, A. C. C., and Phan, R. T. (1987). "Resistance to the flow resistance to the flow of fluids through simple and complex porous-media whose matrices are composed of randomly packed spheres," *Journal of Fluids Engineering*, **109** (3), 268–274.
Fannin, R. J. and Moffat, R. (2006). "Observations on internal stability of cohesionless soils," *Geotechnique*, **56**(7), 497–500.
Federal Emergency Management Agency (FEMA) (2005a). *Technical Manual: Conduits through Embankment Dams*, Washington, DC, USA.
Federal Emergency Management Agency (FEMA) (2005b). *Technical Manual for Dam Owners: Impacts of Animals on Earthen Dams*, Washington, DC, USA.
Federal Emergency Management Agency (FEMA) (2005c). *Technical Manual for Dam Owners: Impacts of Plants on Earthen Dams*, Washington, DC, USA.
Federal Emergency Management Agency (FEMA) (2004). Federal guidelines for dam safety – Hazard potential classification system for dams. Federal Emergency Management Agency, Washington DC, USA.
Federal Energy Regulatory Commission (FERC) (1988). *USA Federal Regulatory Commission – Notice of Revised Emergency Action Plan Guidelines*, Federal Energy Regulatory Commission (FERC), Washington, DC.
Fell, R. (1994). "Landslide risk assessment and acceptable risk," *Canadian Geotechnical Journal*, **31**(2), 261–272.
Fell, R. and Fry, J. J. (2007). "The state of the art of assessing the likelihood of internal erosion of embankment dams, water retaining structures and their foundations," in *Internal Erosion of Dams and their Foundations*, Taylor and Francis, pp. 1–24.
Fell, R. and Hartford, D. (1997). "Landslide risk management" in *Landslide Risk Assessment. Balkema*, Cruden, D. and Fell, R., Eds., Rotterdam, pp. 51–109.
Fell, R., Bowles, D. S., Anderson, L. R., and Bell, G. (2000). "The status of methods for estimation of the probability of failure of dams for use in quantitative risk assessment," in *Proceedings of the 20th Congress on Large Dams*, Question 76, Beijing, pp. 213–236.
Fell, R., Wan, C. F., Cyganiewicz, J., and Foster, M. (2003). "Time for development of internal erosion and piping in embankment dams," *Journal of Geotechnical and Geoenvironmental Engineering*, **129**(4), 307–314.
Fenton, G. A. and Griffiths, D. V. (2008). *Risk Assessment in Geotechnical Engineering*, John Wiley & Sons, Inc., New Jersey.
Feng, Y. W. (2014). "Death toll in Yunnan Ludian earthquake rose to 617 people," *Online* http://news.xinhuanet.com/local/2014–08/09/c_126850338.htm.
FESWMS (2002). *FST2DH User's Manual*, Federal Highway Administration, Publication number FHWA-RD-03-053, Federal Highway Administration, Office of Research, Development, and Technology, Turner-Fairbank Highway Research Center, McLean, Virginia, USA.
Fleshman, M. S. and Rice, J. D. (2014). "Laboratory modeling of the mechanisms of piping erosion initiation," *Journal of Geotechnical and Geoenvironmental Engineering*, **140**(6), 04014017.
FLO-2D Software Inc. (2009). *FLO-2D reference manual*, Nutrioso, Arizona, USA.
Flores-Berrones, R., Ramírez-Reynaga, M., and Macari, E. J. (2011). "Internal erosion and rehabilitation of an earth-rock dam," *Journal of Geotechnical and Geoenvironmental Engineering*, **137**(2), 150–160.
Floris, (2005). "Flood risk and safety in the Netherlands," FLORIS Study-Full Report, Ministerie van Verkeer en Waterstaat, the Netherlands, 134 pp.
Foster, M. and Fell, R. (2001). "Assessing embankment dams filters which do not satisfy design criteria," *Journal of Geotechnical and Geoenvironmental Engineering*, **127**(4), 398–407.
Foster, M., Fell, R., and Spannagle, M. (2000a). "The statistics of embankment dam failures and accidents," *Canadian Geotechnical Journal*, **37**(5), 1000–1024.
Foster, M., Fell, R., and Spannagle, M. (2000b). "A method for assessing the relative likelihood of failure of embankment dams by piping," *Canadian Geotechnical Journal*, **37**(5), 1025–1062.
Fread, D. L. (1977). "The development and testing of a dam-break flood forecasting model" in *Proceedings of the Dam-Break Flood Routing Model Workshop*, Bethesda, Maryland, pp. 164–197.

Fread, D. L. (1988). *BREACH: An Erosion Model for Earthen Dam Failures*, National Weather Service, Office of Hydrology, Silver Spring, MD.

Fread, D. L. and Lewis, J. M. (1998). "NWS FLDWAV Model," Hydrologic Research Laboratory, Office of Hydrology, National Weather Service (NWS), NOAA, Silver Spring, Maryland, USA.

Friedman, D. G. (1975). "Computer simulation in natural hazard assessment," Institute of Behavioral Science, University of Colorado, Boulder, Colorado. 194 pp.

Frieser, B. (2004). *Probabilistic Evacuation Decision Model for River Floods in the Netherlands*, Final report, Delft University of Technology, 138 pp.

Friis-Hansen, A. (2000). *Bayesian Networks as a Decision Support Tool in Marine Applications*, Ph.D. thesis, Technical University of Denmark, Lyngby, Denmark.

Froehlich, D. C. (1995a). "Embankment dam breach parameters revisited," in *Proceedings of the 1995 ASCE Conference on Water Resources Engineering*, Water Resources Engineering, New York, pp. 887–891.

Froehlich, D. C. (1995b). "Peak outflow from breached embankment dam," *Journal of Water Resources Planning and Management Division*, **121**(1), 90–97.

Fujisawa, K., Murakami, A., and Nishimura, S. (2010). "Numerical analysis of the erosion and the transport of fine particles within soils leading to the piping phenomenon," *Soils and Foundations*, **50**(4), 471–482.

Gaskin, S. J., Pieterse, J., Shafie, A. A. I., and Lepage, S. (2003). "Erosion of undisturbed clay samples from the banks of the St Lawrence River," *Canadian Journal of Civil Engineering*, **30**(3), 585–595.

Geo-Slope International Ltd. (2007). *GeoStudio 2007*, Calgary, Alberta, Canada.

Geotechnical Engineering Office (GEO) (1993). *Review of Granular and Geotextile Filters*, GEO Publication No. 1/93, Civil Engineering Department, Hong Kong.

Geotechnical Engineering Office (GEO) (1998). *Landslides and Boulder Falls from Natural Terrain: Interim Risk Guidelines*, GEO Report No. 75, Geotechnical Engineering Office, Hong Kong, 183 pp.

Gilles, D. and Moore, M. (2010). "Review of hydraulic flood modeling software used in Belgium, The Netherlands, and The United Kingdom," in *International Perspectives in Water Resource Management*, IIHR, University of Iowa, College of Engineering, Iowa, USA.

Gimenez, R. and Govers, G. (2002). "Flow detachment by concentrated flow on smooth and irregular beds," *Soil Science Society of America Journal*, **66**, 1475–1483.

Giroud, J. P. (2010). "Development of criteria for geotextiles and granular filters" in *Proceedings 9th International Conference on Geosynthetics*, Terzaghi Lecture.

Glazyrin, G. Y. and Reyzvikh, V. N. (1968). "Computation of the flow hydrograph for the breach of landslide lakes," *Soviet Hydrology*, **5**, 592–596.

Glynn, E. and Kussmaul, J. (2004). *Prediction of Piping Erosion along Middle Mississippi River Levees – an Empirical Model*, US Army Corps of Engineers, Report No. USACE ERDC/GSL TR-04-12, 36 pp.

Goldstein, S., (1929). "The steady flow of viscous fluid past a fixed spherical obstacle at small Reynolds numbers," *Proceedings of Royal Society, London*, **123**, 225–235.

Gomez, B. and Church, M. (1989). "An assessment of bed load sediment transport formulae for gravel bed rivers," *Water Resources Research*, **25**, 1161–1186.

Google Map (2014). "Ludian", *Google.cn*. Retrieved from http://ditu.google.cn/.

Gov.cn (2008). Evacuation decision-making for the Tangjiashan landslide dam. Retrieved from http://www.gov.cn/jrzg/2008–06/01/content_1001962.htm (in Chinese).

Graf, W. H. (1984). *Hydraulics of Sediment Transport*, Water Resources Publications, Vol. 29, Colorado.

Graham, W. J. (1999). *A Procedure for Estimating Loss of Life Caused by Dam Failure*, US Bureau of Reclamation, Dam Safety Office, Denver, USA, Report No. DSO-99-06, 44 pp.

Grêt-Regamey, A. and Straub, D. (2006). "Spatially explicit avalanche risk assessment linking Bayesian networks to a GIS," *Natural Hazards and Earth System Science*, **6**(6), 911–926.

Grissinger, E. H. (1966). "Resistance of selected clay systems to erosion by water," *Water Resources Research*, **2**(1), 132–138.

Guidoux, C., Faure, Y. H., Béguin, R., and Ho, C. C. (2010). "Contact erosion at the interface between granular filter and various base-soils with tangential flow," *Journal of Geotechnical and Geoenvironmental Engineering*, **136**(5), 655–775.

Hagen, V. K. (1982). "Re-evaluation of design floods and dam safety," in *Proceedings of the 14th Congress of International Commission on Large Dams*, International Commission on Large Dams, Paris, pp. 475–491.

Hancox, G. T., McSaveney, M. J., Davies, T. R., and Hodgson, K. (1999). *Mt Adams Rock Avalanche of 6 October 1999 and Subsequent Formation of a Landslide Dam in Poerua River, Westland, New Zealand*, Institute of Geological and Nuclear Sciences, Lower Hutt. Report 19, 22 pp.

Hanson, G. J. (1990). "Surface erodibility of earthen channels at high stresses part I-open channel testing," *Transactions of the ASAE*, **33**(1), 127–131.
Hanson, G. J. and Cook, K. R. (2004). "Apparatus, testing procedures, and analytical methods to measure soil erodibility in situ," *Applied Engineering in Agriculture*, **20**(4), 455–462.
Hanson, G. J. and Hunt, S. L. (2007). "Lessons learned using laboratory Jet method to measure soil erodibility of compacted soils," *Applied Engineering in Agriculture*, **23**(3), 305–312.
Hanson, G. J. and Simon, A. (2001). "Erodibility of cohesive streambeds in the loess area of the midwestern USA," *Hydrological Processes*, **15**(1), 23–38.
Hanson, G. J., Cook, K. R., and Hunt, S. L. (2005a). "Physical modelling of overtopping erosion and breach formation of cohesive embankments," *Transactions of the ASAE*, **48**(5), 1783–1794.
Hanson, G. J., Temple, D. M., Morris, M. W., Hassan, M., and Cook, K. R. (2005b). "Simplified breach analysis model: Part II, Parameter inputs and variable scale model comparisons," in *Proceedings of the 25th Annual United States Society on Dams (USSD) Conference*, Salt Lake City, Utah, pp. 163–174.
Harder, L. F. Jr., Bray, J. D., Volpe, R. L., and Rodda, K. V. (1991). "Performance of Earth Dams during the Loma Prieta Earthquake," in *Proceedings of the Second International Conference on Recent Advances in Geotechnical Earthquake Engineering and Soil Dynamics*, Paper No. LP05.
Harrison, S. (1864). "A complete history of the great flood at Sheffield on March 11 & 12, 1864," *Sheffield Times*, 13, Great Sutton Street, England. Retrieved from http://www2.shu.ac.uk/sfca/extra/A%20Complete%20History%20of%20the%20Great%20Flood%20at%20Sheffield.pdf
Hartford, D. N. D. and Baecher, G. B. (2004). *Risk and Uncertainty in Dam Safety*, Thomas Telford Ltd, London, UK.
Hayashi, T., Ozaki, S., and Ichibashi, T. (1980). "Study on bed-load transport of sediment mixtures," in *Proceedings of the 24th Japanese Conference on Hydraulics*.
He, X. R. (2009). *Summary of the emergency treatment of the Laoyingyan landslide dam in An County*. Proceeding of Survey and Research on Disaster Caused by Wenchuan Earthquake, Beijing, China (in Chinese).
He, B. S., Ding, L. Q., Wang, Y. J., Xu, W. J., and Zhao, Y. F. (2009). "Preliminary evaluation of stability of Xiaojiaqiao dammed lake in Anxian County, Sichuan Province," *Chinese Journal of Rock Mechanics and Engineering*, **28**(2), 3626–3631.
He, B. S., Wang, Y. J., Wei, J. J., and Zhang, S. C. (2010). "Survey on fourteen landslide dams in the areas of the Wenchuan earthquake," *Chinese Culture Pictorial*, **28**(6), 36–42.
Hobeika, A. G. and Jamei, B. (1985). "MASSVAC: A model for calculating evacuation times under natural disaster," in *Proceeding of the Conference on Computer Simulation in Emergency Planning*, Vol. 15, Society of Computer Simulation, La Jolla, Calif. USA.
Hofland, B., Battjes, J. A., and Booij, R. (2005). "Measurement of fluctuating pressures on coarse bed material," *Journal of Hydraulic Engineering*, **131**(9), 770–781.
Honjo, Y., Haque, M. A., and Tsai, K. A. (1996). "Self-filtration behavior of broadly and gap-graded granular soils," in *Proceedings of the 2nd International Conference on Filters and Drainage in Geotechnical and Environmental Engineering*, Geofilters, Montreal, pp. 227–236.
Howad, R. A. and Matheson, J. E. (1984). Influence diagrams. In R. A. Howard and J. E. Matheson, editors. The Principles and Applications of Decision Analysis. Volume II, pages 719–762. Strategic Decision Group. Mento Park, CA, USA.
Howard, I. L., Saucier, C. L., and Tom, J. G. (2009). *Levee Breach Geometries and Algorithms to Simulate Breach Closure*, Civil and Environmental Engineering Department, Mississippi State University. Retrieved from http://www.cee.msstate.edu.
Hromadka T. V. II, and Yen, C. C. (1987). *Diffusive Hydrodynamic Model*, Water Resources Investigations Report 87–4137, USGS, Denver Federal Center, Denver, Coloradan, USA.
HSE (Health and Safety Executive) (2001). *Reducing Risks, Protecting People: HSE's Decision-making Process*, Risk Assessment Policy Unit, HSE Books, Her Majesty's Stationery Office, London, England.
Hu, X. W., Huang, R. Q., Shi, Y. B., Lu, X. P., Zhu, H. Y., and Wang, X. R. (2009). "Analysis of blocking river mechanism of the Tangjiashan landslide and dam-breaking mode of its barrier dam," *Chinese Journal of Rock Mechanics and Engineering*, **28**(1), 181–189.
Huang, M. H., Jin, F., and Yang, W. J. (2008). "Study on the breaching parameters of the Tangjiashan landslide dam," in *Proceeding of Annual meeting of Chinese Hydraulic Engineering Society in 2008*, Haikou, Hainan province, China, pp. 961–966.
Huelman, P. and Corrin, D. (1997). "Drying Walls After a Flood," in *INFO-U of the University of Minnesota Extension Service*, No. 267, Retrieved from http://www.extension.umn.edu/info-u/household/BK267.html.
Hugin Expert A/S. (2009). *Hugin Lite 2004*, Online http://www.hugin.com/Products_Services/Products/Demo/Lite/.
HWRA (2005). *The August 1975 Catastrophic Flood Disaster in Henan*, Henan Water Resources Authority (HWRA), Yellow River Water Conservancy Press, Zhenzhou, China.

Hydrologic Engineering Center (HEC) (2008). *HEC-RAS, River Analysis System, Hydraulic Reference Manual, Version 4.0*, Hydrologic Engineering Center of US Army Corps of Engineers, Washington DC, USA.
HYDROTEC (2001). Hochwasser-Aktionsplan Angerbach. Teil I: Berichte und Anlagen. Studie im Auftrag des StUA Dusseldorf, Aachen (Quoted by Apel et al., 2009).
ICOLD. (2013). "Internal erosion processes and engineering assessment," in *ICOLD Bulletin on Internal Erosion of Existing Dams, Levees and Dikes, and Their Foundations*, Vol. 1, ICOLD, Paris, France.
ICOLD (1998a). *World Register of Dams*, International Commission on Large Dams (ICOLD), Paris.
ICOLD (1998b). *Dam-Break Flood Analysis – Review and Recommendations*. Bulletin 111, International Commission on Large Dams (ICOLD), Paris.
ICOLD (1995). *Dam Failures – Statistical Analysis, Bulletin 99*, International Commission on Large Dams (ICOLD), Paris.
ICOLD. (1994). *Embankment Dams: Granular Filters and Drains*, Bulletin 95, International Commission on Large Dams (ICOLD), Paris.
ICOLD (1988). *World Register of Dams 4th Updating*, International Commission on Large Dams (ICOLD), Paris.
ICPR (2001). *Atlas on the Risk of Flooding and Potential Damage due to Extreme Floods of the Rhine*, International Commission for the Protection of the Rhine (ICPR), Koblenz, Germany.
Indraratna, B. and Raut, A. K. (2006). "Enhanced criterion for base soil retention in embankment," *Journal of Geotechnical and Geoenvironmental Engineering*, **132**(12), 1621–1627.
Indraratna, B. and Radampola, S. (2002). "Analysis of critical hydraulic gradient for particle movement in filtration," *Journal of Geotechnical and Geoenvironmental Engineering*, **128**(4), 347–350.
Indraratna, B., Nguyen, V. T., and Rujikiatkamjorn, C. (2011). "Assessing the potential of internal erosion and suffusion of granular soils," *Journal of Geotechnical and Geoenvironmental Engineering*, **137**(5), 550–554.
Indraratna, B., Raut, A. K., and Khabbaz, H. (2007). "Constriction-based retention criterion for granular filter design," *Journal of Geotechnical and Geoenvironmental Engineering*, **133**(3), 266–276.
Indraratna, B., Vafai, F., and Dilema, E. (1996). "An experimental study of the filtration of a lateritic clay slurry by sand filters," *Proceedings of the Institution of Civil Engineering, Geotechnical Engineering*, **119**(2), 75–83.
Ingles, O. G. and Wood, C. C. (1964). "The contribution of soil and water cations to deflocculation phenomena in earth dams," in *Proceedings of the 37th Congress of Australian and New Zealand Association for the Advancement of Sciences*, Canberra, Australia.
International Electrotechnical Commission (1985). *Analysis Techniques for System Reliability-Procedure for Failure Mode Effects Analysis*, IEC-812, International Electrotechnical Commission, Geneva, Switzerland.
IPET (Interagency Performance Evaluation Task Force) (2006). *Performance evaluation of the New Orleans and Southeast Louisiana hurricane protection system-The consequences*, Vol. 7, Final draft report. US Army Corps of Engineers, Washington.
ISSMGE (2004). *Risk assessment – glossary of terms*, ISSMGE TC32 – Technical Committee on Risk Assessment and Management Glossary of Risk Assessment Terms – Version 1.
Istomina, V. S. (1957). *Filtration Stability of Soils*, Gostroizdat, Moscow, Leningrad.
IWHR (China Institute of Water Resources and Hydropower Research) (2008). *Wenchuan Earthquake Disaster Fighting, Special Subject 6: Landslide Dams*, Special Issue of Technical Abstract Bulletin in IWHR, Beijing, China.
Jansen, R. B. (1988). *Advanced Dam Engineering for Design, Construction, and Rehabilitation*, New York: Van Nostrand Reinhold.
Jennings, D. N., Webby, M. G., and Parkin, D. T. (1993). "Tunawaea landslide dam, King Country, New Zealand," *Landslide News*, **7**, 25–27.
Jensen, F. V. (1996). *An Introduction to Bayesian Networks*, Springer, New York.
Jensen, F. V. (2001). *Bayesian networks and decision graphs*, Springer, New York.
Jia, J. S., Li, N. H., Xu, Z. P., et al. (2014). *Study on Key Technology for the Safety of High CFRDs*, Beijing, China Waterpower Press.
Jiang, L. K., Zhang, S. H., and Huang, L. W. (2009). "Preliminary engineering geological analysis and emergency danger-eliminating actions at the Yibadao barrier lake-Analysis and Investigation on Seismic Damages to Projects Subjected to Wenchuan Earthquake," in *Proceedings of the 2009 Conference of the Chinese Society for Rock Mechanics and Engineering*, pp. 1041–1049.
Johnstone, W. M., Sakamoto, D., Assaf, H., and Bourban, S. (2005). "Architecture, modelling framework and validation of BC Hydro's Virtual Reality Life Safety Model" in Vrijling et al. (Eds.) *Proceedings of the International Symposium on Stochastic Hydraulics*, Nijmegen, the Netherlands.
Jonkman, S. N. (2007). *Loss of Life Estimation in Flood Risk Assessment: Theory and Applications*, Ph. D. thesis, Delft University of Technology, Delft, Netherlands.

Jonkman, S. N., Maaskant, B., Boyd, E., and Levitan, M. L. (2009). "Loss of life caused by the flooding of New Orleans after hurricane Katrina: Analysis of the relationship between flood characteristics and mortality." *Risk Analysis*, **29**(5), 676–698.

Jonkman, S. N. and Penning-Rowsell, E. (2008). "Human instability in flood flows." *Journal of the American Water Resources Association*, **44**(4), 1–11.

Jonkman, S. N., Vrijling, J. K., and Vrouwenvelder, A. C. W. M. (2008). "Methods for the estimation of loss of life due to floods: a literature review and proposal for a new method." *Natural Hazards*, **46**(3), 353–389.

Julian, J. P. and Torres, R. (2006). "Hydraulic erosion of cohesive riverbanks," *Geomorphology*, **76**(1–2), 193–206.

Kamphuis, J. W. and Hall, K. R. (1983). "Cohesive material erosion by unidirectional current." *Journal of Hydraulic Engineering*, **109**(1), 49–61.

Kamphuis, J. W., Gaskin, P. N., and Hoogendoorn, E. (1990)." Erosion tests on four intact Ontario clays," *Canadian Geotechnical Journal*, **27**(5), 692–696.

Karpoff, K. P. (1955). "The use of laboratory tests to develop design criteria for protective filters," in *Proceedings of American Society for Testing and Materials*, pp. 1183–1198.

Kaushik, C. (2006). *Urban flood modeling – A comparative study for 1D and 2D models*, UNESCO-IHE, Delft, The Netherlands.

KECL (Kunming Engineering Corporation Limited) (2014). "New progress in later risk mitigation of the Hongshiyan landslide dam shouldered by our group," *khidi.com*. Online http://www.khidi.com:8083/KHIDI2009/K_Article.asp?ids=61933&DataBaseName=XxzxMainMsg&ListName=%CD%BC%C6%AC%D0%C2%CE%C5.

Kelman, I. (2002). *Physical Flood Vulnerability of Residential Properties in Coastal, Eastern England*, Ph.D. thesis, University of Cambridge, UK.

Kelman, I. and Spence, R. (2004). "An overview of flood actions on buildings," *Engineering Geology*, **73**(3–4), 197–309.

Kenney, T. C. and Lau, D. (1985). "Internal stability of granular filters," *Canadian Geotechnical Journal*, **22**(2), 215–225.

Kenney, T. C., Chahal, R., Chiu, E., Ofoegbu, G. I., Omange, G. N., and Ume, C. A. (1985). "Controlling construction sizes of granular filters," *Canadian Geotechnical Journal*, **22**(1), 32–43.

Kezdi, A. (1979). *Soil Physics-Selected Topics*, Elsevier Scientific Publishing Co., Amsterdam.

King, J., Loveday, I., and Schuster, R. L. (1989). "The 1985 Bairaman landslide dam and resulting debris flow, Papua New Guinea," *Quarterly Journal of Engineering Geology*, **22**(4), 257–270.

Kirkpatrick, G. W. (1977). "Evaluation guidelines for spillway adequacy," in *The Evaluation of Dam Safety, Engineering Foundation Conference*, ASCE, New York, pp. 395–414.

Knapen, A., Poesen, J., Govers, G., Gyssels, G., and Nachtergaele, J. (2007). "Resistance of soils to concentrated flow erosion: A review," *Earth-Science Reviews*, **80**(1–2), 75–109.

Koenders, M. A. and Sellmeijer, J. B. (1992). "Mathematical model for piping," *Journal of Geotechnical Engineering*, **118**(6), 943–946.

Korup, O. (2002). "Recent research on landslide dams – A literature review with special attention to New Zealand," *Progress in Physical Geography*, **26**(2), 206–235.

Korup, O. (2004). "Geomorphometric characteristics of New Zealand landslide dams." *Engineering Geology*, **73**(1–2), 13–35.

Kothyari, U. C. and Jain, R. K. (2008). "Influence of cohesion on the incipient motion condition of sediment mixtures," *Water Resources Research*, **44**(W04410), 1–15.

Kovács, G. (1981). *Seepage Hydraulics. Amsterdam*, Elsevier Scientific Publishing Company.

Koženy, J. (1953). *Hydraulik: Ihre Grundlagen und Praktische Anwendung*, Springer, New York.

Kroup, O. (2002). "Recent research on landslide dams – A literature review with special attention to New Zealand," *Progress in Physical Geography*, **26**(2), 206–235.

Lafleur, J. (1984). "Filter testing of broadly graded cohesionless tills," *Canadian Geotechnical Journal*, **21**(4), 634–643.

Lafleur, J. and Nguyen, P. H. (2007). "Internal stability of particles in dam cores made of granular broadly graded moraines" in *Internal Erosion of Dams and their Foundations*, Taylor & Francis Group, London, pp. 151–158.

Lafleur, J., Mlynarek, J., and Rollin, A. L. (1989). "Filtration of broadly graded granular soils," *Journal of Geotechnical Engineering*, **115**(12), 1747–1769.

Lambe, T. W., Marr, W. A., and Francisco, S. (1981). "Safety of a constructed facility: geotechnical aspects" *Journal of Geotechnical Engineering*, **107**(3), 339–352.

Langford, J. W. (1995). *Logistics: Principles and Applications*, McGraw Hill, New York.

Lee, K. L. and Duncan, J. M. (1975). *Landslide of April 25, 1974, on the Mantaro River, Peru*, Committee on Natural Disasters, Commission on Sociotechnical Systems, National Research Council, 71 pp.

Lee, M. K., Lum, K. Y., and Hartford, D. N. D. (1998). "Calculation of the seismic risk of an earth dam susceptible to liquefaction," *Geotechnical Earthquake Engineering and Soil Dynamics III*, Geotechnical Special Publication No. **75**, Reston, ASCE, 1451–1460.

Lee, R., Hu, P. S., Neal, D. M., Ogles, M. R., Sorensen, J. H., and Trumble, D. A. (1986). *Predicting Loss of Life From Floods*, Oak Ridge National Laboratory Draft Report prepared for the Institute for Water Resources, US Army Corps of Engineers. 131 pp.

Lefebvre, G., Rohan, K., and Douville, S. (1985). "Erosivity of natural intact structured clay: Evaluation," *Canadian Geotechnical Journal*, **22**(4), 508–517.

Lefebvre, G., Rohan, K., and Milette, J. P. (1986). "Erosivity of intact clay: Influence of the natural structure," *Canadian Geotechnical Journal*, **23**(4), 427–434.

Li, D. Q., Yan, L. L., and Shao, D. G. (2007). "Reliability evaluation of embankment dams using Bayesian network." *Engineering Journal of Wuhan University*, **40**(6), 24–29.

Li, G. X., Jie, Y. X., and Li, Q. Y. (2003). "Yangtze dyke and its strengthening." *Lowland Technology International*, **5**(2), 39–46.

Li, L., Wang, R. Z., Sheng, J. B., Wang, Z. S., Peng, X. H., and Zhang, S. C. (2006). *Risk Assessment and Risk Management for Dam*, Water Power Press, Beijing (in Chinese).

Li, M. (2008). *Seepage Induced Instability in Widely Graded Soils*, Ph.D. thesis, The University of British Columbia, Canada.

Li, M. and Fannin, R. J. (2011). "A theoretical envelope for internal instability of cohesionless soil," *Geotechnique*, **62**(1), 77–80.

Li, M. and Fannin, R. J. (2008). "Comparison of two criteria for internal stability of granular soil," *Canadian Geotechnical Journal*, **45**(9), 1303–1309.

Li, P. Y., Zhou, X. Y., Wu, Q. F., and Chen, Y. S. (2008). "Comprehensive introduction to disaster relief and strengthening of earthquake-induced dammed lake," *Journal of Yangtze River Scientific Research Institute*, **25**(6), 52–57.

Li, T. C., Schuster, R. L., and Wu, J. S. (1986). "Landslide dams in south-central China," in *Landslide dams-Processes, Risk, and Mitigation*, Geotechnical Special Publication No. 3, Schuster, R. L., Ed., American Society of Civil Engineers, pp. 146–162.

Liang, J. (2009a). *The Disposal Strategy of the Landslide Dams Induced by the Wenchuan Earthquake*, Report in Sichuan Province Water Resources Department, Sichuan, China.

Liang, J. (2009b). "Analysis and investigation on seismic damages to projects subjected to Wenchuan earthquake-formation of seismic barrier lake and its harnessing scheme," in *Proceedings of the 2009 Conference of the Chinese Society for Rock Mechanics and Engineering*, 1056–1061.

Liang, J. (2009c). "Analysis and investigation on seismic damages to projects subjected to Wenchuan earthquake-overview of Wenchuan earthquake barrier lakes and danger-eliminating actions," in *Proceedings of the 2009 Conference of the Chinese Society for Rock Mechanics and Engineering*, 999–1007.

Lim, S. S. (2006). *Experimental Investigation of Erosion in Variably Saturated Clay Soils*, Ph.D. thesis, The University of New South Wales, Sydney, Australia.

Lim, S. S. and Khalili, N. (2009). "An improved rotating cylinder test design for laboratory measurement of erosion in clayey soils" *Geotechnical Testing Journal*, **32**(3), 1–7.

Lin, B., Wicks, J. M., Falconer, R. A., and Adams, K. (2006). "Integrating 1D and 2D hydrodynamic models for flood simulation," in *Proceedings of the Institution of Civil Engineers, Water Management*, **159**(1), 19–25.

Lin, Z. and Gu, X. M. (2005). "Comments on the piping problem in Shijiao segment of the North River levee system," *Guangdong Water Resources and Hydropower*, **5**, 17–21.

Lind, N., Hartford, D., and Assaf, H. (2004). "Hydrodynamic models of human stability in a flood," *Journal of the American Water Resources Association*, **40**(1), 89–96.

Lindell, M. K. (2008). "EMBLEM2: An empirically based large scale evacuation time estimate model," *Transportation Research Part A: Policy and Practice*, **42**(1), 140–154.

Lindell, M. K. and Perry, R. W. (1992). *Behavioral Foundations of Community Emergency Planning*, Hemisphere Press, Washington DC.

Lindell, M. K., Prater, C. S., and Peacock, W. G. (2007). "Organizational communication and decision-making for hurricane emergencies," *Natural Hazards Review*, **8**(3), 50–60.

Lindell, M. K., Prater, C. S., Perry, R. W., and Wu, J. Y. (2002). *EMBLEM: An Empirically Based Large Scale Evacuation Time Estimate Model*, Texas A&M University Hazard Reduction & Recovery Center, College Station, TX, USA.

Liu, F., Zhang, S. S., Huang, L.W. (2009a) Study on emergency danger-eliminating construction schemes at Xiaogangjian Barrier Lake. Analysis and investigation on seismic damages to projects subjected to Wenchuan earthquake, Proc., 2009 Conference of the Chinese Society for Rock Mechanics and Engineering, pp. 1033–1040 (in Chinese).

Liu, H. and Du, M. (2010). "Disposing the Xiaogangjian landslide dam by the Armed Police Hydropower Troops", chinamil.com.cn, online: http://chn.chinamil.com.cn/xwpdxw/2010- 08/19/content_4282512_3.htm.

Liu, J. (2005). *Seepage Control of Earth-rock Dams: Theoretical Basis, Engineering Experiences and Lessons*, China Waterpower Press, Beijing.

Liu, J., Cui, Y. H., and Xie, D. S. (2008). "Failure mechanism of seepage in uniform sand foundation of river levee," *Chinese Journal of Geotechnical Engineering*, **30**(5), 642–645.

Liu, N. (2014). "Hongshiyan landslide dam danger disposal and coordinated management," *Engineering Science*, **16**(10), 39–46.

Liu, N. (2010). "Practice and cognition for emergency disposal of landslide lakes," *Advances in Water Science*, **21**(4), 541–549.

Liu, N. (2008). "Landslide dams in Wenchuan earthquake and the risk mitigation measures" in *Keynote lecture of Forum on Earthquake Relief vs Science and Technology*, Chinese Academy of Sciences, Chengdu, China.

Liu, N., Chen, Z. Y., Zhang, J. X., Lin, W., Chen, W. Y., and Xu, W. J. (2010). "Draining the Tangjiashan barrier lake," *Journal of Hydraulic Engineering*, **136**(11), 914–923.

Liu, N., Zhang, J. X., Lin, W., Cheng, W. Y., and Chen, Z. Y. (2009b). "Draining Tangjiashan barrier lake after Wenchuan earthquake and the flood propagation after the dam break," *Science in China Series E: Technological Sciences*, **52**(4), 901–809.

Liu, Y. (2009). *An Explicit Risk-Based Approach for Large-Levee Safety Decisions*, MPhil thesis, The Hong Kong University of Science and Technology, Hong Kong.

Lo, K. Y. and Kaniaru, K. (1990). "Hydraulic fracture in earth and rockfill dams," *Canadian Geotechnical Journal*, **27**(4), 496–506.

Locke, M. and Indraratna, B. (2002). "Filtration of broadly graded soils: the reduced PSD method," *Geotechnique*, **52**(4), 285–287.

Lominé, F., Scholtès, L., Sibille, L., and Poullain, P. (2013). "Modeling of fluid–solid interaction in granular media with coupled lattice Boltzmann/discrete element methods: application to piping erosion," *International Journal for Numerical and Analytical Methods in Geomechanics*, **37**(6), 577–596.

Luo, Z. W. (2008). Application of casting blast to emergency treatment of risk-elimination of Ma'anshi landslide dam. Water Resour. Hydropower Eng. **39**(8), 31–33 (in Chinese).

MacDonald, T. C. and Langridge-Monopolis, J. (1984). "Breaching characteristics of dam failures," *Journal of Hydraulic Engineering*, **110**(5), 567–586.

Mansur, C., Postol, G., and Salley, J. R. (2000). "Performance of relief well systems along Mississippi River levees," *Journal of Geotechnical and Geoenvironmental Engineering*, **126**(8), 727–738.

Mao, C. X. (2005). "Study on piping and filters: Part I of piping," Rock and Soil Mechanics, **26**(2), 209–215.

Marot, D., Le, V. D., Garnier, J., Thorel, L., and Audrain, P. (2012). "Study of scale effect in an internal erosion mechanism," *European Journal of Environmental and Civil Engineering*, **16**(1), 1–19.

McCann, M. W., Franzini, J. B., Kavazanjian, E., and Shah, H. C. (1985). *Preliminary Safety Evaluation of Existing Dams, Vol. 1 and 2*, The John A. Blume Earthquake Engineering Center, Report No. 69 and 70, Stanford University, California.

McClelland, D. M. and Bowles, D. S. (2002). *Estimating Life Loss for Dam Safety Risk Assessment – A Review and New Approach*, IWR Report 02-R-3, Institute for Dam Safety Risk Management Utah State University, Logan.

McClelland, D. M. and Bowles, D. S. (1999). "Life-loss estimation: what can we learn from case histories," in *Proc. of the Australian Committee on Large Dams (ANCOLD) Annual Meeting*, Jindabyne, New South Wales, Australia.

Merz, B., Kreibich, H., Schwarze, R., and Thieken, A. (2010). "Review article: Assessment of economic flood damage," *Natural Hazards and Earth System Science*, **10**(8), 1697–1724.

Meyer, W., Schuster, R. L., and Sabol, M. A. (1994). "Potential for seepage erosion of landslide dam," *Journal of Geotechnical Engineering*, **120**(7), 1211–1229.

Meyer-Peter, R. and Müller, R. (1948). "Formulas for bed-load transport," in *Proceedings of the 2nd Meeting International Association of Hydraulic Research*, Stockholm, pp. 39–64.

Meyer-Peter, R., Favre, H., and Einstein, H. A. (1934). "Neuere versuchsresultate tiber den geschiebetrieb," *Schweiz. Bauztg.*, **103**, 147–150.

Mianyang Bureau of Statistics (2009). *Report on the National Economy and Society Development on Mianyang City in 2008*, Mianyang Bureau of Statistics, Sichuan Province, China. Online http://my.gov.cn/bmwz/942947769050464256/20090325/391646.html.

Mileti, D. S. (1995). *Hydrometeorology, Impacts, and Management of Extreme Floods-Factors Related to Flood Warning Response*, Perugia, Italy.

Mileti, D. S. and Peek, L. (2000). "The social psychology of public response to warnings of a nuclear power plant accident," *Journal of Hazardous Materials*, **75**(2–3), 181–194.

Miller, A. C., Merkhofer, M. M., Howard, R. A., Matheson, J. E., and Rice, T. T. (1976). Development of automated aids for decision analysis. Technical report. Stanford Research institute.

Milligan, V. (2003). "Some uncertainties in embankment dam engineering," *Journal of Geotechnical and Geoenvironmental Engineering*, **129**(9), 785–797.

Milligan, V. (1986). "Internal stability of granular filters: Discussion," *Canadian Geotechnical Journal*, **23**(3), 414–418.

Mitchener, H. and Torfs, H. (1996). "Erosion of mud/sand mixtures," *Coastal Engineering*, **29**(1–2), 1–25.

Moffat, R. M. (2005). *Experiments on the Internal Stability of Widely Graded Cohesionless Soils*, Ph.D. thesis, The University of British Columbia, Canada.

Moffat, R. M., Fannin, R. J., and Garner, S. J. (2011). "Spatial and temporal progression of internal erosion in cohesionless soil," *Canadian Geotechnical Journal*, **48**(3), 399–412.

Mohamed, M. (2002). *Embankment Breach Formation and Modeling Methods*, Ph.D. thesis, The Open University, England.

Moore, W. L. and Masch, F. D. (1962). "Experiments on the scour resistance of cohesive sediments," *Journal of Geophysical Research*, **67**(4), 1437–1446.

Mora, S., Madrigal, C., Estrada, J., and Schuster, R. L. (1993). "The 1992 Rio Toro landslide dam, Costa Rica," *Landslide News*, **8**, 19–22.

Morris, M. (2005). *Investigation of Extreme Flood Processes and Uncertainty*, Technical report, IMPACT.

Morris, M. W. (2000). *CADAM Concerted Action on Dam-break Modelling*, Report SR 571, HR Wallingford.

Morris, M. W. (2011). *Breaching of Earth Embankments and Dams*, Ph.D. thesis, Open University, Milton Keynes, UK.

Morris, M. W., Hassan, M. A. A. M., and Vaskinn, K. A. (2005). "Conclusions and recommendations from the IMPACT Project WP2: Breach formation," *Impact Project*. Retrieved from http://www.impact-project.net.

Morris, M., Hassan, M., Buchholzer, Y., and Davies, T. (2008). "HR-BREACH: Developing a practical breach model to meet industry needs," in *Proceedings of the Annual 28th United States Society on Dams (USSD) Conference*, Portland, Oregon, pp. 753–766.

Moss, R. E. S. and Eller, J. M. (2007). "Estimating the probability of failure and associated risk of the California Bay Delta Levee System," in *Probabilistic Applications in Geotechnical Engineering*, Geotechnical Special Publication, No. 170, American Society of Civil Engineers, New York, pp. 215–224.

Munger, D. F., Bowles, D. S., Boyer, D. D., Davis, D. W., Margo, D. A., Moser, D. A., Regan, P. J., and Snorteland, N. (2009). "Interim tolerable risk guidelines for US Army Corps of Engineers dams," in *Proceedings of the US Society on Dams 2009 Annual Lecture*, Nashville, Tennessee.

MURL (2000) Potentielle Hochwasserscha¨den am Rhein in NRW – Ministerium fu¨r Umwelt. Raumordnung und Landwirtschaft des Landes Nordrhein-Westfalen, Du¨sseldorf (quoted by Apel et al., 2009).

MWR (2009). *Standard for Classification of Risk Grade of Landslide Lakes*, Standards in Water Industry of the People's Republic of China, SL 450–2009, Ministry of Water Resources of the People's Republic of China (MWR), Beijing.

MWR (2008). *Special Plan for Rehabilitation of Dangerous Reservoirs*, Ministry of Water Resources of the People's Republic of China (MWR), Beijing (in Chinese).

MWR (1994). *Technical Codes on Earth and Rockfill Dam Safety Monitoring*, Ministry of Water Resources of the People's Republic of China (MWR), Beijing (in Chinese).

MWR. (1993). *National Inventory of Reservoir Dam Failures During 1981–1990*, Ministry of Water Resources of the People's Republic of China (MWR), Beijing (in Chinese).

MWR (1981). *National Inventory of Reservoir Dam Failures*, Ministry of Water Resources of the People's Republic of China (MWR), Beijing (in Chinese).

Nadal, N. C., Zapata, R. E., Pagan, I., Lopez, R., and Agudelo, J. (2010). "Building damage due to riverine and coastal floods," *Journal of Water Resources Planning and Management*, **136**(3), 327–336.

Nagy, L. (2007). *Védekezés az árvízvédelmi gátak suvadása ellen*, Innova-Print, Budapest, Hungary (in Hungarian).

Nagy, L. and Tóth, S. (2005). "WP6: Monitoring and case study, detailed technical report on the collection and analysis of dike breach data with regards to the process and location factors," Impact Project, Retrieved from http://www.impact-project.net.

Nash, T., Bell, D., Davies, T., and Nathan, S. (2008). "Analysis of the formation and failure of Ram Creek landslide dam, South Island, New Zealand", *New Zealand Journal of Geology and Geophysics*, **51**(3), 187–193.

Nash, J. E. (1957). "The form of the instantaneous unit hydrograph," *Hydrol. Sci.*, **45**, 114–121.

Neill, C. R. (1973). *Guide to Bridge Hydraulics*, University of Toronto Press, Toronto.

Nicoletti, P. G. and Parise, M. (2001). "Seven landslide dams of old seismic origin in south-eastern Sicily, Italy," *Geomorphology*, **46**(3–4), 203–222.

Nielsen, N. M., Hartford, D. N. D., and MacDonald, J. J. (1994). "Selection of tolerable risk criteria for dam safety decision-making," in *Proc. 1994 Canadian Dam Safety Conference*, Winnipeg, Manitoba, BiTech Publishers, Vancouver, pp. 355–369.

Niu, Y. G. (2002). *Case Studies on Rehabilitation of Dangerous Reservoirs*, Water Power Press, Beijing.

National Science Foundation (NSF) (2006). "Investigation of the performance of New Orleans flood protection systems in Hurricane Katrina on August 29, 2005, Volume I: Main Text and Executive Summary." Civil and Environmental Engineering, University of California, Berkley. Online http://www.ce.berkeley.edu.

Oak Ridge National Laboratory (1999). *Oak Ridge Evacuation Modeling System (OREMS) User's Guide*, Center for Transportation Analysis, ORNL, Oak Ridge, Tenn.

O'Brien, J. S., Julien, P. Y., and Fullerton, W. T. (1993). "Two-dimensional water flood and mudflow simulation," *Journal of Hydraulic Engineering*, **119**(2), 244–261.

Olmsted, S. M. (1983). On representing and solving decision problems. PhD thesis, Engineering Economics Department, Stanford University.

Osman, A. M. and Thorne, C. R. (1988). "Riverbank stability analysis. I: Theory," *Journal of Hydraulic Research*, **114**(2), 134–150.

Parker, D. J., Green, C. H., and Thompson, P. M. (1987). *Urban Flood Protection Benefits: A Project Appraisal Guide, Gower Technical Press*, Aldershot, Hampshire, UK.

Parker, D. J., Priest, S. M., and Tapsell, S. M. (2009). "Understanding and enhancing the public's behavioral response to flood warning information," *Meteorological Applications*, **16**(1), 103–114.

Parker, G., Kilingeman, P. C., and McLean, D. G. (1982). "Bed-load and size distribution in paved gravel-bed streams," *ASCE Journal of Hydraulic Division*, **108**(4), 544–571.

Patro, S., Chatterjee, C., Mohanty, S., Singh, R., and Raghuwanshi, N. S. (2009). "Flood Inundation Modeling using MIKE FLOOD and Remote Sensing Data," *Journal of the Indian Society of Remote Sensing*, **37**(1), 107–118.

PBS&J (Post, Buckley, Schuh & Jernigan, Inc.) (2000). *Southeast United States Hurricane Evacuation Traffic Study: Evacuation Travel Demand Forecasting System, Technical Memorandum 2*, Final Rep., Tallahassee, Fla.

Peng, M. (2012). *Dam Risk Analysis and Dynamic Decision-making*, Ph.D. thesis, the Hong Kong University of Science and Technology, Hong Kong.

Peng, M. and Zhang, L. M. (2013a). "Dynamic decision-making for dam-break emergency management-Part 1: Theoretical framework," *Natural Hazards and Earth System Science*, **13**(2), 425–437.

Peng, M. and Zhang, L. M. (2013b). "Dynamic decision-making for dam-break emergency management-Part 2: Application to Tangjiashan landslide dam failure," *Natural Hazards and Earth System Science*, **13**(2), 439–454.

Peng, M. and Zhang, L. M. (2012a). "Breaching parameters of landslide dams," *Landslides*, **9**(1), 13–31.

Peng, M. and Zhang, L. M. (2012b). "Analysis of human risks due to dam break floods – part 1: A new model based on Bayesian networks," *Natural Hazards*, **64**(1), 903–933.

Peng, M. and Zhang, L. M. (2012c). "Analysis of human risk due to dam break floods – part 2: Application to Tangjiashan Landslide Dam failure," *Natural Hazards*, **64**(2), 1899–1923.

Peng, M. and Zhang, L. M. (2011). "Breaching of Changkai Levee in June 2010 in Jiangxi Province, China," Geotechnical Special Publication No. 224, pp. 476–483.

Peng, M. and Zhang, L. M. (2010). "Risk analysis of Tangjiashan landslide dam," in *GeoFlorida 2010: Advances in Analysis, Modeling & Design*, Geotechnical Special Publication No. 199, American Society of Civil Engineers, pp. 2192–2201.

Peng, M., Zhang, L. M., Chang, D. S., and Shi, Z. M. (2014). "Engineering risk mitigation measures for the landslide dams induced by the 2008 Wenchuan earthquake," *Engineering Geology*, **180**, 68–84.

Penning-Rowsell, E. C. and Chatterton, J. B. (1977). *The Benefits of Flood Alleviation: A Manual of Assessment Techniques*, Gower, Aldershot, Hampshire.

Penning-Rowsell, E. C., Green, C. H., Thompson, P. M., Coker, A. M., Tunstall, S. M., Richards, C., and Parker, D. J. (1992). *The Economics of Coastal Management: A Manual of Benefit Assessment Techniques*. Belhaven Press, London, UK and Florida, USA.

Penning-Rowsell, E., Johnson, C., Tunstall, S., Tapsell, S., Morris, J., Chatterton, J., and Green, C. (2005). *The Benefits of Flood and Coastal Risk Management: A Manual of Assessment Techniques*, Middlesex University Press, UK.

Petak, W. J. and Atkisson, A. A. (1982). *Natural Hazard Risk Assessment and Public Policy, Anticipating the Unexpected*, Springer-Verlag, Inc., New York.

Peviani, M. A. (1999). *Simulation of Earth-Dams Breaking Processes by Means of A Morphological Numerical Model*, Concerted Action on Dam-Break (CADAM) Meeting, Milano, Italy.

Peyras, L., Royet, P., and Boissier, D. (2006). "Dam ageing diagnosis and risk analysis: Development of methods to support expert judgment," *Canadian Geotechnical Journal*, **43**(2), 169–186.

Phillips, B. C., Yu, S., Thompson, G. R., and de Silva, N. (2005). "1D and 2D modelling of urban drainage systems using XP-SWMM and TUFLOW," in *Proc., 10th Int. Conf. on Urban Storm Drainage*, DTU, Copenhagen, Denmark.

Phoon, K. K. and Ching, J. Y. (2015). *Risk and Reliability in Geotechnical Engineering*, CRC Press, Taylor & Francis Group, Boca Raton, FL.

Pistrika, A. K. and Jonkman, S. N. (2010). "Damage to residential buildings due to flooding of New Orleans after hurricane Katrina." *Natural Hazards*, **54**(2), 413–434.

Plaza-Nieto, G., Yepes, H., and Schuster, R. L. (1990). "Landslide dam on the Pisque river, northern Ecuador," *Landslide News*, **4**, 2–4.

Plaza-Nieto, G. and Zevallos, O. (1993). "The 1993 La Josefina rockslide and Rio Paute landslide dam, Ecuador," *Landslide News*, **8**, 4–6.

Ponce, V. M. and Tsivoglou, A. J. (1981). "Modelling gradual dam breaches," *Journal of Hydraulic Division*," **107**(7), 829–838.

Powledge, G. R., Ralston, D. C., Miller, P., Chen, Y. H., Clopper, P. E., and Temple, D. M. (1989). "Mechanics of overflow erosion on embankments II: Hydraulics and design considerations," *Journal of Hydraulic Research*, **115**(8), 1056–1075.

Priest, S., Wilson, T., Tapsell, S., Penning-Rowsell, E., Viavattene, C., and Fernandez-Bilbao, A. (2007). *Building A Model to Estimate Risk to Life for European Flood Events*, Final Report, Integrated Flood Risk Analysis and Management Methodologies, FLOODsite. Online http://www.floodsite.net/html/partner_area/search_results3b.asp?docID=265.

Prina, E., Bonnard, C., and Vulliet, L. (2004). "Vulnerability and risk assessment of a mountain road crossing landslides," *Rivista Italiana di Geotecnica*, Anno XXXVIII, **2**, 67–79.

QQ News (2014). Photos: *The Ludian Landslide Drainage Completed and the Submerged Village Seen*, Online http://news.qq.com/a/20141007/001967.htm.

Quiroga, C. A. and Singh, V. P. (1987). "A dam-breach erosion model: II. Application," *Water Resources Management*, **1**(3), 199–221.

Radwan, A. E., Hobeika, A. G., and Sivasailam, D. (1985). "A computer simulation model for rural network evacuation under natural disasters," *ITE Journal*, **55**(9), 25–30.

Ramsbottom, D., Floyd, P., and Penning-Rowsell, E. (2003). *Flood Risks to People – Phase 1*, R&D Technical Report FD2317TR, Department for Environment, Food & Rural Affairs – Flood Management Division, London.

Rao, X. J. and Tang, X. L. (2008). "Emergency treatment on cascade landslide dam-lakes of Tangjiashan, Hongyan-Sunjiayuanzi and Guanzipu," *Water Resources and Hydropower Engineering*, **39**(8), 33–35.

Rasmussen, N. C. (1975). *Reactor Safety Study: An Assessment of Accident Risk in US Nuclear Power Plants*, US Nuclear Regulatory Commission.

Raudkivi, A. J. (1990). *Loose Boundary Hydraulics*, Pergamon Press, New York.

Raut, A. K. and Indraratna, B. (2008). "Further advancement in filtration criteria through constriction-based techniques", *Journal of Geotechnical and Geoenvironmental Engineering*, **134**(6), 883–887.

Reddi, L. N., Lee, I. M., and Bonala, V. S. (2000). "Comparison of internal and surface erosion using flow pump tests on a sand-kaolinite mixture," *Geotechnical Testing Journal*, **23**(1), 116–122.

Ren, X. and Dang, Y. P. (2008). "Construction organization and construction management for emergency risk-elimination of landslide dams in Qingchuan County," *Water Resources and Hydropower Engineering*, **39**(8), 27–30.

RESCDAM (2000). *The Use of Physical Models in Dam-Break Flood Analysis: Rescue Actions Based on Dam-Break Flood Analysis*, Final report of Helsinki University of Technology, Helsinki, Finland, 57 pp.

Richards, K. S. and Reddy, K. R. (2012). "Experimental investigation of initiation of backward erosion piping in soils," *Geotechnique*, **62**(10), 933–942.

Richards, K. S. and Reddy, K. R. (2007). "Critical appraisal of piping phenomena in earth dams," *Bulletin of Engineering Geology and the Environment*, **66**(4), 381–402.

Richardson, J. F. and Zaki, W. N. (1954). "Sedimentation and fluidization. Part I," *Transactions of the Institution of Chemical Engineers*, **32**, 35–53.

Rickenmann, D. (1991). "Hyperconcentrated flow and sediment transport at steep slopes," *Journal of Hydraulic Engineering*, **117**(11), 1419–1439.

Ripley, C. F. (1986). "Internal stability of granular filters: Discussion," *Canadian Geotechnical Journal*, **23**(2), 255–258.

Robinson, K. M. and Hanson, G. J. (2001). "Headcut erosion research," in *Proceedings of the Seventh Federal Interagency Sedimentation Conference*, Reno, Nevada.

Rogers, G. O. and Sorensen, J. H. (1989). "Warning and response in two hazardous materials transportation accidents in the US," *Journal of Hazardous Materials*, **22**(1), 57–74.

Rogers, G. O. and Sorensen, J. H. (1991). Diffusion of emergency warning: comparing empirical and simulation results. Risk Analysis, Edited by Zervos, C., Plenum Press, New York,, 117–134.

Rogers, J. D., Boutwell, G. P., Schmitz, D. W., Karadeniz, D., Watkins, C. M., Athanasopoulos-Zekkos, A. G., and Cobos-Roa, D. (2008). "Geologic conditions underlying the 2005 17th Street Canal levee failure in New Orleans," *Journal of Geotechnical and Geoenvironmental Engineering*, **134**(5), 583–601.

Rogers, W. A., Lamson, N., and Rousseau, G. K. (2000). "Warning research: an integrative perspective," *Human Factors*, **42**(1), 102–139.

Roos, I. W. (2003). *Damage to Buildings*. Delft Cluster-publication *DC1–233–9*, Delft, the Netherland.

Rousseeuw, P. J. (1998). "Chapter 17: Robust estimation and identifying outliers" in *Handbook of Statistical Methods for Engineers and Scientists*, 2nd ed., H. M. Wadsworth Jr., Eds., McGraw–Hill, New York, 17.1–17.15.

Ru, N. H. and Niu, Y. G. (2001). *Embankment Dam – Incidents and Safety of Large Dams* Water Power Press, Beijing.

Sagala, S. A. H. (2006). *Analysis of Flood Physical Vulnerability in Residential Areas*, M. S. thesis, International Institute for Geoinformation Science and Earth Observation, Enschede, the Netherlands.

Sager, J. W. and Chambers, D. R. (1986). "Design and construction of the Spirit Lake outlet tunnel, Mount St Helens, Washington," in *Landslide Dams: Processes, Risk, and Mitigation*, Geotechnical Special Publication No. 3, ASCE, Reston, 42–58.

Sasanakul, I., Vanadit-Ellis, W., Sharp, M., Abdoun, T., Ubilla, J., Steedman, S., and Stone, K. (2008). "New Orleans levee system performance during Hurricane Katrina: 17th Street Canal and Orleans Canal North," Journal of Geotechnical and Geoenvironmental Engineering, **134**(5), 657–667.

Sato, H. P. and Harp, E. L. (2009). "Interpretation of earthquake-induced landslides triggered by the 12 May 2008, M7.9 Wenchuan earthquake in the Beichuan area, Sichuan Province, China using satellite imagery and Google Earth," *Landslides*, **6**(2), 153–159.

Sattar, A. and Konagai, K. (2012). "Recent landslide damming events and their hazard mitigation strategies," *Advances in Geotechnical Earthquake Engineering – Soil Liquefaction and Seismic Safety of Dams and Monuments*, Abbas Moustafa (Ed.), InTech, Rijeka, Croatia.

SBC (Sinohydro Bureau 14 CO., LTD), (2014). *The design drawing of drainage tunnel.* Provided by Wang Shuping.

Scawthorn, C., Blais, N., Seligson, H., Tate, E., Mifflin, E., Thomas, W., Murphy, J., and Jones, C. (2006a). "HAZUS-MH flood loss estimation methodology-I: Overview and flood hazard characterization," *Natural Hazards Review*, **7**(2), 60–71.

Scawthorn, C., Flores, P., Blais, N., Seligson, H., Tate, E., Chang, S., Mifflin, E., Thomas, W., Murphy, J., Jones, C., and Lawrence, M. (2006b). "HAZUS-MH flood loss estimation methodology-II: Damage and loss assessment," *Natural Hazards Review*, **7**(2), 72–81.

Schachter, R. D. and Peot, M. A. (1986). "Devaluating influence diagrams." *Operations Research*, **34**(6), 871–882.

Schanze, J., Zeman, E., and Marsalek, J. (2006). *Flood Risk Management: Hazards, Vulnerability and Mitigation Measures*, Springer, Dordrecht, the Netherlands.

Schmitz, S. (2007). *Zur Hydraulischen Kontakterosion Bei Bindigen Basiserdstoffen*, Ph.D. thesis, Universität der Bundeswehr, Munich, Germany.

Schuler, U. (1995). "How to deal with the problem of suffusion," in *Research and Development in the Field of Dams*, SNCLD, Crans-Montana, Switzerland, pp. 145–159.

Schuster, R. L. (2006). "Risk-reduction measures for landslide dams," *Journal of Engineering Geology and Environment*, Special Issue, **1**, 9–13.

Schuster, R. L. (1985). "Landslide dams in the Western United States," in *Proceedings of the IVth International Conference and Field Workshop on Landslides*, Tokyo, 1985, pp. 411–418.

Schuster, R. L. and Evans, S. G. (2011). "Engineering measures for the hazard reduction of landslide dams," in *Natural and Artificial Rockslide Dams*, Lecture Notes in Earth Sciences, Evans, S. G., Hermanns, R. L., Strom, A., and Scarascia-Mugnozza, G., Eds., pp. 77–100.

Schuster, R. L. and Costa, J. E. (1986). "A perspective on landslide dams," in *Landslide dams-Processes, Risk, and Mitigation*, Geotechnical Special Publication No. 3, Schuster, R. L., Ed., American Society of Civil Engineers, pp.1–20.

Seed, H. B. and Duncan, J. M. (1987). "The failure of Teton Dam," *Engineering Geology*, **24**(1–4), 173–205.

Seed, R. B., Bea, R. G., Athanasopoulos-Zekkos, A., Boutwell, G. P., Bray, J. D., Cheung, C., Cobos-Roa, D., Ehrensing, L., Harder, L. F., Pestana, J. M., Riemer, M. F., Rogers, J. D., Storesund, R., Vera-Grunauer, X., and Wartman, J. (2008). "New Orleans and Hurricane Katrina-II: The Central Region and the Lower Ninth Ward," *Journal of Geotechnical and Geoenvironmental Engineering*, **134**(5), 718–739.

Sellmeijer, J. B. (1988). *On the Mechanism of Piping under Impervious Structures*, Ph.D. thesis, TU Delft. The Netherland.

Sellmeijer, J. B. and Koenders, M. A. (1991). "A mathematical model for piping," *Applied Mathematical Modelling*, **115**, 646–661.

Sellmeijer, H., De La Cruz, J. L., Van Beek, V. M., and Knoeff, H. (2011). "Fine-tuning of the piping model through small-scale, medium-scale and IJkdijk experiments," *European Journal of Environmental and Civil Engineering*, **15**(8), 1139–1154.

Shainberg, I., Goldstein, D., and Levy, G. J. (1996). "Rill erosion dependence on soil water content, aging, and temperature," *Soil Science Society of America Journal*, **60**(3), 916–922.

Shang, Y., Yang, Z., Li, L., Liu, D., Liao, Q., and Wang, Y. (2003). "A super-large landslide in Tibet in 2000: Background, occurrence, disaster, and origin," *Geomorphology*, **54**(3–4), 225–243.

Shannon Wilson, Inc. and Wolff, T. F. (1994). *Probability Models for Geotechnical Aspects of Navigation Structures*, Report to the St Louis District, US Army Corps of Engineers.

Sharpe, E. M. (1995). *Capitalism and Calamity; The Mill River Flood of 1874*, Dissertation, University of Delaware, Newark, Delaware, US.

Sheffi, Y., Mahmassani, H., and Powell, W. B. (1982). "A transportation network evacuation model." *Transportation Research Part A*, **16**(3), 209–218.

Sherard, J. L. (1986). "Hydraulic fracturing in embankment dams," *Journal of Geotechnical Engineering*, **112**(10), 905–927.

Sherard, J. L. (1979). "Sinkholes in dams of coarse, broadly graded soils," *Proc. 13th International Congress on Large Dams*, New Delhi, 325–334.

Sherard, J. L. and Dunnigan, L. P. (1989). "Critical filters for impervious soils," *Journal of Geotechnical Engineering*, **115**(7), 927–947.

Sherard, J. L. and Dunnigan, L. P. (1986). "Internal stability of granular filters: Discussion," *Canadian Geotechnical Journal*, **23**(3), 418–420.

Sherard, J. L. and Dunnigan, L. P. (1985). "Filters and leakage control," in *Proceedings Symposium on Seepage and Leakage from Dams and Impoundments*, pp. 1–30.

Sherard, J. L., Decker, R. S., and Dunnigan, L. P. (1976a). "Identification and nature of dispersive soils," *Journal of the Geotechnical Engineering Division*, **102**(4), 287–301.

Sherard, J. L., Decker, R. S., and Ryker, N. L. (1972). "Hydraulic fracturing in low dams of dispersive clay." *Proceedings of Specialty Conference on Performance of Earth and Earth Supported Structures*, pp. 653–689.

Sherard, J. L., Dunnigan, L. P., and Talbot, J. (1984). "Basic properties of sand and gravel filters," *Journal of Geotechnical Engineering*, **110**(6), 684–700.

Sherard, J. L., Dunnigan, L. P., Decker, R. S., and Steel, E. F. (1976b). "Pinhole test for identifying dispersive soils" *Journal of the Geotechnical Engineering Division*, **102**(1), 69–85.

Shields, A. (1936). "Application of the theory of similarity and turbulence research to the bed-load movement," *Mitt. Preuss. Vets. Wasset Schiff*, **26**, 5–24.

Shooman, M. L. (1968). *Probabilistic Reliability: An Engineering Approach*, McGraw-Hill Book Co., New York.

Sina News (2008a). "Emergent evacuations are been exercising in Mianyang City and Jiangyou Town to Tangjiashan landslide dam," Online http://news.sina.com.cn/c/2008–05–28/062015632328.shtml.

Sina News (2008b). "Witnessing the evacuation of 0.2 million people downstream of the Tangjiashan landslide dam," Online http://news.sina.com.cn/c/2008–06–04/110415679694.shtml.

Sinco, E., Xiao, M., Owaidat, L. M., and Smith, L. (2010). "Slurry induced piping progression of a sand," *Geotechnical Special Publication No. 210*, ASCE, Reston, 408–417.

Singh, B. and Varshney, R. S. (1995). *Engineering for Embankment Dams*, Brookfield: A. A. Balkema Publishers.

Singh, V. P. (1996). *Dam Breach Modeling Technology*, Kluwer Academic Publishers, Boston, Massachusetts.

Singh, V. P. and Quiroga, C. A. (1987). "A dam-breach erosion model: I. Formulation," *Water Resources Management*, **1**(3), 177–197.

Singh, V. P. and Scarlatos, P. D. (1988). "Analysis of gradual earth-dam failure," *Journal of Hydraulic Research*, **114**(1), 21–42.

Singh, V. P. and Scarlatos, P. D. (1985). "Breach erosion of earthfill dams and flood routing: BEED model," Research Report, Army Research Office, Battelle, Research Triangle Park, North Carolina.

Singh, V. P. and Snorrason, A. (1984). "Sensitivity of outflow peaks and flood stages to the selection of dam breach parameters and simulation models," *Journal of Hydrology*, **68**(1–4), 295–310.

Skempton, A. W. and Brogan, J. M. (1994). "Experiments on piping in sandy gravels" *Geotechnique*, **44**(3), 449–460.

Smart, G. M. (1984). "Sediment transport formula for steep channels" *Journal of Hydraulic Engineering*, **110**(3), 267–276.

Smerdon, E. T. and Beasley, R. P. (1961). "Critical tractive forces in cohesive soils," *Agricultural Engineering*, **42**(1), 26–29.

Smith, K. and Ward, R. (1998). *Floods: Physical Processes and Human Impacts*, John Wiley & Sons, Chichester.

Smith, M. (2006). "Dam risk analysis using Bayesian networks," in *Proceedings of the 2006 ECI Conference on Geohazards*, Nadim, F., et al., Eds., ECI Symposium Series No. 7, Engineering Conferences International (ECI), New York.

Smith, P. J., Kojiri, T., and Sekii, K. (2006). *Risk-based flood evacuation decision using a distributed rainfall-runoff model*. Annuals of Disas. Prev. Res. Inst., Kyoto Univ., No. 49 B.

Snow, D. T. (1964). *Landslide of Cerro Condor-Seneca*, Department of Ayacucho, Peru, Geological Society of America, pp. 1–6.

Song, C. R., Adhikar, S., Al-Ostaz, A., and Cheng, A. H.-D. (2014). "Re-evaluation of the gap formation in the New Orleans levee system," *Journal of Geotechnical and Geoenvironmental Engineering*, **140**(3), 04013031.

Sorensen, J. H. and Mileti, D. S. (1988). "Warning and evacuation: answering some basic questions," *Industrial Crisis Quarterly* **2**, 195–209.

Stanczak, G. and Oumeraci, H. (2012). "Modeling sea dike breaching induced by wave impact-laboratory experiments and computational model." *Coastal Engineering*, **59**(1), 28–37.

Stanford University. (1994). *National Performance of Dams Program*. Online http://npdp.stanford.edu.

Steedman, R. S. and Sharp, M. K. (2011). "Physical modelling analysis of the New Orleans levee breaches," *Proceedings of Institution of Civil Engineers-Geotechnical Engineering*, **164**(6), 353–372.

Straub, D. and Der Kiureghian, A. (2010a). "Combining Bayesian networks with structural reliability methods: Methodology," *Journal of Engineering Mechanics*, **136**(10), 1248–1258.

Straub, D. and Der Kiureghian, A. (2010b). "Combining Bayesian networks with structural reliability methods: Application," *Journal of Engineering Mechanics*, **136**(10), 1259–1270.

Sun, B. C. B. (1989). *Internal Stability of Clayey to Silty Sands*, Ph.D. thesis, Department of Civil Engineering, University of Michigan.

Swanson, F. J., Oyagi, N., and Tominaga, M. (1986). "Landslide dams in Japan, In Schuster," in *Landslide dams-Processes, Risk, and Mitigation*, Geotechnical Special Publication No. 3, R. L., Ed., American Society of Civil Engineers, pp. 131–145.

Technical Advisory Committee on Water Defences (TAW) (1999). *Guide on Sea and Lake Dikes, TAW*. Retrieved from www.tawinfo.nl (Assessed on 12 Sep, 2014).

Technical Advisory Committee on Water Defences (TAW) (1998). "Fundamentals on water defences," *TAW*. Online www.tawinfo.nl.

Temple, D. M., Hanson, G. J., Neilsen, M. L., and Cook, K. R. (2005). "Simplified breach analysis model: Part I, Background and model components," in *Proceedings of the 25th Annual United States Society on Dams (USSD) Conference*, Salt Lake City, Utah, pp. 151–161.

Terzaghi, K. (1943). *Theoretical Soil Mechanics*, Wiley, New York, 510 pp.

Terzaghi, K. (1939). "Soil mechanics: a new chapter in engineering science," *Journal of the Institution of Civil Engineers*, **12**(7), 106–141.

Terzaghi, K., Peck, R. B., and Mesri, G. (1996). *Soil Mechanics in Engineering Practice*, 3rd ed., John Wiley and Sons, Inc., New York.

The Ministry of Water Resources of People's Republic of China, (1999). *Code for Water Resources and Hydropower Engineering Geological Investigation*. (GB5087-99), Beijing.

Thoman, R. W. and Niezgoda, S. L. (2008). "Determining erodibility, critical shear stress, and allowable discharge estimates for cohesive channels: case study in the Power River Basin of Wyoming," *Journal of Hydraulic Engineering*, **134**(12), 1677–1687.

Tingsanchali, T. and Chinnarasri, C. (2001). "Numerical modelling of dam failure due to flow overtopping." *Hydrological Sciences Journal*, **46**(1), 113–130.

Tomlinson, S. S. and Vaid, Y. P. (2000). "Seepage forces and confining pressure effects on piping erosion," *Canadian Geotechnical Journal*, **37**(1), 1–13.

Tsuchiya, Y. and Yasuda T. (1980) "High tide and life risk, refuge warning in relation to the Ise Bay Typhoon", *Journal of Natural Disaster Science*, **2**(2), 27–60.

Tung, Y. X. (2008). Quantitative analysis for stability of landslide dams. Master's thesis, National Central University, Taiwan.

Tweedie, S. W., Rowland, J. R., Walsh, S., Rohten, R. R., and Hagel, P. L. (1986). "A methodology for estimating emergency evacuation times," *The Social Science Journal*, **21**(2), 189–204.

Ubilla, J., Abdoun, T., Sasanakul, I., Sharp, M., Steedman, S., Vanadit-Ellis, W., and Zimmie, T. (2008). "New Orleans levee system performance during Hurricane Katrina: London Avenue and Orleans Canal South," *Journal of Geotechnical and Geoenvironmental Engineering*, **134**, 668–680.

Urbanik II, T. (2000). "Evacuation time estimates for nuclear power plants," *Journal of Hazardous Materials*, **75**(2–3), 165–180.

Urbanik, T. and Desrosler, A. E. (1981). *An Analysis of Evacuation Time Estimates Around 52 Nuclear Power Plant Sites: An Evacuation*, Rep. No. NUREG/CR 7856, Vol. 1, US Nuclear Regulatory Commission, Rockville, Md.

Urbina, E. and Wolshon, B. (2003). "National review of hurricane evacuation plans and policies: a comparison and contrast of state practices," *Transportation Research Part A: Policy and Practice*, **37**(3), 257–275.

URS (Urban Streams Restoration Program) (2008). "Sunny-day risk analysis, Risk analysis report, Section 9," California Department of Water Resources. Online, http://www.water.ca.gov.

US Army Corps of Engineers (USACE) (2014). *Mississippi River and Tributaries Project-Levee System Evaluation Report for the National Flood Insurance Program*. Washington, DC.
US Army Corps of Engineers (USACE) (2007). *Advanced Measures Report Based on Technical Assistance Investigation, Santa Maria Valley Levees*, County of Santa Barbara. Online, www.countyofsb.org.
US Army Corps of Engineers (USACE) (2000a). *Design and Construction of Levees*, EM 1110–2-1913, Washington, DC.
US Army Corps of Engineers (USACE) (2000b). *Alabama Hurricane Behavioral Analysis*, Final Report, Washington,DC.
US Army Corps of Engineers (USACE) (1995). *Flood Proofing Regulations*, Document EP 1165–2-314, USACE (United States Army Corps of Engineers) National Flood Proofing Committee, Washington,DC.
US Army Corps of Engineers (USACE) (1985). *Business Depth-Damage Analysis Procedure, Research Report 85-R-5*, Engineer Institute for Water Resources, Portland, Oregon.
US Army Corps of Engineers (USACE) (1984). *Shore Protection Manual*, Vol. 2, 4th ed., USACE (United States Army Corps of Engineers) Coastal Engineering Research Center, Washington,DC.
US Army Corps of Engineers (USACE) (1953). *Filter Experiments and Design Criteria*, Waterways Experiment Station, Vicksburg, Technical Memorandum, No. 3-360, Washington, DC.
US Army Corps of Engineers (USACE) (1941). *Investigation of Filter Requirements for Underdrains*, US Army, Waterways Experiment Station, Vicksburg, MS, Technical Memorandum No. 183-1. US Bureau of Reclamation (USBR). (2004). *Design Standards-Embankment Dams*, No. 13, Denver.
US Bureau of Reclamation (USBR) (2012). *Internal Erosion Risks*, Denver, Chapter 26.
US Bureau of Reclamation (USBR) (2011). *Design Standards-Embankment Dams*, No. 13, Denver.
US Bureau of Reclamation (USBR). (2004). *Filters. In Design Standards No. 13: Embankment Dams*, Denver.
US Bureau of Reclamation (USBR). (1982). *Guidelines for defining inundated areas downstream from Bureau of Reclamation dams. Reclamation Planning Instruction No. 82–11*, Denver.
US Bureau of Reclamation (USBR). (1988). *Downstream hazard classification guidelines. ACER Tech. Memorandum No. 11*, US Department of the Interior, Denver.
US Bureau of Reclamation (USBR). (1997). *Guidelines for Achieving Public Protection in Dam Safety Decision-making*. Dam Safety Office, US Bureau of Reclamation, Denver.
US Commission on Large Dams (USCOLD) (1988). *Lessons from Dam Incidents*, USA-II. The Committee on Dam Safety of the USA Committee on Large Dams (USCOLD), ASCE, New York.
US Commission on Large Dams (USCOLD). (1975). *Lessons from Dam Incidents, USA*. The Committee on Dam Safety of the USA Committee on Large Dams (USCOLD), New York.
US Department of Agriculture (USDA) (1994). *Gradation Design of Sand and Gravel Filters*, Natural Resources Conservation Service, US Department of Agriculture, Washington,DC.
US Department of Defense (1980). *Military Standard Procedures for Performing a Failure Modes, Effects, and Criticality Analysis*. US Department of Defense, Washington,DC.
US Department of Transportation (USDOT) (2008). Treatment of Economic Value of a Statistical Life in Departmental Analysis. Memorandum to Secretarial Officers and Modal Administrators from Tyler D. Duval, Assistant Secretary for Transportation Policy. February 5, 2008.
US Society on Dams (USSD) (2011). *Materials for embankment dams*, USSD Committee on Materials for Embankment Dams, Denver.
van Baars, S. (2005). "The horizontal failure mechanism of the Wilnis peat dyke," *Geotechnique*, **55**(4), 319–323.
van Baars, S. and van Kempen, I. M. (2009). "The causes and mechanisms of historical dike failures in the Netherlands," European Water Association. Online, http://www.dwa.de/portale/ewa/ewa.nsf/home?readform&treeid=1.
Van Beek, V. M., Bezuijen, A., Sellmeijer, J. B., and Barends, F. B. J. (2014). "Initiation of backward erosion piping in uniform sands," *Geotechnique*, **64**(12), 927–941.
Van Beek, V. M., Knoeff, J. G., and Sellmeijer, J. B. (2011). "Observations on the process of backward piping by under-seepage in cohesionless soils in small-, medium- and full-scale experiments," *European Journal of Environmental and Civil Engineering*, **15**(8), 1115–1137.
Van Beek, V. M., Knoeff, J. G., Rietdijk, J., Sellmeijer, J. B., and Lopez De La Cruz, J. (2010). "Influence of sand and scale on the piping process – experiments and multivariate analysis" in *Physical Modelling in Geotechnics*, Zürich, Taylor and Francis Group.
Van Emelen, S., Swartenbroekx, C., Zech, Y., and Soares-Frazão, S. (2011). "Numerical modelling of the breaching process in an earthen dike," in *Proceedings of the Fifth International Conference on Advanced Computational Methods in Engineering (ACOMEN 2011)*, Liège, Belgium, pp. 1–10.
Van Genuchten, M. T. (1980). "A closed-form equation for predicting the hydraulic conductivity of unsaturated soils," *Soil Science Society of America Journal*, **44**, 892–898.
Van Manen, S. E. and Vrijling, J. K. (1996). "The problem of the valuation of a human life," in *Proceedings of ESREL 96 – PSAM III Conference*, Crete, Greece.

van Rijn, L. C. (1984a). "Sediment transport. Part I: Bed-load transport," *Journal of Hydraulic Engineering*, **110**(10), 1431–1456.

van Rijn, L. C. (1984b). "Sediment transport. Part II: Suspended load transport," *Journal of Hydraulic Engineering*, **110**(11), 1613–1641.

van Rijn, L. C. (2007). "Unified view of sediment transport by currents and waves. I: initiation of motion, bed roughness, and bed-load transport," *Journal of Hydraulic Engineering*, **133**(6), 649–667.

Vanderkimpen, P., Melger, E., and Peeters, P. (2009). "Flood modeling for risk evaluation – a MIKE FLOOD vs SOBEK 1D2D benchmark study," in *Flood Risk Management: Research and Practice*, Samuels et al. (Eds), Taylor & Francis Group, London, pp. 77–84.

Vaughan, P. R., Kluth, D. J., Leonard, M. W., and Pradourah, H. M. (1970). "Cracking and erosion of the rolled clay core of Balderhead Dam and the remedial works adopted for its repair," in *Transactions of 10th International Congress on Large Dams*. Montrtal, Vol. 1, 73–93.

Verheij, H. (2002). "Time-dependent breach development in cohesive material," Impact Project. Online, http://www.impact-project.net.

Vestad, H. (1976). "Viddalsvatn Dam, a history of leakages and investigations," *Transactions of the 12th International Congress on Large Dams*, Mexico City, **2**, 369–389.

Visser, P. J. (1998). "Breach Growth in Sand-Dikes," Ph.D. thesis, Delft University of Technology, Delft, The Netherlands.

Vogel, A. (1980). *Bibliography of the History of Dam Failures*, Risk Assessment International, Vienna, Austria.

Volk, G. M. (1937). "Method of determination of the degree of dispersion of the day fraction of soils," in *Proceedings of the Soil Science Society of America*, **2**, 561.

Von Thun, J. L. and Gillette, D. R. (1990). *Guidance on breach parameters*. Internal Memorandum, US Dept. of the Interior, Bureau of Reclamation, Denver.

Vrijling, J. K. (2010). *Piping–Realiteit of Rekenfout*, Report TAW/ENW TPN100400, Rijkwaterstaat, Waterdienst, The Netherlands.

Vrijling, J. K. and Van Gelder, P. H. A. J. M. (2000). Policy implications of uncertainty integration in design. In: Eighth International Symposium Stochastic Hydraulics, Beijing, China.

Wahl, T. L. (2004). "Uncertainty of predictions of embankment dam breach parameters," *Journal of Hydraulic Engineering*, **130**(5), 389–397.

Wahl, T. L. (1998). *Prediction of Embankment Dam Breach Parameters – A Literature Review and Needs Assessment*, Dam Safety Rep. No. DSO-98-004, US Dept. of the Interior, Bureau of Reclamation, Denver.

Wahl, T. L., Hanson, G. J., Courivaud, J. R., Morris, M. W., Kahawita, R., McClenathan, J. T., and Gee, D. M. (2008). "Development of next-generation embankment dam breach models," in *Proceedings of the 28th Annual United States Society on Dams (USSD) Conference*, Portland, Oregon, pp. 767–779.

Wahl, T. L., Regazzoni, P., and Erdogan, Z. (2008). "Determining erosion indices of cohesive soils with the hole erosion test and jet erosion test (Hydraulic Laboratory (HL)," No. DSO-08-05). Denver, Colorado, US Department of the Interior, Bureau of Reclamation, Technical Service Center.

Walder, J. S. and O'Connor, J. E. (1997). "Methods for predicting peak discharge of floods caused by failure of natural and constructed earthen dams," *Water Resources Research*, **33**(10), 2337–2348.

Wan, C. F. (2006). *Experimental Investigation of Piping Erosion and Suffusion of Soils in Embankment Dams and Their Foundations*, Ph.D. thesis, The University of New South Wales, Sydney, Australia.

Wan, C. F. and Fell, R. (2008). "Assessing the potential of internal erosion and suffusion in embankment dams and their foundations" *Journal of Geotechnical and Geoenvironmental Engineering*, **134**(3), 410–407.

Wan, C. F. and Fell, R. (2004a). "Investigation of rate of erosion of soils in embankment dams," *Journal of Geotechnical and GeoEnvironmental Engineering*, **130**(4), 373–380.

Wan, C. F. and Fell, R. (2004b). "Laboratory tests on the rate of piping erosion of soils in embankment dams," *Geotechnical Testing Journal*, **27**(3), 295–303.

Wang, G., Liu, F., Fu, X., and Li, T. (2008). "Simulation of dam breach development for emergency treatment of the Tangjiashan Quake Lake in China," *Science in China Series E: Technology Sciences*, **51**, Sup. II, 82–94.

Wang, L. F. and Cao, D. L. (1991). "Study of piping failure in Jingjiang Dike foundation," *Journal of Yangtze River Scientific Research Institute*. **8**(2), 44–51.

Wang, P. and Kahawita, R. (2002). "Modelling the hydraulics and erosion process in breach formation due to overtopping," in *Proceedings of the Symposium on Sedimentation and Sediment Transport*, Monte Verita, Switzerland, pp. 211–220.

Wang, Y. C. and He, B. Q. (2008). "Organization and management of emergency treatment on Shitingjiang Landslide Dam," *Water Resources and Hydropower Engineering*, **39**(8), 52–54 (in Chinese).

Wang, Z. and Bowles, D. S. (2006). "Part 1: Theory and methodology-Three-dimensional non-cohesive earthen dam breach model," *Advances in Water Resources*, **29**(10), 1528–1545.

Watson, D., Serrer, M., and Crookshank, N. (2001). "BC Hydro Life Safety Model (Alpha Version): A two-dimensional model for estimating dam breach flood survival probabilities," Technical report, Canadian Hydraulics Center. 46 pp.

White, C. M. (1940). "The equilibrium of grains on the bed of a stream," in *Proceedings of Royal Society of London*, **A174**(985).

Whitman, R. V. (1984). "Evaluating calculated risk in geotechnical engineering," *Journal of Geotechnical Engineering*, **110**(2), 145–188.

Wiberg, P. L. and Smith, J. D. (1987). "Calculations of the critical shear stress for motion of uniform and heterogeneous sediments," *Water Resources Research*, **23**(8), 1471–1480.

Wolff, T. F. (2008). "Chapter 12: Reliability of levee systems," in *Reliability-based Design in Geotechnical Engineering*, Phoon, K. K. (Ed.), Taylor & Francis, London and New York.

Wolshon, B., Urbina, E., and Levitan, M. (2001). *National Review of Hurricane Evacuation Plans and Policies*, LSU Hurricane Center, Louisiana State University, Baton Rouge.

Wong, H. N., Ho, K. K. S., and Chan, Y. C. (1997). "Assessment of consequence of landslides," in *Proceedings of the International Workshop on Landslide Risk Assessment, Honolulu*, Hawaii, USA, pp. 111–149.

Woo, G. (2008). "Probabilistic criteria for volcano evacuation decisions." *Natural Hazards*, **45**(1): 87–97.

Wood, D. M., Kjaernsli, B., and Høeg, K. (1976). "Thoughts concerning the unusual behaviour of Hyttejuvet Dam," in *Transactions of the 12th International Congress on Large Dams*, Mexico City, **2**, 391–414.

Wörman, A. and Olafsdottir, R. (1992). "Erosion in a granular medium interface," *Journal of Hydraulic Research*, **30**(5), 639–655.

WRD (Water Resources Department of Yunnan Province) (2014a). *The Situations of the 8.03 Earthquake in Ludian, Zhaotong*, Water Resources Department of Yunnan Province, Online, http://www.wcb.yn.gov.cn/news/28094.html.

WRD (Water Resources Department of Yunnan Province) (2014b). *Minister Chen Jian Introduced the Current Situation of 8.03 Earthquake Relief at Press Conference in Ludian, Zhaotong*, Water Resources Department of Yunnan Province, Online, http://www.wcb.yn.gov.cn/news/28138.html.

Wu, L. J. (1980). "Computation of the critical hydraulic gradient for piping of granular soil," *Hydro-Science and Engineering*, **4**, 90–95.

Wu, W. and Wang, S. S. Y. (2007). "One-dimensional modeling of dam-break flow over movable beds," *Journal of Hydraulic Engineering*, **133**(1), 48–58.

Wu, W. and Wang, S. S. Y. (2006). "Formulas for sediment porosity and settling velocity," *Journal of Hydraulic Engineering*, **132**(8), 858–862.

Wu, W., Kang, Y., and Wang, S. S. Y. (2009). "An earthen embankment breach model," in *Proceedings of the 33rd Congress of IAHR C. D.*, Vancouver, Canada.

Wu, W., Wang, S. S. Y., and Jia, Y. (2000). "Non-uniform sediment transport in alluvial rivers," *Journal of Hydraulic Research*, **38**(6), 427–434.

Xinhuanet.com (2008). "The high-risk landslide dam in Qingchuan County was blasting," Retrieved from http://news.xinhuanet.com/local/2008–06/05/content_8315261.htm.

Xu, Q., Fan, X. M., Huang, R. Q., and Westen, C. V. (2009). "Landslide dams triggered by the Wenchuan Earthquake, Sichuan Province, south west China," *Bulletin of Engineering Geology and the Environment*, **68**(3), 373–386.

Xu, T. H. and Zhang, L. M. (2013). "Simulation of piping in earth dams due to concentrated leak erosion," Geo-Congress 2013, ASCE, Reston, 1091–1099.

Xu, Y. (2010). *Analysis of Dam Failures and Diagnosis of Distresses for Dam Rehabilitation*, Ph.D. thesis, the Hong Kong University of Science and Technology, Hong Kong.

Xu, Y. and Zhang, L. M. (2009). "Breaching parameters of earth and rockfill dams," *Journal of Geotechnical and Geoenvironmental Engineering*, **135**(12), 1957–1970.

Xu, Y., Zhang, L. M., and Jia, J. S. (2011). "Diagnosis of embankment dam distresses using Bayesian networks. Part II: Diagnosis of a specific distressed dam," *Canadian Geotechnical Journal*, **48**(11), 1645–1657.

Yager, E. M., Kirchner, J. W., and Dietrich, W. E. (2007). "Calculating bed-load transport in steep boulder bed channels," *Water Resources Research*, **43**(W07418), 1–24.

Yalin, M. S. (1972). *Mechanics of Sediment Transport*, Pergamon Press, Oxford.

Yalin, M. S. (1963). "An expression for bed-load transportation," *Journal of Hydraulic Division*, **89**(3), 221–250.

Yan, R. (2006). *Secondary Disaster and Environmental Effect of Landslides and Collapsed Dams in the Upper Reaches of Minjiang River*, Master thesis, Sichuan University, Chengdu, China.

Yan, Z. W. (2014). "The reason of emergency spillway tunnel water gushing accident of Hongshiyan Landslide Dam in 8.03 Ludian earthquake," *fcbmis.com*. Online, http://www.fcbmis.com/news/jnyw/2014/1011/141011165615B306J3IBCGJ9DEHDG2EB.html.

Yang, C. T. (1973). "Incipient motion and sediment transport," *Journal of Hydraulic Division*, **99**(10), 1679–1704.

Yang, X., Yang, Z., Cao, S., Gao, X., and Li, S. (2010). "Key techniques for the emergency disposal of Quake lakes," *Natural Hazards*, **52**(1), 43–56.

Yao, P. S. (2014). The Ministry of Water Resource: Put the hydraulic risk mitigation of the landslide dam in an important position. chinanews.com. online: http://www.chinanews.com/gn/2014/08–05/6462406.shtml.
YASMG (Yunnan Administration of Surveying, Mapping and Geoinformation) (2014). "The aerial image of the Niulan River landslide lake in Ludian Earthquake," *ynbsm.gov.cn*. Retrieved from http://www.ynbsm.gov.cn/Readinfo.aspx?KindID=fee4165e-8f93–46c4–91dc-97a6011f135e&InfoID=b927fa6a-1dc2–4e5d-99e6–4b08ddad1f77.
Yin, Y. P., Wang, F. W., and Sun, P. (2009). "Landslides hazards triggered by the 2008 Wenchuan earthquake, Sichuan, China," *Landslides*, **6**(2), 139–151.
Yuan, Y. and Whittle, A. J. (2013). "Evaluation and prediction of 17th Street Canal I-Wall stability using numerical limit analyses," *Journal of Geotechnical and Geoenvironmental Engineering*, **139**(6), 841–852.
Zanke, U. (1977). "Berechnung der sinkgeschwindigkeiten von sedimenten," Mitt. Des Franzius-Instituts fuer Wasserbau, Heft 46, Seite 243, Technical University, Hannover, Germany.
Zhai, G., Fukuzono, T., and Ikeda, S. (2006). "An empirical model of fatalities and injuries due to floods in Japan," *Journal of the American Water Resources Association*, **42**(4), 863–875.
Zhang, J. Y., Li, Y., and Xuan, G. X. (2009). "Overtopping breaching of cohesive homogeneous earth dam with different cohesive strength," *Science in China Series E: Technological Sciences*, **52**(10), 3024–3029.
Zhang, J. X. (2009a). *Hydrologic Analysis and Emergency Application of Barrier Lake Breaking*, Ph.D. thesis, Tsinghua University, Beijing, China.
Zhang, L. M. (2009b). "Challenges in multi-hazard risk assessment and management: Geohazard chain in Beichuan Town caused by Great Wenchuan Earthquake," in *Geotechnical Risk and Safety*, Taylor and Francis Group, London, pp. 237–244.
Zhang, L. M. and Chang, D. S. (2012). "Seepage induced soil shear failure under complex stress states," in *Proceedings of the Third International Workshop on Modern Trends in Geomechanics*, IW-MTG3, Nottingham, pp. 93–95.
Zhang, L. M. and Chen, Q. (2006). "Analysis of seepage failure of the Gouhou rockfill dam during reservoir water infiltration," *Soils and Foundations*, **46**(5), 557–568.
Zhang, L. M. and Du, J. C. (1997). "Effects of abutment slope on the performance of high rockfill dams," *Canadian Geotechnical Journal*, **34**(4), 489–497.
Zhang, L. M., Peng, M., and Xu, Y. (2010). "Assessing risks of breaching of earth dams and natural landslide dams," in *Proceedings of Indian Geotechnical Conference*, IGC-2010, Geotrends, Bombay, pp. 81–92.
Zhang, L. M., Xu, Y., and Jia, J. S. (2009). "Analysis of earth dam failures-A database approach," *Georisk*, **3**(3), 184–189.
Zhang, L. M., Xu, Y., and Liu, Y. (2011a). "Assessment of levee breaching risks to the Pearl River Delta," in *Proceedings of the Third International Symposium on Geotechnical Safety and Risk*, Bundesanstalt für Wasserbau, pp. 613–621.
Zhang, L. M., Xu, Y., Jia, J. S., and Zhao, C. (2011b). "Diagnosis of embankment dam distresses using Bayesian networks. Part I. Global-level characteristics based on a dam distress database," *Canadian Geotechnical Journal*, **48**(11), 1630–1644.
Zhang, L. M., Xu, Y., Liu, Y., and Peng, M. (2013). "Assessment of flood risks in Pearl River Delta due to levee breaching," *Georisk*, **7**(2), 122–133.
Zhou, J. W., Yang, X. G., Li, H. T., Zhou, H. W., Cao, P. F., and Qiao, L. (2009). "Engineering Geomechanics analysis of the Barrier Lake in Baisha River at Dujiangyan City induced by Wenchuan Earthquake." *Journal of Sichuan University (Engineering Science Edition)*, **41**(3), 102–108.
Zhou, X., Jie, Y., and Li, G. (2012). "Numerical simulation of the developing course of piping," *Computers and Geotechnics*, **44**, 104–108.
Zhu, Y. (2006). *Breach Growth in Clay Dikes*, Ph.D. thesis, Delft University of Technology, Delft, the Netherlands.
Zobin, V. M., Luhr, J. F., Taran, Y. A., Bréton, M., Cortés, A., De La Cruz-Reyna, S., Domínguez, T., Galindo, I., Gavilanes, J. C., Muñíz, J. J., Navarro, C., Ramírez, J. J., Reyes, G. A., Ursúa, M., Velasco, J., Alatorre, E., and Santiago, H. (2002). "Overview of the 1997–2000 activity of Volcán de Colima, México," *Journal of Volcanology and Geothermal Research*, **117**(1–2), 1–19.

Index

Dam Failure Mechanisms and Risk Assessment, First Edition. Limin Zhang, Ming Peng, Dongsheng Chang, and Yao Xu.

www.ingramcontent.com/pod-product-compliance
Ingram Content Group UK Ltd.
Pitfield, Milton Keynes, MK11 3LW, UK
UKHW050942290926
15470UKWH00005B/1595

9 781118 558515